ABSTRACT

ALGEBRA WITH

APPLICATIONS

IN TWO VOLUMES

VOLUME II

ABSTRACT

ALGEBRA WITH

APPLICATIONS

IN TWO VOLUMES

VOLUME II

RINGS AND FIELDS

KARLHEINZ SPINDLER

Darmstadt, Germany

CRC Press
Taylor & Francis Group
Boca Raton London New York

CRC Press is an imprint of the
Taylor & Francis Group, an **informa** business

CRC Press
Taylor & Francis Group
6000 Broken Sound Parkway NW, Suite 300
Boca Raton, FL 33487-2742

First issued in paperback 2019

ISBN-13: 978-0-8247-9159-9 (hbk)
ISBN-13: 978-0-367-40224-2 (pbk)

Library of Congress Cataloging-in-Publication Data

Spindler, Karlheinz
   Abstract algebra with applications / Karlheinz Spindler.
     p. cm.
   Includes bibliographical references and index.
   Contents: v. 1. Vector spaces and groups -- v. 2. Rings and fields.
   ISBN 0-8247-9144-4 (v. 1). -- ISBN 0-8247-9159-2 (v. 2)
   1. Algebra, Abstract.   I. Title.
QA162.S66 1994
512'.02--dc20
                                        93-32090
                                           CIP

Visit the Taylor & Francis Web site at
http://www.taylorandfrancis.com

and the CRC Press Web site at
http://www.crcpress.com

Meinen Eltern
in Liebe und Dankbarkeit

Preface

THIS volume continues the discussion of abstract algebra and its applications which was begun in Volume I with the treatment of linear algebra and group theory and is now carried on by studying rings and fields. The required set-theoretical background can be found in the Appendix to Volume I; moreover, references to Volume I are given for all facts from linear algebra and group theory which are needed in this volume.† Thus the two volumes together form a self-contained algebra text.

All comments concerning the objectives and the style of this textbook given in the preface of Volume I hold for Volume II as well. This means that I tried to write a text which

- enhances understanding by giving many examples and informal comments;
- offers many exercises which allow the students to check their understanding of the new material and to apply it to concrete problems;
- can be used for an individual self-study program and for reading courses;
- tries to reveal general underlying ideas which guide the students through the technical discussion of the material; and
- also presents the connections between algebra and other mathematical disciplines, thereby demonstrating the pervasive power of algebraic techniques and enabling the students to put the different algebraic theories into a more general framework.

The discussion of rings emphasizes those two topics from which the abstract-structural theory of rings has evolved and by which the introduction of many important concepts is motivated; namely, number theory and algebraic geometry. Ring theory is thereby shown to be rooted in both arithmetic and geometry in the same way in which the theory of vector spaces was traced back to both arithmetical and geometrical considerations in Volume I. Thus the Cartesian program of merging algebra and geometry is extended from linear equations and affine geometry to arbitrary polynomial equations and algebraic geometry. Field theory is presented in two blocks. First, there are three sections dealing with basic properties of algebraic and transcendental field extensions, normality and separability, as these topics are needed in the more advanced sections on ring theory. Second, Galois theory is discussed in some detail. In a short epilogue it is shown that Galois theory and Lie theory can be traced back to the same underlying fundamental idea, namely that of revealing symmetries in the solution set of an equation (be it an algebraic equation or a differential equation) by employing the mathematical concept of a group.

As was already stated in the preface to the first volume, I am greatly indebted to my doctoral thesis advisor, Karl Heinrich Hofmann; my colleagues with whom I have collaborated in preparing lectures and lab sessions in algebra and geometry, namely Benno Artmann, Jürgen Bokowski, Joachim Hilgert, Martin Petschke and Christian Terp; and, above all, to my family for their steadfast support, both practical and moral.

Karlheinz Spindler

---

† References to Volume I include the Roman number I. For example, (I.23.1) refers to item (23.1) in section 23 of Volume I, whereas (23.1) refers to item (23.1) in section 23 of the current volume.

Contents†

Volume I

---

† Headings below section titles refer to topics covered, much like an index, rather than to discrete subsections.

GROUPS

Volume II

1.   Introduction: The Art of Doing Arithmetic

RECALL arithmetical problems that you learned to tackle earlier on in your education: solving linear equations $ax + b = 0$ and quadratic equations $ax^2 + bx + c = 0$, factoring terms like $a^2 - b^2 = (a + b)(a - b)$ or $a^3 + b^3 = (a + b)(a^2 - ab + b^2)$, finding product expansions like $(x + y)^3 = x^3 + 3x^2y + 3xy^2 + y^3$, and so on. All these problems involve the two basic operations of addition and multiplication, and arithmetic can be characterized as the art of handling these two operations. For example, the quadratic equation $ax^2 + bx + c = 0$ is essentially solved by simply rewriting it in the form $a(x + \frac{b}{2a})^2 + c - \frac{b^2}{4a} = 0$ or $(x + \frac{b}{2a})^2 = \frac{b^2 - 4ac}{4a^2}$.

When we talk about doing arithmetic, we usually have in mind the set of integers or the set of rational numbers; when it comes to taking roots also the domains of real or complex numbers. But there are other domains where we can do arithmetic; for example, we can add and multiply functions or matrices. We have to be careful, though, because multiplication in these other domains need not be commutative; for example, there are matrices A, B with $AB \neq BA$ and therefore $A^2 - B^2 \neq (A + B)(A - B)$. Another interesting example of objects that can be added and multiplied are the residue classes introduced in (I.20.14). In this chapter, we want to systematically study domains on which two operations, called addition and multiplication, are defined; these domains are called rings. One main source of ring theory is number theory, and we will see how number-theoretical problems stimulated the development of ring theory.[†]

Why is it that number theory (which is primarily concerned with the properties of natural numbers or integers) should stimulate the investigation of the arithmetical properties of more general domains? The answer is that quite often properties of domains other than the set of integers yield number-theoretical results about the integers themselves. Let us consider a prominent example for this phenomenon, an example which creates a link between the well-known arithmetic of integers and arithmetic in a more "abstract" domain and hence can serve as an "appetizer" for a general theory of rings.

In July 1801, after five years of work, Carl Friedrich Gauss (1777-1855) completed a book entitled *Disquisitiones Arithmeticae* which was, upon publication and ever after, considered as one of the greatest works in the theory of numbers. Why did this book attract such an attention that, for example, Dirichlet carried the French translation with him wherever he went and even slept with the book under his pillow? Even though there were many new things in the book (the first proof of the quadratic reciprocity law, the complete determination of those regular polygons that can be constructed with ruler and compass, extensive studies of binary quadratic forms, primitive roots, and so on), one of Gauss's greatest achievements in this book was the indication of a new way of looking at well-known facts in number theory. Indeed, let us see how Gauss begins his *Disquisitiones Arithmeticae*. After a dedication to his prince, duke Karl Wilhelm Ferdinand of Brunswick and Luneburg, and a short preface, Gauss starts his work with the following phrases: "If a number a divides the difference of two numbers b, c then b and c are called *congruent with respect to a*, otherwise *incongruent*; we call this a the *modulus*. Each of the numbers b, c is called a *residue* of the other in the first case, a *non-*

---

† The other main source of ring theory is algebraic geometry which will be discussed in a separate section later.

residue of the other in the second case."† In other words, Gauss starts his work with the introduction of the residue classes that we encountered in (I.20.14). Let us consider several examples in which the study of residue classes yields number-theoretical results or facilitates computations in the domain of integers.

(1.1) Example. *If the natural numbers a and n are relatively prime, then $a^{\varphi(n)}-1$ is divisible by n (**Euler's Theorem**); here φ denotes Euler's φ-function. In the special case that $n = p$ is a prime this says that $a^{p-1} - 1$ is divisible by p for all $a \in \mathbb{N}$ (**Fermat's Theorem**).*

Proof. The multiplicative group $\mathbb{Z}_n^{\times}$ has $\varphi(n)$ elements, and the residue class $[a]$ of a modulo n is an element of this group because a is relatively prime to n. Hence $[a]^{\varphi(n)} = [1]$ by Lagrange's theorem. But this is the claim. ∎

(1.2) Example. *Let $a_n a_{n-1} \cdots a_1 a_0$ be the decimal representation of a natural number x. Then the following **divisibility rules** hold.*
(a) *x leaves the same remainder when divided by $2(5)$ as its last digit a_0.*
(b) *x leaves the same remainder when divided by $3(9)$ as its digit sum $a_0 + a_1 + \cdots + a_{n-1} + a_n$.*
(c) *x leaves the same remainder when divided by 11 as its alternating digit sum $a_0 - a_1 + a_2 - a_3 + - \cdots$.*
(d) *x leaves the same remainder when divided by $7(11,13)$ as the number $a_2 a_1 a_0 - a_5 a_4 a_3 + a_8 a_7 a_6 - + \cdots$.*
(e) *x leaves the same remainder when divided by 37 as the number $a_2 a_1 a_0 + a_5 a_4 a_3 + a_8 a_7 a_6 + \cdots$.*

Proof. To say that $a_n a_{n-1} \cdots a_1 a_0$ is the decimal representation of x is tantamount to saying $x = \sum_{k=0}^{n} a_k 10^k$.
(a) Let $x' = a_0$. Since $10 = 2 \cdot 5$ we have $10 \equiv 0$ modulo 2 and modulo 5. Therefore, $x - x' = \sum_{k=1}^{n} a_k 10^k \equiv 0 \bmod 2(5)$, which is the claim.
(b) Let $x' = a_0 + a_1 + \cdots + a_{n-1} + a_n$. Since $10 \equiv 1$ modulo 3 and modulo 9, we have $x - x' = \sum_{k=0}^{n} a_k 10^k - \sum_{k=0}^{n} a_k = \sum_{k=0}^{n} a_k (10^k - 1) \equiv 0 \bmod 3(9)$, which is the claim.
(c) Let $x' = a_0 - a_1 + a_2 - + \cdots = \sum_{k=0}^{n} (-1)^k a_k$. Since $10 \equiv -1$ modulo 11, we have $x - x' = \sum_{k=0}^{n} a_k 10^k - \sum_{k=0}^{n} (-1)^k a_k = \sum_{k=0}^{n} a_k \left(10^k - (-1)^k\right) \equiv 0 \bmod 11$, which is the claim.
(d) Let $x' = a_2 a_1 a_0 - a_5 a_4 a_3 + a_8 a_7 a_6 - + \cdots = \sum_{k=0}^{\infty} (-1)^k (a_{3k} + 10 a_{3k+1} + 100 a_{3k+2})$ where we set $a_k := 0$ for $k > n$. Since $7 \cdot 11 \cdot 13 = 1001$ we have $1000 \equiv -1$ modulo 7, 11 and 13. Therefore,

---

† Gauss, like his great models Leibniz and Newton, wrote in Latin; the original text is as follows: "Si numerus a numerorum b, c differentiam metitur, b et c *secundum a congrui* dicuntur, sin minus, *incongrui*: ipsum a *modulum* appellamus. Uterque numerorum b, c priori in casu alterius *residuum*, in posteriori vero *nonresiduum* vocatur."

$$x - x' = \sum_{k=0}^{\infty} 10^{3k}(a_{3k} + 10a_{3k+1} + 100a_{3k+2}) - \sum_{k=0}^{n}(-1)^k(a_{3k} + 10a_{3k+1} + 100a_{3k+2})$$

$$= \sum_{k=0}^{\infty} (1000^k - (-1)^k)(a_{3k} + 10a_{3k+1} + 100a_{3k+2}) \equiv 0$$

mod 7(11, 13), which is the claim.

(e) Let $x' = a_2 a_1 a_0 + a_5 a_4 a_3 + a_8 a_7 a_6 + \cdots = \sum_{k=0}^{\infty}(a_{3k} + 10a_{3k+1} + 100a_{3k+2})$ where we again set $a_k := 0$ for $k > n$. Since $27 \cdot 37 = 999$ we have $1000 \equiv 1$ modulo 37. Therefore, $x - x' = \sum_{k=0}^{\infty} 10^{3k}(a_{3k} + 10a_{3k+1} + 100a_{3k+2}) - \sum_{k=0}^{n}(a_{3k} + 10a_{3k+1} + 100a_{3k+2}) = \sum_{k=0}^{\infty}(1000^k - 1)(a_{3k} + 10a_{3k+1} + 100a_{3k+2}) \equiv 0$ mod 37, which is the claim. ∎

(1.3) Example. *If we add up the squares of four consecutive numbers we can never obtain a square.*

Proof. Suppose $x^2 + (x+1)^2 + (x+2)^2 + (x+3)^2 = y^2$, i.e., $4x^2 + 12x + 14 = y^2$. Modulo 4, this reads $[2] = [y]^2$. But in $\mathbb{Z}_4$ we have $[0]^2 = [2]^2 = [0]$ and $[1]^2 = [3]^2 = [1]$; hence $[2]$ is not a square in $\mathbb{Z}_4$. This is the desired contradiction. ∎

(1.4) Example. *What are the last two digits of the number 7^{99999}?*

Solution. We want to find the residue class of 7^{99999} modulo 100. Now in $\mathbb{Z}_{100}$ we have $[7]^4 = [49^2] = [2401] = [1]$; therefore, $[7]^{99999} = [7]^3 = [343] = [43]$. Hence the two last digits are 43. ∎

(1.5) Example. *What remainder is left if 3^{100} is divided by 34?*

Solution. We want to find the residue class of 3^{100} modulo 34. By Euler's theorem (see (1.1)) we have $[3]^{16} = [1]$ in $\mathbb{Z}_{34}$ because $\varphi(34) = 16$; hence $[3]^{100} = [3]^{16 \cdot 6 + 4} = [3]^4 = [81] = [13]$. Hence the remainder is 13. ∎

(1.6) Example. *Show that there are no integers x and y such that $x^2 - 13y^2 = 275$.*

Solution. If $x^2 - 13y^2 = 275$ then $[x]^2 = [275] = [2]$ in $\mathbb{Z}_{13}$. But in $\mathbb{Z}_{13}$ we have

$$[0]^2 = [0], \quad [\pm 1]^2 = [1], \quad [\pm 2]^2 = [4], \quad [\pm 3]^2 = [9],$$
$$[\pm 4]^2 = [3], \quad [\pm 5]^2 = [12], \quad [\pm 6]^2 = [10];$$

hence the equation $[x]^2 = [2]$ has no solution in $\mathbb{Z}_{13}$. ∎

(1.7) Example. *Pierre de Fermat (1601-1665) claimed in 1640 that all numbers of the form $2^{2^n} + 1$ ($n \in \mathbb{N}_0$) are primes, but he admitted that he did not know a proof. For $n = 0, 1, 2, 3, 4$ Fermat's claim is true, because $3, 5, 17, 257$ and 65537 are primes. However, the great Swiss mathematician Leonhard Euler (1707-1783) proved in 1732 that $2^{32} + 1$ is divisible by 641. Check this fact!*

Solution. We have $2^4 + 5^4 = 16 + 625 = 641$ and $5 \cdot 2^7 = 5 \cdot 128 = 640$; modulo 641 these equations become

$$[5]^4 = -[2]^4 \quad \text{and} \quad [5] \cdot [2]^7 = [-1] \quad \text{in } \mathbb{Z}_{641} .$$

Taking the fourth power of the second equation and plugging in the first we obtain $[1] = [5]^4 [2]^{28} = -[2]^4 [2]^{28} = -[2]^{32}$. But this means $[2^{32} + 1] = [0]$ which is exactly the claim. $\blacksquare$

(1.8) Example. *If $p \in \mathbb{N}$ is a prime number, then $(p-1)! + 1$ is divisible by p* (**Wilson's Theorem**). *Conversely, if $(n-1)! + 1$ is divisible by n, then n is a prime number.*

Proof. Let p be a prime. In the multiplicative group $\mathbb{Z}_p^{\times} = \mathbb{Z}_p \setminus \{0\}$, no element $[x] \neq [\pm 1]$ is its own inverse, because $[x]^2 = [1]$ implies $[0] = [x^2 - 1] = [x - 1] \cdot [x + 1]$ and hence $[x - 1] = [0]$ or $[x + 1] = [0]$ in $\mathbb{Z}_p$. Thus the elements of $\mathbb{Z}_p^{\times}$ other than $\pm [1]$ arise in pairs (ξ, ξ^{-1}). This implies

$$[(p-1)!] = [1] \cdot [2] \cdot [3] \cdots [p-1] = \prod_{\xi \in \mathbb{Z}_p^{\times}} \xi = [1] \cdot [-1] = [-1]$$

which is the claim. (Note that the argument also holds if $p = 2$; in this case $[1] = [-1]$.) Conversely, suppose that $(n-1)! + 1$ is divisible by n, say $1 \cdot 2 \cdot 3 \cdots (n-1) + 1 = k \cdot n$. Obviously, none of the numbers $1, 2, \ldots, n-1$ divides the left-hand side of this equation, hence none of them divides n. This shows that n is prime. $\blacksquare$

(1.9) Example. *Let p be a prime number of the form $4n + 1$. Then there is a natural number x such that $x^2 + 1$ is divisible by p.*

Proof. We claim that $x := 1 \cdot 2 \cdots \frac{p-1}{2}$ has the desired property. Indeed, since $\frac{p-1}{2} = 2n$ is even, we also have $x = (-1)(-2) \cdots (-\frac{p-1}{2})$; therefore, in $\mathbb{Z}_p$ the following equation holds:

$$[x^2] = \underbrace{[-1]}_{= [p-1]} \cdot \underbrace{[-2]}_{= [p-2]} \cdots \underbrace{[-\frac{p-1}{2}]}_{= [\frac{p+1}{2}]} \cdot [1] \cdot [2] \cdots [\frac{p-1}{2}]$$

$$= [1] \cdot [2] \cdots [p-1] = [(p-1)!] = [-1]$$

where the last equation is just Wilson's theorem (see (1.8)). $\blacksquare$

(1.10) Example. *No natural number of the form $4n + 3$ can be written as a sum of two squares.*

Proof. Suppose $4n + 3 = x^2 + y^2$. Modulo 4, this reads $[3] = [x]^2 + [y]^2$. But in $\mathbb{Z}_4$ we have $[0]^2 = [2]^2 = [0]$ and $[1]^2 = [3]^2 = [1]$; hence $[3]$ is not the sum of two squares in $\mathbb{Z}_4$. ∎

THESE examples show that the arithmetic of residue classes can yield results about the integers that could be obtained in a straightforward manner only by very tedious calculations. It should be noted that in some of the above examples this simplification was achieved due to the fact that the transition to residue classes translated a problem involving addition and multiplication into a purely multiplicative problem.† In (1.1) the statement that $a^{\varphi(n)} - 1$ is divisible by n was translated into the purely multiplicative statement $[a]^{\varphi(n)} = [1]$ in $\mathbb{Z}_n$. Similarly, in (1.7) the statement that $2^{32} + 1$ is divisible by 641 was rewritten as $[2]^{32} = [-1]$ in $\mathbb{Z}_{641}$, and in (1.8) the statement that $(p-1)! + 1$ is divisible by p was rewritten as $[(p-1)!] = [-1]$ in $\mathbb{Z}_p$. Finally, in (1.6) the equation $x^2 - 13y^2 = 275$ became $[x]^2 = [2]$ in $\mathbb{Z}_{13}$. But in other problems the use of residue classes (though it gives us some information) does not yield the solution. For example, how should we find all integer solutions of equations like $x^2 - 3y^2 = 1$ or $y^2 + 4 = x^3$? Note that these two equations can be rewritten in a purely multiplicative way if we allow ourselves to use irrational and even complex numbers; namely, we can write $(x + y\sqrt{3})(x - y\sqrt{3}) = 1$ and $(y + 2i)(y - 2i) = x^3$. Since we are only interested in integer solutions we are lead to studying the domains $\mathbb{Z} + \mathbb{Z}\sqrt{3} := \{x + y\sqrt{3} \mid x, y \in \mathbb{Z}\}$ and $\mathbb{Z} + 2i\mathbb{Z} := \{x + 2iy \mid x, y \in \mathbb{Z}\}$. Similarly, a bold (and partly successful!) attempt to solve Fermat's problem is by introducing the number $\varepsilon := e^{2\pi i/n}$ which allows one to rewrite the equation $x^n + y^n = z^n$ in the purely multiplicative form

$$z^n = x^n + y^n = (x + y)(x + \varepsilon y)(x + \varepsilon^2 y) \cdots (x + \varepsilon^{n-1} y) \, .$$

The price we pay for this approach is that we cannot limit our study solely to integers; we want to do arithmetic in a larger domain containing the integers but also the number ε. (See problem 18 below.)

THUS even from the viewpoint of a "pure" number theorist it seems rewarding to do arithmetic in domains other than the integers. It will be the purpose of this chapter to study these domains.

---

† As a matter of fact, it is a characteristic feature of some of the hardest problems in number theory that they intertwine the additive and the multiplicative structure of the number system. The most famous (and most notorious) example is Fermat's Last Theorem, stating that the equation $x^n + y^n = z^n$ has no solution (x, y, z) in the natural numbers for any given exponent $n \geq 3$. Another example is Goldbach's Conjecture that every even natural number $n \geq 4$ can be written as a sum of two primes (where a prime is a number with the simplest possible multiplicative structure, namely a number $p \in \mathbb{N}$ whose only divisors in $\mathbb{N}$ are 1 and p itself). Also, it is not known whether there is only a finite number of primes p such that $p + 2$ is also a prime; i.e., whether there is only a finite number of *twin primes* like $(3, 5)$, $(5, 7)$, $(11, 13)$ or $(17, 19)$.

Exercises

Problem 1. (a) Find the remainder of 2^{47} modulo 30, of 3^{47} modulo 23, of 3^{200} modulo 13, of 94^{200} modulo 13 and of 12^{100} modulo 34.

(b) Find the last digits of $3^{597}, 943^{597}, 7^{123}, 987^{123}$ and 689^{1287}.

(c) Find the last two digits of 2^{400} and 3^{400}.

(d) Find the remainder of the division $2^{1,000,000} : 77$.

(e) What is the remainder if 4^{1000} is divided by 17?

Problem 2. A "magician" asks his audience to choose a number $N < 1000$ and to tell him the remainders r_7, r_{11} and r_{13} of this number modulo 7, 11 and 13. Without any further information, he is able to determine the number N. How does he do that?
Hint. Prove that $715r_7 + 364r_{11} + 924r_{13}$ is congruent to N modulo 1001.

Problem 3. (a) Show that $143^6 + 91^{10} + 77^{12} - 1$ is divisible by 1001.

(b) Show that $2222^{5555} + 5555^{2222}$ is divisible by 7.

(c) Show that $3n^5 + 5n^3 + 7n$ is divisible by 15 whenever $n \in \mathbb{Z}$.

(d) Show that $x^9 - 6x^7 + 9x^5 - 4x^3$ is divisible by 8640 for all $x \in \mathbb{Z}$.

Problem 4. (a) Show that the following congruences do not have solutions.

$$x^2 \equiv 3 \pmod 4, \quad y^2 \equiv 12 \pmod{16}, \quad z^2 \equiv 4444 \pmod{10^4}$$

(b) Show that the sum of the squares of three subsequent natural numbers is never a square itself.

(c) Show that none of the following equations has integer solutions.

$$x^2 + y^2 + z^2 = 8n + 7, \ x^2 + y^2 = 4a + 3, \ x^2 - y^2 = 4b + 2, \ x^2 - 3y^n = 2, \ x^2 - 17m = 855$$

(d) Show that none of the numbers 11, 111, 1111, 11111, ... is a square.
Hint. Consider these numbers modulo 4.

Problem 5. (a) Show that the equations $15x^2 - 7y^2 = 9$ and $x^2 + 3xy - 2y^2 = 122$ do not possess integer solutions.

(b) Find all integer solutions of the equations $5x^2 + 2xy + 2y^2 = 26$ and $x^2 + xy - 2y^2 = 10$.

(c) Show that if a, b, c are integers with $a^2 + b^2 = c^2$ then at least one of these three numbers is divisible by 3.

Problem 6. (a) Show that $x^n - 1$ is divisible by $x - 1$ for any $x \in \mathbb{Z}$ and $n \in \mathbb{N}$.

(b) Let $n \in \mathbb{N}$. Show that $6 \mid n^3 - n$, $30 \mid n^5 - n$, $42 \mid n^7 - n$, $8190 \mid n^{13} - n$, $510 \mid n^{17} - n$, $798 \mid n^{19} - n$, $330 \mid n^{21} - n$ and $383838 \mid n^{37} - n$.

Problem 7. Suppose that $n \in \mathbb{N}$ is odd. Show that $n^2 - 1$ is divisible by 8 and that $n^8 - 1$ is divisible by 32. Can you generalize?

Problem 8. Find all natural numbers n such that $2^n - 1$ is divisible by 31.

Problem 9. Let m and n be natural numbers with $n \geq 2$. Show that $2^m + 1$ is not divisible by $2^n - 1$.

Problem 10. (a) Suppose $m, n \in \mathbb{N}$ are either both even or both odd. Show that $a^m - a^n$ is divisible by 3 for any $a \in \mathbb{Z}$.

(b) Let $p > 3$ be a prime. Show that $a^p - a$ and $a^p b - b^p a$ are divisible by $6p$ for all $a, b \in \mathbb{Z}$.

Problem 11. Let p be a prime number.

(a) Show that $a \equiv 1$ modulo p implies that $a^{p^k} \equiv 1$ modulo p^{k+1} for all $k \in \mathbb{N}_0$.

(b) Show that $a \equiv b \,(\mathrm{mod}\, p^m)$ implies $a^p \equiv b^p \,(\mathrm{mod}\, p^{m+1})$.

Problem 12. (a) Let $p \neq 2, 5$ be a prime number. Show that p divides infinitely many of the numbers $1, 11, 111, 1111, 11111, \ldots$.

Hint.

$$9 \cdot \underbrace{11\ldots1}_{n} = 10^n - 1 .$$

(b) Let S be the set of all natural numbers whose decimal expansions have no digits other than 0 and 1. Show that S is multiplicatively closed, i.e., that if $s_1, s_2 \in S$ then $s_1 s_2 \in S$.

(c) Let $n \in \mathbb{N}$ and $k \in \{1, 2, 3, \ldots, 9\}$. Show that there is a multiple of n whose only digits (in the decimal expansion) are 0 and k.

Hint. It is enough to treat the case $k = 1$ (why?). Note that $10^n - 10$ is a multiple of n which has the form $9s = 1$ with $s \in S$. Now argue that there is an element $x \in \mathbb{N}$ with $9x = 11\ldots1 \in S$.

(d) Show that a natural number divides some number of the form $999\ldots9$ if and only if the last digit of n is $1, 3, 7$ or 9.

Problem 13. Let $p > 3$ be a prime number. Write

$$\frac{1}{1} + \frac{1}{2} + \cdots + \frac{1}{p-1} = \frac{a}{b}, \quad \frac{1}{1^2} + \frac{1}{2^2} + \cdots + \frac{1}{(p-1)^2} = \frac{c}{d} \text{ and } \frac{1}{1^3} + \frac{1}{2^3} + \cdots + \frac{1}{(p-1)^3} = \frac{e}{f}$$

with $a, b, c, d \in \mathbb{N}$. Show that the numerators a, c and e are all divisible by p. Can you even show that a is divisible by p^2?

Hint. Consider the elements $\sum_{k=1}^{p-1} [k]^{-1}$, $\sum_{k=1}^{p-1} [k]^{-2}$ and $\sum_{k=1}^{p-1} [k]^{-3}$ in $\mathbb{Z}_p$ and use the formulas

$$\sum_{k=1}^{n} k = \frac{n(n+1)}{2}, \quad \sum_{k=1}^{n} k^2 = \frac{n(n+1)(2n+1)}{6} \text{ and } \sum_{k=1}^{n} k^3 = \frac{n^2(n+1)^2}{4} .$$

Problem 14. (a) Show that if $p \neq 2$ is a prime then $(p-2)!-1$ and $(p-3)!-\frac{1}{2}(p-1)$ are divisible by p.

(b) For $n = 100, 99, 98, 97, 96$ find the remainder of $n!$ when divided by 101.

(c) For $n = 102, 101, 100, 99, 98$ find the remainder of $n!$ when divided by 103.

(d) Let p be an odd prime. Show that $1^2 3^2 5^2 \cdots (p-2)^2 \equiv (-1)^{\frac{1}{2}(p+1)}$ modulo p.

(e) The fact that n is a prime if and only if $(n-1)!+1$ is divisible by n can be considered as a criterion to check the primality of a given natural number. Can you imagine why this is not a very practical criterion?

Problem 15. Find a single congruence which is equivalent to the two congruences $x \equiv 1 \pmod 4$ and $x \equiv 2 \pmod 3$, i.e., find a congruence whose solutions are exactly those numbers which simultaneously solve the two given congruences.

Problem 16. Solve the equations $x^2 = x$ and $x^3 = x$ in $\mathbb{Z}_6$, in $\mathbb{Z}_{30}$ and in $\mathbb{Z}_q$ where q is the power of an odd prime.

Problem 17. Let p be an odd prime. We denote by $S_p := \{x^2 \mid x \in \mathbb{Z}_p\}$ the set of all squares in $\mathbb{Z}_p$ and by $N_p := \mathbb{Z}_p \setminus S_p$ the set of all non-squares.

(a) Show that $|S_p| = \frac{p+1}{2}$ and conclude that exactly half of the elements in $\mathbb{Z}_p^\times$ are squares.

(b) Show that if $u \in N_p$ then $N_p = uS_p \setminus \{0\}$ and prove the inclusions $Q_pQ_p \subseteq Q_p$, $Q_pN_p \subseteq N_p$ and $N_pN_p \subseteq Q_p$. (Thus the nonzero squares form a subgroup of $(\mathbb{Z}_p^\times, \cdot)$ of index 2.)

(c) Show that for any three elements $a, b, c \in \mathbb{Z}_p^\times$ there are elements $x, y \in \mathbb{Z}_p$ such that $ax^2 + by^2 = c$.

Hint. Use part (a) to show that the sets $\{ax^2 \mid x \in \mathbb{Z}_p\}$ and $\{c - by^2 \mid y \in \mathbb{Z}_p\}$ both have cardinality $(p+1)/2$ and hence have a nonempty intersection.

(d) Show that every element of $\mathbb{Z}_p$ can be written as the sum of two squares.

Problem 18. Let $\varepsilon := e^{2\pi i/3}$ and let $R := \mathbb{Z} + \mathbb{Z}\varepsilon + \mathbb{Z}\varepsilon^2$ be the set of all complex numbers $a + b\varepsilon + c\varepsilon^2$ with $a, b, c \in \mathbb{Z}$. Show that sums and products of elements of R lie again in R.

Problem 19. Suppose that p is a prime of the form $p = 2q + 1$ where q is odd. Using Wilson's theorem, show that $q! \equiv \pm 1$ modulo p.

Problem 20. (a) Show that for each element $x \in \mathbb{Z}_{19}^\times$ there is a number $n \in \mathbb{N}$ such that $x = [2]^n$.

(b) Show that the function

$$\log : \begin{array}{ccc} \mathbb{Z}_{19}^\times & \to & \mathbb{Z}_{18} \\ [2]^n & \mapsto & n \end{array}$$

is well-defined and has the property that $\log(xy) = \log x + \log y$, $\log(x/y) = \log x - \log y$ and $\log x^r = r \log x$ for all $x, y \in \mathbb{Z}_{19}^\times$ and all $r \in \mathbb{Z}$.

(c) List the values of this function, i.e., create a "logarithm table" for $\mathbb{Z}_{19}^\times$ and use this table to determine 30^{14} modulo 19 and 25^{11} modulo 19.

Problem 21. (a) Show that $3x \equiv 4\,(5)$ if and only if $x \equiv 3\,(5)$.
Hint. 3 has a multiplicative inverse modulo 5.
(b) Find a number n such that $8x \equiv 15\,(21)$ if and only if $x \equiv n\,(21)$.
(c) Show that $42x \equiv 259\,(847)$ if and only if $6x \equiv 49\,(121)$.

Problem 22. (a) Show that $3x + 5y$ is divisible by 19 if and only if $13x + 9y$ is.
Hint. We have to show that $3x = -5y$ in $\mathbb{Z}_{19}$ if and only if $13x = -9y$ in $\mathbb{Z}_{19}$. Now 3 and 13 are invertible in $\mathbb{Z}_{19}$.
(b) Find a natural number n such that $2x - 3y$ is divisible by 13 if and only if $5x + ny$ is.

Problem 23. The number $\star\star\star,398,246$ is divisible by $31\star$. Find the missing digits.

Problem 24. Show that the product of four consecutive natural numbers is neither a square nor a cube.

Problem 25. Show that among any 10 consecutive natural numbers there is at least one that is relatively prime to each of the others.

Problem 26. Let $x, y, z \in \mathbb{Z}$.
(a) Show that if $x^3 + y^3 - z^3$ is divisible by 9 then at least one of the numbers x, y, z is divisible by 3.
(b) Show that if $x^5 + y^5 - z^5$ is divisible by 25 then at least one of the numbers x, y, z is divisible by 5.

Problem 27. Show that if n is a square-free natural number then $x^{\varphi(n)+1} - x$ is divisible by n for all $x \in \mathbb{Z}$.

2. Rings and ring homomorphisms

ARITHMETIC is the art of manipulating terms by the basic operations of addition, subtraction and multiplication and also division whenever possible. This art can be practised in different domains; we can not only occupy ourselves with integers or rational numbers, but also with polynomials, matrices or other objects. In all these domains, the used operations satisfy certain arithmetic laws which we usually use without thinking. Now we will axiomatize the notions of addition and multiplication; namely, we will define a ring as a set with two arithmetic operations, called addition and multiplication, which satisfy certain rules (the "arithmetic laws"). This definition will turn out to be "weak" enough to include a variety of diverse examples, but "strong" enough to allow interesting conclusions and non-trivial applications. As guiding examples, we should keep in mind the set $\mathbb{Z}$ of all integers, the set $\mathbb{Z}[x]$ of all polynomials with integer coefficients and the set $\mathbb{Z}^{n \times n}$ of all $n \times n$-matrices with integer entries; each endowed with the usual addition and multiplication.

(2.1) Definitions. (a) *A* **ring** $(R, +, \cdot)$ *is a set* R *with two binary operations*

$$
\begin{array}{ccc}
R \times R & \to & R \\
(x, y) & \mapsto & x + y
\end{array}
\quad and \quad
\begin{array}{ccc}
R \times R & \to & R \\
(x, y) & \mapsto & xy
\end{array},
$$

called addition and multiplication, such that the following conditions hold.
(1) $(R, +)$ *is an abelian group,*
(2) $(xy)z = x(yz)$ *for all* x, y, z (associative law for the multiplication in R),
(3) $x(y + z) = xy + xz$ *and* $(x + y)z = xz + yz$ *for all* x, y, z (distributive laws).
(b) *A ring R is called a* **commutative ring** *if*

$$
xy = yx \qquad for\ all\ x, y \in R \ .
$$

(c) *A ring R is called a* **ring with identity** *or a* **unital ring** *if there is an element* $1 \neq 0$ *in R with*

$$
r \cdot 1 = 1 \cdot r = r \qquad for\ all\ r \in R \ .
$$

WE note that if a ring R possesses an identity element, then this element is uniquely determined; see the argument given in (I.20.3)(a). The condition $1 \neq 0$ serves to exclude the trivial ring $\{0\}$.

It is clear that not all the familiar arithmetic rules which are valid for the integers will hold in arbitrary rings; for example, the formula $a^2 - b^2 = (a + b)(a - b)$ relies on the commutativity of the multiplication and is therefore not true in noncommutative rings. However, some formulas are immediate consequences of the ring axioms and therefore valid in every ring.

(2.2) Proposition. *Let R be a ring and $a, b \in R$. Then*

$$
a \cdot 0 = 0 \cdot a = 0 \ , \quad (-a)b = a(-b) = -ab \ , \quad (-a)(-b) = ab \ .
$$

10

In particular, if R has an identity element 1 then $a(-1) = (-1)a = -a$.

Proof. We have $a \cdot 0 + a \cdot 0 = a \cdot (0 + 0) = a \cdot 0$. Adding $-(a \cdot 0)$ to both sides yields $a \cdot 0 = 0$. Similarly $0 \cdot a = 0$. Also, $(-a)b + ab = ((-a) + a)b = 0 \cdot b = 0$ so that $(-a)b$ is indeed the additive inverse of ab, i.e., $(-a)b = -ab$. Similarly $a(-b) = -ab$. Finally, $(-a)(-b) = -a(-b) = -(-ab) = ab$. ∎

LET $(R, +, \cdot)$ be a ring. Then $(R, +)$ is an abelian group; therefore we can adopt the notation $m \cdot x$ for $m \in \mathbb{Z}$ and $x \in R$ which was introduced after definition (I.25.9).

If R has an identity element 1 we might encounter the phenomenon that the elements $1, 1 + 1, 1 + 1 + 1, \ldots$ are not all pairwise distinct because we might have $1 + 1 + \cdots + 1 = 0$ for a certain number of summands. For example, in $\mathbb{Z}_n$ we have $n \cdot 1 = 0$. To describe this phenomenon, we give the following definition.

(2.3) Definition. *Let R be a ring with identity element 1. The* **characteristic** *char R of R is defined as the smallest number $n \in \mathbb{N}$ such that $n \cdot 1 = 0$; if there is no such number we say that char $R = 0$.*

BEFORE we turn to examples let us give one further definition.

(2.4) Definition. *Let $(R, +, \cdot)$ be a ring. A subset $U \subseteq R$ is called a* **subring** *of R if $(U, +, \cdot)$ is itself a ring. We write $U \leq R$ to express that U is a subring of R and also call $(R : U)$ a* **ring extension**. *If in this situation R has an identity element 1_R such that $1_R \in U$ then we call $(R : U)$ a* **unital ring extension**.

CLEARLY, U is a subring of R if and only if $0 \in U$ and if $x, y \in U$ implies that $-x \in U$, $x + y \in U$ and $xy \in U$. Moreover, if $(R : U)$ is a unital ring extension, then clearly 1_R is an identity element of U. On the other hand, it is possible that a ring extension $(R : U)$ is not unital even though both R and U possess an identity element. (See problem 16 below.)

(2.5) Examples. (a) With the usual addition and multiplication, all the sets in the chain

$$\mathbb{Z} \subseteq \mathbb{Q} \subseteq \mathbb{R} \subseteq \mathbb{C}$$

are commutative rings with identity, each a subring of all of its successors. For any $n \in \mathbb{Z}$, the set $n\mathbb{Z}$ of all multiples of n is a subring of $\mathbb{Z}$. Note that $n\mathbb{Z}$ has no identity unless $n = \pm 1$.

(b) Let $n \in \mathbb{Z} \setminus \{0, 1\}$ be square-free; i.e., $|n|$ is not divisible by the square of a natural number other than 1. Then let $\sqrt{n} := \begin{cases} \sqrt{n}, & \text{if } n > 0; \\ i\sqrt{|n|}, & \text{if } n < 0. \end{cases}$ Then

$$\mathbb{Z}[\sqrt{n}] := \mathbb{Z} + \mathbb{Z}\sqrt{n} = \{x + y\sqrt{n} \mid x, y \in \mathbb{Z}\}$$

is a ring (with addition and multiplication inherited from $\mathbb{C}$), obviously the smallest subring of $\mathbb{C}$ containing $\sqrt{n}$.

(c) Fix a natural number n and consider the set $\mathbb{Z}_n$ of all residue classes modulo n as introduced in (I.20.14) in the group theory chapter. We saw in (I.20.14) that an addition and a multiplication on $\mathbb{Z}_n$ can be defined by

$$[x] + [y] := [x+y] \quad \text{and} \quad [x] \cdot [y] := [xy] .$$

It is easy to see that $(\mathbb{Z}_n, +, \cdot)$ is a commutative ring with identity element $[1]$.

(d) The set $\mathbb{Z}[x]$ of all polynomials with integer coefficients, endowed with the usual addition and multiplication, is a commutative ring with identity. The same holds true for $\mathbb{Z}[x, y]$, the set of all polynomials in two variables x, y with coefficients in $\mathbb{Z}$.

(e) The set $K^{n \times n}$ of all $n \times n$-matrices with entries in the field K is a ring with identity which is not commutative for all $n \geq 2$. The set of all upper triangular matrices is a subring of $K^{n \times n}$.

(f) The set

$$\mathbb{H} := \{ \begin{pmatrix} a & b \\ -\bar{b} & \bar{a} \end{pmatrix} \mid a, b \in \mathbb{C} \} \subseteq \mathbb{C}^{2 \times 2}$$

is a subring of $\mathbb{C}^{2 \times 2}$, called the ring of **quaternions**. (Compare with problem 19 in section [I.21] and with problem 32 in section [I.23].) Writing

$$\mathbf{1} := \begin{pmatrix} 1 & 0 \\ 0 & 1 \end{pmatrix}, \quad I := \begin{pmatrix} 0 & 1 \\ -1 & 0 \end{pmatrix}, \quad J := \begin{pmatrix} 0 & i \\ i & 0 \end{pmatrix}, \quad K := \begin{pmatrix} i & 0 \\ 0 & -i \end{pmatrix}$$

we have

$$\mathbb{H} = \{ a\mathbf{1} + bI + cJ + dK \mid a, b, c, d \in \mathbb{R} \} .$$

(g) Let $(A, +)$ be an arbitrary abelian group. Then A can be made into a commutative ring by defining $x \cdot y := 0$ for all $x, y \in A$; this multiplication is called the **trivial multiplication** on A.

(h) If $(U_i)_{i \in I}$ is a family of subrings of R, then the intersection $\bigcap_{i \in I} U_i$ and the sum

$$\sum_{i \in I} U_i := \{ u_{i_1} + \cdots + u_{i_n} \mid n \in \mathbb{N}, \ i_1, \ldots, i_n \in I, \ u_{i_k} \in U_{i_k} \}$$

are also subrings of R.

(i) An arbitrary ring R has its **center** $C(R) := \{ x \in R \mid xy = yx \text{ for all } y \in R \}$ as a subring.

(j) If R is a subring of S and if $s_1, \ldots, s_n$ are elements of S, then the intersection of all subrings of S containing R and $s_1, \ldots, s_n$ is again a subring of S; it is denoted by $R[s_1, \ldots, s_n]$. † Obviously, $R[s_1, \ldots, s_n]$ is the smallest subring of R containing R and the elements $s_1, \ldots, s_n$; we say that $R[s_1, \ldots, s_n]$ is obtained from R by **adjoining** the elements $s_1, \ldots, s_n$.

IN the sequel we want to show how we can construct many new examples of rings from known rings.

(2.6) Example: Formal power series. Let R be an arbitrary ring. A **formal power series** over R is an expression

$$(\star) \qquad a_0 + a_1 x + a_2 x^2 + a_3 x^3 + \cdots =: \sum_{k=0}^{\infty} a_k x^k$$

---

† This notation was used in part (b) already.

where the coefficients a_k are elements of R and where x is a "symbol", an "indeterminate" or a "variable". This is too vague to be a good definition. Therefore, let us define a power series over R simply as a sequence

$$(\star\star) \qquad\qquad (a_0,\ a_1,\ a_2,\ a_3,\ \cdots)$$

of elements in R. Then $(\star)$ is just a different notation for $(\star\star)$, and the powers of the mysterious object x are merely used to label the positions in the sequence. The advantage of notation $(\star)$ becomes clear when we want to make the set of all power series over R into a ring. In notation $(\star\star)$, we define addition and multiplication by

$$(a_0,\ a_1,\ a_2,\ a_3,\ \cdots) + (b_0,\ b_1,\ b_2,\ b_3,\ \cdots) := (a_0+b_0,\ a_1+b_1,\ a_2+b_2,\ a_3+a_3,\ \cdots)$$

and

$$(a_0,\ a_1,\ a_2,\ a_3,\ \cdots)(b_0,\ b_1,\ b_2,\ b_3,\ \cdots)$$
$$:= (a_0 b_0,\ a_0 b_1 + a_1 b_0,\ a_0 b_2 + a_1 b_1 + a_2 b_0,\ a_0 b_3 + a_1 b_2 + a_2 b_1 + a_3 b_0,\ \cdots)$$
$$= \left(\sum_{i+j=0} a_i b_j,\ \sum_{i+j=1} a_i b_j,\ \sum_{i+j=2} a_i b_j,\ \sum_{i+j=3} a_i b_j,\ \ldots \right).$$

The definition for the multiplication might seem unintelligible at first, but becomes clear if one multiplies two "expressions" $a_0 + a_1 x + a_2 x^2 + \cdots$ and $b_0 + b_1 x + b_2 x^2 + \cdots$ with each other, using the normal arithmetic rules and not worrying too much about what the "symbol" x actually is. The ring of all formal power series over R is denoted by $R[[x]]$.

Observe that for any two ring elements $r, s \in R$ we have

$$(r,0,0,\cdots) + (s,0,0,\cdots) = (r+s,0,0,\cdots) \qquad \text{and}$$
$$(r,0,0,\cdots) \cdot (s,0,0,\cdots) = (rs,0,0,\cdots).$$

This shows that if we identify $r \in R$ with $(r,0,0,\cdots) \in R[x]$, we can consider R as a subring of $R[x]$. In notation $(\star)$ this just means that we view ring elements as "constant power series".

Clearly, $R[[x]]$ is commutative if and only if R is.

(2.7) Example: Polynomials. If R is any ring, then a (formal) **polynomial** over R is a formal power series with only a finite number of nonzero coefficients, i.e., an expression

$$a_0 + a_1 x + a_2 x^2 + \cdots + a_n x^n.$$

The set of all polynomials over R is denoted by $R[x]$ and is clearly a subring of $R[[x]]$. Again, we can identify R with the subring of all "constant" polynomials in $R[x]$.

Now suppose that R has an identity element 1. Then if we write

$$x := (0,\ 1,\ 0,\ 0,\ 0,\ \ldots) \in R[x]$$

it is easy to verify that

$$x^2 = (0,\ 0,\ 1,\ 0,\ 0,\ \ldots),\quad x^3 = (0,\ 0,\ 0,\ 1,\ 0,\ \ldots),\qquad \text{and so on.}$$

Formally defining $x^0 := (1, 0, 0, 0, 0, \ldots)$ we see that

$$(a_0, \; a_1, \; a_2, \; \ldots, \; a_n, \; 0, \; 0, \; \ldots)$$
$$= a_0 x^0 + a_1 x^1 + a_2 x^2 + \cdots + a_n x^n = a_0 + a_1 x + a_2 x^2 + \cdots + a_n x^n$$

under the canonical identification of ring elements with constant polynomials. Hence the notation $a_0 + a_1 x + a_2 x^2 + \cdots + a_n x^n$ for a polynomial is completely justified; the "symbol" x is no longer a mysterious object, but a well-defined element of $R[x]$. Similarly, the power series $\sum_{k=0}^{\infty} a_k x^k$ is the unique power series which coincides in the first n coefficients with the well-defined polynomial $a_0 + a_1 x + \cdots + a_n x^n$, for all $n \in \mathbb{N}_0$. Note however that we do not develop a notion of convergence that would allow us to talk about infinite sums.

IN a similar fashion we can define power series and polynomials in more than one variable.

(2.8) Example: Power series and polynomials in more than one variable.
Let X be an arbitrary non-empty set. For the sake of convenience we choose an indexing $X = \{x_i \mid i \in I\}$ with $0 \notin I$. Then a formal power series over R in the (commuting) variables x_i $(i \in I)$ is an expression

$$a_0 + \sum_{i \in I} a_i x_i + \sum_{i,j \in I} a_{ij} x_i x_j + \sum_{i,j,k \in I} a_{ijk} x_i x_j x_k + \cdots$$

where the coefficients are elements of R and where in each of the sums occurring in this expression only finitely many of the coefficients are different from zero. More formally, we define the power series ring $R[[X]] = R[[(x_i)_{i \in I}]]$ as the set of all sequences

$$\left(a_0, \; (a_i)_{i \in I}, \; (a_{ij})_{i,j \in I}, \; (a_{ijk})_{i,j,k \in I}, \; \ldots \right)$$

where $a_0, a_i, a_{ij}, \ldots \in R$ and where each array $(a_{i_1 i_2 \ldots i_n})_{i_1, i_2, \ldots, i_n \in I}$ occurring in this sequence has only a finite number of nonzero members. The addition in $R[[X]]$ is defined by

$$\left(a_0, \; (a_i)_i, \; (a_{ij})_{i,j}, \; (a_{ijk})_{i,j,k}, \; \ldots \right) + \left(b_0, \; (b_i)_i, \; (b_{ij})_{i,j}, \; (b_{ijk})_{i,j,k}, \; \ldots \right) =$$

$$\left(a_0 + b_0, \; (a_i + b_i)_i, \; (a_{ij} + b_{ij})_{i,j}, \; (a_{ijk} + b_{ijk})_{i,j,k}, \; \ldots \right),$$

the multiplication by

$$\left(a_0, \; (a_i)_i, \; (a_{ij})_{i,j}, \; (a_{ijk})_{i,j,k}, \; \ldots \right) \cdot \left(b_0, \; (b_i)_i, \; (b_{ij})_{i,j}, \; (b_{ijk})_{i,j,k}, \; \ldots \right) =$$

$$\left(a_0 b_0, \; \sum_i (a_0 b_i + a_i b_0), \; \sum_{i,j} (a_0 b_{ij} + a_i b_j + a_{ij} b_0), \; \sum_{i,j,k} (a_0 b_{ijk} + a_i b_{jk} + a_{ij} b_k + a_{ijk} b_0), \; \ldots \right).$$

As above, a polynomial is a power series with only a finite number of nonzero coefficients. The set of all polynomials over R in the variables $(x_i)_{i \in I}$ is denoted by $R[(x_i)_{i \in I}]$ and forms a subring of $R[[(x_i)_{i \in I}]]$. Again, R can be identified as the subring of all constant polynomials.

IN a concrete setting, the above formulas look much less complicated than in their general form. For example, in $\mathbb{Z}[x, y, z]$, the ring of polynomials in 3 variables x, y and z, we have

$$(x^2 y^5 - y)(x^3 - x^2 + xyz) = x^5 y^5 - x^3 y - x^4 y^6 - x^2 y^2 + x^3 y^6 z - xy^2 z \ ,$$

and so on.

THE next example is the most important example of a non-commutative ring.

(2.9) **Example: Matrix rings.** Let R be a ring. Then the set $R^{n \times n}$ of all $n \times n$-matrices with coefficients in R is again a ring if we use the usual operations

$$\begin{pmatrix} a_{11} & \cdots & a_{1n} \\ \vdots & & \vdots \\ a_{n1} & \cdots & a_{nn} \end{pmatrix} + \begin{pmatrix} b_{11} & \cdots & b_{1n} \\ \vdots & & \vdots \\ b_{n1} & \cdots & b_{nn} \end{pmatrix} := \begin{pmatrix} a_{11} + b_{11} & \cdots & a_{1n} + b_{1n} \\ \vdots & & \vdots \\ a_{n1} + b_{n1} & \cdots & a_{nn} + b_{nn} \end{pmatrix}$$

and

$$\begin{pmatrix} a_{11} & \cdots & a_{1n} \\ \vdots & & \vdots \\ a_{n1} & \cdots & a_{nn} \end{pmatrix} \begin{pmatrix} b_{11} & \cdots & b_{1n} \\ \vdots & & \vdots \\ b_{n1} & \cdots & b_{nn} \end{pmatrix} := \sum_{k=1}^{n} \begin{pmatrix} a_{1k} b_{k1} & \cdots & a_{1k} b_{kn} \\ \vdots & & \vdots \\ a_{nk} b_{k1} & \cdots & a_{nk} b_{kn} \end{pmatrix} .$$

If R has an identity element 1, then

$$1 := \begin{pmatrix} 1 & & 0 \\ & \ddots & \\ 0 & & 1 \end{pmatrix}$$

is an identity element of R. If $n \geq 2$, then the ring $R^{n \times n}$ is commutative if and only if the multiplication on R is trivial, i.e., if $xy = 0$ for all $x, y \in R$. Indeed, if there are two elements $x, y \in R$ with $xy \neq 0$, then

$$\begin{pmatrix} x & 0 \\ 0 & 0 \end{pmatrix} \begin{pmatrix} 0 & y \\ 0 & 0 \end{pmatrix} = \begin{pmatrix} 0 & xy \\ 0 & 0 \end{pmatrix} \neq \begin{pmatrix} 0 & 0 \\ 0 & 0 \end{pmatrix} = \begin{pmatrix} 0 & y \\ 0 & 0 \end{pmatrix} \begin{pmatrix} x & 0 \\ 0 & 0 \end{pmatrix} .$$

DEPENDING on the context, there are different ways in which rings of functions can arise. Let us look at three examples.

(2.10) **Example: Rings of functions with pointwise multiplication.** Let R be a ring and $X \neq \emptyset$ be an arbitrary set. Then the set

$$R^X := \{f : X \to R\}$$

of all functions from X to R is a ring with the argumentwise operations

$$(f + g)(x) := f(x) + g(x) \qquad \text{and} \qquad (fg)(x) := f(x)g(x) \ .$$

Clearly, R^X is commutative if and only if R is.

If X is not just a set, but has an additional structure, then we can single out subrings of R^X consisting of functions with special properties. For example, if X is a topological space, then the set $C(X)$ of all real-valued continuous functions on X is a subring of $X^{\mathbb{R}}$.

(2.11) Example: Rings of functions with composition multiplication. If $(A, +)$ is an abelian group, then

$$A^A \;=\; \{f : A \to A\}$$

is a ring with the operations

$$(f + g)(a) \;:=\; f(a) + g(a) \qquad \text{and} \qquad (f \circ g)(a) \;:=\; f\big(g(a)\big) \;.$$

The subset

$$\operatorname{End} A \;:=\; \{f : A \to A \mid f(a + b) = f(a) + f(b) \text{ for all } a, b \in A\}$$

is a subring of $(A^A, +, \circ)$, called the **endomorphism ring** of A.

NOTE that if R is a ring, then we can make R^R into a ring in two different ways, with the pointwise multiplication as in (2.10) or with the composition multiplication as in (2.11). In this case it is important to write down a ring as a triplet $\big($either $(R^R, +, \cdot)$ or $(R^R, +, \circ)\big)$ to make clear what operations are used.

ON some spaces of functions a third type of multiplication can be defined, which is called convolution.

(2.12) Example: Convolution rings. We do not want to give the most general definition of a convolution ring, but instead consider some examples.

The set $C(\mathbb{R}^n)$ of all continuous functions on $\mathbb{R}^n$ or the set $L^2(\mathbb{R}^n)$ of all square-integrable functions on $\mathbb{R}^n$ can be made into a ring with the operations

$$(f + g)(x) \;:=\; f(x) + g(x) \qquad \text{and} \qquad (f \star g)(x) \;:=\; \int_{\mathbb{R}^n} f(x - y) g(y) \, \mathrm{d}\, y \;.$$

For $C[0, \infty)$ we can define a modified version of these operations, namely

$$(f + g)(x) \;:=\; f(x) + g(x) \qquad \text{and} \qquad (f \star g)(x) \;:=\; \int_0^x f(x - t) g(t) \, \mathrm{d}\, t \;.$$

In both cases it is easy to check that the convolution product $\star$ is commutative.

Note that the multiplication of power series is given by a similar pattern. If we identify a power series $(a_0, a_1, a_2, \ldots)$ over R with the function $a : \mathbb{N}_0 \to R$ given by $a(n) = a_n$, then multiplication of power series corresponds to convolution of functions:

$$(a \star b)(n) \;=\; \sum_{k=0}^{n} a(n - k) b(k) \;.$$

LET us now see how new rings can be constructed from a given family of rings.

(2.13) Example: Direct product and direct sum. The **direct product** of a family $(R_i)_{i \in I}$ of rings is defined as

$$\prod_{i \in I} R_i := = \left\{ r : I \to \bigcup_{i \in I} R_i \mid r(i) \in R_i \right\} .$$

This is a ring if we define the operations argumentwise, i.e., by

$$(r + s)(i) := r(i) + s(i) \quad \text{and} \quad (rs)(i) := r(i)s(i) .$$

This means that $\prod_{i \in I} R_i$ is a subring of $(\bigcup_{i \in I} R_i)^I$.

Sometimes it is convenient to write the elements of $\prod_{i \in I} R_i$ as tuples $(r_i)_{i \in I}$; addition and multiplication are then understood componentwise:

$$(r_i)_{i \in I} + (s_i)_{i \in I} := (r_i + s_i)_{i \in I} \quad \text{and} \quad (r_i)_{i \in I}(s_i)_{i \in I} := (r_i s_i)_{i \in I}$$

It is easy to see that $\prod_{i \in I} R_i$ is commutative if and only if all the rings R_i are.

The **direct sum** of a family $(R_i)_{i \in I}$ is the subring of $\prod_{i \in I} R_i$ which consists of all elements r such that $r_i = 0$ for almost all indices i, i.e., $r_i \neq 0$ for only a finite number of indices i. So

$$\bigoplus_{i \in I} R_i := \{(r_i)_{i \in I} \mid r_i = 0 \text{ for almost all } i \in I\} .$$

For a finite family of rings, we also introduce the notation

$$\prod_{i=1}^{n} R_i := R_1 \times \cdots \times R_n \quad \text{and} \quad \bigoplus_{i=1}^{n} R_i := R_1 \oplus \cdots \oplus R_n ;$$

clearly, the direct sum and the direct product coincide in this case.

WE have introduced rings as sets with a certain algebraic structure, and we have seen many examples of rings. Now – as we saw already in our investigation of groups – the study of sets with a certain algebraic structure includes a study of those functions between these sets that preserve the algebraic structure. In the case of rings, this leads to the notion of a ring homomorphism. Since the situation is completely analogous to the group case, we can proceed faster than we did there.

(2.14) Definitions. (a) *A mapping $f : R \to S$ between two rings is called a* **ring homomorphism** *if*

$$f(0) = 0 , \quad f(a + b) = f(a) + f(b) \quad \text{and} \quad f(ab) = f(a)f(b) \text{ for all } a, b \in R .$$

The **kernel** $\ker f$ *and the* **image** $\operatorname{im} f = f(R)$ *are defined by*

$$\ker f := \{a \in R \mid f(a) = 0\} \quad \text{and} \quad \operatorname{im} f := \{f(a) \mid a \in A\} .$$

(b) *A bijective ring homomorphism f is called a* **ring isomorphism**; *in this case f^{-1} is also a ring isomorphism. We say that two rings R and S are* **isomorphic** *and write $R \cong S$ if there is a ring isomorphism $f : R \to S$.*

(c) *An injective ring homomorphism $f : R \to S$ is called an* **embedding**.

AN isomorphism $f : R \to S$ can be thought of as a simple renaming of the elements in R; we call an element $f(r)$ instead of r but nothing else changes. This means that R and S are in some sense simply two realizations of the same ring. Similarly, if an embedding $f : R \to S$ is given we can identify R with its image $f(R)$ and thus treat R as a subring of S. We did this already when we considered R as a subring of $R[x]$ or $R[[x]]$ by identifying a ring element $r \in R$ with the constant polynomial r.

LET us summarize the basic properties of ring homomorphisms.

(2.15) Lemma. *Let $f : R \to S$ be a ring homomorphism.*

(a) *If U is a subring of R, then $f(U)$ is a subring of S. If V is a subring of S, then $f^{-1}(V)$ is a subring of R.*

(b) *If R is commutative, then so is $f(R)$. If R has an identity element 1 and f is not the zero mapping, then $f(R)$ has an identity element, namely $f(1)$.*

(c) *$\ker f$ is a subring of R, and $\operatorname{im} f$ is a subring of S.*

Proof. (a) Let U be a subring of R. Then $0_S = f(0_R) \in f(U)$. Let $u_1, u_2 \in U$. Then $f(u_1) + f(u_2) = f(u_1 + u_2) \in f(U)$, $-f(u_1) = f(-u_1) \in f(U)$ and $f(u_1)f(u_2) = f(u_1 u_2) \in f(U)$. So $f(U) + f(U) \subseteq f(U), -f(U) \subseteq f(U)$ and $f(U)f(U) \subseteq f(U)$. This shows that $f(U)$ is a subring of S.

Now let V be a subring of S. Then $0_R \in f^{-1}(V)$ because $f(0_R) = 0_S \in V$. Let $u_1, u_2 \in f^{-1}(V)$, i.e., $f(u_1), f(u_2) \in V$. Then $f(u_1 + u_2) = f(u_1) + f(u_2) \subseteq V + V \subseteq V$ so that $u_1 + u_2 \in f^{-1}(V)$. Similarly, $-u_1$ and $u_1 u_2$ belong to $f^{-1}(V)$. This shows that $f^{-1}(V)$ is a subring of R.

(b) If $ab = ba$, then $f(a)f(b) = f(ab) = f(ba) = f(b)f(a)$. Let R have an identity element 1 and let $f \not\equiv 0$. Then there is an element $r_0 \in R$ with $0 \neq f(r_0) = f(r_0 \cdot 1) = f(r_0)f(1)$ which implies $f(1) \neq 0$. Moreover, $f(1)f(r) = f(r)f(1) = f(r)$ for all $r \in R$.

(c) $\ker f = f^{-1}(\{0\})$ is a subring of R and $\operatorname{im} f = f(R)$ is a subring of S due to part (a). ∎

(2.16) Examples. (a) If R is any ring the mapping $x \mapsto -x$, i.e., multiplication by -1, is an automorphism.

(b) The complex conjugation $z \mapsto \overline{z}$ defines a ring isomorphism $f : \mathbb{C} \to \mathbb{C}$.

(c) If n is a square-free number then the mapping $\mathbb{Z} + \mathbb{Z}\sqrt{n} \to \mathbb{Z} + \mathbb{Z}\sqrt{n}$ given by $a + b\sqrt{n} \mapsto a - b\sqrt{n}$ is an automorphism.

(d) If $d \mid m$ then we can define a mapping $f : \mathbb{Z}_m \to \mathbb{Z}_d$ by $f([x]_m) = [x]_d$. This mapping is a surjective ring homomorphism.

(e) The mapping $\varphi : \mathbb{Z}_{mn} \to \mathbb{Z}_m \times \mathbb{Z}_n$ given by $[a]_{mn} \mapsto ([a]_m, [a]_n)$ is a well-defined homomorphism. It is easy to see that this mapping is injective (and hence an isomorphism) if and only if m and n are relatively prime. Consequently, if $n = p_1^{r_1} \cdots p_k^{r_k}$ then $\mathbb{Z}_n \cong \mathbb{Z}_{p_1^{r_1}} \times \cdots \times \mathbb{Z}_{p_k^{r_k}}$.

(f) If R is any ring and $r \in R$ an element of R, then the mapping $f : \mathbb{Z} \to R$ given by $f(m) = m \cdot r$ is a ring homomorphism.

(g) For a ring R and a set X, let R^X be the ring of all functions $f : X \to R$ with the pointwise operations. Then for any $x_0 \in X$ the **evaluation map**

$$
\begin{array}{ccc}
R^X & \to & R \\
f & \mapsto & f(x_0)
\end{array}
$$

is a homomorphism. This mapping is obtained by simply "plugging in" the argument x_0 into any given function f.

(h) Let R be a ring and $\varphi \in R[x]$ be a fixed polynomial. Then an endomorphism $\Phi : R[x] \to R[x]$ is defined by $f(x) \mapsto f\big(\varphi(x)\big)$, i.e., by

$$
\Phi\big(a_0 + a_1 x + \cdots + a_n x^n\big) \; := \; a_0 + a_1 \varphi(x) + \cdots + a_n \varphi(x)^n \; .
$$

Thus "plugging in" $\varphi(x)$ for x respects addition and multiplication.

(i) Given a set X and a ring R we can consider the ring R^X of all mappings $f : X \to R$ with the pointwise operations. Then for any subset $Y \subseteq X$ the **restriction map**

$$
\begin{array}{ccc}
R^X & \to & R^Y \\
f & \mapsto & f|_Y
\end{array}
$$

is a ring homomorphism.

(j) Let R be a commutative ring. If $\operatorname{char} R$ is a prime number p then the mapping $f : R \to R$ given by $f(x) = x^p$ is a homomorphism. This stems from the facts that $(xy)^p = x^p y^p$ (due to the commutativity of R) and $(x + y)^p = x^p + y^p + \sum_{k=1}^{p-1} \binom{p}{k} x^{p-k} y^k = x^p + y^p$ (due to the property that $\binom{p}{k}$ is divisible by $p = \operatorname{char} R$ for $1 \le k \le p-1$). One calls $f : R \to R$ the **Frobenius homomorphism** of the ring R.

Exercises

Problem 1. (a) Let x, y be elements of a ring R such that $xy = yx$. Show that

$$(x + y)^n = x^n + \sum_{k=1}^{n-1} \binom{n}{k} x^{n-k} y^k + y^n$$

for all $n \in \mathbb{N}$. (If R has an identity element 1, we formally define $x^0 := 1$ for all $x \in R$; then the above equation becomes $(x + y)^n = \sum_{k=0}^{n} \binom{n}{k} x^{n-k} y^k$.)

(b) Show that if R is a commutative ring of prime characteristic char $R = p$, then $(x + y)^{p^n} = x^{p^n} + y^{p^n}$ for all $x, y \in R$ and all $n \in \mathbb{N}$.

(c) Let R be a commutative ring with identity. Prove the polynomial formula

$$(x_1 + x_2 + \cdots + x_n)^m = \sum_{\substack{m_1, \ldots, m_n \geq 0, \\ m_1 + \cdots + m_n = m}} \frac{m!}{m_1! m_2! \cdots m_n!} x_1^{m_1} x_2^{m_2} \cdots x_n^{m_n}$$

where $x_1, \ldots, x_n \in R$ and $x^0 := 1$. **Hint.** Use induction on m.

Problem 2. Suppose the set R is equipped with two binary operations $+$ and $\cdot$ such that $(R, +)$ is a group, $(R, \cdot)$ is a semigroup with identity element 1, and $x(y + z) = xy + xz$ and $(x + y)z = xz + yz$ for all $x, y, z \in R$. Conclude that $(R, +, \cdot)$ is a ring, i.e., show that the operation $+$ is necessarily commutative.

Hint. Calculate $(a + b)(1 + 1)$ in two ways, using both distributive laws.

Problem 3. Show that the following sets are subrings of $\mathbb{C}$.

(a) $R := \{ \frac{m}{n} \mid m, n \in \mathbb{Z}, \ n$ is not divisible by $p \}$ where $p \in \mathbb{N}$ is a prime number.

(b) $S := \mathbb{Z} + \mathbb{Z}\varepsilon + \mathbb{Z}\varepsilon^2 + \cdots + \mathbb{Z}\varepsilon^{n-1}$ where $\varepsilon := e^{2\pi i/n}$.

Problem 4. (a) Let R be a ring and let S be an arbitrary set. Suppose that there is a bijection $f : R \to S$. Show that then S can be made into a ring (which is isomorphic to R) by defining

$$s_1 + s_2 := f(f^{-1}(s_1) + f^{-1}(s_2)) \quad \text{and} \quad s_1 \cdot s_2 := f(f^{-1}(s_1) \cdot f^{-1}(s_2)) .$$

(b) Given a ring $(R, +, \cdot)$ with identity, define new operations $\oplus$ and $\odot$ on R by $a \oplus b := a + b - 1$ and $a \odot b := a + b - ab$. Show that the zero element of $(R, +, \cdot)$ becomes the identity element of $(R, \oplus, \odot)$ and vice versa.

Problem 5. (a) Let R be a ring whose additive group $(R, +)$ is cyclic. Show that R is commutative.

(b) Let R be a finite ring with p elements where p is a prime number. Show that $R \cong \mathbb{Z}_p$ or $xy = 0$ for all $x, y \in R$.

Problem 6. Let $S = \{a, b, c, d\}$ be a set with 4 elements. Show that S can be made into a non-commutative ring by defining addition and multiplication as follows.

+	a	b	c	d
a	a	b	c	d
b	b	a	d	c
c	c	d	a	b
d	d	c	b	a

·	a	b	c	d
a	a	a	a	a
b	a	a	b	b
c	a	a	c	c
d	a	a	d	d

Hint. Consider the matrices $A = \begin{pmatrix} 0 & 0 \\ 0 & 0 \end{pmatrix}$, $B = \begin{pmatrix} 1 & 1 \\ 1 & 1 \end{pmatrix}$, $C = \begin{pmatrix} 1 & 1 \\ 0 & 0 \end{pmatrix}$ and $D = \begin{pmatrix} 0 & 0 \\ 1 & 1 \end{pmatrix}$ with coefficients in $\mathbb{Z}_2$.

Problem 7. Let R be a commutative ring with identity element 1 and let $R^{\mathbb{N}}$ be the set of all mappings $f : \mathbb{N} \to R$. Define two operations on $R^{\mathbb{N}}$ by

$$(f + g)(n) := f(n) + g(n), \quad (f \star g)(n) := \sum_{xy=n} f(x)g(y)$$

where the sum in the definition of the convolution product $\star$ runs over all pairs $(x, y) \in \mathbb{N} \times \mathbb{N}$ with $xy = n$.

(a) Show that $(R^{\mathbb{N}}, +, \star)$ is a commutative ring which possesses an identity element δ, namely

$$\delta(x) := \begin{cases} 1, & \text{if } n = 1; \\ 0, & \text{if } n \neq 1. \end{cases}$$

(b) A function $f \in R^{\mathbb{N}}$ is called **multiplicative** if $f(mn) = f(m)f(n)$ whenever $m, n \in \mathbb{N}$ are relatively prime. Show that the set of all multiplicative functions is a subring of $R^{\mathbb{N}}$.

(c) Define the **Möbius function** $\mu : \mathbb{N} \to R$ by

$$\mu(n) := \begin{cases} 1, & \text{if } n = 1; \\ (-1)^r, & \text{if } n = p_1 \cdots p_r \text{ with } r \text{ distinct prime factors}; \\ 0, & \text{if } n \text{ has a multiple prime factor.} \end{cases}$$

Show that μ is multiplicative and verify $\mu \star 1 = \delta$ where 1 denotes the constant function with the value 1.

(d) Suppose f and F are elements of $R^{\mathbb{N}}$. Show that $F = \mu \star f$ if and only if $f = \mu \star F$.

(e) Given $f \in R^{\mathbb{N}}$, let $F(n) := \sum_{d|n} f(d)$. Show that f can be reconstructed from F via

$$f(n) = \sum_{d|n} \mu(d) F(\tfrac{n}{d}) \quad \textbf{(Möbius inversion formula)}.$$

Problem 8. Define functions $\sigma, \tau : \mathbb{N} \to \mathbb{Z}$ by

$\tau(n) :=$ number of divisors of n ;

$\sigma(n) :=$ sum of the divisors of n ;

$\varphi(n) :=$ number of natural numbers less than n which are coprime with n .

(Note that φ is Euler's function introduced in (I.21.12) already.)

(a) Show that τ, σ and φ are multiplicative functions which satisfy the equations $\tau = 1 \star 1$, $\sigma = 1 \star \mathrm{id}_\mathbb{N}$ and $\varphi = \mu \star \mathrm{id}_\mathbb{N}$.

(b) Show that if $n = p_1^{r_1} p_2^{r_2} \cdots p_m^{r_m}$ is the prime factorization of a natural number n then

$$\tau(n) = (r_1 + 1)(r_2 + 1)\cdots(r_m + 1),$$

$$\sigma(n) = \frac{p_1^{r_1+1} - 1}{p_1 - 1} \frac{p_2^{r_2+1} - 1}{p_2 - 1} \cdots \frac{p_m^{r_m+1} - 1}{p_m - 1},$$

$$\varphi(n) = (p_1^{r_1} - p_1^{r_1-1})(p_2^{r_2} - p_2^{r_2-1})\cdots(p_m^{r_m} - p_m^{r_m-1}).$$

(c) A natural number n is called **perfect** if it equals the sum of its proper divisors, i.e., if $\sigma(n) = 2n$. Show that if p is a prime number of the form $p = 2^{s+1} - 1$ then the number $n := 2^s p$ is perfect. Find some examples of perfect numbers!

(d) Show that φ takes each of its values only a finite number of times and that $\varphi(n)$ is even for all $n > 2$. Moreover, show that if $d \mid n$ then $\varphi(d) \mid \varphi(n)$.

Problem 9. (a) Let R be a ring with identity such that $x^3 = x$ for all $x \in R$. Show that $6 \cdot x = 0$ and $xy = yx$ for all $x, y \in R$.

(b) Let R be a ring with identity such that $x^4 = x$ for all $x \in R$. Show that R is commutative.

Problem 10. Let R be a ring.

(a) Show that a ring structure on $R \times R$ is given by

$$(a,b) + (c,d) := (a+c, b+d) \quad \text{and} \quad (a,b) \cdot (c,d) := (ac, ad + bc).$$

(b) Let $R \times R$ be the ring with addition and multiplication as in (a). Show that

$$f : \begin{array}{ccc} R \times R & \to & R^{2\times 2} \\ (a,b) & \mapsto & \begin{pmatrix} a & b \\ 0 & a \end{pmatrix} \end{array}$$

is a ring homomorphism.

(c) Can you use the mapping f given in (b) to prove (a) without the need to check all the ring axioms?

Problem 11. Let R be a commutative ring with identity. Its "complexification" is defined as $R \times R$ with the operations

$$(a,b) + (c,d) := (a+c, c+d) \quad \text{and} \quad (a,b)(c,d) := (ac - bd, ad + bc).$$

Considering R is a subring of $R \times R$ via $a \mapsto (a,0)$ and writing $i := (0,1)$, we can express each element of $R \times R$ in the form $a + ib$ with $a, b \in R$ where $i^2 = -1 \in R$. This explains the name "complexification" and the notation $R + iR$ for this ring.

Show that

$$f : \begin{array}{ccc} R + iR & \to & R^{2\times 2} \\ a + ib & \mapsto & \begin{pmatrix} a & b \\ -b & a \end{pmatrix} \end{array}$$

is a ring homomorphism.

Problem 12. Consider the ring $R = K^{n \times n}$ where K is a field.
(a) Find the center of R.
(b) Let U be the subring of R generated by all symmetric matrices. Show that $U = R$.

Problem 13. Let R be a ring and $X \subseteq R$. The **centralizer** of X in R is

$$C(X) := \{a \in R \mid ax = xa \text{ for all } x \in X\} .$$

(a) Show that $C(X)$ is a subring of R.
(b) Show that $X_1 \subseteq X_2$ implies $C(X_1) \supseteq C(X_2)$.
(c) Show that $X \subseteq C(C(X))$ and $C(X) = C\Big(C(C(X))\Big)$.
(d) Let $R = \mathbb{R}^{2 \times 2}$ and $X := \{\begin{pmatrix} x & y \\ 0 & 0 \end{pmatrix} \mid x, y \in \mathbb{R}\}$. Find $C(X)$.
(e) Find $C(X)$ where $X := \{\text{diag}(1, 2)\} \subseteq \mathbb{R}^{2 \times 2}$.

Problem 14. Let R be a ring and $X \subseteq R$. The **(left) annihilator** of X in R is

$$A(X) := \{a \in R \mid ax = 0 \text{ for all } x \in X\} .$$

(a) Show that $A(X)$ is a subring of R.
(b) Show that $X_1 \subseteq X_2$ implies $A(X_1) \supseteq A(X_2)$.
(c) Show that if R is commutative then $X \subseteq A(A(X))$ and $A(X) = A\Big(A(A(X))\Big)$.
(d) Let $R = \mathbb{R}^{2 \times 2}$ and $X := \{\begin{pmatrix} x & y \\ 0 & 0 \end{pmatrix} \mid x, y \in \mathbb{R}\}$. Find $A(X)$.
(e) Find $A(X)$ where $X := \{\text{diag}(1, 2)\} \subseteq \mathbb{R}^{2 \times 2}$.

Problem 15. (a) Let $f : R \to S$ be a ring homomorphism. Show that if R has an identity element 1, then $f(1)$ is an identity element for $f(R)$, but it may happen that S does not possess an identity element.
(b) Let $R = \{\begin{pmatrix} a & 0 \\ 0 & 0 \end{pmatrix} \mid a \in \mathbb{R}\}$ and $S := \mathbb{R}^{2 \times 2}$. Show that both R and S have an identity element, but that the inclusion mapping $f : R \to S$ does *not* satisfy $f(1_R) = 1_S$.

Problem 16. (a) Clearly, $\mathbb{Z}$ is a ring with identity. Find a subring of $\mathbb{Z}$ which has no identity element.

(b) Clearly, $\mathbb{Z}_6$ is a ring with identity. Show that the subring $U := \{[0], [2], [4]\}$ of R also has an identity element, but that the identity element of U is not the same as the identity element of $\mathbb{Z}_6$.

Remark. This phenomenon is further investigated in the next problem. Also compare with problem 15(b).

(c) Define a ring R and a subring $U \leq R$ as follows:

$$R := \{\begin{pmatrix} a & 0 & 0 \\ 0 & 0 & b \\ 0 & 0 & c \end{pmatrix} \mid a, b, c \in \mathbb{R}\}, \quad U := \{\begin{pmatrix} a & 0 & 0 \\ 0 & 0 & 0 \\ 0 & 0 & 0 \end{pmatrix} \mid a \in \mathbb{R}\}.$$

Show that R does not possess an identity element, but U does.

Problem 17. Let U be a subring of $R := \mathbb{Z}_n$. Show that the following conditions are equivalent:

(a) U has an identity element;

(b) there is an element $u \in R$ with $u^2 = u$ and $U = Ru \, (= \mathbb{Z} \cdot u)$.

Problem 18. Let $(R_i)_{i \in I}$ be a family of rings. Under what circumstances do the direct product $\prod_{i \in I} R_i$ and the direct sum $\bigoplus_{i \in I} R_i$ possess an identity element?

Problem 19. (a) Let $R = \{\begin{pmatrix} a & b \\ 0 & 0 \end{pmatrix} \mid a, b \in \mathbb{R}\}$, considered as a subring of $\mathbb{R}^{2 \times 2}$.

Show that R has no identity element but that each element of the form $e = \begin{pmatrix} 1 & \star \\ 0 & 0 \end{pmatrix}$ is a **left-identity**, i.e., satisfies $ex = x$ for all $x \in R$.

(b) Let R be a ring with a *unique* element $e \neq 0$ such that $ex = x$ for all $x \in R$. Show that e is an identity element.

Hint. Calculate $(xe - x + e)y = y$ for $x, y \in R$.

Problem 20. Let R be a ring.

(a) Define operations on $\mathbb{Z} \times R$ as follows:

$$(m, x) + (n, y) := (m + n, x + y), \quad (m, x)(n, y) := (mn, n \cdot x + m \cdot y + xy).$$

Show that $\mathbb{Z} \times R$ is a ring of characteristic 0 with identity element $(1, 0)$ which is commutative if and only if R is. Show that the mapping $x \mapsto (0, x)$ defines an embedding of R into $\mathbb{Z} \times R$. (This shows that every ring can be embedded into a unitary ring.)

(b) Show that if R has the property that $k \cdot x = 0$ for all $x \in x$ then we can do part (a) with $\mathbb{Z}_k \times R$ instead of $\mathbb{Z} \times R$ so that R can be embedded in a unitary ring of characteristic k.

Problem 21. Show that every ring $(R, +, \cdot)$ can be embedded into the endomorphism ring of an abelian group.

Hint. Show that if R has an identity element then

$$R \rightarrow \text{End}(R, +)$$
$$a \mapsto l_a$$

is an embedding where $l_a : R \rightarrow R$ denotes the left-translation $x \mapsto ax$. For the general case, use problem 20.

Problem 22. Let $f : R \rightarrow S$ be a ring homomorphism.
(a) Show that a ring homomorphism $F : R[x] \rightarrow S[x]$ is defined by

$$F(a_0 + a_1 x + \cdots + a_n x^n) := f(a_0) + f(a_1)x + \cdots + f(a_n)x^n .$$

How are $\ker F$ and $\text{im } F$ related to $\ker f$ and $\text{im } f$?
(b) Show that a ring homomorphism $F : R^{n \times n} \rightarrow S^{n \times n}$ is defined by

$$F\big((a_{ij})_{i,j}\big) := \big(f(a_{ij})\big)_{i,j} .$$

How are $\ker F$ and $\text{im } F$ related to $\ker f$ and $\text{im } f$?

Problem 23. (a) Let R be a commutative ring. Show that the translation $l_a : R \rightarrow R$ given by $x \mapsto ax$ is a ring homomorphism if and only if $a^2 = a$.
(b) Find all endomorphisms of $\mathbb{Z}_q$ where q is a prime power.
(c) Find all endomorphisms of the rings $\mathbb{Z}, \mathbb{Z}^n, \mathbb{Q}, \mathbb{Q}[\sqrt{7}]$ and $\mathbb{Q}[\sqrt[3]{2}]$.

Problem 24. Show that the rings $\mathbb{Z}[\sqrt{2}]$ and $\mathbb{Z}[\sqrt{3}]$ are not isomorphic.

Problem 25. (a) Show that the only ring homomorphisms $f : \mathbb{Q} \rightarrow \mathbb{Q}$ are the zero mapping and the identity.
(b) Show that the only ring homomorphisms $f : \mathbb{R} \rightarrow \mathbb{R}$ are the zero mapping and the identity.
(c) Find a ring homomorphism $f : \mathbb{C} \rightarrow \mathbb{C}$ which is neither zero nor the identity.

Problem 26. (a) Show that there is a surjective ring homomorphism $f : \mathbb{Z}_m \rightarrow \mathbb{Z}_n$ if and only if $n \mid m$.
(b) Show that there is an injective ring homomorphism $f : \mathbb{Z}_m \rightarrow \mathbb{Z}_n$ if and only if $m \mid n$ and n/m is relatively prime with m.
(c) Show that if $m \in \mathbb{Z}_n^\times$ then the mapping $f : \mathbb{Z}_n \rightarrow \mathbb{Z}_n$ given by $[k] \mapsto [mk]$ is a bijection.

Problem 27. Let $\{p_1, p_2, p_3, \ldots\}$ the set of all prime numbers.
(a) Show that for each number N the canonical map $\mathbb{Z} \rightarrow \prod_{k=1}^N \mathbb{Z}_{p_k}$ is surjective but not injective.
(b) Show that the canonical map $\mathbb{Z} \rightarrow \prod_{k=1}^\infty \mathbb{Z}_{p_k}$ is injective but not surjective.

Problem 28. Consider the ring $\mathbb{H}$ of quaternions.

(a) Recall that the conjugation of quaternions is defined by $\overline{a + bi + cj + dk} = a - bi - cj - dk$. Compute $x\bar{x}$ and $\bar{x}x$ where $x = a + bi + cj + dk$. Is conjugation a ring homomorphism?

(b) Find all the solutions of the equation $x^2 = x$ in $\mathbb{H}$.

(c) Find all the solutions of the equation $x^2 + 1 = 0$ in $\mathbb{H}$.

Problem 29. Let $f : R \rightarrow S$ be a mapping between rings such that for any two elements $x, y \in R$ the following conditions hold:

(1) $f(x + y) = f(x) + f(y)$;

(2) $f(xy) = f(x)f(y)$ or $f(xy) = f(y)f(x)$.

Show that f is a ring homomorphism or else $f(xy) = f(y)f(x)$ for all $x, y \in R$.

Hint. For all $x \in R$, consider the sets $U_x = \{y \in R \mid f(xy) = f(x)f(y)\}$ and $V_x = \{y \in R \mid f(xy) = f(y)f(x)\}$.

Problem 30. (a) Find all solutions of the equation $x^2 = x$ in $\mathbb{Z}_{10}$, $\mathbb{Z}_{20}$ and $\mathbb{Z}_{30}$.

(b) Find all solutions of the equation $x^2 = x$ in $\mathbb{Z}_{p^k}$ where p is a prime.

Problem 31. An element x in a ring R is called an **idempotent** if $x^2 = x$. A ring R is called a **Boolean ring** if every element of R is an idempotent.

(a) Show that $\mathbb{Z}_2$ is a Boolean ring.

(b) Let $X \neq 0$ be an arbitrary set. Show that the set $\mathbb{Z}_2^X$ of all mappings $f : X \rightarrow \mathbb{Z}_2$ is a Boolean ring if addition and multiplication are defined argumentwise.

(c) Let X be an arbitrary set and $\mathcal{P}(X)$ its power set, i.e., the set of all subsets of X. Show that $\mathcal{P}(X)$ becomes a Boolean ring if we define addition and multiplication by

$$A + B := (A \setminus B) \cup (B \setminus A) \quad \text{and} \quad A \cdot B := A \cap B .$$

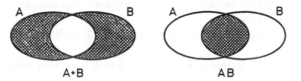

$A+B$ $\qquad\qquad\qquad$ AB

What is the zero element in $(\mathcal{P}(X), +, \cdot)$? Does $\mathcal{P}(X)$ possess an identity element? If X has n elements, how many solutions does the equation $x^2 = x$ have in $\mathcal{P}(X)$?

Hint. Consider the mapping $\mathfrak{P}(X) \rightarrow \mathbb{Z}_2^X$ which assigns to each subset $A \subseteq X$ its characteristic function χ_A which is defined by

$$\chi_A(x) := \begin{cases} 1, & \text{if } x \in A; \\ 0, & \text{if } x \notin A. \end{cases}$$

(d) Let R be an arbitrary commutative ring and let B be the set of all idempotents in R. Define an addition and a multiplication on B by letting $x \oplus y := x + y - 2xy$ and $x \odot y := xy$. Show that $(B, \oplus, \odot)$ is a Boolean ring.

Problem 32. Let R be a Boolean ring.
(a) Show that $xy + yx = 0$ for all $x, y \in R$.
(b) Show that $x + x = 0$ for all $x \in R$. (Hence if $R \neq \{0\}$ then char $R = 2$.)
(c) Show that R is commutative.

Problem 33. Show that every Boolean ring without identity can be embedded into a Boolean ring with identity.

Problem 34. A Boolean algebra is a set B together with two binary operations $\wedge$ and $\vee$ on B, two distinguished elements 0 and 1 in B and a self-mapping $x \mapsto \overline{x}$ [†] such that for all elements $x, y, z \in B$ the following rules:
(1) $x \wedge y = y \wedge x$ and $x \vee y = y \vee x$ (commutative laws);
(2) $x \wedge (y \vee z) = (x \wedge y) \vee (x \wedge z)$ and $x \vee (y \wedge z) = (x \vee y) \wedge (x \vee z)$ (distributive laws);
(3) $x \vee 0 = x$ and $x \wedge 1 = x$ (so that 0 and 1 are neutral elements);
(4) $x \wedge \overline{x} = 0$ and $x \vee \overline{x} = 1$.
(To get some intuition for these axioms see problem 35 for examples.)
(a) Prove that if B is a Boolean algebra and if $x, y, z \in B$ then the following conditions hold:
(5) $x \wedge (y \wedge z) = (x \wedge y) \wedge z$ and $x \vee (y \vee z) = (x \vee y) \vee z$ (associative laws);
(6) $\overline{x \wedge y} = \overline{x} \vee \overline{y}$ and $\overline{x \vee y} = \overline{x} \wedge \overline{y}$ (de Morgan's laws);
(7) $x \wedge x = x$ and $x \vee x = x$ (idempotency laws);
(8) $x \wedge 0 = 0$ and $x \vee 1 = 1$;
(9) $\overline{0} = 1$ and $\overline{1} = 0$;
(10) $\overline{\overline{x}} = x$.
(b) Show that every algebraic identity in a Boolean algebra remains true if one exchanges $\vee$ and $\wedge$ and also exchanges 0 and 1 in this identity.
(c) Let R be a Boolean ring with identity. Show that R becomes a Boolean algebra if we define meet, join and complement by

$$x \wedge y := xy , \quad x \vee y := x + y + xy , \quad \overline{x} = 1 + x .$$

(d) Let R be a Boolean algebra. Show that R becomes a Boolean ring if we define addition and multiplication by

$$x + y := (x \wedge \overline{y}) \vee (\overline{x} \wedge y) \quad \text{and} \quad xy := x \wedge y .$$

Problem 35. In each of the following examples, a set R with two operations $\vee$ and $\wedge$, two distinguished elements 0 and 1 and a self-mapping $x \mapsto \overline{x}$ is given. Decide in each case which of the properties (1) through (10) in problem 34 are satisfied.
(a) Let $R = \{0, 1\}$ be a set with two elements and the following operations.

$$
\begin{array}{c|cc}
\vee & 0 & 1 \\
\hline
0 & 0 & 1 \\
1 & 1 & 1
\end{array}
\qquad
\begin{array}{c|cc}
\wedge & 0 & 1 \\
\hline
0 & 0 & 0 \\
1 & 0 & 1
\end{array}
\qquad
\begin{array}{l}
\overline{0} = 1 \\
\overline{1} = 0
\end{array}
$$

[†] If $x, y \in B$ then $x \wedge y$ and $x \vee y$ are called the **meet** and the **join** of x and y, respectively; moreover, one calls $\overline{x}$ the **complement** of x.

(b) Let $R = \{0, 1, a, b\}$ be a set with four elements and the following operations.

$$
\begin{array}{c|cccc}
\vee & 0 & 1 & a & b \\
\hline
0 & 0 & 1 & a & b \\
1 & 1 & 1 & 1 & 1 \\
a & a & 1 & a & 1 \\
b & b & 1 & 1 & b
\end{array}
\qquad
\begin{array}{c|cccc}
\wedge & 0 & 1 & a & b \\
\hline
0 & 0 & 0 & 0 & 0 \\
1 & 0 & 1 & a & b \\
a & 0 & a & a & 0 \\
b & 0 & b & 0 & b
\end{array}
\qquad
\begin{aligned}
\overline{0} &= 1 \\
\overline{1} &= 0 \\
\overline{a} &= b \\
\overline{b} &= a
\end{aligned}
$$

(c) Let R be the interval $[0, 1]$ and define $x \vee y := \max(x, y)$, $x \wedge y := \min(x, y)$ and $\overline{x} := 1 - x$.

(d) Let N be a given natural number and let R be the set of divisors of N. The special elements $\mathbf{0}$ and $\mathbf{1}$ in R are defined by $\mathbf{0} := 1$ and $\mathbf{1} := N$. Moreover, let $m \vee n := \operatorname{lcm}(m, n)$ and $m \wedge n := \gcd(m, n)$. Finally, $\overline{m} := N/m$.

(e) Let X be a nonempty set and let $R = \mathfrak{P}(X)$ be the power set of X with the operations $A \vee B := A \cup B$, $A \wedge B := A \cap B$ and $\overline{A} := X \setminus A$.

Problem 36. In his work "Investigations of the laws of thought" (1854) the British mathematician George Boole (1815-1864) laid the foundations for an algebraic treatment of logic. In this problem some ideas and consequences of Boole's work are outlined. Let X be a nonempty set.

(a) For any statement A about elements of X we let $T(A) := 1$ if A is true and $T(A) := 0$ if A is false. (This defines a **truth function** on any set of statements about elements of X.) For simplicity we write $T(A) = a$, $T(B) = b$, and so on. Show that

$$
T(A \text{ and } B) = a + b - ab \quad \text{and} \quad T(A \text{ or } B) = ab
$$

for all statements A and B where addition and multiplication are in $\mathbb{Z}_2$. As an example of how the use of T allows one to reduce logical problems to algebraic equations, show that exactly one of four statements A, B, C, D is false (the other three being true) if and only if $abcd = 0$ and $abc + abd + acd + bcd = 1$.

(b) For an element A of an electrical circuit with exactly one entrance and one exit let $T(A) := 1$ if current can flow from the entrace to the exit and $T(A) := 0$ otherwise. (For example, if A is just a piece of wire then $T(A) = 1$ whereas if A is a piece of wire which is disconnected by an open switch then $T(A) = 0$.) For simplicity we write $T(A) = a$, $T(B) = b$, and so on. Denote by $A \vee B$ the parallel connection and by $A \wedge B$ the series connection of A and B. Show that $T(A \vee B) = a + b - ab$ and $T(A \wedge B) = ab$ where addition and multiplication are in $\mathbb{Z}_2$. Find a connection to part (a) and discuss its implications for the possibility of processing information on a digital computer.

(c) Let $\mathfrak{S}(X)$ be all sentential formulas in one variable which become either true or false if an element $x \in X$ is plugged in for the variable. (Typical examples are "x is taller than 6 feet" or "y has brown hair and blue eyes".) We identify two such formulas if for each $x \in X$ they are either both true or both false. Show that $\mathfrak{S}(X)$ is a Boolean algebra if we define meet, join and complement by

$$
A \vee B := A \text{ or } B, \quad A \wedge B := A \text{ and } B, \quad \overline{A} := \neg A := \text{not } A.
$$

(d) Let $\mathfrak{P}(X)$ be equipped with the structure of a Boolean algebra as in part (e) of problem 35. Define a mapping $\theta : \mathfrak{S}(X) \to \mathfrak{P}(X)$ by $\theta(A) := \{x \in X \mid \text{statement } A \text{ is true for element } x\}$. Show that θ is a bijection which respects the Boolean structures on $\mathfrak{P}(X)$ and $\mathfrak{S}(X)$.

3. Integral domains and fields

IT is clear that the arithmetic in an arbitrary ring can differ substantially from the well-known arithmetic in the ring of integers. One example is the possible non-commutativity of multiplication. But even in commutative rings unfamiliar phenomena can occur. We will point out these phenomena and then define a class of rings in which they cannot occur. These rings then resemble the integers much more than arbitrary rings which has the consequence that a reasonable theory of divisibility can be established. Let us start by considering special elements that can occur in a ring.

(3.1) Definitions. *Let R be a ring.*
(a) *An element $r \neq 0$ in R is called a* **zero-divisor** *if there is an element $s \neq 0$ such that $rs = 0$ or $sr = 0$.*
(b) *An element $r \in R$ is called* **nilpotent** *if $r^n = 0$ for some $n \geq 1$; clearly, a nonzero nilpotent element is a zero-divisor.*

(3.2) Examples. (a) In $C[-1, 1]$, the two functions f and g sketched below are zero-divisors because $f \neq 0$, $g \neq 0$, but $fg = 0$. However, the ring $C[-1, 1]$ possesses no nilpotent elements other than 0; if f^n is the zero-function for some $n \geq 1$ then f itself must be the zero-function.

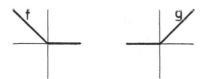

(b) Let $R \neq \{0\}$ be an arbitrary ring. Pick an element $a \neq 0$ in R. Then

$$\begin{pmatrix} 0 & a \\ 0 & 0 \end{pmatrix} \neq \begin{pmatrix} 0 & 0 \\ 0 & 0 \end{pmatrix}, \quad \text{but} \quad \begin{pmatrix} 0 & a \\ 0 & 0 \end{pmatrix}^2 = \begin{pmatrix} 0 & 0 \\ 0 & 0 \end{pmatrix} \quad \text{in } R^{2 \times 2}.$$

Hence the ring $R^{2 \times 2}$ possesses nonzero nilpotent elements.
(c) Suppose m is not divisible by n, but that m and n have a proper common factor a. Then $[m]$ is a zero-divisor in $\mathbb{Z}_n$. Indeed, let $m = xa$ and $n = ya$ where y is a proper factor of n. Then $[m] \neq [0]$ and $[y] \neq [0]$ in $\mathbb{Z}_n$, but $[m] \cdot [y] = [my] = [nx] = [0]$.
(d) Every direct product of two nonzero rings R and S has zero-divisors. In fact, if $r \in R \setminus \{0\}$ and $s \in S \setminus \{0\}$ then $(r, 0)(0, s) = (0, 0)$.
(e) The sum of two commuting nilpotent elements x and y in a ring is nilpotent again. In fact, if $x^m = 0$ and $y^n = 0$ then

$$(x + y)^{m+n-1} = \sum_{k=0}^{m+n-1} \binom{m+n-1}{k} \underbrace{x^k}_{\substack{= 0 \text{ if} \\ k \geq m}} \underbrace{y^{m+n-1-k}}_{\substack{= 0 \text{ if} \\ k < m}} = 0.$$

(f) Let R be a commutative ring. Then a polynomial $f(x) = a_0 + a_1 x + \cdots + a_n x^n$ is nilpotent in $R[x]$ if and only if the coefficients $a_0, \ldots, a_n$ are nilpotent in R. In fact, if $a_0, \ldots, a_n$ are nilpotent in R then the polynomials $f_i(x) := a_i x^i$ are clearly nilpotent in $R[x]$, and hence $f = f_0 + f_1 + \cdots + f_n$ is nilpotent as a sum of commuting nilpotent elements. Let us show conversely by induction on n that if $a_0 + a_1 x + \cdots + a_n x^n$ is nilpotent in $R[x]$ then $a_0, \ldots, a_n$ are nilpotent in R. If $f^N = 0$ then $a_0^N = 0$ because a_0^N is the first coefficient of f^N; this shows that a_0 is nilpotent. Let $g(x) := a_1 + a_2 x + \cdots + a_n x^{n-1}$; then $xg(x) = f(x) + (-a_0)$ is nilpotent as the sum of two commuting nilpotent elements which entails that g is nilpotent. But then $a_1, \ldots, a_n$ are nilpotent by induction hypothesis.

LET us see what it means for a ring to be free of zero-divisors.

(3.3) Proposition. *Let R be a ring. Then the following conditions are equivalent.*
(a) *R has no zero-divisors.*
(b) *The following* **cancellation rules** *hold:*

$$ax = ay, \ a \neq 0 \implies x = y \, ; \qquad xa = ya, \ a \neq 0 \implies x = y \, .$$

(c) *For any given elements $a \in R \setminus \{0\}$ and $b \in R$ each of the equations $ax = b$ and $ya = b$ has at most one solution x or y.*

Proof. Clearly, (c) is simply a restatement of (b); hence it is enough to show the equivalence of (a) and (b). Suppose (a) holds. Then if $a \neq 0$ and $ax = ay$ or $xa = ya$, i.e., $a(x - y) = 0$ or $(x - y)a = 0$, we must have $x = y$; otherwise a and $x - y$ would be zero-divisors. Suppose conversely that (b) holds, but that nevertheless $ab = 0$ with $a \neq 0$ and $b \neq 0$. Then the equation $ax = 0$ has two solutions, namely $x = 0$ and $x = b$, contradicting (b). ∎

(3.4) Proposition. *Let R be a ring with identity and let $n := \operatorname{char} R$ be its characteristic. If R has no zero-divisors then n is zero or a prime number.*

Proof. Suppose that $n = n_1 n_2$ is a composite number. Using the fact that $m \mapsto m \cdot 1$ is a ring homomorphism from $\mathbb{Z}$ into R, we have $0 = (n_1 n_2) \cdot 1 = (n_1 \cdot 1)(n_2 \cdot 1)$. The assumption that R has no zero-divisors implies that $n_1 \cdot 1 = 0$ or $n_2 \cdot 1 = 0$ contradicting the fact that n is the *smallest* number k with $k \cdot 1 = 0$. ∎

IN a ring with identity there is another special class of elements, namely those which possess an inverse.

(3.5) Definition. *Let R be a ring with identity element 1. An element $r \in R$ is called **invertible** or a **unit** in R if there is an element $s \in R$ such that $rs = sr = 1$.*† *This element s is uniquely determined*†† *and denoted by $s = r^{-1}$. It is easy to check that*

$$R^{\times} := \{r \in R \mid r \text{ is invertible}\}$$

*is a group under multiplication; it is called the **group of units** of R.*

(3.6) Examples. (a) Clearly,

$$\boxed{\mathbb{Z}^{\times} = \{1, -1\}.}$$

(b) For all $n \in \mathbb{N}$ we have

$$\boxed{\mathbb{Z}_n^{\times} = \{[m] \mid m \text{ is coprime with } n\}.}$$

Here the inclusion $\subseteq$ is clear because of (3.2)(c) since a zero-divisor cannot be a unit. The converse inclusion was established in (I.20.14).

(c) Let $R = \mathbb{Z} + \mathbb{Z}\sqrt{n}$ where $n \in \mathbb{Z} \setminus \{0, 1\}$ is square-free. To find the units of R we introduce the **norm** $N : \mathbb{Q} + \mathbb{Q}\sqrt{n} \to \mathbb{Q} + \mathbb{Q}\sqrt{n}$ by

$$N(x + y\sqrt{n}) := x^2 - ny^2 \; ;$$

note that if $x, y \in \mathbb{Z}$ then $N(x + y\sqrt{n}) \in \mathbb{Z}$. If we define the **conjugation mapping** of $\mathbb{Z} + \mathbb{Z}\sqrt{n}$ by $\overline{x + y\sqrt{n}} := x - y\sqrt{n}$, we can write

$$N(a) = a\bar{a} \, .$$

Note that $\overline{a + b} = \bar{a} + \bar{b}$ and $\overline{ab} = \bar{a}\bar{b}$ so that

$$N(ab) = ab\overline{ab} = ab\bar{a}\bar{b} = a\bar{a}b\bar{b} = N(a)N(b) \, .$$

Now we are ready to show that

$$\boxed{\begin{aligned} (\mathbb{Z} + \mathbb{Z}\sqrt{n})^{\times} &= \{x + y\sqrt{n} \mid x, y \in \mathbb{Z}, \, x^2 - ny^2 = \pm 1\} \\ &= \{a \in \mathbb{Z} + \mathbb{Z}\sqrt{n} \mid N(a) = \pm 1\} \, . \end{aligned}}$$

Indeed, if $a \in R$ is a unit then there is an element $b \in R$ with $ab = 1$ which implies $1 = N(1) = N(ab) = N(a)N(b)$. Since $N(a)$ and $N(b)$ are integers, this is only possible if $N(a) = N(b) = \pm 1$. Suppose conversely that $N(a) = \pm 1$, i.e., $a\bar{a} = \pm 1$. Then clearly $\pm\bar{a}$ is an inverse of a.

---

† Equivalently, we could define r to be a unit if there are elements $s_1, s_2 \in R$ such that $s_1 r = r s_2 = 1$ for then s_1 and s_2 have to be equal. Indeed, $s_1 = s_1 \cdot 1 = s_1(r s_2) = (s_1 r)s_2 = 1 \cdot s_2 = s_2$.

†† If s and s' are inverse to r, then $s = s \cdot 1 = s(rs') = (sr)s' = 1 \cdot s' = s'$.

(d) Let R be a commutative ring with 1. As in (I.7.5) we can introduce the determinant $\det : R^{n \times n} \to R$. Then

$$\boxed{(R^{n \times n})^\times \;=\; \{A \in R^{n \times n} \mid \det A \in R^\times\}\,.}$$

For example, a matrix with integer entries has an inverse with integer entries if and only if its determinant is ± 1.

To prove this claim, assume first that $A \in (R^{n \times n})^\times$. Then there is a matrix $B \in R^{n \times n}$ such that $AB = BA = 1$. But this implies $(\det A)(\det B) = (\det B)(\det A) = 1$ so that $\det A \in R^\times$. Suppose conversely that $\det A \in R^\times$. Then Cramer's rule gives $A(\operatorname{adj} A) = (\operatorname{adj} A)A = (\det A)1$ so that $(\det A)^{-1}(\operatorname{adj} A)$ is inverse to A.

(e) Let R be a ring with 1 which has no zero-divisors. Then

$$\boxed{R[x]^\times \;=\; R^\times}$$

where we consider R as a subring of $R[x]$. To check this claim, suppose first that $r \in R^\times$. Let $s = r^{-1}$ in R. Then s (considered as a constant polynomial) is also inverse to r in $R[x]$. Conversely, let $f(x) = a_0 + a_1 x + \cdots + a_n x^n \in R[x]^\times$ with $a_n \neq 0$ and let $g(x) = b_0 + b_1 x + \cdots + b_m x^m$ be the inverse of f where $b_m \neq 0$. Then

$$1 \;=\; f(x)g(x) \;=\; a_0 b_0 \;+\; \cdots \;+\; a_n b_m\, x^{n+m}$$

where $a_n b_m \neq 0$ because R has no zero-divisors. We conclude that $n = m = 0$ and $a_0 b_0 = 1$. Also, $1 = g(x)f(x) = b_0 a_0$ so that $b_0 = a_0^{-1}$.

(f) Let R be a commutative ring with identity. Then

$$\boxed{R[x]^\times \;=\; \{a_0 + a_1 x + \cdots + a_n x^n \mid a_0 \in R^\times,\; a_1,\ldots,a_n \text{ nilpotent in } R\}\,.}$$

Let us show the inclusion $\supseteq$ first. If $a_0 \in R^\times$ and if $a_1,\ldots,a_n$ are nilpotent then $g(x) := a_1 x + \cdots + a_n x^n$ is nilpotent by (3.2)(f); hence $f(x) = a_0 + g(x)$ is a sum of a unit and a nilpotent element and therefore a unit. (See problem 3 below.) Let us now prove the converse inclusion $\subseteq$. If $f(x) = a_0 + a_1 x + \cdots + a_n x^n$ is a unit in $R[x]$ then clearly a_0 is a unit in R; without loss of generality we may assume that $a_0 = 1$. Letting $g(x) := -(a_1 + a_2 x + \cdots + a_n x^{n-1})$ we have $f(x) = 1 - x g(x)$, and the inverse of f can be written in the form $1 + x h(x)$. Then $1 = \big(1 - x g(x)\big)\big(1 + x h(x)\big) = 1 - x g(x) + x h(x) - x^2 g(x) h(x)$ which implies that $h = g + x g h$. Substituting the left-hand side of this equation into the right-hand side we find that $h = g + x g(g + x g h) = g + x g^2 + x^2 g^2 h$. Repeating this substitution we find that

$$h \;=\; g + x g^2 + x^2 g^3 + \cdots + x^{N-1} g^N + x^N g^N h$$

for all $N \in \mathbb{N}$. If we had $g^N \neq 0$ for all N this would produce arbitrarily high powers of x on the right-hand side which is, of course, impossible. Thus g is nilpotent so that $f(x) = a_0 + x g(x)$ is a sum of a unit and a nilpotent element and hence is a unit.

(g) Let R be a ring with 1. Then

$$\boxed{R[[x]]^\times \;=\; \{a_0 + a_1 x + a_2 x^2 + a_3 x^3 + \cdots \mid a_0 \in R^\times\}\,.}$$

Let us prove this claim. If $f(x) = \sum_{k=0}^{\infty} a_k x^k$ has an inverse $g(x) = \sum_{k=0}^{\infty} b_k x^k$, then
$1 = f(x)g(x) = a_0 b_0 + (a_0 b_1 + a_1 b_0)x + \cdots$ and $1 = g(x)f(x) = b_0 a_0 + (b_0 a_1 + b_1 a_0)x + \cdots$.
This implies $a_0 b_0 = b_0 a_0 = 1$ so that $a_0 \in R^\times$. Conversely, let $a_0 \in R^\times$. The power
series $g(x) = b_0 + b_1 x + b_2 x^2 + \cdots$ satisfies $fg = 1$ if and only if

$$a_0 b_0 = 1, \quad a_0 b_1 + a_1 b_0 = 0, \quad \ldots, \quad a_0 b_n + a_1 b_{n-1} + \cdots + a_n b_0 = 0, \quad \ldots$$

But the fact that a_0^{-1} exists allows us to solve these equations successively for b_0, b_1, b_2
and so on; this shows that $g \in R[[x]]$ with $fg = 1$ exists. Similarly, we find $h \in R[[x]]$
with $hf = 1$. But then $h = h \cdot 1 = h(fg) = (hf)g = 1 \cdot g = g$ so that $g = f^{-1}$.

(h) Let $\mathbb{H}$ be the ring of quaternions. Then

$$\boxed{\mathbb{H}^\times = \mathbb{H} \setminus \{0\}}$$

which means that every nonzero quaternion is invertible. This follows easily from the
identity

$$\det \begin{pmatrix} a & b \\ -\bar{b} & \bar{a} \end{pmatrix} = |a|^2 + |b|^2 .$$

(i) Let $\operatorname{End} A$ be the endomorphism ring of the abelian group $(A, +)$ as defined in
(2.11). Then

$$\boxed{(\operatorname{End} A)^\times = \operatorname{Aut} A}$$

where $\operatorname{Aut} A$ denotes the automorphism group of A.

LET us study in some more detail the units of the rings $\mathbb{Z} + \mathbb{Z}\sqrt{n}$ since these
rings are of considerable importance in number theory. In fact, the determination of
the units is essentially a number-theoretical problem. This determination is very easy
if $n < 0$, but for positive values of n we need the following non-trivial result.

(3.7) Theorem. *Let $n \in \mathbb{N}$ be a natural number which is not a square.*
(a) *For any given $c \in \mathbb{N}$ there are $x, y \in \mathbb{N}$ with $0 < |x - y\sqrt{n}| < 1/c \leq 1/y$.*
(b) *There are infinitely many pairs $(x, y) \in \mathbb{N}^2$ such that $0 < |x - y\sqrt{n}| < \frac{1}{y}$; each
such pair satisfies $|x^2 - ny^2| < 1 + 2\sqrt{n}$.*
(c) *The equation $x^2 - ny^2 = 1$ (**Pell's equation**)† has a nontrivial solution
$(x, y) \neq (1, 0)$ in $\mathbb{N}^2$.*

Proof. (a) For any $x \in \mathbb{R}$ we denote by $[x]$ the largest integer smaller than x.
Given $c \in \mathbb{N}$, the numbers $k\sqrt{n} - [k\sqrt{n}]$ ($0 \leq k \leq c$) are irrational except for $k = 0$;
hence each of these numbers lies in one of the intervals $[0, \frac{1}{c}), (\frac{1}{c}, \frac{2}{c}), (\frac{2}{c}, \frac{3}{c}), \ldots, (\frac{c-1}{c}, 1)$.
Since we have $c+1$ possibilities for k, but only c intervals, there are at least two different
values $k_1 < k_2$ for k such that both $k_1\sqrt{n} - [k_1\sqrt{n}]$ and $k_2\sqrt{n} - [k_2\sqrt{n}]$ lie in the same
interval. Then

$$\frac{1}{c} > |(k_2\sqrt{n} - [k_2\sqrt{n}]) - (k_1\sqrt{n} - [k_1\sqrt{n}])|$$
$$= |(\underbrace{k_2 - k_1}_{=: \, y \, \in \, \mathbb{N}})\sqrt{n} - (\underbrace{[k_2\sqrt{n}] - [k_1\sqrt{n}]}_{=: \, x \, \in \, \mathbb{N}})| = |x - y\sqrt{n}| ;$$

---

† Named after John Pell (1611-1685).

Note that $x > 0$ because $n \geq 2$. Since $0 < y = k_2 - k_1 \leq k_2 \leq c$ we have $1/c \leq 1/y$ so that the desired inequality holds.

(b) Choose $c_1 \in \mathbb{N}$ and then $x_1, y_1 \in \mathbb{N}$ such that $0 < |x_1 - y_1\sqrt{n}| < 1/c_1 \leq 1/y_1$ as in (a). Then pick $c_2 \in \mathbb{N}$ so large that $1/c_2 < |x_1 - y_1\sqrt{n}|$ and choose x_2, y_2 according to part (a). Continuing in this fashion, we obtain an infinite sequence

$$\frac{1}{c_1} > |x_1 - y_1\sqrt{n}| > \frac{1}{c_2} > |x_2 - y_2\sqrt{n}| > \frac{1}{c_3} > |x_3 - y_3\sqrt{n}| > \cdots$$

where the pairs (x_i, y_i) have the desired property. Furthermore, if $|x - y\sqrt{n}| < 1/y$ then $|x + y\sqrt{n}| \leq |x - y\sqrt{n}| + |2y\sqrt{n}| < (1/y) + 2y\sqrt{n}$ by the triangle inequality and hence

$$|x^2 - ny^2| = |x - y\sqrt{n}|\,|x + y\sqrt{n}| < \frac{1}{y}(\frac{1}{y} + 2y\sqrt{n}) = \frac{1}{y^2} + 2\sqrt{n} \leq 1 + 2\sqrt{n} \ .$$

(c) Due to (b), there must be a natural number $r < 1 + 2\sqrt{n}$ which has an infinite number of representations as $r = |x^2 - ny^2|$ with $x, y \in \mathbb{N}$. Hence for at least one number $\varepsilon \in \{\pm 1\}$ there are infinitely many pairs (x, y) with $x^2 - ny^2 = \varepsilon r$. Since there is only a finite number of remainders modulo r, we can find two different solutions (x_1, y_1) and (x_2, y_2) such that $x_1 \equiv x_2$ and $y_1 \equiv y_2$ modulo r. Then

$$\alpha := \frac{x_1 x_2 - n y_1 y_2}{r} \in \mathbb{Z} \quad \text{and} \quad \beta := \frac{x_2 y_1 - x_1 y_2}{r} \in \mathbb{Z}$$

satisfy

(1) $\quad (x_1 + y_1\sqrt{n})(x_2 - y_2\sqrt{n}) = (x_1 x_2 - n y_1 y_2) + (y_1 x_2 - x_1 y_2)\sqrt{n} = r(\alpha + \beta\sqrt{n}) \ .$

We apply the conjugation mapping $z \mapsto \bar{z}$ of $\mathbb{Q} + \mathbb{Q}\sqrt{n}$ to this equation and obtain

(2) $\qquad\qquad (x_1 - y_1\sqrt{n})(x_2 + y_2\sqrt{n}) = r(\alpha - \beta\sqrt{n}) \ .$

Multiplying (1) and (2), we obtain

$$r^2(\alpha^2 - n\beta^2) = (x_1^2 - n y_1^2)(x_2^2 - n y_2^2) = (\varepsilon r)(\varepsilon r) = \varepsilon^2 r^2 = r^2$$

so that $\alpha^2 - n\beta^2 = 1$. This shows that $(|\alpha|, |\beta|)$ is a solution of Pell's equation. It remains to show that it is not the trivial solution. Suppose $\beta = 0$; then $x_1/x_2 = y_1/y_2 =: \kappa > 0$ so that $\varepsilon r = x_1^2 - n y_1^2 = \kappa^2(x_2^2 - n y_2^2) = \kappa^2 \varepsilon r$ which implies $\kappa^2 = 1$, hence $\kappa = 1$ contradicting the fact that $(x_1, y_1) \neq (x_2, y_2)$. This contradiction shows that $\beta \neq 0$ so that we have indeed found a nontrivial solution. ∎

(3.8) Theorem. *Let $n \in \mathbb{Z} \setminus \{0, 1\}$ be a square-free number.*

(a) *Suppose $n < 0$. Then $(\mathbb{Z} + \mathbb{Z}\sqrt{n})^\times = \{\pm 1\}$ for all $n \geq -2$ whereas $(\mathbb{Z} + i\mathbb{Z})^\times = \{\pm 1, \pm i\}$ for $n = -1$.*

(b) *Suppose $n > 0$ so that $\mathbb{Z} + \mathbb{Z}\sqrt{n} \subseteq \mathbb{R}$. There is a smallest unit u of $\mathbb{Z} + \mathbb{Z}\sqrt{n}$ which is larger than 1, called the **basis unit** of $\mathbb{Z} + \mathbb{Z}\sqrt{n}$, and we have*

$$(\mathbb{Z} + \mathbb{Z}\sqrt{n})^\times = \{\pm u^m \mid m \in \mathbb{Z}\} \ .$$

Proof. (a) The norm of an element $z = x + y\sqrt{n}$ is $N(z) = x^2 - ny^2 = x^2 + |n|y^2 \geq 0$; hence z is a unit if and only if $1 = |N(z)| = N(z) = x^2 + |n|y^2$. For $n = -1$ this equation reads $x^2 + y^2 = 1$ so that the solutions in $\mathbb{Z}^2$ are $(\pm 1, 0)$ and $(0, \pm 1)$. For $|n| \geq 2$, the only solutions of $x^2 + |n|y^2 = 1$ in $\mathbb{Z}^2$ are obviously $(\pm 1, 0)$.

(b) Multiplication by -1 switches positive and negative units, and the inversion map $z \mapsto \frac{1}{z}$ switches the sets of units with $|z| > 1$ and $|z| < 1$, respectively. Hence all units of $\mathbb{Z} + \mathbb{Z}\sqrt{n}$ other than ± 1 are of the form $\pm z$ or $\pm z^{-1}$ where z is a unit with $z > 1$. Now (3.7)(c) guarantees the existence of a unit $z = x + y\sqrt{n}$ with $x, y \in \mathbb{N}$ and hence $z \geq 1 + \sqrt{n} > 1$. † Since $\mathbb{N}$ is discrete, there is clearly a smallest unit u with $u > 1$. Obviously, all powers u^m with $m \in \mathbb{N}$ are again units larger than 1. It remains to show that these powers exhaust the set of units larger than 1. Suppose there is a unit $z > 1$ which is not a power of u. Then there is an exponent M with $u^M < z < u^{M+1}$. Multiplying this inequality by $u^{-M} > 0$, we obtain $1 < zu^{-M} < u$ so that zu^{-M} is a unit lying strictly between 1 and u. This clearly contradicts the choice of u. ∎

SINCE the general notion of a ring is an abstraction of the ring of integers, we can try to carry over the well-known notions of divisibility in $\mathbb{Z}$ to an arbitrary ring R. We could simply define that an element $a \in R$ be a divisor of an element $b \in R$ if there is an element $x \in R$ with $ax = b$. However, if the multiplication is not commutative, this does not imply that also $xa = b$. Hence we will restrict our attention to commutative rings. In $\mathbb{Z}$ we have the notion of the greatest common factor of given numbers. For example, $\gcd(12, 18, 30) = 6$ because 6 is the largest number dividing $12, 18$ and 30 simultaneously. The word 'largest' only has meaning in a ring which possesses an ordering of its elements. Let us try to characterize greatest common divisors in a way that does not refer to the ordering given on $\mathbb{Z}$. For example, the common divisors of $12, 18$ and 30 are $\pm 1, \pm 2, \pm 3, \pm 6$, and the number 6 has the property that it is divisible by all of these. This is a characterization we can use to define a greatest common divisor in arbitrary domains. To define least common multiples, we proceed similarly.

(3.9) Definitions. *Let R be a commutative ring and let $a, b, a_1, \ldots, a_n$ be elements of R.*

(a) *We say that a divides b and write $a \mid b$ if there is an element $r \in R$ such that $b = ra$. Equivalently, we say that b is **divisible** by a and call a a **divisor** or a **factor** of b. (If R has no zero-divisors such an element r is uniquely determined, and we write $r = b/a$.)*

(b) *We call $d \in R$ a **common divisor** of $a_1, \ldots, a_n$ if $d \mid a_k$ for $1 \leq k \leq n$.*

(c) *An element $d \in R$ is called a **greatest common divisor** (gcd) of $a_1, \ldots, a_n$ if d is a common divisor of $a_1, \ldots, a_n$ and if any other common divisor d' of $a_1, \ldots, a_n$ divides d.*

(d) *We call $m \in R$ a **common multiple** of $a_1, \ldots, a_n$ if $a_k \mid m$ for $1 \leq k \leq n$.*

(e) *An element $m \in R$ is called a **least common multiple** (lcm) of $a_1, \ldots, a_n$ if m is a common multiple of $a_1, \ldots, a_n$ and if any other common multiple m' of $a_1, \ldots, a_n$ is a multiple of m.*

---

† Since we studied the equation $x^2 - ny^2 = +1$ in (3.7), we even have $N(z) = 1$, but it may well happen that a unit $z > 1$ satisfies $N(z) = -1$; for example, take $z = 1 + \sqrt{2}$ in $\mathbb{Z} + \mathbb{Z}\sqrt{2}$.

NOTE that two elements in a commutative ring do not necessarily possess a greatest common divisor or a least common multiple. For example, in the ring $R = 2\mathbb{Z}$ of all even integers, the element 6 has no divisors at all in R. It becomes clear at this point that even though we could *define* the basic notions of divisibility in an arbitrary commutative ring, we will have to specialize to a more restricted class of rings to obtain a reasonable theory. For example, if R is a ring without an identity element it is not even true in general that $a \mid a$ for all $a \in R$, and if we allow rings without identity it is not clear how to define that two elements are relatively prime. Also, if R has zero-divisors, we might find that there are two different elements $x_1 \neq x_2$ which satisfy the equation $ax = b$, and it does not make sense to talk about the quotient a/b if a is divisible by b. This is why we now introduce a class of rings which share three characteristic features with the ring of integers; namely commutativity, the existence of an identity element and the absence of zero-divisors; and are thus similar enough to the ring of integers to allow the development of a reasonable theory of divisibility. In fact, one calls these rings integral domains to express their similarity to the ring of integers.

(3.10) Definition. *An* **integral domain** *is a commutative ring with identity which has no zero-divisors.*

(3.11) Examples. (a) $\mathbb{Z}$, $\mathbb{Q}$, $\mathbb{R}$ and $\mathbb{C}$ are all integral domains.

(b) If R is an integral domain and $U \leq R$ a subring with $1 \in U$ then U is itself an integral domain. For example, $\mathbb{Z}[\sqrt{2}] \leq \mathbb{C}$ is an integral domain.

(c) $\mathbb{Z}_n$ is an integral domain if and only if n is a prime number.

(d) If R is an integral domain, then so are the polynomial ring $R[x]$ and the power series ring $R[[x]]$. (See problem 28 below.)

EVEN in an integral domain not all the usual facts known about the integers are true; for example, it is still possible that two elements do not possess a greatest common divisor or a least common multiple.

(3.12) Example. The elements 6 and $2 + 2\sqrt{-5}$ have neither a greatest common divisor nor a least common multiple in the integral domain $\mathbb{Z}[\sqrt{-5}]$.

Indeed, suppose $a + b\sqrt{-5}$ is a common divisor of 6 and $2 + 2\sqrt{-5}$. Taking norms, we see that $a^2 + 5b^2$ is a common divisor of 36 and 24 in $\mathbb{Z}$. Consequently, $a^2 + 5b^2 \mid 12$ which only leaves the possibilities $(\pm 1, 0)$, $(\pm 2, 0)$ and $(\pm 1, \pm 1)$ for (a, b). Since $\pm(1 - \sqrt{-5})$ are not divisors of $1 + \sqrt{-5}$, the common divisors of 6 and $2 + 2\sqrt{-5}$ are the six elements ± 1, ± 2 and $\pm(1 + \sqrt{-5})$. It is easy to see that none of these is divisible by all the others; hence there is no greatest common divisor.

Furthermore, suppose that 6 and $2 + 2\sqrt{-5}$ possess a least common multiple $a + b\sqrt{-5}$. Then $a^2 + 5b^2$ is a common multiple of 36 and 24, hence a multiple of 12. On the other hand, $a + b\sqrt{-5}$ divides every common multiple of 6 and $2 + 2\sqrt{-5}$ so that for example $a + b\sqrt{-5}$ divides 12 and $6(1 + \sqrt{-5})$. But this implies that $a^2 + 5b^2$ divides 144 and 216, hence divides 72. The two conditions together show that $a^2 + 5b^2 = 12, 24, 36$ or 72. This leaves for (a, b) the possibilities $(\pm 2, \pm 2)$, $(\pm 4, \pm 2)$ and $(\pm 6, \pm 0)$. Now 6 is not divisible by $2 + 2\sqrt{-5}$, and $2 \pm 2\sqrt{-5}$ and $4 \pm 2\sqrt{-5}$ are not divisible by 6. So the assumption that $a + b\sqrt{-5}$ is a least common multiple of 6 and $2 + 2\sqrt{-5}$ leads to a contradiction.

NEVERTHELESS, we can transfer some concepts and results from the integers to arbitrary integral domains.

(3.13) Definitions. *Let R be an integral domain.*

(a) *Two elements $a, b \in R$ are called* **associates** *if $a \mid b$ and $b \mid a$ and write $a \sim b$. It is easy to see that $\sim$ is an equivalence relation on R and that $a \sim b$ if and only if there is a unit $u \in R^{\times}$ such that $b = ua$.*

(b) *A factor a of b is called a* **proper factor** *if a is neither a unit nor an associate of b.*

(c) *Elements $a_1, \ldots, a_n$ are called* **relatively prime** *if 1 is a greatest common divisor of $a_1, \ldots, a_n$. For two elements, we also use the word* **coprime** *instead of relatively prime.*

(3.14) Examples. (a) The element $6 \in \mathbb{Z}$ has the proper factors $\pm 2, \pm 3$ and the improper factors $\pm 1, \pm 6$.

(b) -3 and 3 are associates in $\mathbb{Z}$, and $x^2 + 2$ is an associate of $\frac{1}{2}x^2 + 1$ in $\mathbb{Q}[x]$.

(c) The elements $-1 + \sqrt{3}$ and $1 + \sqrt{3}$ are associates in $\mathbb{Z} + \mathbb{Z}\sqrt{3}$.

LET us collect the basic properties of these concepts.

(3.15) Proposition. *Let R be an integral domain.*

(a) *The divisor relation is symmetric and transitive, i.e., $a \mid a$ for all $a \in R$, and if $a \mid b$ and $b \mid c$ then $a \mid c$.*

(b) *Let u be a unit, i.e., $u \mid 1$. The associates and the factors of u are exactly the units in R. If $a \mid b$, then $ua \mid b$.*

(c) *A unit cannot be a zero-divisor.*

(d) *$a \sim b$ if and only if $a \mid b$ and $b \mid a$.*

(e) *Any two greatest common divisors and any two least common multiples of ring elements $a_1, \ldots, a_n$ are associates; conversely, any associate of a greatest common divisor or a least common multiple is again a greatest common divisor or a least common multiple. Moreover, if d is a greatest common divisor of $a_1, \ldots, a_n$ then $\frac{a_1}{d}, \ldots, \frac{a_n}{d}$ are relatively prime.*

(f) *Suppose d is a greatest common divisor of $a_1, \ldots, a_n$. Let $r \neq 0$. Then either rd is a greatest common divisor of $ra_1, \ldots, ra_n$ or else these elements do not possess a greatest common divisor.*

Proof. Exercise. (See problem 27 below.) ∎

(3.16) Proposition. *Let R be an integral domain.*

(a) *If two elements $x, y \in R \setminus \{0\}$ possess a least common multiple $[x, y]$, then they also possess a greatest common divisor (x, y), and we have $xy \sim [x, y] \cdot (x, y)$.*

(b) *It is possible that two elements $x, y \in R \setminus \{0\}$ possess a greatest common divisor, but not a least common multiple.*

(c) *If any two elements in $R \setminus \{0\}$ possess a greatest common divisor, then any two elements in $R \setminus \{0\}$ possess a least common multiple.*

Proof. (a) Suppose $[x, y]$ is a least common multiple of x and y. Since xy is a common multiple of x and y, there is an element $d \in R$ with $xy = d[x, y]$. We are done if we can show that d is a greatest common divisor of x and y.

First of all, there are elements $x', y' \in R$ with $[x, y] = xy' = x'y$, since $[x, y]$ is a common multiple of x and y. Then $xy = d[x, y] = dxy' = dx'y$ so that $y = dy'$ and $x = dx'$; this shows that d is a common divisor of x and y. Now let δ be another common divisor of x and y; say $x = \delta x''$ and $y = \delta y''$. Then $\delta x'' y''$ is a common multiple of x and y and hence a multiple of $[x, y]$, say $\delta x'' y'' = \lambda[x, y]$. Then $d[x, y] = xy = \delta^2 x'' y'' = \delta\lambda[x, y]$ which implies $d = \delta\lambda$, so that δ is a divisor of d. This shows that d is a greatest common divisor of x and y.

(b) We claim that the elements $x := 1 + \sqrt{-5}$ and $y := 2$ of $\mathbb{Z} + \mathbb{Z}\sqrt{-5}$ possess a greatest common divisor, but not a least common multiple.

Let $a + b\sqrt{-5}$ be a common divisor of x and y; taking norms, we see that $a^2 + 5b^2$ is a common divisor of 6 and 4, hence a divisor of 2. Since the equation $a^2 + 5b^2 = 2$ has no solution in $\mathbb{Z}$, we conclude that $a^2 + 5b^2 = 1$ which implies $a + b\sqrt{-5} = \pm 1$. Hence the only divisors of x and y are the units ± 1, which clearly implies the first claim.

Now suppose that $a + b\sqrt{-5}$ is a least common multiple of x and y. Then clearly $a^2 + 5b^2$ is a common multiple of 6 and 4, hence a multiple of 12. On the other hand, $a + b\sqrt{-5}$ divides any common multiple of x and y, hence divides $2 + 2\sqrt{-5}$ and $6 = 2 \cdot 3 = (1 + \sqrt{-5})(1 - \sqrt{-5})$. Taking norms, this shows that $a^2 + 5b^2$ divides 24 and 36, hence divides 12. Thus we obtain $a^2 + 5b^2 = 12$. But this equation does not possess a solution in $\mathbb{Z}$, which is the desired contradiction.

(c) Let $x, y \in R \setminus \{0\}$ and let d be a greatest common divisor of x and y; then $x = dx'$ and $y = dy'$ where x' and y' are relatively prime. Clearly, $dx'y'$ is a common multiple of x and y. We claim that $dx'y'$ is in fact a least common multiple of x and y.

Let m be any other common multiple of x and y. By hypothesis, the elements $dx'y'$ and m possess a greatest common divisor θ. In particular, there is an element α with $dx'y' = \theta\alpha$. Since both x and y are common divisors of $dx'y'$ and m, they are also divisors of θ, say $\theta = rx = sy$. Then $x'y = dx'y' = \theta\alpha = sy\alpha$ (hence $x' = s\alpha$) and $xy' = dx'y' = \theta\alpha = rx\alpha$ (hence $y' = r\alpha$). This shows that α divides x' and y' and hence must be a unit, because x' and y' are relatively prime. Thus $dx'y' = \theta\alpha \sim \theta$ is a divisor of m. This shows that $dx'y'$ is a least common multiple of x and y. ∎

WE now want to consider the situation that division by nonzero elements in an integral domain is always possible.

(3.17) Definition. *An integral domain R is called a **field** if $a \mid b$ for any two elements $a \in R \setminus \{0\}$ and $b \in R$.*

THERE is a non-commutative analogue to the notion of a field. Before we define this notion let us prove the following proposition.

(3.18) Proposition. *Let $R \neq \{0\}$ be a ring. Then the following conditions are equivalent.*

(a) *For all $a, b \in R \setminus \{0\}$ the equation $ax = b$ has a solution $x \in R$.*

(b) *The ring R has an identity element, and $R^\times = R \setminus \{0\}$, i.e., each nonzero element of R is invertible.*

(c) *For all elements $a \in R \setminus \{0\}$ and $b \in R$ the equations $ax = b$ and $ya = b$ have unique solutions $x, y \in R$.*

Proof. The hard part is to show that (a) implies (b). Fix an element $a \in R \setminus \{0\}$. By hypothesis, there is an element $e \in R$ with $ae = a$. Then clearly $e \neq 0$ and $ae^2 = ae = a$ so that $a(e^2 - e) = 0$. We claim that $e^2 = e$. Suppose not; then $e^2 - e \neq 0$, and by hypothesis there is an element e' with $(e^2 - e)e' = e$. But then $0 = 0 \cdot e' = a(e^2 - e)e' = ae = a$ which is the desired contradiction. Hence the claim $e^2 = e$ is established.

We now claim that e is an identity element of R. Let $b \in R$ and suppose $eb \neq b$ so that $eb - b \neq 0$. By hypothesis there is an element $x \in R$ with $(eb - b)x = e$. Then

$$0 = \underbrace{(e^2 - e)}_{=\,0}bx = e\underbrace{(eb - b)x}_{=\,e} = e^2 = e$$

which is a contradiction. We have established that

$(\star)$ $\qquad\qquad\qquad eb = b \qquad$ for all $b \in R$.

Suppose now there is an element $b \in R$ with $be \neq b$, i.e., $be - b \neq 0$. Then by hypothesis, there is an element $y \in R$ with $(be - b)y = a$. Since $ey = y$ by $(\star)$, we obtain $a = (be - b)y = by - by = 0$ which is again a contradiction. Consequently, $be = b$ for all $b \in R$. This establishes that e is an identity element.

Let us now show that $R^\times = R \setminus \{0\}$. Let $r \in R \setminus \{0\}$. By hypothesis, there is an element s with $rs = e$. We claim that s is also a left-inverse of r. Suppose not; then $sr - e \neq 0$ so that there is an element x with $(sr - e)x = e$. This implies $r = re = r(sr - e)x = (rsr - re)x = (er - re)x = 0$ which is the desired contradiction.

Let us show now that (b) implies (c). Given $a \in R \setminus \{0\}$ and $b \in R$, the equations $ax = b$ and $ya = b$ have unique solutions, namely $x = a^{-1}b$ and $y = ba^{-1}$.

Finally, it is trivial that (c) implies (a). ∎

(3.19) Definition. *A ring $R \neq \{0\}$ satisfying the equivalent conditions of proposition (3.18) is called a* **division ring**. *Clearly, a commutative division ring is a field; a non-commutative division ring is called a* **skew-field**.

(3.20) Examples. (a) $\mathbb{Q}$, $\mathbb{R}$ and $\mathbb{C}$ are fields.

(b) $\mathbb{Z}_n$ is a field if and only if n is a prime number.

(c) The ring $\mathbb{H}$ of quaternions is a skew-field.

THE next proposition reveals a connection between integral domains and fields.

(3.21) Proposition. (a) *Every field is an integral domain.*
(b) *Every finite integral domain is a field.*

Proof. (a) Let K be a field; this implies that K is commutative. Also, every nonzero element in K is a unit, hence cannot be a zero-divisor.

(b) Let R be in integral domain with the different elements $a_1, \ldots, a_n$. Let $a \neq 0$. Then the elements $aa_1, \ldots, aa_n$ are pairwise distinct by the cancellation rule, hence are all of the elements of R. In particular, there is an index i such that $aa_i = 1$. This shows that every non-zero a is invertible. A second proof goes as follows. Let $a \neq 0$. The powers $a, a^2, a^3, \ldots$ cannot all be different, because R is finite. Hence there are exponents $m > n$ such that $a^m = a^n$. The cancellation rule yields $a^{m-n} = 1$. This shows that a^{m-n-1} is inverse to a. ∎

FOR an illustration of statement (3.21)(b) compare (3.11)(c) and (3.20)(b). In this context we mention a famous theorem due to Wedderburn stating that every finite division ring is a field; in other words, there are no finite skew-fields. The proof of this statement requires more elaborate techniques and will be given later.

(3.22) Proposition. *Let $K \subseteq R$ be a ring extension such that R is an integral domain and K is a field. If $\dim_K R < \infty$ (where we consider R as a vector space over K) then R is a field.*

Proof. Let $r \in R \backslash K$. Then the left-multiplication $\ell_r : R \to R$ given by $\ell_r(x) = rx$ is injective because R is an integral domain, hence also surjective due to (I.4.32)(a). Consequently, there is an element $s \in R$ with $rs = 1$. This shows that every nonzero element $r \in R$ is invertible. ∎

MIMICKING the construction of the field $\mathbb{Q}$ of rational numbers from the ring $\mathbb{Z}$ of integers by forming fractions, we show in the next theorem that every integral domain can be embedded into a field. The construction exemplifies a procedure which is very common in mathematics: one extends a domain (to obtain more flexibility in handling certain operations) while preserving the characteristic structural properties of the original domain.

(3.23) Theorem. *Let R be a commutative ring without zero-divisors.*
(a) *We define a relation on $R \times (R \setminus 0)$ by $(a, b) \sim (c, d) :\Longleftrightarrow ad = bc$. This is an equivalence relation. The equivalence class of (a, b) is denoted by $\frac{a}{b}$ so that*

$$\frac{a}{b} = \frac{c}{d} \iff ad = bc .$$

(b) *The set $Q(R) := \left\{ \frac{a}{b} \mid a \in R, \, b \in R \setminus \{0\} \right\}$ of equivalence classes, endowed with the operations*

$$\frac{a}{b} + \frac{c}{d} := \frac{ad + bc}{bd} \quad and \quad \frac{a}{b} \cdot \frac{c}{d} := \frac{ac}{bd} ,$$

*is a field, called the **quotient field** of R.*

(c) *Pick any element $u \in Q(R) \setminus \{0\}$. Then*

$$
\begin{array}{rcl}
R & \to & Q(R) \\
r & \mapsto & \frac{ru}{u}
\end{array}
$$

is an embedding so that R can be considered as a subring of $Q(R)$. Since $\frac{ru}{u} = \frac{rv}{v}$ for all $u, v \neq 0$ we can unambiguously write r for this element of $Q(R)$.

Proof. The proof is a straightforward verification and is left as an exercise. (See problem 32 below.) ∎

(3.24) Examples. (a) If $R = \mathbb{Z}$ then $Q(R) = \mathbb{Q}$.
(b) If $R = 2\mathbb{Z}$, then $Q(R) = \{\frac{a}{2^m} \mid a \in \mathbb{Z}, m \in \mathbb{N}_0\}$.
(c) If $R = \mathbb{Z} + \mathbb{Z}\sqrt{n}$ where n is square-free, then $Q(R) = \mathbb{Q} + \mathbb{Q}\sqrt{n}$.
(d) If $R = K[x]$ is the polynomial ring over a field K then Q is the field of all "rational functions" in x, i.e., the set of all quotients $\frac{p}{q}$ where p and $q \neq 0$ are polynomials in x.
(e) Let R be an integral domain. Since every power series $f \in R[[x]]$ can be written in the form $x^n g(x)$ with $n \in \mathbb{N}_0$ and $g \in R[[x]]^\times$, due to (3.6)(g), the quotient field of $R[[x]]$ consists of all elements $x^m \varphi(x)$ with $m \in \mathbb{Z}$ and $\varphi \in R[[x]]^\times$, i.e., of all **Laurent series** $a_{-n} x^{-n} + \cdots + a_{-1} x^{-1} + a_0 + a_1 x + a_2 x^2 + \cdots$.

GENERALIZING on (3.24)(d), we give an example of particular importance.

(3.25) Example. Let R be an integral domain with quotient field K. Then the polynomial ring $R[x_1, \ldots, x_n]$ is again an integral domain; its quotient field is denoted by $R(x_1, \ldots, x_n)$ and consists of all formal expressions $\frac{p(x_1, \ldots, x_n)}{q(x_1, \ldots, x_n)}$ where $p, q \in R[x_1, \ldots, x_n]$ are polynomials with $q \neq 0$. The elements of $R(x_1, \ldots, x_n)$ are called **rational functions** in the variables $x_1, \ldots, x_n$, even though they are not really functions. Note that $R(x_1, \ldots, x_n)$ coincides with the quotient field $K(x_1, \ldots, x_n)$ of $K[x_1, \ldots, x_n]$; this is easily seen by clearing denominators.

WE close this section by an application where in an amazing way the mere introduction of a quotient field helps to solve calculus problems.

(3.26) Application: Mikusinski's Operator Calculus. Consider the ring $C[0, \infty)$ of all complex-valued continuous functions $f : [0, \infty) \to \mathbb{C}$ with the operations

$$
(f + g)(x) = f(x) + g(x) \qquad \text{and} \qquad (f \star g)(x) = \int_0^x f(x - t) g(t) \, \mathrm{d}t \ .
$$

This ring is commutative, and a theorem of Titchmarsh states that $C[0, \infty)$ has no zero-divisors. (The proof of Titchmarsh's theorem, which is purely analytic, has been deferred to the end of this section in order to not interrupt the algebraic argumentation.) Note that $C[0, \infty)$ has no identity element because $(f \star g)(0) = 0$ for all f, g.

By theorem (3.23) we can form the quotient field Q; its elements $\frac{f}{g}$ are called *operators* for reasons that will become clear in the sequel. In order not to mix up the constant function with the value 1 and the identity element in Q, we denote the function with the constant value $\alpha \in \mathbb{C}$ by c_α and let $[\alpha] := \frac{c_\alpha}{c_1} \in Q$. Then $[0]$ is the zero element and $[1]$ the identity element in Q. We note that

$$(1) \qquad\qquad [\alpha] + [\beta] = [\alpha + \beta] \qquad \text{and} \qquad [\alpha] \star [\beta] = [\alpha\beta] \quad \text{in } Q$$

which implies that $\begin{array}{ccc} \mathbb{C} & \to & Q \\ \alpha & \mapsto & [\alpha] \end{array}$ is an embedding. To check (1) we observe that

$$[\alpha] + [\beta] = \frac{c_\alpha}{c_1} + \frac{c_\beta}{c_1} = \frac{c_\alpha + c_\beta}{c_1} = \frac{c_{\alpha+\beta}}{c_1} = [\alpha + \beta] \qquad \text{and}$$

$$[\alpha] \star [\beta] = \frac{c_\alpha}{c_1} \star \frac{c_\beta}{c_1} = \frac{c_\alpha \star c_\beta}{c_1 \star c_1} = \frac{c_{\alpha\beta}}{c_1} = [\alpha\beta] .$$

Here the penultimate equality is equivalent to $c_\alpha \star c_\beta \star c_1 = c_{\alpha\beta} \star c_1 \star c_1$ (by the very definition of Q!) or simply to $c_\alpha \star c_\beta = c_{\alpha\beta} \star c_1$ by the cancellation rule. But this last equation holds true because

$$(c_\alpha \star c_\beta)(x) = \int_0^x \alpha\beta \, dt = \int_0^x (\alpha\beta) \cdot 1 \, dt = (c_{\alpha\beta} \star c_1)(x) \quad \text{for all } x.$$

Moreover, in $C[0,\infty)$ we can multiply a function g by a scalar α to get a new function αg defined by $(\alpha g)(x) = \alpha \cdot g(x)$. Now treating g and αg as elements of Q, we have

$$(2) \qquad\qquad \alpha g = [\alpha] \star g = \frac{c_\alpha}{c_1} \star g .$$

To verify this equation, we have to check that $(\alpha g) \star c_1 = c_\alpha \star g$, but this holds true because

$$((\alpha g) \star c_1)(x) = \int_0^x \alpha \, g(t) \, dt = (c_\alpha \star g)(x) .$$

To avoid one further potential source of misunderstanding, we have to distinguish clearly between a function $f \in C[0,\infty)$ and its value $f(x)$ at x. This is easy for functions which have a fixed name like $\exp, \sin, \tan$ and so on, but we do not want to introduce a new name for each function ("let h be the function defined by $h(x) = 2x - x^3$") and write for short $f = \{f(x)\}$. For example, $\{2x - x^3\}$ is the function h such that $h(x) = 2x - x^3$ for all x.

Now finally, after all the notation is settled, we can talk about the mathematical substance of Q. The key observation is that convolution with the constant function c_1 effects an integration; $c_1 \star f$ is the unique anti-derivative F of f such that $F(0) = 0$ because

$$(c_1 \star f)(x) = \int_0^x f(t) \, dt .$$

This suggests that the inverse s of c_1 (which exists in Q) plays the role of a differential operator (because differentiation and integration are inverse operations). Let us show that this is indeed true! If $f \in C[0,\infty)$ is continuously differentiable with $f' \in C[0,\infty)$ then $f(x) = \int_0^x f'(t) \, dt + f(0)$ for all x so that $f = c_1 \star f' + c_{f(0)} = c_1 \star f' + c_1 \star [f(0)]$. Apply $s := c_1^{-1} = \frac{[1]}{c_1}$ to this equation; the result is $s \star f = f' + [f(0)]$ or

$$(3) \qquad\qquad f' = s \star [f] - [f(0)] .$$

For example,

(4) $$s \star \{e^{\alpha x}\} = \{\alpha e^{\alpha x}\} + [1] = [\alpha] \star \{e^{\alpha x}\} + [1]$$

so that $(s - [\alpha]) \star \{e^{\alpha x}\} = [1]$ or

(5) $$\{e^{\alpha x}\} = \frac{[1]}{s - [\alpha]} \ .$$

Applying s for a second time gives $s \star s \star f = s \star f + s \star [f(0)] = f'' + [f'(0)] + s \star [f(0)]$ so that

(6) $$f'' = s \star s \star f - [f'(0)] - s \star [f(0)] \ .$$

Writing for short $s^n := s \star \cdots \star s$ (n factors), we obtain inductively the general formula

(7) $$f^{(n)} = s^n \star f - [f^{(n-1)}(0)] - s \star [f^{(n-2)}(0)] - \cdots - s^{n-2} \star [f'(0)] - s^{n-1} \star [f(0)] \ .$$

Hence we have transformed the "transcendental" operations of integration and differentiation by algebraic operations in Q. Let us see how this can be exploited to solve the initial value problem

(8) $$y''(x) - 3y'(x) + 2y(x) = e^{3x}, \quad y(0) = 2, \quad y'(0) = 4 \ .$$

We have $y' = s \star y - [2]$ and $y'' = s^2 \star y - [4] - s \star [2]$; so the given differential equation transforms to the algebraic equation

(9) $$s^2 \star y - [4] - s \star [2] - [3] \star s \star y + [6] + [2] \star y = \frac{[1]}{s - [3]}$$

into which the initial conditions are already built in. This algebraic equation can be rewritten as

(10) $$(s - [1]) \star (s - [2]) \star y = [2] \star s - [2] + \frac{[1]}{s - [3]} = \frac{[2] \star s^2 - [8] \star s + [7]}{s - [3]} \ .$$

Solving for y yields

(11) $$y = \frac{[2] \star s^2 - [8] \star s + [7]}{(s - [1])(s - [2])(s - [3])} = [\tfrac{1}{2}] \star \frac{[1]}{s - [1]} + \frac{[1]}{s - [2]} + [\tfrac{1}{2}] \star \frac{[1]}{s - [3]}$$
$$= \{\tfrac{1}{2} e^x\} + \{e^{2x}\} + \{\tfrac{1}{2} e^{3x}\} = \{\tfrac{1}{2} e^x + e^{2x} + \tfrac{1}{2} e^{3x}\} \ .$$

Hence the desired solution is

(12) $$y(x) = \frac{1}{2} e^x + e^{2x} + \frac{1}{2} e^{3x} \ .$$

This example gives us only a glimpse into a whole theory, called operational calculus, that can be developed to treat integral and differential equations. We maintained a rather cumbersome notation to make clear what kind of operations in which domains we used, but usually one drops the multiplication symbol $\star$ for computations in Q and writes α instead of $[\alpha]$, thus treating $\mathbb{C}$ as a subfield of Q. Equation (1) shows that we can do this, and equation (2) shows that the notation αg is unambiguous; therefore, no confusion can arise from this handier notation. (It is understood, of course, that all products in Q are convolution products, where the convolution is the canonical extension of the convolution in $C[0, \infty)$.) For example, equations (9) and (10) can be simply written as

(9')
$$s^2 y - 4 - 2s - 3sy + 6 + 2y = \frac{1}{s-3} \quad \text{and}$$

(10')
$$(s-1)(s-2)y = \frac{2s^2 - 8s + 7}{s-3} .$$

THE remainder of this section is devoted to proving Titchmarsh's theorem. We begin with a lemma.

(3.27) Lemma (Phragmén 1904). *If $g \in C[0, T]$ and if $t \in [0, T)$ is fixed then*

$$\sum_{k=1}^{\infty} \frac{(-1)^{k-1}}{k!} \int_0^T e^{kz(t-\tau)} g(\tau)\, d\tau \ \rightarrow \ \int_0^t g(\tau)\, d\tau \quad as \quad x \to \infty .$$

Proof. Using the fact that summation and integration can be exchanged if the summands are converging uniformly, we see that $\sum_{k=1}^{\infty} \frac{(-1)^{k-1}}{k!} \int_0^T e^{kz(t-\tau)} g(\tau)\, d\tau - \int_0^t g(\tau)\, d\tau$ equals

$$\int_0^T (-g(\tau)) \sum_{k=1}^{\infty} \frac{(-e^{z(t-\tau)})^k}{k!}\, d\tau - \int_0^t g(\tau)\, d\tau$$

$$= \int_0^T g(\tau) \big(1 - \exp(-e^{z(t-\tau)})\big)\, d\tau - \int_0^t g(\tau)\, d\tau$$

$$= -\int_0^t g(\tau) \exp(-e^{z(t-\tau)})\, d\tau + \int_t^T g(\tau) \big(1 - \exp(-e^{z(t-\tau)})\big)\, d\tau .$$

Now we have $\exp(-e^{z(t-\tau)}) \to 0$ as $x \to \infty$ uniformly in τ on each compact subinterval of $(0, t)$ and $\exp(-e^{z(t-\tau)}) \to 1$ as $x \to \infty$ uniformly in τ on each compact subinterval of (t, T). Hence both integrals in the last row of our calculation tend to zero as $x \to \infty$; whence the claim. ∎

(3.28) Theorem. *Let $f \in C[0,T]$.*

(a) *If there is a constant C such that $|\int_0^T e^{nt} f(t) \, \mathrm{d}t| \leq C$ for all $n \in \mathbb{N}$ then $f \equiv 0$.*

(b) *If $\int_0^T t^k f(t) \, \mathrm{d}t = 0$ for all $k \in \mathbb{N}$ then $f \equiv 0$.*

Proof. (a) Let $g(\tau) := f(T - \tau)$ and fix a number $t \in [0, T)$. Then for any natural number x we have

$$
\begin{aligned}
A(x) &:= \left| \sum_{k=1}^{\infty} \frac{(-1)^{k-1}}{k!} \int_0^T e^{kx(t-\tau)} g(\tau) \, \mathrm{d}\tau \right| \\
&= \left| \sum_{k=1}^{\infty} \frac{(-1)^{k-1}}{k!} e^{-kx(T-t)} \int_0^T e^{kx(T-\tau)} f(T - \tau) \, \mathrm{d}\tau \right| \\
&= \left| \sum_{k=1}^{\infty} \frac{(-1)^{k-1}}{k!} e^{-kx(T-t)} \int_0^T e^{kx\theta} f(\theta) \, \mathrm{d}\theta \right| \\
&\leq \sum_{k=1}^{\infty} \frac{1}{k!} e^{-kx(T-t)} \underbrace{\left| \int_0^T e^{kx\theta} f(\theta) \, \mathrm{d}\theta \right|}_{\leq C} \\
&\leq C \sum_{k=1}^{\infty} \frac{1}{k!} e^{-kx(T-t)} = C \left(\exp(e^{-x(T-t)}) - 1 \right)
\end{aligned}
$$

which tends to zero as $x \to \infty$. On the other hand $A(x)$ tends to $|\int_0^t g(\tau) \, \mathrm{d}\tau|$ as $x \to \infty$ due to (3.27). Hence $\int_0^t g(\tau) \, \mathrm{d}\tau = 0$ for all $0 \leq t < T$. Differentiating, we see that $g(t) = 0$ for all $0 < t < T$ which means that $f \equiv 0$ on $(0, T)$. Consequently, $f \equiv 0$ on $[0, T]$ by continuity.

(b) We have

$$
\int_0^T e^{nt} f(t) \, \mathrm{d}t = \int_0^T \sum_{k=0}^{\infty} \frac{(nt)^k}{k!} f(t) \, \mathrm{d}t = \sum_{k=0}^{\infty} \frac{n^k}{k!} \underbrace{\int_0^T t^k f(t) \, \mathrm{d}t}_{= 0 \text{ if } k > 0} = \int_0^T f(t) \, \mathrm{d}t
$$

for all $n \in \mathbb{N}$; hence part (a) applies and yields the claim. $\blacksquare$

(3.29) Proposition. *If $f \in C[0, 2T]$ satisfies $\int_0^t f(t - \tau) f(\tau) \, \mathrm{d}\tau = 0$ for all $0 \leq t \leq 2T$ then $f \equiv 0$ on $[0, T]$.*

Proof. By (3.28)(a) it is enough to show that there is a constant C such that $|\int_0^T e^{nu} f(T - u) \, \mathrm{d}u| \leq C$ for all $n \in \mathbb{N}$. Consider the sets

$$A := \{(u,v) \in \mathbb{R}^2 \mid u \geq -T, v \geq -T, u+v \leq 0\} \quad \text{and}$$
$$B := \{(u,v) \in \mathbb{R}^2 \mid u \leq T, v \leq T, u+v \geq 0\}.$$

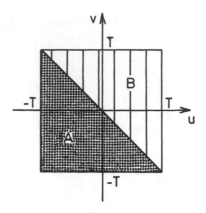

Then

$$\left(\int_{-T}^{T} e^{nu} f(T-u)\,\mathrm{d}u\right)^2 = \left(\int_{-T}^{T} e^{nu} f(T-u)\,\mathrm{d}u\right)\left(\int_{-T}^{T} e^{nv} f(T-v)\,\mathrm{d}v\right)$$

(⋆)
$$= \iint_{A \cup B} e^{n(u+v)} f(T-u)f(T-v)\,\mathrm{d}(u,v)$$

$$= \left(\iint_{A} + \iint_{B}\right) e^{n(u+v)} f(T-u)f(T-v)\,\mathrm{d}(u,v).$$

Let M the maximal value of $|f|$ on $[0, 2T]$. Since $u+v \leq 0$ on A we get the estimate

(1)
$$\left| \iint_{A} e^{n(u+v)} f(T-u)f(T-v)\,\mathrm{d}(u,v) \right| \leq \iint_{A} \underbrace{e^{n(u+v)}}_{\leq 1} \underbrace{|f(T-u)|}_{\leq M} \underbrace{|f(T-v)|}_{\leq M}\,\mathrm{d}(u,v)$$

$$\leq M^2 (\text{area of } A) = 2M^2 T^2 .$$

On the other hand, the substitution $t := 2T - u - v$, $\tau := T - v$ which maps B to $\{(t,\tau) \in \mathbb{R}^2 \mid 0 \leq t \leq 2T, 0 \leq \tau \leq t\}$ shows that

$$\iint_{B} e^{n(u+v)} f(T-u)f(T-v)\,\mathrm{d}(u,v) = \iint_{B} e^{n(2T-t)} f(t-\tau)f(\tau)\,\mathrm{d}(t,\tau)$$

(2)
$$= \int_{0}^{T} e^{n(2T-t)} \underbrace{\left(\int_{0}^{t} f(t-\tau)f(\tau)\,\mathrm{d}\tau\right)}_{= 0 \text{ by hypothesis}}\,\mathrm{d}t = 0 .$$

Substituting (1) and (2) into (⋆), we see that $|\int_{-T}^{T} e^{nu} f(T-u)\,\mathrm{d}u| \leq \sqrt{2}TM$. Consequently,

$$\left| \int_0^T e^{nu} f(T-u)\,\mathrm{d}u \right| = \left| \int_{-T}^T e^{nu} f(T-u)\,\mathrm{d}u - \int_{-T}^0 e^{nu} f(T-u)\,\mathrm{d}u \right|$$

$$\leq \left| \int_{-T}^T e^{nu} f(T-u)\,\mathrm{d}u \right| + \int_{-T}^0 \underbrace{e^{nu}}_{\leq 1} \underbrace{|f(T-u)|}_{\leq M}\,\mathrm{d}u$$

$$\leq \sqrt{2}TM + TM = (1+\sqrt{2})TM .$$

Then (3.28)(a) shows that $f(T-u)$ vanishes for all $u \in [0,T]$ which means that $f \equiv 0$ on $[0,T]$. ∎

(3.30) Corollary. *If $f \in C[0,\infty)$ satisfies $f \star f = 0$ then $f = 0$.*

Proof. By assumption we have $\int_0^t f(t-\tau)f(\tau)\,\mathrm{d}\tau = 0$ for all $t \geq 0$; hence $f \equiv 0$ on each interval $[0,T]$ due to (3.29). ∎

(3.31) Titchmarsh's Theorem. *Let $f,g \in C[0,\infty)$. If $f \star g = 0$ then $f = 0$ or $g = 0$.*

Proof. Let $F(t) := t \cdot f(t)$ and $G(t) := t \cdot g(t)$; then

$$(F \star g + f \star G)(t) = \int_0^t (t-\tau)f(t-\tau)g(\tau)\,\mathrm{d}\tau + \int_0^t f(t-\tau)\tau g(\tau)\,\mathrm{d}\tau$$

$$= t\int_0^t f(t-\tau)g(\tau)\,\mathrm{d}\tau = t \cdot (f \star g)(t) = 0$$

for all $t \geq 0$ so that $F \star g + f \star G = 0$. Multiplying both sides of this equation by $f \star G$ and using the commutativity and associativity of the convolution product, we see that

$$0 = \underbrace{(f \star g)}_{=0} \star (F \star G) + (f \star G) \star (f \star G) = (f \star G) \star (f \star G) ;$$

hence $f \star G = 0$ by (3.30). We have shown that $f \star g = 0$ implies $f \star (tg(t)) = 0$, and inductively we see that $f \star (t^n g(t)) = 0$ for all $n \in \mathbb{N}$. This means that $\int_0^t f(t-\tau)\tau^n g(\tau)\,\mathrm{d}\tau = 0$ for all $n \in \mathbb{N}$ and all $t \geq 0$. But then (3.28)(b) implies that $f(t-\tau)g(\tau) = 0$ whenever $0 \leq \tau \leq t < \infty$. Now suppose that $f \neq 0$ and $g \neq 0$, say $f(t_1) \neq 0$ and $g(t_2) \neq 0$. Letting $t := t_1 + t_2$ and $\tau := t_2$, this gives $0 \neq f(t_1)g(t_2) = f(t-\tau)g(\tau) = 0$ which is absurd. This contraction shows that we must have $f = 0$ or $g = 0$. ∎

Exercises

Problem 1. (a) Show that a zero-divisor in a ring cannot be a unit.

(b) Let a, b be elements of a commutative ring. Show that if neither a nor b is a zero-divisor then ab is not a zero-divisor either.

(c) Let R be a ring with identity element 1. Show that all elements of R which are no zero-divisors have the same order in the abelian group $(R, +)$.

Problem 2. Let R be a ring. Show that 0 is the only nilpotent element of R if and only if 0 is the only element $x \in R$ with $x^2 = 0$.

Problem 3. Let R be a ring with identity element 1.

(a) Show that if $x \in R$ is nilpotent then $1 - x$ is a unit.

Hint. Mimick the geometric series to obtain $(1 - x)^{-1}$.

(b) Show that if u is a unit and if $x_1, \ldots, x_n \in R$ are nilpotent elements which commute with each other and with u then $u + x_1 + \cdots + x_n$ is a unit.

(c) Show that $1 - xy$ is a unit if and only if $1 - yx$ is.

Hint. To get an idea, mimick the geometric series.

Problem 4. Let R and S be rings with identity.

(a) Show that $(R \times S)^\times = R^\times \times S^\times$.

(b) Apply part (a) in the situation $R = \mathbb{Z}_m$ and $S = \mathbb{Z}_n$ to show that if m and n are coprime then $\varphi(mn) = \varphi(m)\varphi(n)$.

Problem 5. Let $R = R_1 \times \cdots \times R_n$ be a direct product of rings. How are the zero-divisors and the nilpotent elements of R related to the zero-divisors and the nilpotent elements in the factors R_i? If each R_i is a unital ring, how are the units of R related to the units of the factors R_i?

Problem 6. (a) Find an example for a ring homomorphism which does not map zero-divisors to zero-divisors.

(b) Show that an injective ring homomorphism maps zero-divisors to zero-divisors.

(c) Let $f : R \to S$ be a homomorphism between rings with identity such that $f(1) = 1$. Show that f maps units to units.

Problem 7. (a) Find all units in the polynomial ring $\mathbb{Z}_4[x]$.

(b) Find all units in the matrix ring $\{ \begin{pmatrix} a & 0 \\ b & c \end{pmatrix} \mid a, b, c \in \mathbb{Z}_6 \}$.

(c) Let R be a commutative ring with identity and let $R + iR$ be its complexification as defined in problem 11 of section 2. Show that

$$(R + iR)^\times = \{ a + ib \mid a^2 + b^2 \in R^\times \} .$$

What are the units of $\mathbb{Z} + i\mathbb{Z}$?

Problem 8. Let $\{A_{ij} \mid 1 \leq i, j \leq n\} \subseteq K^{m \times m}$ be a commuting family of matrices over a field K and let $R := K[A_{11}, \ldots, A_{nn}]$ be the subring of $K^{m \times m}$ generated by these A_{ij}. Then the block matrix

$$A := \begin{pmatrix} A_{11} & \cdots & A_{1n} \\ \vdots & & \vdots \\ A_{n1} & \cdots & A_{nn} \end{pmatrix}$$

can be interpreted both as an element of $K^{mn \times mn}$ or as an element of $R^{n \times n}$. Show that $\det_K(A) = \det_K(\det_R(A))$.

Hint. Use problem 19 in section [I.7].

Problem 9. (a) Let $n \geq 2$. Show that every nonzero element of $\mathbb{Z}_n$ is either a unit or a zero-divisor.

(b) Show that $\mathbb{Z}_n$ has no nonzero nilpotent elements if and only if n is square-free.

Problem 10. Let $\text{End}(V)$ be the endomorphism ring of a vector space. Show that if $\dim V < \infty$ then every nonzero element of R is either a unit or a zero divisor. Show that if $\dim V = \infty$ then there are nonzero elements in R which are neither a unit nor a zero divisor.

Problem 11. Let $X \neq \emptyset$ be a non-empty set and let $\mathcal{P}(X)$ be the Boolean ring introduced in problem 31(c) of section 2. Show that the identity element X of $\mathcal{P}(X)$ is the only unit in $\mathcal{P}(X)$ whereas all other elements $Y \neq X$ of $\mathcal{P}(X)$ are zero-divisors.

Problem 12. Let $R = C(\mathbb{R})$ be the ring of all continuous functions $f : \mathbb{R} \to \mathbb{R}$.

(a) Show that $f \in R$ is a unit if and only if $f(x) \neq 0$ for all $x \in R$.

(b) Show that $f \in R$ is a zero-divisor if and only if the set $\{x \in \mathbb{R} \mid f(x) = 0\}$ has non-empty interior.

(c) Find elements of R which are neither units nor zero-divisors.

Problem 13. Let $A \in \mathbb{Z}^{2 \times 2}$ be a nonzero matrix such that $A^2 = \lambda A$ for some $\lambda \in \mathbb{Z} \setminus \{0\}$; then $R := \{mA \mid m \in \mathbb{Z}\}$ is a commutative subring of $\mathbb{Z}^{2 \times 2}$.

(a) Show that R has an identity element if and only if $\lambda = \pm 1$; determine $R^\times$ in this case.

(b) Show that R has no zero-divisors.

(c) Show that $mA \mid mA$ if and only if $\lambda = n = \pm 1$. Which elements r of R have the property that $r \mid r_1 r_2$ implies that $r \mid r_1$ or $r \mid r_2$?

(d) Suppose mA possesses proper factors if and only if $m = \lambda m'$ where m' is not a prime.

Problem 14. Consider the ring $(R^{\mathbb{N}}, +, \star)$ as in problem 7 of section 2.

(a) Show that f is a unit in $R^{\mathbb{N}}$ if and only if $f(1)$ is a unit in R.

(b) Suppose $f \in R^{\mathbb{N}}$ is multiplicative with $f(1) \in R^\times$. Show that f^{-1} is also multiplicative.

Problem 15. Let R be a commutative ring with identity element 1. Show that a polynomial $f \in R[x]$ is a zero-divisor in $R[x]$ if and only if $af = 0$ for some $a \in R \setminus \{0\}$.

Hint. Let $f(x) = a_0 + a_1 x + \cdots + a_n x^n$ and let $m \geq 0$ be the minimal number such that there is a polynomial $g(x) = b_0 + b_1 x + \cdots + b_m x^m$ with $b_m \neq 0$ and $fg = 0$. (We call m the degree of g.) Then $a_n b_m = 0$ so that $g_0 := a_n g$ has a smaller degree than g; hence $fg_0 = 0$ implies $g_0 = 0$, i.e., $a_n g = 0$. Consequently, $(a_0 + a_1 x + \cdots + a_{n-1} x^{n-1})g(x) = 0$ which implies that $a_{n-1} b_m = 0$ so that $g_1 := a_{n-1} g$ has smaller degree than g; hence $fg_1 = 0$ implies $g_1 = 0$, i.e., $a_{n-1} g = 0$. Continuing in this way, show that $a_k g = 0$ for $0 \leq k \leq n$.

Problem 16. Suppose R is a finite ring and $a \in R$ is not a zero-divisor. Show that R has an identity element and that a is a unit.

Problem 17. For each of the values $n = 6, 7, 8$, find three solutions in natural numbers of the equation $x^2 - ny^2 = 1$.

Problem 18. Show that there are infinitely many right triangles such that the difference of the lengths of the two small legs is 1.

Problem 19. Let $n \neq 1$ be a square-free number with $n \equiv 1$ modulo 4. Let $R_n := \mathbb{Z} + \mathbb{Z}\sqrt{n}$ and $\mathcal{O}_n := \mathbb{Z} + \mathbb{Z}(\frac{1 + \sqrt{n}}{2})$.

(a) Show that each element of $\mathcal{O}_n$ has a unique representation $z = \frac{1}{2}(a + b\sqrt{n})$ with $a, b \in \mathbb{Z}$ either both even and both odd. Conclude that R_n is a subring of $\mathcal{O}_n$.

(b) Using the norm $N(z) = z\bar{z}$ as defined in (3.6)(c), show that

$$\mathcal{O}_n^{\times} = \{z \in \mathcal{O}_n \mid N(z) = \pm 1\} = \{\frac{a + b\sqrt{n}}{2} \in \mathcal{O}_n \mid a^2 - nb^2 = \pm 4\} .$$

Conclude that $\mathcal{O}_n^{\times} = R_n^{\times}$ if $n \equiv 1$ modulo 8.

(c) Suppose that $n < 0$. Show that $\mathcal{O}_n^{\times} = \{\pm 1\}$ if $|n| > -3$ whereas $\mathcal{O}_{-3}^{\times}$ is the group of all sixth roots of unity, i.e.,

$$\mathcal{O}_{-3}^{\times} = \{ \pm 1, \frac{-1 \pm \sqrt{-3}}{2}, \frac{1 \pm \sqrt{-3}}{2} \} .$$

(d) Suppose now that $n > 0$ so that $\mathcal{O}_n \subseteq \mathbb{R}$. Show that there is a smallest unit $u > 1$ (called the **basis unit** of $\mathcal{O}_n$) and that $\mathcal{O}_n^{\times} = \{\pm u^m \mid m \in \mathbb{Z}\}$.

(e) Suppose that $z \in \mathcal{O}_n \setminus R_n$. Show that if $n \equiv 1$ modulo 8 then $z^N \in \mathcal{O}_n \setminus R_n$ for all $N \in \mathbb{N}$ whereas if $n \equiv 5$ modulo 8 then $z^N \in \mathcal{O}_n \setminus R_n$ if and only if N is not divisible by 3. In the case that $n \equiv 5$ modulo 8 and that N divisible by 3, we have $z^N = x + y\sqrt{n}$ with $x \in \mathbb{Z}$ odd and $y \in \mathbb{Z}$ even.

Hint. Write $z = \frac{1}{2}(a + b\sqrt{n})$ with a, b both odd and use the fact that $a^2 \equiv b^2 \equiv 1$ modulo 8.

Problem 20. Let $n \in \mathbb{Z} \setminus \{0, 1\}$ be a square-free number and let

$$\mathcal{O}_n := \begin{cases} \mathbb{Z} + \mathbb{Z}\sqrt{n}, & \text{if } n \not\equiv 1 \text{ modulo 4;} \\ \mathbb{Z} + \mathbb{Z}(\frac{1+\sqrt{n}}{2}), & \text{if } n \equiv 1 \text{ modulo 4.} \end{cases}$$

(a) Show that the basis units of $\mathcal{O}_2$, $\mathcal{O}_3$, $\mathcal{O}_5$, $\mathcal{O}_{13}$ and $\mathcal{O}_{21}$ are $1 + \sqrt{2}$, $2 + \sqrt{3}$, $\frac{1}{2}(1+\sqrt{5})$, $\frac{1}{2}(3+\sqrt{13})$ and $\frac{1}{2}(5+\sqrt{21})$, respectively. Use these facts to find for each number $n \in \{2,3,5,13,21\}$ the smallest solution $(x,y) \in \mathbb{N}^2$ of the equation $x^2 - ny^2 = 1$.

Hint. If (x,y) is this solution, then $z := x + y\sqrt{n}$ must be a power of the basis unit of $\mathcal{O}_n$.

(b) Suppose $n = x^2 + 1 \neq 5$. Show that $z := x + \sqrt{n}$ is the basis unit of $\mathcal{O}_n$. (For example, $6 + \sqrt{37}$ is the basis unit of $\mathcal{O}_{37}$.) Conclude that $\mathcal{O}_n^{\times} = (\mathbb{Z} + \mathbb{Z}\sqrt{n})^{\times}$.

Hint. Let u be the basis unit of $\mathcal{O}_n$. Since $N(z) = -1$, the element z must be an odd power of u; hence if $z \neq u$ then $z = u^r$ with $r \geq 3$ so that $u \leq \sqrt[3]{z}$. Derive a contradiction from this inequality.

Problem 21. A field $\mathbb{Q} \subsetneq K \subseteq \mathbb{C}$ is called a **quadratic number field** if each element $x \in K$ satisfies a quadratic equation $ax^2 + bx + c = 0$ with coefficients $a, b, c \in \mathbb{Q}$. Show that the quadratic number fields are exactly the sets $K_n := \mathbb{Q} + \mathbb{Q}\sqrt{n}$ where $n \in \mathbb{Z} \setminus \{0,1\}$ is a square-free number and that $K_n \neq K_m$ if $n \neq m$. Moreover, show that if K and K' are different quadratic number fields then $K \cap K' = \mathbb{Q}$.

Problem 22. If a, b are elements of a ring with identity such that $ab = 1$ then b is called a **right-inverse** of a whereas a is called a **left-inverse** of b.

(a) By A we denote the abelian group of all sequences $(x_0, x_1, x_2, \ldots)$ with coefficients in $\mathbb{Z}_2$, equipped with the coordinatewise operations. Let $R = \operatorname{End} A$ be the endomorphism ring of A and define $\alpha \in R$ by

$$\alpha(x_0, x_1, x_2, \ldots) := (x_1, x_2, x_3, \ldots) .$$

Find all right-inverses of α, i.e., all elements $\beta \in \operatorname{End} A$ with $\alpha \circ \beta = \operatorname{id}$.

(b) Let R be a ring with identity. Show that if an element $a \in R$ possesses two different right-inverses, then it possesses infinitely many right-inverses.

Hint. Let S be the set of all right-inverses of a and let $b \in S$ be a fixed element. Show that

$$
\begin{array}{ccc}
S & \to & S \\
b' & \mapsto & b'a + b - 1
\end{array}
$$

is a well-defined injective mapping. Assuming S finite, obtain a contradiction.

(c) Let R be a ring with identity element 1. Show that if $a \in R$ possesses a right-inverse then either a is a zero-divisor or a unit.

Problem 23. Let $(R, +, \cdot)$ be a ring (not necessarily unitary). An element x is called a **quasi-unit** if there are elements $y_1, y_2 \in R$ with $x + y_1 = xy_1$ and $x + y_2 = y_2 x$.

(a) Define an operation on R by

$$x \circ y := x + y - xy .$$

Show that $(R, \circ)$ is a semigroup with identity element 0 which is commutative if and only if $(R, \cdot)$ is. Show that the quasi-units of R are exactly the invertible elements in $(R, \circ)$.

(b) Show that x is a quasi-unit if and only if there is an element $y \in R$ with $x + y = xy = yx$; this element is uniquely determined and is called the **quasi-inverse** of x.

(c) Show that every nilpotent element of R is a quasi-unit.

Hint. Consider $y = -(x + x^2 + x^3 + \cdots + x^n)$.

(d) Suppose now that R has an identity element 1. Show that $x \in R$ is a quasi-unit (with quasi-inverse y) if and only if $1 - x$ is a unit (with inverse $1 - y$).

Problem 24. Show that up to isomorphism there is a unique integral domain R with 4 elements. Determine its addition and multiplication table.

Hint. Let $R = \{0, 1, a, b\}$. Show that char $R = 2$, $b = a + 1$ and $ab = 1$.

Problem 25. Let R be the ring of all entire functions, i.e., all holomorphic functions $f : \mathbb{C} \to \mathbb{C}$. Show that R is an integral domain with $R^\times = \{f \in R \mid f(z) \neq 0$ for all $z \in \mathbb{C}\} = \{e^F \mid F \in R\}$.

Hint. If $f \in R^\times$ then f has no zero so that

$$F(z) := \int_0^z \frac{f'(\zeta)}{f(\zeta)} \, d\zeta$$

is also an element of R. Now show that $f = f(0)e^F$ by verifying that the function fe^{-F} has zero derivative.

Problem 26. Let R be an integral domain with char $R \neq 2$. Show that $U := \{x^2 \mid x \in R^\times\}$ form a subgroup of $R^\times$ of index 2.

Hint. The mapping $\varphi : R^\times \to R^\times$ given by $x \mapsto x^2$ is a group homomorphism.

Problem 27. Prove proposition (3.15).

Problem 28. Let R be a commutative ring and let $X = \{x_i \mid i \in I\}$ be a non-empty set of variables. Prove that the following conditions are equivalent.

(a) R is an integral domain.

(b) $R[X]$ is an integral domain.

(c) $R[[X]]$ is an integral domain.

Problem 29. (a) Let R be an integral domain with identity element 1_R and let $U \neq \{0\}$ be a subring of R. Show that if U has an identity element 1_U then $1_U = 1_R$.

(b) Let R be a ring with identity element 1_R and let $f : R \to S$ be a ring homomorphism into an integral domain S. Show that either $f(1_R) = 1_S$ or else $f \equiv 0$.

Remark. Compare with problems 15 and 16 in section 2.

Problem 30. Let R be an integral domain of characteristic $p > 0$. Show that $m \cdot a = n \cdot a$ if and only if $m - n$ is divisible by p.

Problem 31. Let R be an integral domain. Show that the automorphisms of the polynomial ring $R[x]$ are exactly the mappings $f(x) \mapsto f(ax+b)$ with $a \in R^\times$ and $b \in R$.

Problem 32. Prove theorem (3.23).

Problem 33. Find two elements in $\mathbb{Z} + \mathbb{Z}\sqrt{-3}$ which have a greatest common divisor, but not a least common multiple.

Problem 34. Let R be an integral domain with quotient field K.
(a) Show that each embedding $f : R \to L$ into a field L can be uniquely extended to an embedding $F : K \to L$.
(b) Show that each automorphism of R can be uniquely extended to an automorphism of K.

Problem 35. Let $R \neq \{0\}$ be a ring. Suppose that for each element $a \neq 0$ there is a unique element b with $aba = a$. Show that R is a division ring.
Hint. Show first that R has no zero-divisors and that the condition $aba = a$ in the hypothesis implies $bab = b$.

Problem 36. Show that a finite ring $R \neq \{0\}$ without zero-divisors is a division ring.

Problem 37. Let $f : R \to S$ by a nonzero ring homomorphism. Show that if R is a division ring then f is injective.

Problem 38. Let $R \neq \{0\}$ be a commutative ring without zero-divisors. Show that each of the following conditions implies that R is a field.
(1) Every proper subring of R is finite.
(2) R has only a finite number of ideals.
Here an ideal is a subring of R with the special property that multiplying an element of this subring with any element of R yields an element of the subring again.
Hint. If (a) holds, show that $Rx = R$ for all $x \neq 0$. If (b) holds, show that for any $x \neq 0$ there are natural numbers $i < j$ with $Rx^i = Rx^j$ and conclude that $x^j \in x^j R$.

Problem 39. (a) Let K be an arbitrary field. Define a relation on K by letting $a \sim b$ if and only if ab can be written as a sum of two squares in K. Show that $\sim$ is an equivalence relation.
(b) Let $K = \mathbb{R}(x)$ be the rational function field over $\mathbb{R}$. Each element $f \in \mathbb{R}(x)$ can be identified with a real-valued function which is defined on $\mathbb{R}$ with a finite number of elements removed. Show that f can be written as a sum of two squares in $\mathbb{R}(x)$ if and only if $f \geq 0$ on the domain of f.
Hint. Use the fact that every polynomial $p \in \mathbb{R}[x]$ can be factorized into polynomials of degree 1 and 2 in $\mathbb{R}[x]$.
(c) Show that if K is a finite field then every element of K can be written as a sum of two squares.

Problem 40. (a) Show that there is no field K such that $(K, +) \cong (K^\times, \cdot)$.
(b) Show that there are fields K and L such that $(K, +) \cong (L^\times, \cdot)$.

4. Polynomial and power series rings

POLYNOMIAL rings are probably the most important examples of rings. The main purpose of this section is to introduce some terminology and basic results which are necessary to discuss these rings. Since polynomials are special power series, it is natural to also include power series rings in the discussion. This section has a preparatory character; deeper results will be obtained only in subsequent sections, as we dwell into the general theory in greater depth. Let us start by defining the degree of a polynomial.

(4.1) Definition. *Let R be an arbitrary ring. If $f(x) = a_0 + a_1 x + \cdots + a_n x^n$ with $a_n \neq 0$ then we call n the* **degree** *of f and write $n = \deg f$. Moreover, we call a_n the* **leading coefficient** *and $a_n x^n$ the* **leading term** *of f. If R is a unitary ring and $a_n = 1$, then f is called* **monic**. *Formally, the degree of the zero polynomial is defined as $\deg 0 := -\infty$.*

OCCASIONALLY we will adopt the common terminology of calling a polynomial f constant, linear, quadratic, or cubic if $\deg f \leq 0$, $\deg f = 1$, $\deg f = 2$, or $\deg f = 3$, respectively. There are two rather obvious properties of the degree of a polynomial.

(4.2) Theorem. *Let R be a ring and let $f, g \in R[x]$.*
(a) $\deg(f + g) \leq \max(\deg f, \deg g)$ with equality if and only if either f and g have different degrees or else the sum of the leading coefficients of f and g is not zero.
(b) $\deg(fg) \leq \deg(f)\deg(g)$ with equality if and only if the product of the leading coefficients of f and g is not zero.

Proof. Trivial. ∎

WE now want to generalize the notion of degree to polynomials in several variables. Before we do so, it is appropriate to introduce a simplifying notation.

(4.3) Terminology. Recall that a power series over a ring R in n (commuting) variables $x_1, \ldots, x_n$ is an expression of the form

$$(\star) \qquad f(x_1, \ldots, x_n) = \sum_{i_1, \ldots, i_n \geq 0} a_{i_1, \ldots, i_n} x_1^{i_1} \cdots x_n^{i_n}$$

where $a_{i_1, \ldots, i_n} \in R$. We now combine the variables $x_1, \ldots, x_n$ into a vector $x = (x_1, \ldots, x_n)$ and any numbers $i_1, \ldots, i_n \in \mathbb{N}_0$ into a **multi-index** $I = (i_1, \ldots, i_n)$; then we can simply write $x^I := x_1^{i_1} \cdots x_n^{i_n}$. Every polynomial of the special form x^I is called a **monomial** or **power product**. With the use of multi-indices, the power series $(\star)$ simply reads

$$(\star\star) \qquad f(x) = \sum_{I \in \mathbb{N}_0^n} a_I x^I .$$

Moreover, addition and multiplication of power series (and hence polynomials) are in this notation given by

$$\sum_I a_I x^I + \sum_I b_I x^I \;=\; \sum_I (a_I + b_I) x^I \,, \quad \Big(\sum_I a_I x^I\Big)\Big(\sum_I b_I x^I\Big) \;=\; \sum_I \Big(\sum_{I_1 + I_2 = I} a_{I_1} b_{I_2}\Big) x^I \,.$$

Finally, the **length** of a multi-index $I = (i_1, \ldots, i_n)$ is defined as $|I| = i_1 + \cdots + i_n$.

LET us now turn to defining the degree of a polynomial in several variables and some related concepts.

(4.4) Definition. *Let $f(x) = \sum_I a_I x^I \in R[x_1, \ldots, x_n]$ be a power series in n commuting variables $x_1, \ldots, x_n$ over a ring R.*

(a) *If $f \neq 0$, then the **subdegree** $\operatorname{subdeg} f$ of f is the smallest length of a multi-index I which occurs in f, i.e., for which $a_I \neq 0$. Formally, we define the subdegree of the zero power series as $\operatorname{subdeg} 0 := \infty$.*

(b) *If $f \neq 0$ is a polynomial, then there are only finitely many multi-indices I with $a_I \neq 0$. Then we can define the **degree** $\deg f$ of f as the largest length of a multi-index I which occurs in f. Formally, we define the degree of the zero polynomial as $\deg 0 := -\infty$.*

(c) *The polynomial f is called **homogeneous** of degree k if $|I| = k$ for all occurring monomials x^I. This is clearly the case if and only if $\operatorname{subdeg} f = \deg f = k$.*

(4.5) Example. If $f(x, y, z) = 3x^2 y z + 4x y^2 z + 5 x^3 y^4 z^5 - x^7 z^6$ then $\operatorname{subdeg} f = 4$ and $\deg f = 13$.

WE then have the following properties which are easily verified.

(4.6) Proposition. *Let R be a ring.*

(a) *If $f, g \in R[[x_1, \ldots, x_n]]$ are power series over R, then $\operatorname{subdeg}(f + g) \leq \operatorname{subdeg} f + \operatorname{subdeg} g$. Moreover, $\operatorname{subdeg}(fg) \geq (\operatorname{subdeg} f)(\operatorname{subdeg} g)$ with equality if R has no zero-divisors.*

(b) *If $f, g \in R[x_1, \ldots, x_n]$ are polynomials over R, then $\deg(f + g) \leq \deg f + \deg g$. Moreover, $\deg(fg) \leq (\deg f)(\deg g)$ with equality if R has no zero-divisors.*

Proof. Exercise. (See problem 1 below.) ∎

SOMETIMES it is convenient to write down the multi-indices occurring in a polynomial in a specified order. The following remark lists some convenient possibilities to do so.

(4.7) Remark. There are several ways in which we can arrange the terms of a power series $f \in R[x_1, \ldots, x_n]$.

(a) We can write $f(x) = \sum_I a_I x^I$ where the occurring monomials are arranged to ascend with respect to the lexicographic ordering on the set of all multi-indices.

(b) We may pick one variable, say x_n, and order everything in terms of x_n so that

$$f(x_1,\ldots,x_n) \;=\; \sum_{i=0}^{\infty} c_i(x_1,\ldots,x_{n-1})x_n^i$$

where $c_i \in R[[x_1,\ldots,x_{n-1}]]$. Doing this, we treat the power series f as an element of $\big(R[[x_1,\ldots,x_{n-1}]]\big)([[x_n]])$; this quite often allows one to transfer results for power series in one variable to power series in several variables.

If f is a polynomial then there is a largest number s such that $c_s \neq 0$; this number is called the **degree of f in the variable x_n** and is denoted by $\deg_{x_n} f$.

(c) We may also order a power series by its homogeneous components:

$$f(x) \;=\; f_0(x) + f_1(x) + f_2(x) + \cdots$$

where $f_k(x) = \sum_{|I|=k} a_I x^I$.

THE next point to discuss is the fact that every polynomial $p \in R[x_1,\ldots,x_n]$ over a *commutative* ring R gives rise to a function $R^n \to R$ as follows. Given a polynomial $p(x) = \sum_I a_I x^I$ and ring elements $(r_1,\ldots,r_n)$ we can "plug in" the ring elements $r_1,\ldots,r_n$ for the variables $x_1,\ldots,x_n$ to obtain the well-defined ring element $p(r_1,\ldots,r_n) := \sum_I a_I r^I$ where $r^I := r_1^{i_1}\cdots r_n^{i_n}$. Clearly, the assignment $(r_1,\ldots,r_n) \mapsto p(r_1,\ldots,r_n)$ is a well-defined function $R^n \to R$ which we call the polynomial function associated with p.† An important property is as follows.

(4.8) Proposition. *Let R be a commutative ring. Then for any elements $r_1,\ldots,r_n \in R$ the* **evaluation map**

$$\begin{array}{ccc} R[x_1,\ldots,x_n] & \mapsto & R \\ \sum_I a_I x^I & \mapsto & \sum_I a_I r^I \end{array}\;, \qquad \text{for short} \qquad \begin{array}{ccc} R[x_1,\ldots,x_n] & \to & R \\ p & \mapsto & p(r_1,\ldots,r_n) \end{array}\;,$$

is a homomorphism.

Proof. Exercise. (See problem 1 below.) ∎

WE should note at this point that it is perfectly possible that two different polynomials p and q define the same polynomial function; for example, all polynomials $p(x) = x^n + x \in \mathbb{Z}_2[x]$ with $n \geq 1$ define the zero function on $\mathbb{Z}_2$. Thus it is important to distinguish between a polynomial p and the polynomial function induced by p. It will always be clear from the context whether we are talking about polynomials or polynomial functions. Occasionally, we will use capital letters to denote variables and small letters to denote ring elements to make the distinction between a polynomial

---

† Since we defined only polynomial rings in *commuting* variables, this definition does not make sense if $n \geq 2$ and R is not commutative. In the case $n = 1$, the definition $p(r) := a_0 + a_1 r + \cdots + a_n r^n$ makes sense even if R is not commutative, but we do not have $(pq)(r) = p(r)q(r)$ for all $p,q \in R[x]$ and $r \in R$ in this case.

$p(X_1, \ldots, X_n) \in R[X_1, \ldots, X_n]$ in the variables $X_1, \ldots, X_n$ and the polynomial function $(x_1, \ldots, x_n) \mapsto p(x_1, \ldots, x_n)$ with ring elements $x_1, \ldots, x_n$ very clear.

(4.9) Definition. *Let R be a commutative ring and let $f \in R[x_1, \ldots, x_n]$. If $f(r_1, \ldots, r_n) = 0$ with elements $r_1, \ldots, r_n \in R$, then $(r_1, \ldots, r_n)$ is called a **root** or a **zero** of f.*

LET us first discuss roots of polynomials in one variable. The first theorem shows that if r is a root of f then we can "split apart" the factor $(x - r)$. Since in the case of one variable the assignment $r \mapsto f(r)$ is well-defined even if R is not commutative, we do not assume commutativity for this theorem.

(4.10) Theorem. *Let R be a ring with identity element 1 and let $f, g \in R[x]$.*

*(a) (**Division with remainder.**) Suppose the leading coefficient of g is a unit in R. Then there are unique polynomials $q, r \in R[x]$ such that*

$$f = qg + r \quad and \quad \deg r < \deg g .$$

(b) For any $c \in R$, there is a unique polynomial $q \in R[x]$ such that

$$f(x) = q(x) \cdot (x - c) + f(c) .$$

Consequently, $f(c) = 0$ if and only if there is a polynomial q such that $f(x) = q(x)(x - c)$.

(c) Let $c \in R$ and $m \geq 1$. Then the following conditions are equivalent:

(1) there is a polynomial $q \in R[x]$ with $f(x) = q(x)(x - c)^m$ and $q(c) \neq 0$;

(2) there is a polynomial g with $f(x) = g(x)(x - c)^m$, but there is no polynomial h with $f(x) = g(x)(x - c)^{m+1}$.

Proof. (a) Let us first prove the *existence* of q and r. If $\deg g > \deg f$, let $q := 0$ and $r := f$. Suppose $\deg g \leq \deg f$. Then we have $f(x) = \sum_{k=0}^{n} a_k x^k$ and $g(x) = \sum_{k=0}^{m} b_k x^k$ where $a_n \neq 0$, $b_m \in R^\times$ and $m \leq n$. We proceed by induction on n. If $n = 0$, then $m = 0$ so that $f = a_0$ and $g = b_0$. In this case, we can take $q := a_0 b_0^{-1}$ and $r := 0$. Let $n \geq 1$. Then $a_n b_m^{-1} x^{n-m} g(x)$ has the same leading term as $f(x)$ so that

$$F(x) := f(x) - a_n b_m^{-1} x^{n-m} g(x)$$

has a degree less than n. By induction hypothesis, we can find $q', r \in R[x]$ such that $F = q'q + r$ and $\deg r < \deg g$. Now let

$$q(x) := a_n b_m^{-1} x^{n-m} + q'(x) ;$$

then $f = qg + r$. Next, we show the *uniqueness of q and r*. Suppose $q_1 g + r_1 = q_2 g + r_2$ where r_1 and r_2 have smaller degree than g. Then $(q_1 - q_2)g = r_2 - r_1$. This implies

$$\deg g > \deg(r_2 - r_1) = \deg(q_1 - q_2)g = \deg(q_1 - q_2) + \deg g$$

where the equality $\deg(q_1 - q_2)g = \deg(q_1 - q_2) + \deg g$ holds because the leading coefficient of g is a unit. But this shows that $\deg(q_1 - q_2) < 0$ which is only possible if $\deg(q_1 - q_2) = -\infty$, i.e., $q_1 = q_2$ and consequently $(q_1 - q_2)g = 0$, whence $q_1 = q_2$.

(b) We apply (a) with $g(x) := x - c$. Then $f(x) = q(x)(x - c) + r(x)$ where $\deg r < \deg g = 1$ so that $r(x) = r_0$ is a constant polynomial (possibly 0). We want to see that $r_0 = f(c)$. Write† $q(x) = \sum_{k=0}^{n-1} b_k x^k$. Then

$$f(x) = q(x)(x - c) + r_0 = \sum_{k=0}^{n-1} b_k x^{k+1} - \sum_{k=0}^{n-1} b_k c x^k + r_0$$

so that

$$f(c) = \sum_{k=0}^{n-1} b_k c^{k+1} - \sum_{k=0}^{n-1} b_k c^{k+1} + r_0 = r_0 .$$

If $f(c) = 0$ then the existence of q follows immediately from the above equation. Conversely suppose that $f(x) = q(x)(x - c)$ for some $q \in R[x]$. On the other hand, $f(x) = q_1(x)(x - c) + f(c)$ for some q_1 by the above equation. The uniqueness claim in (a) gives $f(c) = 0$ (and $q_1 = q$).

(c) Suppose (1) holds. Let $g := q$ and suppose that $f(x) = h(x)(x - c)^{m+1}$ for some h. Then the uniqueness claim in (a) gives $g(x) = h(x)(x - c)$ which implies $g(c) = 0$ (plugging in is allowed here!). This is a contradiction; hence (2) holds. Suppose conversely that (2) holds. Let $q := g$. Assume that $q(c) = 0$; then $q(x) = q_1(x)(x - c)$ for some q_1. But this implies $f(x) = q_1(x)(x - c)^{m+1}$ which contradicts (2). Hence $q(c) \neq 0$, and (1) is proved. ∎

NOTE that for a field or skew-field where every nonzero element is a unit the hypothesis in (4.10)(a) is automatically satisfied.

SUPPOSE R is a commutative ring with identity. Then theorem (4.10)(b) tells us that $c \in R$ is a root of a polynomial $f \in R[x]$ if and only if $x - c$ divides f, i.e., if and only if we can "split apart" the linear factor $x - c$ to write $f(x) = (x - c)g(x)$ where g has lower degree than f (and is thus easier to handle). It is natural to try continuing this process of splitting apart linear factors. In view of theorem (4.10) the following definitions make sense.

(4.11) Definition. *Let R be a commutative ring with identity and let $f \in R[x]$. An element $c \in R$ is called a root of f with* **multiplicity** *m if $(x - \alpha)^m$ divides $f(x)$, but $(x - \alpha)^{m+1}$ does not. Equivalently, c is a root with multiplicity m if and only if there is a polynomial g such that $f(x) = g(x)(x - c)^m$ and $g(c) \neq 0$.*

(4.12) Definition. *Let R be a commutative ring and let $f \in R[x]$. We say that f* **splits** *over R if there are elements $a, c_1, \ldots, c_n \in R$ such that $f(x) = a(x - c_1) \cdots (x - c_n)$, i.e., if f can be completely decomposed into linear factors.*

---

† One is tempted to simply plug in $x := c$ in the equation $f(x) = q(x)(x - c) + r_0$ but this is in general only allowed if R is commutative. We have to *verify* that it is allowed in our case.

WE note that it is possible that a polynomial $f \in R[x]$ does not split over R but splits over a ring extension S. For example, $f(x) = x^2 + 1 \in \mathbb{R}[x]$ does not split over $\mathbb{R}$, but can be considered as an element of $\mathbb{C}[x]$ and splits over $\mathbb{C}$ because $f(x) = (x - i)(x + i)$. In general, there is not much one can say about the number of roots of a given polynomial. For example, the polynomial $p(x) = x^2 + 1$ has no root over $\mathbb{R}$, two roots over $\mathbb{C}$ and an infinite number of roots over the ring $\mathbb{H}$ of quaternions.†
Also, if X is a set with n elements and if $R := \mathcal{P}(X)$ is the ring introduced in problem 31(c) of section 2, then the polynomial $q(x) = x^2 - x$ has 2^n roots over R. However, if R is an integral domain, the following important theorem holds.

(4.13) Proposition. *Let R be an integral domain and $f \in R[x]$ a polynomial of degree n. Then f has at most n disctinct roots in R.*

Proof. Suppose $c_1, \ldots, c_m$ are distinct roots of f. Then $f(x) = q_1(x)(x - c_1)$ for some q_1. Commutativity ensures that $0 = f(c_2) = q_1(c_2)(c_2 - c_1)$. Since $c_1 \neq c_2$ this implies $q_1(c_2) = 0$; therefore, $q_1(x) = q_2(x)(x - c_2)$ for some q_2 so that $f(x) = q_2(x)(x - c_1)(x - c_2)$. Inductively, we obtain

$$f(x) = q_m(x)(x - c_1) \cdots (x - c_m)$$

for some $q_m \in R[x]$. But then f has a factor of degree m, namely $(x - c_1) \cdots (x - c_m)$. This is only possible if $m \leq n = \deg f$. ∎

AS a corollary of this proposition, we will now show that the multiplicative group of a finite field is always cyclic (in fact, we will prove a bit more). In the proof of this statement we will use polynomials even though the statement itself has nothing to do with polynomials; this gives us an idea of the power of the theory of polynomials.

(4.14) Theorem. *Let R be an integral domain and let G be a finite subgroup of $R^\times$. Then G is cyclic.*

Proof. Suppose $|G| = n$. Let m be the maximal order of an element of G; then $a^m = 1$ for all $a \in G$, due to (I.21.10)(b). But this means that the polynomial $x^m - 1$ has at least n roots in K, namely the elements of G. By (4.13) this is only possible if $m \geq n$. On the other hand, we have $m \mid n$ by Lagrange's theorem; consequently, $m = n = |G|$. Hence there is an element in G of order $|G|$; this is the claim. ∎

THE discussion of the multiplicity of roots is eased by the introduction of the derivative of a polynomial.

---

† See problem 28 in section 2.

(4.15) Definition. *Let R be a ring. The* **(formal) derivative** *of a polynomial $f \in R[x]$ is defined as follows:*

If $f(x) = a_0 + a_1 x + \cdots + a_n x^n$ then $f'(x) := a_1 + 2a_2 x + \cdots + na_n x^{n-1} \in R[x]$.

The higher derivatives of f are defined inductively by $f^{(0)} := f$ and $f^{(n+1)} := (f^{(n)})'$.

IT is easy to check that the familiar rules for taking derivatives in calculus also hold for this purely formal derivative.

(4.16) Lemma. *The formal derivative in a polynomial ring $R[x]$ satisfies the following conditions:*

$$(cf)' = cf', \quad (f+g)' = f'+g', \quad (fg)' = f'g+fg', \quad (f^n)' = nf^{n-1}f'.$$

Also, **Leibniz' rule** *$(fg)^{(n)} = \sum_{k=0}^{n} \binom{n}{k} f^{(k)} g^{(n-k)}$ holds.*

Proof. Exercise. (See problem 27 below.) ∎

(4.17) Proposition. *Let R be an integral domain and $f \in R[x]$.*
(a) If $\deg f \geq 1$ then $f' = 0$ if and only if $\operatorname{char} R$ is a prime p and $f(x) = g(x^p)$ for some $g \in R[x]$, i.e., if f has the form

$$f(x) = a_0 + a_p x^p + a_{2p} x^{2p} + \cdots + a_{np} x^{np}.$$

(b) $(x-c)^2$ divides f if and only if $x-c$ divides f and f'. This implies that c is a multiple root of f if and only if $f(c) = f'(c) = 0$.

Proof. (a) Let $f(x) = \sum_{k=0}^{n} a_k x^k$. If $f'(x) = \sum_{k=0}^{n} ka_k x^{k-1}$ is the zero-polynomial then $ka_k = 0$ for all k which means that $k = 0$ whenever $a_k \neq 0$.
(b) If $(x-c)^2$ divides f, then $f(x) = (x-c)^2 g(x)$ for some g which implies $f'(x) = (x-c)^2 g'(x) + 2(x-c)g(x)$; hence $x-c$ divides f and f'. Suppose conversely that $f(x) = (x-c)g(x)$ and $f'(x) = g(x) + (x-c)g'(x)$ are divisible by $x-c$. Clearly, this implies that g is divisible by $x-c$, say $g(x) = (x-c)g_1(x)$. But then $f(x) = (x-c)^2 g_1(x)$ is divisible by $(x-c)^2$. ∎

THE notion of a derivative can be extended to polynomials in several variables.

(4.18) Definition. *Let $f(x_1, \ldots, x_n) = \sum_{i_1,\ldots,i_n} a_{i_1,\ldots,i_n} x_1^{i_1} \cdots x_n^{i_n}$ be a polynomial in n variables $x_1, \ldots, x_n$. The* **partial derivative** *of f with respect to the variable x_k is defined as*

$$\frac{\partial f}{\partial x_k}(x_1, \ldots, x_n) := \sum_{i_1,\ldots,i_n} i_k \, a_{i_1,\ldots,i_n} x_1^{i_1} \cdots x_k^{i_k-1} \cdots x_n^{i_n}.$$

CLEARLY this is just the ordinary derivative of f if we consider f as an element of $K[x_1, \ldots, x_{k-1}, x_{k+1}, \ldots, x_n][x_k]$. It should be obvious how the higher partial derivatives can be defined inductively.

LET us show how one can use results for polynomials in one variable to obtain some information on the roots of polynomials in several variables.

(4.19) Theorem. *Let R be an infinite integral domain and $f \in R[x_1, \ldots, x_n]$. If $f \neq 0$, there is an infinite number of n-tuples $(r_1, \ldots, r_n) \in R^n$ which satisfy $f(r_1, \ldots, r_n) \neq 0$, i.e., which are not roots of f.*

Proof. We proceed by induction on n. The case $n = 1$ is clear, because the number of roots of f is at most $\deg f$ in this case, due to (4.13). Suppose the claim is true for $n - 1$. Then we write $f \in R[x_1, \ldots, x_n]$ in the form

$$f(x_1, \ldots, x_n) = \varphi_0 x_n^d + \varphi_1 x_n^{d-1} + \cdots + \varphi_{d-1} x_n + \varphi_d$$

where $\varphi_k \in R[x_1, \ldots, x_{n-1}]$. Let i be the first index with $\varphi_i \neq 0$. By induction hypothesis, there is an infinite number of $(n-1)$-tuples $(r_1, \ldots, r_{n-1})$ such that $\varphi_i(r_1, \ldots, r_{n-1}) \neq 0$. For each such tuple we have

$$f(r_1, \ldots, r_{n-1}, x_n) = \varphi_i(r_1, \ldots, r_{n-1}) x_n^{d-i} + \sum_{k < d-i} \varphi_{d-k}(r_1, \ldots, r_{n-1}) x_n^k \ .$$

This is a polynomial in x_n of degree $d - i$ over R and hence has at most $d - i$ distinct roots by (4.13). Thus for each of the infinitely many $(n-1)$-tuples $(r_1, \ldots, r_{n-1})$ we find an infinite number of ring elements r_n with $f(r_1, \ldots, r_n) \neq 0$. ∎

WE now turn to a very special class of polynomials in several variables. Let R be an integral domain. Then the symmetric group Sym_n acts on the polynomial ring $R[x_1, \ldots, x_n]$ and also on its quotient field $R(x_1, \ldots, x_n)$ via

$$(\sigma \star f)(x_1, \ldots, x_n) = f(x_{\sigma^{-1}(1)}, \ldots, x_{\sigma^{-1}(n)}) \ ,$$

i.e., by permuting the variables $x_1, \ldots, x_n$. Now an element $f \in K(x_1, \ldots, x_n)$ is called **symmetric** if it is invariant under this action.

(4.20) Definition. *Let R be an integral domain and let $f \in R(x_1, \ldots, x_n)$ be a rational function in the variables $x_1, \ldots, x_n$ over R. Then f is called **symmetric** if*

$$f(x_{\sigma(1)}, \ldots, x_{\sigma(n)}) = f(x_1, \ldots, x_n) \quad \text{for all permutations } \sigma \in \mathrm{Sym}_n.$$

*Moreover, we define the **elementary symmetric polynomials** $s_1, \ldots, s_n$ in n variables $x_1, \ldots, x_n$ by*

$$\boxed{s_k(x_1, \ldots, x_n) = \sum_{1 \le i_1 < \cdots < i_k \le n} x_{i_1} \cdots x_{i_k} \ .}$$

THUS

$$s_1(x_1,\ldots,x_n) \;=\; x_1 + x_2 + \cdots + x_n,$$
$$s_2(x_1,\ldots,x_n) \;=\; \textstyle\sum_{i<j} x_i x_j,$$
$$\cdots$$
$$s_n(x_1,\ldots,x_n) \;=\; x_1 x_2 \cdots x_n.$$

For example, $s_2(x,y,z) = xy + xz + yz$ and $s_2(a,b,c,d) = ab + ac + ad + bc + bd + cd$.

THE purpose of studying symmetric functions is to exhibit relations between the coefficients of a polynomial and its roots. To solve a polynomial equation $p(x) = 0$ means to express the roots of p as functions of its coefficients. Now we will ask the converse question: How can the coefficients of a polynomial be expressed by its roots? Suppose R is an integral domain and $p(x) = x^n + a_{n-1}x^{n-1} + \cdots + a_1 x + a_0 \in R[x]$ is a monic polynomial which has n roots $\xi_1,\ldots,\xi_n$ in some extension S of R. Then we can write

$$x^n + a_{n-1}x^{n-1} + \cdots + a_1 x + a_0 \;=\; (x - \xi_1)(x - \xi_2)\cdots(x - \xi_n) \;.$$

Multiplying out the right-hand side yields

$$a_{n-k} = (-1)^k s_k(\xi_1,\ldots,\xi_n) \qquad \text{for } 1 \le k \le n \;;$$

hence the coefficients of p are up to sign the values of the elementary symmetric functions at its roots. For a quadratic polynomial $x^2 + px + q = (x - \xi_1)(x - \xi_2)$, this just says that $p = -(\xi_1 + \xi_2)$ and $q = \xi_1\xi_2$ which are Vieta's formulas.

WE are now going to prove that every symmetric function can be expressed in terms of the elementary symmetric polynomials (which explains the name of the latter). Let us give some examples before we turn to a precise formulation and a proof of this statement.

(4.21) Examples. (a) Let us express some symmetric polynomials in 3 variables in terms of the elementary symmetric functions s_1, s_2, s_3. We have

$$x^2 + y^2 + z^2 \;=\; (x + y + z)^2 - 2(xy + xz + yz) \;=\; s_1^2 - 2s_2,$$
$$x^3 + y^3 + z^3 \;=\; (x + y + z)^3 - 3(x + y + z)(xy + xz + yz) + 3xyz$$
$$=\; s_1^3 - 3s_1 s_2 + 3s_3,$$
$$x^2 y^2 + x^2 z^2 + y^2 z^2 \;=\; (xy + xz + yz)^2 - 2(xy + xz + yz) \;=\; s_2^2 - 2s_2.$$

Finally, $x^2 y + x^2 z + xy^2 + xz^2 + y^2 z + yz^2$ equals $(x + y + z)(xy + xz + yz) - 3xyz = s_1 s_2 - 3s_3$.

(b) Suppose that $p(x) = x^3 - 7x^2 - 8x + 9 \in \mathbb{C}[x]$ has the (unknown) roots $\alpha_1, \alpha_2, \alpha_3$. We want to find the monic cubic polynomial q with the roots α_1^2, α_2^2 and α_3^2. Now we know that the coefficients of a polynomial are the values of the elementary symmetric functions at its roots. For p, this gives

$$s_1 = \alpha_1 + \alpha_2 + \alpha_3 = 7,$$
$$s_2 = \alpha_1\alpha_2 + \alpha_2\alpha_3 + \alpha_3\alpha_1 = -8,$$
$$s_3 = \alpha_1\alpha_2\alpha_3 = -9.$$

On the other hand, we obtain $q(x) = x^3 - \sigma_1 x^2 + \sigma_2 x - \sigma_3$ where

$$\sigma_1 = \alpha_1^2 + \alpha_2^2 + \alpha_3^2,$$
$$\sigma_2 = \alpha_1^2\alpha_2^2 + \alpha_1^2\alpha_3^2 + \alpha_2^2\alpha_3^2,$$
$$\sigma_3 = \alpha_1^2\alpha_2^2\alpha_3^2.$$

Using the results from part (a), we obtain $\sigma_1 = s_1^2 - 2s_2 = 65$, $\sigma_2 = s_2^2 - 2s_2 = 80$, and $\sigma_3 = s_3^2 = 81$. So

$$q(x) = x^3 - 65x^2 + 80x - 81 .$$

Note that we were able to find the polynomial q without explicitly knowing the roots of p.

BEFORE we can prove the main theorem on symmetric functions, we need an auxiliary result.

(4.22) **Lemma.** *Let R be an integral domain. We define the **lexicographic degree**[†] lexdeg p of a polynomial $p = \sum a_I x^I \in R[x_1, \ldots, x_n]$ as the maximal occurring multi-index I where the maximum is taken with respect to the lexicographic ordering on $\mathbb{N}_0^n$. Then we call $a_I x^I$ the leading term and a_I the leading coefficient of p. Formally, we define the lexicographic degree of the zero polynomial as lexdeg $0 := (-\infty, \ldots, -\infty)$.*

(a) If $p, q \in R[x_1, \ldots, x_n]$ then lexdeg$(p + q) \leq$ max(lexdeg p, lexdeg q) *and* lexdeg$(pq) =$ lexdeg $p +$ lexdeg q.

(b) If $p \in R[x_1, \ldots, x_n]$ is symmetric and lexdeg $p = (i_1, \ldots, i_n)$ *then $i_1 \geq i_2 \geq \cdots \geq i_n$.*

Proof. The proof of (a) is straightforward. For (b), we observe that if $ax_1^{j_1} \cdots x_n^{j_n}$ ($a \neq 0$) occurs as a monomial in p, then so does $ax_{\sigma^{-1}(1)}^{j_1} \cdots x_{\sigma^{-1}(n)}^{j_n}$ for any $\sigma \in \text{Sym}_n$, by symmetry. In other words: If $(j_1, \ldots, j_n)$ occurs as the index degree of a monomial, then all index degrees $(j_{\sigma(1)}, \ldots, j_{\sigma(n)})$ ($\sigma \in \text{Sym}_n$) occur. But the greatest of these degrees (with respect to the lexicographic ordering on $\mathbb{N}_0^n$) is the one with entries in non-increasing order. ∎

FOR example, the lexicographic of the k-th elementary symmetric polynomial s_k is lexdeg $s_k = (1, 1, \ldots, 1, 0, \ldots, 0)$ with 1's in the first k entries and 0's in the remaining $n - k$ entries.

NOW we can turn to the main theorem on symmetric functions.

---

† The notion of the lexicographic degree is an *ad hoc* definition and not in general use; associating this degree with a polynomial does not treat all the variables on an equal footing but depends on the (arbitrary) numbering of the variables.

(4.23) Main theorem on symmetric polynomials. *Let R be an integral domain.*

(a) *If $p \in R[x_1, \ldots, x_n]$ is a symmetric polynomial, then there is a unique polynomial $\varphi \in R[x_1, \ldots, x_n]$ such that $p = \varphi(s_1, \ldots, s_n)$. Thus every symmetric polynomial can be expressed by the elementary symmetric polynomials.*

(b) *If $f \in R(x_1, \ldots, x_n)$ is a symmetric rational function, then f can be represented as a quotient $f = p/q$ with symmetric polynomials $p, q \in R[x_1, \ldots, x_n]$.*

Proof. (a) *Existence.* We proceed by induction on the lexicographic degree of p. (See problem 34 below.) Let p be symmetric with lexdeg $p = (i_1, \ldots, i_n)$. Then $i_1 \geq i_2 \geq \cdots \geq i_n$ by (4.22)(b). Hence we can form the symmetric polynomial

$$f = s_1^{i_1 - i_2} s_2^{i_2 - i_3} \cdots s_{n-1}^{i_{n-1} - i_n} s_n^{i_n}$$

which has lexicographic degree

$$
\begin{aligned}
\text{lexdeg } f &= (i_1 - i_2)\,\text{lexdeg } s_1 + (i_2 - i_3)\,\text{lexdeg } s_2 + \cdots + i_n \,\text{lexdeg } s_n \\
&= (i_1 - i_2, 0, \ldots, 0) + (i_2 - i_3, i_2 - i_3, \ldots, 0) + \cdots + (i_n, i_n, \ldots, i_n) \\
&= (i_1, i_2, \ldots, i_n) = \text{lexdeg } p .
\end{aligned}
$$

Obviously, the leading coefficient of f is 1. So if $a \neq 0$ is the leading coefficient of p, then af and p have the same leading term. Consequently, the polynomial

$$p_1 := p - af$$

is of lower lexicographic degree than p (possibly $p_1 = 0$) and is obviously symmetric (since p and f are). By induction hypothesis, p_1 can be expressed as a polynomial in $s_1, \ldots, s_n$; but then the same holds for

$$p = p_1 + af = p_1 + a s_1^{i_1 - i_2} s_2^{i_2 - i_3} \cdots s_n^{i_n} .$$

Uniqueness. Suppose $\varphi(s_1, \ldots, s_n) = \psi(s_1, \ldots, s_n)$ and let $h := \varphi - \psi$; then $h(s_1, \ldots, s_n) = 0$. If $h \neq 0$ and $a\, x_1^{i_1} \cdots x_n^{i_n}$ is a term occurring in $h(x_1, \ldots, x_n)$ then $h(s_1, \ldots, s_n)$ possesses the term $a\, s_1^{i_1} \cdots s_n^{i_n}$ which has lexicographic degree

$$i_1\, \text{lexdeg } s_1 + \cdots + i_n\, \text{lexdeg } s_n = (i_1 + \cdots + i_n,\ i_2 + \cdots + i_n,\ \ldots,\ i_{n-1} + i_n,\ i_n) .$$

But this lexicographic degree is different from $(0, \ldots, 0)$ since $(i_1, \ldots, i_n)$ is. Consequently, $h(s_1, \ldots, s_n) \neq 0$ which is a contradiction. (The fact that $h(s_1, \ldots, s_n) = 0$ implies $h = 0$ is expressed by saying that $s_1, \ldots, s_n$ are algebraically independent.)

(b) Let $f = \frac{P}{Q}$ be an arbitrary representation of f as a quotient of two polynomials. If Q is symmetric, then so is $P = fQ$ (as a product of symmetric functions), and we are done. Otherwise let $Q = Q_1, Q_2, \ldots, Q_r$ be the various distinct polynomials obtained from Q by permuting the indeterminates. Then $q := Q_1 Q_2 \cdots Q_r$ is obviously symmetric (any permutation of the indeterminates merely permutes the factors). Let $p := P Q_2 \cdots Q_r$. Then $f = \frac{p}{q}$, and $p = fq$ is symmetric (as a product of symmetric functions). ∎

WE will now define an important example of a symmetric polynomial.

(4.24) Definition. *The polynomial*

$$D(x_1, \ldots, x_n) \; := \; \prod_{1 \le i < j \le n} (x_i - x_j)^2$$

in n variables $x_1, \ldots, x_n$ is called the **discriminant** *of $x_1, \ldots, x_n$.*

(4.25) Proposition. *The discriminant is a symmetric polynomial.*

Proof. The effect of applying any permutation $\sigma \in \mathrm{Sym}_n$ to D is a mere permutation of the factors $(x_i - x_j)^2$; hence $\sigma \star D = D$ for all $\sigma \in \mathrm{Sym}_n$. ∎

BEING a symmetric polynomial, the discriminant is expressible by the elementary symmetric polynomials by (4.23). Let us determine this representation in low dimensions.

(4.26) Examples. (a) For $n = 2$ we have $D(x, y) = (x - y)^2 = (x + y)^2 - 4xy = s_1^2 - 4s_2$.

(b) In the case $n = 3$ we let $\Delta(x, y, z) = (x - y)(x - z)(y - z)$; then $D = \Delta^2$. Now $\Delta(x, y, z) = A - B$ where $A := x^2 y + y^2 z + z^2 x$ and $B := xy^2 + yz^2 + zx^2$. Hence $D(x, y, z) = (A - B)^2 = (A + B)^2 - 4AB$. Now in (4.21)(a) we saw that $A + B = s_1 s_2 - 3s_3$, and another straightforward computation yields $AB = s_1^3 s_3 + s_2^3 - 6s_1 s_2 s_3 + 9s_3^2$. (See problem 35 below.) Thus we obtain

$$\begin{aligned} D(x, y, z) &= (s_1 s_2 - 3s_3)^2 - 4(s_1^3 s_3 + s_2^3 - 6s_1 s_2 s_3 + 9s_3^2) \\ &= s_2^2 s_1^2 + 18 s_1 s_2 s_3 - 27 s_3^2 - 4 s_1^3 s_3 - 4 s_2^3 \, . \end{aligned}$$

THE meaning of the discriminant can be described as follows. Let $R \subseteq S$ be integral domains and suppose that a monic polynomial $p(x) = x^n + a_{n-1} x^{n-1} + \cdots + a_1 x + a_0 \in R[x]$ splits over S so that

$$x^n + a_{n-1} x^{n-1} + \cdots + a_1 x + a_0 \; = \; (x - x_1)(x - x_2) \cdots (x - x_n)$$

with elements $x_i \in S$. Plugging the ring elements $x_1, \ldots, x_n \in S$ into the discriminant $D(X_1, \ldots, X_n) = \prod_{i < j} (X_i - X_j)^2$, we see that $D(x_1, \ldots, x_n) = 0$ if and only if $x_i = x_j$ for some indices $i \ne j$, i.e., if and only if f has a multiple root. On the other hand, $D(x_1, \ldots, x_n)$ can be written as a polynomial expression in $s_k(x_1, \ldots, x_n) = (-1)^k a_{n-k}$, i.e., depends only on the coefficients of f and hence can be calculated without knowing the roots. This makes the following definition of the discriminant of a polynomial possible.

(4.27) Definition. *Let R be an integral domain and let $f \in R[x]$ be a polynomial which splits over some extension S of R. Let $\alpha_1, \ldots, \alpha_n \in S$ be the roots of f (counted with multiplicity). If D is the discriminant in n variables then $D(\alpha_1, \ldots, \alpha_n)$ is called the **discriminant** of the polynomial f and is denoted by $D(f)$.*

NOTE that $D(\alpha_1, \ldots, \alpha_n)$, being a symmetric function in $\alpha_1, \ldots, \alpha_n$, can be expressed by the coefficients of f without prior knowledge of the roots of f; in particular, $D(f)$ is an element of R (and not just of S).

(4.28) Examples. (a) The discriminant of a quadratic polynomial $f(x) = x^2 + px + q$ is $D(f) = p^2 - 4q$. This is an immediate consequence of (4.26)(a).
(b) The discriminant of a cubic polynomial $f(x) = x^3 + ax^2 + bx + c$ is $D(f) = a^2b^2 + 18abc - 27c^2 - 4a^3c - 4b^3$. This is an immediate consequence of (4.26)(b).
(c) As a special case of part (b), we see that the discriminant of $f(x) = x^3 + px + q$ is $D(f) = -27q^2 - 4p^3$.

SINCE the discriminant decides whether or not a polynomial has multiple roots we can find out, without actually knowing the roots of a polynomial $f \in R[x]$, whether or not f has multiple roots – if we know offhand that f splits in some extension of R. This is, for example, the case for all polynomial with real coefficients because such a polynomial can be considered as an element of $\mathbb{C}[x]$ and hence splits over $\mathbb{C}$ because $\mathbb{C}$ is algebraically closed. We conclude this section by providing some useful information about the discriminant of real polynomials.

(4.29) Proposition. *Let $f \in \mathbb{R}[x]$.*
(a) *If all roots of f are real then $D(f) \geq 0$.*
(b) *Let $\deg f = 2$ or 3. If $D(f) \geq 0$ then all roots of f are real.*
(c) *If f has the non-real roots $x_1, \overline{x_1}, \ldots, x_r, \overline{x_r}$ and the real roots $y_1, \ldots, y_s$ and if these roots are pairwise distinct then $D(f) > 0$ if and only if r is even, i.e., if the number of non-real roots is divisible by 4.*

Proof. (a) If the roots of f are $x_1, \ldots, x_n \in \mathbb{R}$ then $D(f) = \prod_{i<j}(x_i - x_j)^2$ is a product of squares of real number and hence is nonnegative.
(b) Let f be a real quadratic polynomial with a nonreal root x_1. Then the second root of f is the complex conjugate $x_2 = \overline{x_1}$. Letting $\alpha := x_1 - \overline{x_1}$ we see that the discriminant of f is $D(f) = (x_1 - x_2)^2 = (x_1 - \overline{x_1})^2 = -(x_1 - \overline{x_1})(\overline{x_1} - x_1) = -\alpha\overline{\alpha} = -|\alpha|^2 < 0$. Analogously, let f be a real cubic polynomial with a nonreal root x_1. Then $x_2 := \overline{x_1}$ is also a root. Finally, the third root x_3 must be real because every real cubic polynomial has at least one real root. (See problem 22 below.) Let $\Delta := (x_1 - x_2)(x_1 - x_3)(x_2 - x_3)$; then $\overline{\Delta} = (x_2 - x_1)(x_2 - x_3)(x_1 - x_3) = -\Delta$. Hence the discriminant of f is $D(f) = \Delta^2 = -\Delta\overline{\Delta} = -|\Delta|^2 < 0$.
(c) In the product by which the discriminant of f is given all the terms $(x_i - y_k)^2(\overline{x_i} - y_k)^2 = |x_i - y_k|^4$, $(y_k - y_l)^2$ and $(x_i - x_j)^2(\overline{x_i} - x_j)^2(x_i - \overline{x_j})^2(\overline{x_i} - \overline{x_j})^2 = |x_i - x_j|^4|\overline{x_i} - x_j|^4$ are positive. Consequently, the sign of $D(f)$ equals the sign of

$$(x_1 - \overline{x_1})^2 \cdots (x_r - \overline{x_r})^2 = (2i\,\mathrm{Im}\,x_1)^2 \cdots (2i\,\mathrm{Im}\,x_r)^2 = 4^r(-1)^r(\mathrm{Im}\,x_1)^2 \cdots (\mathrm{Im}\,x_r)^2$$

which is $(-1)^r$; whence the claim. ∎

Exercises

Problem 1. Prove the propositions (4.6) and (4.8).

Problem 2. Let R be a ring. Show that R is commutative if and only if all the evaluation maps $R[x] \to R$ given by $p \mapsto p(r)$ with $r \in R$ are homomorphisms.

Problem 3. (a) If you multiply out the product $(1 - 3x + 3x^2)^{743}(1 + 3x - 3x^2)^{744}$ to obtain a polynomial $p(x) = \sum_{k=0}^{n} a_k x^k \in \mathbb{Z}[x]$, which value has the sum $a_0 + \cdots + a_n$ of the coefficients of p?

(b) Show that by multiplying out the product
$$(1 - x + x^2 - x^3 + \cdots - x^{99} + x^{100})(1 + x + x^2 + x^3 + \cdots + x^{99} + x^{100})$$
you obtain a polynomial in which only even powers of x occur.

(c) Let $f(x) = a_4 x^4 + a_3 x^3 + a_2 x^2 + a_1 x + a_0$ where $a_0, \ldots, a_4$ are the numbers $1, -2, 3, 4, -6$ in any order. Show that f has a rational root.

Problem 4. Let p be a prime number and let $f(x) = a_n x^n + a_{n-1} x^{n-1} + \cdots + a_1 x + a_0 \in \mathbb{Z}_p[x]$. Determine f^{p^N}.

Problem 5. (a) Find all roots of the polynomials $f(x) = x^2 - x$ and $g(x) = x^3 - x$ in $\mathbb{Z}_{29}$, $\mathbb{Z}_{30}$ and $\mathbb{Z}_{31}$. Moreover, find all roots of the polynomials $f(x) = x^3 - 4$ and $g(x) = x^2 + 3x + 2$ in the ring $\mathbb{Z}_{210}$.

(b) Let p be a prime. Show that $f(x) = x^p - x^2 - x + a$ has a root in $\mathbb{Z}_p$ if and only if a is a square in $\mathbb{Z}_p$.

Problem 6. Show that if $f \in \mathbb{C}[x_1, \ldots, x_n]$ vanishes on $\mathbb{Z}^n$ then $f \equiv 0$.

Problem 7. (a) Show that if $f \in \mathbb{Z}[x]$ then $f(b) - f(a)$ is divisible by $b - a$ whenever $a \neq b$ are integers.

(b) Let $n \in \mathbb{N}$. Find a polynomial $p \in \mathbb{Z}[x]$ such that $x^n - 1 = p(x)(x - 1)$. Conclude that $k^n - 1$ is divisible by $k - 1$ for all integers $k \neq 1$.

(c) Let $n \in \mathbb{N}$ be odd. Find a polynomial $q \in \mathbb{Z}[x]$ such that $x^n + 1 = q(x)(x + 1)$. Conclude that $k^n + 1$ is divisible by $k + 1$ for all integers $k \neq -1$.

Problem 8. (a) Let $f(x) = x^2 + x + 41$. Show that $f(n)$ is a prime number for $1 \leq n \leq 39$.

(b) Show that there is no polynomial function $f \in \mathbb{Z}[x]$ such that $f(n)$ is a prime number for all $n \in \mathbb{N}$.

Hint. If $f(n)$ is a prime number p choose a number n' with $n' \equiv n \ (p)$ with $|f(n')| > p$. (Why is this possible?) Show that then $f(n')$ is not a prime number.

(c) Let $f \in \mathbb{Z}[x]$ be a nonconstant polynomial. Show that there is an infinite number of primes p such that the congruence $f(x) \equiv 0$ modulo p has a solution in $\mathbb{Z}$.

Problem 9. Show that there is no nonzero polynomial function $f : \mathbb{C} \to \mathbb{C}$ such that $f(x+2) - 2f(x+1) = f(x)$ for all $x \in \mathbb{C}$.

Problem 10. (a) Find all polynomials $p \in \mathbb{Z}_2[x]$ whose associated polynomial function $\mathbb{Z}_2 \to \mathbb{Z}_2$ is the zero function.
(b) Let p be a prime number so that $K := \mathbb{Z}_p$ is a field. Show that the homogeneus polynomial $f(x, y) = x^p y - y^p x \in K[x, y]$ induces the zero function on $\mathbb{Z}_p \times \mathbb{Z}_p$.
(c) Show that if K is an infinite field, then only the zero polynomial induces the zero function.

Problem 11. Let K be a field and let $f \in K[x]$ be a polynomial whose associated function satisfies $f(a+b) = f(a)f(b)$ for all $a, b \in K$. Show that if K is infinite, then $f = 0$ or $f = 1$. Show that this conclusion is not correct if K is a finite field.

Problem 12. If f and g are polynomials over a commutative ring define a new polynomial $f \circ g$ by $(f \circ g)(x) = f(g(x))$.
(a) Show that $(f \circ g) \circ h = f \circ (g \circ h)$.
(b) Show that if $e(x) = x$ then $f \circ e = f$ and $e \circ g = g$ for all f and g.
(c) Show that in general $f \circ g \neq g \circ f$.
(d) Show that $f_1 \circ g = f_2 \circ g$ implies $f_1 = f_2$.
(e) Show that $f \circ g_1 = f \circ g_2$ does not imply $g_1 = g_2$.
(f) Show that $(f \circ g)' = (f' \circ g) \cdot g'$.
(g) Show that $f \circ g_1 - f \circ g_2$ is a multiple of $g_1 - g_2$.
(h) Show that if $p(x) := f(x) - x$ does not have a multiple root then neither does $q(x) := f(f(x)) - x$. How are the roots of q related to those of p?

Problem 13. Let K be a finite field with q elements. Let $f : K \to K$ be an arbitrary function. Show that there is a unique polynomial $p \in K[X]$ of degree less than q such that $f(x) = p(x)$ for all $x \in K$.
Hint. Show first that the function $p_a(x) := (x - a)^{q-1}$ satisfies $p_a(a) = 0$ and $p_a(x) = 1$ if $x \neq a$.

Problem 14. Let K be a finite field with q elements and let $f \in K[x_1, \ldots, x_n]$ be a polynomial of degree $d < n$ such that $f(0, \ldots, 0) = 0$. Show that there are elements $a_1, \ldots, a_n \in K$, not all zero, such that $f(a_1, \ldots, a_n) = 0$.
Hint. Suppose $f(a) \neq 0$ for all $a \neq 0$ in K^n. Then

$$g(x_1, \ldots, x_n) = 1 - f(x_1, \ldots, x_n)^{q-1} \quad \text{and}$$
$$h(x_1, \ldots, x_n) = (1 - x_1^{q-1}) \cdots (1 - x_n^{q-1})$$

induce the same polynomial function. Now compare the degrees of g and h!

Problem 15. Suppose that K is a field which is not algebraically closed. Show that there is a polynomial $f \in K[x_1, \ldots, x_n]$ which has $(0, \ldots, 0)$ as its only root.
Hint. Proceed by induction on n, the case $n = 1$ being trivial.

Problem 16. Let K be a field and let $f \in K[x_1, \ldots, x_n]$ be a nonzero polynomial. Suppose that $\alpha_1, \alpha_2, \ldots$ is an infinite sequence of pairwise different elements of K. Show that one can find indices $i_1, \ldots, i_n$ such that $f(\alpha_{i_1}, \ldots, \alpha_{i_n}) \neq 0$.

Hint. Proceed by induction on n, the case $n = 1$ being trivial.

Problem 17. Let K be a field and let $p \in K[x_1, \ldots, x_n]$. Show that each of the following statements implies the next.

(1) p is homogenous of degree m.

(2) $p(tx_1, \ldots, tx_n) = t^m p(x_1, \ldots, x_n)$ in the polynomial ring $K[x_1, \ldots, x_n, t]$.

(3) For $1 \leq \mu \leq m$ we have

$$\sum_{m_1 + \cdots + m_n = \mu} \frac{x_1^{m_1} \cdots x_n^{m_n}}{m_1! \cdots m_n!} \frac{\partial^{m_1 + \cdots + m_n} f}{\partial x_1^{m_1} \cdots \partial x_n^{m_n}}(x_1, \ldots, x_n) = \left(\binom{m}{\mu} \right) f(x_1, \ldots, x_n).$$

Remark. In multi-index notation, this simply reads $\sum_{|I| = \mu} (x^I / I!)(\partial^I f)(x) = \left(\binom{m}{\mu} \right) f(x)$ for $1 \leq \mu \leq m$.

(4) The following formula (**Euler's identity**) holds:

$$x_1 \frac{\partial f}{\partial x_1} + x_2 \frac{\partial f}{\partial x_2} + \cdots + x_n \frac{\partial f}{\partial x_n} = mf.$$

Moreover, show that if char $K = 0$ then (4) implies (1) so that all four statements are equivalent in this case. This is no longer true for arbitrary fields; to see an example, choose $K = \mathbb{Z}_p$ and $f(x, y) = xy + x^{p+1} y^{p+1}$.

Problem 18. Let K be an infinite field and let $p \in K[x_1, \ldots, x_n]$ be homogeneous of degree m. We want to show that p can be brought into a certain normal form by applying a suitable linear coordinate transformation.

(a) Show that if $x_i = \sum_k a_{ik} y_k$ then

$$x_1^{m_1} x_2^{m_2} \cdots x_n^{m_n} = \sum_{k=1}^{n} a_{1k}^{m_1} a_{2k}^{m_2} \cdots a_{nk}^{m_n} y_k^{m_1 + \cdots + m_n} + \text{ mixed terms}$$

where mixed terms are terms containing products $y_i y_j$ with $i \neq j$.

(b) Conclude that $p(x_1, \ldots, x_n) = \sum_{k=1}^{n} p(a_{1k}, a_{2k}, \ldots, a_{nk}) y_k^m + \text{ mixed terms}$.

(c) Show that the coefficients a_{ik} can be chosen such that

$$p(x_1, \ldots, x_n) = c_1 y_1^m + c_2 y_2^m + \cdots + c_n y_n^m + \text{ mixed terms}$$

where $c_1, c_2, \ldots, c_n$ are nonzero elements of K.

Problem 19. (Interpolation.) Let K be a field.

(a) Given $n + 1$ points $(x_0, y_0), \ldots, (x_n, y_n) \in K \times K$, show that there is a unique polynomial $f \in K[x]$ of degree $\leq n$ such that $f(x_i) = y_i$ for $0 \leq i \leq n$.

Hint. For the existence proof try f in the form

$$f(x) = a_0 + a_1(x - x_0) + a_2(x - x_0)(x - x_1) + \cdots + a_n(x - x_0)(x - x_1) \cdots (x - x_n)$$

and determine $a_0, a_1, \ldots, a_n$ inductively. For the uniqueness proof, assume there are two such polynomials f_1, f_2. Then $f_1 - f_2$ has the $n + 1$ distinct roots $x_0, x_1, \ldots, x_n$.

(b) Conclude from (a) that if K is a finite field then every function $f : K \to K$ is a polynomial function. (Compare with problem 13 above.)

(c) Suppose char $K = 0$ or char $K > 5$. Find the polynomial of least possible degree that passes through the points $(0,0), (1,0), (2,1), (3,7), (4,22)$ and $(5,50)$.

Problem 20. Let R be a commutative ring with identity and let $p \in R[x]$ be a polynomial of degree $d \geq 1$ whose leading coefficient is a unit. Show that for any polynomial $f \in R[x]$ there are unique polynomials $f_0, \ldots, f_m$ of degree $< d$ such that

$$f = f_0 + f_1 p + \cdots + f_m p^m .$$

Hint. If the statement is not true, there is a counterexample f of minimal degree. Now write $f = gp + r$ with $\deg r < d$ according to (4.10)(a); then $\deg g < \deg f$ so that g is not a counterexample.

Problem 21. Let R be a commutative ring. Given a polynomial $p(x) = a_0 + a_1 x + \cdots + a_n x^n \in R[x]$ of degree n and an element $x_0 \in R$, we can try to expand p in powers of $x - x_0$, i.e., to find coefficients $b_0, \ldots, b_n \in R$ such that $p(x) = b_0 + b_1(x - x_0) + \cdots + b_n(x - x_0)^n$. The following procedure to find $b_0, \ldots, b_n$ was devised in 1819 by the English mathematician William George Horner (1786-1837) and is therefore called **Horner's scheme.**

(a) Determine coefficients $a'_n, \ldots, a'_1, a'_0$ by the scheme

p	a_n	a_{n-1}	a_{n-2}	$\cdots$	a_1	a_0
x_0	0	$a'_n x_0$	$a'_{n-1} x_0$	$\cdots$	$a'_2 x_0$	$a'_1 x_0$
	$\downarrow$	$\nearrow \quad \downarrow$	$\nearrow \quad \downarrow$	$\nearrow \cdots \nearrow$	$\downarrow$	$\nearrow \quad \downarrow$
p_{n-1}	$a'_n = a_n$	a'_{n-1}	a'_{n-2}	$\cdots$	a'_1	a'_0

where a downward vertical arrow $\downarrow$ stands for an addition of the first two entries in the corresponding column and where a diagonal arrow $\nearrow$ stands for a multiplication by x_0. Show that $p(x_0) = a'_0$. Moreover, show that if $p_{n-1}(x) := a'_n x^{n-1} + a'_{n-1} x^{n-2} + \cdots + a'_2 x + a'_1$ then $p(x) = (x - x_0)p_{n-1}(x) + p(x_0)$.

Apply the scheme with $p(x) = 2x^5 - x^4 - 5x^2 + 3x - 9 \in \mathbb{Z}[x]$ and $x_0 = 3$.

(b) Applying this scheme to p_{n-1} instead of p, then to p_{n-2} instead of p_{n-1}, and so on, you obtain polynomials $p_{n-1}, p_{n-2}, \ldots, p_0$. Show that

$$(\star) \qquad \begin{aligned} p(x) = {} & p(x_0) + p_{n-1}(x_0)(x - x_0) + p_{n-2}(x_0)(x - x_0)^2 + \cdots \\ & + p_1(x_0)(x - x_0)^{n-1} + p_0(x_0)(x - x_0)^n \end{aligned}$$

so that the coefficients b_k sought for are given by $b_k = p_{n-k}(x_0)$, i.e., by the last elements in the rows of Horner's scheme. Moreover, conclude from $(\star)$ that $p^{(k)}(x_0) = k! b_k$. Hence if R is a field with char $R = 0$ (or char $R > n$), then $p(x) = \sum_{k=0}^{n} \frac{p^{(k)}(x_0)}{k!}(x - x_0)^k$ (**Taylor's formula**).

(c) Expand $x^4 - 2x^3 + x^2 - 7x + 3$ in powers of $x - 3$.

(d) Expand $x^4 + 2x^3 - x + 5$ in powers of $x - 2$.

Problem 22. Using the intermediate value theorem from calculus, show that every polynomial $f \in \mathbb{R}[x]$ of odd degree has a root in $\mathbb{R}$.

Problem 23. For $f(x) = \sum_k a_k x^k \in \mathbb{C}[x]$ define $\overline{f}(x) = \sum_k \overline{a_k} x^k$.

(a) Show that the assignment $f \mapsto \overline{f}$ defines an automorphism of $\mathbb{C}[x]$.

Remark. Compare with problem 22(a) in section 2.

(b) Show that if α is a root of f then $\overline{\alpha}$ is a root of $\overline{f}$ with the same multiplicity.

(c) Show that $f = \overline{f}$ if and only if $f \in \mathbb{R}[x]$. Conclude that if a real polynomial has a root $\alpha \in \mathbb{C}$ then $\overline{\alpha}$ is also a root.

Problem 24. Let $f(x) = a_n x^n + a_{n-1} x^{n-1} + \cdots + a_1 x + a_0 \in \mathbb{C}[x]$ be a polynomial of degree n.

(a) Show that if $\alpha \in \mathbb{C}$ is a root of f then $|\alpha| \leq nm$ where

$$m := \max\{|\frac{a_0}{a_n}|, \ |\frac{a_1}{a_n}|, \ \ldots, \ |\frac{a_{n-1}}{a_n}|, \ |\frac{a_n}{a_n}|\} \ .$$

Hint. If α is a root of f then $-\alpha^n = \frac{a_{n-1}}{a_n}\alpha^{n-1} + \cdots + \frac{a_1}{a_n}\alpha + \frac{a_0}{a_n}$. Suppose that $|\alpha| > nm \, (\geq 1)$ and use the triangle inequality to obtain a contradiction.

(b) Suppose that $a_0 \neq 0$ and that f has n real roots (counted with multiplicity). Show that then $n \cdot |a_0| \leq |a_{n-1}|$.

Hint. The geometric mean of nonnegative real numbers α_i cannot exceed their arithmetic mean; i.e., $n \cdot \sqrt[n]{\alpha_1 \cdots \alpha_n} \leq \alpha_1 + \cdots + \alpha_n$.

Problem 25. Let R be an integral domain with characteristic $\neq 2$ such that $R^\times$ has $2n$ elements.

(a) Show that each element of $R^\times$ is a root of $x^{2n} - 1$.

(b) Show that the set $\{y^2 \mid y \in R^\times\}$ has n elements and that each of these is a root of $x^n - 1$.

(c) Show that there are n elements in $R^\times$ which are not squares and that each of these is a root of $x^n + 1$.

(d) Show that -1 is a square in $R^\times$ if and only if n is even.

(e) Let p be an odd prime. Show that the congruence $x^2 \equiv -1$ modulo p has a solution if and only if $p \equiv 1$ modulo 4.

Hint. Apply (d) to $R := \mathbb{Z}_p$.

Problem 26. If p is a prime number, then $\mathbb{Z}_p^\times$ is cyclic due to (4.14). In this problem, we ask whether $\mathbb{Z}_q^\times$ is also cyclic if q is a prime power.

(a) Let $n \in \mathbb{N}$ and let $p \in \mathbb{N}$ be a prime number. Show that $p^n \| y - 1$ implies $p^{n+1} \| y^p - 1$ except in the single case that $p = 2$ and $n = 1$; here $p^n \| a$ means that $p^n \mid a$, but $p^{n+1} \nmid a$.

Hint. If $p^n \| y - 1$, say $y = 1 + up^n$ with $p \nmid u$, then $y^p - 1 = \sum_{k=1}^{p} \binom{p}{k} p^{nk} u^k$. Check how often each term in this sum is divisible by p.

(b) Let p be odd and let $x \in \mathbb{Z}$ be a generator of $\mathbb{Z}_p^\times$. Show that x is a generator of $\mathbb{Z}_{p^n}$ for all $n \in \mathbb{N}$. In particular, this establishes that $\mathbb{Z}_q^\times$ is cyclic whenever q is a power of an odd prime.

Hint. We have to show that the smallest natural number r with $x^r \equiv 1(p^n)$ is $\varphi(p^n) = p^n - p^{n-1}$. Now if $x^r \equiv 1(p^n)$ then $x^r \equiv 1(p)$ so that r must be a multiple of $p - 1$. Applying (a) to $y := x^{p-1}$, show that $p^k \| y^{p^{k-1}} - 1$ for $1 \leq k \leq n$ so that $y^m \not\equiv 1(p^n)$ for $1 \leq m < p^{n-1}$.

(c) Let $n \geq 3$. Show that $\mathbb{Z}_{2^n}^{\times}$ is not cyclic, but that

$$\mathbb{Z}_{2^n}^{\times} = \{\pm[5]^k \mid k \in \mathbb{Z}\} \cong \{\pm 1\} \times \langle [5] \rangle$$

so that $\mathbb{Z}_{2^n}^{\times}$ is isomorphic to a direct product of a cyclic group of order $\frac{1}{2}\varphi(2^n) = 2^{n-2}$ and a cyclic group of order 2.

Hint. Using (a), show that the order of 5 modulo 2^n is 2^{n-2} for all $n \geq 2$.

Problem 27. Prove lemma (4.16).

Problem 28. (a) Let R be a ring and let $p \in R[x]$. Show that $p(-x) = p(x)$ if and only if p has only even powers and $p(-x) = -p(-x)$ if and only if p has only odd powers. (**Hint.** Use induction on $\deg p$, taking a derivative for the induction step.)
(b) Look at problem 3(b) again.

Problem 29. (a) Let R be a commutative ring and let $f \in R[x]$ be a polynomial of degree d. Show that $f(x + h) = f(x) + \sum_{i=1}^{d} \varphi_i(x) h^i$ where $i! \, \varphi_i(x) = f^{(i)}(x)$. Replacing x by x_0 and h by $x - x_0$, this reads

$$f(x) = f(x_0) + \sum_{i=1}^{d} \varphi_i(x_0)(x - x_0)^i$$

which is Taylor's formula. (Compare with problem 21 above.)
(b) Generalizing to several variables, let R be a commutative ring again and let $f \in R[x_1, \ldots, x_n]$ be a polynomial of degree d. Show that

$$f(x_1 + h_1, \ldots, x_n + h_n) = \sum_{\substack{m_1, \ldots, m_n \geq 0, \\ 0 \leq m_1 + \cdots + m_n \leq d}} \frac{\partial^{m_1 + \cdots + m_n} f}{\partial x_1^{m_1} \cdots \partial x_n^{m_n}}(x_1, \ldots, x_n) \frac{h_1^{m_1} \cdots h_n^{m_n}}{m_1! \cdots m_n!} \,.$$

In this case, Taylor's formula reads

$$f(x_1, \ldots, x_n) = \sum_{0 \leq m_1 + \cdots + m_n \leq d} \frac{\partial^{m_1 + \cdots + m_n} f}{\partial x_1^{m_1} \cdots \partial x_n^{m_n}}(x_1^{(0)}, \ldots, x_n^{(0)}) \frac{h_1^{m_1} \cdots h_n^{m_n}}{m_1! \cdots m_n!}$$

where $h_i = x_i - x_i^{(0)}$. This is $f(x) = \sum_{0 \leq |I| \leq d} (\partial^I) f(x^{(0)})(x - x^{(0)})^I / I!$ in multi-index notation.

Problem 30. Let f be a polynomial with $f' \neq 0$. Show that $f / \gcd(f, f')$ has the same roots as f but only with multiplicity 1.

Problem 31. Let f be a polynomial over a field of characteristic p. Show that $f'' = 0$ if and only if there are polynomials $f_1, f_2 \in K[x]$ such that $f(x) = f_1(x^p) + x \cdot f_2(x^p)$. Describe all polynomials in $K[x]$ whose m-th derivative vanishes.

Problem 32. Let R be an integral domain.

(a) Show that if $f_1, \ldots, f_n \in R[x]$ then

$$\frac{(f_1 \cdots f_n)'}{f_1 \cdots f_n} = \frac{f_1'}{f_1} + \cdots + \frac{f_n'}{f_n}.$$

(b) Let $R(x)$ be the quotient field of $R[x]$. Show that the derivation map $R(x) \to R(x)$ given by

$$\left(\frac{f}{g}\right)' := \frac{f'g - fg'}{g^2}$$

is well-defined and satisfies the usual rules.

Problem 33. Let R be an integral domain of characteristic 0. Show that there are no polynomials $p, q, r \in R[x]$ with $p^n + q^n = r^n$ whenever $n \geq 3$.

Hint. Suppose $p^n + q^n = r^n$. Take the derivative on both sides of this equation to see that the following system of equations is satisfied:

$$\begin{pmatrix} p & q & -r \\ p' & q' & -r' \end{pmatrix} \begin{pmatrix} p^{n-1} \\ q^{n-1} \\ r^{n-1} \end{pmatrix} = \begin{pmatrix} 0 \\ 0 \end{pmatrix}.$$

Conclude that there is a polynomial $\lambda \in R[x]$ with $\lambda(p^{n-1}, q^{n-1}, r^{n-1}) = (rq' - qr', pr' - rp', pq' - qp')$. Show that this is impossible by considering the degrees of p, q and r.

Problem 34. Let $\leq$ be the lexicographic ordering on $\mathbb{N}_0^n$. Prove that there is no strictly decreasing series $v_1 > v_2 > v_3 > \cdots$ where $v_i \in \mathbb{N}_0^n$. Then use this fact to prove the following induction principle: If a statement holds for the vector $(0, 0, \ldots, 0)$ and if the validity of this statement for all $v < v_0$ implies the validity for v_0, then this statement holds for all $v \in \mathbb{N}_0^n$.

Problem 35. Express the following polynomials in terms of the elementary symmetric polynomials s_1, s_2, s_3.

(a) $x^2yz + xy^2z + xyz^2 + xyz$

(b) $x^3y^3 + y^3z^3 + x^3z^3 + xyz^4 + yzx^4 + xzy^4$

(c) $(x^2y + y^2z + z^2x)(xy^2 + yz^2 + zx^2)$

Problem 36. Express the following symmetric polynomials in four variables x_1, x_2, x_3, x_4 as polynomials in the elementary symmetric polynomials s_1, s_2, s_3, s_4.

(a) $x_1^2x_2^2 + x_1^2x_3^2 + x_1^2x_4^2 + x_2^2x_3^2 + x_2^2x_4^2 + x_3^2x_4^2$

(b) $\sum x_i^2(x_jx_j + x_jx_l + x_kx_l)$ where the sum ranges over all quadruplets (i, j, k, l) with i, j, k, l being pairwise different.

Problem 37. Let $p(x_1, \ldots, x_n) := \sum x_i^2 x_j^2 x_k$ where the sum runs over all triples (i, j, k) with i, j, k being pairwise distinct. Express p in terms of the elementary symmetric polynomials $s_1, \ldots, s_n$.

Problem 38. Find all numbers x such that $x + y + z = 6$, $xy + yz + zx = 11$ and $xyz = 6$.

Problem 39. (a) Find a quadratic polynomial whose roots are the cubes of the roots of $x^2 + ax + b$.
(b) Find all quadratic polynomials $x^2 + ax + b$ such that the set of the squares of the roots coincides with the set of the roots.

Problem 40. Suppose that $f(x) = x^3 + px + q$ has the roots a, b and c.
(a) Find a cubic polynomial with roots $(a - b)^2, (a - c)^2$ and $(b - c)^2$.
(b) Find a polynomial which has the roots $(b + c)/a^2$, $(c + a)/b^2$ and $a + b)/c^2$.

Problem 41. For all $k \in \mathbb{N}$, we can define a symmetric polynomial σ_k in n variables $x_1, \ldots, x_n$ by
$$\sigma_k := x_1^k + \cdots + x_n^k .$$
Show that if $s_1, \ldots, s_n$ are the elementary symmetric polynomials in the variables $x_1, \ldots, x_n$ and if $s_j := 0$ for $j > n$ then
$$\sigma_k - \sigma_{k-1} s_1 + \sigma_{k-2} s_2 - + \cdots + (-1)^{k-1} \sigma_1 s_{k-1} + (-1)^k \cdot s_k = 0$$
for all $k \in \mathbb{N}$. Thus
$$\sigma_k = \begin{cases} \sum_{i=1}^{k-1} (-1)^{i+1} s_i \sigma_{k-i} + (-1)^{k+1} k s_k, & k \leq n; \\ \sum_{i=1}^{n} (-1)^{i+1} s_i \sigma_{k-i}, & k > n. \end{cases}$$

This formula allows one to inductively express all power sums of the roots of a polynomial without explicitly knowing these roots. For example,
$$\sigma_2 = s_1^2 - 2s_2,$$
$$\sigma_3 = s_1^3 - 3s_1 s_2 + 3s_3,$$
$$\sigma_4 = s_1^4 - 4s_1^2 s_2 + 4s_1 s_3 + 2s_2^2 - 4s_4.$$

The formulas were first published by Isaac Newton in his "Arithmetica Universalis".

Problem 42. Let $s_1, \ldots, s_n$ be the elementary symmetric polynomials in n variables $x_1, \ldots, x_n$. Show that an arbitrary polynomial in the variables $x_1, \ldots, x_n$ can be written as
$$\sum_{i_1, \ldots, i_n} a_{i_1, \ldots, i_n}(s_1, \ldots, s_n) x_1^{i_1} x_2^{i_2} \cdots x_n^{i_n}$$
where $0 \leq i_1 \leq n - 1$, $0 \leq i_2 \leq n - 2$, $0 \leq i_3 \leq n - 3$, and so on.

Problem 43. (a) Let $D = \prod_{i<j}(x_i - x_j)^2$ be the discriminant and let $\sigma_k = x_1^k + \cdots + x_n^k$ be as in problem 41. Show that

$$D(x_1, \ldots, x_n) = \det \begin{pmatrix} \sigma_0 & \sigma_1 & \cdots & \sigma_{n-1} \\ \sigma_1 & \sigma_2 & \cdots & \sigma_n \\ \vdots & \vdots & & \vdots \\ \sigma_{n-1} & \sigma_n & \cdots & \sigma_{2n-2} \end{pmatrix}$$

Hint. Recall the **Vandermonde determinant**

$$\det V = \det \begin{pmatrix} 1 & 1 & \cdots & 1 \\ x_1 & x_2 & \cdots & x_n \\ \vdots & \vdots & \ddots & \vdots \\ x_1^{n-1} & x_2^{n-1} & \cdots & x_n^{n-1} \end{pmatrix} = \prod_{i>j}(x_j - x_i) .$$

Now use the fact that $D^2 = (\det V)^2 = \det(VV^T)$.

(b) Use part (a) to show that the discriminant of $f(x) = x^n + px + q$ is $D(f) = (-1)^{n(n-1)/2}\big((1-n)^{n-1}p^n + n^n q^{n-1}\big)$.

Problem 44. Define a polynomial in n variables by $\Delta(x_1, \ldots, x_n) := \prod_{i>j}(x_i - x_j)$. Show that $\sigma \star \Delta = (\text{sign}\,\sigma)\Delta$ for all $\sigma \in \text{Sym}_n$. Why does this imply that the discriminant of $x_1, \ldots, x_n$ is a symmetric polynomial?

Problem 45. Find the discriminants of the polynomials $x^3 - 6x^2 + 7x + 9$, $x^4 - 5x^3 + 6x^2 + 4x - 8$, $x^5 + 5x + 5$ and $x^5 + 6x^3 + 3x + 4$. Which of these polynomials do possess a multiple root?

5. Ideals and quotient rings

IDEALS play the same role in a ring that normal subgroups play in a group; they arise as kernels of homomorphisms and allow the construction of quotient structures. Quite often one is in the situation that a certain ideal in a ring consists of elements which are "uninteresting" with respect to a certain question; in this case, one factorizes the ring modulo this ideal (modulo the "uninteresting" objects) and does only the essential computations. A special case of this procedure was already encountered in section 1, where we passed from calculations in $\mathbb{Z}$ to calculations in the residue-class rings $\mathbb{Z}_n$.

(5.1) Definition. *A subring I of a ring R is called an* **ideal** *of R if $xy \in I$ and $yx \in I$ whenever $x \in I$ and $y \in R$; i.e., if multiplication of elements of I by arbitrary elements from either side does not lead outside I. We write $I \trianglelefteq R$ to express that I is an ideal of R.*

(5.2) Examples. (a) For any $n \in \mathbb{Z}$, the set $n\mathbb{Z}$ of all multiples of n is an ideal of $\mathbb{Z}$. These are in fact the only ideals of $\mathbb{Z}$. (See problem 1(a) below.)

(b) Let R be a commutative ring and let $r_1, \ldots, r_n$ be elements of R. Then

$$\{p \in R[x_1, \ldots, x_n] \mid f(r_1, \ldots, r_n) = 0\}$$

is an ideal of the polynomial ring $R[x_1, \ldots, x_n]$.

(c) Let R be a ring and $X \neq \emptyset$ be any set. The ring of all functions $f : X \to R$ (with the pointwise operations) is denoted by R^X. Then for any subset $Y \subseteq X$ the set $\{f : X \to R \mid f|_Y \equiv 0\}$ of all functions vanishing on Y is an ideal of R^X.

(d) If $f : R \to S$ is a ring homomorphism then $\ker f$ is an ideal of R. Indeed, if $x \in \ker f$ and $r \in R$ then $f(rx) = f(r)f(x) = f(r) \cdot 0 = 0$ so that $rx \in \ker f$; similarly, $xr \in \ker f$.

(e) If $(R_i)_{i \in I}$ is a family of rings, then the direct sum $\bigoplus_{i \in I} R_i$ is an ideal of the direct product $\prod_{i \in I} R_i$.

(f) If $(I_\lambda)_{\lambda \in L}$ is a family of ideals of a ring R, then the intersection $\bigcap_{\lambda \in L} I_\lambda$ and the sum $\sum_{\lambda \in L} I_\lambda$ are also ideals of R. Here the sum of ideals is defined as

$$\sum_{\lambda \in L} I_\lambda := \{x_{\lambda_1} + \cdots + x_{\lambda_n} \mid n \in \mathbb{N}, \ \lambda_1, \ldots, \lambda_n \in L, \ x_{\lambda_k} \in I_{\lambda_k}\} \ .$$

LET us observe that one can develop a certain calculus on the set of all ideals of a ring R.

(5.3) Proposition. *Let R be an arbitrary ring and let $\mathcal{I}(R)$ be the set of all ideals of R. If we define addition and multiplication on $\mathcal{I}(R)$ by*

$$I + J := \{x + y \mid x \in I, y \in J\} \ and \ IJ := \{\sum_{i=1}^n x_i y_i \mid n \in \mathbb{N}, x_i \in I, y_j \in J\}$$

we obtain an algebraic structure $(\mathcal{I}(\mathcal{R}), +, \cdot)$ with the following properties.

(a) *The addition in $\mathcal{I}(\mathcal{R})$ is commutative and associative and possesses a neutral element, namely $\{0\}$.*

(b) *The multiplication in $\mathcal{I}(\mathcal{R})$ is associative; it is commutative if the multiplication in R is. If R has an identity element 1, then R is an identity element for $\mathcal{I}(R)$ because in this case $RI = IR = I$ for all $I \trianglelefteq R$.*

(c) *The distributive laws $I(J + K) = IJ + IK$ and $(I + J)K = IK + JK$ hold.*

(d) *$IJ \subseteq I \cap J \subseteq I + J$ for all $I, J \in \mathcal{I}(\mathcal{R})$.*

Proof. Exercise. (See problem 8 below.) ∎

HENCE $\mathcal{I}(\mathcal{R})$ is "almost a ring"; the only axiom which is not satisfied is the existence of additive inverses. Clearly, if $I \neq \{0\}$ is a nonzero element of a ring R then there is no ideal $J \trianglelefteq R$ with $I + J = \{0\}$.

NOW let us see how we can construct an ideal starting with an arbitrary subset of a ring.

(5.4) Definition. *If X is a nonempty subset of a ring R, then*

$$\langle X \rangle := \bigcap_{I \trianglelefteq R,\ I \supseteq X} I = \text{intersection of all ideals containing } X$$

*is called the ideal **generated** by X. Obviously, $\langle X \rangle$ is the smallest ideal of R containing X. It is easy to see that $\langle X \rangle$ consists of all finite sums of elements of the form †*

$$n \cdot x + ax' + x''b + cx'''d \quad \text{where} \quad x, x', x'', x''' \in X,\ n \in \mathbb{Z},\ a, b, c, d \in R .$$

In a commutative ring with 1, this simplifies to

$$\langle X \rangle = \{a_1 x_1 + \cdots + a_n x_n \mid a_i \in R, x_i \in X\} = Ra_1 + \cdots + Ra_n .$$

Note that $\langle X \rangle + \langle Y \rangle = \langle \{x + y \mid x \in X, y \in Y\} \rangle$ and $\langle X \rangle \langle Y \rangle = \langle \{xy \mid x \in X, y \in Y\} \rangle$.

THE case in which the generating set is finite is particularly important.

(5.5) Definitions. *Let R be a ring. An ideal $I \trianglelefteq R$ is called **finitely generated** if I can be generated by a finite number of elements, i.e., if there are elements $a_1, \ldots, a_n \in R$ such that $I = \langle a_1, \ldots, a_n \rangle$. An ideal $I \trianglelefteq R$ is called a **principal ideal** if it can be generated by a single element.*

---

† Recall that $n \cdot x$ means $x + \cdots + x$ if $n > 0$ and $(-x) + \cdots + (-x)$ if $n < 0$ (in each case there are $|n|$ summands); this is the same notation as the one for abelian groups.

(5.6) Examples. (a) Let $I := \langle 4, 6 \rangle$ in $\mathbb{Z}$. Since $2 = 6 - 4 \in I$, all even numbers belong to I. On the other hand, every element of I is a linear combination of 4 and 6, hence is divisible by 2. This reveals I as a principal ideal, namely $I = \langle 2 \rangle$.

(b) Let R be a commutative ring and let $I := \{f \in R[x,y] \mid f(0,0) = 0\} \trianglelefteq R[x,y]$ be the ideal of all polynomials in two variables x, y with a zero constant term. Then $I = \langle x, y \rangle$.

SOME rings have the remarkable property that every ideal is a principal ideal. For example, the ring $\mathbb{Z}$ of integers has this property. Since we will encounter rings of this type of rings more often, we give the following definition.

(5.7) Definition. *A ring in which every ideal is a principal ideal is called a* **principal ideal ring**. *A principal ideal ring which is also an integral domain is called a* **principal ideal domain**.

CLEARLY, we have always two ideals in an arbitrary ring R, namely $\{0\}$ and R itself. Sometimes there are no other ideals; in this case the ring R is called *simple*.

(5.8) Definition. *A ring R is called* **simple** *if it has no ideals other than $\{0\}$ and R.*

(5.9) Examples. (a) Every division ring is simple. Indeed, if R is a division ring and if $I \trianglelefteq R$ is a nonzero ideal, then I must contain a unit of R because $R^{\times} = R \setminus \{0\}$. But an ideal I which contains a unit $u \in R^{\times}$ must be all of R, because if $r \in R$ then $r = r \cdot 1 = (ru^{-1})u \in I$.

(b) If K is any field then the matrix ring $K^{n \times n}$ is simple. To see this, let I be an ideal of $K^{n \times n}$ which contains a nonzero element $A \neq 0$. Suppose $a_{i_0 j_0} \neq 0$. Then I contains the element $E_{i i_0} A E_{j_0 j} = a_{i_0 j_0} E_{ij}$, hence also $(a_{i_0 j_0}^{-1} 1)(a_{i_0 j_0} E_{ij}) = E_{ij}$; here E_{ij} denotes the matrix with (i,j)-entry 1 and all other entries equal to zero. This shows that I contains all the matrices E_{ij}, but then also all of $K^{n \times n}$.

LET us now study the behavior of ideals under ring homomorphisms.

(5.10) Proposition. *Suppose that $f : R \to S$ is a ring homomorphism.*
(a) *If I is an ideal of R, then $f(I)$ is an ideal of $f(R)$ (but not necessarily of S).*
(b) *If J is an ideal of S, then $f^{-1}(J)$ is an ideal of R.*

Proof. (a) Take arbitrary elements in $f(R)$ and $f(I)$, say $f(r)$ and $f(x)$ where $r \in R$ and $x \in I$. Then $rx \in I$ and $xr \in I$ so that $f(r)f(x) = f(rx) \in f(I)$ and $f(x)f(x) = f(xr) \in f(I)$. This shows that $f(R)f(I) \subseteq f(I)$ and $f(I)f(R) \subseteq f(I)$ so that $f(I)$ is an ideal of $f(R)$.

(b) Let J be an ideal of S. If $r \in R$ and $x \in f^{-1}(J)$ then $f(rx) = f(r)f(x) \in SJ \subseteq J$ so that $rx \in f^{-1}(J)$. Similarly, $xr \in f^{-1}(J)$. This shows that the subring $f^{-1}(I)$ of R is in fact an ideal. ∎

TAKING $J := \{0\}$ in (5.10)(b) we see that the kernel of a ring homomorphism is automatically an ideal; this was already observed in (5.2)(d). We now show conversely that every ideal is the kernel of some ring homomorphism. This is done by introducing the notion of a quotient ring modulo an ideal. Since the construction is completely analogous to the construction of a quotient group modulo a normal subgroup, we proceed without further comments.

(5.11) Definition. *Let I be an ideal of the ring R.*

(a) *The definition $x \overset{I}{\equiv} y :\Longleftrightarrow x - y \in I$ gives an equivalence relation on R. The equivalence class of an element $x \in R$ is given by*

$$[x] := x + I = \{x + a \mid a \in I\} .$$

(b) *The set $R/I := \{\, [x] \mid x \in R \,\}$ of all equivalence classes becomes a ring with the operations*

$$[x] + [y] := [x+y] \qquad and \qquad [x] \cdot [y] := [xy] .$$

*This ring is called the **quotient ring** or the **factor ring** of R modulo I.*

(c) *The **canonical projection** of R onto R/I given by*

$$\pi : \begin{array}{ccc} R & \to & R/I \\ x & \mapsto & [x] \end{array}$$

is a ring homomorphism with $\ker \pi = I$.

Proof. (a) Write for short $x \equiv y$ instead of $x \overset{I}{\equiv} y$. Then we have $x \equiv x$ for all $x \in R$ because $0 \in I$. Also, $x \equiv y$ implies $y \equiv x$ because $-I = I$. Furthermore, if $x \equiv y$ and $y \equiv z$, then $x \equiv z$ because $I + I \subseteq I$. Notice that we did not use the fact that I is an ideal; it is enough if $(I, +)$ is a subgroup of $(R, +)$.

(b) We have only to check that the operations are well-defined; the validity of the arithmetic laws is then inherited from R to R/I. We have to show that the conditions $[x] = [x']$ and $[y] = [y']$ imply that $[x+y] = [x'+y']$ and $[xy] = [x'y']$; i.e., from $x - x' \in I$ and $y - y' \in I$ we have to conclude that $(x + y) - (x' + y') \in I$ and $xy - x'y' \in I$. Now

$$(x + y) - (x' + y') = (x - x') + (y - y') \in I + I \subseteq I \quad \text{and}$$
$$xy - x'y' = (x - x')y + x'(y - y') \in IR + RI \subseteq I .$$

(c) We have $\pi(x + y) = [x + y] = [x] + [y] = \pi(x) + \pi(y)$ and $\pi(xy) = [xy] = [x] \cdot [y] = \pi(x)\pi(y)$ so that π is a homomorphism. An element $x \in R$ belongs to $\ker \pi$ if and only if $[x] = [0]$, i.e., if and only if $x \in I$. ∎

(5.12) Examples. (a) Taking $R = \mathbb{Z}$ and $I = n\mathbb{Z} = \langle n \rangle$, we obtain the ring

$$\mathbb{Z}_n := \frac{\mathbb{Z}}{n\mathbb{Z}} = \{[x]_n \mid x \in \mathbb{Z}\}$$

of all residue classes modulo n.

(b) Let K be a field. Taking $R = K[x]$ and $I = f \cdot K[x] = \langle f \rangle$, we obtain the ring

$$K[x]_f := K[x]/\langle f \rangle = \{[p]_f \mid p \in K[x]\}$$

of all polynomials modulo f. Let us look at the example $\mathbb{Z}_2[x]_{x^2+x+1}$ to see how arithmetic modulo f works. Writing

$$0 := [0], \quad 1 := [1], \quad \alpha := [x], \quad \beta := [x+1]$$

and using the fact that $[x^2 + x + 1] = [0]$, we obtain the following tables for addition and multiplication in $\mathbb{Z}_2[x]_f = \mathbb{Z}_2[x]/\langle f \rangle$.

$$
\begin{array}{c|cccc}
+ & 0 & 1 & \alpha & \beta \\
\hline
0 & 0 & 1 & \alpha & \beta \\
1 & 1 & 0 & \beta & \alpha \\
\alpha & \alpha & \beta & 0 & 1 \\
\beta & \beta & \alpha & 1 & 0
\end{array}
\qquad
\begin{array}{c|cccc}
\cdot & 0 & 1 & \alpha & \beta \\
\hline
0 & 0 & 0 & 0 & 0 \\
1 & 0 & 1 & \alpha & \beta \\
\alpha & 0 & \alpha & \beta & 1 \\
\beta & 0 & \beta & 1 & \alpha
\end{array}
$$

For example, $\alpha\beta = [x] \cdot [x+1] = [x^2 + x] = [-1] = [1] = 1$. Observe that every nonzero element in R/I is invertible. This means that R/I is a field! We will see in (6.14)(d) that this is not an accident, but follows from a general fact. Also, a funny application in which the field $\mathbb{Z}_2[x]_{x^2+x+1}$ is used is given in problem 33 below.

(c) Let us find the remainder of $x^4 - 2x^3 + x^2$ when divided by $x^2 + 1$ in $\mathbb{Q}[x]$. Let $I := \langle x^2 + 1 \rangle \trianglelefteq \mathbb{Q}[x]$. We have $x^2 \equiv -1$ modulo I and hence $x^4 - 2x^3 + x^2 \equiv (-1)^4 - 2(-1)x + (-1) = 2x$; thus the remainder is $2x$. Note that we did not have to perform a polynomial division to obtain this result!

(d) We claim that $x^{10000} - x^6 - 2$ is divisible by $x^2 + 1$. Indeed, modulo $I := \langle x^2 + 1 \rangle$ we have $x^2 \equiv -1$ and hence $x^{10000} - x^6 - 2 \equiv (-1)^{5000} - (-1)^3 - 2 = 0$. Similarly, we see that $x^N - 1$ is divisible by $x^n - 1$ whenever N is divisible by n.

(5.13) Isomorphism theorems. (a) *Let $f : R \to S$ be a ring homomorphism. Then*

$$\Phi : \begin{array}{ccc} R/\ker f & \to & \operatorname{im} f \\ [x] & \mapsto & f(x) \end{array}$$

is well-defined and an isomorphism.

(b) *If $S \leq R$ and $I \trianglelefteq R$, then $(S + I)/I \cong S/(S \cap I)$.*

(c) *Let $I \trianglelefteq R$. Then the mapping*

$$\begin{array}{ccc} \{\text{ ideals of } R \text{ containing } I \} & \to & \{\text{ ideals of } R/I \} \\ J & \mapsto & J/I \end{array}$$

is a bijection. Moreover, if $I \subseteq J \trianglelefteq R$ then $(R/I)/(J/I) \cong R/J$.

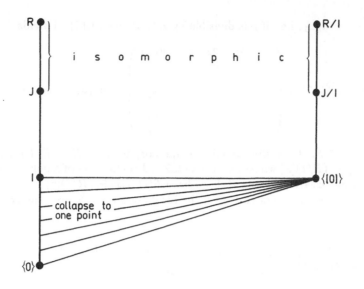

Proof. (a) $[x] = [x']$ if and only if $x - x' \in \ker f$, i.e., if $f(x) = f(x')$. This shows that Φ is well-defined and one-to-one. Obviously, Φ is onto. It is straightforward from the definitions to check that Φ is a homomorphism.

(b) We only have to observe that the mapping $f : S \to (S+I)/I$ given by $x \mapsto x+I$ is a surjective homomorphism; the claim then follows from (a).

(c) Let $\pi : R \to R/I$ be the canonical projection. If J is an ideal of R with $J \supseteq I$, then $\pi(J) = J/I$ is an ideal of $\pi(R) = R/I$. Conversely, if $\overline{J}$ is an ideal of R/I, then $J := \pi^{-1}(\overline{J})$ is an ideal of R with $J/I = \pi(J) = \overline{J}$ and $J \supseteq I$. Finally, if $I \subseteq J \trianglelefteq R$, then

$$f : \begin{array}{ccc} R & \mapsto & \frac{R/I}{J/I} \\ r & \mapsto & (r+I) + J/I \end{array}$$

is a surjective homomorphism with $\ker f = J$. Indeed, $r \in \ker f$ if and only if $r + I \in J/I$, i.e., if $r \in J + I = J$. ∎

(5.14) Example. (a) Let K be a field. Then we define a homomorphism $f : K[x] \to K^{n \times n}$ by

$$\sum_{k=0}^{m} a_k x^k \mapsto \begin{pmatrix} a_0 & a_1 & a_2 & \cdots & a_{n-1} \\ & \ddots & \ddots & \ddots & \vdots \\ & & \ddots & \ddots & a_2 \\ & & & \ddots & a_1 \\ & & & & a_0 \end{pmatrix} ;$$

obviously, a polynomial $p(x) = \sum_k a_k x^k$ belongs to the kernel of f if and only if

$a_0 = a_1 = \cdots = a_{n-1}$, i.e., if p is divisible by x^{10}. Hence (5.13)(a) yields

$$K[x]/\langle x^{10} \rangle \cong \left\{ \begin{pmatrix} a_0 & a_1 & a_2 & \cdots & a_{n-1} \\ & \ddots & \ddots & \ddots & \vdots \\ & & \ddots & \ddots & a_2 \\ & & & \ddots & a_1 \\ & & & & a_0 \end{pmatrix} \;\middle|\; a_0, a_1, \ldots, a_{n-1} \in K \right\}.$$

(b) Let $I \trianglelefteq R$. Then the canonical homomorphism $\varphi : R \to R/I$ induces a mapping $\Phi : R[x] \to (R/I)[x]$ whose effect is to take all coefficients of a polynomial $p \in R[x]$ modulo I. Since clearly $\ker \Phi = I[x]$, we have $R[x]/I[x] \cong (R/I)[x]$.

(c) As a consequence of (5.13)(c), the ideals of $\mathbb{Z}_n = \frac{\mathbb{Z}}{n\mathbb{Z}}$ are exactly the sets $\frac{m\mathbb{Z}}{n\mathbb{Z}}$ where $m \mid n$.

(5.15) Definition. *Let R be a ring. An ideal $I \neq R$ of R is called a* **maximal ideal** *if there is no ideal $J \trianglelefteq R$ with $I \subsetneqq J \subsetneqq R$.*

THE following proposition characterizes maximal ideals.

(5.16) Proposition. *Let R be a ring and let $I \neq R$ an ideal of R. Then the following conditions are equivalent.*
  (a) *I is a maximal ideal.*
  (b) *$\langle I \cup \{x\} \rangle = I + \langle x \rangle = R$ for all $x \in R \setminus I$.*
  (c) *R/I is a simple ring.*

Proof. If $x \in R \setminus I$ then $\langle I \cup \{x\} \rangle$ is an ideal strictly larger than I, hence must be R if I is maximal. This shows that (a) implies (b). Conversely, if $\langle I \cup \{x\} \rangle = R$ for all $R \setminus I$ then no proper ideal of R contains I properly. But this means that I is maximal. Finally, the equivalence of (a) and (c) is an immediate consequence of (5.13)(c). $\blacksquare$

IN a commutative ring with identity, we simply have $\langle I \cup \{x\} \rangle = I + Rx$; this will allow us to say much more about maximal ideals in the next section where we focus on commutative rings.

(5.17) Examples. (a) The maximal ideals of $\mathbb{Z}$ are exactly the ideals $\langle p \rangle = p\mathbb{Z}$ where p is a prime number.

(b) Let $C(\mathbb{R})$ be the ring of all continuous functions $f : \mathbb{R} \to \mathbb{R}$. For each $x \in \mathbb{R}$, the set $I_x = \{f \in C(\mathbb{R}) \mid f(x) = 0\}$ is a maximal ideal of $C(\mathbb{R})$. Indeed, the evaluation map $f \mapsto f(x)$ is a surjective homomorphism $C(\mathbb{R}) \to \mathbb{R}$ with kernel I_x. Hence $C(\mathbb{R})/I_x \cong \mathbb{R}$ is a field and therefore a simple ring which implies that I_x is a maximal ideal, due to (5.16).

THE following result guarantees a big supply of maximal ideals in rings with identity.

(5.18) Proposition. *Let R be a ring with identity and $I \neq R$ an ideal of R. Then I is contained in a maximal ideal of R.*

Proof. Let

$$\mathcal{M} := \{J \unlhd R \mid I \subseteq J \subsetneq R\} \,.$$

This set is not empty because $I \in \mathcal{M}$. We show that every chain in $\mathcal{M}$ possesses an upper bound in $\mathcal{M}$; then Zorn's lemma will yield the claim. Given a chain $\mathcal{C}$ in $\mathcal{M}$, let $J_0 := \bigcup_{J \in \mathcal{C}} J$. This is again an ideal of R containing I; moreover, $J_0 \neq R$ because $1 \notin J_0$. Consequently, J_0 belongs to $\mathcal{M}$ and is clearly an upper bound for $\mathcal{C}$. ∎

(5.19) Definition. *Two ideals I and J in a ring R are called* **comaximal** *if $I + J = R$.*

(5.20) Chinese remainder theorem. *Let R be a ring and let $I_1, \ldots, I_n$ be ideals of R such that $R^2 + I_k = R$ for all k.* † *If $I_1, \ldots, I_n$ are pairwise comaximal, i.e., if $I_r + I_s = R$ for all $r \neq s$, then*

$$\Phi : \begin{array}{ccc} R & \to & R/I_1 \times \cdots \times R/I_n \\ r & \mapsto & (r + I_1, \ \ldots, \ r + I_n) \end{array}$$

is a surjective homomorphism with $\ker \Phi = I_1 \cap \cdots \cap I_n$.

Proof. It is clear that Φ is a homomorphism with kernel $I_1 \cap \cdots \cap I_n$; only the surjectivity has to be proved. We claim that

$$R = I_1 + \bigcap_{k \neq 1} I_k \,.$$

Indeed,

$$\begin{aligned}
R = I_1 + R^2 &= I_1 + (I_1 + I_2)(I_1 + I_3) \\
&\subseteq \underbrace{I_1 + I_1^2 + I_1 I_3 + I_2 I_1}_{\subseteq I_1} + \underbrace{I_2 I_3}_{\subseteq I_2 \cap I_3} \ \subseteq I_1 + (I_2 \cap I_3) \,.
\end{aligned}$$

But then also

$$\begin{aligned}
R = I_1 + R^2 &= I_1 + \underbrace{(I_1 + I_2 \cap I_3)}_{= R} \underbrace{(I_1 + I_4)}_{= R} \\
&= \underbrace{I_1 + I_1 I_4 + (I_2 \cap I_3)I_4}_{\subseteq I_1} + \underbrace{(I_2 \cap I_3)I_4}_{\subseteq I_2 \cap I_3 \cap I_4} \ \subseteq I_1 + (I_2 \cap I_3 \cap I_4) \,.
\end{aligned}$$

Continuing in this fashion, we obtain $R = I_1 + (I_2 \cap I_3 \cap I_4 \cap \cdots \cap I_n)$ as claimed. Similarly,

$$R + I_2 + \bigcap_{k \neq 2} I_k = I_3 + \bigcap_{k \neq 3} I_k = \cdots = I_n + \bigcap_{k \neq n} I_k \,.$$

---

† This condition is automatically satisfied if R has an identity; in this case even $R^2 = R$.

Now let $r_1, \ldots, r_n$ be arbitrarily given. Then we can find elements $a_k \in I_k$ and $b_k \in \bigcap_{\nu \neq k} I_\nu$ such that $r_k = a_k + b_k$. Then

$$r := b_1 + \cdots + b_n \equiv b_k \equiv r_k \quad \text{modulo } I_k$$

which is just another way of saying that

$$\Phi(r) = (r_1 + I_1, \, r_2 + I_2, \, \ldots, \, r_n + I_n) \, .$$

$\blacksquare$

THE last result certainly looks quite abstract, but when we specialize it to certain ideals in $\mathbb{Z}$, we obtain an important result of elementary number theory concerning the simultaneous solution of a finite number of congruences. This result, stated and proved in (5.21), was known to Chinese mathematicians in the first century A.D., a fact which gave the abstract version (5.20) its name.

(5.21) Example. *Suppose that the natural numbers $a_1, \ldots, a_n$ are relatively prime. Given arbitrary integers $r_1, \ldots, r_n$ we can find an integer m such that*

$(\star)$ $\qquad\qquad m \equiv r_1 \bmod a_1, \quad m \equiv r_2 \bmod a_2, \quad \ldots, \quad m \equiv r_n \bmod a_n \, .$

Moreover, if m and m' are two solutions of the simultaneous congruences $(\star)$ then

$(\star\star)$ $\qquad\qquad\qquad\qquad m \equiv m' \bmod a_1 a_2 \cdots a_n \, .$

Proof. Just apply (5.20) with $R = \mathbb{Z}$ and $I_k = a_k \mathbb{Z}$ to get $(\star)$. To obtain $(\star\star)$ we observe that

$$I_1 \cap \cdots \cap I_n = a_1 \mathbb{Z} \cap a_2 \mathbb{Z} \cap \cdots \cap a_n \mathbb{Z} = (a_1 a_2 \ldots a_n)\mathbb{Z}$$

because $a_1, \ldots, a_n$ are relatively prime. $\blacksquare$

(5.22) Example. Let us find all integers $x \in \mathbb{Z}$ with $x \equiv 1 \,(\mathrm{mod}\,2)$, $x \equiv 2 \,(\mathrm{mod}\,3)$ and $x \equiv 3 \,(\mathrm{mod}\,5)$. The ideals involved are $I_1 = 2\,\mathbb{Z}$, $I_2 = 3\,\mathbb{Z}$ and $I_3 = 5\,\mathbb{Z}$. Following the proof of (5.20), we have to find numbers $a_i, \, b_i \in \mathbb{Z}$ such that

$$
\begin{aligned}
1 &= a_1 + b_1 \quad \text{with } a_1 \in I_1 = 2\,\mathbb{Z} \text{ and } b_1 \in I_2 \cap I_3 = 15\,\mathbb{Z}, \\
2 &= a_2 + b_2 \quad \text{with } a_2 \in I_2 = 3\,\mathbb{Z} \text{ and } b_2 \in I_1 \cap I_3 = 10\,\mathbb{Z} \quad \text{and} \\
3 &= a_3 + b_3 \quad \text{with } a_3 \in I_3 = 5\,\mathbb{Z} \text{ and } b_3 \in I_1 \cap I_2 = 6\,\mathbb{Z} \, .
\end{aligned}
$$

It is easy to see[†] that we can take $(a_1, b_1) = (16, -15)$, $(a_2, b_2) = (-18, 20)$ and $(a_3, b_3) = (-15, 18)$. Then $x := b_1 + b_2 + b_3 = 23$ is a solution of our problem,

---

† A systematic way of solving these congruences is the Euclidean algorithm to be presented in section 7.

and all other solutions can differ from x only by a multiple of $2 \cdot 3 \cdot 5 = 30$. Hence the solutions of our problem are exactly the numbers $23 + 30k$ with $k \in \mathbb{Z}$.

WE conclude this section with another result about comaximal ideals.

(5.23) Theorem. *Let R be a commutative ring with identity element 1. If $I_1, \ldots, I_n$ are pairwise comaximal, then*

$$I_1 \cap I_2 \cap \cdots \cap I_n = I_1 I_2 \cdots I_n \ .$$

Proof. We proceed by induction on n, the case $n = 1$ being trivial. In the case $n = 2$, we have $I_1 + I_2 = R$, hence

$$I_1 \cap I_2 = R(I_1 \cap I_2) = (I_1 + I_2)(I_1 \cap I_2) \subseteq I_1 I_2 \ ;$$

the converse inclusion $I_1 I_2 \subseteq I_1 \cap I_2$ is trivial. Now let $n \geq 3$ and suppose that the claim is true for $n - 1$ so that $I_1 \cap I_2 \cap \cdots \cap I_{n-1} = I_1 I_2 \cdots I_{n-1} =: J$. We claim that $I_n + J = R$. Indeed, since $I_n + I_k = R$, we find $x_k \in I_k$ and $y_k \in I_n$ with $x_k + y_k = 1$. Hence $J \ni \prod_{k=1}^{n-1} x_k = \prod_{k=1}^{n-1} (1 - y_k) \in 1 + I_n$ so that $1 \in J + I_n$. Then using the case $n = 2$ we have $I_1 I_2 \cdots I_{n-1} I_n = J I_n = J \cap I_n = I_1 \cap I_2 \cap \cdots \cap I_{n-1} \cap I_n$. $\blacksquare$

Exercises

Problem 1. (a) Show that if $R \neq \{0\}$ is a subring of $\mathbb{Z}$ then there is a smallest natural number n in R, and this number satisfies $R = n\mathbb{Z} = \langle n \rangle$. (This shows that every subring of $\mathbb{Z}$ is already a principal ideal.)

(b) Show that the property specified in (a) is inherited to all homomorphic images of $\mathbb{Z}$. What can you conclude about the subrings of $\mathbb{Z}_n$?

(c) Show that $\langle [a] \rangle = \langle [b] \rangle$ in $\mathbb{Z}_n$ if and only if $\gcd(a, n) = \gcd(b, n)$. Use this fact to find all subrings of $\mathbb{Z}_{24}$, $\mathbb{Z}_{36}$ and $\mathbb{Z}_{63}$.

(d) Show that $\mathbb{Z} + 2i\mathbb{Z}$ is a subring of $\mathbb{Z} + i\mathbb{Z}$ but not an ideal.

Problem 2. (a) Let K be a field. Show that $\langle x, y \rangle$ is not a principal ideal of $K[x, y]$.

(b) Consider the set I of all polynomials in $\mathbb{Z}[x]$ whose constant term is divisible by 5. Show that I is an ideal, but not a principal ideal of $\mathbb{Z}[x]$.

(c) Let $n \equiv 1$ modulo 4. Show that $\langle 2, \sqrt{n} \rangle$ is a principal ideal in $\mathbb{Z} + \mathbb{Z}\frac{1+\sqrt{n}}{2}$.

(d) Let $n \equiv 1$ modulo 4. Show that $\langle 2, \sqrt{n} \rangle$ is not a principal ideal in $\mathbb{Z} + \mathbb{Z}\sqrt{n}$.

Problem 3. Let K be a field. Show that the nonzero ideals of the power series ring $K[[x]]$ are exactly the sets $\langle x^n \rangle$ with $n \in \mathbb{N}_0$; in particular, $K[[x]]$ is a principal ideal domain.

Problem 4. Let $R = C^\infty(-1, 1)$ be the ring of all functions $f : (-1, 1) \to \mathbb{R}$ which are infinitely often differentiable. Show that for each fixed value of n the set

$$I := \{ f \in C^\infty(-1, 1) \mid f^{(k)}(0) = 0 \text{ for } 0 \le k \le n \}$$

is an ideal of R.

Problem 5. Find the ideal of $\mathbb{Z} \times \mathbb{Z}$ generated by $(4, 9)$ and $(7, 15)$.

Problem 6. Let X be a finite set and let $\mathfrak{P}(X)$ be the Boolean ring introduced in problem 31(c) of section 2. Find all ideals in $\mathfrak{P}(X)$.

Problem 7. Let $R = \mathbb{Z} + i\mathbb{Z}$. Sketch the ideal $\langle 2 + i \rangle$ as a subset of $\mathbb{C}$.

Problem 8. Prove proposition (5.3).

Problem 9. Let R be the ring of all strictly upper triangular matrices with coefficients in a field K. Determine the ideals R^2, R^3, $R^4, \ldots$ of R.

Problem 10. Let I, J and K be ideals of a ring R.
(a) Show that $I + J = I$ if and only if $J \subseteq I$.
(b) Show that $I \cap J = I$ if and only if $I \subseteq J$.

(c) Show that $(I \cap K) + (J \cap K) \subseteq (I + J) \cap K$. Show that if $I \subseteq K$ or $J \subseteq K$ then even equality holds, but use the example $R = K[x, y]$, $I = \langle x \rangle$, $J = \langle y \rangle$, $K = \langle x + y \rangle$ to show that the inclusion may be strict in general.

Problem 11. Let $K^{n \times n}$ be the ring of all $(n \times n)$-matrices over a field K.
(a) For $1 \le k \le n$ let

$$U_k := \{ \begin{pmatrix} | & & | & & | \\ 0 & \cdots & v & \cdots & 0 \\ | & & | & & | \end{pmatrix} \mid v \in K^n \}$$

be the set of all matrices whose entries are zero except in the k-th column. Show that U_k is a subring of $K^{n \times n}$ with the property that $AB \in U_k$ whenever $B \in U_k$, but is not an ideal. (One says that U_k is a **left-ideal** of $K^{n \times n}$ because U_k is invariant under multiplication from the left.) Moreover, show that

$$K^{n \times n} = U_1 \oplus \cdots \oplus U_n$$

is a decomposition of $K^{n \times n}$ into minimal left-ideals.
(b) Find a decomposition of $K^{n \times n}$ into minimal right-ideals.

Problem 12. Let R be a ring whose only left-ideals are $\{0\}$ and R. Show that either R is a division ring or else that R is a ring with a prime number of elements such that $xy = 0$ for all $x, y \subset R$.
Hint. Show that either $Rx = R$ for all $x \ne 0$ or else $Rx = \{0\}$ for all $x \in R$.

Problem 13. Let $\mathbb{H}^*$ be the set of all quaternions $a + bi + cj + dk \in \mathbb{H}$ such that either $a, b, c, d \in \mathbb{Z}$ or else a, b, c, d are all halves of odd integers.
(a) Show that $\mathbb{H}^*$ is a subring of $\mathbb{H}$.
(b) Show that $|\alpha| \in \mathbb{Z}$ for all $\alpha \in \mathbb{H}^*$.
(c) Show that every left-ideal of $\mathbb{H}^*$ has the form $\mathbb{H}^* \alpha$ for some $\alpha \in \mathbb{H}^*$, i.e., is a principal left-ideal.

Problem 14. (a) Show that if $f : R \to S$ is a surjective ring homomorphism the ideals of S are exactly the sets $f(I)$ where I is an ideal of R.
(b) Find all ideals of $\mathbb{Z}_n$.

Problem 15. Let R be a ring.
(a) Show that if $I \trianglelefteq R$ then $I^{n \times n} \trianglelefteq R^{n \times n}$.
(b) Suppose R has an identity element and let $J \trianglelefteq R^{n \times n}$. Show that

$$I := \{r \in R \mid r \text{ occurs as a coefficient of a matrix in } J\}$$

is an ideal of R with $I^{n \times n}$.
(c) Find all ideals in $\mathbb{Z}_k^{n \times n}$.

Problem 16. Let R be a ring, $U \leq R$ a subring and $I, J \trianglelefteq R$ ideals of R. Show that if $U \subseteq I \cup J$ then $U \subseteq I$ or $U \subseteq J$.

Problem 17. Let $R \subseteq S$ be commutative rings.
(a) Show that there is a unique maximal set $I \subseteq R$ with the property that I is an ideal of R as well as an ideal of S, namely $I = \{x \in R \mid xS \subseteq R\}$. This ideal is called the **leading ideal** of R in S.
(b) Let $R := \mathbb{Z} + \mathbb{Z}\sqrt{-3}$ and $S := \mathbb{Z} + \mathbb{Z}(\frac{1+\sqrt{-3}}{2})$. Find the leading ideal of R in S.

Problem 18. (a) Let R be a commutative ring with identity. Show that if R is simple then R is a field.
(b) Let p be an odd prime number and let $R := \{a + bi + cj + dk \mid a, b, c, d \in \mathbb{Z}_p\}$ be the ring of all "quaternions modulo p". Show that the ring R is simple, but not a division ring.

Problem 19. Let R be a field or a skew-field and let $f : R \rightarrow S$ be a ring homomorphism. Show that f is either one-to-one or the zero mapping.

Problem 20. (a) Let $R = R_1 \oplus \cdots \oplus R_n$ be a direct sum of a finite number of rings with identity. Show that the ideals of R are exactly the sets $I_1 \oplus \cdots \oplus I_n$ where I_k is an ideal of R_k for $1 \leq k \leq R_n$. Characterize the maximal ideals of R.
(b) Let $R = K_1 \oplus \cdots \oplus K_n$ be a direct sum of fields. List all the ideals of R. (How many ideals are there?)
(c) Let $R = 2\mathbb{Z}$ be the ring of all even integers. Find an ideal of $R \oplus R$ which is not of the form $I \oplus J$ with $I, J \trianglelefteq R$.

Problem 21. Give an example for a ring homomorphism $f : R \rightarrow S$ and an ideal $I \trianglelefteq R$ such that $f(I)$ is not an ideal of S.

Problem 22. Let $f : R \rightarrow S$ be a surjective ring homomorphism and let $I \trianglelefteq R$ and $J \trianglelefteq S$. Show that $f(I) \cap J = f(I \cap f^{-1}(J))$.

Problem 23. Let $f : R \rightarrow S$ be a ring homomorphism. We denote by $\mathcal{I}(R)$ and $\mathcal{I}(S)$ the set of ideals of R and S, respectively, and define mappings $\mathcal{E} : \mathcal{I}(R) \rightarrow \mathcal{I}(S)$ ("extension") and $\mathcal{C} : \mathcal{I}(S) \rightarrow \mathcal{I}(R)$ ("contraction") as follows:

$$\mathcal{E}(I) := \langle f(I) \rangle \trianglelefteq S ; \qquad \mathcal{C}(J) := f^{-1}(J) \trianglelefteq R .$$

Prove the following statements!
(a) $I \subseteq \mathcal{C} \circ \mathcal{E}(I)$ for all $I \trianglelefteq R$ and $J \supseteq \mathcal{E} \circ \mathcal{C}(J)$ for all $J \trianglelefteq S$.
(b) $\mathcal{C} = \mathcal{C} \circ \mathcal{E} \circ \mathcal{C}$ and $\mathcal{E} = \mathcal{E} \circ \mathcal{C} \circ \mathcal{E}$.
(c) Investigate how sums, products and intersections behave under the operators $\mathcal{E}$ and $\mathcal{C}$!

Problem 24. (a) Let $f : R \to S$ be a ring homomorphism and let $J \trianglelefteq S$ be an ideal. Show that the mapping

$$\Phi : \begin{array}{ccc} R/f^{-1}(J) & \to & S/J \\ x + f^{-1}(J) & \mapsto & f(x) + J \end{array}$$

is a well-defined injective ring homomorphism.

(b) Suppose that $f : R \to S$ is a surjective ring homomorphism between commutative rings with identity. Deduce from (a) that J is maximal in S/J if and only if $f^{-1}(J)$ is maximal in $R/f^{-1}(J)$.

Problem 25. (a) Find all ideals and all homomorphic images (up to isomorphism) of

$$R = \{ \begin{pmatrix} a & b \\ 0 & c \end{pmatrix} \mid a, b, c \in \mathbb{Z} \} \quad \text{and} \quad S = \{ \begin{pmatrix} a & b \\ 0 & c \end{pmatrix} \mid a, b, c \in \mathbb{R} \} .$$

(b) Find all units, all ideals and all homomorphic images of the ring

$$R = \{ \begin{pmatrix} a & b & c \\ 0 & a & b \\ 0 & 0 & a \end{pmatrix} \mid a, b, c \in \mathbb{Q} \} \cong \mathbb{Q}[x]/\langle x^3 \rangle .$$

Problem 26. Let

$$R := \{ \begin{pmatrix} a & b \\ 0 & c \end{pmatrix} \mid a, b, c \in \mathbb{Z} \} \quad \text{and} \quad I := \{ \begin{pmatrix} 6p & q \\ 0 & 8r \end{pmatrix} \mid p, q, r \in \mathbb{Z} \} .$$

Show that I is an ideal of R and find all units in the quotient ring R/I.

Problem 27. Show that $\mathbb{R}[x]/\langle x^2 + 1 \rangle \cong \mathbb{C}$.

Problem 28. Let $I := \langle x_1, \ldots, x_n \rangle \trianglelefteq K[x_1, \ldots, x_n]$. Show that I is a maximal ideal of $K[x_1, \ldots, x_n]$ and describe the powers I^m $(m \in \mathbb{N})$ of I.

Problem 29. For any prime $p \in \mathbb{N}$ we denote by $\mathbb{Z}_{(p)}$ the ring of all rational numbers $m/n \in \mathbb{Q}$ with $p \nmid n$.

(a) Show that $\mathbb{Z}_{(p)}$ is a principal ideal domain.

(b) Show that if p and q are different primes then there is no nonzero ring homomorphism $\mathbb{Z}_{(p)} \to \mathbb{Z}_{(q)}$.

(c) Show that if p and q are different primes then $\mathbb{Z}_{(p)} \cap \mathbb{Z}_{(q)}$ is a principal ideal domain with exactly two maximal ideals.

Problem 30. Let X be a compact Hausdorff space (for example the interval $[0, 1]$) and let $C(X)$ be the ring of all continuous functions $f : X \to \mathbb{R}$ with the pointwise operations. For any $x \in X$ let

$$I_x := \{ f \in C(X) \mid f(x) = 0 \} .$$

(a) Let $x \in X$. Show that I_x is a maximal ideal of $C(X)$.

(b) Let M be a maximal ideal of $C(X)$. Show that $M = I_x$ for some $x \in X$.

Hint. Suppose $M \neq I_x$ for all x. Then for each $x \in X$ there is a function $f_x \in M$ with $f_x(x) \neq 0$ and hence $f_x \neq 0$ on an open neighborhood U_x of x. Pick $x_1, \ldots, x_n \in X$ such that $\{U_{x_1}, \ldots, U_{x_n}\}$ is a covering of X. Then $f := f_{x_1}^2 + \cdots + f_{x_n}^2 \in M$ takes only positive values. Why is this a contradiction?

Remark. By Urysohn's lemma we have $I_x \neq I_y$ for $x \neq y$. Hence there is a 1-1-correspondence between the points of X and the maximal ideals of $C(X)$.

Problem 31. (a) Find the remainders of the divisions $(x^{1000} - 7) : (x + 1)$ and $(x^{1000} - 7) : (x^2 - 1)$ in $\mathbb{Z}[x]$.

(b) Let p be a prime number. Show that all coefficients of the polynomial $x^{p-1} + \cdots + x + 1 - (x - 1)^{p-1}$ are divisible by p.

Problem 32. (a) Verify that $\mathbb{Q}[x]/\langle x^3 - 2 \rangle$ is a field by finding for each element $[a + bx + cx^2]$ modulo $x^3 - 2$ its inverse.

(b) Show that $\mathbb{Z}_2[x]/\langle x^2 + 1 \rangle$ and $\mathbb{Z}_2[x]/\langle x^2 + x + 4 \rangle$ are isomorphic fields. What is the number of elements in these fields?

Problem 33. (This problem is adapted from the very nice book *Concepts of Modern Mathematics* by Ian Stewart, Penguin Books 1975, pp. 91-94.]) The game "solitaire" is played on a board with 33 holes as shown below.

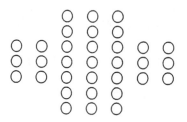

Initially, there is a peg in each of these holes except the the hole in the middle. The object of the game is to transform this initial configuration into the configuration with only one peg in the middle by admissible moves. Here, an admissible move consists of jumping vertically or horizontally with one peg across a neighboring peg into a free hole and then removing the crossed peg. We now introduce coordinates for the 33 holes as follows.

$$
\begin{pmatrix}
& & (-1,3) & (0,3) & (1,3) & & \\
& & (-1,2) & (0,2) & (1,2) & & \\
(-3,1) & (-2,1) & (-1,1) & (0,1) & (1,1) & (2,1) & (3,1) \\
(-3,0) & (-2,0) & (-1,0) & (0,0) & (1,0) & (2,0) & (3,0) \\
(-3,-1) & (-2,-1) & (-1,-1) & (0,-1) & (1,-1) & (2,-1) & (3,-1) \\
& & (-1,-2) & (0,-2) & (1,-2) & & \\
& & (-1,-3) & (0,-3) & (1,-3) & &
\end{pmatrix}
$$

Then each configuration S, i.e., each set of occupied holes, can be identified with a set of pairs of numbers, namely the coordinates of the positions which are occupied. Let

$\mathfrak{S}$ be the set of all configurations and let $K = \{0, 1, \alpha, \beta\}$ be the field introduced in (5.12)(b). We define a mapping $\varphi : \mathfrak{S} \to K \times K$ by

$$\varphi(S) := \left(\sum_{(k,l) \in S} \alpha^{k+l} , \sum_{(k,l) \in S} \alpha^{k-l} \right).$$

(a) Show that $\varphi(S) = (\alpha, \beta)$ for the configuration S shown below.

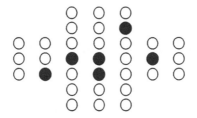

(b) Show that if the configuration S' is obtained from the configuration S by an admissible move then $\varphi(S') = \varphi(S)$.

(c) Show that there are only 5 possible final configurations with only one single peg.

Problem 34. Let $K = \{0, 1, \alpha, \beta\}$ be the field introduced in (5.12)(b). Find the addition table and the multiplication table for the endomorphism ring of K.

Problem 35. Let $K := \mathbb{Z}_3$ and let I be the ideal of $K[x]$ which is generated by the polynomials $f(x) = x^4 + x^3 + x + 2$ and $g(x) = x^4 + 2x^3 + 2x + 2$.
(a) Show that $x^4 + 2$ is an element of I.
(b) Show that $L := K[x]/I$ is a field.
(c) Find the prime factorizations of $x^4 + 2$ over K and over L.

Problem 36. Show that $K[x, y]/\langle x^2 - y^3 \rangle$ is isomorphic to the ring of all polynomials of the form $a_0 + a_2 t^2 + a_3 t^3 + \cdots + a_n t^n$ with coefficients $a_i \in K$.
Hint. Consider the homomorphism $K[x, y] \to K[t]$ given by $f(x, y) \mapsto f(t^3, t^2)$.

Problem 37. Let $I = \langle p \rangle$ be the ideal of $\mathbb{Q}[x]$ generated by $p(x) = x^3 + 3x - 2$ and let $\overline{x} = x + I \in \mathbb{Q}[x]/I$. Express the elements

$$(2\overline{x}^2 + \overline{x} - 3)(3\overline{x}^2 - 4\overline{x} + 1) \qquad \text{and} \qquad (2\overline{x}^2 + 4\overline{x} - 5)^{-1}$$

of $\mathbb{Q}[x]/I$ in the form $a\overline{x}^2 + b\overline{x} + c$.

Problem 38. Let I be the ideal of $R := \mathbb{Z}[x]$ which is generated by the polynomials $f(x) = x^4 - 2x^3 + x^2$ and $g(x) = x^6 - 2x^4 + x^2 - 2$. Show that R/I is finite and determine the number of elements of R/I. Moreover, determine all nilpotent elements, all units and all ideals in R/I.

Problem 39. Let R be a ring with identity element 1.
(a) Show that

$$J(R) := \{x \in R \mid 1 - rx \text{ is a unit for any } r \in R\}$$

is an ideal of R; it is called the **Jacobson ideal** of R.

Hint. Show first that if $x \in J(R)$ and $a \in R$ then $ax \in J(R)$ and $xa \in J(R)$, using problem 3(c) in section 3. To show that $x, y \in J(R)$ implies $x + y \in J(R)$ note that

$$1 - r(x + y) = \left(1 - rx(1 - ry)^{-1}\right)(1 - ry) \;;$$

use problem 3(c) in section 3 again to conclude that the right-hand side is a product of units.
(b) Show that if $xrx = 0$ for all $r \in R$ then $x \in J(R)$.
(c) Show that $J\big(R/J(R)\big) = \{0\}$.
(d) Show that $J(R[[x]]) = \{a_0 + a_1 x + a_2 x^2 + \cdots \mid a_0 \in J(R)\}$.

Problem 40. Let K be a field. Show that if $f \in K[x]$ has a non-constant multiple factor then the quotient ring $K[x]/\langle f \rangle$ has nonzero nilpotent elements.

Problem 41. Let R be an infinite integral domain which possesses only a finite number of units. Show that R possesses an infinite number of maximal ideals.

Problem 42. Let R be a commutative ring with identity. Show that two ideals I and J of R are comaximal if and only if $(x + I) \cap (y + J) \neq \emptyset$ for all $x, y \in R$.

Problem 43. Let R be a commutative ring with identity. Show that if I and J are comaximal ideals in R then so are I^2 and J^2.

Problem 44. Let $a_1, \ldots, a_n, b \in \mathbb{Z}$. Show that the equation $a_1 x_1 + \cdots + a_n x_n = b$ possesses a solution $(x_1, \ldots, x_n) \in \mathbb{Z}^n$ if and only if the congruence $a_1 x_1 + \cdots + a_n x_n \equiv b$ modulo m is solvable for any $m \in \mathbb{N}$.

Problem 45. Solve the following systems of congruences!
(a) $x \equiv 2 \,(\mathrm{mod}\,3)$, $x \equiv 3 \,(\mathrm{mod}\,5)$, $x \equiv 5 \,(\mathrm{mod}\,2)$
(b) $x \equiv 1 \,(\mathrm{mod}\,4)$, $x \equiv 0 \,(\mathrm{mod}\,3)$, $x \equiv 5 \,(\mathrm{mod}\,7)$
(c) $x \equiv 7 \,(\mathrm{mod}\,9)$, $x \equiv 6 \,(\mathrm{mod}\,7)$, $x \equiv 2 \,(\mathrm{mod}\,5)$
(d) $x \equiv 0 \,(\mathrm{mod}\,10)$, $x \equiv -2 \,(\mathrm{mod}\,26)$, $x \equiv -1 \,(\mathrm{mod}\,231)$

Problem 46. Find the two smallest natural numbers n with the property that $n \equiv 1, 2, 3, 4, 5 \bmod 3, 4, 5, 6, 7$.

Problem 47. Fix a natural number k. Show that for each $n \in \mathbb{N}$ one can find n consecutive natural numbers each of which is divisible by some k-th power.

Problem 48. Let $m, n \in \mathbb{N}$ be relatively prime.

(a) Use the Chinese remainder theorem to show that

$$
\begin{array}{ccc}
\mathbb{Z}_{mn} & \to & \mathbb{Z}_m \times \mathbb{Z}_n \\
[x]_{mn} & \mapsto & ([x]_m, [x]_n)
\end{array}
$$

is a ring isomorphism. Verify this explicitly for $m = 2$ and $n = 3$.

(b) Suppose n is odd and $\mathbb{Z}_n^\times$ is cyclic†, say $\mathbb{Z}_n^\times = \langle [a]_n \rangle$. Show that $\mathbb{Z}_{2n}^\times$ is also cyclic and generated by $[a]_{2n}$ or $[a+n]_{2n}$. Use this result to find generators of $\mathbb{Z}_{11}^\times$ and $\mathbb{Z}_{22}^\times$.

(c) Recall that for any $n \in \mathbb{N}$, the value $\varphi(n)$ of Euler's φ-function is defined as the number of natural numbers k with $1 \le k \le n$ which are relatively prime to n. Show that $\varphi(mn) = \varphi(m)\varphi(n)$ whenever m and n are relatively prime.

Hint. $\varphi(n)$ is the cardinality of $\mathbb{Z}_n^\times$; use part (a).

Problem 49. Show that if $m \mid n$ then the canonical mapping $\mathbb{Z}_n^\times \to \mathbb{Z}_m^\times$ is surjective.

Hint. First treat the case that n is a prime power; then invoke the Chinese remainder theorem for the general case.

---

† If n is a prime this is satisfied by theorem (4.14).

6. Ideals in commutative rings

IN order to study the ideals of a ring, it is natural to try to decompose arbitrary ideals into special ones which can more easily be investigated. As in the theory of factorization in rings (whose rudiments were developed in section 3), a reasonable theory can be developed only for commutative rings to which we therefore restrict our attention. We start by defining two types of ideals which will play an important role as the smallest building-blocks into which we can decompose arbitrary ideals.

(6.1) Definition. *Let R be a commutative ring and let $I \neq R$ be an ideal of R.*
(a) *We call I a **prime ideal** if $ab \in I$ implies that $a \in I$ or $b \in I$.*
(b) *We call I a **primary ideal** if $ab \in I$ and $a \notin I$ implies $b^n \in I$ for some $n \in \mathbb{N}$.*

(6.2) Examples. (a) Let R be a commutative ring. The ideal $\{0\}$ is a prime ideal if and only if R has no zero-divisors, and a primary ideal if and only if R has no nonzero nilpotent elements.

(b) The prime ideals of $\mathbb{Z}$ are exactly the ideals $\{0\}$ and $\langle p \rangle = p\mathbb{Z}$ where p is a prime number. The primary ideals of $\mathbb{Z}$ are exactly the ideals $\{0\}$ and $\langle p^n \rangle = p^n \mathbb{Z}$ where p^n is a prime power. Since $\langle p^n \rangle = \langle p \rangle^n$ we see that the primary ideals of $\mathbb{Z}$ are exactly the powers of prime ideals.

(c) Let K be a field and let $R = K[x_1, \ldots, x_n]$ be the ring of polynomials in n variables over K. For any subset $\{i_1, \ldots, i_r\}$ of $\{1, \ldots, n\}$ the ideal $I = \langle x_{i_1}, \ldots, x_{i_r} \rangle$ is prime, and all powers I^n are primary.

(d) Let $R = \mathbb{Z}[x]$ and let p be a prime number. Then the ideal $\langle p, x \rangle$, consisting of all polynomials $\sum_k a_k x^k$ with a_0 divisible by p, and the ideal $\langle p \rangle$, consisting of all polynomials $\sum_k a_k x^k$ with all coefficients a_k are divisible by p, are prime ideals of R.

(e) Clearly, each prime ideal is primary.

THE following proposition characterizes prime ideals.

(6.3) Proposition. *Let R be a commutative ring and $I \neq R$ an ideal of R. Then the following conditions are equivalent:*
(1) *I is a prime ideal;*
(2) *$R \setminus I$ is multiplicatively closed;*
(3) *if $a_1, \ldots, a_n \in R$ are elements of R with $a_1 \cdots a_n \in I$ then $a_k \in I$ for some k;*
(4) *if $I_1, I_2 \trianglelefteq R$ are ideals of R with $I_1 I_2 \subseteq I$ then $I_1 \subseteq I$ or $I_2 \subseteq I$;*
(5) *if $I_1, \ldots, I_n \trianglelefteq R$ are ideals of R with $I_1 \cdots I_n \subseteq I$ then $I_k \subseteq I$ for some k.*

Proof. (2) is merely a restatement of (1), and the equivalences (1)$\Longleftrightarrow$(3) and (4)$\Longleftrightarrow$(5) can easily be obtained by induction in one direction and specialization $n = 2$ in the other direction. Thus the only nontrivial part of the proof is to show that (1) and (4) are equivalent.

Suppose that (1) holds and assume that $I_1 \not\subseteq I$ and $I_2 \not\subseteq I$; we have to show that in this case $I_1 I_2 \not\subseteq I$. By assumption, there are elements $x_1 \in I_1 \setminus I$ and $x_2 \in I_2 \setminus I$; since I is prime we have $x_1 x_2 \notin I$ and hence $x_1 x_2 \in I_1 I_2 \setminus I$ which implies $I_1 I_2 \not\subseteq I$. Thus (1) implies (4). (Note that we did not use the commutativity of R.)

Suppose conversely that (4) holds and let $ab \in I$, hence $\langle ab \rangle \subseteq I$. Using the commutativity of R, we may write $\langle ab \rangle = \langle a \rangle \langle b \rangle$; hence $\langle a \rangle \langle b \rangle \subseteq I$. By property (4), this implies $\langle a \rangle \subseteq I$ or $\langle b \rangle \subseteq I$, i.e., $a \in I$ and $b \in I$. Thus (4) implies (1). $\blacksquare$

THE next result yields some information about the existence of prime ideals.

(6.4) Proposition. *Let R be a commutative ring.*
(a) *Let I be an ideal of R and $S \subseteq R$ a multiplicatively closed set with $S \cap I = \emptyset$. Then the set*

$$\mathcal{M} := \{ J \trianglelefteq R \mid I \subseteq J, \ S \cap J = \emptyset \}$$

possesses maximal elements. Each such element is a prime ideal of R.
(b) *If $R^2 = R$ (which is automatically satisfied if R has an identity element) then every maximal ideal in R is prime.*

Proof. (a) The set $\mathcal{M}$ is not empty because $I \in \mathcal{M}$. We show that every chain in $\mathcal{M}$ possesses an upper bound in $\mathcal{M}$; then Zorn's lemma will give the claim. Given a chain $\mathcal{C}$ in $\mathcal{M}$, let $J_0 := \bigcup_{J \in \mathcal{C}} J$. This is again an ideal of R which contains I and is disjoint with S; moreover $J_0 \neq R$ because $J_0 \cap S = \emptyset$. Consequently, J_0 belongs to $\mathcal{M}$ and is clearly an upper bound for $\mathcal{C}$. This shows that $\mathcal{M}$ has maximal elements.
Let J be such a maximal element. To show that J is prime, we assume that $a \notin J$ and $b \notin J$. By the maximality of J, the ideals $\langle J \cup \{a\} \rangle$ and $\langle J \cup \{b\} \rangle$ are not disjoint with S; hence there are elements $s_1, s_2 \in S$ with $s_1 \in \langle J \cup \{a\} \rangle$ and $s_2 \langle J \cup \{a\} \rangle$. Using the commutativity of R, we can write $s_1 = x_1 + r_1 a + m_1 \cdot a$ and $s_2 = x_2 + r_2 b + m_2 \cdot b$ with $x_i \in J$, $r_i \in R$ and $m_i \in \mathbb{Z}$. Using $ab \in J$ this implies $s_1 s_2 \in J$; on the other hand $s_1 s_2 \in S$ because S is multiplicatively closed. Hence $s_1 s_2 \in S \cap J$ which contradicts the fact that J belongs to $\mathcal{M}$. This is the desired contradiction.
(b) Let I be a maximal ideal of R and suppose that $ab \in I$ but $a \notin I$ and $b \notin I$. Since I is maximal with $a, b \notin I$ we have $R = \langle I \cup \{a\} \rangle = I + Ra + \mathbb{Z} \cdot a$ and $R = \langle I \cup \{b\} \rangle = I + Rb + \mathbb{Z} \cdot b$. Consequently, $R = R^2 = (I + Ra + \mathbb{Z} \cdot a)(I + Rb + \mathbb{Z} \cdot b) \subseteq I$ where the last inclusion holds because $ab \in I$. This is the desired contradiction. $\blacksquare$

WE will now show that prime and primary ideals behave better under homomorphisms than maximal ideals. Note that if $f : R \to S$ is a homomorphism and $J \trianglelefteq S$ is an ideal of S, then $f^{-1}(J)$ is an ideal of R. It may happen that J is maximal in S but $f^{-1}(J)$ is not maximal in R; as an example, take f to be the embedding of $R := \mathbb{Z}$ into $S := \mathbb{Q}$ and let $J := \{0\}$. However, we will now show that if J is prime (primary) then so is $f^{-1}(J)$, i.e., that pre-images of prime (primary) ideals are again prime (primary) ideals.

(6.5) Proposition. *Let $f : R \to S$ be a ring homomorphism between commutative rings and let $J \trianglelefteq S$ be a prime (primary) ideal of S. Then $f^{-1}(J)$ is a prime (primary) ideal of R.*

Proof. Assume $xy \in f^{-1}(J)$ but $x \notin f^{-1}(J)$. Then $f(x)f(y) = f(xy) \in J$, but $f(x) \notin J$. If J is prime, this implies $f(y) \in J$, i.e., $y \in f^{-1}(J)$. If J is primary, this implies $f(y^n) = f(y)^n \in J$, i.e., $y^n \in f^{-1}(J)$ for some $n \in \mathbb{N}$. ∎

TO proceed further in the study of ideals in a commutative ring, we extend the calculus of ideals whose rudiments were developed in (5.3). Recall that in an arbitrary ring we could define sums and products of ideals; in a commutative ring, we can also define quotients and roots of ideals.

(6.6) Definition. *Let $I, J \trianglelefteq R$ be ideals of the commutative ring R. Then*

$$(I : J) := \{x \in R \mid xJ \subseteq I\}$$

is called the **ideal quotient** *or the* **residual quotient** *of I and J.*

THE basic properties of ideal quotients are as follows.

(6.7) Proposition. *Let R be a commutative ring and let I, J, K, I_α and J_α be ideals of R.*
  (a) $(I : J)$ is an ideal of R, in fact the largest ideal K of R such that $KJ \subseteq I$. In particular, $I \subseteq (I : J)$.
  (b) If $A \neq \emptyset$ is any index set then $\left(\bigcap_{\alpha \in A} I_\alpha\right) : J = \bigcap_{\alpha \in A}(I_\alpha : J)$.
  (c) If $A \neq \emptyset$ is any index set then $I : \left(\sum_{\alpha \in A} J_\alpha\right) = \bigcap_{\alpha \in A}(I : J_\alpha)$.
  (d) $(I : J) : K = I : (JK) = (I : K) : J$.
  (e) $I : J = I : (I + J)$.

Proof. (a) Let $x, y \in (I : J)$ and $r \in R$. Then $(x + y)J \subseteq xJ + yJ \subseteq I + I = I$ and $(rx)J = r(xJ) \subseteq rI \subseteq I$; hence $x + y$ and rx both belong to $I : J$. This shows that $I : J$ is an ideal of R. The remaining claims are clear.
  (b) $x \in \bigcap_\alpha(I_\alpha : J)$ if and only if $xJ \subseteq I_\alpha$ for all α, i.e., if $xJ \subseteq \bigcap_\alpha I_\alpha$. But this means exactly $x \in \left(\bigcap_\alpha I_\alpha\right) : J$.
  (c) $x \in \bigcap_\alpha(I : J_\alpha)$ if and only if $xJ_\alpha \subseteq I$ for all α. But this is the case if and only if $x\left(\sum_\alpha J_\alpha\right) \subseteq I$, i.e., if $x \in I : \left(\sum_\alpha J_\alpha\right)$.
  (d) $x \in (I : J) : K \Longleftrightarrow xK \subseteq (I : J) \Longleftrightarrow (xK)J \subseteq I \Longleftrightarrow x(JK) \subseteq I \Longleftrightarrow x \in I : (JK)$. The second equation follows by exchanging the roles of J and K.
  (e) This follows immediately from (c) and the trivial fact that $I : I = R$. ∎

WE now define the radical of an ideal as the set of all elements which have a "root" in this ideal.

(6.8) Definition. *Let R be a commutative ring and $I \trianglelefteq R$ and ideal of R. Then*

$$\sqrt{I} := \operatorname{Rad} I := \{x \in R \mid x^n \in I \text{ for some } n = n_x \in \mathbb{N}\}$$

is called the **radical** *of I. The ideal I is called a* **radical ideal** *if $I = \sqrt{I}$. Moreover, $\sqrt{\{0\}}$ is called the* **nilradical** *of R.*

CLEARLY, $\sqrt{R}$ is all of R and $\sqrt{\{0\}}$ is the set of all nilpotent elements of R. Let us collect the basic properties of the formation of radicals.

(6.9) Theorem. *Let $I, J, I_1, \ldots, I_n$ be ideals of a commutative ring R.*
(a) $\sqrt{I}$ *is itself an ideal of R with $I \subseteq \sqrt{I} \subseteq R$.*
(b) *If $I \neq R$ then $\sqrt{I}$ is the intersection of all prime ideals of R containing I.*
(c) *If $I \subseteq J$ then $\sqrt{I} \subseteq \sqrt{J}$.*
(d) $\sqrt{\sqrt{I}} = \sqrt{I}$.
(e) $\sqrt{I_1 \cdots I_n + J} = \sqrt{(I_1 \cap \cdots \cap I_n) + J} = \sqrt{I_1 + J} \cap \cdots \cap \sqrt{I_n + J}$; *in particular,* $\sqrt{I_1 \cdots I_n} = \sqrt{I_1 \cap \cdots \cap I_n} = \sqrt{I_1} \cap \cdots \cap \sqrt{I_n}$ *and* $\sqrt{I^n} = \sqrt{I}$ *for all $n \in \mathbb{N}$.*
(f) $\sqrt{I + J} = \sqrt{\sqrt{I} + \sqrt{J}}$.
(g) *If $R^2 = R$ (which is automatically the case if R has an identity element) then $\sqrt{I} = R$ if and only if $I = R$. In this case, two ideals I and J are comaximal if and only if $\sqrt{I}$ and $\sqrt{J}$ are.*
(h) *If $\sqrt{I}$ is finitely generated, there is a natural number $n \in \mathbb{N}$ with $(\sqrt{I})^n \subseteq I$.*

Proof. (a) It is clear that $I \subseteq \sqrt{I} \subseteq R$. It remains to show that $\sqrt{I}$ is itself an ideal. Let $a, b \in \sqrt{I}$; then there are natural numbers $m, n \in \mathbb{N}$ such that $a^m \in I$ and $b^n \in I$. Using the commutativity of the ring R, we obtain

$$(a+b)^{m+n-1} = \sum_{k=0}^{m+n-1} \binom{m+n-1}{k} a^{m+n-1-k} b^k$$

$$= \sum_{k=0}^{n-1} \binom{m+n-1}{k} \underbrace{a^{m+n-1-k}}_{\in I} b^k + \sum_{k=n}^{m+n-1} \binom{m+n-1}{k} a^{m+n-1-k} \underbrace{b^k}_{\in I} \in I.$$

This shows that $a + b \in \sqrt{I}$. Also, if $x \in R$ is an arbitrary ring element, then $(xa)^m = x^m a^m \in x^m I \subseteq I$; this shows that $xa \in \sqrt{I}$.

(b) Let P be any prime ideal containing I and let $a \in \sqrt{I}$. Then $a^n \in I \subseteq P$ for some $n \in \mathbb{N}$; since P is a prime ideal, this implies $a \in P$ due to (6.3)(c). This shows that $\sqrt{I}$ is contained in every prime ideal $P \supseteq I$.
Conversely, let $a \notin \sqrt{I}$. Then $a^n \notin I$ for all $n \in \mathbb{N}$ so that the set $S := \{a^n \mid n \in \mathbb{N}\} + I$ is multiplicatively closed and disjoint with I. By (6.4)(a) there is a prime ideal $P \supseteq I$ disjoint with S so that in particular $a \notin P$.

(c) This is trivial.

(d) Since $I \subseteq \sqrt{I}$ we have $\sqrt{I} \subseteq \sqrt{\sqrt{I}}$ due to part (d). Conversely, let $a \in \sqrt{\sqrt{I}}$; then $a^m \in \sqrt{I}$ for some $m \in \mathbb{N}$, hence $(a^m)^n = a^{mn} \in I$ for some $n \in \mathbb{N}$. But this implies $a \in \sqrt{I}$.

(e) We have $(I_1 \cdots I_n) + J \subseteq (I_1 \cap \cdots \cap I_n) + J \subseteq I_1 + J, \ldots, I_n + J$; hence the inclusions $\sqrt{(I_1 \cdots I_n) + J} \subseteq \sqrt{(I_1 \cap \cdots \cap I_n) + J} \subseteq \sqrt{I_1 + J} \cap \cdots \cap \sqrt{I_n + J}$ are immediate consequences of part (d). Suppose conversely that $a \in \sqrt{I_1 + J} \cap \cdots \cap \sqrt{I_n + J}$. Then there are natural numbers $k_1, \ldots, k_n$ with $a^{k_1} \in I_1 + J, \ldots, a^{k_n} \in I_n + J$. Consequently, $a^{k_1 + \cdots + k_n} = a^{k_1} \cdots a^{k_n} \in (I_1 + J) \cdots (I_n + J) \subseteq (I_1 \cdots I_n) + J$ so that $a \in \sqrt{(I_1 \cdots I_n) + J}$. This shows the inclusion $\sqrt{I_1 + J} \cap \cdots \cap \sqrt{I_n + J} \subseteq \sqrt{(I_1 \cdots I_n) + J}$.
The second statement follows immediately by taking $J := \{0\}$, the third statement by further specializing to $I_1 = I_2 = \cdots = I_n = I$.

(f) We have $I + J \subseteq \sqrt{I} + \sqrt{J}$ by part (a) and hence $\sqrt{I + J} \subseteq \sqrt{\sqrt{I} + \sqrt{J}}$ by part (d). Conversely, $\sqrt{I} + \sqrt{J} = \sqrt{I \cap J}$ by part (f) and hence $\sqrt{\sqrt{I} + \sqrt{J}} = \sqrt{\sqrt{I \cap J}} = \sqrt{I \cap J} \subseteq \sqrt{I + J}$ by parts (e) and (d).

(g) Suppose $I \neq R$. Then there is a maximal ideal $M \trianglelefteq R$ with $I \subseteq M$. But then M is prime due to (6.4)(b); hence $\sqrt{I} \subseteq M$ by part (b). Consequently, $\sqrt{I} \neq R$. Then the second claim follows immediately, because two ideals are comaximal if and only if their sum equals R.

(h) By hypothesis, the ideal $\sqrt{I}$ is finitely generated, say $I = \langle x_1, \ldots, x_k \rangle$. By the definition of $\sqrt{I}$, there are natural numbers $n_1, \ldots, n_k$ with $x_i^{n_i} \in I$. Let $n := n_1 + \cdots + n_k + k - 1$. Then $\sqrt{I}^n$ is generated by the products $x_1^{r_1} \cdots x_k^{r_k}$ with $r_1 + \cdots + r_k = n$. Now if $r_i < n_i$ for all indices i then $n = r_1 + \cdots + r_k < n_1 + \cdots + n_k = n - k + 1$, hence $k < 1$ which is absurd; hence $r_i \geq n_i$ for at least one index i. Hence each of the products $x_1^{r_1} \cdots x_k^{r_k}$ contains a factor in I, hence lies in I itself. This shows that $(\sqrt{I})^n \subseteq I$. ∎

AS a corollary of (6.9), we show that there is a close relation between prime ideals and primary ideals.

(6.10) Proposition. *Let R be a commutative ring. If $I \trianglelefteq R$ is a primary ideal then $\sqrt{I}$ is a prime ideal, in fact the smallest prime ideal containing I.*

Proof. Suppose $ab \in \sqrt{I}$ and $a \notin \sqrt{I}$. Then there is a natural number $n \in \mathbb{N}$ with $(ab)^n = a^n b^n \in I$. Since $a^n \notin I$ and since I is primary this implies that $(b^n)^m = b^{nm} \in I$ for some $m \in \mathbb{N}$. Consequently, $b \in \sqrt{I}$. ∎

WE can now give further characterizations of prime and primary ideals.

(6.11) Proposition. *Let R be a commutative ring. An ideal $I \neq R$ of R is prime if and only if I is radical and primary.*

Proof. Suppose I is prime. Then trivially I is primary. Moreover, let $a \in \sqrt{I}$. Then $a^n \in I$ for some $n \in \mathbb{N}$; whence $a \in I$ by (6.3)(c). This shows that $\sqrt{I} \subseteq I$ so that I is also radical.

Suppose conversely that I is a radical and primary ideal. Then $I = \sqrt{I}$ is prime by (6.10). ∎

(6.12) Proposition. *Let R be a commutative ring and $I \neq R$ an ideal of R. Then the following conditions are equivalent:*
  (a) *I is a primary ideal;*
  (b) *if $a_1, \ldots, a_n \in R$ are such that $a_1 \cdots a_n \in I$ then $a_k \in \sqrt{I}$ for some k;*
  (c) *if $I_1, I_2 \trianglelefteq R$ are ideals of R with $I_1 I_2 \subseteq I$ then $I_1 \subseteq \sqrt{I}$ or $I_2 \subseteq \sqrt{I}$;*
  (d) *if $I_1, \ldots, I_n \trianglelefteq R$ are ideals of R with $I_1 \cdots I_n \subseteq I$ then $I_k \subseteq \sqrt{I}$ for some k.*

Proof. Exercise. (See problem 22 below.) ∎

(6.13) Remark. It is possible that I is a primary ideal of a commutative ring R and that I_1, I_2 are ideals of R with $I_1 I_2 \subseteq I$, $I_1 \nsubseteq I$ and $I_2^N \nsubseteq I$ for any $N \in \mathbb{N}$. (See problem 23 below.)

IN a commutative ring R such that $R^2 = R$ we know from (6.4)(b) that every maximal ideal is prime. For such a ring, the correlations between maximal, prime, primary and radical ideals can be summarized in the following diagram.

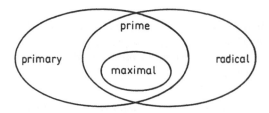

THE following theorem characterizes the various types of ideals by properties of the corresponding quotient rings. Clearly, (6.11) can also be obtained as a corollary of the following result.

(6.14) Theorem. *Let R be a commutative ring and let $I \neq R$ be an ideal of R.*
(a) *I is prime if and only if R/I has no zero-divisors.*
(b) *I is radical if and only if R/I has no nonzero nilpotent elements.*
(c) *I is primary if and only if every zero-divisor of R/I is nilpotent.*
(d) *If R has an identity element then I is maximal if and only if R/I is a field.*

Proof. (a) For any $x \in R$ we write for short $[x]$ for the coset $x + I$. Then R/I has no zero-divisors if and only if $[a][b] = [0]$ implies $[a] = [0]$ or $[b] = [0]$, i.e., if $ab \in I$ implies $a \in I$ or $b \in I$. By definition, this means exactly that I is a prime ideal.

(b) The nonzero nilpotent elements of R/I are exactly the elements $[a] \neq [0]$ with $[a]^n = [0]$ for some $n \in \mathbb{N}$, i.e., the cosets $[a]$ where $a \notin I$ but $a^n \in I$ for some $n \in \mathbb{N}$ which means $a \in \sqrt{I} \setminus I$. Thus there are no such elements if and only if $\sqrt{I} \setminus I$ is empty, i.e., if $\sqrt{I} \subseteq I$.

(c) Let I be primary and let $[a] \neq [0]$ be a zero-divisor in R/I. Then there is an element $[b] \neq [0]$ in R/I with $[a][b] = [0]$. But this means $ab \in I$ and $b \notin I$. Since I is primary this implies $a^n \in I$ for some $n \in \mathbb{N}$, i.e., $[a]^n = [0]$.

Conversely, let $ab \in I$ and $a \notin I$. Then $[a][b] = [0]$ and $[a] \neq [0]$. Hence $[b]$ is zero or a zero-divisor in R/I; by assumption, this implies $[b]^n = [0]$ for some $n \in \mathbb{N}$, i.e., $b^n \in I$.

(d) Suppose I is maximal and $[a] \neq [0]$ in R/I, i.e., $a \notin I$. By the maximality of I, this implies $R = \langle I \cup \{a\} \rangle = I + Ra$. In particular, there is an element $r \in R$ with

$1 \in I + ra$. But this means $[1] = [r][a]$ in R/I; hence every nonzero element in R/I is invertible. Suppose conversely that I is not maximal. Then there is an element $a \in R$ with $I \subsetneq I + Ra \subsetneq R$ so that $1 \notin I + Ra$. But then $[a]$ is not invertible in R/I so that R/I is not a field. ∎

THIS classification of different types of ideals can be very helpful. For example, let R be a commutative ring and $I \trianglelefteq R$ and ideal of R. Recall from (5.13)(c) the 1-1-correspondence between ideals of R containing I and ideals of R/I. As an immediate corollary of (6.14), we now prove that under this correspondence prime ideals (radical, primary, maximal ideals) correspond to prime ideals (radical, primary, maximal ideals).

(6.15) Proposition. *Let R be a commutative ring and let $I \trianglelefteq R$ be an ideal of R. Then an ideal J of R with $I \subseteq J \subseteq R$ is prime (radical, primary, maximal) if and only if J/I is. (As far as the statement on maximal ideal is concerned we assume that R has an identity element.)*

Proof. The rings R/J and $\frac{R/I}{J/I}$ are isomorphic, due to (5.13)(c). Hence one of them satisfies any of the conditions listed in (6.14) if and only if the other does. ∎

ANOTHER corollary of the characterization theorem (6.14) is as follows.

(6.16) Proposition. *Let R be a finite commutative ring with identity. Then every prime ideal of R is maximal.*

Proof. An ideal $I \neq R$ of R is prime if and only if R/I is an integral domain and is maximal if and only if R/I is a field. In (3.2)(b) we proved that every finite integral domain is a field; whence the claim. ∎

IT was observed in (6.10) that if I is primary then $\sqrt{I}$ is prime. To further explore the correspondence between primary and prime ideals we give the following definition.

(6.17) Definition. *Let R be a commutative ring and let Q, P be ideals of R. We say that Q is **P-primary** if Q is primary with $\sqrt{Q} = P$.*

FOR example, if $P = \langle x, y \rangle \trianglelefteq \mathbb{Z}[x, y]$ then all ideals $Q = \langle x^m, y^n \rangle$ with $m, n \in \mathbb{N}$ are P-primary. We now characterize those pairs (Q, P) of ideals such that Q is P-primary.

(6.18) Proposition. *Let R be a commutative ring and let $P, Q \trianglelefteq R$ be ideals of R. Then the following conditions are equivalent:*

(1) *Q is P-primary;*
(2) *$Q \subseteq P \subseteq \sqrt{Q}$, and the conditions $I_1 I_2 \subseteq Q$ and $I_1 \not\subseteq Q$ imply $I_2 \subseteq P$;*
(3) *$Q \subseteq P \subseteq \sqrt{Q}$, and the conditions $ab \in Q$ and $a \notin Q$ imply $b \in P$;*
(4) *$Q \subseteq P \subseteq \sqrt{Q}$, and the conditions $ab \in Q$ and $a \notin P$ imply $b \in Q$.*

Proof. (1)$\Longrightarrow$(2). Suppose Q is P-primary; then $Q \subseteq \sqrt{Q} = P$. Moreover, if I_1 and I_2 are ideals of R with $I_1 I_2 \subseteq Q$ and $I_1 \not\subseteq Q$, pick an element $y \in Q \setminus I_1$. Then for any $x \in I_2$ we have $xy \in I_2 Q \subseteq Q$ but $y \notin Q$. Since Q is P-primary this implies $x \in \sqrt{Q} = P$. Since $x \in I_2$ was arbitrary, this shows $I_2 \subseteq P$.

(2)$\Longrightarrow$(3). If $ab \in Q$ and $a \notin Q$ then $\langle a \rangle \langle b \rangle = \langle ab \rangle \subseteq Q$ and $\langle a \rangle \not\subseteq Q$. Hence $\langle b \rangle \subseteq P$ due to (2); in particular, $b \in P$.

(3)$\Longrightarrow$(1). By assumption, $ab \in Q$ and $a \notin Q$ implies $b \in P \subseteq \sqrt{Q}$; hence Q is primary. To prove (1) it remains to show that $\sqrt{Q} = P$. By assumption we have $P \subseteq \sqrt{Q}$. Conversely, let $x \in \sqrt{Q}$ and let $n \in \mathbb{N}$ be the smallest natural number such that $x^n \in Q$. If $n = 1$ then $x \in Q \subseteq P$. If $n > 1$ then $x^{n-1} x \in Q$ but $x^{n-1} \notin Q$, hence $x \in P$ by assumption (3). Since $x \in \sqrt{Q}$ was arbitrary this shows $\sqrt{Q} \subseteq P$.

(3)$\Longleftrightarrow$(4). This equivalence is easily verified by changing the roles of a and b. $\blacksquare$

AS a next step, we show that under certain circumstances an ideal can inherit the property of being P-primary to other ideals.

(6.19) Proposition. *Let $Q_1, \ldots, Q_n, P, I$ be ideals of a commutative ring R.*
(a) *If $Q_1, \ldots, Q_n$ are P-primary then so is $Q_1 \cap \cdots \cap Q_n$.*
(b) *If Q is P-primary and if $I \not\subseteq Q$ then $Q : I$ is also P-primary.*
(c) *If $I \not\subseteq P$ then $Q : I = Q$. If $I \subseteq Q$ then $Q : I = R$.*

Proof. (a) We apply (6.18)(3) to $Q := Q_1 \cap \cdots \cap Q_n$ and P. First of all, $Q \subseteq \sqrt{Q} = \sqrt{Q_1 \cap \cdots \cap Q_n} = \sqrt{Q_1} \cap \cdots \cap \sqrt{Q_n} = P \cap \cdots \cap P = P$. Second, if $ab \in Q$ and $a \notin Q$ then $ab \in Q_i$ and $a \notin Q_i$ for some i; hence $b \in \sqrt{Q_i} = P$. This gives the claim.

(b) Suppose first $I \not\subseteq Q$. To show that $Q : I$ is P-primary, we apply (6.18)(2) to $Q : I$ and P. Let $x \in (Q : I)$ and pick $y \in I \setminus Q$. Then $xy \in (Q : I)I \subseteq Q$ but $y \notin Q$; therefore, $x \in \sqrt{Q} = P$. Now let $z \in P = \sqrt{Q}$. Then there is an $n \in \mathbb{N}$ with $z^n \in Q \subseteq (Q : I)$; therefore, $z \in \sqrt{(Q : I)}$. Since $x \in (Q : I)$ and $z \in P$ were arbitrary, we have proved that $(Q : I) \subseteq P \subseteq \sqrt{(Q : I)}$ which is the first condition in (6.18)(2). To prove the second condition in (6.18)(2), suppose $I_1 I_2 \subseteq (Q : I)$ and $I_1 \not\subseteq (Q : I)$. This means $I_1 I_2 I \subseteq Q$ (i.e., $(I_1 I)I_2 \subseteq Q$) and $I_1 I \not\subseteq Q$. Hence (6.18)(2), applied to Q and P, yields $I_2 \subseteq P$.

(c) Suppose $I \not\subseteq P$; we want to show that $Q = (Q : I)$. Trivially, $Q \subseteq (Q : I)$. Assume $Q \subsetneqq (Q : I)$ and pick an element $x \in (Q : I) \setminus Q$. Then $xI \subseteq Q$ but $x \notin Q$. Since Q is P-primary this implies $I \subseteq P$. Thus if $I \not\subseteq P$ then $(Q : I) = Q$.

On the other hand, if $I \subseteq Q$ then $RI \subseteq I \subseteq Q$ so that $R \subseteq (Q : I)$, hence $R = (Q : I)$. $\blacksquare$

WE now turn to the important problem of "decomposing" an ideal into "simpler" ideals. To get an idea of what this could mean, note that the fundamental theorem of arithmetic, stating that every natural number possesses a unique prime factorization, is equivalent to a purely ideal-theoretical statement, namely, that every nonzero ideal of $\mathbb{Z}$ can be uniquely written as a product of prime ideals. Indeed, every nonzero ideal I of $\mathbb{Z}$ is a principal ideal, say $I = \langle m \rangle$ with $m \in \mathbb{N}$, and if $m = p_1^{k_1} \cdots p_n^{k_n}$ is the prime factorization of m then

$$ I = \langle p_1^{k_1} \cdots p_n^{k_n} \rangle = \langle p_1 \rangle^{k_1} \cdots \langle p_n \rangle^{k_n} $$

which is a product of prime ideals. Thus it is a natural attempt to "decompose" an ideal by writing it as a finite product of prime ideals. This was indeed the historical approach, and, as a matter of fact, the very notion of an ideal was introduced to obtain unique factorization results for certain types of rings. However, only a very limited class of rings has the property that every nonzero ideal can be written as a product of prime ideals; these rings will be studied in a separate section later.

The unique factorization property in $\mathbb{Z}$ can be restated ideal-theoretically in a second way which can be generalized to a much larger class of rings. Namely, if $m = p_1^{k_1} \cdots p_n^{k_n}$ and $I = \langle m \rangle$ as above, then

$$ I = \langle p_1^{k_1} \cdots p_n^{k_n} \rangle = \langle p_1^{k_1} \rangle \cdots \langle p_n^{k_n} \rangle = \langle p_1^{k_1} \rangle \cap \cdots \cap \langle p_n^{k_n} \rangle \; ; $$

hence every nonzero ideal in $\mathbb{Z}$ can be written as a product as well as an intersection of primary ideals.† Now in a setting as general as possible, it is more natural to seek for a decomposition via set-theoretical intersection rather than a multiplicative decomposition, because the largest common subideal of two ideals I and J is $I \cap J$ and not IJ. Thus we are led to the following definition.

(6.20) **Definition.** *Let R be a commutative ring. We say that an ideal I has a* **primary decomposition** *if there are primary ideals $Q_1, \ldots, Q_m$ of R such that*

$$ I = Q_1 \cap \cdots \cap Q_m \; . $$

Such a decomposition is called **reduced** *if no Q_i is contained in the intersection of the remaining Q_j and if the prime ideals $P_i = \sqrt{Q_i}$ $(1 \le i \le m)$ are pairwise distint.*

(6.21) **Example.** Let $R = K[x, y]$ where K is a field. Then

$$ \langle x^2, xy \rangle = \langle x \rangle \cap \langle x^2, xy, y^2 \rangle = \langle x \rangle \cap \langle x^2, y \rangle = \langle x \rangle \cap \langle x^2, x + y \rangle $$

are three different reduced primary decompositions of $\langle x^2, xy \rangle$.

IT is clear that every primary decomposition $I = Q_1 \cap \cdots \cap Q_m$ can be reduced. Indeed, suppose first that $P_1, \ldots, P_m$ are not pairwise different but that, say, $Q_{i_1}, \ldots, Q_{i_r}$ belong to the same prime ideal P. But then $Q := Q_{i_1} \cap \cdots \cap Q_{i_r}$ is also P-primary

---

† The fact that the product coincides with the intersection follows from (5.23) because if r and s are coprime integers, then $\langle r \rangle$ and $\langle s \rangle$ are comaximal ideals of $\mathbb{Z}$, due to (I.21.10).

by (6.19)(a); hence we can replace $Q_{i_1}, \ldots, Q_{i_r}$ in the decomposition by the single primary ideal Q. Doing this for all occurring P, we achieve that all of the occurring prime ideals are pairwise distinct. Then, if some Q_i contains the intersection of the remaining Q_j, it may be left out altogether; thus, dropping redundant terms one by one, we obtain a reduced primary decomposition. Now example (6.21) shows already that even a reduced primary decomposition of a given ideal need not be unique, i.e., the individual components of a given (reduced) primary decomposition depend on the special choice of the decomposition. However, there are certain features of primary decompositions that are uniquely determined. Our first result concerning uniqueness is the next theorem.

(6.22) Theorem. *Let I be an ideal of a commutative ring R and let $I = Q_1 \cap \cdots \cap Q_n$ be a reduced primary decomposition of I where Q_i is P_i-primary. Then $P_1, \ldots, P_n$ are exactly those ideals of the form $\sqrt{I : \langle x \rangle}$ $(x \in R)$ which are prime. In particular, the number n and the prime ideals $P_1, \ldots, P_n$ depend only on I and not on the special choice of the decomposition; i.e., if $I = Q'_1 \cap \cdots \cap Q'_m$ is another reduced primary decomposition of I where Q'_i is P'_i-primary, then $m = n$ and (after reordering if necessary) $P_i = P'_i$ for all i.*

Proof. For any $x \in R$ we have $I : \langle x \rangle = (\bigcap_{i=1}^{n} Q_i) : \langle x \rangle = \bigcap_{i=1}^{n}(Q_i : \langle x \rangle)$ and therefore $\sqrt{I : \langle x \rangle} = \bigcap_{i=1}^{n} \sqrt{Q_i : \langle x \rangle}$. Now if $x \notin Q_i$ then $Q_i : \langle x \rangle$ is P_i-primary due to (6.19)(b) so that $\sqrt{Q_i : \langle x \rangle} = P_i$. On the other hand, if $x \in Q_i$ then $Q_i : \langle x \rangle = R$ due to (6.19)(c) and hence $\sqrt{Q_i : \langle x \rangle} = R$. This shows

$$(\star) \qquad \qquad \sqrt{I : \langle x \rangle} \; = \; \bigcap_{x \notin Q_i} P_i \; .$$

Now suppose that $\sqrt{I : \langle x \rangle}$ is prime. Then $\prod_{x \notin Q_i} P_i \subseteq \bigcap_{x \notin Q_i} P_i = \sqrt{I : \langle x \rangle}$ implies $P_{i_0} \subseteq \sqrt{I : \langle x \rangle} = \bigcap_{x \notin Q_i} P_i \subseteq P_{i_0}$ for some index i_0 with $x \notin Q_{i_0}$ and hence $\sqrt{I : \langle x \rangle} = P_{i_0}$ so that $\sqrt{I : \langle x \rangle}$ is one of the ideals $P_1, \ldots, P_n$.

Conversely, let $1 \le i_0 \le n$. Since the decomposition $I = \bigcap_{i=1}^{n} Q_i$ is reduced, there is an element $x_{i_0} \in (\bigcap_{i \ne i_0} Q_i) \setminus Q_{i_0}$. Then $\sqrt{I : \langle x_{i_0} \rangle} = P_{i_0}$ due to $(\star)$. This shows that each of the prime ideals P_i is of the form $\sqrt{I : \langle x \rangle}$ for some $x \in R$. ∎

AS was remarked before, it is not true that the primary components $Q_1, \ldots, Q_n$ of an ideal I are uniquely determined. However, we will next show that certain intersection of the ideals Q_i are uniquely determined.

(6.23) Proposition. *Suppose an ideal I in a commutative ring R possesses a primary decomposition $I = Q_1 \cap \cdots \cap Q_n$ where Q_i is P_i-primary. Let S be a multiplicatively closed subset of R. We assume that the indices $1, \ldots, n$ are arranged such that $P_i \cap S = \emptyset$ for $1 \le i \le m$ and $P_i \cap S \ne \emptyset$ for $m + 1 \le i \le n$. Then*

$$Q_1 \cap \cdots \cap Q_m \; = \; \{x \in R \mid xs \in I \text{ for some } s \in S\} \; =: \; I_S \; ;$$

in particular, $Q_1 \cap \cdots \cap Q_m$ does not depend on the particular choice of the primary decomposition, because I_S does not.†

Proof. Let $x \in I_S$. Then there is an element $s \in S$ with $sx \in I = Q_1 \cap \cdots \cap Q_n$. Let $1 \le i \le m$. Then $sx \in Q_i \subseteq P_i$ and $s \notin P_i$; hence $x \in Q_i$ by (6.18)(4). This shows $x \in Q_1 \cap \cdots \cap Q_m$. Thus the inclusion $I_S \subseteq Q_1 \cap \cdots \cap Q_m$ is proved.

Suppose conversely that $x \in Q_1 \cap \cdots \cap Q_m$. For each index $i > m$ pick an element $s_i \in S \cap P_i$ and let $t \in s_{m+1} \cdots s_n$; note that $t \in S$ because S is multiplicatively closed. If $N \in \mathbb{N}$ is chosen large enough, then $t^N = s_{m+1}^N \cdots s_n^N \in Q_{m+1} \cdots Q_n$, hence

$$x \underbrace{t^N}_{\in S} \in (Q_1 \cap \cdots \cap Q_m)Q_{m+1} \cdots Q_n \subseteq Q_1 \cap \cdots \cap Q_n = I$$

so that $x \in I_S$. This shows $Q_1 \cap \cdots \cap Q_m \subseteq I_S$. ∎

TO exploit (6.23) in a concrete situation, we have to make a choice for the multiplicatively closed set S. Now note that if P is a prime ideal then the complement $R \setminus P$ is multiplicatively closed. This observation will help us to apply (6.23).

(6.24) Definition. *Let $I = Q_1 \cap \cdots \cap Q_n$ be a decomposition of I where Q_i is P_i-primary. A family $\mathcal{P} \subseteq \{P_1, \ldots, P_n\}$ is called **isolated** if none of the prime ideals P_i outside $\mathcal{P}$ is contained in one the prime ideals belonging to $\mathcal{P}$.*

NOTE that if P is a minimal element of $\{P_1, \ldots, P_n\}$ with respect to inclusion then $\{P\}$ is an isolated set. We now claim that if $\{P_{i_1}, \cdots, P_{i_r}\}$ is isolated then $Q_{i_1} \cap \cdots \cap Q_{i_r}$ depends only on $P_{i_1}, \ldots, P_{i_r}$ and not on the particular choice of the decomposition. The proof is an application of (6.23), but also involves the following lemma.

(6.25) Lemma. *Let $U \le R$ be a subring of a commutative ring R. If $P_1, \ldots, P_n$ are prime ideals of R with $U \subseteq P_1 \cup \cdots \cup P_n$ then $U \subseteq P_k$ for some index k.* ††

Proof. We proceed by induction on n, the case $n = 1$ being trivial. Suppose the claim is true for $n - 1$. To get a contradiction, suppose that $U \subseteq P_1 \cup \cdots \cup P_n$ but $U \subsetneq P_k$ for all k. By induction hypothesis, we have $U \not\subseteq \bigcup_{i \ne k} P_i$ for all k; hence we can pick elements

$$x_k \in U \setminus \bigcup_{i \ne k} P_i \subseteq \bigcup_i P_i \setminus \bigcup_{i \ne k} P_i = P_k \setminus \bigcup_{i \ne k} P_i$$

---

† It is easy to check that I_S is an ideal of R with $I \subseteq I_S \subseteq R$. Both extremes can occur: if $S = \{0\}$ then $I_S = R$, if $S = \{1\}$ (in a ring with identity) then $I_S = I$. Sometimes one calls I_S the S-component of I.

†† Compare with problem 16 in section 5.

for $1 \leq k \leq n$. Then $y_k := x_1 \cdots x_{k-1} x_{k+1} \cdots x_n \in U$ does not belong to P_k (because P_k is prime and $x_1, \ldots, x_{k-1}, x_{k+1}, \ldots, x_n \notin P_k$), but belongs to P_i if $i \neq k$ (because $x_i \in P_i$ is a factor of y_k). Therefore, the element $y := y_1 + \cdots + y_n \in U$ does not belong to $\bigcup_{k=1}^{n} P_k$, contradicting our assumption. $\blacksquare$

(6.26) Theorem. *Let $I = Q_1 \cap \cdots \cap Q_n$ where Q_i is P_i-primary. If $\mathcal{P} = \{P_{i_1}, \cdots, P_{i_r}\}$ is isolated, then $Q_{i_1} \cap \cdots \cap Q_{i_r}$ does not depend on the particular choice of the decomposition.*

Proof. The set $S := (R \setminus P_{i_1}) \cap \cdots \cap (R \setminus P_{i_r})$ is multiplicatively closed, and by (6.23) we are done if we can show that $I_S = Q_{i_1} \cap \cdots \cap Q_{i_r}$. To do so, it suffices to show that if $P_j \notin \mathcal{P}$ then $P_j \cap S \neq \emptyset$. But $P_j \notin \mathcal{P}$ implies $P_j \not\subseteq P_{i_1}, \ldots, P_j \not\subseteq P_{i_r}$ (since $\mathfrak{P}$ is isolated) and hence $P_j \not\subseteq P_{i_1} \cup \cdots \cup P_{i_r}$ by lemma (6.25). But this means exactly $P_j \cap S \neq \emptyset$. $\blacksquare$

CONSIDERING the examples given in (6.2)(b) and (c), one might be tempted to think that powers of prime ideals are primary. This is not the case in general; counterexamples are provided in the exercises. However, we can obtain some information on prime powers. To do so, let us introduce "symbolic prime powers".

(6.27) Definition. *Let P be a prime ideal of a commutative ring R and let $m \in \mathbb{N}$. Then*
$$P^{(m)} := \{x \in R \mid xs \in P^m \text{ for some } s \in R \setminus P\}$$
*is called the m-th **symbolic power** of P.*

THE role of symbolic prime powers is exhibited in the next proposition.

(6.28) Proposition. *Let P be a prime ideal of a commutative ring and let $m \in \mathbb{N}$.*
(a) $P^m \subseteq P^{(m)} \subseteq P$.
(b) $P^{(m)}$ is a P-primary ideal.
(c) *If P^m has a reduced primary decomposition $P^m = Q_1 \cap \cdots \cap Q_n$ where Q_i is P_i-primary, then P is a minimal element of $\{P_1, \ldots, P_n\}$, and the corresponding primary component is just $P^{(m)}$.*

Proof. (a) If $x \in P^m$ then $xs \in P^m$ for any $s \in R$. Also, if $xs \in P^m \subseteq P$ with $s \notin P$ then $x \in P$ because P is prime. This proves the two inclusions.

(b) It is straightforward to check that $P^{(m)}$ is an ideal of R. To verify that it is P-primary, we use the criterion (6.18)(4). By part (a), we have $P^{(m)} \subseteq P \subseteq \sqrt{P^{(m)}}$. Moreover, if $xy \in P^{(m)}$, say $(xy)s \in P^m$ with $s \in R \setminus P$, and $y \notin P$ then $x(ys) \in P^m$ with $ys \in R \setminus P$ (since $R \setminus P$ is multiplicatively closed) so that $x \in P^{(m)}$.

(c) Taking the radical on both sides of the equation $P^m = Q_1 \cap \cdots \cap Q_n$ we obtain $P = P_1 \cap \cdots \cap P_n$ which implies $P_1 \cdots P_n \subseteq P$ and hence $P_{i_0} \subseteq P$ for some index i_0 because P is prime. On the other hand $P = P_1 \cap \cdots \cap P_n \subseteq P_i$ for all i; this implies

$P = P_{i_0}$ and (since $P_1, \ldots, P_n$ are pairwise different) $P \subsetneq P_i$ for $i \neq i_0$. This shows that P is a minimal prime for P^m; the corresponding primary component is $P^{(m)}$ due to (6.23) with $S := R \setminus P$. ∎

UP to now we have considered an ideal which was assumed to possess a primary decomposition, but we have not developed criteria which would ensure the existence of such a decomposition. There are in fact examples of ideals which do not possess a primary decomposition at all. (See problem 39 below.) However, in section 11 below we will see that a large class of commutative rings (including all polynomial rings over fields, for example) has the property that *every* ideal possesses a primary decomposition.

Exercises

Problem 1. (a) Let $R = \mathbb{Z}$ and $I = \{0\}$. Show that I is a prime ideal of R which is not maximal.

(b) Let $R = 2\mathbb{Z}$ and $I = 4\mathbb{Z}$. Show that I is a maximal ideal of R which is not prime. Show that the multiplication on R/I is trivial. (In particular, R/I is not a field even though I is maximal.)

(c) Let $R = \mathbb{R}^{2 \times 2}$ be the (non-commutative) ring of all real 2×2-matrices and let $I = \{0\}$. Show that there are elements $a, b \in R \setminus I$ with $ab \in I$, but that $I_1 I_2 \subseteq I$ with ideals $I_1, I_2 \trianglelefteq R$ implies $I_1 \subseteq I$ or $I_2 \subseteq I$.

Problem 2. Show that $I := \{f \in \mathbb{Z}[x] \mid f(0) = 0\}$ is a principal ideal of $\mathbb{Z}[x]$. Is I a prime ideal? A maximal ideal?

Problem 3. Let $n = p_1^{k_1} \cdots p_r^{k_r}$ be the prime factorization of a natural number n. Show that the prime ideals of $\mathbb{Z}_n$ are exactly the sets $(p_i \mathbb{Z})/(n\mathbb{Z})$ where $1 \leq i \leq r$ and that all of these are in fact maximal ideals.

Problem 4. Let R be an integral domain and let $a_1, \ldots, a_n \in R$. Show that $\langle x_1 - a_1, \ldots, x_n - a_n \rangle$ is a prime ideal of $R[x_1, \ldots, x_n]$.

Problem 5. Let R be a commutative ring and let $I \trianglelefteq R$. Using the isomorphism $R[x]/I[x] \cong (R/I)[x]$, show that $I[x]$ is a prime ideal if and only if I is.

Problem 6. Let R be a commutative ring and let $I \neq R$ be an ideal of R. Show that I is prime if and only if $I = \ker f$ for some surjective homomorphism $f : R \to S$ onto a commutative ring without zero-divisors.

Problem 7. Let K be a field and let $R = K[x, y, z]/\langle xy, xz \rangle$. For any polynomial $f \in K[x, y, z]$ we write for short $\overline{f} := f + \langle xy, xz \rangle$. Show that $\langle \overline{x}, \overline{y} \rangle$, $\langle \overline{x}, \overline{z} \rangle$ and $\langle \overline{y}, \overline{z} \rangle$ are prime ideals of R. Are any of the ideals $\langle \overline{x} \rangle$, $\langle \overline{y} \rangle$, $\langle \overline{z} \rangle$ prime?

Problem 8. (a) Show that if $x, y \in \mathbb{Z} \setminus \{0\}$ then $\langle x \rangle + \langle y \rangle = \langle \gcd(x, y) \rangle$, $\langle x \rangle \langle y \rangle = \langle xy \rangle$, $\langle x \rangle \cap \langle y \rangle = \langle \mathrm{lcm}(x, y) \rangle$ and $\langle x \rangle : \langle y \rangle = \langle x/\gcd(x, y) \rangle$.

(b) Let $p_1, \ldots, p_k \in \mathbb{N}$ be pairwise different prime numbers and let $n_1, \ldots, n_k \geq 1$ be natural numbers. Show that $\sqrt{\langle p_1^{n_1} \cdots p_k^{n_k} \rangle} = \langle p_1 \cdots p_k \rangle$.

Problem 9. Let I, J, K be ideals of a commutative ring R. Prove that $(I \cap J) \cdot (I + J) \subseteq IJ$. Moreover, show that if $I \subseteq J$ then $I : K \subseteq J : K$ and $K : I \supseteq K : J$.

Problem 10. Let $K(x)$ be the quotient field of $K[x]$. For $\alpha \in K$ let R_α be the set of all $\varphi \in K(x)$ which can be expressed in the form $\varphi = f/g$ with $f, g \in K[x]$ and $g(\alpha) \neq 0$. Show that R_α is a ring and that the assignment $\varphi \mapsto \varphi(\alpha)$ is a well-defined homomorphism $R_\alpha \to K$ whose kernel is a prime ideal of R_α.

Problem 11. Let $R = R_1 \oplus \cdots \oplus R_n$ be a direct sum of commutative rings with identity. Show that the prime (primary) ideals of R are exactly the sets

$$R_1 \oplus \cdots \oplus R_{k-1} \oplus I_k \oplus R_{k+1} \oplus \cdots \oplus R_n$$

where $1 \leq k \leq n$ and where I_k is a prime (primary) ideal of R_k.

Problem 12. Let R be a commutative ring.
(a) Show that a set $S \subseteq R$ with $0 \notin S$ is multiplicatively closed and contains with any element $x \in R$ also all divisors of x if and only if $R \setminus S$ is a union of prime ideals.
(b) Check that the following sets S satisfy the conditions stated in (a)!
(1) The complement of the set of all zero-divisors.
(2) The set of all units of R (if R has an identity).
(3) The set of all products of prime elements, where an element $p \in R$ is called prime if p is neither zero nor a unit and if $p \mid ab$ implies that $p \mid a$ or $p \mid b$.

Problem 13. Let $R \neq \{0\}$ be a commutative ring.
(a) Show that if $(P_i)_{i \in I}$ is a chain of prime ideals (so that $P_i \subseteq P_j$ or $P_j \subseteq P_i$ for all $i, j \in I$) then $\bigcap_{i \in I} P_i$ and $\bigcup_{i \in I} P_i$ are also prime ideals. Conclude that the set of prime ideals possesses minimal elements with respect to inclusion.
Hint. An ideal $I \neq R$ is a prime ideal if and only if $R \setminus I$ is multiplicatively closed.
(b) Show that if $P \subsetneq Q$ are prime ideals of R, then there are distinct prime ideals P_1, Q_1 with $P \subseteq P_1 \subsetneq Q_1 \subseteq Q$ such that there is no prime ideal properly between P_1 and Q_1.
Hint. Use Zorn's lemma to find a maximal chain $\mathcal{P}$ of prime ideals between P and Q. Then if $x \in Q \setminus P$, define $Q_1 := \bigcap_{\substack{P \in \mathcal{P} \\ x \in P}} P$ and $P_1 := \bigcap_{\substack{P \in \mathcal{P} \\ x \notin P}} P$.

Problem 14. Let R be a commutative ring with identity. Show that if every principal ideal $I \neq R$ of R is prime, then R is a field.
Hint. First of all, R is an integral domain since $\{0\}$ is a prime ideal. Suppose $x \neq 0$ is not invertible so that $Rx \neq R$. Then Rx^2 is a prime ideal.

Problem 15. Let R be a commutative ring.
(a) Let $\mathcal{M}$ be the set of those ideals of R all of whose elements are zero-divisors. Show that $\mathcal{M}$ has maximal elements and that every such maximal element is a prime ideal.
(b) Conclude that the set of all zero-divisors in R is a union of prime ideals.

Problem 16. Let R be a commutative unitary ring with nilradical $N(R)$ and Jacobson radical $J(R)$. (See problem 39 in section 5.)
(a) Show that $J(R)$ is the intersection of all maximal ideals of R and conclude that $N(R) \subseteq J(R)$.
(b) Show that $N\big(R/N(R)\big) = \{0\}$.
(c) Show that $N(R[x]) = J(R[x])$.
(d) Show that $N(\mathbb{Z}_n) = J(\mathbb{Z}_n)$ for all $n \in \mathbb{N}$. (You can use problem 3.)

Problem 17. Let R be a commutative ring with identity and let $N(R)$ be the nilradical of R. Show that the following conditions are equivalent:
(1) R has exactly one prime ideal;
(2) every element of R is either a unit or nilpotent;
(3) $R/N(R)$ is a field.

Problem 18. Let R be a commutative ring with identity. Suppose for each element $x \in R$ there is a natural number $n > 1$ with $x^n = x$. Show that every prime ideal of R is maximal. **Hint.** Use theorem (6.14).

Problem 19. Let R be the set of all Cauchy sequences $(a_1, a_2, a_3, \ldots)$ of rational numbers and let $I \subseteq R$ be the subset of all null sequences, i.e., all sequences $(a_1, a_2, a_3, \ldots)$ with $a_n \to 0$ as $n \to \infty$. Show that R is a ring with the coordinatewise operations and that I is a maximal ideal of R. Conclude that R/I is a field. (The field of real numbers can be defined as $\mathbb{R} := R/I$.) Moreover, show that the mapping $\mathbb{Q} \to R/I$ given by $a \mapsto [(a, a, a, \ldots)]$ is an embedding.

Problem 20. Show that an ideal in a commutative ring is a radical ideal if and only if it is an intersection of prime ideals.

Problem 21. Let R be a commutative ring. Prove the following statements.
(a) If an ideal $I \trianglelefteq R$ is such that $\sqrt{I}$ is maximal then I is primary.
(b) If $M \trianglelefteq R$ is maximal then every power M^k is primary.

Problem 22. Prove proposition (6.12).

Problem 23. Let $R := K[x_1, x_2, x_3, \ldots]$ be a polynomial ring in a countably infinite number of variables and consider the following ideals of R.

$$P := \langle x_1, x_2, x_3, \ldots \rangle \ , \ Q_1 := \langle x_1^2, x_2, x_3, \ldots \rangle \ , \ Q_2 := \langle x_1^2, x_2^2, x_3^2, \ldots \rangle \ .$$

(a) Show that P is a maximal ideal of R and that Q_1 and Q_2 are P-primary.
(b) Show that $P^2 \subseteq Q_1$ but that $P^n \not\subseteq Q_2$ for all $n \in \mathbb{N}$.

Problem 24. (a) Let $R = K[x_1, x_2, x_3, \ldots]$ as in problem 23 and let $Q := \langle x_i x_j \mid i, j \in \mathbb{N} \rangle$. Show that Q is a primary ideal of R. Moreover, if $I := \langle x_1 \rangle$ and $J := \langle x_1, x_2, x_3, \ldots \rangle$, then $IJ \subseteq Q$, $I \not\subseteq Q$ and $J^n \not\subseteq Q$ for all $n \in \mathbb{N}$.
(b) Let Q be a primary ideal and suppose $IJ \subseteq Q$, but $I \not\subseteq Q$. Show that if J is finitely generated, then $J^n \subseteq Q$ for some $n \in \mathbb{N}$.

Problem 25. Let R be a commutative ring with identity. Show that if there is an ideal $I \trianglelefteq R$ which is not a principal ideal, there is also a *prime* ideal $P \trianglelefteq R$ which is not principal.
Hint. Show that the set $\mathcal{M} := \{J \trianglelefteq R \mid I \subseteq J, J \text{ is not principal }\}$ possesses maximal elements with respect to inclusion and that each such maximal element is a prime ideal of R.

Problem 26. An ideal I in a ring R is called **irreducible** if the equation $I = I_1 \cap I_2$ with ideals I_1, $I_2 \trianglelefteq R$ implies that $I = I_1$ or $I = I_2$; otherwise I is called **reducible**.

(a) Show that $I = \langle x^2 \rangle \trianglelefteq K[x]$ is irreducible and not prime.

(b) Show that $I = \langle x, y \rangle^2 \trianglelefteq K[x, y]$ is primary and reducible.

(c) Let $R = \mathbb{Z}[x]$ and $I = \langle 4, 2x, x^2 \rangle$. Show that I is primary and reducible; in fact, $I = \langle 4, x \rangle \cap \langle 2, x^2 \rangle$.

Problem 27. Consider the ring $K[x, y]$ where K is a field. Show that $I_1 = \langle x \rangle$ and $I_2 = \langle x - y^2 \rangle$ are radical ideals of $K[x, y]$ but that $I_1 + I_2$ is not. Moreover, show that $I := \langle x^2, xy \rangle$ is not primary but $\sqrt{I}$ is prime.

Problem 28. (a) Show that if $P^n \subseteq Q \subseteq P$ where P is prime and Q is primary, then Q is P-primary.

(b) In $K[x, y]$, let $P := \langle x \rangle$ and $Q := \langle x^2, xy \rangle$. Show that P is prime, $P^2 \subseteq Q \subseteq P$, but Q is not primary.

Problem 29. Let K be a field. Consider the two ideals of $K[x, y]$ given by

$$Q := \langle x^2, y \rangle \qquad \text{and} \qquad P := \langle x, y \rangle .$$

(a) Show that P is a maximal ideal and the only prime ideal containing Q.

(b) Show that P is Q-primary.

(c) Show that $P^2 \subsetneq Q \subsetneq P$ and conclude that Q is not the power of any prime ideal.

Problem 30. Show that in each of the following cases that P is a prime ideal of R but P^2 is not primary.

(a) $R := K[x, y, z]/\langle xy - z^2 \rangle$ where K is a field; $P := \langle \overline{x}, \overline{z} \rangle$;

(b) $R := \{a_0 + a_1 x + \cdots + a_n x^n \in \mathbb{Z}[x] \mid a_1 \text{ is divisible by } 3\}$; $P := \langle 3x, x^2, x^3 \rangle$.

Problem 31. Let $R = \mathbb{Z}[x]$.

(a) Show that $\langle 2, x \rangle$ is a maximal ideal of R and that $\langle 4, x \rangle$ is $\langle 2, x \rangle$-primary but not a power of $\langle 2, x \rangle$.

(b) Show that $\langle x^2, 2x \rangle$ is not primary but contains the prime power $\langle x \rangle^2$.

(c) Find a reduced primary decomposition of the ideal $\langle 9, 3x + 3 \rangle$.

Problem 32. Find reduced primary decompositions of the ideals $\langle 4, 2x, x^2 \rangle$ and $\langle x^2, xy, 2 \rangle$ in $\mathbb{Z}[x, y]$.

Problem 33. Let K be a field. Consider the polynomials $f(x, y, z) = y^2 - xz$, $g(x, y, z) = yz - x^3$ and $h(x, y, z) = z^2 - x^2 y$ and the ideal $P = \{f \in K[x, y, z] \mid f(t^3, t^4, t^5) = 0 \text{ as a polynomial in } t\}$.

(a) Show that P is a prime ideal and that $P = \langle f, g, h \rangle$.

(b) Check that $x(x^5 - 3x^2 yz + xy^3 + z^3) = g^2 - fh \in P^2$ and prove that $x \notin P$ and that $x^5 - 3x^2 yz + xy^3 + z^3 \notin P^2$. Conclude that P^2 is not a primary ideal.

Problem 34. Let R be a subring of a commutative ring S. Show that if $Q \unlhd S$ is P-primary in S, then $R \cap Q$ is $(R \cap P)$-primary in R.

Problem 35. Let R be a commutative ring.
(a) Show that if Q is P-primary in R then $Q[x]$ is $P[x]$-primary in $R[x]$. (In particular, if P is a prime ideal of R then $P[x]$ is a prime ideal of $R[x]$.)
(b) Show that if $I = \bigcap_{i=1}^{n} Q_i$ is a reduced primary decomposition of $I \unlhd R$ then $I[x] = \bigcap_{i=1}^{n} Q_i[x]$ is a reduced primary decomposition of $I[x] \unlhd R[x]$. If P is a minimal prime for I, then $P[x]$ is a minimal prime for $I[x]$.

Problem 36. Show that $M = 2\mathbb{Z}$ is a maximal ideal of $\mathbb{Z}$ but that $M[x]$ is not a maximal ideal of $\mathbb{Z}[x]$.

Problem 37. Let K be a field and let $I := \langle x^2, xy \rangle$.
(a) For $\alpha \in K$ let $I_\alpha := \langle x^2, y + \alpha x \rangle$. Show that the ideals I_α are pairwise different and that $I = \langle x \rangle \cap I_\alpha$ is a reduced primary decomposition of I for any $\alpha \in K$.
(b) Show that $I = \langle x \rangle \cap \langle x^2, xy, y^2 \rangle$ is still another reduced primary decomposition.
(c) Is it an accident that all reduced primary decompositions of I given in (a) and (b) contain $\langle x \rangle$ as one of their factors?

Problem 38. Consider the ideals $P_1 = \langle x, y \rangle$, $P_2 = \langle x, z \rangle$ and $M = \langle x, y, z \rangle$ of the polynomial ring $K[x, y, z]$ where K is a field. Let $I := P_1 P_2$ and show that $I = P_1 \cap P_2 \cap M^2$ is a reduced primary decomposition of I. Which of the three factors P_1, P_2 and M^2 must occur in *every* reduced primary decomposition of I?

Problem 39. Let R be the set of all sequences with coefficients in $\mathbb{Z}_2$ which become eventually constant. (This means that a sequence $a = (a_1, a_2, a_3, \ldots)$ belongs to R if and only if there is an index N such that $a_n = a_N$ for all $n \geq N$.) For $k \in \mathbb{N}$ let $P_k := \{a \in R \mid a_k = 0\}$; moreover, let $P_0 := \{(a_1, \ldots, a_n, 0, 0, \ldots) \mid a_i \in \mathbb{Z}_2\}$.
(a) Show that the prime ideals of R other than $\{0\}$ are exactly the ideals P_k ($k \geq 0$).
(b) Show that an ideal of R is primary if and only if it is prime.
(c) Show that $\{0\}$ cannot be written as a finite intersection of primary ideals.

Problem 40. Let P be a prime ideal of a commutative ring R. Show that if the ideal $P^{(m)} P^{(n)}$ has a primary decomposition then $P^{(m+n)}$ is its P-primary component.

Problem 41. Show that if Q_1 and Q_2 are primary ideals with $\sqrt{Q_1} \neq \sqrt{Q_2}$ then $Q_1 \cap Q_2$ is not primary unless $Q_1 \subseteq Q_2$ or $Q_2 \subseteq Q_1$.

Problem 42. Let X be a compact Hausdorff space with more than one element and let $R := C(X)$ be the ring of all continuous functions $f : X \to \mathbb{R}$. Show that for each maximal ideal M of R there is a prime ideal $P \unlhd R$ with $P \subsetneq M$.
Hint. According to problem 30 in section 5 the ideal M must be one of the ideals $I_x = \{f \in R \mid f(x) = 0\}$ with $x \in X$. Pick any point $y \in X$ with $y \neq x$; then the set $S := \{f \in R \mid f(x)f(y) \neq 0\}$ is multiplicatively closed.

7. Factorization in integral domains

IN section 3 we carried over the well-known notions of divisibility from the ring $\mathbb{Z}$ of integers to arbitrary integral domains. Now one of the key features of the ring $\mathbb{Z}$ is that every number $m \in \mathbb{Z} \setminus \{0, \pm 1\}$ possesses a prime factorization $m = \pm p_1^{k_1} \cdots p_n^{k_n}$ which is unique up to the order of the factors.[†] We want to explore, in general, under what circumstances an integral domain has this unique factorization property.[††] The first step is to define, for an arbitrary integral domain R, the analogue of the prime numbers in $\mathbb{Z}$. Thus we want to consider those elements in R which can serve as last multiplicative building blocks. It turns out that there are two possible ways to generalize the notion of a prime number; these are given in the following definitions.

(7.1) Definitions. *Let R be an integral domain.*
(a) *The set*

$$\boxed{R^\heartsuit := \{x \in R \mid x \text{ is neither zero nor a unit}\,\}}$$

is called the **heart** *of the domain R.*
(b) *An element $p \in R$ is called* **irreducible** *if it belongs to $R^\heartsuit$ and has no proper divisors. The last property can be restated as follows:*

$$\boxed{\text{if } a \mid p \text{ then } a \text{ is a unit or an associate of } p.}$$

(c) *An element $p \in R$ is called* **prime** *if it belongs to $R^\heartsuit$ and has the following property:*

$$\boxed{\text{if } p \mid ab \text{ then } p \mid a \text{ or } p \mid b.}$$

(7.2) Examples. (a) An element of $\mathbb{Z}$ is irreducible if and only if it is prime if and only if it is of the form $\pm p$ where $p \in \mathbb{N}$ is an ordinary prime number.
(b) The polynomial $x^2 + 1$ is irreducible in $\mathbb{Q}[x]$ and $\mathbb{R}[x]$, but not in $\mathbb{C}[x]$ because of the decomposition $x^2 + 1 = (x + i)(x - i)$.
(c) The polynomial $2x + 2$ is irreducible in $\mathbb{Q}[x]$, but reducible in $\mathbb{Z}[x]$. Indeed, $2x + 2 = 2(x + 2)$ is a proper factorization in $\mathbb{Z}[x]$ because 2 is not a unit in $\mathbb{Z}$.
(d) The polynomial $x^2 - x - 2$ is reducible in $\mathbb{Z}[x]$, but not in $\mathbb{Z}[[x]]$. Indeed,

---

[†] We assume this as known for the time being; a proof – in a more general setting – will be given in (7.10) below.

[††] If we allow rings without identity, there is an easy example where such a property does not hold. Namely, in the ring $R := 2\mathbb{Z}$ of all even integers, all numbers of the form $2k$ with k odd are "indecomposable", i.e., cannot be factored in R, whereas all numbers of the form $2k$ with k even are "composite", i.e., can be factored in R like $8 = 2 \cdot 4$. Then the number 60 possesses two different "prime factorizations", namely, $60 = 2 \cdot 30 = 6 \cdot 10$.

$x^2 - x - 2 = (x+1)(x-2)$ is a proper factorization in $\mathbb{Z}[x]$, but not in $\mathbb{Z}[[x]]$; note that $(1+x)(1-x+x^2-x^3+-\cdots) = 1$ so that $1+x$ is a unit in $\mathbb{Z}[[x]]$. Then $x^2 - x - 2$ is irreducible in $\mathbb{Z}[[x]]$, because, obviously, $x-2$ is.

(e) The polynomial $x^4 + 4$ is reducible in $\mathbb{Z}[x]$ because

$$x^4 + 4 = (x^2 + 2x + 2)(x^2 - 2x + 2) .$$

(f) Every polynomial of the form $x^4 + a^4$ with $a \in \mathbb{R}$ is reducible in $\mathbb{R}[x]$ because

$$x^4 + a^4 = (x^2 + a^2)^2 - 2a^2 x^2 = (x^2 + a^2 + \sqrt{2}ax)(x^2 + a^2 - \sqrt{2}ax) .$$

(g) The irreducible elements of $\mathbb{C}[x]$ are exactly the linear polynomials; this is a restatement of the fundamental theorem of algebra. Consequently, the irreducible elements of $\mathbb{R}[x]$ are the linear polynomials and those quadratic polynomials which do not possess a real root, i.e., those polynomials $ax^2 + bx + c \in \mathbb{R}[x]$ with $b^2 - 4ac < 0$.

(h) The elements $2, 3, 1 + \sqrt{-5}$ and $1 - \sqrt{-5}$ are irreducible, but not prime in $\mathbb{Z}[\sqrt{-5}]$. Let us prove this for the element $1 - \sqrt{-5}$; for the other elements, one proceeds in a completely analogous manner. The irreducibility can be established by taking norms. If $a + b\sqrt{-5}$ divides $1 + \sqrt{-5}$, then $N(a + b\sqrt{-5})$ divides $N(1 + \sqrt{-5})$; i.e., $a^2 + 5b^2$ divides 6. This is only possible if $a = \pm 1, b = 0$ or $a = \pm 1, b = \pm 1$. Now check that $1 - \sqrt{-5}$ is not a divisor of $1 + \sqrt{-5}$; thus the only divisors are ± 1 and $\pm(1 + \sqrt{-5})$, i.e., units or associates of $1 + \sqrt{-5}$. This shows that $1 + \sqrt{-5}$ has no proper divisors, i.e., is irreducible.

Also, the equation $6 = 2 \cdot 3 = (1 + \sqrt{5})(1 - \sqrt{5})$ shows that $1 + \sqrt{-5}$ divides $2 \cdot 3$. On the other hand, $1 + \sqrt{-5}$ divides neither 2 nor 3, because the elements $\frac{2}{1+\sqrt{-5}} = \frac{1+\sqrt{-5}}{3}$ and $\frac{3}{1+\sqrt{-5}} = \frac{1+\sqrt{-5}}{2}$ of $\mathbb{Q} + \mathbb{Q}\sqrt{-5}$ do not belong to $\mathbb{Z} + \mathbb{Z}\sqrt{-5}$. This shows that $1 + \sqrt{-5}$ is not prime in $\mathbb{Z} + \mathbb{Z}\sqrt{-5}$.

(7.3) Lemma. *Let R be an integral domain.*

(a) *Every associate of a prime element (an irreducible element) is again prime (irreducible).*

(b) *If p is prime and $p \mid a_1 \cdots a_n$ then $p \mid a_i$ for some i.*

(c) *If $p_1 \cdots p_m = q_1 \cdots q_n$ where the elements p_i and q_j are primes, then $m = n$ and there is a permutation $\sigma \in \mathrm{Sym}_n$ such that q_i is an associate of $p_{\sigma(i)}$. This means that a decomposition into primes is unique up to a rearrangement of the factors or multiplication of the factors by units.*

(d) *Every prime element is irreducible.*

Proof. (a) This follows immediately from the definitions.

(b) We proceed by induction on n. The case $n = 1$ is trivial; let $n \geq 2$. Now $p \mid (a_1 \cdots a_{n-1})a_n$ implies $p \mid a_1 \cdots a_{n-1}$ or $p \mid a_n$ because p is prime. If $p \mid a_n$ we are done; if $p \mid a_1 \cdots a_{n-1}$ then $p \mid a_k$ for some $1 \leq k \leq n-1$ by induction hypothesis.

(c) We proceed by induction on $m + n$. If $m + n = 2$, then $m = n = 1$ and $p_1 = q_1$. Suppose the claim is proved for some value of $m + n$ and assume that $p_1 \cdots p_m p_{m+1} = q_1 \cdots q_n$. (If n is raised by 1, the argument is analogous.) Then p_{m+1} divides the right-hand side, hence must divide one of the factors, say q_i. Then $q_i = r p_{m+1}$ for some $r \in R$. Since q_i is prime this implies $q_i \mid r$ or $q_i \mid p_{m+1}$. But $q_i \mid r$ is impossible; otherwise there would exist an $s \in R$ such that $r = s q_i$, hence $q_i = s q_i p_{m+1}$, hence $1 = s p_{m+1}$

whereas p_{m+1} is no unit. So $q_i \mid p_{m+1}$ which shows that q_i and p_{m+1} are associates; i.e., $q_i = u p_{m+1}$ for some $u \in R^\times$. Divide the equation $p_1 \cdots p_m p_{m+1} = q_1 \cdots q_n$ by p_{m+1} to obtain $p_1 \ldots p_m = (u q_1) q_2 \cdots \widehat{q_i} \cdots q_n$. By induction hypothesis, $m = n - 1$ and $q_j \sim q_{\sigma(j)}$ for some bijection $\sigma : \{1, 2, \ldots, i-1, i+1, \ldots, n\} \to \{1, \ldots, m\}$. Define $\sigma(i) := m + 1$ to obtain the claim.

(d) Let a be divisor of p so that $p = ab$ for some $b \in R$. Then $p \mid ab$ so that $p \mid a$ or $p \mid b$ because p is prime. Now also $a \mid p$ and $b \mid p$. If $p \mid a$, then a is an associate of p. If $p \mid b$, then b is an associate of p which implies that a must be a unit. ∎

IT is an important fact that the key notions of divisibility can be expressed in terms of ideals. This is done in the next proposition.

(7.4) Proposition. *Let R be an integral domain.*
(a) *$a \mid b$ if and only if $\langle a \rangle \supseteq \langle b \rangle$.*
(b) *a and b are associates if and only if $\langle a \rangle = \langle b \rangle$.*
(c) *$x = 0$ if and only if $\langle x \rangle = \{0\}$.*
(d) *x is a unit if and only if $\langle x \rangle = R$.*
(e) *x is irreducible if and only if $\langle x \rangle$ is maximal amongst the proper principal ideals of R.*
(f) *x is prime if and only if $\langle x \rangle$ is a prime ideal.*
(g) *If x is an associate of a prime power then $\langle x \rangle$ is a primary ideal. If x possesses a factorization into primes and if $\langle x \rangle$ is a primary ideal, then x is an associate of a prime power.†*
(h) *x is a common multiple of $a_1, \ldots, a_n$ if and only if $\langle x \rangle \subseteq \langle a_1 \rangle \cap \cdots \cap \langle a_n \rangle$.*
(i) *x is a least common multiple of $a_1, \ldots, a_n$ if and only if $\langle x \rangle = \langle a_1 \rangle \cap \cdots \cap \langle a_n \rangle$.*
(j) *x is a common divisor of $a_1, \ldots, a_n$ if and only if $\langle a_1 \rangle + \cdots + \langle a_n \rangle \subseteq \langle x \rangle$.*
(k) *If $\langle a_1 \rangle + \cdots + \langle a_n \rangle = \langle x \rangle$ then x is a greatest common divisor of $a_1, \ldots, a_n$.*

Proof. (a) $a \mid b$ if and only if $b = ra$ for some $r \in R$ if and only if $b \in \langle a \rangle$ if and only if $\langle b \rangle \subseteq \langle a \rangle$.

(b) a and b are associates if and only if $a \mid b$ and $b \mid a$. By part (a), this means exactly that $\langle a \rangle = \langle b \rangle$.

(c) This is trivial.

(d) Let x be a unit and $r \in R$ an arbitrary element. Then $r = (rx^{-1})x \in Rx = \langle x \rangle$ so that $R = \langle x \rangle$. Suppose conversely that $\langle x \rangle = R$, i.e., $Rx = R$. Then in particular $1 \in xR$; thus there is an element $y \in R$ with $xy = yx = 1$. This shows that x is invertible.

(e) Let x be irreducible and suppose that $\langle x \rangle \subseteq \langle a \rangle$ for some $a \in R$. Then $a \mid x$. But x has no proper divisors! So either a is an associate of x (which means $\langle a \rangle = \langle x \rangle$) or a unit (which means $\langle a \rangle = R$). We see that $\langle x \rangle$ is maximal among the proper principal ideals of R. Suppose conversely that this condition holds; we want to show that x is irreducible. Let r be a divisor of x. Then $\langle x \rangle \subseteq \langle r \rangle$. By maximality, either $\langle r \rangle = \langle x \rangle$ (so that r is an associate of r) or $\langle r \rangle = R$ (so that r is a unit). We see that r has no proper divisors.

---

† It may happen that $\langle x \rangle$ is primary but that x is not an associate of a prime power. (See problem 19 below.)

(f) Let x be prime and suppose that $ab \in \langle x \rangle$. Then $x \mid ab$, hence $x \mid a$ or $x \mid b$, i.e., $a \in \langle x \rangle$ or $b \in \langle x \rangle$. This shows that $\langle x \rangle$ is a prime ideal. Suppose conversely that $\langle x \rangle$ is a prime ideal and assume $x \mid ab$. Then $ab \in \langle x \rangle$ which implies $a \in \langle x \rangle$ or $b \in \langle x \rangle$, i.e., $x \mid a$ or $x \mid b$. This shows that x is a prime element.

(g) Let x be an associate of a prime power, say $ux = p^k$ with $u \in R^\times$, and suppose $ab \in \langle x \rangle$, but $a \notin \langle x \rangle$. Then there are an element $r \in R$ with $ab = rp^n$ and an integer $m \in \mathbb{Z}$ with $0 \le m < n$ such that $p^m \mid a$ but $p^{m+1} \nmid a$. Hence we can write $a = sp^m$ with $p \nmid s$. Now $rp^n = ab = sp^m b$ which implies $bs = rp^{n-m}$. Then p divides the right-hand side; since p is prime and $p \nmid s$ this implies $p \mid b$. Hence $b \in \langle p \rangle$ and $b^n \in \langle p^n \rangle = \langle x \rangle$. This shows that $\langle x \rangle$ is primary.

Suppose now that $\langle x \rangle$ is primary and x possesses a prime factorization. Then, if x is not an associate of a prime power, there are two non-associate prime factors p, q of x. Let p^m and q^n be the highest powers dividing x so that $x = rp^m q^n$ with $p \nmid r$ and $q \nmid r$. Then $p^m q^n \in \langle x \rangle$ and $p^m \notin \langle x \rangle$. Since $\langle x \rangle$ is primary this implies $q^N \in \langle x \rangle$ for some multiple N of n, say $q^N = sx$. But then $q^N = sx = rsp^m q^n$ which implies $p \mid q^N$, hence $p \mid q$ which is the desired contradiction.

(h) $\langle x \rangle \subseteq \bigcap_i \langle a_i \rangle$ if and only if $x \in \bigcap_i \langle a_i \rangle$ if and only if there are elements $r_i \in R$ with $x = r_1 a_1 = \cdots = r_n a_n$. But this means exactly that x is a common multiple of $a_1, \ldots, a_n$.

(i) Suppose $\langle x \rangle = \bigcap_i \langle a_i \rangle$. Then, due to part (h), x is a common multiple of $a_1, \ldots, a_n$, and every other common multiple of $a_1, \ldots, a_n$ is contained in $\bigcap_i \langle a_i \rangle = \langle x \rangle$, hence is a multiple of x. This shows that x is a least common multiple of $a_1, \ldots, a_n$.

Suppose conversely that x is a least common multiple of $a_1, \ldots, a_n$. Then $\langle x \rangle \subseteq \bigcap_i \langle a_i \rangle$ due to (h). To prove the converse inclusion, let $y \in \bigcap_i \langle a_i \rangle$. Then y is a common multiple of $a_1, \ldots, a_n$, hence must be a multiple of x which means $y \in \langle x \rangle$. Since y was chosen arbitrarily, this means $\bigcap_i \langle a_i \rangle \subseteq \langle x \rangle$.

(j) $\sum_i \langle a_i \rangle \subseteq \langle x \rangle$ if and only if $\langle a_i \rangle \subseteq \langle x \rangle$ for all i if and only if $a_i \in \langle x \rangle$ for all i. But this means exactly that x divides all a_i, i.e., is a common divisor of $a_1, \ldots, a_n$.

(k) If $\sum_i \langle a_i \rangle = \langle x \rangle$ then x is a common divisor of $a_1, \ldots, a_n$, due to part (j). Let y be another common divisor. Then $\langle y \rangle \supseteq \sum_i \langle a_i \rangle = \langle x \rangle$ which means that $x \in \langle y \rangle$, i.e., that y divides x. This shows that x is a greatest common divisor of $a_1, \ldots, a_n$. ∎

NOW let R be an integral domain and let $x \in R^\heartsuit$. It seems obvious that we can write x as a product of irreducible factors. Why? Well, if x itself is irreducible – fine! If x is reducible, we can write down a nontrivial factorization $x = x_1 x_2$. Then we apply the same argument to x_1 and x_2, and so on. Now in $\mathbb{Z}$ this procedure will terminate after a finite number of steps because the absolute values of the factors become smaller and smaller, and this cannot continue indefinitely. In a general domain, however, it is possible that the process never stops (and we will see examples for this phenomenon).

But even if we can always find a factorization into irreducible elements, this factorization need not be unique. Now at this point we should remark that even in $\mathbb{Z}$ the factorization of a number into irreducible ones need not be unique. For example,

$$30 = 2 \cdot 3 \cdot 5 = 2 \cdot 5 \cdot 3 = (-2) \cdot 3 \cdot (-5), \quad \text{and so on.}$$

However, all these factorizations are essentially the same; the factors differ only up to units and in the order in which they occur. This is in fact the utmost of uniqueness that we may expect from a factorization into irreducible elements. We can always change the order of the factors or multiply them by units without changing the product. But as we will see, there can be "substantially" different factorizations of elements of a domain. To have a convenient terminology available, we give the following definitions.

(7.5) Definitions. *An integral domain R is called a* **factorization domain** *if any element $r \in R^\heartsuit$ possesses a factorization into irreducible elements. We call such a factorization essentially unique, if an equation $r = p_1 \cdots p_m = q_1 \cdots q_n$ with irreducible elements p_i, q_j implies that $n = m$ and $q_j \sim p_{\sigma(j)}$ for some permutation $\sigma \in \mathrm{Sym}_n$; i.e., if it is unique up to multiplication with units and a rearrangement of the factors. An integral domain R is called a* **unique factorization domain** *if each element $r \in R^\heartsuit$ possesses an essentially unique factorization.*

(7.6) Examples. (a) $\mathbb{Z}$ is a unique factorization domain. As stated in the introduction of this section, a proof of this statement in a more general setting will be given in (7.10) below.

(b) $\mathbb{Z}(\sqrt{-5})$ is not a unique factorization domain because the element 6 has two essentially different factorizations, namely

$$6 = 2 \cdot 3 = (1 + \sqrt{-5})(1 - \sqrt{-5}) \, .$$

(c) Let R be the set of formal expressions $a_1 x^{q_1} + \cdots + a_n x^{q_n}$ with $a_i, q_i \in \mathbb{Q}$ and $q_i \geq 0$ with the obvious addition and the multiplication determined by $x^\alpha x^\beta = x^{\alpha+\beta}$; clearly, R is an integral domain with identity element $x^0 = 1$. Then $x = x^1$ is a non-unit which cannot be factored into irreducible elements. Thus R is not even a factorization domain.

(d) In problem 6 below it will be shown that the set of all complex numbers z which satisfy a monic equation $z^n + a_{n-1} z^{n-1} + \cdots + a_1 z + a_0 = 0$ with coefficients $a_i \in \mathbb{Z}$ form a ring R whose only units are ± 1. In R we can write

$$2 = \sqrt{2} \cdot \sqrt{2} = \sqrt[4]{2} \cdot \sqrt[4]{2} \cdot \sqrt[4]{2} \cdot \sqrt[4]{2} = \cdots$$

so that the number 2 cannot be factored into irreducible elements. Hence R is not a factorization domain.

LET us return to our discussion preceding (7.4). What goes wrong if an integral domain is not a factorization domain? If $x \in R^\heartsuit$ does not possess a factorization into irreducible elements, then there is a chain of divisors

$$x_1 \mid x \, , \quad x_2 \mid x_1 \, , \quad x_3 \mid x_2 \, , \quad \ldots$$

such that none of the quotients x_i / x_{i+1} is a unit. Using (7.4)(a),(b) this can be expressed ideal-theoretically as

$$\langle x \rangle \subsetneq \langle x_1 \rangle \subsetneq \langle x_2 \rangle \subsetneq \langle x_3 \rangle \subsetneq \cdots \, .$$

This observation proves the following criterion for the existence of factorizations.

(7.7) Proposition. *Let R be an integral domain. Then the following conditions are equivalent.*

(a) *R is a factorization domain.*

(b) *There is no strictly ascending sequence of principal ideals in R.* ■

IT was Emmy Noether's idea to study the more general class of rings which do not possess a strictly ascending chain of ideals $I_1 \subsetneq I_2 \subsetneq \cdots$ whatsoever (not necessarily principal ideals). We will discuss these rings in section 11 below. Let us now find criteria for the uniqueness of decomposition into irreducible elements.

(7.8) Proposition. *Let R be a factorization domain. Then the following conditions are equivalent:*

(1) *R is a unique factorization domain;*

(2) *any n elements $a_1, \ldots, a_n \in R$ possess a greatest common divisor and a least common multiple;*

(3) *any n elements $a_1, \ldots, a_n \in R$ possess a greatest common divisor;*

(4) *whenever n elements $a_1, \ldots, a_n \in R$ possess a gcd g then gb is a gcd of $a_1 b, \ldots, a_n b$;*

(5) *every irreducible element if prime.*

Proof. $(1) \Longrightarrow (2)$. There are units $u_1, \ldots, u_k$, pairwise non-associate irreducible elements $p_1, \ldots, p_k$ and integers $n_1, \ldots, n_k \geq 0$ such that

$$a_1 = u_1 p_1^{n_{11}} p_2^{n_{12}} \cdots p_k^{n_{1k}}, \quad a_2 = u_2 p_1^{n_{21}} p_2^{n_{22}} \cdots p_k^{n_{2k}}, \quad \ldots, \quad a_k = u_k p_1^{n_{k1}} p_2^{n_{k2}} \cdots p_k^{n_{kk}}.$$

Let $n_i := \min\{n_{1i}, n_{2i}, \ldots, n_{ki}\}$ and $N_i := \max\{n_{1i}, n_{2i}, \ldots, n_{ki}\}$. It is straightforward to check that $d := p_1^{n_1} p_2^{n_2} \cdots p_k^{n_k}$ is a greatest common divisor and $m := p_1^{N_1} p_2^{N_2} \cdots p_k^{N_k}$ is a least common multiple of $a_1, \ldots, a_k$.

$(2) \Longrightarrow (3)$. This is trivial.

$(3) \Longrightarrow (4)$. Let g be a gcd of $a_1, \ldots, a_n$ and let $b \in R$. By assumption, the elements $a_1 b, \ldots, a_n b$ possess a gcd, say g'. Now gb divides $a_1 b, \ldots, a_n b$, hence divides g' so that $g' = xgb$ for some $x \in R$. Since g' divides $a_1 b, \ldots, a_n b$; the element xg divides $a_1, \ldots, a_n$, hence also divides the gcd of these elements, namely g. Hence there is an element $y \in R$ with $xyg = g$. This shows that x is a unit. Since g' and gb differ only by a unit, we conclude that gb is a gcd of $a_1 b, \ldots, a_n b$.

$(4) \Longrightarrow (5)$. Suppose p is irreducible with $p \mid ab$ and $p \nmid a$; we have to show that $p \mid b$. Now 1 is a gcd of p and a because p is irreducible and $p \nmid a$. Thus (d) yields that b is a gcd of pb and ab. But then p is a common divisor of pb and ab, hence divides the greatest common divisor of these elements which is b.

$(e) \Longrightarrow (a)$. Suppose an element has two decompositions

$$u_1 p_1 \cdots p_m = u_2 q_1 \cdots q_n$$

with units u_1, u_2 and irreducible elements p_i, q_j. Since all irreducible elements are prime by hypothesis, we can invoke (7.3)(c) which gives the claim. ■

(7.9) Theorem. *Let R be an integral domain. Then the following conditions are equivalent:*

(1) *R is a unique factorization domain;*

(2) *every nonzero prime ideal $P \trianglelefteq R$ contains a prime element;*

(3) *every element $x \in R^\heartsuit$ can be written as a product of prime elements.*

Proof. (1)$\Longrightarrow$(2). Let $P \neq \{0\}$ be a prime ideal and let $x \neq 0$ be a nonzero element of P; then $x \in R^\heartsuit$. Since R is a unique factorization domain, we can write $x = p_1 \cdots p_n$ where each element p_i is irreducible and hence prime due to (7.8)(e). But $p_1 \cdots p_n \in P$ implies $p_k \in P$ for some index k, since P is a prime ideal.

(2)$\Longrightarrow$(3). Let S be the set of all products of prime elements. Suppose there is an element $x \in R^\heartsuit$ with $x \notin S$. Then even $Rx \cap S = \emptyset$. In fact, suppose $rx \in S$, say $rx = p_1 \cdots p_n$. Then each p_i divides either r or s; after renumbering if necessary, we can assume that $p_1 \cdots p_m \mid r$ and $p_{m+1} \cdots p_n \mid x$, say $r = p_1 \cdots p_m u$ and $x = p_{m+1} \cdots p_n v$. Since $rx = p_1 \cdots p_n$, we conclude that $uv = 1$ so that u and v are units. This implies that $x = p_{m+1} \cdots (p_n v)$assumption that $x \notin S$. Hence we established that $Rx \cap S = \emptyset$. Then (6.4)(a) shows that there is a prime ideal P with $P \cap S = \emptyset$ and $Rx \subseteq P$ (hence $P \neq \{0\}$), contradicting our hypothesis (2).

(3)$\Longrightarrow$(1). This is clear since every prime element is irreducible by (7.3)(d) and since any decomposition into prime elements is essentially unique by (7.3)(c). ∎

AS an important consequence of expressing notions of divisibility in terms of ideals, we can show that every principal ideal domain is automatically a unique factorization domain. In particular, the ring $\mathbb{Z}$ of integers is a unique factorization domain.

(7.10) Theorem. *Let R be a principal ideal domain.*

(a) *Any n elements $a_1, \ldots, a_n$ of R possess a greatest common divisor, and an element $d \in R$ is a greatest common divisor of $a_1, \ldots, a_n$ if and only if*

$$\langle d \rangle = \langle a_1, \ldots, a_n \rangle .$$

This implies in particular that we have a linear representation $d = r_1 a_1 + \cdots + r_n a_n$ with elements $r_i \in R$ for the greatest common divisor d of $a_1, \ldots, a_n$.

(b) *R is a unique factorization domain.*

Proof. (a) The ideal $\langle a_1 \rangle + \cdots + \langle a_n \rangle$ is a principal ideal; hence there is an element $d \in R$ with $\langle a_1 \rangle + \cdots + \langle a_n \rangle = \langle d \rangle$. Then (7.4)(k) shows that d is a greatest common divisor of $a_1, \ldots, a_n$. Since any two greatest common divisors of $a_1, \ldots, a_n$ must be associates of each other, this condition already characterizes the greatest common divisors of $a_1, \ldots, a_n$.

(b) We apply proposition (7.7) to show that R is a factorization domain. Let $\langle x_1 \rangle \subseteq \langle x_2 \rangle \subseteq \langle x_3 \rangle \subseteq \cdots$ be an ascending chain of principal ideals. Then $I := \bigcup_{n=1}^{\infty} \langle x_i \rangle$ is again an ideal. Since R is a principal ideal domain there is an element x with $I = \langle x \rangle$. But then $x \in \langle x_N \rangle$ for some $N \in \mathbb{N}$ which clearly implies $\langle x_n \rangle = \langle x_N \rangle = \langle x \rangle$ for all $n \geq N$. This shows that R does not possess a strictly ascending chain of principal ideals and hence is a factorization domain. Then (7.8)(c) and part (a) of this theorem imply immediately that R is a unique factorization domain. ∎

THUS the desirable unique factorization property in an integral domain R can be established by proving that R is a principal ideal domain. However, the only principal ideal domains we know as yet are $\mathbb{Z}$ and (trivially) all fields. We now want to introduce a class of rings which encompasses many examples of practical interest and for which we can show rather easily that they are all principal ideal domains. Let us start by giving a characterization of principal ideal domains.

(7.11) Theorem. *Let R be an integral domain. Then R is a principal ideal domain if and only if there is a function $\beta : R \setminus \{0\} \to \mathbb{N}$ such that whenever $a, b \in R \setminus \{0\}$ with $a \nmid b$ there are elements $x, y \in R$ with $xa + yb \neq 0$ and $\beta(xa + yb) < \beta(a)$.*

Proof. Suppose first that R is a principal ideal domain. Let P be a set of prime elements such that each prime element of R is an associate of exactly one element of P.† Since R is a unique factorization domain, we can define a function $\beta : R \setminus \{0\} \to \mathbb{N}$ by

$$\beta(a) := 2^{\sum_{p \in P} k(p)} \qquad \text{if} \qquad a = \varepsilon \prod_{p \in P} p^{k(p)} \text{ with } \varepsilon \in R^{\times} \ ;$$

then clearly $\beta(a) = 1$ if and only if a is a unit and $\beta(ab) = \beta(a)\beta(b)$ for all $a, b \in R^{\heartsuit}$. This shows that if d is a proper divisor of b then $\beta(d) < \beta(b)$. Now let us verify that β has the property stated in the theorem. Let $a, b \in R \setminus \{0\}$ be elements with $a \nmid b$. By (7.10)(a), we can find a greatest common divisor d of a and b as a linear combination $d = xa + yb$; then $a \nmid b$ implies that d is a proper divisor of b and hence satisfies $\beta(d) < \beta(b)$, i.e., $\beta(xa + yb) < \beta(b)$.

Suppose conversely that a function $\beta : R \setminus \{0\} \to \mathbb{N}$ with the indicated property is given. Let $I \neq \{0\}$ be an ideal of R and pick an element $a \in I \setminus \{0\}$ for which β takes a minimal value; i.e., such that $\beta(a) \leq \beta(b)$ for all $b \in I \setminus \{0\}$. We claim that $I = \langle a \rangle = Rb$. Suppose not. Then there is an element $b \in I \setminus \{0\}$ with $a \nmid b$. By hypothesis there are elements $x, y \in R$ with $xa + yb \neq 0$ (hence $xa + yb \in I \setminus \{0\}$) with $\beta(xa + yb) < \beta(a)$ which contradicts the choice of a. ∎

THIS necessary and sufficient criterion is too hard to check in general to be of great practical value. Therefore, we introduce a class of rings for which the existence of a suitable function β is easier to establish.

(7.12) Definition. *An integral domain with a mapping $\beta : R \to \mathbb{N}_0$ is called a* **Euclidean domain** *if the following conditions are satisfied:*
(1) $\beta(a) = 0$ *if and only if* $a = 0$;
(2) *if* $a \mid b$ *then* $\beta(a) \leq \beta(b)$;
(3) *given any two elements* $a, b \in R \setminus \{0\}$, *there are elements* $q, r \in R$ *with*

$$b = qa + r \qquad \text{and} \qquad \beta(r) < \beta(a) \quad \textbf{(division with remainder)} \ .$$

---

† This means that we pick exactly one member from each class of associate prime elements.

(7.13) Remark. Considering the proof of (7.11), we see that if $I \neq \{0\}$ is an ideal of a Euclidean domain R and if $x \in I \setminus \{0\}$ is an element for which β takes its minimal value on $I \setminus \{0\}$, then $I = Rx$. In particular, every Euclidean domain is a principal ideal domain and hence a unique factorization domain.

(7.14) Examples. (a) $R = \mathbb{Z}$ with the function $\beta(a) := |a|$ is a Euclidean domain.
(b) An arbitrary field becomes a Euclidean domain if we define

$$\beta(a) \;:=\; \begin{cases} 1, & \text{if } a \neq 0; \\ 0, & \text{if } a = 0. \end{cases}$$

(c) If K is a field then $R = K[x]$ is a Euclidean domain with the function $\beta(p) := 2^{\deg p}$; this follows easily from (4.10)(a). (Here $2^{-\infty} := 0$.)

NOTE that in all these examples the "degree-function" β could be chosen as a multiplicative function, i.e., a function such that $\beta(ab) = \beta(a)\beta(b)$. Suppose now that R is an integral domain with quotient field K and that $\beta : R \to \mathbb{N}_0$ is a multiplicative function with $\beta(a) = 0$ if and only if $a = 0$. Then clearly β satisfies conditions (1) and (2) in (7.12). Moreover, β can unambiguously be extended to a function $\beta : K \to \mathbb{Q}_0^+$ via

$$\beta\left(\frac{a}{b}\right) \;:=\; \frac{\beta(a)}{\beta(b)} \;.$$

Then condition (7.12)(3), stating that for all $a, b \in R \setminus \{0\}$ there is an element $q \in R$ with $\beta(b - qa) < \beta(a)$ or $\beta(\frac{b}{a} - q) < \beta(1) = 1$, can be rephrased as follows: For all $x \in K$, there is an element $q \in R$ with $\beta(x - q) < 1$. Thereby, we have obtained the following result.

(7.15) Proposition. *Let R be an integral domain with quotient field K. Suppose that there is a function $\beta : K \to \mathbb{Q}_0^+$ with the following properties:*
(1) $\beta(0) = 0$ and $\beta(a) \in \mathbb{N}$ for all $a \in R$;
(2) $\beta(xy) = \beta(x)\beta(y)$ for all $x, y \in K$;
(3) for all $x \in K$ there is an element $q \in R$ with $\beta(x - q) < 1$.
Then R is a Euclidean domain with degree function β. ∎

(7.16) Remark. The last condition in (7.15) can be formulated geometrically. Indeed, for each $x \in K$ we can consider the set

$$U_x \;:=\; \{y \in K \mid \beta(x - y) < 1\}$$

of all "points" with "distance" less than 1 from x. Let us call this set the "zone" around x. Then condition (3) means that each zone around a point x contains at least one element of R. This is tantamount to saying that $\bigcup_{q \in R} U_q$ covers all of K.

WE now want to show that some of the rings $\mathbb{Z} + \mathbb{Z}\sqrt{n}$ with $n \in \mathbb{Z} \setminus \{0, 1\}$ square-free are Euclidean with degree-function $\beta(x + y\sqrt{n}) := |x^2 - ny^2|$. Note that this just means $\beta(a) = |N(a)|$ where N is the norm function introduced in (3.6)(c). Since N is multiplicative, it is clear that β satisfies the conditions (1) and (2) in proposition (7.15). Thus what remains is to check the third condition. It will be helpful to prove a purely technical lemma first before we verify this third condition.

(7.17) Lemma. *Let $n \geq 2$ be a square-free natural number and let*

$$N := \begin{cases} n, & \text{if } n \not\equiv 1 \bmod 4; \\ n/4, & \text{if } n \equiv 1 \bmod 4. \end{cases}$$

Suppose that $r, s \in \mathbb{Q}$ are rational numbers with $0 \leq r, s \leq \frac{1}{2}$ such that $|(r - a)^2 - Ns^2| \geq 1$ for all $a \in \{0, 1, -1\}$. Then $N \geq 8$.

Proof. Let us write down what the indicated condition means explicitly in the cases $a = 0$, $a = 1$ and $a = -1$.

$a = 0:$ (1a) $r^2 \geq 1 + Ns^2$ or   (1b) $Ns^2 \geq 1 + r^2$

$a = 1:$ (2a) $(1 - r)^2 \geq 1 + Ns^2$ or   (2b) $Ns^2 \geq 1 + (1 - r)^2$

$a = -1:$ (3a) $(1 + r)^2 \geq 1 + Ns^2$ or   (3b) $Ns^2 \geq 1 + (1 + r)^2$

Now (1a) and (2a) do not hold because $0 \leq r \leq \frac{1}{2}$ and $N > 0$; hence (1b) and (2b) hold. Suppose (3a) holds. Then

$$1 + 2r + r^2 = (1 + r)^2 \geq 1 + Ns^2 \overset{(2b)}{\geq} 2 + (1 - r)^2 = 3 - 2r + r^2 \quad \text{so that} \quad r \geq \frac{1}{2}.$$

Since $r \leq 1/2$ by assumption, this implies $r = 1/2$. Now (2b) reads $Ns^2 \geq 5/4$, and (3a) says $1 + Ns^2 \leq 9/4$; these conditions together yield $Ns^2 = 5/4$. Write $s = p/q$ with coprime natural numbers p, q. Then $4Np^2 = 5q^2$ which implies $p^2 \mid 5$ and hence $p = 1$; this gives $4N = 5q^2$. If $n \not\equiv 1 \bmod 4$ this reads $4n = 5q^2$, and since n is square-free this implies $q = 2$ and hence $n = 5$ which is a contradiction. If $n \equiv 1 \bmod 4$ we have $4N = n$ and hence $n = 5q^2$ which is impossible since n is square-free. Thus the assumption that (3a) is true leads to a contradiction in any case; hence (3b) must be true. Taking into account that $0 \leq s \leq 1/2$ and hence $0 \leq s^2 \leq 1/4$, this shows that

$$\frac{N}{4} \geq Ns^2 \overset{(3b)}{\geq} 1 + (1 + r)^2 \geq 2, \quad \text{i.e., } N \geq 8.$$

∎

(7.18) Proposition. (a) *If $n \leq -3$ then $\mathbb{Z} + \mathbb{Z}\sqrt{n}$ is not Euclidean with the degree function $\beta = |N|$.*

(b) *If $n \in \{-2, -1, 2, 3, 6, 7\}$ then $\mathbb{Z} + \mathbb{Z}\sqrt{n}$ is a Euclidean domain with the degree function $\beta = |N|$.*

Proof. We only have to check the third condition in (7.15). In our case, this condition reads as follows.

$(\star)$ If $|r^2 - ns^2| \geq 1$ with $r, s \in \mathbb{Q}$ then there are integers $a, b \in \mathbb{Z}$
with $|(a - r)^2 - n(b - s)^2| < 1$.

(a) Suppose $n \leq -3$. Let $r = s = 1/2$; then $|r^2 - ns^2| = (1/4) - (n/4) = (|n| + 1)/4 \geq 1$. Now if $a, b \in \mathbb{Z}$ are arbitrary integers then $|(a - r)^2 - n(b - s)^2| = (a - \frac{1}{2})^2 + |n|(b - \frac{1}{2})^2 \geq (1/2)^2 + |n|(1/2)^2 = (|n| + 1)/4 \geq 1$; hence condition $(\star)$ is violated.

(b) Suppose first that $n \in \{-2, -1, 2, 3\}$. Given $r, s \in \mathbb{Q}$, we can choose $a, b \in \mathbb{Z}$ with $|a - r| \leq 1/2$ and $|b - s| \leq 1/2$; i.e., we choose $a, b \in \mathbb{Z}$ as close to $r, s \in \mathbb{Q}$ as possible. If $|n| \leq 2$, this gives

$$|(a - r)^2 - n(b - s)^2| \leq \underbrace{(a - r)^2}_{\leq 1/4} + \underbrace{|n|}_{\leq 1/2} \underbrace{(b - s)^2}_{\leq 1/4} \leq \frac{3}{4} < 1 .$$

For $n = 3$, we have to be a little bit more careful with our estimate. We obtain

$$-\frac{3}{4} \leq (a - r)^2 - \frac{3}{4} \leq (a - r)^2 - 3(b - s)^2 \leq \frac{1}{4} - 3(b - s)^2 \leq \frac{1}{4} .$$

Let us now treat the cases $n = 6$ and $n = 7$. Suppose $(\star)$ does *not* hold to get a contradiction. Then there are elements $r, s \in \mathbb{Q}$ such that

$$|r^2 - ns^2| \geq 1 \quad \text{and} \quad |(a - r)^2 - n(b - s)^2| \geq 1 \quad \text{for all } a, b \in \mathbb{Z} ;$$

obviously it is no loss of generality if we assume that $0 \leq r \leq 1/2$ and $0 \leq s \leq 1/2$. Considering the second inequality for the special choices $a \in \{0, \pm 1\}$ and $b = 0$ and using lemma (7.17), we obtain $n \geq 8$ which is the desired contradiction. ∎

WE will now show that $\mathbb{Z} + \mathbb{Z}\sqrt{n}$ cannot be Euclidean if $n \equiv 1$ modulo 4. To this end we present a criterion for finding the roots of a polynomial with coefficients in a unique factorization domain.

(7.19) Theorem. *Let R be a unique factorization domain with quotient field Q. If $f(x) = a_0 + a_1 x + \cdots + a_n x^n$ has a root $r/s \in Q$ where $r, s \in R$ are coprime then $r \mid a_0$ and $s \mid a_n$.*

Proof. Multiply the equation

$$0 = f\left(\frac{r}{s}\right) = a_0 + a_1 \frac{r}{s} + a_2 \frac{r^2}{s^2} + \cdots + a_{n-1} \frac{r^{n-1}}{s^{n-1}} + a_n \frac{r^n}{s^n}$$

by s^n to obtain

$$0 = \overbrace{a_0 s^n + a_1 r s^{n-1} + \underbrace{a_2 r^2 s^{n-2} + \cdots + a_{n-1} r^{n-1} s + a_n r^n}_{\text{divisible by } s}}^{\text{divisible by } r} .$$

We see that $r \mid a_0 s^n$ and $s \mid a_n r^n$. Since r and s are coprime this implies that $r \mid a_0$ and $s \mid a_n$. ∎

(7.20) Example. To find the rational roots of $f(x) = 12x^4 - 4x^3 + 6x^2 + x - 1 \in \mathbb{Z}[x]$, we have to try only the numbers $\frac{r}{s}$ with $r \mid -1$ and $s \mid 12$, i.e., the numbers $1, \frac{1}{2}, \frac{1}{3}, \frac{1}{4}, \frac{1}{6}$ and $\frac{1}{12}$ and their negatives. Only $\frac{1}{3}$ works so that we obtain the factorization $f(x) = (3x - 1)(4x^3 + 2x + 1)$.

WE now note that theorem (7.19) is not just a useful criterion for finding roots of polynomials, but also imposes a necessary condition for an integral domain to be a unique factorization domain.

(7.21) Proposition. *Let R be a unique factorization domain with quotient field Q. If $x \in Q$ satisfies a monic equation*

$$x^n + a_{n-1}x^{n-1} + \cdots + a_1 x + a_0 = 0$$

with coefficients $a_i \in R$ then $x \in R$.

Proof. Let $x = r/s$ with coprime elements $r, s \in R$. Applying (7.19) with $a_n = 1$ we see that $s \mid 1$, i.e., s is a unit. But then $x = rs^{-1} \in R$. ∎

THIS result is exploited in the following proposition.

(7.22) Proposition. *Let $n \in \mathbb{Z} \setminus \{0, 1\}$ be a square-free number with $n \equiv 1$ modulo 4. Then $\mathbb{Z} + \mathbb{Z}\sqrt{n}$ is not a unique factorization domain.*

Proof. We can write $n = 4k + 1$ with $k \in \mathbb{Z}$. The quotient field of $R := \mathbb{Z} + \mathbb{Z}\sqrt{n}$ is $Q = \mathbb{Q} + \mathbb{Q}\sqrt{n}$. The element $x := \frac{1+\sqrt{n}}{2}$ belongs to $Q \setminus R$ and satisfies

$$x^2 = \frac{1 + 2\sqrt{n} + n}{4} = \frac{4k + 2 + 2\sqrt{n}}{4} = k + \frac{1 + \sqrt{n}}{2} = k + x \, ,$$

i.e., $x^2 - x - k = 0$. This is a monic equation with coefficients in $\mathbb{Z} \subseteq R$. Hence (7.21) shows that R cannot be a unique factorization domain. ∎

HOWEVER, in the case $n \equiv 1$ modulo 4 the set $\mathbb{Z} + \mathbb{Z}\frac{1+\sqrt{n}}{2}$ is a ring, and it turns out that some of these rings *are* Euclidean with the degree function $\beta = |N|$. This is proved in the following proposition.

(7.23) Proposition. *Let $n \neq 1$ be square-free with $n \equiv 1$ modulo 4 and let*
$\mathcal{O}_n := \mathbb{Z} + \mathbb{Z}\frac{1+\sqrt{n}}{2}$.

    (a) If $n < 0$ then $\mathcal{O}_n$ is Euclidean with degree function $\beta = |N|$ if and only if $n = -3, -7$ or -11.

    (b) If $n \in \{5, 13, 17, 21, 29\}$ then $\mathcal{O}_n$ is Euclidean with degree function $\beta = |N|$.

Proof. $(\mathcal{O}_n, \beta)$ is Euclidean if and only if the third condition in (7.15) holds which now reads as follows:

$(\star)$
    If $|r^2 - ns^2| \geq 1$ with $r, s \in \mathbb{Q}$, then there are integers $a, b \in \mathbb{Z}$
    with $|(r - a - \frac{b}{2})^2 - n(s - \frac{b}{2})^2| < 1$.

    (a) Let us treat the case $n < 0$. Suppose first that $|n| \geq 15$ and choose $r := s := 1/4$. Then

$$\left| (r - a - \frac{b}{2})^2 - n(s - \frac{b}{2})^2 \right| = \underbrace{\left(\frac{1}{4} - (a + \frac{b}{2}) \right)^2}_{\geq (1/4)^2} + |n| \underbrace{\left(\frac{1}{4} - \frac{b}{2} \right)^2}_{\geq (1/4)^2} \geq \frac{1 + |n|}{16} \geq 1$$

so that $(\star)$ is violated. Since $n \equiv 1$ modulo 4 and $n < 0$ this leaves only the possibilities $n = -3, -7, -11$ for $\mathcal{O}_n$ to be Euclidean. Let us show that $\mathcal{O}_n$ is indeed Euclidean for these three values. Given r, s with $|r^2 - ns^2| \geq 1$ we pick an integer $b \in \mathbb{Z}$ with $|s - (b/2)| \leq (1/4)$ and then an integer $a \in \mathbb{Z}$ with $|a - (-(b/2) + r)| \leq 1/2$. Then

$$\left| (r - a - \frac{b}{2})^2 - n(s - \frac{b}{2})^2 \right| = \underbrace{(a + \frac{b}{2} - r)^2}_{\leq (1/4)^2} + \underbrace{|n|}_{\leq 11} \underbrace{(s - \frac{b}{2})^2}_{\leq (1/16)} \leq \frac{1}{4} + \frac{11}{16} = \frac{15}{16} < 1 .$$

    (b) Now suppose $n > 0$. We will show that if $\mathcal{O}_n$ is *not* Euclidean then $n \geq 32$; this will clearly imply the claim. Suppose $\mathcal{O}_n$ is not Euclidean. Then condition $(\star)$ does not hold; hence there are rational numbers $r, s \in \mathbb{Q}$ such that $| (r - a - (b/2))^2 - n(s - (b/2))^2 | \geq 1$ for all $a, b \in \mathbb{Z}$. Substituting $N := n/4$ and $\sigma := 2s$, this reads

$(\star\star)$
$$| (r - a - \frac{b}{2})^2 - N(\sigma - b)^2 | \geq 1 \quad \text{for all } a, b \in \mathbb{Z} .$$

Clearly, we can assume $0 \leq r \leq (1/2)$ and $0 \leq \sigma \leq (1/2)$.[†] Evaluating $(\star\star)$ with $a \in \{0, \pm 1\}$ and $b = 0$ and using lemma (7.17), we obtain $N \geq 8$, i.e., $n \geq 32$. ∎

---

    † Let X be the set of all pairs $(r, \sigma) \in \mathbb{Q}^2$ satisfying $(\star\star)$. If $(r, \sigma) \in X$ then $(r + m, \sigma) \in X$ and $(-r + m, -\sigma) \in X$ for all $m \in \mathbb{Z}$; hence we may assume $0 \leq r \leq (1/2)$. Also, if $(r, \sigma) \in X$ then $(r, \pm \sigma + 2m) \in X$ for all $m \in \mathbb{Z}$; hence we may assume $0 \leq \sigma \leq 1$. Now if $0 \leq \sigma \leq (1/2)$ we are done; otherwise we observe that if $(r, \sigma) \in X$ then $(r', \sigma') := ((1/2) - r, 1 - \sigma) \in X$. Since $0 \leq r \leq (1/2)$ and $(1/2) \leq \sigma \leq \frac{1}{2}$ we have $0 \leq r' \leq (1/2)$ and $0 \leq \sigma' \leq (1/2)$, as desired.

(7.24) Remark. Suppose that $n \in \mathbb{Z} \setminus \{0,1\}$ is a square-free number. Letting $\mathcal{O}_n = \mathbb{Z} + \mathbb{Z}(\frac{1+\sqrt{n}}{2})$ if $n \equiv 1$ modulo 4 and $\mathcal{O}_n = \mathbb{Z} + \mathbb{Z}\sqrt{n}$ if $n \equiv 2,3$ modulo 4, one can show that $\mathcal{O}_n$ is norm-Euclidean if and only if

$$n \in \{-11, -7, -3, -2, -1, 1, 2, 3, 5, 6, 7, 11, 13, 17, 19, 21, 29, 33, 37, 41, 57, 73\} .$$

For references, see *An Introduction to the Theory of Numbers* by Hardy and Wright, Oxford University Press 1954.

WE have collected several examples of Euclidean rings. Let us now introduce an important algorithm which allows us to find the greatest common divisor of two elements in a very effective way.

(7.25) Euclidean algorithm. *Given two elements* a_1, a_2 *in a Euclidean ring such that* $0 < \beta(a_2) \leq \beta(a_1)$, *we can successively perform division with remainder:*

$$
\begin{array}{llll}
(1) & a_1 = q_1 a_2 + a_3 & \text{where} & \beta(a_3) < \beta(a_2) ; \\
(2) & a_2 = q_2 a_3 + a_4 & \text{where} & \beta(a_4) < \beta(a_3) ; \\
(3) & a_3 = q_3 a_4 + a_5 & \text{where} & \beta(a_5) < \beta(a_4) ; \\
& \qquad \ldots\ldots
\end{array}
$$

Since the sequence $\beta(a_1) > \beta(a_2) > \beta(a_3) > \ldots$ *is a strictly decreasing sequence of natural numbers, it must be finite; hence the procedure has to terminate after some time, say after the n-th step:*

$$
\begin{array}{llll}
(n-1) & a_{n-1} = q_{n-1} a_n + a_{n+1} & \text{where} & \beta(a_{n+1}) < \beta(a_n) ; \\
(n) & a_n = q_n a_{n+1} .
\end{array}
$$

Then the following statements hold.

(a) a_{n+1} *is a greatest common divisor of* a_1 *and* a_2.

(b) *We can express* a_{n+1} *as a linear combination of* a_1 *and* a_2; *i.e., there are elements* $x, y \in R$ *such that*

$$xa_1 + ya_2 = a_{n+1} \quad \textbf{(linear representation of the gcd) .}$$

Proof. (a) a_{n+1} divides a_n by equation (n); but then also a_{n-1} by equation $(n-1)$; ...; but then also a_3 by equation (3); but then also a_2 by equation (2); but then also a_1 by equation (1). This shows that a_{n+1} is a common divisor of all the emerging elements a_k, in particular of a_1 and a_2.

Now let d be any common divisor of a_1 and a_2; we want to show that d divides a_{n+1}. Now d divides a_1 and a_2; but then also a_3 by equation (1); but then also a_4 by equation (2); but then also a_5 by equation (3); ...; but then also a_{n+1} by equation $(n-1)$.

(b) Using equation (1), we can write a_3 as a linear combination of a_1 and a_2, namely

$$(1') \qquad\qquad a_3 = a_1 - q_1 a_2 .$$

Using (2), we write a_4 as a linear combination of a_2 and a_3, but then with (1') also as a linear combination of a_1 and a_2, as follows:

$$(2') \qquad a_4 \; = \; a_2 - q_2 a_3 \; = \; a_2 - q_2(a_1 - q_1 a_2) \; = \; -q_2 a_1 + (1 + q_1 q_2)a_2 \; .$$

Continuing this process, we clearly can write a_{n+1} as a linear combination of a_1 and a_2. ∎

(7.26) Examples. (a) The Euclidean algorithm in $\mathbb{Z}$. Let us find the gcd of 243 and 51 and express it as a linear combination of these numbers. We obtain the following iterated divisions with remainder:

$$243 \; = \; 4 \cdot 51 + 39, \quad 51 \; = \; 1 \cdot 39 + 12, \quad 39 \; = \; 3 \cdot 12 + 3, \quad 12 \; = \; 4 \cdot 3 \; .$$

Hence $\gcd(243, 51) = 3$, and reading backwards, we obtain

$$\begin{aligned} 3 \; &= \; 39 - 3 \cdot 12 \; = \; 39 - 3 \cdot (51 - 1 \cdot 39) \; = \; 4 \cdot 39 - 3 \cdot 51 \\ &= \; 4 \cdot (243 - 4 \cdot 51) - 3 \cdot 51 \; = \; 4 \cdot 243 - 19 \cdot 51 \; . \end{aligned}$$

(b) The Euclidean algorithm in $\mathbb{Q}[x]$. Let us find the gcd of $f(x) = x^4 + 3x^2 + 2$ and $g(x) = x^3 + 5x^2 + x + 5$. We obtain the divisions

$$x^4 + 3x^2 + 2 \; = \; (x - 5)(x^3 + 5x^2 + x + 5) + (27x^2 + 27) \; ,$$

$$x^3 + 5x^2 + x + 5 \; = \; \frac{1}{27}(x + 5)(27x^2 + 27) \; .$$

Hence $x^2 + 1$ (or $27(x^2 + 1)$) is a gcd of f and g, and we have $27(x^2 + 1) = f(x) - (x - 5)g(x)$.

(c) The Euclidean algorithm in $\mathbb{Z}(i)$. Recall from the proof of (7.17)(b) that we obtain a number $q \in \mathbb{Z}(i)$ with $\beta(b - qa) < \beta(a)$ if we choose $q = m + ni$ such that the coefficients $m, n \in \mathbb{Z}$ are as close to the coefficients of $(b/a) = r + si \in \mathbb{Q}(i)$ as possible. Let us find a gcd of $a_1 := 18 - i$ and $a_2 := 11 + 7i$. We obtain the following results.

$$\frac{a_1}{a_2} \; = \; \frac{191}{170} - \frac{137}{170}i \; ; \qquad \text{hence let } q_1 := 1 - i \text{ and } a_3 := a_1 - q_1 a_2 \; = \; 3i \; .$$

$$\frac{a_2}{a_3} \; = \; \frac{7}{3} - \frac{11}{3}i \; ; \qquad \text{hence let } q_2 := 2 - 4i \text{ and } a_4 := a_2 - q_2 a_3 \; = \; -1 + i \; .$$

$$\frac{a_3}{a_4} \; = \; \frac{3}{2} - \frac{3}{2}i \; ; \qquad \text{hence let } q_3 := 2 - i \text{ and } a_5 := a_3 - q_3 a_4 \; = \; 1 \; .$$

(Instead of $q_3 = 2 - i$ we could choose $q_3 = 2 - 2i$, $1 - 2i$ or $1 - i$ as well.)

$$\frac{a_4}{a_5} \; = \; -1 + i \; =: \; q_4 \; ; \qquad \text{then} \quad a_4 \; = \; q_4 a_5 \; .$$

We see that $\gcd(a_1, a_2) = 1$ so that the two elements are coprime. Reading backwards, we obtain

$$\begin{aligned} 1 \; &= \; a_3 - q_3 a_4 \; = \; a_3 - q_3(a_2 - q_2 a_3) \; = \; (1 + q_2)a_3 - q_3 a_2 \\ &= \; (1 + q_2)(a_1 - q_1 a_2) - q_3 a_2 \; = \; (1 + q_2)a_1 - (q_1 + q_1 q_2 - q_3)a_2 \; . \end{aligned}$$

Plugging in $q_1 = 1 - i$, $q_2 = 2 - 4i$ and $q_3 = 2 - i$, this gives $1 = (3 - 4i)a_1 + (3 + 6i)a_2$.

(7.27) Application: Inversion in $\mathbb{Z}_n$. We know that $\mathbb{Z}_n^\times = \{[a] \mid \gcd(a, n) = 1\}$, but we do not yet know a method of finding $[a]^{-1}$ if $[a] \in \mathbb{Z}_n^\times$. Now if a and n are relatively prime, the Euclidean algorithm allows us to find $x, y \in \mathbb{Z}$ with $xa + yn = 1$. But then $[x] \cdot [a] = [1]$ in $\mathbb{Z}_n$ which shows that $[x]$ is the desired inverse of $[a]$ in $\mathbb{Z}_n$.

For example, the element $[55]$ is invertible in $\mathbb{Z}_{63}$ because 55 and 63 are coprime in $\mathbb{Z}$. To find $[55]^{-1}$ we apply the Euclidean algorithm:

$$63 = 1 \cdot 55 + 8, \qquad 55 = 6 \cdot 8 + 7, \qquad 8 = 1 \cdot 7 + 1 \,.$$

Therefore, $1 = 8 - 7 = 8 - (55 - 6 \cdot 8) = 7 \cdot 8 - 55 = 7 \cdot (63 - 55) - 55 = 7 \cdot 63 - 8 \cdot 55$. Modulo 63 this reads $[1] = [-8] \cdot [55]$ so that $[55]^{-1} = [-8] = [55]$. Hence $[55]$ is its own inverse in $\mathbb{Z}_{63}$.

THE reader might well have wondered why we took such pains to establish the unique factorization property of rather "fancy" rings as in (7.18) and (7.23). The reason is that these rings have important applications in number theory some of which will be presented in section 9 below. We close this section with a criterion that can sometimes be used to show that a given integral domain is *not* Euclidean.

(7.28) Proposition. *Let R be a Euclidean domain. Then there is an element $u \in R^\heartsuit$ with the property that for any $x \in R$ there is an element $r \in R^\times \cup \{0\}$ with $u \mid x - r$.*

Proof. Let β be the degree function of the Euclidean domain R. Choose $u \in R^\heartsuit$ with the property that $\beta(u) \leq \beta(v)$ for all $v \in R^\heartsuit$. Now let $x \in R$. Then there are elements $q, r \in R$ with $x = qu + r$ with $\beta(r) < \beta(u)$. The choice of u then forces that $r \notin R^\heartsuit$, i.e., $r \in R^\times \cup \{0\}$. ∎

AS an application we now show that none of the rings $\mathcal{O}_{-n}$ with $n > 11$ is Euclidean. (Up to this point we only know that the norm function is not a Euclidean degree function for these rings.)

(7.29) Proposition. *Let $n \geq 13$ be a square-free natural number. Then $\mathcal{O}_{-n}$ is not a Euclidean domain.*

Proof. Suppose that $R := \mathcal{O}_{-n}$ is Euclidean; then there is an element $u \in R^\heartsuit$ as in (7.28). Considering norms it is easy to see that $R^\times = \{\pm 1\}$; hence if x is an arbitrary element of R then one of the three conditions $u \mid x$, $u \mid x - 1$ or $u \mid x + 1$ must hold. Taking $x := 2$ we see that either $u \mid 2$ or $u \mid 3$, say $uv = 2$ or $uv = 3$. If we had $u \in R \setminus \mathbb{Z}$ then also $v \in R \setminus \mathbb{Z}$ and hence $N(u) > 3$ and $N(v) > 3$. (This is where the condition $n \geq 13$ enters.) But then we would have $N(uv) > 9$ which contradicts the fact that $uv = 2$ or 3. Hence $u \in \mathbb{Z}$ and consequently $u = \pm 2$ or $u = \pm 3$. Now let $x := 1 + \sqrt{-n}$ if $n \not\equiv 1$ modulo 4 and $x := (1 + \sqrt{-n})/2$ if $n \equiv 1$ modulo 4. Then u divides none of the three numbers x, $x + 1$ and $x - 1$ in R, contradicting the choice of u. This is the desired contradiction. ∎

Exercises

Problem 1. (a) Suppose $a = rb \neq 0$ in an integral domain R. Show that r is a unit if and only if $a \sim b$.

(b) Suppose $a \mid bc$ in a unique factorization domain R. Show that if a and c are coprime then $a \mid b$.

(c) Suppose a and b are coprime elements in a unique factorization domain R. Show that if $a \mid c$ and $b \mid c$ then $ab \mid c$. Also, show that if $ax = by$ then $a \mid y$ and $b \mid x$.

Problem 2. (a) Show that $\gcd(a, b, c) = \gcd(\gcd(a, b), c)$. Hence the greatest common divisor of a finite number of elements can be found by successively finding the greatest common divisor of two elements at a time.

(b) Suppose that R is an integral domain in which any two elements a_1, a_2 possess a gcd. Show that any n elements $a_1, \ldots, a_n \in R$ possess a gcd.

(c) Suppose R is a domain in which any two elements a_1, a_2 possess a gcd which can be written as an R-linear combination $r_1 a_1 + r_2 a_2$. Show that any n elements $a_1, \ldots, a_n \in R$ possess a gcd which can be written as an R-linear combination $r_1 a_1 + \cdots + r_n a_n$.

Problem 3. (a) Let $R := \mathbb{Z} + \mathbb{Z}\sqrt{-5}$. Show that the elements 3 and $2 + \sqrt{-5}$ possess a gcd d in R, but that d cannot be written as an R-linear combination of 3 and $2 + \sqrt{-5}$.

(b) Find all units in $\mathbb{Z} + \mathbb{Z}\sqrt{5}$.

Problem 4. Let R be a subring of S where R is a principal ideal domain and where S is an integral domain. Show that if d is a greatest common divisor of a and b in R then d is also a greatest common divisor of a and b in S.

Problem 5. (a) Let R be an integral domain and let $p \in R$ be a prime element. Show that

$$\{a_0 + a_1 x + a_2 x^2 + \cdots + a_n x^n \mid a_0, a_1, \ldots, a_n \text{ are divisible by } p\} \quad \text{and}$$
$$\{a_0 + a_1 x + a_2 x^2 + \cdots + a_n x^n \mid a_0 \text{ is divisible by } p\}$$

are prime ideals of $R[x]$.

(b) Describe all prime ideals of $R[x]$ where $R = \mathbb{Z}, \mathbb{Q}, \mathbb{R}$ and $\mathbb{C}$.

Problem 6. (a) Let $\alpha, \beta \in \mathbb{C}$ be complex numbers and let $p, q \in \mathbb{Z}[x]$ be monic polynomials such that $p(\alpha) = q(\beta) = 0$. Let $x_1, \ldots, x_n \in \mathbb{C}$ be the roots of pq (counted with multiplicity) and let

$$f(x) := \prod_{i<j}(x - (x_i + x_j)), \quad g(x) := \prod_{i<j}(x - (x_i x_j)), \quad h(x) := \prod_{i<j}(x + x_j).$$

Show that f, g, h belong to $\mathbb{Z}[x]$.

Hint. Show that the coefficients of f, g and h are symmetric polynomials in

$x_1, \ldots, x_n$, hence can be expressed in terms of the elementary symmetric polynomials in $x_1, \ldots, x_n$.

(b) Let R be the set of all complex numbers $z \in \mathbb{C}$ which satisfy a monic equation with coefficients in $\mathbb{Z}$. Show that R is a ring whose only units are ± 1.

Problem 7. Verify that each of the following equations is an example for non-unique factorization into irreducible elements.

(a) $9 = 3 \cdot 3 = (2 + \sqrt{-5})(2 - \sqrt{-5})$ in $\mathbb{Z} + \mathbb{Z}\sqrt{-5}$

(b) $4 = 2 \cdot 2 = (3 + \sqrt{5})(3 - \sqrt{5})$ in $\mathbb{Z} + \mathbb{Z}\sqrt{5}$

(c) $4 = 2 \cdot 2 = (1 + \sqrt{-3})(1 - \sqrt{-3})$ in $\mathbb{Z} + \mathbb{Z}\sqrt{-3}$

(d) $6 = 2 \cdot 3 = (4 + \sqrt{10})(4 - \sqrt{10})$ in $\mathbb{Z} + \mathbb{Z}\sqrt{10}$

(e) $6 = 2 \cdot 3 = (\sqrt{-6})(-\sqrt{-6})$ in $\mathbb{Z} + \mathbb{Z}\sqrt{-6}$

(f) $6 = 2 \cdot 3 = \sqrt{6} \cdot \sqrt{6}$ in $\mathbb{Z} + \mathbb{Z}\sqrt{6}$

(g) $55 = 5 \cdot 11 = (7 + \sqrt{-6})(7 - \sqrt{-6})$ in $\mathbb{Z} + \mathbb{Z}\sqrt{-6}$

Problem 8. Show that

$$21 = 3 \cdot 7 = (1 + 2\sqrt{-5})(1 - 2\sqrt{-5}) = (4 + \sqrt{-5})(4 - \sqrt{-5})$$

are three pairwise non-equivalent factorizations of 21 in $\mathbb{Z} + \mathbb{Z}\sqrt{-5}$.

Problem 9. (a) Show that $1 + \sqrt{3}$ and $-1 + \sqrt{3}$ are associates in $\mathbb{Z} + \mathbb{Z}\sqrt{3}$.

(b) Show that $1 + i$ is a prime in $\mathbb{Z} + i\mathbb{Z}$.

(c) Show that $11 + 2\sqrt{6}$ is a prime in $\mathbb{Z} + \mathbb{Z}\sqrt{6}$.

(d) Show that 3 is a prime in $\mathbb{Z} + i\mathbb{Z}$ but not in $\mathbb{Z} + \mathbb{Z}\sqrt{6}$.

Problem 10. Let $n \in \mathbb{Z} \setminus \{0, 1\}$ be a square-free number and let $\alpha \in \mathbb{Z} + \mathbb{Z}\sqrt{n}$. Show that if $N(\alpha)$ is a prime element of $\mathbb{Z}$, then α is irreducible in $\mathbb{Z} + \mathbb{Z}\sqrt{n}$. The same statement holds for $\mathbb{Z} + \mathbb{Z}\frac{1 + \sqrt{n}}{2}$ if $n \equiv 1$ modulo 4.

Problem 11. In which of the rings $\mathbb{Z}[x]$, $\mathbb{Q}[x]$, $\mathbb{R}[x]$ and $\mathbb{C}[x]$ is the polynomial $f(x) = x^3 - 5x^2 + 25x + 10$ reducible?

Problem 12. (a) Let K be a field in which $2 \neq 0$. Show that for the polynomial $q(x, y) = ax^2 + bxy + cy^2 \in K[x, y]$ the following conditions are equivalent:

(1) q is reducible;

(2) $D := b^2 - 4ac$ is a square in K;

(3) $D = 0$, or else q has a root $(\xi, \eta) \neq (0, 0)$ in K^2.

(b) Find all solutions of the equation $x^2 + xy + 4y^2 = 0$ over $\mathbb{Z}_5$ and all solutions of the equation $x^2 + 2xy + 3y^2 = 0$ over $\mathbb{Z}_{11}$.

Problem 13. Consider the polynomial $f(x_1, \ldots, x_n) = x_1^3 + \cdots + x_n^3$ over a field K with char $K \neq 3$. Show that f is irreducible in $K[x_1, \ldots, x_n]$ if and only if $n \geq 3$.

Problem 14. (a) Let K be an arbitrary field and let $\alpha \in K$. Show that if $\alpha \neq 0$ then $f_\alpha(x,y) = x^2 + y^2 + \alpha$ is irreducible in $K[x,y]$. Show that if $\alpha = 0$ then f_α may or may not be irreducible, depending on the field K.

(b) Show that $f(x,y,z) = xy + yz + xz$ is irreducible over $\mathbb{Z}_2$, whereas $g(x,y,z) = x^2 + y^2 + z^2$ is reducible.

Problem 15. Let $f(x) = \sum_{k=0}^{\infty} a_k x^k$ be an element of $R[[x]]$ where R is a commutative ring with identity. Show that if a_0 is irreducible in R, then f is irreducible in $R[[x]]$.

Problem 16. (a) Find all divisors of 6 in $\mathbb{Z}[\sqrt{-6}]$ and of 21 in $\mathbb{Z}[\sqrt{-5}]$.

(b) Find all units in $\mathbb{Z} + i\mathbb{Z}$. Which elements of $\mathbb{Z} + i\mathbb{Z}$ have the property that they are associates of their complex conjugates?

Problem 17. Show that $R = \{m/n \in \mathbb{Q} \mid m,n \in \mathbb{Z}, n \text{ is odd }\}$ is an integral domain. Find all units and all prime elements in R.

Problem 18. Decompose the following elements into prime factors!
(a) $2,3,5$ in $\mathbb{Z} + i\mathbb{Z}$
(b) $3,5,7$ in $\mathbb{Z} + \mathbb{Z}\sqrt{2}$
(c) $x^4 - 3$ in $\mathbb{Z}_5[x]$
(d) $x^5 - 11x + 6$ in $\mathbb{Z}_2[x]$, $\mathbb{Z}_3[x]$ and $\mathbb{Z}_5[x]$
(e) $x^4 + 8x^3 + x^2 + 2x + 5$ in $\mathbb{Q}[x]$
(f) $x^4 + x^2 + 1$ in $\mathbb{Q}[x]$
(g) $(-1 + 3i)x^2 - (3 + i)x + (4 + 2i)$ in $(\mathbb{Z} + i\mathbb{Z})[x]$

Problem 19. Let R be the ring of all formal expressions $a_1 x^{q_1} + \cdots + a_n x^{q_n}$ with $q_i, q_i \in \mathbb{Q}$ and $q_i \geq 0$ as in example (7.6)(c).

(a) Show that no element of the form x^α with $\alpha > 0$ is prime.

(b) Show that $I := \langle x^{1/2} \rangle$ is a primary ideal but that $x^{1/2}$ is not an associate of a prime power.

Hint. Show that $I = \{\sum_i a_i x^{q_i} \mid q_i \geq \frac{1}{2}\}$ and $\sqrt{I} = \{\sum_i a_i x^{q_i} \mid q_i > 0\}$.

Problem 20. Let K be an arbitrary field.
(a) Show that $p(x) = x + x^2$ is irreducible in $K[[x]]$, but reducible in $K[x]$.
(b) Find elements $f, g \in K[x]$ which are coprime in $K[[x]]$, but not in $K[x]$.

Problem 21. Let $I, J \trianglelefteq \mathbb{Z}$ be ideals of $\mathbb{Z}$ with $\langle 21 \rangle \subsetneq I \subsetneq J$. Show that $J = \mathbb{Z}$.

Problem 22. Let R be an integral domain and let $P \subseteq R \setminus \{0\}$. Suppose that each element $a \neq 0$ in R has a unique representation $a = u \prod_{p \in P} p^{k(p)}$ with $u \in R^\times$ and $k(p) \geq 0$ for all p with $k(p) > 0$ only for a finite number of elements $p \in P$. Show that R is a unique factorization domain and that each prime element of R is an associate of exactly one element of P.

Give an example for such a set P in the cases $R = \mathbb{Z}$ and $R = K[x]$ where K is a field.

Problem 23. Let R be the ring of all entire functions $f : \mathbb{C} \to \mathbb{C}$.
(a) Show that R is not a factorization domain.
Hint. Prove that a prime element of R has only a finite number of zeros in $\mathbb{C}$. Conclude that $f(z) = \sin z \in R$ cannot be written as a (finite) product of primes.
(b) Show that every irreducible element of R is prime.
Hint. If f is irreducible, then $f(a) = 0$ for some $a \in \mathbb{C}$. Now write $f(z) = (z - a)g(z)$.

Problem 24. Let R be a factorization domain. Show that R is a unique factorization domain if and only the intersection of any two principal ideals of R is again a principal ideal.

Problem 25. Suppose R is a unique factorization domain such that $\langle p \rangle$ is a maximal ideal whenever p is a prime element. Show that R is a principal ideal domain.

Problem 26. Let R be a factorization domain such that every ideal generated by two elements is principal. Show that R is a principal ideal domain.
Hint. Show first that R is a unique factorization domain by establishing that every irreducible element $p \in R$ is prime. (If $p \mid ab$ and $p \nmid a$, consider the ideal $\langle p, a \rangle$.) Then pick a complete system $(p_i)_{i \in I}$ of pairwise non-associate prime elements. If $J \neq \{0\}$ is an ideal, choose an element $x = u \prod_{i \in I} p_i^{k_i}$ in J such that $\sum_{i \in I} k_i$ is minimal. To prove that $J = \langle x \rangle$, suppose that $\langle x \rangle \subsetneq J$ to obtain a contradiction.

Problem 27. Suppose $f(x) = a_n x^n + \cdots + a_1 x + a_0 \in \mathbb{Z}[x]$. Show that $\alpha \in \mathbb{Q}$ is a root of f then α can be written in the form $\alpha = y/(a_n^n)$ where $y \in \mathbb{Z}$ divides $a_0 a_n^{n-1}$.
Hint. Substitute $y := a_n x$.

Problem 28. Let $R = K[[x]]$ where K is a field.
(a) Show that if $\text{subdeg} f = n$ if and only if $f(x) \sim x^n$ and conclude that $f \mid g$ in R if and only if $\text{subdeg} f \leq \text{subdeg} g$.
(b) Show that the function $\beta(f) := \text{subdeg} f$ satisfies the condition in theorem (7.11), but not condition (3) in (7.12).

Problem 29. Let R be a Euclidean domain. Show that $R^\times = \{x \in R \setminus \{0\} \mid \beta(x) = \beta(1)\}$.

Problem 30. Show that $R = \mathcal{O}_{-19} = \mathbb{Z} + \mathbb{Z}(\frac{1+\sqrt{-19}}{2})$ is a principal ideal domain but not a Euclidean domain.

Hint. Show that the function $\beta = |N|$ satisfies the conditions of theorem (7.11).

Problem 31. Show that $\mathcal{O}_{23} = \mathbb{Z} + \mathbb{Z}\sqrt{23}$ is not Euclidean with $\beta = |N|$.

Hint. Choose $r = 0$ and $s = 7/23$ to show that condition $(\star)$ in the proof of (7.17) is violated.

Problem 32. (a) Are the equations

$$6 = 2 \cdot 3 = \frac{1 + \sqrt{-23}}{2} \frac{1 - \sqrt{-23}}{2} \quad \text{and} \quad 27 = 3 \cdot 3 \cdot 3 = (2 + \sqrt{-23})(2 - \sqrt{-23})$$

examples for non-unique factorization in $\mathcal{O}_{-23}$?

(b) Find all prime factorizations of 8 in $\mathcal{O}_{-23}$.

Problem 33. (a) Show that $4 = 2 \cdot 2 = (1 + \sqrt{-3})(1 - \sqrt{-3})$ is an example for non-unique factorization on $\mathbb{Z} + \mathbb{Z}\sqrt{3}$ but not in $\mathcal{O}_{-3} := \mathbb{Z} + \mathbb{Z}(\frac{1+\sqrt{-3}}{2})$.

(b) Why are the equations

$$13 = \frac{7 + \sqrt{-3}}{2} \frac{7 - \sqrt{-3}}{2} = (1 + 2\sqrt{-3})(1 - 2\sqrt{-3})$$

not in conflict with the unique factorization property in $\mathcal{O}_{-3}$?

(c) Find all units in $\mathcal{O}_{-3}$.

Problem 34. Let R be a commutative ring with a mapping $\beta : R \to \mathbb{N}_0$ such that the following conditions are satisfied.

(1) $\beta(x) = 0$ if and only if $x = 0$;

(2) $\beta(xy) = \beta(x)\beta(y)$ for all $x, y \in R$;

(3) if $0 < \beta(a) \leq \beta(b)$ there is an element $q \in R$ with $\beta(b - qa) < \beta(a)$.

Show that R has an identity element and is free of zero-divisors, hence is a Euclidean domain. Moreover, show that the group of units is $R^{\times} = \{r \in R \mid \beta(r) = 1\}$.

Problem 35. (a) Find a greatest common divisor of $78, 130$ and 195 and represent it as a $\mathbb{Z}$-linear combination of these three numbers.

(b) Find a greatest common divisor of $a = 1,148,408,683$ and $b = 326,987$ and represent it as a linear combination of a and b.

Problem 36. Perform the following polynomial divisions with remainder.

(a) $(x^6 + 13x^5 + 5x^4 + 52x^3 + 3x^2 + 3x - 13) : (x^2 + 4)$ in $\mathbb{Z}[x]$

(b) $x^3 : (x^2 + 4x + 3)$ in $\mathbb{Z}_7[x]$

(c) $(x^6 + 3x^5 + 2x^3 + 3x^2 + 3x) : (x^2 - 1)$ in $\mathbb{Z}_5[x]$

Problem 37. In each of the following cases, determine a greatest common divisor of the given elements of R and express it as a linear combination of these elements.
  (a) $R = \mathbb{Z}$, $\alpha = 816$, $\beta = 294$
  (b) $R = \mathbb{Z}$, $\alpha = 3587$, $\beta = 1819$
  (c) $R = \mathbb{Z} + \mathbb{Z}i$, $\alpha = 4 + 9i$, $\beta = 2 + 7i$
  (d) $R = \mathbb{Z} + \mathbb{Z}i$, $\alpha = 3 + 4i$, $\beta = 4 - 3i$
  (e) $R = \mathbb{Z} + \mathbb{Z}\sqrt{2}$, $\alpha = -5 + 2\sqrt{-2}$, $\beta = 1 + 5\sqrt{-2}$
  (f) $R = \mathbb{Z}_5[x]$, $f(x) = x^5 + 4x$, $g(x) = x^6 + x^5 + 3x^4 + x^3 + x^2 + x + 2$
  (g) $R = \mathbb{Z}_2[x]$, $f(x) = x^5 + x^4 + 1$, $g(x) = x^5 + x + 1$
  (h) $R = \mathbb{Z}_2[x]$, $f(x) = x^4 + x^3 + x^2 + 1$, $g(x) = x^3 + 1$

Problem 38. In each of the following cases, express the greatest common divisors of the given polynomials $f, g \in \mathbb{Q}[x]$ as a linear combination of f and g.
  (a) $f(x) = x^2 + x + 1$, $g(x) = x - 1$
  (b) $f(x) = x^5 + 2x^4 + x^3 + 2x^2 + x + 2$, $g(x) = x^4 - 4$
  (c) $f(x) = x^6 - 1$, $g(x) = x^4 + 2x^3 + 3x^2 + 2x + 1$
  (d) $f(x) = 15x^4 + 2x^3 + 15x^2 + 4$, $g(x) = 12x^3 - 11x^2 + 3x - 10$
  (e) $f(x) = 5x^3 + 9x + 1$, $g(x) = x^2 + 2$
  (f) $f(x) = x^3 - 6x^2 + x + 4$, $g(x) = x^2 + 1$
  (g) $f(x) = x^6 + x^3 + x + 1$, $g(x) = x^5 - 6x + 1$
  (h) $f(x) = x^5 + 2x^3 + x^2 + x + 1$, $g(x) = x^4 - 1$
  (i) $f(x) = x^5 + x^3 - x^2 - 1$, $g(x) = x^4 + x^3 + x^2 + x$
  (j) $f(x) = 3x^3 + 2x^2 - 12x + 6$, $g(x) = x^2 + 2x - 2$
  (k) $f(x) = x^3 + 2x^2 + 2x + 1$, $g(x) = x^2 + x + 2$
  (l) $f(x) = x^4 + x^3 - x^2 + x + 2$, $g(x) = x^3 + 2x^2 + 2x + 1$

Problem 39. (a) Find the inverse of 20 in the group of units of $\mathbb{Z}_{107}$.
  (b) Find the inverse of 365 in $\mathbb{Z}_{1876}^{\times}$.

Problem 40. Consider the elements $f(x) = x^4 + 1$, $g(x) = x^2 - i\sqrt{2}x - 1$ and $h(x) = x^2 - 1$ of the polynomial ring $\mathbb{C}[x]$. Find a greatest common divisor and a least common multiple of f, g and h and express the greatest common divisor as a linear combination of these three elements.

Problem 41. Let K be a field.
  (a) Show that if $f, g \in K[x_1, \ldots, x_n]$ are relatively prime then there are polynomials $p, q \in K[x_1, \ldots, x_n]$ and $d \in K[x_1, \ldots, x_{n-1}]$ such that $pf + qg = d \neq 0$.
  (b) Show that two coprime polynomials $f \in K[x, y]$ and $g \in K[x, y]$ cannot have infinitely many common roots in K^2.

Problem 42. Find a single polynomial $f \in \mathbb{Q}[x]$ which generates the ideal $\langle x^4 + 4x^3 + x^2 - 2, \, 2x^4 - 2x^3 - x^2 + 6x - 7, \, x^3 + 3x^2 - x - x \rangle$.

Problem 43. Solve the following system of equations with coefficients in $\mathbb{Z}_{19}$.

$$\begin{pmatrix} 2 & 7 \\ 9 & 4 \end{pmatrix} \begin{pmatrix} x \\ y \end{pmatrix} = \begin{pmatrix} 4 \\ 1 \end{pmatrix}$$

Problem 44. Which of the following equations are solvable in $\mathbb{Z}$? Find a solution, if possible!

$$81x + 3y = 5 , \quad 175x - 13y = 1 , \quad 14x + 217y = 35$$

Problem 45. (a) Show that the integer solutions of the equation $803x + 154y = 0$ are exactly the pairs $(14t, -73t)$ with $t \in \mathbb{Z}$.

(b) Show that the integer solutions of the equation $803x + 154y = 11$ are exactly the pairs $(5 + 14t, -26 - 73t)$ with $t \in \mathbb{Z}$.

Problem 46. Let R be a Euclidean ring.

(a) Show that every matrix in $R^{m \times n}$ can be transformed into a matrix

$(\star) \qquad C = \begin{pmatrix} c_1 & & & \\ & \ddots & & 0 \\ & & c_r & \\ \hline & 0 & & 0 \end{pmatrix}$ with $c_1 \cdots c_r \neq 0$ and $c_1 \mid c_2 \mid \cdots \mid c_r$

by applying the following transformations:
- addition of a multiple of one row [column] to another row [column];
- exchange of two rows [columns].

(b) Define an equivalence relation on $R^{m \times n}$ by declaring A and B equivalent if there are invertible matrices $P \in (R^{m \times m})^{\times}$ and $Q \in (R^{n \times n})^{\times}$ such that $B = P^{-1}AQ$. Show that every equivalence class contains an element of the form $(\star)$. Here the number r is uniquely determined, and the elements $c_1, \ldots, c_r \in R$ are unique up to multiplication by units.

(c) Show that part (b) remains true if R is a general principal ideal domain.
Hint. Mimic the proof of (I.10.29).

8. Factorization in polynomial and power series rings

AFTER having discussed the general theory of factorizations in integral domains, we now want to apply this theory to polynomial rings. The only class of examples of unique factorization domains we know as yet is the class of principal ideal domains. This class comprises the polynomial rings $K[x]$ where K is a field because these rings are Euclidean. On the other hand, it is true that $\mathbb{Z}[x]$ is a unique factorization domain even though it is not a principal ideal domain. (For example, the ideal $\langle 2, x \rangle$ is not a principal ideal.) The fact that $\mathbb{Z}[x]$ is in fact a unique factorization domain is contained in a general result which we are going to establish now, namely, that if R is a unique factorization domain then so is the polynomial ring $R[x]$. For the proof of this statement, it is convenient to introduce the following terminology.

(8.1) Definition. *Let R be a unique factorization domain. A polynomial in $R[x]$ is called* **primitive** *if its coefficients are relatively prime.*

FOR example, the polynomial $f(x) = 6x^2 - 10x + 15 \in \mathbb{Z}[x]$ is primitive whereas $g(x) = 4x^5 - 6$ is not.

(8.2) Theorem. *Let R be a unique factorization domain and let K be its quotient field.*

*(a) (**Gauß's Lemma.**) Two polynomials $f, g \in R[x]$ are primitive if and only if their product fg is.*

(b) Let $f \in K[x]$. Then there are an element $c \in K$ and a primitive polynomial $f^* \in R[x]$ such that $f = cf^*$. This decomposition is essentially unique, i.e., if also $f = c'f'$ with $c' \in K$ and as primitive polynomial $f' \in R[x]$, then there is a unit $u \in R^\times$ such that $c' = uc$ and $f' = u^{-1}f^*$.

(c) Let $f \in R[x]$ be a nonconstant polynomial. If there are polynomials $f_1, f_2 \in K[x]$ with $f = f_1 f_2$, then there is an element $c \in K^\times$ such that cf_1 and $c^{-1}f_2$ belong to $R[x]$.

(d) Let $f \in R[x]$. If f is irreducible in $R[x]$, then f is also irreducible in $K[x]$ (unless $\deg f = 0$ in which case f is a unit in $K[x]$). If f is primitive and irreducible in $K[x]$, then f is also irreducible in $R[x]$.

(e) Let $f \in R[x]$. If f is prime in $R[x]$, then f is also prime in $K[x]$. If f is primitive and prime in $K[x]$, then f is also prime in $R[x]$.

(f) $R[x]$ is a unique factorization domain.

(g) $R[x_1, \ldots, x_n]$ is a unique factorization domain.

Proof. (a) If fg is primitive, then f and g are. Indeed, any common divisor of the coefficients of f also divides all the coefficients of fg, hence must be a unit; this shows that f is primitive. The same argument applies to g.

Suppose conversely that f and g are primitive. To obtain a contradiction, suppose that the coefficients of fg are *not* relatively prime. Then there is an irreducible element p dividing all of them; p is also prime, by (7.8)(e). Now p can neither divide all the coefficients of f nor all the coefficients of g because f and g are primitive. Write

135

$f(x) = \sum_i a_i x^i$ and $g(x) = \sum_j b_j x^j$. Let k be the first index with $p \nmid a_k$ and l the first index with $p \nmid b_l$; then $p \nmid a_k b_l$ since p is prime. On the other hand, p divides the coefficient of x^{k+l} in the product, and this coefficient is

$$\underbrace{a_0 b_{k+l} + a_1 b_{k+l-1} + \cdots + a_{k-1} b^{l+1}}_{\text{divisible by } p \text{ because } p \mid a_i \text{ for } i < k} + a_k b_l + \underbrace{a_{k+1} b_{l-1} + \cdots + a_{k+l-1} b_1 + a_{k+l} b_0}_{\text{divisible by } p \text{ because } p \mid b_j \text{ for } j < l} .$$

But then p has to divide $a_k b_l$, and this is the desired contradiction.

(b) Write

$$f(x) = \frac{a_0}{b_0} + \frac{a_1}{b_1} x + \frac{a_2}{b_2} x^2 + \cdots + \frac{a_n}{b_n} x^n \quad \text{with} \quad a_i, b_i \in R, \ b_i \neq 0 .$$

Putting $b := b_0 b_1 b_2 \cdots b_n$ we have $f = b^{-1}\varphi$ with $\varphi \in R[x]$. Now let d be a greatest common divisor of the coefficients of φ and let $f^* := d^{-1}\varphi$; then f^* is primitive, and $f = db^{-1}f^* =: cf^*$.

Now let $f = cf^* = c'f'$ with $c, c' \in K$ and f^*, f' primitive. We can write $c = c_1/c_2$ and $c' = c_1'/c_2'$ with $c_i, c_i' \in R$; then

$$c_1 c_2' f^* = c_1' c_2 f' .$$

Since f^* and f' are primitive, the coefficients of the polynomial on the left-hand side have $c_1 c_2'$ as a gcd, whereas $c_1' c_2$ is a gcd of the coefficients on the right-hand side. But the two sides are equal; hence $c_1 c_2'$ and $c_1' c_2$ can differ only by a unit u of R, say $c_1' c_2 = u c_1 c_2'$. But then $c' = uc$.

(c) Let d be a gcd of the coefficients of f. Then $f = df^*$ where f^* is primitive. By part (b), we can write $f_1 = c_1 f_1^*$ and $f_2 = c_2 f_2^*$ with $c_i \in K$ and f_i^* primitive. Then

$$d \underbrace{f^*}_{\text{primitive}} = f = f_1 f_2 = c_1 c_2 \underbrace{f_1^* f_2^*}_{\text{primitive by (a)}} .$$

The uniqueness statement in (b) shows that $c_1 c_2 = ud \in R$ for some unit $u \in R^\times$. So letting $c := c_2$, we have $cf_1 = c_1 c_2 f_1^* \in R[x]$ and $c^{-1} f_2 = f_2^* \in R[x]$.

(d) Suppose f is irreducible in $R[x]$, but reducible in $K[x]$. Then there is a proper factorization $f = f_1 f_2$ with $f_i \in K[x]$. According to part (c), we can find an element $c \in K^\times$ such that $F_1 := cf_1$ and $F_2 := c^{-1} f_2$ belong to $R[x]$. But then $f = F_1 F_2$ is a proper factorization in $R[x]$. This is the desired contradiction.

Suppose conversely that $f \in R[x]$ is primitive and irreducible in $K[x]$, but reducible in $R[x]$. Then there is a proper factorization $f = f_1 f_2$ in $R[x]$. But this factorization must be trivial in $K[x]$ so that $\deg f_1 = 0$ or $\deg f_2 = 0$; say $\deg f_1 = 0$. Then $f_1 \equiv c_1$ where $c_1 \in R$ is a constant dividing all the coefficients of f. But these coefficients are relatively prime because f is primitive. Thus c_1 must be a unit in R, hence f_1 a unit in $R[x]$, contradicting the assumption that $f = f_1 f_2$ is a proper factorization in $R[x]$.

(e) Let f be prime in $R[x]$ and suppose that $f \mid f_1 f_2$ in $K[x]$. Then there is a polynomial $g \in K[x]$ with $f_1 f_2 = gf$. According to part (b), there are elements $c_1, c_2, d \in K, c \in R$ and primitive polynomials $f^*, f_1^*, f_2\star, g^* \in R[x]$ such that $f = cf^*, f_1 = c_1 f_1^*, f_2 = c_2 f_2^*$ and $g = dg^*$. This implies

$$c_1 c_2 \underbrace{f_1^* f_2^*}_{\text{primitive by (a)}} = f_1 f_2 = gf = cd \underbrace{g^* f^*}_{\text{primitive by (a)}} .$$

The uniqueness claim in (b) shows that $uc_1c_2 = cd$ for some $u \in R^\times$. Consequently, $f_1^* f_2^* = ug^* f^*$, i.e., $(cf_1^*)f_2^* = ug^* f^*$. Now f^* divides the left-hand side and is prime in $R[x]$, hence divides $cf_1^* f_2^*$. But this implies that f divides f_1 or f_2 in $K[x]$.

Conversely, suppose that $f \in R[x]$ is primitive and prime in $K[x]$. Assume $f \mid f_1 f_2$ in $R[x]$. Then $f \mid f_1$ or $f \mid f_2$ in $K[x]$, say $f_1 = fg$ where $g \in K[x]$. According to (b), there are elements $c \in R, c_1 \in K$ and primitive elements $f_1^*, g^* \in R[x]$ such that $f_1 = cf_1^*$ and $g = cg^*$. Then

$$ c_1 f_1^* \ = \ f_1 \ = \ fg \ = \ c(fg^*) \ . $$

The uniqueness statement in (b) yields that $c = c_1 u \in R$ so that $g \in R[x]$. But this shows that $f \mid f_1$ in $R[x]$.

(f) Let $f \in R[x]$ be neither zero nor a unit. Let g be a gcd of the coefficients of f. Then $f = df^*$ where $f^* \in R[x]$ is primitive. Since R is a unique factorization domain and since $R[x]$ is a polynomial ring, we have decompositions

$$ d \ = \ d_1 \cdots d_m \ , \quad f^* \ = \ p_1 \cdots p_n $$

where d_i is irreducible in R (hence in $R[x]$) and p_j irreducible in $R[x]$. (These decompositions may degenerate if d or f^* is a unit, but not both at the same time, because then f would be a unit.) This shows that $R[x]$ is a factorization domain.

To prove that factorizations are essentially unique, it is enough to show that every prime element in $R[x]$ is irreducible. Now if f is prime in $R[x]$, then also in $K[x]$, by part (e). But then f is even irreducible in $K[x]$, because $K[x]$ is a Euclidean ring, hence a unique factorization domain. Now part (d) implies that f is also irreducible in $R[x]$, which is what we wanted to show.

(g) In part (f) we proved that if R is a unique factorization domain then so is $R[x]$. The claim then follows easily by induction on n. ∎

THE main part of Gauß's theorem is that if R is a unique factorization domain then so is $R[x]$. As a matter of fact, the converse of this statement is also true, as we now show.

(8.3) Theorem. *Let R be an integral domain. Then $R[x]$ is a factorization domain if and only if R is.*

Proof. In (8.2) we proved already that if R is a unique factorization domain then so is $R[x]$. Suppose conversely that $R[x]$ is a unique factorization domain and let $a \in R^\heartsuit$. Considering a as a constant polynomial, we have a factorization $a = up_1(x) \cdots p_n(x)$ with a unit $u \in R[x]^\times = R^\times$ and prime elements $p_i \in R[x]$. Considering degrees, we see that each p_i must be a constant polynomial and hence a prime element of R. Thus every element $a \in R^\heartsuit$ possesses a prime factorization. By (7.9)(3), this shows that R is a unique factorization domain. ∎

AS an application let us show how the fact that $\mathbb{Z}[x_1, \ldots, x_n]$ is a unique factorization domain can be exploited to derive an identity for determinants. (Cf. (I.7.16).)

(8.4) Application: The Vandermonde determinant. *Let $a_1, \ldots, a_n$ be elements of an arbitrary commutative ring with identity. Then*

$$\det \begin{pmatrix} 1 & 1 & \cdots & 1 \\ a_1 & a_2 & \cdots & a_n \\ a_1^2 & a_2^2 & \cdots & a_n^2 \\ \vdots & \vdots & & \vdots \\ a_1^{n-1} & a_2^{n-1} & \cdots & a_n^{n-1} \end{pmatrix} = \prod_{i>j} (a_i - a_j) \, .$$

Proof. Let

$$p(x_1, \ldots, x_n) := \det \begin{pmatrix} 1 & 1 & \cdots & 1 \\ x_1 & x_2 & \cdots & x_n \\ x_1^2 & x_2^2 & \cdots & x_n^2 \\ \vdots & \vdots & & \vdots \\ x_1^{n-1} & x_2^{n-1} & \cdots & x_n^{n-1} \end{pmatrix} \in \mathbb{Z}[x_1, \ldots, x_n] \, .$$

Consider $\mathbb{Z}[x_1, \ldots, x_n]$ as $\left(\mathbb{Z}[x_1, \ldots, \widehat{x_i}, \ldots, x_n] \right)[x_i]$ and take any $j \neq i$. Then x_j is a root of p; indeed, if one plugs in $x_i = x_j$ then the determinant has two identical columns, hence vanishes. This means that $x_i - x_j$ is a factor of p (which is obviously irreducible). Since this holds true for all $i \neq j$ we see that $\prod_{i>j}(x_i - x_j)$ is a factor of p in $\mathbb{Z}[x_1, \ldots, x_n]$. If we imagine p being multiplied out and grouped by powers of x_n then clearly the coefficient of x_n^{n-1} is 1; this shows that in fact $p(x_1, \ldots, x_n) = \prod_{i>j}(x_i - x_j)$. ∎

WE have proved that whenever R is a unique factorization domain then so is the polynomial ring $R[x_1, \ldots, x_n]$. The same statement for the power series ring $R[[x_1, \ldots, x_n]]$ is false as was first shown by Pierre Samuel, but we are going to show that $K[[x_1, \ldots, x_n]]$ is a unique factorization domain whenever K is an infinite field. We need a preparatory result which helps to transform a given power series into a more suitable form. In fact, by applying a coordinate transformation and a multiplication by a unit, we can arrange that any given power series has finite degree in one of the variables and hence can be treated as a polynomial in this variable. This will allow us to apply facts on polynomial rings to power series rings.

(8.5) Proposition. *Let K be an infinite field.*

(a) **(Normalization Theorem.)** *Given a finite family $\mathfrak{P}$ of nonzero homogeneous polynomials $p \in K[x_1, \ldots, x_n]$ there is a transformation of the form $y_i := x_i + \alpha_i x_n$ for $1 \leq i \leq n-1$, $y_n := \alpha_n x_n$ with $\alpha_i \in R$ such that all $p \in \mathfrak{P}$ take the form*

$$p(y_1, \ldots, y_n) = c y_n^m + \text{lower terms in } y_n$$

where $c \in K^\times$ is a nonzero constant (depending on p).

(b) **(Weierstraß's Preparation Theorem.)** *Given a power series $p \in K[[x_1, \ldots, x_n]]$ of subdegree $m \geq 1$ whose lowest homogeneous component takes the form*

$$p_m(x_1, \ldots, x_n) = cx_n^m + \text{lower terms in } x_n \qquad \text{where } c \in K^\times,$$

there is a unit $u \in R[[x_1, \ldots, x_n]]^\times$ such that

$$pu = x_n^m + A_1 x_n^{m-1} + A_2 x_n^{m-1} + \cdots + A_m$$

where $A_i \in K[[x_1, \ldots, x_{n-1}]]$. (Note that necessarily subdeg $A_i \geq i$.)

Proof. (a) By (4.19) we can find elements $\alpha_1, \ldots, \alpha_n \in K$ such that $p(\alpha_1, \ldots, \alpha_n) \neq 0$ for all $p \in \mathfrak{P}$. Then the indicated transformation $y_i = x_i + \alpha_i x_n$ yields

$$
\begin{aligned}
y_1^{m_1} \cdots y_n^{m_n} &= (x_1 + \alpha_1 x_n)^{m_1} (x_2 + \alpha_2 x_n)^{m_2} \cdots (\alpha_n x_n)^{m_n} \\
&= \alpha_1^{m_1} \alpha_2^{m_2} \cdots \alpha_n^{m_n} x_n^{m_1 + m_2 + \cdots + m_n} + \text{lower terms in } x_n.
\end{aligned}
$$

Hence for every homogeneous polynomial p of degree m, we have

$$p(y_1, \ldots, y_n) = p(\alpha_1, \ldots, \alpha_n) x_n^m + \text{lower terms in } x_n .$$

The choice of $\alpha_1, \ldots, \alpha_n$ yields the claim.

(b) Trying $u = u_0 + u_1 + u_2 + \cdots$ (which is not necessarily the decomposition of u into its homogeneous components!), we obtain $pu = r_m + r_{m+1} + r_{m+2} \cdots$ where

$$r_m = p_m u_0, \qquad r_{m+1} = p_{m+1} u_0 + p_m u_1, \qquad r_{m+2} = p_{m+2} u_0 + p_{m+1} u_1 + p_m u_2$$

and so on; in general, $r_{m+k} = \sum_{j=0}^{k} p_{m+k-j} u_j$. Now we try to determine $u_0, u_1, u_2, \ldots$ to meet our objective. First of all, we choose $u_0 := c^{-1} \in K^\times$. Next, we treat p_{m+1} and p_m as polynomials in x_n over the ring $R[x_1, \ldots, x_{n-1}]$ and apply (4.10)(a) to find a unique polynomial $u_1 \in R[x_1, \ldots, x_{n-1}, x_n]$ such that $\deg_{x_n}(p_{m+1}u_0 + p_m u_1) < \deg_{x_n}(p_m) = m$; this inequality forces subdeg $u_1 \geq 1$. Applying (4.10)(a) again, we find a unique polynomial $u_2 \in R[x_1, \ldots, x_n]$ such that $\deg_{x_n}(p_{m+2}u_0 + p_{m+1}u_1 + p_m u_2) < \deg_{x_n}(p_m) = m$; again, this condition implies subdeg $u_2 \geq 1$. Proceeding in this way, we obtain polynomials $u_0, u_1, u_2, \ldots$ such that $p(u_0 + u_1 + u_2 + \cdots)$ has the desired form (this is easily seen by collecting like powers of x_n). It remains to note that $u = u_0 + u_1 + u_2 + \cdots$ is indeed a unit in $R[[x_1, \ldots, x_n]]^\times$; but this follows from the fact that $u_0 = c^{-1} \in K^\times$ and subdeg$(u_1 + u_2 + \cdots) \geq 1$ because subdeg $u_i \geq 1$ for all i. ■

(8.6) Theorem. *If K is an infinite field, then the power series ring $K[[x_1, \ldots, x_n]]$ is a unique factorization domain.*

Proof. We have to show that every irreducible element $f \in K[[x_1, \ldots, x_n]]$ is prime. To do so, we proceed by induction on n. The case $n = 1$ is rather trivial. Indeed, if $f \in K[[x]]$ is irreducible, then f is not a unit, hence has the form $f(x) = x^m u(x)$

where $m \geq 1$ is the subdegree of f and u is a unit in $K[[x]]$. Since x^m is irreducible if and only if $m = 1$, the only irreducible element in $K[[x]]$ (up to multiplication with units) is x, and this element is clearly prime.

Suppose by induction hypothesis that $K[[x_1, \ldots, x_{n-1}]]$ is a unique factorization domain; then so is $K[[x_1, \ldots, x_{n-1}]][x_n]$, due to (8.2)(f). Now let $f \in K[[x_1, \ldots, x_n]]$ be irreducible and suppose that $f \mid ab$; we have to show that $f \mid a$ or $f \mid b$. If a or b is a unit, we are done. Otherwise f, a and b have positive subdegrees. After a coordinate transformation as in (8.5)(a), which does not change any divisibility properties, we can assume that f, a and b have the form $cx_n^m +$lower terms in x_n (where $c \in K^\times$ and $m \geq 1$ depend on f, a and b). Now by (8.5)(b) we can find units $u, v, w \in K[[x_1, \ldots, x_n]]$ such that $F := fu$, $A := av$ and $B := bw$ are not only elements of $K[[x_1, \ldots, x_n]]$, but even of $K[[x_1, \ldots, x_{n-1}]][x_n]$ which is a unique factorization domain. Now $f \mid ab$ in $K[[x_1, \ldots, x_n]]$, say $ab = fh$, implies $F \mid AB$ in $K[[x_1, \ldots, x_{n-1}]][x_n]$ because $AB = vwab = vwfh = (u^{-1}vwh)F$, and the fact that A, B and F have finite degree in x_n implies that $u^{-1}vwh$ does. Once we know that F is irreducible in the unique factorization domain $K[[x_1, \ldots, x_{n-1}]][x_n]$ (hence prime!) we can conclude that $F \mid A$ or $F \mid B$ in $K[[x_1, \ldots, x_{n-1}]][x_n]$ which clearly implies $f \mid a$ or $f \mid b$ in $K[[x_1, \ldots, x_{n-1}, x_n]]$.

Now it is possible that an element of $K[[x_1, \ldots, x_{n-1}]][x_n]$ is irreducible in $K[[x_1, \ldots, x_{n-1}, x_n]]$, but reducible in $K[[x_1, \ldots, x_{n-1}]][x_n]$ (see example (7.2)(d) above); however, the special form of F implies that this does not happen for F. Indeed, we have $F = x_n^m + A_{m-1}x_n^{m-1} + \cdots + A_1 x_n + A_0$ where $A_{m-i} \in K[[x_1, \ldots, x_{n-1}]]$ has subdegree $\geq i$. Now suppose we have a factorization $F = GH$ in $K[[x_1, \ldots, x_{n-1}]][x_n]$, say with

$$G = B_k x_n^k + B_{k-1} x_n^{k-1} + \cdots + B_1 x_n + B_0 \quad \text{and}$$
$$H = C_l x_n^l + C_{l-1} x_n^{l-1} + \cdots + C_1 x_n + C_0$$

where $A_i, B_i \in K[[x_1, \ldots, x_{n-1}]]$ and $k + l = m$. Since this must be a trivial factorization in $K[[x_1, \ldots, x_n]]$ where F is irreducible, at least one of F or G must be a power series unit, i.e., must have subdegree 0. Suppose subdeg $G = 0$ so that subdeg $B_0 = 0$. Multiplying out the product $F = GH$ and comparing powers of x_n, we obtain

$$A_0 = B_0 C_0, \quad A_1 = B_1 C_0 + B_0 C_1, \quad A_2 = B_2 C_0 + B_1 C_1 + B_0 C_2, \quad \text{and so on.}$$

Since subdeg $A_{m-i} \geq i$ and subdeg $B_0 = 0$, we obtain successively subdeg $C_0 \geq m$, subdeg $C_1 \geq m - 1$, ..., subdeg $C_i \geq m - i$; finally, we have subdeg $C_l \geq m - l = k$ and consequently subdeg$(B_k C_l) \geq k$. But $B_k C_l = 1$ (the leading coefficient of F); hence we must have $k = 0$. This shows that the factorization $F = GH$ must be a trivial factorization in $K[[x_1, \ldots, x_{n-1}]][x_n]$. ∎

FROM a purely economical viewpoint, we could have omitted (8.5) and (8.6) because we are now going to prove a more general result than (8.6) with less effort than was necessary there. However, (8.5) is interesting in its own right, and the proof of (8.6) is instructive in that it shows how results on polynomial rings can be used to obtain results on power series rings. We are now going to prove that if R is a principal ideal domain, then $R[x]$ is a unique factorization domain. For this, we need the following preparatory result, also interesting in its own right.

(8.7) Proposition. *Suppose that R is a commutative ring and that P is a prime ideal of $R[[x]]$. Let $P_0 \trianglelefteq R$ be the ideal of all elements of R which occur as the a_0-term of a power series in P.*

(a) If P is finitely generated, then so is P_0.

(b) If P_0 is finitely generated, then so is P. More precisely, if $P_0 = \langle a_1, \ldots, a_n \rangle$ and if $f_1, \ldots, f_n \in R[[x]]$ are power series leading off with $a_1, \ldots, a_n$, then

$$
P = \begin{cases} \langle a_1, \ldots, a_n, x \rangle, & \text{if } x \in P; \\ \langle f_1, \ldots, f_n \rangle, & \text{if } x \notin P. \end{cases}
$$

Proof. (a) This is clear, because P_0 is a homomorphic image of P; namely, $P_0 = \varphi(P)$ where $\varphi : R[[x]] \to R$ is given by $\varphi(\sum_k a_k x^k) = a_0$.

(b) If $x \in P$, then clearly $P = \langle a_1, \ldots, a_n, x \rangle$. Suppose $x \notin P$ and let $g \in P$. Then the initial coefficient of g can be written as $\sum_{i=1}^n r_i a_i$ with $r_i \in R$ so that $g - \sum_{i=1}^n r_i f_i$ has initial coefficient zero and hence can be written as $x g_1$ with $g_1 \in R[[x]]$. Then $x g_1 = g - \sum_{i=1}^n r_i f_i \in P$ and $x \notin P$; since P is a prime ideal, this implies $g_1 \in P$. Repeating the procedure, we obtain a representation $g_1 = \sum_{i=1}^n r_i' f_i + x g_2$ with $g_2 \in P$. Continuing in this fashion, we obtain

$$
\begin{aligned}
g &= \sum_{i=1}^n r_i f_i + x \sum_{i=1}^n r_i' f_i + x^2 \sum_{i=1}^n r_i'' f_i + \cdots \\
&= \sum_{i=1}^n (r_i + r_i' x + r_i'' x^2 + \cdots) f_i \in \langle f_1, \ldots, f_n \rangle.
\end{aligned}
$$

Since g was an arbitrary element of P, the claim is established. ∎

(8.8) Theorem. *If R is a principal ideal domain, then $R[[x]]$ is a unique factorization domain.*

Proof. By (7.9)(2) it is enough to show that every nonzero prime ideal $P \trianglelefteq R[[x]]$ contains a prime element. If $x \in P$, then we can take x as our prime element. Suppose $x \notin P$; then the ideal P_0 as defined in (8.7) is different from $\{0\}$ and is generated by a single element, since R is a principal ideal domain. But then (8.7)(b) implies that P is also generated by a single element, say $P = \langle f \rangle$. Then (7.4)(f) shows that f is a prime element. ∎

WE have seen that if R is a unique factorization domain then so is $R[x]$, but merely *knowing* that a given polynomial has an essentially unique factorization into irreducible polynomials does not mean that we can actually *find* this decomposition. It turns out, however, that there are factorization algorithms for all practically important polynomial rings. Since these algorithms are rather cumbersome and rarely used in practice, however, the reader may wish to skip the discussion and continue with the irreducibility criteria for polynomials presented in (8.18). To present various factorization algorithms for polynomials (all due to Kronecker) let us start by stating precisely what we mean by a factorization algorithm.

(8.9) Definition. *Let R be a unique factorization domain. A* **factorization algorithm** *for R is a method which, when applied to any element $r \in R^{\heartsuit}$, produces in a finite number of steps a prime factorization of R.*

Equivalently, a factorization algorithm allows in a finite number of steps to find all divisors of r (up to units); in fact, if $r = p_1 \cdots p_m$ is a prime factorization of r then all factors can be found by taking all products $p_{i_1} \cdots p_{i_k}$ with $i_1, \ldots, i_k \in \{1, \ldots, m\}$.

(8.10) Example. *Let (R, β) be a Euclidean domain such that for all $n \in \mathbb{N}$ the set $R_n := \{r \in R \mid \beta(r) \leq n\}$ is finite. (For example, one can take $R = \mathbb{Z}$ and $\beta(r) = |r|$ or $R = K[x]$ and $\beta(p) := 2^{\deg p}$ where K is a finite field.) Then simple trial and error may serve as a factorization algorithm for R.*

In fact, given $b \in R^{\heartsuit}$, every divisor a of b satisfies $\beta(a) \leq \beta(b)$ and hence lies in the finite set $R_{\beta(b)}$. Thus the prime factors of b can be found by simply checking all elements of $R_{\beta(b)}$.

TURNING to polynomial rings, we now establish the existence of a factorization for $\mathbb{Z}[x]$.

(8.11) Theorem. *There is a factorization algorithm for $\mathbb{Z}[x]$. More precisely, given a polynomial $f \in \mathbb{Z}[x]^{\heartsuit} = \mathbb{Z}[x] \setminus \{0, \pm 1\}$, the following steps lead to either a proper factor of f or the proof that f is irreducible.*

(a) If $\deg f = 0$ then simply apply the factorization algorithm for $\mathbb{Z}$, identifying constant polynomials with integers.

(b) If $\deg f = n > 0$ let m be the largest integer $\leq n/2$ and calculate $f(0), f(1), \ldots, f(m)$. If $f(i) = 0$ for some i then $x - i$ divides f; hence either $x - i$ is a proper factor of f or else $f(x) = \pm(x - i)$ in which case f is irreducible. Hence if $f(i) = 0$ for some $1 \leq i \leq m$ then we are done.

(c) If $f(i) \neq 0$ for $1 \leq i \leq m$ write down all $(m+1)$-tuples $d = (d_0, d_1, \ldots, d_m)$ such that d_i is a divisor of $f(i)$ in $\mathbb{Z}$; there is only a finite number of such $(m+1)$-tuples. For each of these $(m+1)$-tuples d there is a unique polynomial $g = g_d \in \mathbb{Q}[x]$ of degree $\leq m$ with $g(0) = d_0, g(1) = d_1, \ldots, g(m) = d_m$.

(d) If there is an $(m+1)$-tuple d such that g_d is a proper factor of f in $\mathbb{Z}[x]$, then we are done; otherwise f is irreducible (and we are also done).

Proof. (a) Clearly, the factorization of polynomials of degree 0 is ordinary factorization of integers for which an algorithm exists by (8.10).

(b) This step is self-explanatory.

(c) Let us quickly prove the existence and uniqueness of the **interpolation polynomial** g_d for a given $(m+1)$-tuple g_d. *Uniqueness:* If g_1 and g_2 have degree $\leq m$ and take the same values on $0, \ldots, m$ then the polynomial $g_1 - g_2$ (which has a degree $\leq m$) has the $m+1$ different roots $0, \ldots, m$ and hence must be the zero polynomial. *Existence:* For $0 \leq i \leq m$ define $p_i \in \mathbb{Q}[x]$ by

$$p_i(x) = \frac{(x-0) \cdots \big(x - (i-1)\big)\big(x - (i+1)\big) \cdots (x - m)}{(i-0) \cdots \big(i - (i-1)\big)\big(i - (i+1)\big) \cdots (i - m)} .$$

Since clearly $p_i(i) = 1$ and $p_i(j) = 0$ if $i \neq j$, the polynomial $g(x) := g_d(x) := d_0 \cdot p_0(x) + d_1 \cdot p_1(x) + \cdots + d_m \cdot p_m(x)$ has the desired property.

(d) If f has a proper factorization $f = gh$ in $\mathbb{Z}[x]$, then either g or h must have a degree $\leq m$; assume without loss of generality that $\deg g \leq m$. Since $f = gh$ implies that $f(i) = g(i)h(i)$ for $1 \leq i \leq m$ so that $g(i)$ is a divisor of $f(i)$ for all i, we see that g must be one of the polynomials g_d. Thus all the proper divisors of f in $\mathbb{Z}[x]$ can be found among the polynomials g_d, and since there are only finitely many of these one can simply check all of them. ∎

THE existence of a factorization algorith for $\mathbb{Z}[x]$ implies the existence of one for $\mathbb{Q}[x]$; this is an immediate consequence of the following result.

(8.12) Theorem. *Suppose that R is a unique factorization domain with quotient field K. If there is a factorization algorithm for $R[x]$ then there is also one for $K[x]$.*

Proof. Let $f \in K[x]^{\heartsuit}$. Then $\deg f > 0$, and we can write

$$f(x) = \frac{r_n}{s_n}x^n + \frac{r_{n-1}}{s_{n-1}}x^{n-1} + \cdots + \frac{r_1}{s_1}x + \frac{r_0}{s_0}$$

with $r_i, s_i \in R$. Choose an element $d \in R$ which is divisible by all of the denominators s_i (for example, take $d = s_0 s_1 \cdots s_n$ or let d be a least common multiple of $s_0, \ldots, s_n$). Then we can apply the factorization algorithm for $R[x]$ to $F(x) := d \cdot f(x)$ to obtain a decomposition $F(x) = F_1(x) \cdots F_m(x)$ where each F_i is irreducible in $R[x]$. Then either F_i is a unit in $K[x]$ (if $\deg F_i = 0$) or else is irreducible in $K[x]$ (if $\deg F_i > 0$), due to Gauß's lemma (8.2)(a). Since d is a unit in $K[x]$, we have the prime decomposition $f(x) = d^{-1} F_1(x) \cdots F_m(x)$ of f in $K[x]$. ∎

(8.13) Examples. (a) There is a factorization algorithm for $\mathbb{Z}[x]$ by (8.11); hence (8.12) implies that there is also one for $\mathbb{Q}[x]$.

(b) Let K be a field. If there is a factorization algorithm for $K[x, y]$, then there is also one for $K(y)[x]$. To see this, simply apply (8.12) with $R = K[y]$.

NEXT we consider polynomial rings in several variables.

(8.14) Theorem. *Let K be a field. If there is a factorization algorithm for $K[x]$ then there is also one for $K[x, y]$. More precisely, given a polynomial $f \in K[x, y]$, the following steps lead to either a proper factor of f or the proof that f is irreducible:*

(a) Let $n := \deg_y f$ be the degree of f in the second variable y. If $n = 0$ then f is a polynomial in x alone, and we can apply the algorithm for $K[x]$ to factor f.

(b) If $n > 0$, pick any number $N > n$ and let $\overline{f}(x) := f(x^N, x)$. Then each factorization $f = gh$ yields a factorization $\overline{f} = \overline{g}\overline{h}$. Apply the factorization algorithm for $K[x]$ to find all proper factors θ of $\overline{f}$. For each such factor one can determine a polynomial $g \in K[x, y]$ with $\theta = \overline{g}$. If g divides f in $K[x, y]$ (which can be checked by a simple division) then we have found a factor of f; if none of the polynomials g divides f then f is irreducible.

Factorization in polynomial and power series rings / 143

Proof. The only claim that needs to be proved is that every divisor θ of $\overline{f}$ is of the form $\overline{g}$ and that g can be effectively reconstructed from $\overline{g}$. Write $\theta(t) = \sum_i a_i t^i$. Now each $i \in \mathbb{N}_0$ can be uniquely written as $i = rN + s$ where $r \geq 0$ and $0 \leq s < N$. Then it is easy to check that $\theta = \overline{g}$ where $g(x,y) = \sum_{r,s} a_{rN+s} y^r x^s$. $\blacksquare$

WE now turn to the question whether or not for a field extension $(L : K)$ the existence of a factorization algorithm for $K[x]$ implies the existence of one for $L[x]$. This is indeed true if $(L : K)$ is a purely transcendental extension of finite transcendence degree, as the following proposition shows.†

(8.15) Proposition. *Let $L = K(\alpha_1, \ldots, \alpha_n)$ where $\alpha_1, \ldots, \alpha_n)$ are algebraically independent over K. If there is a factorization algorithm for $K[x]$, then there is also one for $L[x]$.*

Proof. We may assume $n = 1$; if this special case is established then the general case will follow by induction. Introduce a new variable y. If there is a factorization algorithm for $K[x]$, there is also one for $K[x,y]$ by (8.14) and then also one for $K(y)[x]$ by (8.13)(b); but $K(y)[x] \cong K(\alpha)[x] = L[x]$. $\blacksquare$

TURNING to algebraic extension, we have the following result.

(8.16) Theorem. *Suppose that the field L is obtained from a field K by adjoining a separable algebraic element α so that $L = K(\alpha)$. If there is a factorization algorithm for $K[x]$, then there is also one for $L[x]$. Namely, the following steps lead to a factorization of a given polynomial $f \in L[x]$.*

(a) *Determine the minimal polynomial $\Phi \in K[x]$ of α over K and find a splitting field M of Φ.*

(b) *Introduce a new variable u, define $L \in K[x,u]$ by $F(x,u) := f(x + u\alpha)$ and then let*
$$(NF)(x,u) := \prod_{\sigma \in G_K^M} (\sigma \star F)(x,u) = \prod_{\sigma \in G_K^M} f(x + u\,\sigma(\alpha)) .$$

This polynomial has coefficients in K, and since there is a factorization algorithm for $K[x,u]$ by (8.14) we can effectively find a decomposition $(NF)(x,u) = G_1(x,u) \cdots G_m(x,u)$ of NF into irreducible factors.

(c) *For $1 \leq i \leq m$, use the Euclidean algorithm for $L(u)[x]$ to find a greatest common divisor $d_i \in L(u)[x]$ of F and G_i in $L(u)[x]$; then we can write $F = d_i e_i$ with $e_i \in L(u)[x]$. The coefficients of d_i and e_i are rational functions in u over L; clearing denominators in the equation $F = d_i e_i$ we obtain an equation*

$(\star)$ $\qquad\qquad H_i(u)F(x,u) = H_i f(x + u\alpha) = D_i(x,u)E_i(x,u)$

---

† The next three theorems can only be understood with some knowledge on field extensions as developed in the field theory sections of this book.

with $H_i \in L[x]$ and $D_i, E_i \in L[x, u]$. We can arrange that $H_i(0) \neq 0$; then setting $u = 0$ in $(\star)$ and dividing by $H_i(0) \neq 0$ yields

$(\star\star)$ $$f(x) = H_i(0)^{-1} D_i(x, 0) E_i(x, 0)$$

which is a factorization of f in $L[x]$.

(d) If $(\star\star)$ happens to be a proper factorization for some $1 \leq i \leq m$, then we are done. Otherwise f is irreducible in $L[x]$ (and we are done too).

Proof. (a) We note that $(M : K)$ is a normal and separable field extension, hence a Galois extension.

(b) The key idea of the proof is to associate with $f \in L[x]$ the polynomial $(Nf)(x) := \prod_{\sigma \in G_K^M} (\sigma \star f)$ which has coefficients in $F(G_K^M) = K$ and hence can be decomposed into irreducible factors using the existing factorization algorithm for $K[x]$, say $Nf = g_1 g_2 \cdots g_m$. Then for $1 \leq i \leq m$ we can use the Euclidean algorithm to find a greatest common divisor d_i of f and g_i in $L[x]$; if d_i happens to be a proper factor of f, then we are done. However, it may happen that none of $d_1, \ldots, d_m$ is a proper factor of f in $L[x]$, but that f is still not irreducible in $L[x]$.†

In this case we might apply the same technique to $f(x + \alpha)$ instead of $f(x)$; if this yields a proper factorization $f(x + \alpha) = g(x)h(x)$ for $f(x + \alpha)$, then also a proper factorization $f(x) = g(x - \alpha)h(x - \alpha)$ for f itself. Of course, we might try as well $f(x + 2\alpha)$, $f(x - \alpha)$, and so on. Now instead of trying $f(x + m\alpha)$ for various choices of m, we consider the polynomial $f(x + u\alpha)$ where u is a new variable. This explains the idea of part (b) of the proof.

(c) If $H_i(u)$ should be divisible by some power u^t, then there are exponents r, s with $r + s = t$ such that u^r divides D_i and u^s divides E_i, and the factor u^t can be cancelled on both sides. Hence we may in fact assume that $H_i(0) \neq 0$.

(d) We have to show that if none of the greatest common divisors of $F(x + u\alpha)$ and the irreducible factors of $(NF)(x + u\alpha)$ yields a proper factor of f in $L[x]$, then f is irreducible in $L[x]$. Note that the corresponding conclusion was wrong if applied to $(Nf)(x)$ instead of $(Nf)(x + u\alpha)$; this is exactly the reason why we had to modify the original idea as described in part (b) of the proof.

Using the Euclidean algorithm for $L(u)[x]$, we find polynomials $A_i, B_i \in L(u)[x]$ with $d_i(x) = A_i(x) \cdot f(x + u\alpha) + B_i(x) \cdot G_i(x, u)$. Hence the product $d_1 \cdots d_m$ has the form

(1) $$d_1(x) \cdots d_m(x) = f(x + u\alpha) \cdot A(x) + B_1(x) \cdots B_m(x) \cdot G_1(x, u) \cdots G_m(x, u) = $$
$$f(x + u\alpha) \cdot A(x) + B(x) \cdot (NF)(x, u) = f(x + u\alpha) \cdot A(x) + B(x) \cdot (Nf)(x + u\alpha).$$

Suppose now that none of the factorizations $(\star\star)$ is proper. Then $\deg D_i(x, 0) = \deg_x D_i = \deg d_i$ is always either 0 or $\deg f$. If we had $\deg d_i = 0$ for all i, then the left-hand side of (1) would be an element of $L(u)$ so that (1) would imply that $f(x + u\alpha)$ and $(Nf)(x + u\alpha)$ are coprime in $L(u)[x]$. But this is impossible because f divides Nf (by the very definition of Nf).

Thus there is at least one index i_0 with $\deg d_{i_0} = \deg f$. We wish to show that this implies the irreducibility of f over L. Suppose that $f = gh$ in $L[x]$. Then

---

† For example, if f lies in $K[x]$ and is irreducible over K, but reducible over L, then $Nf = f^n$ and $d_i = f$ for $1 \leq i \leq m$ so that none of the d_i is a proper factor of f.

$f(x + u\alpha) = g(x + u\alpha)h(x + u\alpha)$ and $G_1(x,u) \cdots G_m(x,u) = (Nf)(x + u\alpha) = (Ng)(x + u\alpha) \cdot (Nh)(x + u\alpha)$. Since $K[x,u]$ is a unique factorization domain and since G_{i_0} is irreducible in $K[x,u]$, we conclude that $G_{i_0}(x,u)$ divides $(Ng)(x + u\alpha)$ or $(Nh)(x + u\alpha)$; let us assume without loss of generality that $G_{i_0}(x,u)$ divides $(Ng)(x + u\alpha)$. Since d_{i_0} divides both $f(x + u\alpha)$ and $G_{i_0}(x,u)$ in $L(u)[x]$ and since $\deg d_{i_0} = \deg f$, we see that $f(x + u\alpha)$ divides $G_{i_0}(x,u)$ and hence also $(Ng)(x + u\alpha)$ in $L(u)[x]$, say $(Ng)(x + u\alpha) = f(x + u\alpha) \cdot \varphi(x)$ with $\varphi \in L(u)[x]$. Clearing the denominators which occur in the coefficients of φ, this becomes

(2) $H(u) \cdot (Ng)(x + u\alpha) = Q(x,u) \cdot f(x + u\alpha)$ with $H \in L[x]$ and $Q \in L[x,u]$.

The term with the highest power of $x + u\alpha$ on the right-hand side is $c \cdot Q(x,u) \cdot (x + u\alpha)^{\deg f}$ with some $c \in L^\times$, whereas the corresponding term on the left-hand side has the form

$$d \cdot u^r \cdot \prod_{\sigma \in G_K^M} (x + u\sigma(\alpha))^{\deg g} = d \cdot u^r \cdot N((x + u\alpha)^{\deg g}) = d \cdot u^r \cdot N(x + u\alpha)^{\deg g}$$

with $d \in L^\times$. Since these two terms must be equal, we conclude that $(x + u\alpha)^{\deg f}$ divides $u^r \cdot N(x + u\alpha)^{\deg g}$. Plugging in $u = -1$, we see that

(3) $$(x - \alpha)^{\deg f} \text{ divides } N(x - \alpha)^{\deg g} .$$

On the other hand $x - \alpha$ divides $N(x - \alpha)$ in $L[x]$, say $N(x - \alpha) = (x - \alpha)q(x)$ and hence

(4) $$N(x - \alpha)^{\deg g} = (x - \alpha)^{\deg g} q(x)^{\deg g} .$$

As a consequence of (3) and (4) we see that $(x - \alpha)^{\deg f - \deg g}$ divides $q(x)^{\deg g}$. Let us suppose that g is a proper divisor of f so that $\deg f > \deg g$; then $x - \alpha$ divides $q(x)^{\deg g}$ and hence $q(x)$. But then

(5) $$(x - \alpha)^2 \text{ divides } (x - \alpha)q(x) = N(x - \alpha) \text{ in } L[x] .$$

Now consider again the minimal polynomial Φ of α over K. Clearly, there is a factorization $\Phi(x) = (x - \alpha)\varphi(x)$ in $L[x]$; taking the norm on both sides we obtain $\Phi(x)^n = (N\Phi)(x) = N(x - \alpha)(N\varphi)(x)$. Since $K[x]$ is a unique factorization domain and since Φ is irreducible, this implies that $N(x - \alpha)$ is a power of Φ (up to a unit of $K[x]$), say $N(x - \alpha) = \lambda\Phi(x)^r$ with $\lambda \in K^\times$. Since $r \deg \Phi = \deg \Phi^r = \deg N(x - \alpha) = [K(\alpha):K] = \deg \Phi$ we conclude that $r = 1$ so that $N(x - \alpha) = \lambda\Phi(x)$. Hence (5) shows that $(x - \alpha)^2$ divides $\Phi(x)$ so that α is a multiple root of its minimal polynomial Φ. This clearly contradicts the separability of α; hence our assumption that g is a proper factor of f was wrong. Consequently, f has no proper factors which is what we wanted to show. ∎

OUR results can be summarized as follows.

(8.17) Theorem. *Let L be obtained from $\mathbb{Q}$ by adjoining a finite number of (transcendental or algebraic) elements. Then there is a factorization algorithm for $L[x]$.*

Proof. Let $Z = \mathbb{Q}(X)$ where X is a transcendence basis of L over $\mathbb{Q}$ (the hypothesis implies that X must be finite). Then there is a factorization algorithm for $Z[x]$ by (8.11) and (8.15). Now L is obtained from Z by adjoining a finite number of algebraic elements (which are automatically separable because $\operatorname{char} \mathbb{Q} = 0$). Hence (8.16) implies that there is also a factorization algorithm for $L[x]$. ∎

THE last result exhibits a large class of fields K such that $K[x]$ allows a factorization algorithm. However, since the indicated algorithms in general involve lengthy and tedious computations, we should not conceal that our results are more of theoretical than of practical value.

LET us now find criteria for the irreducibility of polynomials.

(8.18) Theorem. *Let R be an integral domain and $f \in R[x]$, $f \neq 0$.*
(a) *If $\deg f = 2$ or 3, then f is reducible in $R[x]$ if and only if f has a root in R.*
(b) *Let $c \in R^{\times}$ and $d \in R$. Then f is reducible if and only if $g(x) := f(cx + d)$ is reducible.*
(c) *Let S be a commutative ring and suppose that $\varphi : R \to S$ is a homomorphism which does not map the leading coefficient of f to zero. If $f(x) = a_0 + a_1 x + \cdots + a_n x^n \in R[x]$, let*

$$\overline{f}(x) := \varphi(a_0) + \varphi(a_1)x + \cdots + \varphi(a_n)x^n \in S[x] .$$

If $\overline{f}$ is irreducible then so is f.
(d) *Let $f(0) \neq 0$. Then $f(x) = a_0 + a_1 x + \cdots + a_n x^n$ is reducible if and only if the* **reciprocal polynomial**

$$\widehat{f}(x) := a_0 x^n + a_1 x^{n-1} + \cdots + a_{n-1} x + a_n$$

is reducible.
(e) **(Eisenstein's criterion.)** *Let R be a unique factorization domain with quotient field Q. Let $f(x) = a_0 + a_1 x + \cdots + a_n x^n$ with $n = \deg f \geq 1$ such that the nonzero coefficients of f are relatively prime. If there is a prime $p \in R$ with*

$$p \mid a_0, \quad p \mid a_1, \quad p \mid a_{n-1}, \quad p \nmid a_n, \quad p^2 \nmid a_0$$

then f is irreducible in $R[x]$.

Proof. (a) If $f = gh$ then $\deg g + \deg h = \deg f = 2$ or 3; hence any proper factorization of f must involve a linear factor.
(b) If $f = f_1 f_2$ then $g = g_1 g_2$ where $g_i := f_i(cx + d)$ is a polynomial with the same degree as f_i. Conversely, if $g = g_1 g_2$ then $f = f_1 f_2$ where $f_i(x) := g_i(c^{-1}x - c^{-1}d)$.
(c) If $f = gh$ is reducible, then so is $\overline{f} = \overline{g}\,\overline{h}$; the condition on φ ensures that the transition from f to $\overline{f}$ and from g to $\overline{g}$ does not lower the degree.
(d) Simply observe that $\widehat{f}(x) = x^n f(\frac{1}{x})$; thus $f = gh$ if and only if $\widehat{f} = \widehat{g}\widehat{h}$. The

condition $a_0 \neq 0$ ensures that the transition from f to $\widehat{f}$ and from g to $\widehat{g}$ does not lower the degree.

(e) Suppose f possesses a proper factorization $f = gh$ in $R[x]$, say $g(x) = b_0 + b_1 x + \cdots + b_r x^r$ and $h(x) = c_0 + c_1 x + \cdots + c_s x^s$. By assumption, p divides $a_0 = b_0 c_0$; since p is prime, this implies $p \mid b_0$ or $p \mid c_0$, say $p \mid b_0$. But then $p \nmid c_0$ because $p^2 \nmid a_0$. Also $p \nmid a_n$ by assumption; hence p divides neither b_r nor c_s. Let i_0 be the smallest index i such that $p \nmid b_i$. Then

$$a_{i_0} = \underbrace{b_0 c_{i_0} + b_1 c_{i_0-1} + b_2 c_{i_0-2} + \cdots + b_{i_0-1} c_1}_{\text{divisible by } p} + \underbrace{b_{i_0} c_0}_{\text{not divisible by } p}$$

so that $p \nmid a_{i_0}$. By assumption this implies $i_0 = n$ so that $r \geq n$, which is only possible if $r = n$ and $s = 0$. So h is a constant polynomial, since the coefficients of f are relatively prime, and hence even an element of $R^\times = R[x]^\times$. This clearly contradicts our assumption that $f = gh$ is a proper factorization. ∎

UNFORTUNATELY, the above criteria encompass about all one can say in general about the irreducibility of polynomials. The examples below give an idea of how one can proceed to establish the irreducibility of a polynomial and of the trickery that can be useful in doing so.

(8.19) Examples. (a) The polynomials $f(x) = 3x^6 + 7x^4 + 21x^3 + 14$ and $g(x) = x^{2222} + 2x^{2220} + 4x^{2218} + \cdots + 2220x^2 + 2222$ are irreducible in $\mathbb{Z}[x]$ and $\mathbb{Q}[x]$; just apply Eisenstein's criterion with $p = 7$ for f and $p = 2$ for g.

(b) Let $p \in \mathbb{N}$ be a prime number. Then for any $n \in \mathbb{N}$ the polynomial $x^n - p$ is irreducible in $\mathbb{Z}[x]$ and $\mathbb{Q}[x]$, again by Eisenstein's criterion.

(c) Let $p \in \mathbb{N}$ be a prime number. Then the polynomial

$$f(x) = x^{p-1} + x^{p-2} + \cdots + x + 1$$

is irreducible in $\mathbb{Z}[x]$. The clue is to observe that $f(x) = \frac{x^p - 1}{x - 1}$. Now due to (8.18)(b), the polynomial f is irreducible if and only if $g(x) := f(x+1) = \sum_{k=0}^{p-1} \binom{p}{k+1} x^k$ is. But the irreducibility of g follows immediately from Eisenstein's criterion.

(d) Let $g(x) = x^5 + x^2 + 1 \in \mathbb{Z}_2[x]$. Since g has obviously no root in $\mathbb{Z}_2$, a possible factorization in $\mathbb{Z}_2[x]$ must involve a quadratic and a cubic factor. By the "brute force method" of trying all possible factorizations, we write

$$x^5 + x^2 + 1 = (x^3 + ax^2 + bx + 1)(x^2 + cx + 1) ;$$

multiplying out and trying to solve for a, b, c, we easily see that there is no such factorization.

(e) The reciprocal polynomial of $f(x) = 2x^4 + 4x^2 + 4x + 1$ is $g(x) = x^4 + 4x^3 + 4x^2 + 2$. Now g is irreducible by Eisenstein's criterion; hence f is irreducible by (8.18)(d).

(8.20) Example. A typical application of (8.18)(c) is to prove the irreducibility of a polynomial $f \in \mathbb{Z}[x]$ by reading its coefficients modulo some number $n \in \mathbb{N}$, using the homomorphism $\varphi : \mathbb{Z} \to \mathbb{Z}_n$. For example, let us show that for any odd number α the polynomial

$$f(x) = 3x^5 - 4x^4 + 2x^3 + x^2 + 18x + \alpha$$

is irreducible in $\mathbb{Z}[x]$. Reading the coefficients modulo 2, we obtain the polynomial

$$\overline{f}(x) = x^5 + x^2 + 1 \in \mathbb{Z}_2[x]$$

which was shown to be irreducible in (8.19)(d). Hence $f \in \mathbb{Z}[x]$ is irreducible due to (8.18)(c).

(8.21) Example. Let R be a unique factorization domain; then $\overline{R} := R[x]$ is also a unique factorization domain, due to (8.2). We claim that

$$f(x,y) := y^3 + x^m y^2 + x^n y + x \in R[x,y]$$

is irreducible for any $m, n \geq 1$. Indeed, $x \in \overline{R}$ is irreducible in $\overline{R}$, hence prime because $\overline{R}$ is a unique factorization domain. But then we can apply Eisenstein's criterion with $p := x$ to

$$F(y) := y^3 + x^m y^2 + x^n y + x \in \overline{R}[y] .$$

(8.22) Example. Let us give two proofs that the polynomial $p(x) = 5x^7 + 4 \in \mathbb{Q}[x]$ is irreducible. For the first proof observe that the reciprocal polynomial of p is $q(x) = 4x^7 + 5$. Now q is irreducible by Eisenstein's criterion, hence p is irreducible by (8.18)(d). For the second proof, we want to find a number $\alpha \in \mathbb{Z}$ such that $r(x) := p(x + \alpha) = 5(x + \alpha)^7 + 4$ is irreducible; then (8.18)(d) will yield the claim. Multiplying out, we have $r(x) = 5x^7 + 7(\cdots) + 5\alpha^7 + 4$; if we can choose α such that $5\alpha^7 + 4$ is divisible by 7, but not by 49, then Eisenstein's criterion will establish the irreducibility of r. Now Fermat's theorem yields $5\alpha^7 + 4 \equiv 5\alpha + 4$ modulo 7; this leads us to try $\alpha = 2$, a choice which does work.

(8.23) Example. Let us give two proofs for the irreducibility of $p(x) := 4x^3 - 3x - \frac{1}{2} \in \mathbb{Q}[x]$. First of all, p is irreducible if and only if $f(x) := 2p(x) = 8x^3 - 6x - 1$ is. But f has no rational root (the only possibilities being ± 1 due to (7.19)) and hence is irreducible due to (8.18)(a). Second, p is irreducible if and only if $g(x) := p(\frac{x+1}{2}) = x^3 + 3x^2 - 3$ is. But g is irreducible by Eisenstein's criterion.

FOR the fields $\mathbb{R}$ and $\mathbb{C}$ one can classify all irreducible polynomials in one variable.

(8.24) Theorem. (a) *The irreducible polynomials in $\mathbb{C}[x]$ are exactly the constant and the linear polynomials.*

(b) *The irreducible polynomials in $\mathbb{R}[x]$ are exactly the constant and linear polynomials and those quadratic polynomials which have no real root.*

Proof. (a) $\mathbb{C}$ is algebraically closed; hence every polynomial $f \in \mathbb{C}[x]$ splits into linear factors.

(b) Let $f \in \mathbb{R}[x]$ be irreducible with $\deg f > 1$. Then we can consider f as an element of $\mathbb{C}[x]$ and split apart a linear factor $x - \alpha$ with $\alpha \in \mathbb{C} \setminus \mathbb{R}$. Then $\bar{\alpha} \in \mathbb{C} \setminus \mathbb{R}$ is a second root of f so that f contains the factor $(x-\alpha)(x-\bar{\alpha}) = x^2 - (2\operatorname{Re}\alpha)x + |\alpha|^2 \in \mathbb{R}[x]$. Since f is irreducible, this implies that f is an associate of $x^2 - (2\operatorname{Re}\alpha)x + x^2$ which gives the claim. ∎

WE will now establish a criterion that allows one to decide whether or not two given polynomials have a common factor. Note that if we know offhand that the polynomials in question split (possibly in some extension of their coefficient ring) then it is clear that such a criterion must exist. In fact, suppose that a polynomial f has the roots $\alpha_1, \ldots, \alpha_n$. Then a polynomial g has a factor in common with f if and only if it has a root in common with f which is tantamount to saying that $g(\alpha_1) \cdots g(\alpha_n) = 0$. But this equation is symmetric in $\alpha_1, \ldots, \alpha_n$, hence is expressible as a polynomial equation in the elementary symmetric polynomials in $\alpha_1, \ldots, \alpha_n$ which are up to sign just the coefficients of f.

(8.25) Definition. *Let* $f(x) = a_0 + a_1 x + \cdots + a_n x^n$ *and* $g(x) = b_0 + b_1 x + \cdots + b_m x^m$ *be polynomials over a commutative ring* R *such that* $a_n \neq 0$ *and* $b_m \neq 0$. *The* **resultant** $\operatorname{Res}(f, g)$ *of* f *and* g *is the element of* R *which is defined as the determinant of the following* $(m + n) \times (m + n)$*-matrix:*

$$
M(f,g) := \begin{pmatrix}
a_0 & a_1 & a_2 & \cdots & \cdots & a_{n-1} & a_n & & & & 0 \\
 & a_0 & a_1 & a_2 & \cdots & \cdots & a_{n-1} & a_n & & & \\
 & & \ddots & \ddots & \ddots & & & & \ddots & \ddots & \\
0 & & & a_0 & a_1 & a_2 & \cdots & \cdots & a_{n-1} & a_n \\
b_0 & b_1 & \cdots & \cdots & b_m & & & & & & 0 \\
 & b_0 & b_1 & \cdots & \cdots & b_m & & & & & \\
 & & \ddots & \ddots & & & & \ddots & & & \\
0 & & & b_0 & b_1 & \cdots & & \cdots & b_m &
\end{pmatrix}.
$$

(8.26) Examples. (a) If $f(x) = ax + 1$ and $g(x) = bx + 1$ then

$$
\operatorname{Res}(f, g) = \det \begin{pmatrix} a & 1 \\ b & 1 \end{pmatrix} = a - b .
$$

(b) If $f(x) = x^2 + ax + 1$ and $g(x) = x^2 + bx + 1$, then

$$
\operatorname{Res}(f, g) = \det \begin{pmatrix} 1 & a & 1 & 0 \\ 0 & 1 & a & 1 \\ 1 & b & 1 & 0 \\ 0 & 1 & b & 1 \end{pmatrix} = (a - b)^2 .
$$

(c) If $f(x) = ax^2 + bx + c$ and $g(x) = px + q$, then

$$\operatorname{Res}(f,g) \;=\; \det \begin{pmatrix} a & b & c \\ p & q & 0 \\ 0 & p & q \end{pmatrix} \;=\; aq^2 - bpq + cp^2 \;=\; p^2 \cdot f\left(-\frac{q}{p}\right).$$

(8.27) Theorem. *Let R be a unique factorization domain. For two polynomials $f, g \in R[x]$, the following conditions are equivalent.*
   (1) *f and g have a non-constant common factor in $R[x]$.*
   (2) *There are polynomials $u, v \in R[x]$ with $\deg u < \deg g$ and $\deg v < \deg f$ such that $fu = gv$.*
   (3) *$\operatorname{Res}(f,g) = 0$.*
If f and g split over some extension S of R, then condition (1) is equivalent to
   (1') *f and g have a common root in S.*

Proof. Suppose (1) holds so that f and g have a non-constant factor $h \in R[x]$. Then we can write $f = hv$ and $g = hu$ where $\deg v < \deg f$ and $\deg u < \deg g$ and obtain $fu = huv = gv$ so that (2) holds. Suppose conversely that (2) holds. Since $R[x]$ is a unique factorization domain, due to (8.2), we have a decomposition $f = p_1 \cdots p_r$ which implies $p_1 \cdots p_r u = fu = gv$. Since $\deg v < \deg f$, one of the polynomials p_i cannot divide v, hence must divide g and is hence a common factor of f and g. So the equivalence of (1) and (2) is proved.

To show the equivalence of (2) and (3), let $M := M(f,g)$ and $r := \operatorname{Res}(f,g)$. By the definition of M, we have

$$M \begin{pmatrix} 1 \\ x \\ x^2 \\ \vdots \\ x^{m-1} \\ x^m \\ \vdots \\ x^{m+n-1} \end{pmatrix} = \begin{pmatrix} f(x) \\ xf(x) \\ x^2 f(x) \\ \vdots \\ x^{m-1} f(x) \\ g(x) \\ \vdots \\ x^{n-1} g(x) \end{pmatrix}.$$

Using $\operatorname{adj}(M)M = (\det M)\mathbf{1}$ we obtain

$$\operatorname{adj}(M) \begin{pmatrix} f(x) \\ xf(x) \\ x^2 f(x) \\ \vdots \\ x^{m-1} f(x) \\ g(x) \\ \vdots \\ x^{n-1} g(x) \end{pmatrix} = (\det M) \begin{pmatrix} 1 \\ x \\ x^2 \\ \vdots \\ x^{m-1} \\ x^m \\ \vdots \\ x^{m+n-1} \end{pmatrix}.$$

The first component of this equality gives an equation of the form

(⋆) $$U(x)f(x) + V(x)g(x) \;=\; \det M \;=\; r$$

where $\deg U < m$ and $\deg V < n$ (just imagine that the matrix multiplication on the left-hand side is really performed).

Now ($\star$) shows clearly that (3) implies (2); just choose $u := U$ and $v := -V$. Conversely, suppose that $fu = gv$ holds as in (3). Then ($\star$) implies that $fuV = gvV = rv - fUv$ so that

$$f(uV + Uv) = rv .$$

Now if r were nonzero, then the left-hand side would be nonzero, hence a polynomial of degree $\geq \deg f$, whereas the degree of the right-hand side would be $\deg v < \deg f$. This shows that $r = 0$ so that (3) holds. ∎

SUPPOSE that a polynomial f splits over some integral domain R. By (4.17)(b), the polynomial f has a multiple root if and only if f and f' have a common root. By (8.27) this is the case if and only if $\mathrm{Res}(f, f') = 0$. Comparing with our discussion at the end of section 4 this strongly suggests that there is a connection between the resultant of f and f' and the discriminant of f as defined in (4.27). The key step in revealing this connection is the following result. (See problem 52 below for a further discussion.)

(8.28) Proposition. *Let R be a commutative ring and let $f, g \in R[x]$ be polynomials which split over some extension S of R † so that $f(x) = a_n(x - x_1) \cdots (x - x_n)$ and $g(x) = b_m(x - y_1) \cdots (x - y_m)$. Then*

$$\mathrm{Res}(f, g) = a_n^m b_m^n \prod_{\substack{1 \leq i \leq n \\ 1 \leq j \leq m}} (x_i - y_j) = a_n^m \prod_{i=1}^n g(x_i) = (-1)^{mn} b_m^n \prod_{j=1}^m f(y_j) .$$

Proof. Let us introduce the polynomials $F(x, X_1, \ldots, X_n) := a_n(x - X_1) \cdots (x - X_n)$ and $G(x, Y_1, \ldots, Y_m) := b_m(x - Y_1) \cdots (x - Y_m)$ where $X_1, \ldots, X_n, Y_1, \ldots, Y_m$ are new variables. Then the resultant $\mathrm{Res}(F, G)$ of F and G with respect to the variable x is a polynomial in $X_1, \ldots, X_n, Y_1, \ldots, Y_m$, say $p(X_1, \ldots, X_n, Y_1, \ldots, Y_m) \in R[X_1, \ldots, X_n, Y_1, \ldots, Y_m]$. Now whenever we plug in $X_{i_0} = Y_{j_0}$ for some indices $1 \leq i \leq n$ and $1 \leq j_0 \leq m$ then F and G have a common root so that $P = \mathrm{Res}(F, G)$ vanishes; this implies that $X_{i_0} - Y_{j_0}$ is a factor of P. By considering degrees we see that this implies $P = c \cdot \prod_{i,j}(X_i - Y_j)$ with some constant c. This constant is easily recognized as $a_n^m b_m^n$ by finding the coefficient of $x_1^m x_2^m \cdots x_n^m$ in the expansion of the determinant by which $\mathrm{Res}(F, G)$ is defined. Thus

$$\mathrm{Res}(F, G) = a_n^m b_m^n \prod_{i,j}(X_i - Y_j) .$$

Now the claim follows easily by plugging in the ring elements x_i and y_j for the variables X_i and Y_j. ∎

---

† We will see later that if R is an integral domain then there is always a field L over which f and g split.

NOTE that the resultant can be used to find simultaneous solutions of polynomial equations. Suppose we want to find all pairs (x, y) with $f(x, y) = g(x, y) = 0$ where f and g are given polynomials, say

$$f(x, y) = a_0(y) + a_1(y)x + \cdots + a_{n-1}(y)x^{n-1} + a_n(y)x^n \quad \text{and}$$
$$g(x, y) = b_0(y) + b_1(y)x + \cdots + b_m(y)x^m.$$

If y is such that there is some x such that (x, y) is a solution then necessarily $\text{Res}(f(\cdot, y), g(\cdot y)) = 0$ which means that the determinant of the matrix

$$\begin{bmatrix} a_0(y) & a_1(y) & a_2(y) & \cdots & & a_{n-1}(y) & a_n(y) & & & 0 \\ & a_0(y) & a_1(y) & a_2(y) & \cdots & & a_{n-1}(y) & a_n(y) & & \\ & \ddots & \ddots & \ddots & & & & \ddots & \ddots & \\ 0 & & & a_0(y) & a_1(y) & a_2(y) & \cdots & & a_{n-1}(y) & a_n(y) \\ b_0(y) & b_1(y) & \cdots & \cdots & b_m(y) & & & & & 0 \\ & b_0(y) & b_1(y) & \cdots & \cdots & b_m(y) & & & & \\ & \ddots & \ddots & & & & \ddots & & & \\ 0 & & & b_0(y) & b_1(y) & \cdots & & \cdots & & b_m(y) \end{bmatrix}$$

vanishes. Unless this determinant is the zero polynomial in y we get a finite number of roots y, and for each of these we can find all possible values for x such that (x, y) is a solution of the given system of two polynomial equations.

(8.29) Application. Suppose a curve in $\mathbb{R}^2$ is given by the parametric representation

$$x(t) = t^2 + t, \qquad y(t) = t^2 - t + 1.$$

We want to use the concept of the resultant of two polynomials to write down the equation for the curve in implicit form $F(x, y) = 0$. To do so, we observe that a point (x, y) lies on the curve if and only if the two polynomials

$$f(t) := t^2 + t - x \in \mathbb{R}[t] \quad \text{and} \quad g(t) := t^2 - t + 1 - y \in \mathbb{R}[t]$$

have a common root. Since f and g are quadratic polynomials and neither is a multiple of the other, this happens exactly if f and g have a non-constant common factor, i.e., if $\text{Res}(f, g) = 0$. Now

$$\text{Res}(f, g) = \det \begin{pmatrix} -x & 1 & 1 & 0 \\ 0 & -x & 1 & 1 \\ 1-y & -1 & 1 & 0 \\ 0 & 1-y & -1 & 1 \end{pmatrix} = (x - y)^2 - 4y + 3$$

so that the implicit equation for the curve is

$$(x - y)^2 - 4y + 3 = 0.$$

WE close this section with a result which is sometimes useful to find factorizations of polynomials in several variables.

(8.30) Proposition. *Let R be an integral domain and let $f \in R[x_1, \ldots, x_n]$ be homogeneous. If $f = gh$ with $g, h \in R[x_1, \ldots, x_n]$, then g and h are also homogeneous.*

Proof. Decompose the polynomials g and h into their homogeneous components, say $g = \sum_{i=m}^{M} g_i$ and $h = \sum_{j=n}^{N} h_j$ with g_m, g_M, h_n and h_N all different from zero. Then $g_m h_n$ and $g_M h_N$ are homogeneous components of $gh = f$ of degrees $m + n$ and $M + N$, respectively. Since f is homogeneous, this implies $m + n = M + N$; but this is only possible if $m = M$ and $n = N$, since $m \leq M$ and $n \leq N$. We conclude that $g = g_m = g_M$ and $h = h_n = h_N$, which is the claim. ∎

Exercises

Problem 1. Let $f, g \in \mathbb{Q}[x]$ such that $fg \in \mathbb{Z}[x]$.
(a) Show that any product of a coefficient of f with a coefficient of g is an integer.
(b) Show that if f and g are monic then they both lie in $\mathbb{Z}[x]$.
(c) Verify in the example $f(x) = 3x + \frac{9}{2}$ and $g(x) = 4x^2 - \frac{2}{3}x - \frac{4}{3}$ that fg can be factorized in $\mathbb{Z}[x]$.

Problem 2. Let $K \subseteq L$ be fields. Show that if $f, g \in K[x]$ and if h is a monic greatest common divisor of f and g in $L[x]$ then actually $h \in K[x]$.

Problem 3. Let R be a unique factorization domain with quotient field K and let $f_1, \ldots, f_n \in R[x]$. Show that for a primitive polynomial $\Phi \in R[x]$ the following conditions are equivalent.
(1) Φ is a greatest common divisor of $f_1, \ldots, f_n$ in $K[x]$;
(2) Φ is a common divisor of $f_1, \ldots, f_n$ in $R[x]$, and there is an element $r \in R \setminus \{0\}$ with $r\Phi \in \langle f_1, \ldots, f_n \rangle \unlhd R[x]$.
Show that Φ is not necessarily a greatest common divisor of $f_1, \ldots, f_n$ in $R[x]$.

Problem 4. Let K be a field.
(a) Suppose $f, g \in K[x_1, \ldots, x_n]$ are relatively prime. Show that there are polynomials $p, q \in K[x_1, \ldots, x_n]$ and $d \in K[x_1, \ldots, x_{n-1}]$ with $pf + qg = d$. (I.e., we can choose a linear combination of f and g which does not depend on one of the variables.)
(b) Show that if $f, g \in K[x, y]$ are relatively prime, there are only finitely many points $(x, y) \in K^2$ with $f(x, y) = g(x, y) = 0$.

Problem 5. Let K be a field. Show that the polynomial ring $K[x_1, x_2, x_3, \ldots]$ (with an infinite number of variables) is a unique factorization domain.

Problem 6. (a) Let R be an integral domain which is not a field (so that $R^\heartsuit \neq \emptyset$). Show that $R[x]$ is not a principal ideal domain.
Hint. If R is not even a factorization domain, the claim is an immediate consequence of (8.2). Otherwise there is an irreducible element $c \in R$. Show that in this case the ideal $\langle c, x \rangle \unlhd R[x]$ is not a principal ideal.
(b) Show that $\mathbb{Z}[x]$ is not a principal ideal domain. Also, if K is a field, show that $K[x_1, \ldots, x_n]$ is not a principal ideal domain whenever $n \geq 2$.

Problem 7. (a) Let $f(x, y)$ be a polynomial in two variables. Show that if $f(x, x)$ is the zero polynomial then $f(x, y) = (x - y)g(x, y)$ for some polynomial g.
(b) Let $f(x_0, x_1, \ldots, x_n)$ be a polynomial in $n + 1$ variables $x_0, x_1, \ldots, x_n$. Show that if there is an index $1 \leq i \leq n$ such that $f(x_i, x_1, \ldots, x_n)$ is the zero polynomial then f is divisible by $x_i - x_0$.

Problem 8. Factorize the following polynomials in three variables.

$$
\det \begin{pmatrix} 1 & x & x^3 \\ 1 & y & y^3 \\ 1 & z & z^3 \end{pmatrix} \qquad
\det \begin{pmatrix} 1 & x & x^4 \\ 1 & y & y^4 \\ 1 & z & z^4 \end{pmatrix} \qquad
\det \begin{pmatrix} 1 & x & yz \\ 1 & y & xz \\ 1 & z & xy \end{pmatrix}
$$

Problem 9. Let $a_1, \ldots, a_n \in \mathbb{C}$ and let $\varepsilon_k := e^{2k\pi i/n}$. Show that

$$
\det \begin{pmatrix}
a_1 & a_2 & a_3 & \cdots & a_n \\
a_n & a_1 & a_2 & \cdots & a_{n-1} \\
a_{n-1} & a_n & a_1 & \cdots & a_{n-2} \\
\vdots & \vdots & \vdots & & \vdots \\
a_2 & a_3 & a_4 & \cdots & a_1
\end{pmatrix} = \Pi_{k=1}^n (a_1 + \varepsilon_k a_2 + \varepsilon_k^2 a_3 + \cdots + \varepsilon_k^{n-1} a_n) \ .
$$

(This determinant is called the **circulant** of $a_1, \ldots, a_n$.)

Problem 10. Apply the algorithm presented in (8.11) to find a prime factorization of $f(x) = 2x^4 + 8x^3 + 9x^2 + 2x - 3$ in $\mathbb{Z}[x]$.

Problem 11. Let K be a finite field. Show that there is a factorization algorithm for $K[x]$.

Problem 12. Let K be a field. Show that the following statements are equivalent:
(1) there is a factorization algorithm for $K[x]$;
(2) there is an method which exhibits in a finite number of steps an irreducible factor of any polynomial $p \in K[x]^\heartsuit$.

Problem 13. Let $L = K(\alpha)$ where α is algebraic over K with minimal polynomial $\Phi \in K[x]$. Let $n := \deg \Phi$.
(a) Show that every element $f \in L[x]$ ben be uniquely written in the form $f(x) = f_0(x) + f_1(x)\alpha + \cdots + f_{n-1}(x)\alpha^{n-1}$.
(b) Let $f \in L[x]$. Show that for any index $0 \le i \le n-1$ there are uniquely determined polynomials $f_{ij} \in K[x]$ such that

$$
f(x) \cdot \alpha^i = \sum_{j=0}^{n-1} f_{ij}(x) \cdot \alpha^j \ .
$$

Show that $(Nf)(x) = \det \big(f_{ij}(x) \big)_{i,j}$.
(c) Deduce from (b) the basic properties of the norm mapping $N : L[x] \to K[x]$, namely, that $N(fg) = (Nf)N(g)$ for all $f, g \in L[x]$ and that $N(f) = f^n$ for all $f \in K[x]$.

Problem 14. Check the following polynomials for irreducibility!

$$
x^2 + x + 1 \in \mathbb{Z}_2[x] \ , \quad x^2 + 1 \in \mathbb{Z}_7[x] \ , \quad x^3 - 9 \in \mathbb{Z}_{11}[x] \ , \quad x^3 - p \in \mathbb{Z}_{31}[x]
$$

Problem 15. Which of the following cubic polynomials are irreducible over $\mathbb{Q}$?
(a) $x^3 - 2$
(b) $x^3 - x + 1$
(c) $x^3 - x - 1$
(d) $8x^3 - 6x - 1$
(e) $x^3 + x^2 - 2x - 1$
(f) $x^3 + 2x + 10$
(g) $x^3 - 2x^2 + x + 15$
(h) $3x^3 + x - 5$
(i) $x^3 - 5x^2 + 1$
(j) $8x^3 - 6x - 1$
(k) $x^3 + x^2 - 2x - 1$
(l) $x^3 + 39x^2 - 4x + 8$
(m) $x^3 + x^2 + x + 1$

Problem 16. Which of the following quartic polynomials are irreducible over $\mathbb{Q}$?
(a) $x^4 - x + 1$
(b) $x^4 + 1$
(c) $x^4 + 2$
(d) $x^4 + 4$
(e) $x^4 - 2x^2 + 8x + 6$
(f) $2x^4 - 4x + 3$
(g) $x^4 + 2x + 2$
(h) $x^4 - 2x^2 + 2$
(i) $x^4 + 3x + 7$
(j) $x^4 + 15x^3 + 7$
(k) $2x^4 - 8x^2 + 1$
(l) $x^4 + 3x^3 + x^2 - 2x + 1$
(m) $3x^4 + 6x^2 - 12x + 10$
(n) $7x^4 + 4x^3 + 40x^2 + 48x + 96$
(o) $x^4 + 5x^3 + 10x^2 + 10x + 5$
(p) $x^4 + x^3 + x^2 + x + 1$
(q) $x^4 - x^2 + 1$

Problem 17. Which of the following polynomials are irreducible over $\mathbb{Q}$?
(a) $x^5 - 4x^3 - 2$
(b) $x^5 + 4x^4 + 2x^3 + 3x^2 - x + 5$
(c) $x^5 - 2x^4 + 6x + 10$
(d) $x^6 + x^3 + 1$
(e) $x^6 - 72$
(f) $x^6 + x^3 + 1$
(g) $x^6 + x^3 + 1$
(h) $x^7 - 21x^6 + 14x^3 + 6x - 15$
(i) $x^{12} - 27$
(j) $x^{13} + 1$
(k) $x^n - p$

Problem 18. Which of the following polynomials are irreducible over $\mathbb{Z}_3$?
(a) $x^2 + x + 2$
(b) $x^3 + x + 1$
(c) $x^3 + 2x - 1$
(d) $x^4 + 2x^2 + 1$
(e) $x^4 + x^3 + x^2 + 1$
(f) $x^3 - x - 1$
(g) $2x^5 + 3x^4 + 9x^3 + 15$
(h) $x^3 + x^2 + x + 1$
(i) $x^4 - x + 1$
(j) $x^4 + 15x^3 + 7$

Problem 19. Find the prime factorizations of the following polynomials in $\mathbb{Q}[x]$.
(a) $x^4 - 1$
(b) $x^4 + 4$
(c) $x^5 + 2x^3 + x^2 + x + 1$
(d) $x^5 - x^4 - 3x^2 + 9x - 6$
(e) $x^5 + 4x^3 + 4x$
(f) $x^5 + 2x^4 + x^3 + 3x^2 + 6x + 3$
(g) $x^4 + x^2 + 1$
(h) $3x^3 + x - 5$
(i) $2x^4 + x^3 + x^2 + x - 1$

Problem 20. Find the prime factorizations of the following polynomials in $\mathbb{Q}[x]$.
(a) $f(x) = x^7 - x^5 + x^3 - x$
(b) $f(x) = x^4 + 2x^3 + 2x^2 + (1/4)$
(c) $f(x) = x^9 - 1$
(d) $f(x) = x^5 + x^3 - 2x^2 - 2$
(e) $f(x) = 2x^4 + 8x^3 + 9x^2 + 2x - 3$
(f) $f(x) = x^4 + 8x^3 + x^2 + 2x + 5$
(g) $f(x) = x^8 - 1$
(h) $f(x) = 2x^2 - 3x + 4$

Problem 21. Let $f(x) := x^4 + x + 1$.
(a) Show that f is irreducible over $\mathbb{Q}$ by using reduction modulo 2.
(b) Find the prime decomposition of f over $\mathbb{R}$.
(c) Show that if $a + ib \in \mathbb{C}$ is a root of f then $4a^2$ is a root of $g(x) = x^3 - 4x - 1$.

Problem 22. Consider the polynomial $f(x) = x^4 + 2ax^2 + b^2$ with $a, b \in \mathbb{Z}$.
(a) Show that if p is a prime then f is reducible over $\mathbb{Z}_p$ for all choices of a and b.
   Hint. Since $f(x) = (x^2 + a)^2 - (a^2 - b^2) = (x^2 + b)^2 - 2(b - a)x^2 = (x^2 - b)^2 - 2(-a - b)x^2$ it is enough to show that one of the numbers $a^2 - b^2$, $2(b - a)$ and $2(-a - b)$ is necessarily a square in $\mathbb{Z}_p$.
(b) Find conditions for a and b which ensure that f is irreducible over $\mathbb{Q}$.
   Remark. This shows that there are polynomials whose irreducibility cannot be checked by reduction modulo a prime.

Problem 23. (a) Find the prime factorizations of $x^4 + 1$ in $\mathbb{C}[x]$, $\mathbb{R}[x]$ and $\mathbb{Q}[x]$.
(b) Find the prime decompositions of $x^5 + x + 1$ in $\mathbb{Q}[x]$, $\mathbb{Z}_2[x]$, $\mathbb{Z}_5[x]$ and $\mathbb{Z}_{19}[x]$.
(c) Find the prime factorization of $x^4 - x^3 + x^2 - x$ in $\mathbb{Z}_3[x]$.
(d) Find the prime factorizations of $x^3 - 2$ in $\mathbb{R}[x]$, $\mathbb{C}[x]$ and $\mathbb{Z}_5[x]$.
(e) Find the prime factorizations of $x^9 - x$ and $x^{27} - x$ in $\mathbb{Z}_3[x]$.
(f) Find the prime decomposition of $x^{10} + x^5 + 1$ in $\mathbb{Z}_5[x]$.
(g) Find the prime factorizations of $x^5 - 11x + 6$ in $\mathbb{Z}_2[x]$, $\mathbb{Z}_3[x]$ and $\mathbb{Z}_5[x]$.
(h) Find the prime factorization of $(-1 + 3i)x^2 - (3 + i)x + (4 + 2i)$ in $(\mathbb{Z} + i\mathbb{Z})[x]$.
(i) Find the prime factorizations of $x^5 - x^4 - 6x^3 + 6x^2 - 3x + 3$ in $\mathbb{Q}[x]$, $\mathbb{Z}_5[x]$ and $\mathbb{Z}_{13}[x]$.

Problem 24. Find all irreducible polynomials in $\mathbb{Z}_2[x]$ of degree ≤ 4.

Problem 25. Let $f \in \mathbb{R}[x]$ be a monic quadratic polynomial. Show that f is irreducible in $\mathbb{R}[x]$ if and only if there are $a, b \in \mathbb{R}$ with $b \neq 0$ such that $f(x) = (x - a)^2 + b^2$.

Problem 26. Let $a \in \mathbb{Q} \setminus \{0\}$. Show that $x^4 + a^4$ is irreducible over $\mathbb{Q}$ but reducible over $\mathbb{Q}(\sqrt{2})$.

Problem 27. (a) Show that $x^{n-1} + x^{n-2} + \cdots + x + 1$ is irreducible in $\mathbb{Q}[x]$ if and only if n is a prime number.
(b) Show that $x^{n+1} - n(x - 1) - x$ is divisible by $(x - 1)^2$ for all $n \in \mathbb{N}$.
(c) Show that $x^{2^n} + 1$ is irreducible in $\mathbb{Q}[x]$ for all $n \in \mathbb{N}$.
(d) Show that every polynomial of the form $x^{2m+2} + x^{2n+1} + 1$ is reducible.

Problem 28. (a) For which $n \in \mathbb{Z}$ is $f(x) = x^3 + x^2 + nx + 2$ irreducible over $\mathbb{Q}$?
(b) For which values of $k \in \mathbb{Z}$ is $g(x) = x^5 - kx + 1$ irreducible over $\mathbb{Q}$?

Problem 29. (a) Show that $x^3 + mx + n$ is irreducible in $\mathbb{Z}[x]$ whenever m and n are odd numbers.
(b) Show that $x^3 + x^2 + 5ax - 6$ is irreducible in $\mathbb{Z}[x]$ for any choice of $a \in \mathbb{Z}$.

Problem 30. Suppose that $f \in \mathbb{Z}[x]$ satisfies $f(0) \neq 0$ and has the property that $n - 1$ of its roots in $\mathbb{C}$ have an absolute value less than 1. Show that f is irreducible in $\mathbb{Z}[x]$.

Problem 31. (a) Show that if $a_1, \ldots, a_n \in \mathbb{Z}$ are pairwise different then $p(x) := (x - a_1) \cdots (x - a_n) - 1$ is irreducible in $\mathbb{Z}[x]$.
Hint. If $f = gh$ in $\mathbb{Z}[x]$ then $g(a_i) = h(a_i) = \pm 1$ for $1 \leq i \leq n$.
(b) Suppose that $f(x) = (x - a_1)^2 \cdots (x - a_n)^2 + 1$ where $a_1, \ldots, a_n \in \mathbb{Z}$ are pairwise different. Show that f is irreducible in $\mathbb{Z}[x]$.
(c) Let $f \in \mathbb{Z}[x]$ be a polynomial of odd degree m. Show that if f takes the value 1 at least m times then f is irreducible.

Problem 32. Find all integers a with the property that $x^2 - x + a$ divides a polynomial of the form $x^m + x + n$ where $m, n \in \mathbb{N}$ and where m is odd.

Problem 33. Show that a monic polynomial $f \in \mathbb{Z}[x]$ has at most three different rational roots if $f(0)$ is a prime number.

Problem 34. Find all complex roots of the following polynomials.
(a) $6x^4 - 31x^3 + 25x^2 + 33x + 7$
(b) $2x^4 + 8x^3 + 9x^2 + 2x - 3$
(c) $x^4 - 4x^3 - 6x^2 - 12x + 9$
(d) $x^4 - 10x^3 + 26x^2 - 5x - 2$
(e) $x^4 + 4x^3 + 6x^2 + 4x + 1$
(f) $x^4 - x^3 + x^2 - 3x + 2$
(g) $x^4 + x^3 - 3x^2 - 5x - 2$
(h) $x^4 - 2x^3 + 3x^2 - 2x + 2$

Problem 35. Let $f \in \mathbb{Z}[x]$.
(a) Show that $a \equiv b$ modulo m implies $f(a) \equiv f(b)$ modulo m.
(b) Show that if $f(0)$ and $f(1)$ are both odd then f has no root in $\mathbb{Z}$. If in addition the leading coefficient of f is odd, then f has no root in $\mathbb{Q}$.

Problem 36. Show that if $d + 1$ is the greatest common divisor of $m + 1$ and $n + 1$ in $\mathbb{N}$ then $1 + x + \cdots + x^d$ is a greatest common divisor of $1 + x + \cdots + x^n$ and $1 + x + \cdots + x^m$ in $\mathbb{Q}[x]$.

Problem 37. Show that $x^2 - 1 = (x - 1)(x - 14) = (x - 4)(x - 11)$ in $\mathbb{Z}_{15}[x]$.

Problem 38. Show that for each number $t \in [-1, 1]$ the polynomial $4x^3 - 3x - t$ has a real root. (**Hint.** There is a trigonometric formula which expresses $\cos(3\theta)$ as a function of $\cos\theta$).

Problem 39. Let $n \in \mathbb{Z} \setminus \{0, 1\}$ be a square-free number and let $f \in \mathbb{Q}[x]$.
(a) Show that if $a + b\sqrt{n}$ is a root of f with $a, b \in \mathbb{Q}$ then so is $a - b\sqrt{n}$.
(b) Suppose f is cubic polynomial. Show that if f has a root in $\mathbb{Q} + \mathbb{Q}\sqrt{n}$ then f has a root in $\mathbb{Q}$.

Problem 40. Suppose that $f(x) = (x - a)(x - b)(x - c)(x - d) - p^2$ where $a, b, c, d \in \mathbb{Z}$ are pairwise distinct and where p is a prime number. Show that if f has a root in $\mathbb{Z}$ then this root must be $(a + b + c + d)/4$.

Problem 41. (a) Suppose that $p(x) = a_n x^n + a_{n-1} x^{n-1} + \cdots + a_1 x + a_0$ where $a_0 \neq 0$ and $a_{n-k} = (-1)^{n-k} a_k$ for $1 \leq k \leq n$. Show that if α is a root of p then so is $-1/\alpha$.
(b) Find the roots of $x^4 + x^3 - x^2 - x + 1$.

Problem 42. (a) Let $f(x) = x^4 + ax^3 + bx^2 + cx + d$. Show that if $c^2 = da^2$ then there is a factorization $f(x) = (x^2 + px + q)(x^2 + rx + q)$.

(b) Find all roots of the polynomial $x^4 + x^3 - x^2 - x + 1$.

Problem 43. Let K be a field with an element ε such that $\varepsilon^3 = 1$; for example $K = \mathbb{Q}, \varepsilon = 1$, $K = \mathbb{C}, \varepsilon = e^{2\pi i/3}$ or $K = \mathbb{Z}_7, \varepsilon = [2]$. Show that $x^3 - 3abx + a^3 + b^3$ is divisible by $x + a\varepsilon + b\varepsilon^2$.

Problem 44. Let $p, q \in \mathbb{C}[x]$ be nonconstant polynomials. Suppose that p and q have the same roots and that $p + 1$ and $q + 1$ have the same roots (possibly with different multiplicities). Show that $p = q$.

Problem 45. Show that for a polynomial $p \in \mathbb{C}[x]$ the following conditions are equivalent.

(1) Some power of p' divides some power of p.
(2) p' divides some power of p.
(3) p' divides p.
(4) $p(x) = c(x - x_0)^n$.

Problem 46. Let $K \subseteq L$ be fields and let $I = \langle q \rangle \trianglelefteq L[x]$ be the ideal generated by a polynomial $q \in L[x]$. Show that if there is an element $p \in I \cap K[x]$ which has no multiple prime factors then q also has no multiple prime factors.

Hint. A polynomial has no multiple prime factors if and only if it is coprime with its derivative.

Problem 47. Let K be a field and let $g \in K[x]$ be a polynomial with a decomposition $g(x) = c \prod_{i=1}^{n} g_i(x)^{d_i}$ where $c \in K$ and where $g_1, \ldots, g_n$ are monic irreducible polynomials. Show that if f is a polynomial with $\deg f < \deg g$ then there are polynomials f_i with $\deg f_i < \deg g_i^{d_i}$ such that

$$\frac{f}{g} = \frac{f_1}{g_1^{d_1}} + \cdots + \frac{f_n}{g_1^{d_n}}.$$

Problem 48. (a) Does the polynomial $f(x) = x^9 + x^7 + 8x^4 + 5x^3 + 1 \in \mathbb{C}[x]$ have a multiple root?

(b) Suppose there is a field $L \supseteq \mathbb{Z}_3$ such that $g(x) = x^9 + x^6 + x^3 + 1 \in \mathbb{Z}_3[x]$ splits over L. Does g have a multiple root in L?

(c) Show that the polynomials $p(x) = x^5 - ix^4 + 5x^3 - 5ix^2 + 6x - 6i$ and $q(x) = 2x^3 - ix^2 + 6x - 3i$ have a common root in $\mathbb{C}$.

Problem 49. Find the resultant of $f(x) = x^3 + 2x^2 + 3x + 5$ and $g(x) = 6x^2 + 8x + 9$.

Problem 50. Show without any calculation that

$$\det \begin{pmatrix} 1 & 0 & -1 & 0 & 0 \\ 0 & 1 & 0 & -1 & 0 \\ 0 & 0 & 1 & 0 & -1 \\ 1 & 3 & 3 & 1 & 0 \\ 0 & 1 & 3 & 3 & 1 \end{pmatrix} = 0 .$$

Problem 51. Let R be a unique factorization domain.
(a) Find $\mathrm{Res}(a_n x^n + \cdots + a_1 x + a_0, ax + b)$, $\mathrm{Res}(x^2 + px + q, x^2 + ax + b)$ and $\mathrm{Res}(x^3 + px + q, x^2 + ax + b)$.
(b) Prove that $\mathrm{Res}(f + rg, g) = \mathrm{Res}(f, g)$ and $\mathrm{Res}(f_1 f_2, g) = \mathrm{Res}(f_1, g)\,\mathrm{Res}(f_2, g)$ for all $f, f_1, f_2, g \in R[x]$ and $r \in R$.

Problem 52. Let R be a unique factorization domain and let $f(x) = a_n x^n + \cdots + a_1 x + a_0 = a_n(x - x_1) \cdots (x - x_n) \in R[x]$. The **discriminant** of f is defined as

$$\mathrm{Dis}(f) := \frac{(-1)^{\frac{1}{2}n(n-1)}}{a_n} \,\mathrm{Res}(f, f') .$$

(a) Show that

$$\mathrm{Dis}(f) = \frac{(-1)^{\frac{1}{2}n(n-1)}}{a_n^{n-2}} \prod_{i=1}^{n} f'(x_i) = a_n^{2n-2} \prod_{i<k} (x_i - x_k)^2$$

and relate this definition of the discriminant of a polynomial to the slightly different one given in (4.27).
(b) Verify that $\mathrm{Dis}(fg) = \mathrm{Dis}(f)\,\mathrm{Dis}(g)\big(\mathrm{Res}(f,g)\big)^2$ and that $\mathrm{Dis}(f_c) = \mathrm{Dis}(f)$ where $f_c(x) := f(x + c)$.

Problem 53. Find the discriminant of a pure polynomial $x^n - a$. Find the resultant of two pure polynomials $x^m - a$ and $x^n - b$.

Problem 54. Show that the discriminant of $f(x) = x^5 + px + q$ is $4^4 p^5 + 5^5 q^4$.

Problem 55. Let R be an integral domain. A polynomial $f \in R[x_1, \ldots x_n]$ is called **weight-homogeneous** of degree d with weight $\alpha = (\alpha_1, \ldots, \alpha_n) \in \mathbb{N}^n$ if all monomials $x_1^{i_1} \cdots x_n^{i_n}$ occurring in f satisfy $\alpha_1 i_1 + \cdots + \alpha_n i_n = d$.
(a) Show that if f is weight-homogeneous with weight α and if $f = gh$ is a factorization of f, then g and h are also weight-homogeneous with weight α.
(b) Let p and q be coprime natural numbers. Show that $f(x, y) = x^p - y^q$ is irreducible in $R[x, y]$.

9. Number-theoretical applications of unique factorization

IN this section we want to show how nontrivial number-theoretical results can be obtained by using arithmetic in seemingly "fancy" rings introduced in previous sections. Our first application deals with the representability of prime numbers by quadratic forms.

(9.1) Theorem. *Let $n \in \mathbb{Z} \setminus \{0,1\}$ be square-free and let*

$$\mathcal{O}_n := \begin{cases} \mathbb{Z} + \mathbb{Z}\sqrt{n}, & \text{if } n \not\equiv 1 \mod 4; \\ \mathbb{Z} + \mathbb{Z}\frac{1+\sqrt{n}}{2}, & \text{if } n \equiv 1 \mod 4 \end{cases}$$

and

$$Q_n(x,y) := \begin{cases} x^2 - ny^2, & \text{if } n \not\equiv 1 \mod 4; \\ x^2 + xy + \frac{1}{4}y^2(1-n), & \text{if } n \equiv 1 \mod 4. \end{cases}$$

Suppose that $\mathcal{O}_n$ is a unique factorization domain. Then for an odd prime number $p \in \mathbb{N}$ the following conditions are equivalent.

(1) The congruence $z^2 \equiv n$ modulo p possesses a solution $z \in \mathbb{Z}$.
(2) p is not a prime element in $\mathcal{O}_n$.
(3) p is a product of two prime elements in $\mathcal{O}_n$, each with norm $\pm p$.
(4) There is an element $\pi \in \mathcal{O}_n$ with $N(\pi) = \pm p$.
(5) There are integers $x, y \in \mathbb{Z}$ such that $p = \pm Q_n(x,y)$.

Proof. Before we start let us motivate the definition of Q_n. Depending on whether $n \not\equiv 1$ or $n \equiv 1 \mod 4$, an element π of $\mathcal{O}_n$ can be written as $\pi = x + y\sqrt{n}$ or $\pi = x + y\frac{1+\sqrt{n}}{2}$ with $x, y \in \mathbb{Z}$. Then the norm of π is given by

$$(\star) \quad N(\pi) = \pi\overline{\pi} = \begin{cases} x^2 - ny^2, & \text{if } n \not\equiv 1 \mod 4; \\ (x + \frac{y}{2})^2 - n(\frac{y}{2})^2 = x^2 + xy + \frac{y^2}{4}(1-n), & \text{if } n \equiv 1 \mod 4. \end{cases}$$

This shows that the quadratic form Q_n is just a suitable expression of the norm on $\mathcal{O}_n$.

(1)$\Longrightarrow$(2). By hypothesis, there is a number $z \in \mathbb{Z}$ such that p divides $z^2 - n = (z + \sqrt{n})(z - \sqrt{n})$ in $\mathbb{Z}$ and hence also in $\mathcal{O}_n$. On the other hand, p divides none of the factors $z \pm \sqrt{n}$ in $\mathcal{O}_n$ because $\frac{z}{p} \pm \frac{1}{p}\sqrt{n}$ are obviously not elements of $\mathcal{O}_n$. (Here the assumption $p \neq 2$ is used.) This shows that p is not a prime element of $\mathcal{O}_n$.

(2)$\Longrightarrow$(3). Since $N(p) = p^2 \neq \pm 1$ the element p is not a unit in $\mathcal{O}_n$, hence possesses a prime factor π, say $p = \pi\alpha$. Here α is not a unit, because p is not a prime element. Taking norms, we see that $p^2 = N(\pi)N(\alpha)$. Since $N(\pi)$ and $N(\alpha)$ are different from ± 1, this leaves only the possibilities $N(\pi) = N(\alpha) = \pm p$ (which implies that α is irreducible and hence prime).

(3)$\Longrightarrow$(4). This is trivial.

(4)$\Longrightarrow$(5). This follows immediately from our initial remark that if $\pi = x + y\sqrt{n}$ or $\pi = x + y\frac{1+\sqrt{n}}{2}$, respectively, then $N(\pi) = Q_n(x,y)$.

(5)$\Longrightarrow$(1). Suppose $p = \pm Q_n(x,y)$ with $x, y \in \mathbb{Z}$. Then $p \nmid y$. Indeed, if $p \mid y$ then $p \mid Q_n(x,y)$ implies $p \mid x^2$, hence $p \mid x$ and consequently $p^2 \mid Q_n(x,y)$ which contradicts

163

the hypothesis that $p = \pm Q_n(x, y)$. Now $p|y$ implies that the congruence $y\eta \equiv 1$ modulo p possesses a solution $\eta \in \mathbb{Z}$. Now let us treat the two cases $n \not\equiv 1$ and $n \equiv 1$ modulo 4 separately. In the first case, we multiply the equation $\pm p = Q_n(x, y) = x^2 - ny^2$ by η^2 to obtain $\pm p\eta^2 = \eta^2 x^2 - n\eta^2 y^2$; modulo p this reads $0 \equiv \eta^2 x^2 - n$ which is a congruence of the desired form (take $z = \eta x$). In the second case, we have $\pm 4p = 4Q_n(x, y) = 4x^2 + 4xy + y^2(1 - n) = (2x + y)^2 - ny^2$; multiplying this equation by η^2, we obtain $\pm 4p\eta^2 = (2x + y)^2 \eta^2 - ny^2 \eta^2$. Modulo p, this reads $0 \equiv (2x + y)^2 \eta^2 - n$ which is again a congruence of the desired form; just take $z = (2x + y)\eta$. Note that this implication did not invoke the unique factorization property. ∎

AS a corollary, let us prove a famous theorem first found by Fermat.

(9.2) Theorem (Fermat). *An odd prime number p can be represented as a sum of two squares if and only if p is of the form $4n + 1$.* [†]

Proof. The ring $\mathbb{Z} + i\mathbb{Z} = \mathcal{O}_{-1}$ is a unique factorization domain due to (7.18)(b); hence we can apply (9.1) with $n = -1$. Since $Q_n(x, y) = x^2 + y^2$ this shows that an odd prime p can be represented as a sum of two squares if and only if the equation $[z]^2 = [-1]$ is solvable in $\mathbb{Z}_p$.

If $p = 4n + 1$ then the congruence $z^2 \equiv -1$ modulo p has a solution, due to (1.9); hence p can be written as a sum of two squares. On the other hand, no number of the form $4n + 3$ (prime or not) can be written as the sum of two squares (see (1.10)). Note by the way that, as a consequence of (9.1), the congruence $z^2 \equiv 1 \bmod p$ is not solvable if $p = 4n + 3$. ∎

USUALLY theorem (9.1) is applied by establishing condition (5) from the knowledge of condition (1); one knows that a certain prime is a quadratic residue modulo n and then concludes that p or $-p$ can be represented by the quadratic form Q_n. To be able to apply (9.1) in a concrete situation, we must be able to find out whether or not a given prime is a square modulo n. Statements about quadratic reciprocity are most easily formulated with the aid of Legendre symbols which we now define.

(9.3) Definition. *Let p be a prime number and let $a \in \mathbb{Z}_p \setminus \{0\}$. One calls a a* **quadratic residue** *modulo p if the equation $x^2 = a$ has a solution in $\mathbb{Z}_p$; otherwise p is called a* **quadratic nonresidue** *modulo p. Then the* **Legendre symbol** *of a modulo p is defined as*

$$\left(\frac{a}{p}\right) := \begin{cases} 1, & \text{if } a \text{ is a quadratic residue modulo } p; \\ -1, & \text{if } a \text{ is a quadratic nonresidue modulo } p. \end{cases}$$

(Here $\{\pm 1\}$ should be considered as a subgroup of $\mathbb{Z}_p^\times$.)

[†] If such a representation exists it is essentially unique. (See problem 5 below.)

(9.4) Proposition. *The Legendre symbol modulo p defines a group homomorphism* $\mathbb{Z}_p^\times \to \{\pm 1\}$.

Proof. The claim is that $(\frac{ab}{p}) = (\frac{a}{p})(\frac{b}{p})$ for all $a, b \in \mathbb{Z}_p^\times$ which is tantamount to saying that the product of two quadratic residues and the product of two quadratic non-residues is always a quadratic residue whereas the product of a residue and a nonresidue is always a nonresidue. This fact follows immediately from problem 17 in section 1 or from problem 26 in section 3. ∎

WE now formulate a rather general criterion which exploits the fact that the group $\mathbb{Z}_p^\times$ is cyclic, a fact which was proved in (4.14) above.

(9.5) Euler's criterion. *Let p be an odd prime and let* $a \in \mathbb{Z}_p \setminus \{0\}$. *Let* α *be a generator of* $\mathbb{Z}_p^\times$ *and write* $a = \alpha^r$. *Moreover, let* $n \in \mathbb{N}$ *be a natural number and let* $d := \gcd(n, p-1)$. *Then the following conditions are equivalent:*
(1) *the equation* $x^n = a$ *has a solution x in* $\mathbb{Z}_p$;
(2) $d \mid r$;
(3) $r(p-1)/d$ *is divisible by* $p-1$;
(4) $a^{(p-1)/d} = 1$ *in* $\mathbb{Z}_p$.
Consequently, the equation $x^2 = a$ *is solvable in* $\mathbb{Z}_p$ *if and only if* $a^{(p-1)/2} = 1$ *in* $\mathbb{Z}_p$. *Hence*

$$\left(\frac{a}{p}\right) = a^{(p-1)/2} \bmod p \ .$$

Proof. Since any element $x \in \mathbb{Z}_p \setminus \{0\}$ can be written in the form α^s, condition (1) is equivalent to the existence of a number $s \in \mathbb{Z}$ with $\alpha^{sn} = \alpha^r$, i.e., $\alpha^{r-sn} = 1$. But $\alpha^{r-sn} = 1$ if and only if $r - sn$ is a multiple of $\operatorname{ord}\alpha = p - 1$, i.e., if $r \in sn + \mathbb{Z}(p-1)$. Hence (1) holds if and only if $r \in \mathbb{Z}n + \mathbb{Z}(p-1) = \mathbb{Z}d$ which means $d \mid r$; this shows the equivalence of (1) and (2). Trivially, (2) and (3) are equivalent. Since $\operatorname{ord}\alpha = p - 1$, condition (3) is equivalent to the equation $1 = \alpha^{\frac{r(p-1)}{d}} = a^{\frac{p-1}{d}}$ in $\mathbb{Z}_p$ which is condition (4).

To prove the last claim just use the equivalence of (1) and (4) with $n := 2$. (Note that $x := a^{(p-1)/2} \in \mathbb{Z}_p$ satisfies $x^2 = a^{p-1} = 1$ due to Fermat's theorem (1.1); hence there are only the two possibilities $x = \pm 1$.) ∎

TO be able to exploit Euler's criterion, we introduce the following notion.

(9.6) Definition. *Let p be an odd prime number. A subset* $S \subseteq \mathbb{Z}_p \setminus \{0\}$ *is called a* **half-system** *for p if* $|S| = \frac{p-1}{2}$ *(i.e., if S contains half of the elements of* $\mathbb{Z}_p \setminus \{0\}$*) and if* $S \cup (-S) = \mathbb{Z}_p \setminus \{0\}$ *(i.e., if each element in* $\mathbb{Z}_p \setminus \{0\}$ *can be written as* $\pm s$ *with* $s \in S$*). In other words, a half-system is obtained by picking exactly one element from each pair* $(x, -x)$ *with* $x \in \mathbb{Z}_p \setminus \{0\}$.

(9.7) Gauß's Lemma. *Let p be an odd prime and let $a \in \mathbb{Z}_p \setminus \{0\}$. Let S be any half-system for p. Then $aS = \{as \mid s \in S\}$ is again a half-system for p. Let ω be the number of elements in $aS \cap (-S)$, i.e., the number of those elements in S which fall into the negative half-system $-S$ when being multiplied by a. Then a is a square in $\mathbb{Z}_p$ if and only if ω is even; i.e.,*

$$\left(\frac{a}{p}\right) = (-1)^{\omega} .$$

Proof. Let $S = \{s_1, \ldots, s_{\frac{p-1}{2}}\}$. Then each element as_i is exactly one of the elements $\pm s_j$ where the number of negative signs is ω. Hence

$$a^{\frac{p-1}{2}} s_1 s_2 \cdots s_{\frac{p-1}{2}} = (as_1)(as_2) \cdots (as_{\frac{p-1}{2}}) = (-1)^{\omega} s_1 s_2 \cdots s_{\frac{p-1}{2}}$$

so that $a^{\frac{p-1}{2}} = (-1)^{\omega}$. Now a comparison with Euler's criterion (9.5)(b) immediately yields the claim. ∎

WE now establish the fundamental facts about the properties of the Legendre symbol. These facts allow one in a very convenient way to determine whether or not a given number is a square modulo some given prime in any concrete case.

(9.8) Theorem. (a) *If p is an odd prime then*

$$\left(\frac{-1}{p}\right) = \begin{cases} 1, & \text{if } p \equiv 1\,(4); \\ -1, & \text{if } p \equiv 3\,(4); \end{cases} \qquad \left(\frac{2}{p}\right) = \begin{cases} 1, & \text{if } p \equiv \pm 1\,(8); \\ -1, & \text{if } p \equiv \pm 5\,(8). \end{cases}$$

(b) **(Quadratic Reciprocity Law.)** *If $p \neq q$ are odd primes then*

$$\left(\frac{p}{q}\right) = \left(\frac{q}{p}\right) \cdot (-1)^{\frac{p-1}{2} \cdot \frac{q-1}{2}} = \begin{cases} \left(\frac{q}{p}\right), & \text{if } p \equiv q \equiv 3\,(4); \\ -\left(\frac{q}{p}\right), & \text{if } p \equiv 1\,(4) \text{ or } q \equiv 1\,(4). \end{cases}$$

Thus if $p \equiv q \equiv 3\,(4)$ then q is a quadratic residue modulo p if and only if p is a quadratic nonresidue modulo q, and if $p \equiv 1\,(4)$ or $q \equiv 1\,(4)$ then q is a quadratic residue modulo p if and only if p is a quadratic residue modulo q.

Proof. (a) For the first claim we observe that if S is any half-system for p then the set $(-1)S \cap (-S) = -S$ has $\frac{p-1}{2}$ elements. Hence the number ω in (9.7) is $(p-1)/2$, and this number is even if and only if $p = 4n + 1$.

For the second claim let us choose the half-system $S = \{1, 3, 5, 7, \ldots, p - 2\}$ and then determine the cardinality of $2S \cap (-S)$. If $s \in S$ satisfies $s < \frac{p}{2}$, then $2s < p$ is even and hence belongs to $-S$; on the other hand, if $s > \frac{p}{2}$ then $2s \equiv 2s - p$ is odd and hence belongs to S. Consequently, ω is the number of odd natural numbers less than $\frac{p}{2}$. If $p = 8n + 1\,(8n + 3,\ 8n + 5,\ 8n + 7)$, then $\omega = 2n\,(2n + 1,\ 2n + 1,\ 2n + 2)$, respectively; whence the claim.

(b) Let Γ be the set of all pairs $(x, y) \in \mathbb{N}^2$ with $1 \leq x \leq \frac{p-1}{2}$ and $1 \leq y \leq \frac{q-1}{2}$. Applying (9.7) to the half-system $\{1, 2, \ldots, \frac{p-1}{2}\}$ for p, we see that $\left(\frac{q}{p}\right) = (-1)^{\omega}$ where

ω is the number of those elements $x \in \{1, 2, \dots, \frac{p-1}{2}\}$ for which $qx \in \{\frac{p+1}{2}, \dots, p-1\}$ modulo p which is just the number of pairs $(x, y) \in \Gamma$ with $-\frac{p}{2} < qx - py \le 0$. Similarly, $(\frac{p}{q}) = (-1)^\nu$ where ν is the number of pairs $(x, y) \in \Gamma$ with $0 \le qx - py < \frac{q}{2}$. Thus $(\frac{q}{p})(\frac{p}{q}) = (-1)^{\omega + \nu}$ where $\omega + \nu$ is the number of pairs $(x, y) \in \Gamma$ inside the sketched strip.

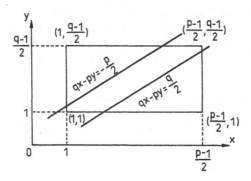

By symmetry, this number $\omega + \nu$ differs from the cardinality $|\Gamma| = \frac{p-1}{2} \cdot \frac{q-1}{2}$ only by an even number; whence the claim. ∎

THE quadratic reciprocity law was first formulated by Euler in 1744 and again, fully developed, in 1783; then independently also by Legendre (1785) and Gauß (1801) who gave the first proof in his *Disquisitiones Arithmeticae*. Let us give an example which shows how useful the quadratic reciprocity law is from an algorithmic viewpoint.

(9.9) Example. Does the equation $x^2 + 43y = 29$ have an integer solution? This is the case if and only if 29 is a quadratic residue modulo 43, i.e., if $(\frac{29}{43}) = 1$. Using the quadratic reciprocity law we find that

$$\left(\frac{29}{43}\right) = \left(\frac{43}{29}\right) = \left(\frac{14}{29}\right) = \left(\frac{2}{29}\right)\left(\frac{7}{29}\right) = -\left(\frac{7}{29}\right) = -\left(\frac{29}{7}\right) = -\left(\frac{1}{7}\right) = -1$$

which shows that the equation in question does not possess a solution.

(9.10) Example. Is 221 a quadratic residue modulo 383? Using the facts that $221 = 13 \cdot 17$ and that 383 is a prime, we see that

$$\left(\frac{221}{383}\right) = \left(\frac{13}{383}\right)\left(\frac{17}{383}\right) = \left(\frac{383}{13}\right)\left(\frac{383}{17}\right) = \left(\frac{6}{13}\right)\underbrace{\left(\frac{9}{17}\right)}_{=1}$$

$$= \left(\frac{6}{13}\right) = \underbrace{\left(\frac{2}{13}\right)}_{= -1}\left(\frac{3}{13}\right) = -\left(\frac{13}{3}\right) = -\left(\frac{1}{13}\right) = -1 \, .$$

Hence the congruence $x^2 \equiv 221 \bmod 383$ does not possess a solution.

(9.11) Example. Let us show that if p is an odd prime then

$$\left(\frac{-2}{p}\right) = \begin{cases} 1, & \text{if } p = 8n + 1 \text{ or } p = 8n + 3; \\ -1, & \text{if } p = 8n + 5 \text{ or } p = 8n + 7. \end{cases}$$

Since $\left(\frac{-2}{p}\right) = \left(\frac{-1}{p}\right)\left(\frac{2}{p}\right)$ there are two possibilities that $\left(\frac{-2}{p}\right) = 1$.

• First possibility: $\left(\frac{-1}{p}\right) = \left(\frac{2}{p}\right) = 1$. Due to (9.8)(a) this is the case if both $p \equiv 1$ mod 4 and $p \equiv 1$ or 7 mod 8 which is tantamount to saying that $p \equiv 1$ mod 8.

• Second possibility: $\left(\frac{-1}{p}\right) = \left(\frac{2}{p}\right) = -1$. Due to (9.8)(a) this is the case if both $p \equiv 3$ mod 4 and $p \equiv 3$ or 5 mod 8 which is tantamount to saying that $p \equiv 3$ mod 8. Thus $\left(\frac{-2}{p}\right) = 1$ if and only if $p \equiv 1$ or 3 mod 8 which is the claim.

TO obtain statements as in (9.8)(a) or (9.11) we can, in principle, apply Gauß's lemma; however, this becomes more and more cumbersome for larger numbers. Let us obtain one more result in this direction without invoking Gauß' lemma.

(9.12) Proposition. *Let $p > 3$ be a prime number. Then*

$$\left(\frac{-3}{p}\right) = \begin{cases} 1, & \text{if } p = 3n + 1; \\ -1, & \text{if } p = 3n + 2. \end{cases}$$

Proof. Suppose first that there is an element $x \in \mathbb{Z}_p$ with $x^2 + 3 = 0$. Then $y := (x - 1) \cdot 2^{-1}$ satisfies the equation $y^2 + y + 1 = 0$ and hence $y^3 = 1$. If $y = 1$, we have $0 = y^2 + y + 1 = 3$ and hence $p = 3$. If $y \neq 1$, the equations $y^3 = 1$ and $y^{p-1} = 1$ (the latter due to Lagrange's theorem) imply that $3 \mid p - 1$, i.e., $p \equiv 1$ modulo 3.

Suppose conversely that $p \equiv 1$ modulo 3, say $p = 3n + 1$, and let α be a generator of $\mathbb{Z}_p^\times$ so that $1 = \alpha^{p-1} = \alpha^{3n}$. Then the element $y := \alpha^n$ satisfies $y^3 = 1$, but $y \neq 1$. Since $(y^3 - 1) = (y - 1)(y^2 + y + 1)$, this implies $y^2 + y + 1 = 0$ and consequently $0 = 4(y^2 + y + 1) = (2y + 1)^2 + 3$. Hence the element $x := 2y + 1 \in \mathbb{Z}_p$ satisfies $x^2 = -3$. ∎

THE multiplicativity of the Legendre symbol enables us to immediately obtain some associated statements.

(9.13) Proposition. *Let $p > 3$ be a prime. Then*

$$\left(\frac{3}{p}\right) = \begin{cases} 1, & \text{if } p = 12n + 1 \text{ or } p = 12n + 11; \\ -1, & \text{if } p = 12n + 5 \text{ or } p = 12n + 7. \end{cases}$$

$$\left(\frac{6}{p}\right) = \begin{cases} 1, & \text{if } p = 24n + k \text{ with } k \in \{1, 5, 19, 23\}; \\ -1, & \text{if } p = 24n + k \text{ with } k \in \{7, 11, 13, 17\}. \end{cases}$$

$$\left(\frac{-6}{p}\right) = \begin{cases} 1, & \text{if } p = 24n + k \text{ with } k \in \{1, 5, 7, 11\}; \\ -1, & \text{if } p = 24n + k \text{ with } k \in \{13, 17, 19, 23\}. \end{cases}$$

Proof. This can be obtained from (9.8)(a), (9.11) and (9.12) in the same way as (9.11) was obtained; we leave the proof as an exercise. (See problem 13 below.)

(9.14) Proposition. *For a prime number p the following equivalences hold.*
(a) *p can be written as $p = x^2 + y^2$ if and only if $p = 2$ or $p = 4n + 1$.*
(b) *p can be written as $p = x^2 + 2y^2$ if and only if $p = 2$ or $p = 8n+1$ or $p = 8n+3$.*
(c) *p can be written as $p = x^2 + 3y^2$ if and only if $p = 3$ or $p = 3n + 1$.*
(d) *p can be written as $p = |x^2 - 2y^2|$ if and only if $p = 2$ or $p = 8n \pm 1$.*
(e) *p can be written as $p = |x^2 - 3y^2|$ if and only if $p = 2$ or $p = 12n \pm 1$.*
(f) *p can be written as $p = |x^2 - 6y^2|$ if and only if $p = 2$, $p = 3$ or $p = 24n + k$
with $k \in \{1, 5, 19, 23\}$.*

Proof. These results follow immediately from the equivalence (1)$\Longleftrightarrow$(5) in theorem (9.1) if we take into account (9.8)(a) and (9.11)-(9.13); the cases $p = 2$ and $p = 3$ have to be treated separately. ∎

NOTE that we did not make a statement about the representability of prime numbers in the form $p = x^2 + 6y^2$. Here theorem (9.1) is not applicable because $\mathcal{O}_{-6}$ is not a unique factorization domain. (See problem 6 below.) Some additional information on quadratic residues will be provided in the exercises. Let us now turn to other applications of unique factorization. We start with the following lemma.

(9.15) Lemma. *Let R be a unique factorization domain. If $a, b \in R^{\heartsuit}$ are coprime and if*

$$ab = c^n$$

for some $c \in R$, then there is a unit $u \in R^{\times}$ such that ua and $u^{-1}b$ are n-th powers in R. If all units in R happen to be n-th powers themselves, then so are a and b.

Proof. Let $c = p_1 \cdots p_k$ be a decomposition into irreducible elements. Then $ab = p_1^n \cdots p_k^n$ is the (essentially unique) decomposition of ab into irreducible factors. The factor p_1 must appear on the left, hence must divide a or b; say $p_1 \mid a$. But since a and b are coprime, this gives $p_1^n \mid a$. The same arguments applies to the other elements p_i. So we obtain

$$ a = \alpha p_{i_1}^n \cdots p_{i_r}^n \qquad \text{and} \qquad b = \beta p_{j_1}^n \cdots p_{j_s}^n $$

where $\{i_1, \ldots, i_r, j_1, \ldots, j_s\} = \{1, \ldots, k\}$ and where α, β are units in R. But then we must have $\alpha\beta = 1$, and the claim follows. ∎

(9.16) Theorem (Fermat). *The only integer solutions of the equation $y^2 + 2 = x^3$ are $y = \pm 5$, $x = 3$.*

Proof. If y were even, then the right-hand side of the given equation would be divisible by 8, but the left-hand side only by 2. Hence y must be odd. Rewrite the equation as

$$(y + \sqrt{-2})(y - \sqrt{-2}) = x^3$$

and treat it is an equation in the unique factorization domain $\mathbb{Z}[\sqrt{-2}]$. We will show in a moment that $y + \sqrt{-2}$ and $y - \sqrt{-2}$ are coprime in $\mathbb{Z}[\sqrt{-2}]$. Since their product is a cube, they have to be cubes themselves, due to (9.15) and the fact that all the units in $\mathbb{Z}[\sqrt{-2}]$, namely ± 1, are cubes. Hence there are integers $a, b \in \mathbb{Z}$ such that

$$y + \sqrt{-2} = (a + b\sqrt{-2})^3 = (a^3 - 6ab^2) + (3a^2 b - 2b^3)\sqrt{-2}.$$

Comparing imaginary parts, we have $1 = 3a^2 b - 2b^3 = b(3a^2 - 2b^2)$. This yields $b = 1, a = \pm 1$ so that $y = a^3 - 6ab^2 = \pm 5$ which implies $x = 3$.

It remains to check that $y + \sqrt{-2}$ and $y - \sqrt{-2}$ are coprime in $\mathbb{Z}[\sqrt{-2}]$. Let $a + b\sqrt{-2}$ $(a, b \in \mathbb{Z})$ be a common divisor of these two elements. Then $a + b\sqrt{-2}$ divides also their sum $2y$ and their difference $2\sqrt{-2}$; hence there are integers $m, n, r, s \in \mathbb{Z}$ such that

$$2y = (a + b\sqrt{-2})(m + n\sqrt{-2}) \quad \text{and} \quad 2\sqrt{-2} = (a + b\sqrt{-2})(r + s\sqrt{-2}).$$

In each of these two equations, apply the Euclidean norm d on both sides to obtain

$$4y^2 = (a^2 + 2b^2)(m^2 + 2n^2) \quad \text{and} \quad 8 = (a^2 + 2b^2)(r^2 + 2s^2).$$

We see that $a^2 + 2b^2$ divides $4y^2$ and 8, hence divides 4 because y is odd. This implies that $a^2 + 2b^2 = 1, 2$ or 4 which leaves for (a, b) only the possibilities $(\pm 1, 0)$, $(0, \pm 1)$ and $(\pm 2, 0)$. Thus

$$a + b\sqrt{-2} \in \{\pm 1, \ \pm\sqrt{-2}, \ \pm 2\}.$$

But of the six elements in this set, only ± 1 are divisors of $y + \sqrt{-2}$ in $\mathbb{Z}[\sqrt{-2}]$, due to the fact that y is odd (check!). We have shown that the only common divisors of $y + \sqrt{-2}$ and $y - \sqrt{-2}$ are 1 and -1. ∎

(9.17) Theorem (Fermat). *The only integer solutions of the equation $y^2 + 1 = 2x^3$ are $y = \pm 1, x = 1$.*

Proof. We note first that y must be odd. Then we consider the given equation as an equation in $\mathbb{Z}[i]$ and rewrite it as

$$(y + i)(y - i) = 2x^3.$$

Any common divisor of $y + i$ and $y - i$ must also divide $(y + i) - (y - i) = 2i = (1 + i)^2$, hence must be $1, 1 + i$ or $(1 + i)^2$ (up to multiplication by a unit). Now the fact that y is odd shows that $1 + i$ divides $y + i$ and $y - i$ but $(1 + i)^2 = 2i$ does not. Hence $1 + i$ is a gcd of $y + i$ and $y - i$ so that we can write

$$y + i = (1 + i)(a + ib) \quad \text{and} \quad y - i = (1 + i)(c + id)$$

where $a + ib$ and $c + id$ are coprime. Then

$$2x^3 \;=\; (y+i)(y-i) \;=\; (1+i)^2(a+bi)(c+di) \;=\; 2i(a+bi)(c+di)$$

so that $(a+bi)(c+di) = \frac{1}{i}x^3 = -ix^3 = (ix)^3$. By lemma (9.15) the numbers $a + bi$ and $c + di$ must be cubes in $\mathbb{Z}[i]$; note that all the units ± 1, $\pm i$ in $\mathbb{Z}[i]$ are cubes. So there exist integers $\alpha, \beta \in \mathbb{Z}$ with

$$y + i \;=\; (1+i)(\alpha + \beta i)^3 \;=\; (\alpha^3 - 3\alpha\beta^2 - 3\alpha^2\beta + \beta^3) \;+\; i(\alpha^3 - 3\alpha\beta^2 + 3\alpha^2\beta - \beta^3)\,.$$

This breaks up into the two equations

$$
\begin{aligned}
y &= \alpha^3 - 3\alpha\beta^2 - 3\alpha^2\beta + \beta^3 \;=\; (\alpha+\beta)(\alpha^2 - 4\alpha\beta + \beta^2) \quad \text{and} \\
1 &= \alpha^3 - 3\alpha\beta^2 + 3\alpha^2\beta - \beta^3 \;=\; (\alpha-\beta)(\alpha^2 + 4\alpha\beta + \beta^2)\,.
\end{aligned}
$$

The last equation is only satisfied for $\alpha = 0$, $\beta = -1$ and $\alpha = 1$, $\beta = 0$. Plugging these possibilities into the first equation gives $y = \pm 1$, and we are done. ∎

(9.18) Theorem (Fermat). *The only integer solutions of the equation $y^2 + 4 = x^3$ are $y = \pm 11$, $x = 5$ and $y = \pm 2$, $x = 2$.*

Proof. If y is even, then so is x, and we can write $y = 2Y$ and $x = 2X$. This gives

$$Y^2 + 1 \;=\; 2X^3$$

which has only the solutions $Y = \pm 1$, $X = 1$ by (9.17). So $y = \pm 2$ and $x = 2$. Now suppose that y is odd. Treating y and x as elements of $\mathbb{Z}[i]$, we can rewrite the given equation as

$$(2+iy)(2-iy) \;=\; x^3\,.$$

We claim that $2 + iy$ and $2 - iy$ must be coprime. Once this is established, we conclude as above that $2 + iy$ and $2 - iy$ must be cubes in $\mathbb{Z}[i]$. Write

$$2 + iy \;=\; (a+ib)^3 \;=\; (a^3 - 3ab^2) \;+\; i(3a^2b - b^3)\,.$$

This gives

$$2 \;=\; a^3 - 3ab^2 \;=\; a(a^2 - 3b^2) \quad \text{and} \quad y \;=\; 3a^2b - b^3 \;=\; b(3a^2 - b^2)\,.$$

The first equation has only the solutions $a = -1$, $b = \pm 1$ and $a = 2$, $b = \pm 1$. Then the second equation gives $y = \pm 1$ or $y = \pm 11$. Since $y = \pm 1$ does not give a solution, we see that y must be ± 11.

Now let us verify that $2 + iy$ and $2 - iy$ must be coprime in $\mathbb{Z}[i]$. A common factor $a + bi$ of these two elements must also divide their sum 4 and their difference $2iy$. Taking norms, this implies $a^2 + b^2 \mid 16$ and $a^2 + b^2 \mid 4y^2$, hence $a^2 + b^2 \mid 4$ because y is odd. This gives $a^2 + b^2 = 1, 2$ or 4 so that

$$(a,b) \;\in\; \{(\pm 1, 0),\; (0, \pm 1),\; (\pm 1, \pm 1),\; (\pm 2, 0),\; (0, \pm 2)\}\,.$$

But neither $\pm 1 \pm i$ nor ± 2 nor $\pm 2i$ are divisors of $2 + iy$ because y is odd. Hence $a + bi$ is ± 1 or $\pm i$, i.e., a unit. This shows that $2 + iy$ and $2 - iy$ are coprime. ∎

THE next result presents Nagell's proof of a conjecture due to Srinivasa Ramanujan (1887-1920). Our exposition follows that found in *Algebraic Number Theory* by Ian Stewart and David Tall, Chapman and Hall 1979, pp. 99-102.

(9.19) Theorem (Ramanujan-Nagell). *The only integer solutions of the equation*

$$x^2 + 7 = 2^n$$

are $(x, n) = (\pm 1, 3)$, $(\pm 3, 4)$, $(\pm 5, 5)$, $(\pm 11, 7)$ *and* $(\pm 181, 15)$.

Proof. Let (x, n) be a solution. Then clearly x is odd. Without loss of generality let $x > 0$. Suppose first that n is even, say $n = 2N$. Then the given equation reads

$$7 = 2^n - x^2 = 2^{2N} - x^2 = (2^N + x)(2^N - x) .$$

This is only possible if $2^N + x = 7$ and $2^N - x = 1$ which leads to $x = 3$ and $N = 2$. So the only solutions (x, n) with n even are $(\pm 3, 4)$.

Suppose now that n is odd. We consider the given equation as an equation in the Euclidean domain $\mathcal{O}_{-7} = \mathbb{Z} + \mathbb{Z}\frac{1+\sqrt{-7}}{2}$. Let

$$\alpha := \frac{1 + \sqrt{-7}}{2} \quad \text{and} \quad \beta := \frac{1 - \sqrt{-7}}{2} ;$$

these are prime elements in $\mathcal{O}_{-7}$ and satisfy the equations

$$\alpha + \beta = 1, \quad \alpha - \beta = \sqrt{-7}, \quad \alpha\beta = 2, \quad \alpha^2 = \alpha - 2 \quad \text{and} \quad \beta^2 = \beta - 2 .$$

Now x is odd, say $x = 2k + 1$. Then $x^2 + 7 = 4k^2 + 4k + 8 = 4(k^2 + k + 2)$. Let $m := n - 2$. Then the given equation reads $k^2 + k + 2 = 2^{n-2} = 2^m$ or

$$(\star) \qquad\qquad (k + \alpha)(k + \beta) = \alpha^m \beta^m .$$

Now any common divisor z of $k + \alpha$ and $k + \beta$ also divides $(k + \alpha) - (k + \beta) = \alpha - \beta = \sqrt{-7}$ which implies that $N(z)$ divides $N(\sqrt{-7}) = 7$. Consequently, neither α nor β is a common divisor of $k + \alpha$ and $k + \beta$ because $N(\alpha) = N(\beta) = 2$. This fact and the fact that the only units in $\mathcal{O}_{-7}$ are ± 1 show that $(\star)$ implies

$$\begin{bmatrix} k + \alpha = \pm\alpha^m \\ k + \beta = \pm\beta^m \end{bmatrix} \quad \text{or} \quad \begin{bmatrix} k + \alpha = \pm\beta^m \\ k + \beta = \pm\alpha^m \end{bmatrix} .$$

In either case we obtain

$$(\star\star) \qquad\qquad \alpha^m - \beta^m = \pm(\alpha - \beta) .$$

We claim that $(\star\star)$ holds if and only if $m = 3, 5, 13$; the theorem follows easily once this is established.

Assume first that $\alpha^m - \beta^m = \alpha - \beta$. Then $\alpha^m - \alpha = \beta^m - \beta$. Since m is odd, the left-hand side of this equation is divisible by $-(1+\alpha) = \beta^2$ whereas the right-hand side is not (β^m is divisible by β^2, but β is not). So we must have $\alpha^m - \beta^m = \beta - \alpha$, i.e.,

$$\left(\frac{1+\sqrt{-7}}{2}\right)^m - \left(\frac{1-\sqrt{-7}}{2}\right)^m = -\sqrt{-7}.$$

We multiply this equation by 2^m and expand by the binomial theorem to obtain

$$-2^m\sqrt{-7} = (1+\sqrt{-7})^m - (1-\sqrt{-7})^m = \sum_{k=0}^{m}\binom{m}{k}(\sqrt{-7})^k\underbrace{(1-(-1)^k)}_{=\begin{cases}2, & k \text{ odd,} \\ 0, & k \text{ even}\end{cases}}$$

$$= \sum_{\substack{0 \le k \le m, \\ k \text{ odd}}}\binom{m}{k}(\sqrt{-7})^k.$$

This means

$$-2^{m-1} = \binom{m}{1} - 7\binom{m}{3} + 7^2\binom{m}{5} - + \cdots \pm 7^{\frac{m-1}{2}}\binom{m}{m}$$

which implies that

$$(\star\star\star) \qquad\qquad -2^{m-1} \stackrel{(7)}{\equiv} m.$$

Using the fact that $2^6 \stackrel{(7)}{\equiv} 1$, one concludes easily that the validity of $(\star\star\star)$ depends only on the residue class of m modulo 42, i.e., if $m \stackrel{(42)}{\equiv} m'$, then m solves $(\star\star\star)$ if and only if m' does. Now checking the numbers $1 \le m \le 42$ reveals exactly 3 solutions, namely 3, 5 and 13, and these indeed yield solutions of our original equation. The only other possible solutions must be of the form $3 + 42k$, $5 + 42k$ or $13 + 42k$ with $k > 1$. We are going to show that none of these numbers can satisfy equation $(\star\star)$.

The equations $\alpha^2 = \alpha - 2$ and $\beta^2 = \beta - 2$ imply that the powers of α and β are given by the following formulas.

$$\begin{array}{ll}\alpha^n = r_n\alpha + s_n \\ \beta^n = r_n\beta + s_n\end{array} \quad \text{where} \quad \binom{r_0}{s_0} = \binom{0}{1}$$

and

$$\binom{r_{n+1}}{s_{n+1}} = \begin{pmatrix}1 & 1 \\ -2 & 0\end{pmatrix}\binom{r_n}{s_n}, \quad \text{i.e.,} \quad \binom{r_n}{s_n} = \begin{pmatrix}1 & 1 \\ -2 & 0\end{pmatrix}^n\binom{0}{1}.$$

Now equation $(\star\star)$ is equivalent to $\beta - \alpha = \alpha^n - \beta^n = r_n(\alpha - \beta)$; i.e., $r_n = -1$. We have

$$\binom{r_3}{s_3} = \binom{-1}{-2}, \quad \binom{r_5}{s_5} = \binom{-1}{6} \quad \text{and} \quad \binom{r_{13}}{s_{13}} = \binom{-1}{90}.$$

We want to show that $r_{n+42k} \neq -1$ for $n = 3, 5, 13$ and $k > 1$. We will even prove that $r_{n+42k} \not\equiv -1 \bmod 3$ in these cases. It is easy to check that $\begin{pmatrix} 1 & 1 \\ -2 & 0 \end{pmatrix}^{42k} \equiv \begin{pmatrix} -1 & -1 \\ -1 & -1 \end{pmatrix}$ mod 3. Therefore, we obtain the following equations modulo 3.

$$\begin{pmatrix} r_{3+42k} \\ s_{3+42k} \end{pmatrix} = \begin{pmatrix} 1 & 1 \\ -2 & 0 \end{pmatrix}^{42k} \begin{pmatrix} r_3 \\ s_3 \end{pmatrix} = \begin{pmatrix} -1 & -1 \\ -1 & -1 \end{pmatrix} \begin{pmatrix} -1 \\ -2 \end{pmatrix} = \begin{pmatrix} 0 \\ 0 \end{pmatrix},$$

$$\begin{pmatrix} r_{5+42k} \\ s_{5+42k} \end{pmatrix} = \begin{pmatrix} 1 & 1 \\ -2 & 0 \end{pmatrix}^{42k} \begin{pmatrix} r_5 \\ s_5 \end{pmatrix} = \begin{pmatrix} -1 & -1 \\ -1 & -1 \end{pmatrix} \begin{pmatrix} -1 \\ 0 \end{pmatrix} = \begin{pmatrix} 1 \\ 1 \end{pmatrix},$$

$$\begin{pmatrix} r_{13+42k} \\ s_{13+42k} \end{pmatrix} = \begin{pmatrix} 1 & 1 \\ -2 & 0 \end{pmatrix}^{42k} \begin{pmatrix} r_{13} \\ s_{13} \end{pmatrix} = \begin{pmatrix} -1 & -1 \\ -1 & -1 \end{pmatrix} \begin{pmatrix} -1 \\ 0 \end{pmatrix} = \begin{pmatrix} 1 \\ 1 \end{pmatrix}.$$

In each of the cases, $r_{n+42k} \neq -1$ (modulo 3). ∎

OUR next aim is to prove a special case of Fermat's Last Theorem, namely, we will prove that there is no solution to the equation $x^3 + y^3 = z^3$ in nonzero integers. The main idea of the proof will be to introduce the element $\varepsilon := e^{2\pi i/3} = \frac{1}{2}(-1+\sqrt{-3})$ and to rewrite the equation $x^3 + y^3 = z^3$ as

$$(-z)^3 = x^3 + y^3 = (x+y)(x+\varepsilon y)(x+\varepsilon^2 y)$$

which is an equation in the ring $\mathbb{Z}[\varepsilon] = \mathbb{Z} + \mathbb{Z}\varepsilon = \mathbb{Z} + \mathbb{Z}(\frac{-1+\sqrt{-3}}{2})$ which is just the unique factorization domain $\mathcal{O}_{-3}$. Recall that the units of $\mathcal{O}_{-3}$ are ± 1, $\pm \varepsilon$ and $\pm \varepsilon^2$. An important element in the proof will be played by the element $\theta := \sqrt{-3} = 2\varepsilon + 1 = \varepsilon - \varepsilon^2$. Since $N(\theta) = 3$, this element is irreducible; its associates in $\mathcal{O}_{-3}$ are $\pm(\varepsilon - \varepsilon^2)$, $\pm(1-\varepsilon)$ and $\pm(1 - \varepsilon^2)$.

(9.20) Theorem. *Let $\theta := \sqrt{-3} \in \mathcal{O}_{-3}$.*
(a) *If $\alpha \in \mathcal{O}_{-3}$ then either $\alpha \equiv 0\,(\theta)$, $\alpha \equiv 1\,(\theta)$ or $\alpha \equiv -1\,(\theta)$. Consequently, $\alpha^3 \equiv \alpha\,(\theta)$.*
(b) *If $\alpha \equiv 1\,(\theta)$ then $\alpha^3 \equiv 1\,(\theta^4)$; if $\alpha \equiv -1\,(\theta)$ then $\alpha^3 \equiv -1\,(\theta^4)$.*
(c) *If $\alpha, \beta, \gamma \in \mathcal{O}_{-3}$ are relatively prime with $\alpha^3 + \beta^3 + \gamma^3 = 0$ then θ divides exactly one of the numbers α, β, γ.*
(d) *If $\alpha^3 + u\beta^3 + v(\theta^r \gamma)^3 = 0$ with $\alpha, \beta, \gamma \in \mathcal{O}_{-3}$ not divisible by θ, $u, v \in \mathcal{O}_{-3}^{\times}$ and $r > 0$, then $u = \pm 1$ and $r \geq 2$.*
(e) *It is not possible to choose $\alpha, \beta, \gamma \in \mathcal{O}_{-3} \setminus \{0\}$, $v \in \mathcal{O}_{-3}^{\times}$ and $r \in \mathbb{N}$ such that $\alpha^3 + \beta^3 + v(\theta^r \gamma)^3 = 0$.*
(f) *There are no nonzero elements $\alpha, \beta, \gamma \in \mathcal{O}_{-3} \setminus \{0\}$ such that $\alpha^3 + \beta^3 + \gamma^3 = 0$.*

Proof. (a) Write $\alpha = \frac{a+b\theta}{2}$ with $a, b \in \mathbb{Z}$ either both even or both odd. Then one of the congruences $2a \equiv 0, \pm 1\,(3)$ must hold in $\mathbb{Z}$, hence even more so in $\mathcal{O}_{-3}$ since 3 is divisible by θ. But

$$\alpha = \underbrace{\frac{b+a\theta}{2}}_{\in \mathcal{O}_{-3}} \cdot \theta + 2a \overset{(\theta)}{\equiv} 2a \overset{(\theta)}{\equiv} 0, \pm 1 .$$

(b) Suppose that $\alpha \equiv 1\,(\theta)$, say $\alpha = 1 + \beta\theta$ with $\beta \in \mathcal{O}_{-3}$. Then

$$\alpha^3 = 1 + \underbrace{3\beta\theta}_{=\,-\beta\theta^3} + \underbrace{3\beta^2\theta^2}_{=\,-\theta^4\beta^2} + \beta^3\theta^3 = 1 - \theta^4\beta^2 + \theta^3(\underbrace{\beta^3 - \beta}_{\text{divisible by }\theta}) \stackrel{(\theta^4)}{\equiv} 1\,.$$

Suppose now that $\alpha \equiv -1\,(\theta)$. Then $-\alpha \equiv 1\,(\theta)$, so $(-\alpha)^3 \equiv 1\,(\theta^4)$ by what we just proved. Since $(-\alpha)^3 = -\alpha^3$, this means $\alpha^3 \equiv -1\,(\theta^4)$.

(c) Suppose none of the elements α, β, γ is divisible by θ. Then $\alpha, \beta, \gamma \equiv \pm 1\,(\theta)$ and hence $0 = \alpha^3 + \beta^3 + \gamma^3 \equiv \pm 1 \pm 1 \pm 1\,(\theta^4)$. Considering all possible combinations of signs, we see that one of the numbers ± 1 and ± 3 is divisible by θ^4. But this is impossible since $N(\theta^4) = N(\theta)^4 = 81$. Hence at least one of the elements α, β, γ is divisible by θ. But if two of them are, then so is the third, which contradicts our hypothesis that α, β, γ are relatively prime. Hence exactly one of these elements is divisible by θ.

(d) Since $r > 0$, we have $\alpha^3 + u\beta^3 \equiv 0\,(\theta^3)$. On the other hand, we have $\alpha^3 + u\beta^3 \equiv \pm 1 \pm u\,(\theta^4)$ due to part (b), hence even more $\alpha^3 + u\beta^3 \equiv \pm 1 \pm u\,(\theta^3)$. Now the elements $\pm 1 \pm u$ with $u \in \mathcal{O}_{-3}^\times$ are exactly the elements $0, \pm 2, \pm(1 \pm \varepsilon), \pm(1 \pm \varepsilon^2)$. None of these except 0 is divisible by θ^3, since $1 + \varepsilon = -\varepsilon^2$ and $1 + \varepsilon^2 = -\varepsilon$ are units, $1 - \varepsilon$ and $1 - \varepsilon^2$ are associates of θ, and $N(\pm 2) = 4$ whereas $N(\theta^3) = N(\theta)^3 = 27$. Hence the congruence $\pm 1 \pm u \equiv 0\,(\theta^3)$ implies $\pm 1 \pm u = 0$, hence $u = \pm 1$.

Moreover, $\alpha^3 + u\beta^3 \equiv 0\,(\theta^3)$ implies that $\alpha^3 + u\beta^3 \equiv 0\,(\theta)$ and hence $\alpha + u\beta \equiv 0\,(\theta)$ due to (a). Now $\alpha, \beta \equiv \pm 1\,(\theta)$ and therefore $\alpha^3, \beta^3 \equiv \pm 1\,(\theta^4)$ by (b). Since $u = \pm 1$, this implies $\alpha^3 + u\beta^3 \equiv 0\,(\theta^4)$. Hence $v(\theta^r\gamma)^3$ is divisible by θ^4 which implies $r \geq 2$.

(e) If there is an equation $\alpha^3 + \beta^3 + v(\theta^r\gamma)^3 = 0$ of the indicated form, we can also find one with minimal value for r. Dividing out a gcd of α, β, γ we can assume that α, β, γ are relatively prime and hence – due to the special form of the equation – pairwise coprime. By (d), we have $r \geq 2$ and hence $\alpha^3 + \beta^3 \equiv 0\,(\theta^6)$. Now

$$\alpha^3 + \beta^3 = (\alpha + \beta)(\alpha + \varepsilon\beta)(\alpha + \varepsilon^2\beta)\,.$$

We claim that θ is a gcd of any two of the three factors on the right. First, θ divides the right-hand side, hence divides one of the factors since θ is a prime element. But $\alpha + \beta \equiv \alpha + \varepsilon\beta \equiv \alpha + \varepsilon^2\beta\,(\theta)$ since $1 \equiv \varepsilon \equiv \varepsilon^2\,(\theta)$; hence θ divides all three factors. On the other hand, a common divisor of $\alpha + \beta$ and $\alpha + \varepsilon\beta$ also divides $(\alpha + \varepsilon\beta) - (\alpha + \beta) = (\varepsilon - 1)\beta$ and $\varepsilon(\alpha + \beta) - (\alpha + \varepsilon\beta) = (\varepsilon - 1)\alpha$, hence – since α and β are coprime – divides $\varepsilon - 1$ which is an associate of θ. Similarly, every common divisor of $\alpha + \beta$ and $\alpha + \varepsilon^2\beta$ and of $\alpha + \varepsilon\beta$ and $\alpha + \varepsilon^2\beta$ must divide θ. This shows that θ divides two of the factors exactly once and the third factor exactly $3r - 2$ times. Thus the equation $-u\theta^{3r}\gamma^3 = \alpha^3 + \beta^3$ becomes

$$-u\gamma^3 = \frac{\alpha + \beta}{\theta^a} \cdot \frac{\alpha + \varepsilon\beta}{\theta^b} \cdot \frac{\alpha + \varepsilon^2\beta}{\theta^c}$$

where the numbers a, b, c are $1, 1, 3r - 2$ in some order. Since the factors on the right are pairwise relatively prime, each of them must be an associate of a cube, say

$$\frac{\alpha + \beta}{\theta^a} = u_1 A^3\,, \qquad \frac{\alpha + \varepsilon\beta}{\theta^b} = u_2 B^3\,, \qquad \frac{\alpha + \varepsilon^2\beta}{\theta^c} = u_3 C^3$$

with $u_1, u_2, u_3 \in \mathcal{O}_{-3}$ and $A, B, C \in \mathcal{O}_{-3}$ not divisible by θ and relatively prime. Then

$(\star)$
$$u_1\theta^a A^3 + (\varepsilon u_2)\theta^b B^3 + (\varepsilon^2 u_3)\theta^c C^3 = (\alpha + \beta) + \varepsilon(\alpha + \varepsilon\beta) + \varepsilon^2(\alpha + \varepsilon^2\beta)$$
$$= (\alpha + \beta)(1 + \varepsilon + \varepsilon^2) = 0\,.$$

Now suppose that $a = b = 1$ and $c = 3r - 2$ (the other two possibilities are treated in a completely analogous way). Plugging into ($\star$) and dividing by $u_1 \theta$, we obtain

$$A^3 + \frac{\varepsilon u_2}{u_1} B^3 + \frac{\varepsilon^2 u_3}{u_1} (\theta^{r-1} C)^3 = 0 .$$

Then part (d) implies that $\frac{\varepsilon u_2}{u_1} = \pm 1$; hence this last equation reads

$$A^3 + (\pm B)^3 + \overline{u}(\theta^{r-1} C)^3 = 0 .$$

This clearly contradicts the fact that our initial equation was chosen with the minimal possible value for r.

(f) Suppose there are nonzero elements $\alpha, \beta, \gamma \in \mathcal{O}_{-3}$ with $\alpha^3 + \beta^3 + \gamma^3 = 0$. Dividing out a greatest common divisor, we can assume that α, β and γ are relatively prime. Then part (c) implies that θ divides exactly one of the numbers α, β, γ; say $\gamma = \theta^r \gamma'$ with $r > 0$. But then $\alpha^3 + \beta^3 + (\theta^r \gamma')^3 = 0$ which is impossible by (e). ∎

RECALL that Fermat's Last Theorem states that the equation $x^n + y^n = z^n$ has no solution in natural numbers for any $n \geq 3$. We just established the case $n = 3$ of this theorem; the case $n = 4$ can be settled with elementary methods. (See problem 24 below.) Note that to establish Fermat's theorem it is enough to prove the claim for all prime exponents, because if $x^n + y^n = z^n$ where $n = pq$ then $(x^q)^p + (y^q)^p = (z^q)^p$. Historically, mathematicians were able to treat several exponents, one after the other, thereby establishing Fermat's theorem in various individual cases. We will now close this section by proving part of a celebrated result due to Ernst Eduard Kummer which, for the first time, did not just establish Fermat's Last Theorem for an individual exponent but for a whole class of exponents at once. The idea of Kummer's approach was outlined at the end of section 1 already; namely, the equation $x^p + y^p = z^p$ in $\mathbb{Z}$ is rewritten in purely multiplicative form as

$$z^p = (x + y)(x - \varepsilon y)(x + \varepsilon^2 y) \cdots (x + \varepsilon^{p-1} y)$$

where $\varepsilon = e^{2\pi i/p}$. In the proof below we will assume that $Z[\varepsilon]$ is a unique factorization domain and then show that the factors of the right-hand side are pairwise coprime in $\mathbb{Z}[\varepsilon]$; we can then invoke (9.15) to conclude that each of the factors must be an associate of a p-th power in $\mathbb{Z}[\varepsilon]$. This will eventually lead to a contradiction to the assumption that the above equation has a solution. Let us begin with the following lemma.

(9.21) Lemma. *Let $\varepsilon = e^{2\pi i/p}$ where p is an odd prime and let $R := \mathbb{Z}[\varepsilon]$.*

*(a) Every element of R possesses a unique representation $x = a_0 + a_1 \varepsilon + a_2 \varepsilon^2 + \cdots + \varepsilon^{p-2}$. (This is called the **canonical representation** of $x \in R$.)*

(b) The elements $1 - \varepsilon^k$ with $1 \leq k \leq p - 1$ are pairwise associate non-units in R whose product is p.

(c) Let $x, y \in \mathbb{Z}$ and $1 \leq r \neq s \leq p - 1$. Then $x + \varepsilon^r y$ and $x + \varepsilon^s y$ are coprime in $\mathbb{Z}[\varepsilon]$ if and only if x and y are coprime in $\mathbb{Z}$ and if $x + y$ is not divisible by p.

Proof. (a) Let $f(x) := x^{p-1} + \cdots + x + 1$. Since $f(\varepsilon) = (\varepsilon^p - 1)/(\varepsilon - 1) = 0$ we have $\varepsilon^{p-1} = -\sum_{k=0}^{p-2} \varepsilon^k$ so that all powers of ε can be expressed by $1, \varepsilon, \ldots, \varepsilon^{p-2}$.

Consequently, every element of R possesses a representation as indicated. As for the uniqueness, suppose that an element $x \in R$ has two different representations of this form. Taking the difference, we see that in this case ε is the root of a polynomial $g \neq 0$ in $\mathbb{Z}[x]$ of degree $\leq p - 2$. Choose such a g with minimal degree and divide f by g with remainder, say $f(x) = q(x)g(x) + r(x)$ with $\deg r < \deg g$. Plugging in $x = \varepsilon$ shows that $r(\varepsilon) = 0$ which implies $r = 0$ due to the minimality of g; hence g divides f. But this is impossible because f is irreducible, due to (8.19)(c).

(b) Let $1 \leq r \neq s \leq p - 1$ and let $m \in \mathbb{N}$ be such that $ms \equiv r$ modulo p (i.e., $m = s^{-1}r$ in the multiplicative group $\mathbb{Z}_p^{\times}$). Plugging in $x = \varepsilon^s$ in the identity

$$\frac{1 - x^m}{1 - x} = 1 + x + x^2 + \cdots + x^{m-1}$$

we see that $1 - \varepsilon^{ms} = 1 - \varepsilon^r$ is divisible by $1 - \varepsilon^s$ in R. This shows that all quotients $(1 - \varepsilon^r)/(1 - \varepsilon^s)$ are units in R which clearly implies that all elements $1 - \varepsilon^k$ with $1 \leq k \leq p - 1$ are associates. Thus to prove that none of them is a unit it is enough to show that $1 - \varepsilon$ is not a unit, i.e., that $1/(1 - \varepsilon)$ is not an element of R. This is an easy task which is left as an exercise. (See problem 27 below.) Finally, we have

$$p = x^{p-1} + x^{p-2} + \cdots + x + 1|_{x=1}$$
$$= (x - \varepsilon)(x - \varepsilon^2) \cdots (x - \varepsilon)^{p-1}|_{x=1} = (1 - \varepsilon)(1 - \varepsilon^2) \cdots (1 - \varepsilon^{p-1}) .$$

(c) We show first that if x and y are not coprime in $\mathbb{Z}$ or if $x + y$ is divisible by p then $x + \varepsilon^r y$ and $x + \varepsilon^s y$ cannot be coprime in R. If x and y have a proper common divisor m in $\mathbb{Z}$ then m is also a proper common divisor of $x + \varepsilon^r y$ and $x + \varepsilon^s y$ in R. If $x + y$ is divisible by p in $\mathbb{Z}$ then $x + y$ is divisible by $1 - \varepsilon$ in R because p is an associate of $(1 - \varepsilon)^{p-1}$ by part (b). Hence $1 - \varepsilon$ is a common divisor of $x + \varepsilon^r y = (x + y) + (\varepsilon^r - 1)y$ and $x + \varepsilon^s y = (x + y) + (\varepsilon^s - 1)y$.

Conversely, assume that $x + y$ are coprime in $\mathbb{Z}$ and that $x + y$ is not divisible by p. We will show that in this case the ideal

$$I := \langle x + \varepsilon^r y, \, x + \varepsilon^s y \rangle \trianglelefteq R$$

contains the element 1 which clearly implies that $x + \varepsilon^r y$ and $x + \varepsilon^s y$ are coprime in R. First of all, the ideal I contains $(x + \varepsilon^r y) - (x + \varepsilon^s y) = \varepsilon^r (1 - \varepsilon^{r-s}y) \sim (1 - \varepsilon)y$ and also $(x + \varepsilon^r y)\varepsilon^s - (x + \varepsilon^s y)\varepsilon^r = \varepsilon^s(1 - \varepsilon^{r-s})x \sim (1 - \varepsilon)x$. Since x and y are coprime in $\mathbb{Z}$ there are integers $m, n \in \mathbb{Z}$ with $mx + ny = 1$. Therefore, the ideal I contains

$$m\underbrace{(1 - \varepsilon)x}_{\in I} + n\underbrace{(1 - \varepsilon)y}_{\in I} = (1 - \varepsilon)(mx + ny) = 1 - \varepsilon ,$$

hence also all elements $1 - \varepsilon^k$ with $1 \leq k \leq p - 1$ which are associates of $1 - \varepsilon$ and then also p which is the product of these associates. Moreover, I contains the element

$$\underbrace{(x + \varepsilon^r y)}_{\in I} + \underbrace{(1 - \varepsilon^r)}_{\in I} y = x + y .$$

Since $x + y$ is not divisible by p in $\mathbb{Z}$ there are integers $m', n' \in \mathbb{Z}$ with $m'(x+y) + n'p = 1$, and since $x + y$ and p are both elements of I this implies that $1 \in I$ as desired. ∎

WE now need a technical result whose proof can be given only later because it requires some field theory.

(9.22) Kummer's lemma. Let $\varepsilon = e^{2\pi i/p}$ where p is an odd prime. Then every unit in the ring $\mathbb{Z}[\varepsilon]$ can be written as a product of a real unit and a power of ε.

ASSUMING this lemma for the time being we can now prove Kummer's celebrated theorem.

(9.23) Kummer's theorem. Let $p \geq 5$ be a prime number and let $\varepsilon = e^{2\pi i/p}$. If $\mathbb{Z}[\varepsilon]$ is a unique factorization domain then there are no natural numbers x, y, z, all coprime with p, such that $x^p + y^p = z^p$.

Proof. If there were such numbers then (by dividing out the greatest common divisor of these numbers) we could also find such numbers with the additional property of being relatively prime and hence pairwise coprime. Thus suppose that there are natural numbers x, y, z, none divisible by p, with

$$z^p = x^p + y^p = (x - y)(x + \varepsilon y)(x + \varepsilon^2 y) \cdots (x + \varepsilon^{p-1} y) .$$

By Fermat's theorem (1.1) we have $x + y \equiv x^p + y^p = z^p \equiv z \not\equiv 0$ modulo p so that $x + y$ is not divisible by p. Moreover, x and y are coprime by assumption. Hence lemma (9.21) implies that the elements $x + \varepsilon^r y$ $(0 \leq r \leq p - 1)$ are pairwise coprime. Then due to (9.15) each of these elements must be an associate of a p-th power in $\mathbb{Z}[\varepsilon]$. In particular there are a unit u and an element a in $\mathbb{Z}[\varepsilon]$ with

$$(1) \qquad\qquad x + \varepsilon y = u \cdot a^p .$$

Rewriting the equation $x^p + y^p = z^p$ in the form $x^p + (-z)^p = (-y)^p$ and repeating the argument we see that there are also a unit v and an element b in $\mathbb{Z}[\varepsilon]$ with

$$(2) \qquad\qquad x - \varepsilon z = v \cdot b^p .$$

Now write $a = a_0 + a_1 \varepsilon + \cdots + a_{p-2}\varepsilon^{p-2}$ with coefficients $a_i \in \mathbb{Z}$; then

$$a^p \equiv a_0^p + a^p \varepsilon^p + \cdots + a_{p-2}\varepsilon^{p(p-2)} \equiv a_0 + a_1 + \cdots + a_{p-2} =: N \bmod p .$$

According to Kummer's lemma (9.22) we can find a real unit r such that $u = \varepsilon^k r$. Then the equation $x + \varepsilon y = ua^p$ implies that $\varepsilon^{-k}(x + \varepsilon y) = ra^p \equiv rN =: \rho \in \mathbb{R}$ modulo p. Taking complex conjugates in this equation (which preserves congruences modulo p) we also see that $\varepsilon^k(x + \varepsilon^{-1}y) \equiv \rho$ modulo p; hence

$$\varepsilon^{-k}(x + \varepsilon y) \equiv \varepsilon^k(x + \varepsilon^{-1}y) \text{ modulo } p .$$

This means that

$$(\star\star) \qquad \omega := x\varepsilon^k + y\varepsilon^{k-1} - x\varepsilon^{-k}y - y\varepsilon^{1-k} \text{ is divisible by } p \text{ in } \mathbb{Z}[\varepsilon] .$$

Now if an element $x \in \mathbb{Z}[\varepsilon]$ is written in the canonical form $x = a_0 + a_1\varepsilon + \cdots + a_{p-2}\varepsilon^{p-2}$ then clearly x is divisible by p in $\mathbb{Z}[\varepsilon]$, i.e., $x/p \in \mathbb{Z}[\varepsilon]$, if and only if all coefficients a_i are divisible by p in $\mathbb{Z}$. Now we distinguish three cases.

First case: The four numbers k, $k-1$, $-k$ and $1-k$ are pairwise incongruent modulo p, and none of these numbers is congruent with $p-1$ modulo p. In this case $(\star\star)$ is already in canonical form, and we conclude that both x and y are divisible by p, contradicting our assumption.

Second case: One of the numbers k, $k-1$, $-k$, $1-k$ is congruent with $p-1$ modulo p. Let us study the four possibilities.

- If $k \equiv p-1\,(p)$ then $\varepsilon^k = \varepsilon^{-1}$ so that $\omega = x\varepsilon^{-1} + y\varepsilon^{-2} - x\varepsilon - y\varepsilon^2$. Since ω is divisible by p then so is $\varepsilon^2\omega = x\varepsilon + y - x\varepsilon^3 - y\varepsilon^4$ which is written in canonical form already; hence the same argument as before shows that x and y are divisible by p.
- If $k-1 \equiv p-1\,(p)$ then $\varepsilon^k = 1$ so that $\omega = x + y\varepsilon^{-1} - x - y\varepsilon$. Then $\varepsilon\omega = y - y\varepsilon^2$ is divisible by p in $\mathbb{Z}[\varepsilon]$ which implies that y is divisible by p.
- If $-k \equiv p-1\,(p)$ then $\varepsilon^k = \varepsilon$ so that $\omega = x + y\varepsilon^{-1} - x - y - y\varepsilon$. Then $\varepsilon\omega = (y-x) + x\varepsilon^2 - y\varepsilon^3$ is divisible by p in $\mathbb{Z}[\varepsilon]$, hence x and y are divisible by p in $\mathbb{Z}$.
- If $1-k \equiv p-1\,(p)$ then $\varepsilon^k = \varepsilon^2$ so that $\omega = x\varepsilon^2 + y\varepsilon - x\varepsilon^{-2} - y\varepsilon^{-1}$. Then $\varepsilon^2\omega = x\varepsilon^4 + y\varepsilon^3 - x - y\varepsilon$ is divisible by p in $\mathbb{Z}[\varepsilon]$, hence x and y are divisible by p in $\mathbb{Z}$.

Third possibility: Two of the numbers k, $k-1$, $-k$ and $1-k$ are congruent modulo p. The congruence $k \equiv -k$ leads to $k \equiv 0$, and this case was covered in part (b) already. The congruence $k-1 \equiv 1-k$ leads to $k \equiv 1$ which was also treated under (b). Finally, the congruences $k \equiv 1-k$ and $k-1 \equiv -k$ lead to $2k \equiv 1$. In this case we have $\varepsilon^{-k} = \varepsilon \cdot \varepsilon^k$ and therefore $\varepsilon^k\omega = (y-x) + \varepsilon(x-y)$. In this case we can only conclude that $x-y$ is divisible by p. However, we can now invoke the second equation above, namely $x - \varepsilon z = vb^p$, and apply exactly the same argument. Again, there are three cases, and the least we can conclude is that $x - (-z)$ is divisible by p. (If one of the first two cases occurs for the second equation we can even conclude that x or z is divisible by p.) Hence we know that $x-y$ and $x+z$ are divisible by p so that $y \equiv x$ and $z \equiv -x$ modulo p. Consequently,

$$-x \equiv z \equiv z^p = x^p + y^p \equiv x+y \equiv 2x$$

modulo p, i.e., $3x \equiv 0$ and hence $x \equiv 0$. This contradicts our hypothesis that none of the numbers x, y and z is divisible by p. ∎

THIS does not establish Fermat's Last Theorem for the exponents in question, but only a part of it; namely, the equation $x^p + y^p = z^p$ has no solution in natural numbers coprime with p. Using deeper properties of the ring $\mathbb{Z}[e^{2\pi i/p}]$, Kummer was able to prove that if $\mathbb{Z}[e^{2\pi i/p}]$ is a unique factorization domain then the equation $x^p + y^p = z^p$ has no solution in natural numbers at all. Thereby he established Fermat's theorem for all prime exponents p with the property that $\mathbb{Z}[e^{2\pi i/p}]$ is a unique factorization domain.

Exercises

Problem 1. Let $n \in \mathbb{N}$ be a square-free integer such that $\mathcal{O}_n$ is a unique factorization domain.

(a) Show that for every prime element π in $\mathcal{O}_n$ there is a unique prime number $p \in \mathbb{N}$ with $\pi \mid p$ in $\mathcal{O}_n$.

Hint. Since $\pi \mid N(\pi)$ the set of natural numbers n with $\pi \mid n$ in $\mathcal{O}_n$ is not empty and hence possesses a smallest element.

(b) Show that if $p \in \mathbb{N}$ is an odd prime number then p is

$$\begin{cases} \text{a prime element in } \mathcal{O}_n, & \text{if } p \nmid n \text{ and } \left(\frac{n}{p}\right) = -1; \\ \text{a product of two non-associate prime elements in } \mathcal{O}_n, & \text{if } p \nmid n \text{ and } \left(\frac{n}{p}\right) = 1; \\ \text{an associate of a square of a prime element in } \mathcal{O}_n, & \text{if } p \mid n. \end{cases}$$

(c) Show that 2 is

$$\begin{cases} \text{a prime element in } \mathcal{O}_n, & \text{if } n \equiv 5 \bmod 8; \\ \text{a product of two non-associate prime elements in } \mathcal{O}_n, & \text{if } n \equiv 1 \bmod 8; \\ \text{an associate of a square of a prime element in } \mathcal{O}_n, & \text{if } n \equiv 3 \bmod 4. \end{cases}$$

(d) Show that if an odd prime number $p \in \mathbb{N}$ can be written as a product of two non-associate prime elements π_1 and π_2 in $\mathcal{O}_n$ π_1 and $\overline{\pi_2}$ are associates, and so are $\overline{\pi_1}$ and π_2.

Problem 2. Let $n < -1$ be a square-free number such that $|n|$ is not a prime number. Show that $\mathcal{O}_n$ is not a unique factorization domain.

Hint. Suppose that $\mathcal{O}_n$ is a unique factorization domain and let p is a prime divisor of n in $\mathbb{Z}$. Then p is an associate of a square of a prime in $\mathcal{O}_n$, due to part (b) of problem 1. Now take norms.

Problem 3. Using problem 1, describe the set of all prime elements in $\mathcal{O}_n$ for $n \in \{-3, -2, -1, 2, 3\}$.

Problem 4. (a) Show that if $m, n \in \mathbb{N}$ are coprime numbers which can both be represented as a sum of two squares, then mn can also be represented as a sum of two squares.

(b) Characterize the natural numbers which can be written as a sum of two squares.

Problem 5. Let a and b be natural numbers. Suppose that a number $n \in \mathbb{N}$ possesses two representations $n = ax^2 + by^2 = aX^2 + bY^2$ with $x, y, X, Y \in \mathbb{N}$.

(a) Show that $ab(Xy - xY)(Xy + xY)$ is divisible by n.

(b) Show that $|xY \pm Xy| \leq n$ where the inequality is strict except if $a = b = 1$ and the sign is "+". To do so, use the estimate

$$n^2 = (ax^2 + by^2)(aX^2 + bY^2) = (axX \pm byY)^2 + ab(xY \mp Xy)^2 \geq ab(xY \mp Xy)^2 .$$

(c) Suppose now that n is a prime number. Show that $xY - Xy$ must be zero (where X and Y may have to be replaced in the case $a = b = 1$). Conclude that $(x, y) = (X, Y)$ or $(x, y) = (Y, X)$ so that there is essentially only one representation $n = ax^2 + by^2$.

(d) Verify the equations

$$2537 = 1 \cdot 15^2 + 2 \cdot 34^2 = 1 \cdot 45^2 + 2 \cdot 16^2,$$
$$8003 = 2 \cdot 10^2 + 3 \cdot 51^2 = 2 \cdot 20^2 + 3 \cdot 49^2,$$
$$14017 = 1 \cdot 49^2 + 6 \cdot 44^2 = 1 \cdot 79^2 + 6 \cdot 36^2.$$

Conclude that the numbers 2537, 8003 and 14017 are not prime and find their prime factorizations.

Problem 6. The number -6 is a quadratic residue modulo 5; nevertheless, 5 cannot be written in the form $x^2 + 6y^2$. Why doesn't this contradict the equivalence of (1) and (5) in theorem (9.1)?

Problem 7. Show by direct calculating all squares in $\mathbb{Z}_{11}^{\times}$ that an element $x \in \mathbb{Z}_{11}^{\times}$ satisfies $\left(\frac{x}{11}\right)$ if and only if $x \in \{1, 3, 4, 5, 9\}$.

Problem 8. Find all prime numbers p such that -2 is a square modulo p
(a) by considering the half-system $\{1, 3, 5, \ldots, p - 2\}$;
(b) by considering the half-system $\{1, 2, 3, \ldots, \frac{p-1}{2}\}$.

Problem 9. Suppose that a natural number n can be written as $n = ax^2 + bxy + cy^2$.

(a) Show that if p is a prime divisor of n then either p is a common divisor of x and y or $b^2 - 4ac$ is a square modulo p (possibly divisible by p).

Hint. Use problem 12 in section 7.

(b) Suppose that a number $n \in \mathbb{N}$ has the property that all prime numbers $p \leq \sqrt{n}$ are neither common divisors of x and y nor divisors of $b^2 - 4ac$ and that $\left(\frac{b^2-4ac}{p}\right) = -1$ for all these p. Conclude that n is a prime number.

Problem 10. (a) Show that the equation $x^2 = 93$ has a solution in $\mathbb{Z}_{137}$.
(b) Is 14993 a quadratic residue modulo 65537?
(c) Is 1801 a quadratic residue modulo 8191?
(d) Does the equation $x^2 + 391y = 7$ have an integral solution?
(e) Does the equation $1001x^2 + 1000y = 999$ have an integral solution?

Problem 11. (a) For which prime numbers p are the polynomials $f(x) = x^2 + 1$ and $g(x) = x^2 + x + 1$ irreducible over $\mathbb{Z}_p$?
(b) Show that for any prime number p there is a natural number x such that $(x^2 + 1)(x^4 - 4) \equiv 0$ modulo p even though the equation $(x^2 + 1)(x^4 - 4) = 0$ has no solution in $\mathbb{Z}$.

Problem 12. (a) Let $a \neq 0$ and b be coprime natural numbers where b is odd. If $b = p_1^{r_1} p_2^{r_2} \cdots p_n^{r_n}$ is the prime factorization of b then the **Jacobi symbol** $(\frac{a}{b})$ is defined as

$$(\frac{a}{b}) := (\frac{a}{p_1})^{r_1} (\frac{a}{p_2})^{r_2} \cdots (\frac{a}{p_n})^{r_n}$$

where the symbols on the right-hand side are Legendre symbols. † Show that that Jacobi symbol has the following properties.

$$(\frac{-1}{b}) = (-1)^{(b-1)/2}, \quad (\frac{2}{b}) = (-1)^{(b^2-1)/8}, \quad (\frac{b}{a}) = (\frac{a}{b})(-1)^{\frac{a-1}{2} \cdot \frac{b-1}{2}} \quad (a \text{ odd}).$$

(b) Suppose you want to decide whether or not a number is a square modulo some other number. Explain why the use of the Jacobi symbol is more convenient than the use of the Legendre symbol.

(c) Give an example in which a Jacobi symbol $(\frac{a}{n})$ equals 1 even though a is not a square modulo n.

(d) Is 1363 a quadratic residue modulo 65537?

Hint. $1363 = 29 \cdot 47$, and 65537 is a prime.

(e) Is $20,002$ a quadratic residue modulo $134,353$?

Hint. $134,353$ is a prime.

Problem 13. Prove proposition (9.13).

Problem 14. Let $p \in \mathbb{N}$ be a prime number.

(a) Show that p can be written in the form $p = x^2 - 3y^2$ if and only if $p \equiv 1\,(12)$ and in the form $p = -x^2 + 3y^2$ if and only if $p = 2, 3$ or $p \equiv -1\,(12)$.

(c) Show that p can be written in the form $p = x^2 - 6y^2$ if and only if $p = 3$ or $p \equiv 1, -5\,(24)$ and in the form $p = -x^2 + 6y^2$ if and only if $p = 2$ or $p \equiv -1, 5\,(24)$.

Problem 15. Let p be a prime of the form $p = 4k + 3$. Show that $2p + 1$ is a prime number if and only if $2p + 1$ divides $2^p - 1$.

Problem 16. (a) Let $p > 5$ be a prime. Show that

$$(\frac{5}{p}) = \begin{cases} 1, & \text{if } p \equiv 1, 4\,(5); \\ -1, & \text{if } p \equiv 2, 3\,(5). \end{cases}$$

$$(\frac{-5}{p}) = \begin{cases} 1, & \text{if } p \equiv 1, 3, 7, 9\,(20); \\ -1, & \text{if } p \equiv 11, 13, 17, 19\,(20). \end{cases}$$

(b) Show that a prime number $p > 5$ can be written in the form $p = |x^2 - 5y^2|$ if and only if its last digit is 1 or 9.

(c) Show that p can be written in the form $p = x^2 + xy - y^2$ if and only if $p = 5$ or $p \equiv \pm 1\,(5)$.

---

† If b is a prime then the Jacobi symbol coincides with the Legendre symbol; hence there is no ambiguity in notation.

Problem 17. Let $p > 7$ be a prime. Show that

$$\left(\frac{7}{p}\right) = \begin{cases} 1, & \text{if } p \equiv 1,3,9,19,25,27 \, (28); \\ -1, & \text{if } p \equiv 5,11,13,15,17,23 \, (28). \end{cases}$$

$$\left(\frac{-7}{p}\right) = \begin{cases} 1, & \text{if } p \equiv 1,2,4 \, (7); \\ -1, & \text{if } p \equiv 3,5,6 \, (7). \end{cases}$$

Which prime numbers are representable in the form $p = x^2 + 7y^2$ or $p = |x^2 - 7y^2|$?

Problem 18. (a) Show that if p is a prime number of the form $p = 6n + 1$ then there are integers $x, y \in \mathbb{Z}$ with $4p = x^2 + 27y^2$.
Hint. We can write p in the form $p = a^2 + 3b^2$. Then $4p = (a \pm 3b)^2 + 3(a \mp b)^2$.
(b) Show that every prime number of the form $p = 3n + 1$ can be written as $p^2 = a^2 - ab + b^2$ with $a, b \in \mathbb{Z}$.

Problem 19. Let n be a prime number with $n \equiv 3 \bmod 4$ such that $\mathcal{O}_{-n}$ is a unique factorization domain.
(a) Consider the polynomial $f(x) := x^2 + x + m$ where $m := (n+1)/4$. Show that $f(x)$ is a prime number for all $x < (n-3)/4$.†
(b) Show that all prime numbers $p < n/4$ are quadratic nonresidues modulo n.
Hints. (a) Suppose that $f(x)$ is not prime for some $x < (n-3)/4$. Then there is a prime factor $p \leq \sqrt{f(x)} < m$. Conclude that p must be irreducible in $\mathcal{O}_{-n}$ by estimating the norm of a possible prime factor of p in $\mathcal{O}_{-n}$. On the other hand, use the fact that

$$p \mid x^2 + x + m = \frac{1}{4}((2x+1)^2 + n) = \frac{2x+1+\sqrt{-n}}{2} \cdot \frac{2x+1-\sqrt{-n}}{2}$$

to show that p is not a prime element in $\mathcal{O}_{-n}$.
(b) Suppose that some prime $p < n/4$ is a quadratic residue modulo p. Use the quadratic reciprocity law to show that then $-n$ is a quadratic residue modulo p, say $k^2 \equiv -n \bmod p$. After replacing k by $k + p$, if necessary, we can assume that k is odd, say $k = 2x + 1$. Now consider $f(x)$.

Problem 20. (a) Show that $\mathcal{O}_{-79}$ is not a unique factorization domain by either considering the polynomial $x^2 + x + 20$ or by finding a prime $p < 20$ with $\left(\frac{p}{79}\right) = 1$.
Hint. Apply problem 19.
(b) Find other primes n for which $\mathcal{O}_{-n}$ is not a unique factorization domain.

Problem 21. Which prime numbers p can be written in the form $p = x^2 + 5y^2$ or $p = 3x^2 + 2xy$?

Problem 22. Show that the equations $13x^2 + 34xy + 22y^2 = 23$ and $5x^2 + 16xy + 13y^2 = 23$ do not possess integer solutions.

---

† Assuming the nontrivial fact that $\mathcal{O}_{-163}$ is a unique factorization domain this explains the remarkable fact that $x^2 + x + 41$ is a prime number for $1 \leq x \leq 39$. (See problem 8 in section 4.)

Problem 23. Find all integer solutions of the following equations.

$$x^2 + 3 = y^2 \qquad\qquad 4x^2 + 4x + 4 = y^2 \qquad\qquad x^2 + x + 1 = y^2$$
$$x^2 - x + 1 = y^2 \qquad\qquad x^3 - 1 = y^2 \qquad\qquad x^3 + 1 = y^2$$
$$5x^2 + 2xy + 2y^2 = 26 \qquad 5x^2 - 2y^2 = 3 \qquad\qquad 80x^2 - y^2 = 16$$

Problem 24. A triple $(x, y, z) \in \mathbb{N}^3$ is called a **primitive Pythagorean triple** if $x^2 + y^2 = z^2$ such that x, y, z are relatively prime and x is even.

(a) Show that every solution $(a, b, c) \in \mathbb{N}^3$ of the equation $a^2 + b^2 = c^2$ can be obtained from a primitive Pythagorean triple (x, y, z) by a permutation of x, y, z and a multiplication of x, y and z by some natural number.

(b) Show that if $a > b$ are coprime natural numbers, one of them even, the other odd, then $(x, y, z) = (2ab, a^2 - b^2, a^2 + b^2)$ is a primitive Pythagorean triple. Conversely, show that every primitive Pythagorean triple can be obtained in this way.

(c) Show that the equation $x^4 + y^4 = z^2$ has no solution (x, y, z) in natural numbers.

Hint. If there is a solution, there is also one with a minimal value for z. Then $\gcd(x, y, z) = 1$, and after a possible exchange of x and y we can assume that x is even so that (x^2, y^2, z) is a primitive Pythagorean triple, say $(x^2, y^2, z) = (2ab, a^2 - b^2, a^2 + b^2)$. Conclude that (b, y, a) is a primitive triple, say $(b, y, a) = (2cd, c^2 - d^2, c^2 + d^2)$ so that $x^2 = 4cd(c^2 + d^2)$. Deduce that c, d and $c^2 + d^2$ are squares, say $c = e^2$, $d = f^2$ and $c^2 + d^2 = g^2$. Then $e^4 + f^4 = g^2$ and $g < z$ which contradicts the minimality of z.

(d) Show that the equation $x^4 + 4y^4 = z^2$ has no solution in natural numbers.

Problem 25. Use arithmetic in the ring $\mathbb{Z} + \mathbb{Z}\sqrt{-2}$ to find all integer solutions of the equation $y^2 = x^3 - 2$.

Remark. It is interesting to note that there are rational solutions $(x, y) \in \mathbb{Q}^2$ which are not integral, for example

$$(x, y) = (\frac{129}{100}, \frac{383}{1000}) \quad \text{and} \quad (x, y) = (\frac{97723}{9^2 19^2}, \frac{29,719,175}{9^3 19^3}) \, .$$

Problem 26. (a) Show that the equation $x^3 + y^3 = 3z^3$ has no solution in nonzero integers.

(b) Show that the equation $x^3 + y^3 = 5z^3$ has no solution in nonzero integers.

Problem 27. Let $\varepsilon = e^{2\pi i/p}$ where p is a prime number. Show that $1 - \varepsilon$ is not a unit in $\mathbb{Z}[\varepsilon]$.

Hint. If it were one could find integers $a_0, \ldots, a_{p-2} \in \mathbb{Z}$ with $1/(1 - \varepsilon) = a_0 + a_1\varepsilon + \cdots + a_{p-2}\varepsilon^{p-2}$.

Problem 28. Let a be a natural number.

(a) Show that there are infinitely many primes p with $\left(\frac{a}{p}\right) = 1$.

(b) Show that if a is not a square then there are infinitely many primes p with $\left(\frac{a}{p}\right) = -1$.

10. Modules and integral ring extensions

ROUGHLY speaking, a module over a ring R is an abelian group on which R acts linearly. Let us give the precise definition and some examples before we explain why we are introducing the concept of a module at this point.

(10.1) Definition. *Let R be a ring. An abelian group $(A, +)$ is called a **module over R** or simply an **R-module** if there is an operation $R \times A \to A$ denoted by $(r, a) \mapsto r \star a$ such that for all elements $a, b \in A$ and $r, s \in R$ the following conditions hold:*

(1) $r \star (a + b) = r \star a + r \star b$;

(2) $(r + s) \star a = r \star a + s \star a$;

(3) $(rs) \star a = r \star (s \star a)$.

If R has an identity element 1 such that

(4) $1 \star a = a$ for all $a \in A$

*then A is called a **unitary R-module**.*

(10.2) Remarks. (a) If A is an R-module we can define a group homomorphism $\varphi : (R, +) \to (\operatorname{End} A, \circ)$ by $\varphi(r)a := r \star a$. Conversely, given such a homomorphism φ, we can define a module action of R on A by $r \star a := \varphi(r)a$.

(b) Let A be an R-module. Let us denote the zero element of the abelian group A by 0_A and the zero element of the ring R by 0_R. It is easy to derive from the definition of a module that we have

$$0_R \star a = 0_A, \quad r \star 0_A = 0_A, \quad (-r) \star a = -(r \star a) = r \star (-a) \text{ and } m \cdot (r \star a) = r \star (m \cdot a)$$

for all $a \in A$, $r \in R$ and $m \in \mathbb{Z}$. Henceforth we will simply write 0 for both 0_A and 0_R; it will always be clear from the context which zero element is meant.

(10.3) Examples. (a) If K is a field then a unitary K-module is the same as a vector space over K.

(b) A unitary $\mathbb{Z}$-module is the same as an abelian group, the action of $\mathbb{Z}$ being given by $n \star a = a + \cdots + a$ if $n > 0$.

(c) Any abelian group A is a unitary $(\operatorname{End} A)$-module via $f \star a := f(a)$.

(d) Every abelian group A can be made into a module over an arbitrary ring R with the trivial action $r \star a := 0$ for all $r \in R$ and all $a \in A$.

(e) Let $X \neq \emptyset$ be an arbitrary set and R a ring. Let $R^{(X)}$ be the set of all formal sums $\sum_{x \in X} r_x x$ with coefficients $r_x \in R$ such that $r_x = 0$ for all but a finite number of elements $x \in X$. Then $R^{(X)}$ becomes an R-module via

$$\sum_{x \in X} r_x x + \sum_{x \in X} s_x x := \sum_{x \in X} (r_x + s_x)x \quad \text{and} \quad r \star (\sum_{x \in X} r_x x) := \sum_{x \in X} (rr_x)x \,.$$

Equipped with these operations, $R^{(X)}$ is called the **free R-module on X**.

THE reason we introduced the concept of a module is that this concept encompasses several different situations which occur in ring theory and which can be treated on an equal footing by using the notion of a module.

(10.4) Examples. Let R be a ring.

(a) If I is an ideal of R, then I is an R-module via $r \star x := rx$ (ring multiplication in R).

(b) If I is an ideal of R, then the quotient ring R/I is an R-module via $r \star (y+I) := ry + I$.

(c) If $S \supseteq R$ is a ring extension of R then S is an R-module via $r \star s := rs$ (ring multiplication in S).

BEFORE we use the notion of a module in "pure" ring theory, let us develop some general theory of modules, thus treating modules as algebraic objects in their own right. We will define submodules, quotient modules, module homomorphisms, direct sums and products and so on in much the same way as we proceeded in the theory of groups and rings. In fact, this formal similarity led to the general concept of a category as a class of objects with a certain structure (in our case an algebraic structure). Then in category theory one tries to establish in a very general setting to derive statements about substructures, quotient structures, structure-preserving maps, and so on, thereby capturing common features of different theories.

(10.5) Definitions. *Let A be an R-module.*

(a) *A subgroup $U \leq A$ is called an* **R-submodule** *of A if $r \star u \in U$ for all $r \in R$ and all $u \in U$, i.e., if U is itself a module under the given action of R.*

(b) *If $U \leq A$ is an R-submodule then the quotient group A/U becomes itself an R-module via $r \star (a + U) := (r \star a) + U$: this module is called the* **quotient module** *of A modulo U.*

(10.6) Examples. (a) Let K be a field and V a unitary K-module, i.e., a vector space over K. Then a K-submodule of K is the same as a vector subspace.

(b) If $(U_i)_{i \in I}$ is a family of R-submodules of an R-module A, then the intersection $\bigcap_{i \in I} U_i$ and the sum $\sum_{i \in I} U_i$ are also R-submodules of A.

(c) Let A be an R-module and $X \subseteq A$ a subset of A. By part (b), the intersection of all submodules U of A with $U \supseteq X$ is again a submodule of A, obviously the smallest submodule of A containing X. It is called the submodule **generated** or **spanned** by X; we can easily verify that it is given by

$$\{\sum_{i=1}^m r_i \star x_i + \sum_{j=1}^n k_j \cdot x_j' \mid m, n \in \mathbb{N}, x_i, x_j' \in X, r_i \in R, k_j \in \mathbb{Z}\} .$$

If U is generated by a finite set X then U is called **finitely generated**.

(10.7) Remark. Let A be a unitary R-module. It is clear that every submodule U of A is automatically unitary. Moreover, if X is a subset of A, then the submodule generated by X is

$$\{\sum_{i=1}^m r_i \star x_i \mid m \in \mathbb{N}, x_i \in X, r_i \in R\}$$

because in a unitary module we have $k \cdot y = (k \cdot 1_R) \star y \in R \star y$ where $R \star y := \{r \star y \mid r \in R\}$. In particular, if X is a finite set $X = \{x_1, \ldots, x_n\}$ then the module generated by X is

$$R \star x_1 + \cdots + R \star x_n .$$

AS in the case of groups and rings, we can define the direct product and the direct sum of modules.

(10.8) Definition. *Let $(A_i)_{i \in I}$ be a family of R-modules. Then the (group-theoretical) direct product $\prod_{i \in I} A_i$ becomes an R-module with the componentwise action $r \star (a_i)_{i \in I} := (r \star a_i)_{i \in I}$. Clearly, the (group-theoretical) direct sum $\bigoplus_{i \in I} A_i$ is a submodule of $\prod_{i \in I} A_i$ under this action. We call $\prod_{i \in I} A_i$ and $\bigoplus_{i \in I} A_i$ the* **direct product** *and the* **direct sum** *of the modules A_i.*

WE have observed already that it is useful for the study of algebraic objects to also study the mappings between these objects which preserve the given algebraic structure. For modules, this leads to the notion of a module homomorphism.

(10.9) Definition. *Let A and B be R-modules. A mapping $f : A \rightarrow B$ is called an* **R-module homomorphism** *if*

$$f(a + a') \;=\; f(a) + f(a') \quad and \quad f(r \star a) \;=\; r \star f(a)$$

for all $a, a' \in A$ and all $r \in R$. In other words, an R-module homomorphism is a group homomorphism which is **R-linear** *in the sense that it commutes with the action of R.*

A bijective module homomorphism f is called a **module isomorphism**; *in this case, the inverse mapping f^{-1} is again a module homomorphism.*

(10.10) Example. (a) Let $(A_i)_{i \in I}$ be a family of R-modules. Then the projections $p_{i_0} : \prod_{i \in I} A_i \rightarrow A_{i_0}$ are surjective R-module homomorphisms.

(b) Let R be a ring, $X \neq \emptyset$ a set and $f : X \rightarrow A$ a mapping from X into an R-module A. Then there is a unique R-module homomorphism $F : R^{(X)} \rightarrow A$ such that $F(\sum_{x \in X} r_x x) = \sum_{x \in X} r_x f(x)$. (If R possesses an identity element we can consider X as a subset of $R^{(X)}$ by identifying $x \in X$ with $1 \cdot x \in R^{(X)}$. In this case, F is the unique extension of f to an R-module homomorphism.)

IT is straightforward to check that if $f : A \rightarrow B$ is a module homomorphism then ker f is a submodule of A and im f is a submodule of B. Also, the various isomorphism theorems from group theory and ring theory can be easily carried over to modules. (See problem 5 below.)

WE now introduce two important classes of modules.

(10.11) Definition. *Let A be an R-module. Then A is called* **Noetherian** *if every ascending chain $A_1 \subseteq A_2 \subseteq A_3 \subseteq \cdots$ of submodules becomes stationary, and* **Artinian** *if every descending chain $A_1 \supseteq A_2 \supseteq A_3 \supseteq \cdots$ of submodules becomes stationary.*

HERE a sequence $(A_1, A_2, A_3, \ldots)$ of objects is said to become *stationary* if there is an index N such that $A_n = A_N$ for all $n \geq N$. The terminology was chosen to honor Emmy Noether (1882-1935) and Emil Artin (1898-1962), two prominent figures in the history of algebra. Emmy Noether's father, Max Noether (1844-1921), also contributed to the subject.

(10.12) Examples. (a) Let A be a module with only finitely many elements; for example $A = \mathbb{Z}_n$, considered as a $\mathbb{Z}$-module. Then clearly A is both Noetherian and Artinian.

(b) Let $A = \mathbb{Z}$, considered as a $\mathbb{Z}$-module so that a submodules is the same as an ideal. Then A is Noetherian; this follows easily from the fact that every ideal of $\mathbb{Z}$ has the form $\mathbb{Z}m$, and $\mathbb{Z}m \subseteq \mathbb{Z}n$ if and only if n divides m. On the other hand, A is not Artinian; namely, if $a \neq 0$, then $\langle a \rangle \supsetneq \langle a^2 \rangle \supsetneq \langle a^3 \rangle \supsetneq \cdots$ is a strictly decreasing chain of ideals.

(c) Let $p \in \mathbb{N}$ be a prime and let

$$A := \{ \frac{m}{p^n} + \mathbb{Z} \mid m \in \mathbb{Z}, n \in \mathbb{N}_0 \} \subseteq \frac{\mathbb{Q}}{\mathbb{Z}},$$

considered as a unitary $\mathbb{Z}$-module. Then a submodule is the same as a subgroup. Now it is easy to check that the only subgroups of A are the sets

$$A_n := \{ \frac{m}{p^n} + \mathbb{Z} \mid m \in \mathbb{Z} \} \qquad (n \in \mathbb{N}_0) \,.$$

Since $A_0 \subsetneq A_1 \subsetneq A_2 \subsetneq A_3 \subsetneq \cdots$ we see that A is Artinian, but not Noetherian.

(d) Let V be a vector space over K, considered as a unitary K-module so that a submodule is the same as a vector subspace. If V is finite-dimensional then V is both Noetherian and Artinian, if V is infinite-dimensional then V is neither Noetherian nor Artinian.

Let us quickly prove these statements. If V is finite-dimensional, then every strictly ascending chain $A_1 \subsetneq A_2 \subsetneq \cdots$ and every strictly descending chain $B_1 \supsetneq B_2 \supsetneq \cdots$ must be finite, because $\dim A_1 < \dim A_2 < \cdots \leq \dim V$ and $\dim B_1 > \dim B_2 > \cdots \geq 0$. Conversely, if V is infinite-dimensional, there is an infinite sequence $(x_1, x_2, \ldots)$ of linearly independent elements of V. For $n \in \mathbb{N}$ let $A_n := \langle x_1, \ldots, x_n \rangle$ and $B_n := \langle x_n, x_{n+1}, x_{n+2}, \ldots \rangle$. Then $A_1 \subsetneq A_2 \subsetneq \cdots$ and $B_1 \supsetneq B_2 \supsetneq \cdots$.

THE next proposition gives characterizations of Noetherian and Artinian modules. Note that there is an asymmetry between Noetherian and Artinian modules, despite the formal similarity in the definitions of these two types of modules.

(10.13) Proposition. *Let A be an R-module.*
(a) *The following conditions are equivalent:*
(1) *A is Noetherian* (**ascending chain condition**);
(2) *every non-empty set of submodules has a maximal element* (**maximum condition**);
(3) *every submodule of A is finitely generated* (**finiteness condition**).

(b) *The following conditions are equivalent:*

(1) *A is Artinian* (**descending chain condition**);

(2) *every non-empty set of submodules has a minimal element* (**minimum condition**).

Proof. (a) Let us show first that (1) implies (2). Suppose there is a set $\mathcal{M} \neq \emptyset$ of submodules which does not possess a maximal element. Pick $A_1 \in \mathcal{M}$. Since A_1 is not maximal, there is $A_2 \in \mathcal{M}$ with $A_1 \subsetneq A_2$. Since A_2 is not maximal, there is $A_3 \in \mathcal{M}$ with $A_2 \subsetneq A_3$. Continuing in this fashion, we obtain a strictly increasing sequence of submodules, contradicting the hypothesis that A is Noetherian.

Let us show now that (2) implies (3). Let U be a submodule of A. Then the set $\mathcal{M}$ of all finitely generated submodules of U is not empty (since $\{0\} \in \mathcal{M}$), hence possesses a maximal element U_0. Let $x \in U$. Then $U_0 + Rx$ is finitely generated, hence must be U_0 due to the maximality of U_0. This implies $x \in U_0$. Since $x \in U$ was arbitrary, we see that $U = U_0$; hence U is finitely generated.

Let us finally show that (3) implies (1). Let $A_1 \subseteq A_2 \subseteq \cdots$ be an ascending sequence of submodules of A. Then $\bigcup_{n=1}^{\infty} A_n$ is again a submodule of A, hence by hypothesis finitely generated, say by $x_1, \ldots, x_r$ where $x_i \in A_{n_i}$. Let $N := \max\{n_1, \ldots, n_r\}$; then $x_1, \ldots, x_r \in A_N$ and consequently $\bigcup_{n=1}^{\infty} A_n = A_N$ which implies $A_n = A_N$ for all $n \geq N$.

(b) The proof that (1) implies (2) is completely analogous to the corresponding proof in part (a); let us show that (2) implies (1). Let $A_1 \supseteq A_2 \supseteq \cdots$ be a descending sequence of submodules of A. By hypothesis, the set $\{A_1, A_2, \ldots\}$ possesses a minimal element, say A_N. But then $A_n = A_N$ for all $n \geq N$. ∎

THERE is a very convenient criterion which allows one to verify that a given module is Noetherian or Artinian. To formulate this criterion, we need the concept of an exact sequence of modules.

(10.14) Definition. *Let R be a ring. A sequence*

$$\cdots \to A_{n-1} \overset{f_{n-1}}{\to} A_n \overset{f_n}{\to} A_{n+1} \to \cdots$$

of R-modules and R-module homomorphisms is called an **exact sequence** *if* $\operatorname{im} f_{n-1} = \ker f_n$ *for all n.*

An exact sequence of the special form

$$\{0\} \to A \overset{\alpha}{\to} B \overset{\beta}{\to} C \to \{0\}$$

is called a **short exact sequence;** *here the exactness means that α is injective, β is surjective and $\operatorname{im} \alpha = \ker \beta$.*

(10.15) Examples. (a) Let A, C be R-modules. Then

$$\{0\} \to A \to A \oplus C \to C \to \{0\}$$

is a short exact sequence.

(b) Let U be a submodule of the R-module V. Then

$$\{0\} \ \to \ U \ \to \ V \ \to \ V/W \ \to \ \{0\}$$

is a short exact sequence.

NOW we can present the promised criterion for verifying that a given module is Noetherian or Artinian.

(10.16) Theorem. *Let $\{0\} \to A \xrightarrow{\alpha} B \xrightarrow{\beta} C \to \{0\}$ be a short exact sequence of R-modules.*

(a) *B is Noetherian if and only if A and C are Noetherian.*

(b) *B is Artinian if and only if A and C are Artinian.*

Proof. We will only prove (a); the proof of (b) is completely analogous.

Suppose first that B is Noetherian. let $A_1 \subseteq A_2 \subseteq \cdots$ be an ascending sequence of submodules of A. Then $\alpha(A_1) \subseteq \alpha(A_2) \subseteq \cdots$ is an ascending sequence of submodules of B, hence becomes stationary; since α is injective, this implies that the original sequence $A_1 \subseteq A_2 \subseteq \cdots$ becomes stationary. This shows that A is Noetherian. Now let $C_1 \subseteq C_2 \subseteq \cdots$ be an ascending sequence of submodules of C. Then $\beta^{-1}(C_1) \subseteq \beta^{-1}(C_2) \subseteq \cdots$ is an ascending sequence of submodules of B, hence becomes stationary; since β is surjective (so that $C_k = \beta\big(\beta^{-1}(C_k)\big)$ for all k) this implies that the original sequence $C_1 \subseteq C_2 \subseteq \cdots$ becomes stationary. This shows that C is Noetherian.

Suppose conversely that A and C are Noetherian and let $B_1 \subseteq B_2 \subseteq \cdots$ be an ascending sequence of submodules of B. Then the sequences $\alpha^{-1}(B_1) \subseteq \alpha^{-1}(B_2) \subseteq \cdots$ and $\beta(B_1) \subseteq \beta(B_n) \subseteq \cdots$ become stationary; say $\alpha^{-1}(B_n) = \alpha^{-1}(B_N)$ and $\beta(B_n) = \beta(B_N)$ for all $n \geq N$. We claim that $B_n = B_N$ for all $n \geq N$. Let $x \in B_n$. Since $\beta(B_n) = \beta(B_N)$ there is an element $y \in B_N$ with $\beta(x) = \beta(y)$ so that $x - y \in \ker \beta = \operatorname{im} \alpha$, say $x - y = \alpha(z)$. Then $z \in \alpha^{-1}(B_n) = \alpha^{-1}(B_N)$ so that $x - y = \alpha(z) \in B_N$ which implies $x \in y + B_N = B_N$. Since $x \in B_n$ was arbitrary, this shows $B_n \subseteq B_N$, hence $B_n = B_N$. ∎

AS an application of this theorem let us prove that finite direct sums of Noetherian (Artinian) modules are again Noetherian (Artinian). Note that this statement can also be directly derived from the definition of Noetherian and Artinian modules.

(10.17) Proposition. *If $A_1, \ldots, A_n$ are Noetherian (Artinian) R-modules, then $A_1 \oplus \cdots \oplus A_n$ is also Noetherian (Artinian).*

Proof. This follows easily by induction on n. The step from $n-1$ to n is achieved by applying theorem (10.16) to the short exact sequence

$$\{0\} \ \to \ A_n \ \to \ \bigoplus_{i=1}^{n} A_i \ \to \ \bigoplus_{i=1}^{n-1} A_i \ \to \ \{0\} \ .$$

∎

WE will now end our general discussion of modules and rather apply the concept of a module in the study of rings *per se*. If R is a subring of a ring S then S is not simply a ring, but at the same time an R-module. This coupling of a ring structure and a module structure occurs quite often and hence is given a name of its own.

(10.18) Definition. *Let R be a commutative ring. A commutative ring which is at the same time an R-module is called an* **R-algebra**. *The* **subalgebra** *of S generated by elements $s_1, \ldots, s_n$ is the set of all sums $\sum_{i_1, \ldots, i_n} r_{i_1, \ldots, i_n} \star (s_1^{i_1} s_2^{i_2} \cdots s_n^{i_n})$ with $r_{i_1, \ldots, i_n} \in R$ and $i_k \geq 0$.*

IN the case that S is a ring extension of R, the subalgebra of S generated by $s_1, \ldots, s_n$ is clearly just the ring $R[s_1, \ldots, s_n] = \{f(s_1, \ldots, s_n) \mid f \in R[x_1, \ldots, x_n]\}$. It will turn out that such a ring extension of R (which is a finitely generated R-algebra) has particularly nice properties if it is also finitely generated as an R-module. (This is generally a much stronger condition; clearly, if $S = Rs_1 + \cdots + Rs_n$ then $S = R[s_1, \ldots, s_n]$.) Let us see what this condition means in a simple (yet important) special case. Suppose that S is obtained from R by ring-adjunction of a single element s, say $S = R[s]$. Assuming that R has an identity element 1, we have

$$S = R[s] = \{r_0 + r_1 s + \cdots + r_n s^n \mid n \in \mathbb{N}, r_i \in R\} = \sum_{k \geq 0} Rs^k$$

where $s^0 := 1$. From this equation it is clear that S is finitely generated as an R-module if some power s^n can be expressed as an R-linear combination of lower powers of s, because then $S = \sum_{k=0}^{n-1} Rs^k$. Let us introduce the following terminology.

(10.19) Definition. *Let $R \subseteq S$ be commutative rings.*
   (a) *An element $s \in S$ is called* **integral** *over R if there are elements $r_0, \ldots, r_{n-1} \in R$ such that*

$$s^n + r_{n-1} s^{n-1} + \cdots + r_1 s + r_0 = 0 \,.$$

We say that S is integral over R if every element $s \in S$ is integral over R; in this case $(S : R)$ is called an **integral ring extension**.
   (b) *An element $s \in S$ is called* **algebraic** *over R if there are elements $r_0, \ldots, r_{n-1}, r_n \in R$ with $r_n \neq 0$ such that*

$$r_n s^n + r_{n-1} s^{n-1} + \cdots + r_1 s + r_0 = 0 \,.$$

We say that S is algebraic over R if every element $s \in S$ is algebraic over R; in this case $(S : R)$ is called an **algebraic ring extension**.
   (c) *An element $s \in S$ is called* **transcendental** *over R if it is not algebraic over R. If such an element exists in S then $(S : R)$ is called a* **transcendental ring extension**.

(10.20) Examples. (a) The number $3 + \sqrt{2}$ is a root of $p(x) = x^2 - 6x + 7 \in \mathbb{Z}[x]$ and hence is integral over $\mathbb{Z}$. Similarly, the number $\frac{1}{2}(\sqrt{5} - 1)$ is integral over $\mathbb{Z}$ because it is a root of the monic polynomial $p(x) = x^2 + x - 1 \in \mathbb{Z}[x]$.

(b) Let $R \subseteq S$ be commutative rings. Clearly, every element $s \in S$ which is integral over R is also algebraic over R. (This is true even if R does not possess an identity element; to see this, just multiply an equation of the form $s^n + r_{n-1}s^{n-1} + \cdots + r_1 s + r_0 = 0$ by an arbitrary element $r \in R$.) If R is a field, the converse is also true: every element on S which is algebraic over R is also integral over R.

(c) The number $\frac{3}{4} \in \mathbb{Q}$ is a root of the polynomial $p(x) = 4x - 3 \in \mathbb{Z}[x]$ and hence is algebraic over $\mathbb{Z}$. However, (7.19) shows that there is no *monic* polynomial in $\mathbb{Z}[x]$ which has $\frac{3}{4}$ as a root; hence $\frac{3}{4}$ is not integral over $\mathbb{Z}$.

(d) The numbers $\frac{1}{2}(1 \pm \sqrt{3})$ are roots of $p(x) = 2x^2 - 2x - 1 \in \mathbb{Z}[x]$ and hence are algebraic over $\mathbb{Z}$. However, they are not roots of a *monic* polynomial in $\mathbb{Z}[x]$ and hence are not integral over $\mathbb{Z}$. (See problem 24 below.)

(e) All real numbers of the form $\sqrt[n]{a}$ where $a > 0$ is a rational number are integral over $\mathbb{Q}$. Indeed, $\sqrt[n]{a}$ is a root of the monic polynomial $x^n - a \in \mathbb{Q}[x]$.

(f) The numbers e and π are transcendental over $\mathbb{Q}$, i.e., they do not arise as roots of a polynomial with rational coefficients. This is hard to prove. (Proofs can be found in *Algebra* by Serge Lang, pp. 493-499. The first proofs were given by Hermite in 1873 for the number e and by Lindemann in 1882 for the number π.)

(g) Let K be a field and let $R = K[X,Y]/I$ where $I := \langle X^2 - Y^2 - XY^2 \rangle$. Writing $x := X + I$ and $y := y + I$, we have $R = K[x,y]$ with $x^2 - y^2 - xy^2 = 0$. Let S be the quotient field of R. Then the element $\alpha = x/y \in S$ is integral over R since $\alpha^2 - (1 + x) = 0$. Note that α also satisfies linear equations with coefficients in R, namely $y\alpha - x = 0$ and $x\alpha - y - xy = 0$. Hence to establish that α is algebraic over R it is enough to consider linear equations, whereas we need a quadratic equation to establish that α is integral over R.

(h) Let K be a field and let $S = K[X_1, \ldots, X_n]$ be the polynomial ring in n variables over K. Consider the subring $R = K[s_1, \ldots, s_n]$ where s_i is the i-th symmetric polynomial in $X_1, \ldots, X_n$. (We know by (4.23)(a) that R consists of *all* symmetric polynomials in $X_1, \ldots, X_n$, but this fact is not needed here.) Since each X_i is a root of the polynomial

$$\prod_{i=1}^{n}(X - X_i) = X^n - s_1 X^{n-1} + \cdots + (-1)^n s_n \in R[X]$$

we see that S is an integral extension of R.

WE will now concentrate on the case that S has an identity element 1 and that $1 \in R$. Our main goal will be to show that if S is finitely generated as an R-algebra, say $S = R[s_1, \ldots, s_n]$, then S is finitely generated as an R-module if and only if S is integral over R. We will prove this fact in the case $n = 1$ first and then proceed by induction. One new definition is needed.

(10.21) Definition. *Let R be a commutative ring and let A be an R-module. For any subset $X \subseteq A$, we define the **annihilator** of X in R as*

$$\mathrm{Ann}_R(X) := \{r \in R \mid r \star x = 0 \text{ for all } x \in X\} .$$

It is easy to see (using the commutativity of R) that $\mathrm{Ann}_R(X)$ is an ideal of R.

NOW we can characterize integral elements in module-theoretical terms.

(10.22) Theorem. *Let S be a commutative ring with identity element 1 and $R \leq S$ a subring with $1 \in R$. Then for any $s \in S$, the following statements are equivalent.*

(1) *s is integral over R;*

(2) *$R[s]$ is finitely generated as an R-module;*

(3) *there is an intermediate ring $R[s] \subseteq Z \subseteq S$ which is finitely generated as an R-module;*

(4) *there is an $R[s]$-module A which is finitely generated as an R-module and whose annihilator in $R[s]$ is zero.*

Proof. (1)$\Longrightarrow$(2). By assumption, there are elements $a_0, \ldots, a_{n-1} \in R$ such that $s^n = -(a_0 + a_1 s + \cdots + a_{n-1} s^{n-1})$ which implies $s^{n+k} = -(a_0 s^k + a_1 s^{k+1} + \cdots + a_{n-1} s^{n+k-1})$ for all $k \geq 0$. We see inductively that every power s^m with $m \geq n$ can be expressed as an R-linear combination of $1, s, \ldots, s^{n-1}$. Hence the n elements $1, s, \ldots, s^{n-1}$ generate $R[s]$ as an R-module.

(2)$\Longrightarrow$(3). Take $Z := R[s]$.

(3)$\Longrightarrow$(4). Take $A := Z$. Then A is finitely generated by assumption, and if $x \in R[s]$ annihilates Z then (since $1 \in Z$) we have $0 = x \cdot 1 = x$.

(4)$\Longrightarrow$(1). Let $A = Ra_1 + \cdots + Ra_n$. Since A is an $R[s]$-module we have $sA \subseteq A$. So there are elements $t_{ij} \in R$ with

$$sa_1 = t_{11}a_1 + \cdots + t_{1n}a_n,$$
$$\cdots$$
$$sa_n = t_{n1}a_1 + \cdots + t_{nn}a_n.$$

This may be rewritten as a matrix equation:

$$\begin{pmatrix} t_{11} - s & t_{12} & \cdots & t_{1n} \\ t_{21} & t_{22} - s & \cdots & t_{2n} \\ \vdots & \vdots & \ddots & \vdots \\ t_{n1} & t_{n2} & \cdots & t_{nn} - s \end{pmatrix} \begin{pmatrix} a_1 \\ a_2 \\ \vdots \\ a_n \end{pmatrix} = \begin{pmatrix} 0 \\ 0 \\ \vdots \\ 0 \end{pmatrix}$$

or shorter $(T - s\mathbf{1})a = 0$. Then $\det(T - s\mathbf{1})\mathbf{1} = \mathrm{adj}(T - s\mathbf{1})(T - s\mathbf{1})$ maps a to zero, so that $\det(T - s\mathbf{1}) \in R[s]$ annihilates $a_1, \ldots, a_n$, hence all of A. By hypothesis this implies $\det(T - s\mathbf{1}) = 0$. But this equation is a polynomial equation with leading coefficient ± 1. $\blacksquare$

THEOREM (10.22) dealt with the situation that one single element is adjoined to a ring. Before we turn to the situation that several elements are adjoined, let us prove two general facts about extension rings which are finitely generated as modules over the base ring. Note that in part (b) of the following proposition the "determinant trick" occurring in the proof of (10.22) is used again.

(10.23) Lemma. *Let $R \leq S \leq T$ be commutative rings with the same identity element $1 \in R$.*

(a) *If S is finitely generated as an R-module and T is finitely generated as an S-module, then T is finitely generated as an R-module.*

(b) *Assume that $\operatorname{Ann}_R(S) = \{r \in R \mid rs = 0 \text{ for all } s \in S\}$ is $\{0\}$. If S is finitely generated as an R-module, then the only ideal $I \trianglelefteq R$ with $IS = S$ is $I = R$.*

Proof. (a) If $S = \sum_{i=1}^m Rs_i$ and $T = \sum_{j=1}^n St_j$ then $T = \sum_{j=1}^n (\sum_{i=1}^m Rs_i)t_j = \sum_{i=1}^m \sum_{j=1}^n Rs_i t_j$.

(b) Suppose $S = Rs_1 + \cdots + Rs_n$. Then each s_i lies in $S = IS = Is_1 + \cdots + Is_n$; hence there are elements $a_{ij} \in I$ with $s_i = \sum_j a_{ij} s_j$ so that $\sum_j (a_{ij} - \delta_{ij})s_j = 0$. Introducing the matrix $A := (a_{ij})$, this reads $(A - 1)\vec{s} = \vec{0}$. Multiplying this equation by the adjunct $\operatorname{adj}(A - 1)$, we see that $\det(A - 1)\vec{s} = \vec{0}$, i.e., that $\det(A - 1)s_i = 0$ for all i so that $\det(A - 1)$ lies in $\operatorname{Ann}_R(S)$. By hypothesis, this means $\det(A - 1) = 0$; on the other hand, expansion of the determinant shows that $\det(A - 1) = (-1)^n + x$ with $x \in I$. Hence $1 \in I$ which shows that $I = R$. ∎

WE are now ready to prove that a ring S of the form $R[s_1, \ldots, s_n]$ is finitely generated as an R-module if and only if the elements s_i are integral over R and that in this case $(S : R)$ is an integral ring extension.

(10.24) Theorem. *Let $R \leq S \leq T$ be commutative rings with the same identity element $1 \in R$.*

(a) *If S is finitely generated as an R-module, then S is an integral extension of R.*

(b) *If the elements $s_1, \ldots, s_n \in S$ are integral over R, then the ring $R[s_1, \ldots, s_n]$ is finitely generated as an R-module.*

(c) *If T is integral over S and S is integral over R, then T is integral over R.*

(d) *The* **integral closure of R in S**, *defined as*

$$\overline{R} := \{s \in S \mid s \text{ is integral over } R\} ,$$

is a ring with $R \subseteq \overline{R} \subseteq S$. If $\overline{R} = R$, we say that R is **integrally closed** *in S.*

(e) *If S is the integral closure of R in T, then S is integrally closed in T.*

Proof. (a) Let $s \in S$. Letting $Z := S$ in (10.22)(3), we see that s is integral over R, due to theorem (10.22).

(b) We proceed by induction on n. The case $n = 1$ was settled in (10.22). Let $n > 1$. By induction hypothesis, the ring $T := R[s_1, \ldots, s_{n-1}]$ is finitely generated as an R-module. Moreover, $R[s_1, \ldots, s_n] = T[s_n]$ is finitely generated as a T-module by (10.22) because s_n is integral over R, the more so over T. But then (10.23)(a) implies that $R[s_1, \ldots, s_n]$ is finitely generated as an R-module.

(c) Let $t \in T$. Since T is integral over S, there are elements $s_i \in S$ with $t^n + s_{n-1}t^{n-1} + \cdots + s_1 t + s_0 = 0$. Then t is integral over $S' := R[s_0, \ldots, s_{n-1}]$ so that $S'[t]$ is a finitely generated S'-module due to (10.22). But S' is a finitely generated R-module due to part (b); consequently, (10.23)(a) implies that $S'[t]$ is also a finitely generated R-module. By (10.22)(3) this implies that t is integral over R. Since $t \in T$

was arbitrarily chosen, the claim is established.

(d) The inclusions $R \subseteq \overline{R} \subseteq S$ are clear; for $R \subseteq \overline{R}$ just observe that an element $r \in R$ is the root of the monic polynomial $p(x) = x - r \in R[x]$. To exhibit $\overline{R}$ as a ring, it suffices to show that if $x, y \in S$ are integral over R, then so are $x \pm y$ and xy. Now if x and y are integral over R, then the ring $R[x, y]$ (which contains $x \pm y$ and xy) is finitely generated as an R-module, due to part (b). But then part (a) implies that $x \pm y$ and xy are integral over R.

(e) Let $t \in T$ be integral over S. By part (c), t is also integral over R, hence lies in the integral closure of R in T which is S. $\blacksquare$

(10.25) Example. Let R be a unique factorization domain with quotient field K. Then the integral closure of R in K is R itself; this is merely a restatement of proposition (7.19).

NOTE that (10.24)(d) says that sums and products of integral elements are integral again, a fact which is not immediately clear from the definition of integrality.†

WE will have a little more to say about integral ring extensions once we know more about field extensions. In fact, many important examples of integral ring extensions $(S : R)$ arise in the following way: Let R be an integral domain with quotient field K and let L be a field extension of K; then let S be the integral closure of R in L so that we have the following constellation.

In this constellation we can hope to deduce information on the ring extension $(S : R)$ from information on the field extension $(L : K)$, and we will see in later sections that this hope is not in vain.

---

† However, see problem 6 in section 7.

Exercises

Problem 1. Let R be a ring and let A, B be R-modules. Show that the set $\text{Hom}_R(A, B)$ of all module homomorphisms $f : A \to B$ becomes itself a module via

$$(f + g)(a) := f(a) + g(a) \quad \text{and} \quad (r \star f)(a) := r \star f(a) .$$

Problem 2. Suppose R is a ring with identity and A is an R-module, but not a unitary R-module. Show that there is an element $a \in A$ with $a \neq 0$ such that $r \star a = 0$ for all $r \in R$.

Problem 3. Suppose A, B, C are R-submodules of an R-module M such that $A \subseteq C$. Show that $A + (B \cap C) = (A + B) \cap C$ (**modular law for submodules**).

Problem 4. Let R be a commutative ring.
(a) Consider R in a canonical way as an R-module, so that all the left-multiplications $l_a : x \mapsto ax$ are module homomorphisms. Show that l_a is a ring homomorphism if $a^2 = a$.
(b) Let $g \in R[x]$ be a fixed polynomial and consider the mapping $\varphi_g : R[x] \to R[x]$ given by $f \mapsto gf$. Show that φ_g is an R-module homomorphism. Under what conditions on g is φ_g a ring homomorphism?

Problem 5. (**Isomorphism theorems for modules.**) (a) Let $\varphi : A \to B$ be a homomorphism of R-modules. Show that $\text{im}\,\varphi \cong A/\ker\varphi$ (isomorphism of R-modules).
(b) Let $A \subseteq B \subseteq C$ be submodules of some R-module. Show that $\frac{C/A}{B/A} \cong C/B$ (isomorphism of R-modules).
(c) Let A, B be submodules of an R-module C. Then $\frac{A+B}{B} \cong \frac{A}{A \cap B}$ (isomorphism of R-modules).

Problem 6. Let A be an R-module. For $a \in A$, let $I_a := \{r \in R \mid r \star a = 0\}$. Show that R/I_a is an R-module (where the action is multiplication from the left) which is isomorphic to a submodule of A.

Problem 7. If A is an R-module, we can consider the set

$$\text{End}_R(A) := \{f : A \to A \mid f(a + a') = f(a) + f(a') \text{ and } f(r \star a) = r \star f(a)$$
$$\text{for all } a, a' \in A, r \in R\}$$

of all module endomorphisms of A.
(a) Show that $\text{End}_R(A)$ is a ring with pointwise addition and composition of mappings and that A becomes an $\text{End}_R(A)$-module via $f \star a := f(a)$.
(b) Let $s \in S$. Show that the "left-multiplication" $l_s : A \to A$, given by $a \mapsto s \star a$, belongs to $\text{End}_R(A)$ if and only if s belongs to the center of R.
(c) Suppose that A is a **simple** R-module in the sense that A has no submodules

other than $\{0\}$ and A. Show that in this case $\mathrm{End}_R(A)$ is a division ring.

Hint. If $f \in \mathrm{End}_R(A)$, consider $\ker f$ and $\mathrm{im}\, f$.

Problem 8. (a) Let A be an R-module and let $\mathrm{End}_R(A)$ be as in problem 7. Suppose there is a subring $S \leq \mathrm{End}_R(A)$ such that A is simple as an S-module. Show that

$$D_S := \{T \in \mathrm{End}(R, +) \mid TS = ST \text{ for all } S \in S\}$$

is a division ring **(Schur's lemma)**.

Hint. Show that $D_S = \mathrm{End}_S(A)$ and apply part (c) of problem 7.

(b) Suppose R is a commutative ring and A is a simple R-module. Show that $S := \{l_r \mid r \in R\}$ satisfies the hypothesis of part (a); here $l_r : A \to A$ denotes the mapping $a \mapsto r \star a$.

(c) Consider the actions of $\mathbb{R}$ on $\mathbb{R}$, $\mathbb{R}$ on $\mathbb{R}^2$ and $\mathbb{C}$ on $\mathbb{R}^4$ given as follows:

$$a \star t := at \qquad (t \in \mathbb{R}),$$

$$a \star v := \begin{pmatrix} 0 & a \\ -a & 0 \end{pmatrix} v \qquad (v \in \mathbb{R}^2),$$

$$(a + ib) \star v := \begin{pmatrix} 0 & a & 0 & b \\ -a & 0 & b & 0 \\ 0 & -b & 0 & a \\ -b & 0 & -a & 0 \end{pmatrix} v \qquad (v \in \mathbb{R}^4).$$

Show in each case that the module in question is simple and verify Schur's lemma with S as in (b). Show that D_S is isomorphic to $\mathbb{R}$, $\mathbb{C}$ and $\mathbb{H}$, respectively.

Problem 9. Let $B := \{\frac{m}{p^n} \mid m \in \mathbb{Z}, n \in \mathbb{N}_0\} \subseteq \mathbb{Q}$, considered as a unitary $\mathbb{Z}$-module. Show that B is neither Noetherian nor Artinian.

Hint. Let $A := \mathbb{Z}$ and let C be the module from example (10.12)(c); then there is an exact sequence $\{0\} \to A \to B \to C \to \{0\}$.

Problem 10. Suppose A is an R-module such that every non-empty set of finitely generated submodules has a maximal element. Show that A is Noetherian.

Problem 11. Let A be an R-module and let $f : A \to A$ be a module homomorphism.

(a) Show that if $f \circ f = f$ then $A = \ker f \oplus \mathrm{im}\, f$.

(b) Show that if A is Noetherian and f is surjective then f is an isomorphism.

Hint. Consider the submodules $\ker f \subseteq \ker f^2 \subseteq \ker f^3 \subseteq \cdots$ of A. Conclude that $(f^n)^{-1}(f^{-1}(0)) = (f^n)^{-1}(0)$ for some $n \in \mathbb{N}$ and apply f^n to this equation.

(c) Show that if A is Artinian and f is injective then f is an isomorphism.

Hint. Consider the submodules $\mathrm{im}\, f \supseteq \mathrm{im}\, f^2 \supseteq \mathrm{im}\, f^3 \supseteq \cdots$ of A.

(d) Let $R := \mathbb{Z}_2[x]$, considered as an R-module. Show that R is Noetherian and that the mapping $f : R \to R$ given by $p \mapsto p^2$ is an injective ring homomorphism which is not surjective.

Problem 12. Let A be a Noetherian R-module and let $\varphi : A \to A$ be a module homomorphism which is neither nilpotent nor an isomorphism. Find submodules $A_1, A_2 \neq \{0\}$ with $A_1 \cap A_2 = \{0\}$.

Problem 13. Let $\{0\} \to M_1 \to M_2 \to M_3 \to \{0\}$ be an exact sequence of modules over a commutative ring with identity. Show that if M_1 and M_3 are finitely generated then so is M_2.

Problem 14. (a) Let $0 \to A \xrightarrow{\alpha} B \xrightarrow{\beta} C \to 0$ be a short exact sequence. Show that the following conditions are equivalent.

(1) $\operatorname{im} \alpha$ is a direct summand of B; i.e., there is a submodule U of B such that $B = (\operatorname{im} \alpha) \oplus U$.

(2) The homomorphism β possesses a right inverse; i.e., there is a module homomorphism $\psi : C \to B$ with $\beta \circ \psi = \operatorname{id}_C$.

(3) The homomorphism α possesses a left-inverse; i.e., there is a module homomorphism $\varphi : B \to A$ such that $\varphi \circ \alpha = \operatorname{id}_A$.

A short exact sequence satisfying these conditions is called **split exact**.

$$\{0\} \longrightarrow A \underset{\varphi}{\overset{\alpha}{\rightleftarrows}} B \underset{\psi}{\overset{\beta}{\rightleftarrows}} C \longrightarrow \{0\}$$

(b) Show that for any two modules A, C there is a split-exact sequence $0 \to A \to M \to C \to 0$.

(c) Show that if U is a vector subspace of a vector space V then the sequence $0 \to U \to V \to V/U \to 0$ is split-exact.

(d) Construct an exact sequence $0 \to \mathbb{Z}_2 \to \mathbb{Z}_4 \to \mathbb{Z}_2 \to 0$ which is not split-exact.

Problem 15. Show that if $0 \to A \to B \to C$ is an exact sequence of R-modules, then so is $0 \to \operatorname{Hom}_R(X, A) \to \operatorname{Hom}_R(X, B) \to \operatorname{Hom}_R(X, C)$.

Problem 16.† Consider the following commuting diagram of modules and module homomorphisms. Suppose the two rows are exact.

$$
\begin{array}{ccccccccc}
\{0\} & \longrightarrow & A & \xrightarrow{f} & B & \xrightarrow{g} & C & \longrightarrow & \{0\} \\
 & & \alpha \downarrow & & \beta \downarrow & & \gamma \downarrow & & \\
\{0\} & \longrightarrow & A' & \xrightarrow{\varphi} & B' & \xrightarrow{\psi} & C' & \longrightarrow & \{0\}
\end{array}
$$

Show that the sequence

$$\{0\} \longrightarrow \ker \alpha \xrightarrow{f|} \ker \beta \xrightarrow{g|} \ker \gamma \xrightarrow{\partial} A'/\operatorname{im}\alpha \xrightarrow{\overline{\varphi}} B'/\operatorname{im}\beta \xrightarrow{\overline{\psi}} C'/\operatorname{im}\gamma \longrightarrow \{0\}$$

is exact where $f \mid$ and $g \mid$ are the restrictions of f and g, $\overline{\varphi}$ and $\overline{\psi}$ are induced by φ and ψ, and the **connecting homomorphism** ∂ is defined as follows. If $c = g(b) \in \ker \gamma$ then $0 = \gamma\big(g(b)\big) = \psi\big(\beta(b)\big)$ so that $\beta(b) \in \ker \psi = \operatorname{im}\varphi$, say $\beta(b) = \varphi(a')$. Define $\partial(c) := a' + \operatorname{im}\alpha$. (Check that ∂ is well-defined, i.e., independent of the choices made for b and a'.)

---

† The problems 16 and 17 are exercises in "diagram-chasing", a term that denotes the procedure of deducing properties of mappings in a diagram from properties of other mappings occurring in that diagram. Diagram-chasing is applied quite often in algebraic topology, for example.

Remark. As a way of memorizing the definition of ∂, we can simply think $\partial = \varphi^{-1} \circ \beta \circ g^{-1}$, but this is not a proper definition, of course.

Problem 17. Consider the following commuting diagram of modules and module homomorphisms. Suppose the two rows are exact.

$$
\begin{array}{ccccccccc}
A_1 & \xrightarrow{\alpha_1} & A_2 & \xrightarrow{\alpha_2} & A_3 & \xrightarrow{\alpha_3} & A_4 & \xrightarrow{\alpha_4} & A_5 \\
f_1 \downarrow & & f_2 \downarrow & & f_3 \downarrow & & f_4 \downarrow & & f_5 \downarrow \\
B_1 & \xrightarrow{\beta_1} & B_2 & \xrightarrow{\beta_2} & B_3 & \xrightarrow{\beta_3} & B_4 & \xrightarrow{\beta_4} & B_5
\end{array}
$$

(a) Show that if f_1 is surjective and f_2, f_4 are injective, then f_3 is injective.

(b) Show that if f_5 is injective and f_2, f_4 are surjective, then f_3 is surjective.

(c) Prove the **Five Lemma**: If f_1, f_2, f_4, f_5 are all isomorphisms, then f_3 is also an isomorphism.

Problem 18. Let R be a commutative ring with identity and let A, B be R-modules. A mapping $f : A \times B \to C$ into an R-module C is called **R-bilinear** if $f(\sum_i r_i a_i, \sum_j s_j b_j) = \sum_{i,j} r_i s_j f(a_i, b_j)$ for all $a_i \in A$, $b_j \in B$ and r_i, $s_j \in R$.

Let $R^{(A \times B)}$ be the free R-module on $A \times B$, i.e., the set of all formal sums $\sum_{i=1}^{n} r_i (a_i, b_i)$. Let U be the submodule of $R^{(A \times B)}$ generated by all elements

$$
\begin{aligned}
&(a + a', b) - (a, b) - (a', b) , \\
&(a, b + b') - (a, b) - (a, b') , \\
&(ra, b) - r(a, b) \quad \text{and} \\
&(a, rb) - r(a, b) .
\end{aligned}
$$

Then the quotient module $R^{(A \times B)}/U$ is called the **tensor product** of A and B over R and is denoted by $A \otimes_R B$. We write $a \otimes b$ for the element $(a, b) + U$.

(a) Show that the mapping $\otimes : A \times B \to A \otimes_R B$ which maps (a, b) to $a \otimes b$ is R-bilinear.

(b) Show that for every R-bilinear mapping $f : A \times B \to C$ into an R-module C there is a unique R-linear mapping $F : A \otimes_R B \to C$ with $F(a \otimes b) = f(a, b)$ for all $a \in A$ and $b \in B$.

(c) Suppose T is an R-module and $\beta : A \times B \to T$ is a bilinear mapping. Suppose that for every R-bilinear mapping $f : A \times B \to C$ into an R-module C there is a unique R-linear mapping $F : T \to C$ with $F(\beta(a, b)) = f(a, b)$ for all $a \in A$ and $b \in B$. Show that there is a unique isomorphism $\Phi : A \otimes_R B \to T$ such that $\Phi \circ \otimes = \beta$.

Remark. Part (b) reveals the important *universal property of the tensor product*, namely, that there is a 1-1-correspondence between bilinear mappings on $A \times B$ and linear mappings on $A \otimes B$. Part (c) simply says that $A \otimes_R B$ is uniquely determined by this property.

Problem 19. (a) Let A be an abelian group. Show that $\mathbb{Z}_n \otimes_{\mathbb{Z}} A \cong A/(nA)$.
(b) Show that $\mathbb{Z}_m \otimes_{\mathbb{Z}} \mathbb{Z}_n \cong \mathbb{Z}_{\gcd(m,n)}$.
(c) Show that $\mathbb{Q} \otimes_{\mathbb{Z}} \mathbb{Z}_n = \{0\}$.
(d) Show that if A is a finite abelian group then $A \otimes_{\mathbb{Z}} \mathbb{Q} = \{0\}$.
(e) Show that $\mathbb{Q} \otimes_{\mathbb{Z}} \mathbb{Q} \cong \mathbb{Q}$.

Problem 20. Let R be an integral domain and let $A := R[x_1, \ldots, x_n]$ be the polynomial ring in n variables over R, considered as an R-module. For each $d \in \mathbb{N}_0$ let A_d be the submodule of all homogeneous polynomials of degree d. Show that $A = \bigoplus_{d \in \mathbb{N}_0} A_d$ as a direct sum of R-modules and that $A_d A_e \subseteq A_{d+e}$ for all $d, e \in \mathbb{N}_0$.

Problem 21. Let R be a ring with identity and let A be a unitary R-module. A subset $B \subseteq A$ is called a **basis** of A over R if every element $a \in A$ possesses a unique representation as a finite sum $a = \sum_{b \in B} r_b \star b$ (which means that only a finite number of the ring elements r_b are nonzero).
(a) Show that B is a basis of R if and only if A is isomorphic to the free module $R^{(B)}$. (Hence a module which possesses a basis is called a **free module**.)
(b) Let A be the abelian group of all finite sequences $a = (a_1, \ldots, a_n, 0, 0, 0, \ldots)$ of real numbers and let $R = \operatorname{End}(A)$ be the endomorphism ring of A. Consider R as a module over itself (the module action just being ring multiplication from the left). Pick any natural number $n \in \mathbb{N}$ and define elements $f_1, \ldots, f_n \in R$ as follows.

$$f_1(a) := (a_n, a_{2n}, a_{3n}, \ldots),$$
$$f_2(a) := (a_{n-1}, a_{2n-1}, a_{3n-1}, \ldots),$$
$$\ldots$$
$$f_n(a) := (a_1, a_{n+1}, a_{2n+1}, \ldots).$$

Show that $\{f_1, \ldots, f_n\}$ is a basis of R, for any $n \in \mathbb{N}$.
(c) Show that if A has an infinite basis then all bases of A have the same cardinality.
Hint. Show first that all bases of A must be infinite; then mimick the proof of (I.25.14).

Problem 22. Let R be a ring with identity and let A be an R-module. For any ideal $I \trianglelefteq R$ we let $I \star A := \{r_1 \star a_1 + \cdots r_n \star a_n \mid n \in \mathbb{N}, r_i \in I, a_i \in A\}$.
(a) Show that $I \star A$ is a submodule of A.
(b) Show that the quotient module $\overline{A} := A/(I \star A)$ can be made into a module over $\overline{R} := (R/I)$ by defining $(r + I) \star (a + I \star A) := (r \star a) + (I \star A)$. (In short notation this simply reads $\overline{r} \star \overline{a} := \overline{r \star a}$.)
(c) Show that if B is a basis of the R-module A then $\overline{B} := \{b + (I \star A) \mid b \in B\}$ is a basis of the $\overline{R}$-module $\overline{B}$ and that $|\overline{B}| = |B|$.

Problem 23. Let R be a commutative ring with identity. Show that if A is a unitary R-module which possesses a basis then all bases of A over R have the same cardinality. (The cardinality of a basis is called the **rank** of A.)
Hint. Pick a prime ideal P and apply part (c) of problem 22 with $I := P$. Then observe that R/P is a field so that every unitary module over R/P is in fact a vector space over R/P.

Problem 24. Let $\alpha \in \mathbb{C}$ be a complex number and suppose that there is an irreducible polynomial $p \in \mathbb{Z}[x]$ such that $p(\alpha) = 0$. Show that if p is not monic then α is not integral over $\mathbb{Z}$.

Hint. Show first that there is no polynomial in $\mathbb{Q}[x]$ of smaller degree than p which has α as a root. Then conclude that any polynomial in $\mathbb{Q}[x]$ which has α as a root must be divisible by p.

Problem 25. Is the following statement true or false? If $a, b \in \mathbb{R}$ are real numbers, then $a + ib \in \mathbb{C}$ is algebraic (integral) over $\mathbb{Q}$ if and only if a and b are algebraic (integral) over $\mathbb{Q}$. (Proof or counterexample!)

Problem 26. Show that a complex number $\alpha \in \mathbb{C}$ is algebraic over $\mathbb{Q}$ if and only if there is a natural number $m \in \mathbb{N}$ such that $m\alpha$ is integral over $\mathbb{Z}$. Moreover, show that if m is chosen as small as possible then a number $n\alpha$ with $n \in \mathbb{N}$ is integral over $\mathbb{Z}$ if and only if n is a multiple of m.

Problem 27. (a) Suppose that $r \in \mathbb{Q}$ is a rational number such that $\sin r\pi$ and $\cos r\pi$ are also rational numbers. Show that there is an integer m such that $r = \frac{m}{2}$.

Hint. If r is rational then $\cos r\pi + i \sin r\pi = e^{ir\pi}$ is integral over $\mathbb{Z}$, the more so over $\mathbb{Z} + i\mathbb{Z}$.

(b) Find all numbers $n \in \mathbb{N}$ such that $\cos \frac{2\pi}{n}$ is a rational number.

Problem 28. Let $R \subseteq S$ be commutative rings. Show that an element $\alpha \in S$ is algebraic over R if and only if there is a polynomial $p \in R[x]$ such that $p(\alpha)$ is algebraic over R.

Problem 29. Let $R \subseteq S$ be commutative rings with the same identity element 1 and let $s_1, \ldots, s_n \in S$. Show that $R[s_1, \ldots, s_n]$ is a finitely generated R-module if and only if s_k is integral over $R[s_1, \ldots, s_{k-1}]$ for $1 \leq k \leq n$.

Problem 30. Let R be a commutative ring with identity and let $G \leq \operatorname{Aut} R$ be a finite group of automorphisms of R. Show that R is integral over

$$R^G := \{x \in R \mid \sigma x = x \text{ for all } \sigma \in G\} .$$

Hint. For any $x \in R$, consider the polynomial $p(t) := \prod_{\sigma \in G}(t - \sigma(x))$.

Problem 31. Let $A \subseteq B$ be commutative rings. Show that if $B \setminus A$ is multiplicatively closed then A is integrally closed in B.

Problem 32. (a) Let R be an integral domain which is integrally closed in its quotient field K. Show that $R[x]$ is integrally closed in its quotient field $K(x)$.

(b) Show that any field K is integrally closed in its rational function field $K(x)$.

Problem 33. Let $A \subseteq B$ be commutative rings with the same identity element such that B is integral over A.

(a) Show that if $x \in A$ is a unit in B then x is a unit in A.

(b) Show that $J(A) = J(B) \cap A$ where J denotes the Jacobson radical.

Problem 34. Let $R \subseteq S$ be commutative rings and let $\overline{R}$ be the integral closure of R in S.

(a) Show that if f, g are monic polynomials in $S[x]$ satisfying $fg \in \overline{R}[x]$ then $f, g \in \overline{R}[x]$.

(b) Show that $\overline{R[x]} = \overline{R}[x]$.

Hint. If $f \in S[x]$ is integral over $R[x]$, then there are polynomials $g_i \in R[x]$ with $f^n + g_{n-1}f^{n-1} + \cdots + g_1 f + g_0 = 0$. Choose $r > \max(n, \deg g_0, \ldots, \deg g_{n-1})$ and let $F = f - x^r$ so that

$$
\begin{aligned}
0 &= (F + x^r)^n + g_{n-1}(F + x^r)^{n-1} + \cdots + g_1(F + x^r) + g_0 \\
&= F^n + h_{n-1}F^{n-1} + \cdots + h_1 F + h_0
\end{aligned}
$$

with $h_i \in R[x]$. Now apply part (a) to $-F$ and $F^{n-1} + h_{n-1}F^{n-2} + \cdots + h_1$.

11. Noetherian rings

IN section 7 we observed that an integral domain R is a factorization domain if and only if there is no strictly ascending chain of principal ideals in R. In this section, we want to study those rings which do not allow a strictly ascending chain of ideals whatsoever. Let us first prove a statement which is analogous to (10.13)(a); in fact, if R is *commutative* then an ideal of R is the same as a submodule of R, so that the content of the next theorem follows immediately from (10.13)(a) in this case.

(11.1) Theorem. *Let R be a ring. Then the following conditions are equivalent:*
(1) *every ideal of R is finitely generated* (**finiteness condition**);
(2) *every ascending sequence $I_1 \subseteq I_2 \subseteq I_3 \subseteq \cdots$ of ideals of R becomes stationary* (**ascending chain condition**);
(3) *every set M of ideals contains a maximal element, i.e., an element $J \in M$ which is not properly contained in any $I \in M$* (**maximum condition**).

Proof. (1)$\Longrightarrow$(2). Let $I_1 \subseteq I_2 \subseteq I_3 \subseteq \cdots$ be an ascending sequence of ideals. Then $J := \bigcup_{k=1}^{\infty} I_k$ is an ideal of R (check!), hence finitely generated. Let $J = \langle a_1, \ldots, a_r \rangle$. Each generator a_i is contained in one of the ideals of the chain, say in I_{n_i}. Let $n := \max\{n_1, \ldots, n_r\}$. Then all the generators a_i are contained in I_n. This clearly implies that $I_n = I_{n+1} = I_{n+2} = \cdots$.

(2)$\Longrightarrow$(3). Suppose there is a set $\mathfrak{M}$ of ideals without a maximal element. Pick any $I_1 \in \mathfrak{M}$. Since I_1 is not maximal, there is an ideal $I_2 \in \mathfrak{M}$ with $I_1 \subsetneq I_2$. But I_2 is also not maximal; thus $I_2 \subsetneq I_3$ for some $I_3 \in M$. Clearly, this gives an ascending chain $I_1 \subsetneq I_2 \subsetneq I_3 \subsetneq \cdots$ which does not become stationary.

(3)$\Longrightarrow$(1). Let I be any ideal of R and set

$$M := \{ J \trianglelefteq R \mid J \subseteq I, \ J \ \text{finitely generated} \} .$$

This set is not empty because $\{0\} \in M$. By (3) there is a maximal element $J_0 = \langle a_1, \ldots, a_n \rangle \subseteq I$ in M. Let $a \in I$ be arbitrary. Then $\langle a_0, \ldots, a_n, a \rangle$ lies in M and contains J_0, hence must be equal to J_0 because J_0 is maximal in M. This shows $a \in J_0$. Since $a \in I$ was arbitrary and $J_0 \subseteq I$, this gives $I = J_0$ so that I is finitely generated. ∎

(11.2) Definition. *A ring satisfying the equivalent conditions of theorem (11.1) is called a* **Noetherian ring**.

(11.3) Examples. (a) Every principal ideal ring is Noetherian; in particular, every Euclidean domain is Noetherian.

(b) Every homomorphic image of a Noetherian ring is again Noetherian. Indeed, let R be a Noetherian ring and let $f : R \to S$ be a surjective ring homomorphism. If $J \trianglelefteq S$ is an ideal of S, then $J = f(I)$ where $I := f^{-1}(J)$ is an ideal of R. Since R is Noetherian, the ideal I is finitely generated, say $I = \langle r_1, \ldots, r_n \rangle$. But then $J = f(I) = \langle f(r_1), \ldots, f(r_n) \rangle$ is also finitely generated.

(c) Let K be a field. Then the polynomial ring $K[x_1, x_2, x_3, \ldots]$ in a countably infinite number of variables is not Noetherian, because $\{0\} \subsetneq \langle x_1 \rangle \subsetneq \langle x_1, x_2 \rangle \subsetneq \langle x_1, x_2, x_3 \rangle \cdots$ is a strictly ascending chain of ideals.

(d) Let $R = \{(a_1, a_2, a_3, \ldots) \mid a_i \in \mathbb{Z}\}$ be the ring of all integer sequences with the coordinatewise operations and let $I_n := \{(a_1, \ldots, a_n, 0, 0, \ldots) \mid a_i \in \mathbb{Z}\}$. Then $I_1 \subsetneq I_2 \subsetneq I_3 \subsetneq \cdots$ is a strictly ascending chain of ideals; hence R is not Noetherian.

(e) In view of (7.7), every Noetherian integral domain is a factorization domain. For example, the ring R introduced in (7.6)(d), consisting of those complex numbers which satisfy a monic equation with coefficients in $\mathbb{Z}$, cannot be Noetherian. In fact, $\{0\} \subsetneq \langle \sqrt{2} \rangle \subsetneq \langle \sqrt[4]{2} \rangle \subsetneq \langle \sqrt[8]{2} \rangle \subsetneq \cdots$ is a strictly ascending chain of ideals in R.

THE next theorems reveal new examples of Noetherian rings.

(11.4) Theorem. *Let $n \in \mathbb{Z} \setminus \{0, 1\}$ be a square-free number.*

(a) *Every ideal in $\mathbb{Z} + \mathbb{Z}\sqrt{n}$ can be generated by two elements. In particular, $\mathbb{Z} + \mathbb{Z}\sqrt{n}$ is a Noetherian ring.*

(b) *Suppose $n \equiv 1$ modulo 4 so that $\mathcal{O}_n := \mathbb{Z} + \mathbb{Z}\frac{1+\sqrt{n}}{2}$ is a ring. Then every ideal of $\mathcal{O}_n$ can be generated by two elements. In particular, $\mathcal{O}_n$ is a Noetherian ring.*

Proof. (a) Let $I \trianglelefteq \mathbb{Z} + \mathbb{Z}\sqrt{n}$ be an arbitrary ideal of $\mathbb{Z} + \mathbb{Z}\sqrt{n}$. Then one easily verifies that

$$I_0 := \{a \in \mathbb{Z} \mid a + b\sqrt{n} \in I \text{ for some } b \in \mathbb{Z}\} \quad \text{and} \quad I_1 := \{b \in \mathbb{Z} \mid b\sqrt{n} \in I\}$$

are ideals of $\mathbb{Z}$. Since every ideal of $\mathbb{Z}$ is a principal ideal, there are elements $a_0, b_1 \in \mathbb{Z}$ with $I_0 = \mathbb{Z}a_0$ and $I_1 = \mathbb{Z}b_1$. By the definition of these ideals, we can find an element $b_0 \in \mathbb{Z}$ with $x := a_0 + b_0\sqrt{n} \in I$, and we have $y := b_1\sqrt{n} \in I$. We claim that I is generated by x and y. To prove our claim, take an arbitrary element $a + b\sqrt{n}$ in I. Then $a \in I_0$ so that there is an element $\lambda \in \mathbb{Z}$ with $a = \lambda a_0$. Then

$$I \ni (a + b\sqrt{n}) - \lambda x = (\lambda a_0 + b\sqrt{n}) - \lambda(a_0 + b_0\sqrt{n}) = (b - \lambda b_0)\sqrt{n} \ .$$

Hence $b - \lambda b_0 \in I_1$ so that we can find $\mu \in \mathbb{Z}$ with $b - \lambda b_0 = \mu b_1$. But then

$$a + b\sqrt{n} = \lambda a_0 + (\lambda b_0 + \mu b_1)\sqrt{n} = \lambda(a_0 + b_0\sqrt{n}) + \mu b_1\sqrt{n} = \lambda x + \mu y \ .$$

(b) The proof is completely analogous to the one in part (a). Observe that $\mathcal{O}_n$ consists of all elements of the form $\frac{1}{2}(a + b\sqrt{n})$ where $a, b \in \mathbb{Z}$ have the same parity, i.e., are either both even or both odd. Given an ideal $I \trianglelefteq \mathcal{O}_n$, the sets

$$I_0 := \{a \in \mathbb{Z} \mid \frac{1}{2}(a + b\sqrt{n}) \in I \text{ for some } b \in \mathbb{Z}\} \quad \text{and} \quad I_1 := \{b \in \mathbb{Z} \mid \frac{1}{2}b\sqrt{n} \in I\}$$

are ideals of $\mathbb{Z}$ so that $I_0 \doteq \mathbb{Z}a_0$ and $I_1 = \mathbb{Z}b_1$ with suitable elements $a_0, b_1 \in \mathbb{Z}$. By the definition of these ideals, we can find an element $b_0 \in \mathbb{Z}$ with $x := \frac{1}{2}(a_0 + b_0\sqrt{n}) \in I$, and we have $y := \frac{1}{2}b_1\sqrt{n} \in I$. As above, one can show that I is generated by x and y. $\blacksquare$

(11.5) (Hilbert's Basis Theorem.) *Let R be a commutative ring with identity element 1. If R is Noetherian, then so is the polynomial ring $R[x_1, \ldots, x_n]$.*

Proof. Clearly, it is enough to prove that $R[x]$ is Noetherian; then we can proceed by induction on n. Suppose there is an ideal $I \unlhd R[x]$ which is not finitely generated. Once this assumption leads to a contradiction we know that $R[x]$ is Noetherian. We choose succesively polynomials in I as follows:

Pick $f_1 \in I$ with $n_1 := \deg f_1$ minimal.

Pick $f_2 \in I \setminus \langle f_1 \rangle$ with $n_2 := \deg f_2$ minimal.

$\cdots$

Pick $f_{k+1} \in I \setminus \langle f_1, \ldots, f_k \rangle$ with $n_{k+1} := \deg f_{k+1}$ minimal.

Let $a_k \in R$ be the leading coefficient of f_k. We claim that

$$\langle a_1, \ldots, a_k \rangle \subsetneqq \langle a_1, \ldots, a_k, a_{k+1} \rangle \qquad \text{for all } k \text{ ;}$$

clearly, this will give us an ascending chain of ideals in R which does not become stationary, which is the desired contradiction.

Suppose that $\langle a_1, \ldots, a_k \rangle = \langle a_1, \ldots, a_{k+1} \rangle$. Then there are elements $r_1, \ldots, r_k \in R$ with $a_{k+1} = \sum_{j=1}^{k} r_j a_j$. (Here we use the fact that R is commutative and has an identity element!) Then

$$g(x) := \sum_{j=1}^{k} r_j f_j(x) \cdot x^{n_{k+1} - n_j} = a_{k+1} x^{n_{k+1}} + \text{lower order terms}$$

belongs to $\langle f_1, \ldots, f_n \rangle$. Hence $f_{k+1} - g$ does not belong to $\langle f_1, \ldots, f_n \rangle$. Now observe that f_{k+1} and g have the same leading term so that $\deg(f_{k+1} - g) < \deg f_{k+1}$. But this contradicts the choice of f_{k+1}. ∎

WE will now prove that if a commutative ring R with identity is Noetherian then so is the power series ring $R[[x_1, \ldots, x_n]]$. In fact, we will offer two proofs; one is straightforward, the other one very elegant, but not as direct as the first one. As a preparation of this elegant proof, we need the following result.

(11.6) Cohen's Theorem. *Let R be a commutative ring with identity and let $\mathcal{M}$ be the set of all ideals of R which are not finitely generated.*

(a) *Suppose that $\mathcal{M}$ is not empty so that R is not Noetherian. Then $\mathcal{M}$ possesses maximal elements, and each such maximal element is a prime ideal of R.*

(b) *R is Noetherian if and only if every prime ideal of R is finitely generated.*

Proof. (a) The existence of maximal elements is an immediate consequence of Zorn's lemma. Suppose P is such a maximal element, but not a prime ideal. Then there are elements $a, b \in R$ with $ab \in P$, but $a \notin P$ and $b \notin P$. Then the ideals $P + Ra$ (which contains a) and $P : \langle a \rangle$ (which contains b) are both strictly larger than P, hence are finitely generated due to the maximality of P. Suppose $P+Ra = \langle p_1 + r_1 a, \ldots, p_n + r_n a \rangle$ with $p_i \in P$ and $r_i \in R$. Then $P_0 := \langle p_1, \ldots, p_n \rangle \subseteq P$ clearly satisfies $P_0 + Ra = P + Ra$. Moreover,

$$(\star) \qquad\qquad P = P_0 + a(P : \langle a \rangle).$$

Indeed, the inclusion $\supseteq$ is trivial, and to obtain the converse inclusion, take any element $p \in P$. Since $P \subseteq P + Ra = P_0 + Ra$, we can write $p = p_0 + ra$ with $p_0 \in P_0$ and $r \in R$. Then $ra = p - p_0 \in P$ so that $r \in (P : \langle a \rangle)$; this shows $p = p_0 + ra \in P_0 + a(P : \langle a \rangle)$. Since both P_0 and $P : \langle a \rangle$ are finitely generated, the equation $(\star)$ shows that P is also finitely generated, which is clearly a contradiction. Hence P must be a prime ideal.

(b) This is an immediate consequence of (a). In fact, if R is not Noetherian, then (a) states that there is a prime ideal of P which is not finitely generated. ∎

(11.7) Theorem. *Let R be a commutative ring with identity element 1. If R is Noetherian, then so is the power series ring $R[[x_1, \ldots, x_n]]$.*

Proof. Again, it is enough to prove that $R[[x]]$ is Noetherian; then we can use induction on n. Now since R is Noetherian, (8.7)(b) implies that every prime ideal of $R[[x]]$ is finitely generated; hence $R[[x]]$ is Noetherian, due to (11.6)(b). ∎

(11.8) Remark. As stated before we will give a second proof of this theorem, more tedious, but also more straightforward. For simplicity of notation, we will write $o(k)$ for a power series which has a subdegree strictly greater than k. Let I be an ideal of $R[[x]]$. Then for each $k \in \mathbb{N}_0$ the set

$$I_k := \{a \in R \mid \text{ there is an element in } I \text{ which has the form } aX_k + o(k)\}$$

is an ideal of R, and clearly $I_k \subseteq I_{k+1}$ for all k. Since R is Noetherian, there is an index m with $I_m = I_{m+1} = I_{m+2} = \cdots$. Also, the ideals $I_0, \ldots, I_m$ are finitely generated, and for the sake of convenience, we take the same number of generators, say s, for each of these ideals. Hence we can write

$$I_0 = \langle a_1^{(0)}, \ldots, a_s^{(0)} \rangle, \quad \ldots, \quad I_m = \langle a_1^{(m)}, \ldots, a_s^{(m)} \rangle$$

with suitable elements $a_i^{(k)} \in R$ ($0 \le k \le m, 1 \le i \le s$). By the definition of these ideals, there are power series $\varphi_i^{(k)} \in I$ with

$$\varphi_i^{(k)} = a_i^{(k)} X^k + o(k+1).$$

We claim that

$$I = \langle \varphi_i^{(k)} \mid 0 \le k \le m, 1 \le i \le s \rangle$$

so that I is finitely generated. To prove this claim, take an arbitrary element $\varphi \in I$, say $\varphi = a_0 + o(1)$. Then $a_0 \in I_0 = \langle a_1^{(0)}, \ldots, a_s^{(0)} \rangle$; hence we can find elements $r_i^{(0)} \in R$ with $a_0 = \sum_{i=1}^{s} r_i^{(0)} a_i^{(0)}$. This implies

$$\varphi - \sum_{i=1}^{s} r_i^{(0)} \varphi_i^{(0)} = a_1 X + o(1).$$

Since the left-hand side is in I, we have $a_1 \in I_1 = \langle a_1^{(1)}, \ldots, a_s^{(1)} \rangle$; hence we can find elements $r_i^{(1)} \in R$ with $a_1 = \sum_{i=1}^{s} r_i^{(1)} a_i^{(1)}$. This implies

$$\varphi - \sum_{i=1}^{s} r_i^{(0)} \varphi_i^{(0)} - \sum_{i=1}^{s} r_i^{(1)} \varphi_i^{(1)} = a_2 X^2 + o(2) .$$

Proceeding in this way, we obtain a representation

$$\psi := \varphi - \sum_{i=1}^{s} r_i^{(0)} \varphi_i^{(0)} - \sum_{i=1}^{s} r_i^{(1)} \varphi_i^{(1)} - \cdots - \sum_{i=1}^{s} r_i^{(m-1)} \varphi_i^{(m-1)} = a_m X^m + o(m) .$$

We are done if we can show that ψ lies in the ideal spanned by $\varphi_1^{(m)}, \ldots, \varphi_s^{(m)}$. To do so, we continue our procedure and obtain

$$\psi - \sum_{i=1}^{s} r_i^{(m)} \varphi_i^{(m)} = a_{m+1} X^{m+1} + o(m+1) .$$

Now $a_{m+1} \in I_{m+1} = I_m$; hence we can find elements $r_1^{m+1}, \ldots, r_s^{(m+1)}$ with

$$\psi - \sum_{i=1}^{s} r_i^{(m)} \varphi_i^{(m)} - \sum_{i=1}^{s} r_i^{(m+1)} X \varphi_i^{(m)} = a_{m+2} X^{m+2} + o(m+2) .$$

Since $a_{m+2} \in I_{m+2} = I_m$, we can find elements $r_1^{m+2}, \ldots, r_s^{(m+2)}$ with

$$\psi - \sum_{i=1}^{s} r_i^{(m)} \varphi_i^{(m)} - \sum_{i=1}^{s} r_i^{(m+1)} X \varphi_i^{(m)} - \sum_{i=1}^{s} r_i^{(m+2)} X^2 \varphi_i^{(m)} = a_{m+3} X^{m+3} + o(m+3) .$$

Continuing in this fashion, we obtain sequence $\left(r_i^{(m)}, r_i^{(m+1)}, r_i^{(m+2)}, \ldots \right)$ $(1 \leq i \leq r)$ such that

$$\psi - \sum_{k=0}^{n} \sum_{i=1}^{s} r_i^{(m+k)} X^k \varphi_i^{(m)} \, a_{m+n+1} X^{m+n+1} + o(m+n+1) = o(m+n) .$$

For $1 \leq i \leq s$ let $\theta_i = \sum_{k=0}^{\infty} r_i^{(m+k)} X^k \in R[[X]]$. Then

$$\psi - \sum_{i=1}^{s} \theta_i \varphi_i^{(m)} = \psi - \sum_{k=0}^{n} \sum_{i=1}^{s} r_i^{(m+k)} X^k \varphi_i^{(m)} - \sum_{i=1}^{s} \sum_{k=n+1}^{\infty} r_i^{(m+k)} X^k \varphi_i^{(m)}$$

$$= o(m+n) + o(m+n) = o(m+n) .$$

(Note that $\varphi_i^{(m)} = a_i^{(m)} X^m + o(m) = o(m-1)$.) Now this equation holds for an arbitrary n which is possible only if the right-hand side is zero. This shows $\psi = \sum_{i=1}^{s} \theta_i \varphi_i^{(m)}$. ∎

CONSIDER a subset X of a Noetherian ring and let $\langle X \rangle$ be the ideal generated by this set. We know that $\langle X \rangle$ is finitely generated, but we will now prove the more precise statement that we can choose a finite set of generators of $\langle X \rangle$ inside the initial set X.

(11.9) Proposition. *Let R be a Noetherian ring. For any subset $X \neq \emptyset$ of R, there are finitely many elements $x_1, \ldots, x_n \in X$ such that $\langle X \rangle = \langle x_1, \ldots, x_n \rangle$.*

Proof. Suppose the claim is false for some set X. Pick any element $x_1 \in X$. Since $\langle x_1 \rangle \subsetneq \langle X \rangle$ there is an element $x_2 \in X$ with $x_2 \notin \langle x_1 \rangle$. Hence $\langle x_1 \rangle \subsetneq \langle x_1, x_2 \rangle$. Since $\langle x_1, x_2 \rangle \subsetneq \langle X \rangle$ there is an element $x_3 \in X$ with $x_3 \notin \langle x_1, x_2 \rangle$. Hence $\langle x_1, x_2 \rangle \subsetneq \langle x_1, x_2, x_3 \rangle$. Continuing in this way, we obtain a strictly increasing sequence

$$\langle x_1 \rangle \subsetneq \langle x_1, x_2 \rangle \subsetneq \langle x_1, x_2, x_3 \rangle \subsetneq \cdots$$

which is impossible since R is Noetherian. ∎

LET us now present a peculiar strategy of proof which can be applied in Noetherian rings, the so-called "Noetherian induction". If one wants to show that ideals with a certain property do not exist, one simply considers the set S of all these ideals and assumes that $S \neq \emptyset$. Then S possesses a maximal element by (11.3)(c), and quite often the maximality of this element can be used to obtain the desired contradiction. Let us illustrate this method by proving a decomposition theorem for ideals.

(11.10) Definition. *An ideal I in a ring R is called* **irreducible** *if the equation $I = I_1 \cap I_2$ with ideals $I_1, I_2 \trianglelefteq R$ implies that $I = I_1$ or $I = I_2$; otherwise, I is called* **reducible**.

(11.11) Proposition. *Every ideal I in a Noetherian ring R can be written as an intersection of a finite number of irreducible ideals.*

Proof. Let S be the set of all ideals which cannot be written as a finite intersection of irreducible ideals; we want to show that $S = \emptyset$. Suppose $S \neq \emptyset$; then there is a maximal element I of S since R is Noetherian. Then I cannot be irreducible; hence $I = I_1 \cap I_2$ with $I \subsetneq I_1$ and $I \subsetneq I_2$. Now if both I_1 and I_2 could be written as finite intersections of irreducible ideals, then so could I which is not the case. Hence I_1 and I_2 both belong to S, contradicting the choice of I as a maximal element of S. ∎

RECALL that in section 6 we studied the situation in which an ideal in an arbitrary commutative ring possesses a primary decomposition, but we did not develop criteria which would ensure the existence of such a decomposition. We will now prove that in a commutative Noetherian ring with identity *every* ideal possesses a primary decomposition. Due to proposition (11.11), it suffices to show that in such a ring every irreducible ideal is primary.

(11.12) Proposition. (a) *Every prime ideal in a commutative ring is irreducible.*
(b) *Every irreducible ideal in a Noetherian commutative ring with identity is primary.*

Proof. (a) Let I be a prime ideal with $I = I_1 \cap I_2$. Suppose $I \subsetneq I_1$ and $I \subsetneq I_2$. Then there are elements $x_1 \in I_1 \setminus I$ and $x_2 \in I_2 \setminus I$. Hence $x_1 x_2 \in I_1 I_2 \subseteq I_1 \cap I_2 = I$, so that $x_1 x_2 \in I$ but $x_1 \notin I$ and $x_2 \notin I$. This contradicts the hypothesis that I is prime.

(b) Let I be an irreducible ideal in a Noetherian commutative ring R. Suppose $xy \in I$, but $x \notin I$; we have to find a natural number N with $y^N \in I$. For $n \in \mathbb{N}$, let $I_n := \{r \in R \mid ry^n \in I\}$; then $I \subseteq I_1 \subseteq I_2 \subseteq I_3 \subseteq \cdots$ and $I \neq I_1$ because $x \in I_1 \setminus I$. Since R is Noetherian, this sequence becomes stationary at some index N so that $I_N = I_{N+1} = \cdots$. We claim that

$$(\star) \qquad\qquad (Ry^N + I) \cap (Rx + I) \;=\; I \;;$$

once this is established, we have $I = Ry^N + I$ or $I = Rx + I$ by irreducibility, and since $x \notin I$ this implies that $I = Ry^N + I$ so that $y^N \in I$ as desired. Let us prove $(\star)$. The inclusion $\supseteq$ is trivial. Suppose conversely that $\alpha \in (Ry^N + I) \cap (Rx + I)$. Then there are elements $r, s \in R$ such that $\alpha \equiv ry^N \equiv sx$ modulo I. Since $xy \in I$ we have $\alpha y \equiv 0$ modulo I. Therefore, $ry^{N+1} \equiv \alpha y \equiv 0$ so that $r \in I_{N+1} = I_N$ and hence $\alpha \equiv ry^N \equiv 0$ modulo I. But this means $\alpha \in I$. ∎

(11.13) Remark. In a commutative Noetherian ring, the set of prime ideals is contained in the set of all irreducible ideals as well as in the set of all primary ideals. Both inclusions can be proper, as the following examples show.

(a) Let $p \in \mathbb{N}$ be a prime number and let $n \geq 2$. Then $\langle p^n \rangle = \mathbb{Z}p^n$ is a primary ideal in the Noetherian ring $\mathbb{Z}$, but not a prime ideal.

(b) Let K be a field and let $R = K[x,y]$. Then $\langle x^2, xy, y^2 \rangle$ is primary (more precisely, $\langle x, y \rangle$-primary), but not irreducible since $\langle x^2, xy, y^2 \rangle = \langle x^2, y \rangle \cap \langle x, y^2 \rangle$.

AS stated before, this yields the following result.

(11.14) Theorem. *Let R be a Noetherian commutative ring with identity. Then every ideal $I \neq R$ possesses a primary decomposition.*

Proof. This is an immediate consequence of (11.11) and (11.12)(b). ∎

AS d'Alembert once said, algebra is generous; she often gives more than is asked of her. Not only does the Noetherian property guarantee the existence of primary decomposition, but it also yields primary decompositions which special properties. These properties are due to the fact that if Q is a P-primary ideal in a Noetherian commutative ring then some power of P is contained in Q; this statement an immediate consequence of (6.9)(h). To derive these special properties of primary decompositions in Noetherian rings, let us start with a definition.

(11.15) Definition. *Let R be a Noetherian commutative ring with identity. Suppose Q is a P-primary ideal of R. Then the smallest natural number $k \in \mathbb{N}$ such that $P^k \subseteq Q \subseteq P$ is called the **exponent** of Q.*

THE following lemma complements proposition (6.19)(c).

(11.16) Lemma. *Let R be a commutative Noetherian ring with identity. Let $Q \trianglelefteq R$ be P-primary and let $I \trianglelefteq R$ be an ideal with $I \subseteq P$, but $I \not\subseteq Q$. Then $Q \subsetneq (Q : I) \subsetneq R$.*

Proof. Let n be the exponent of Q so that $P^n \subseteq Q \subseteq P$ but $P^{n-1} \not\subseteq Q$. Since $I \subseteq P$ we have $IP^{n-1} \subseteq P^n \subseteq Q$ so that $P^{n-1} \subseteq (Q : I)$. Since $P^{n-1} \not\subseteq Q$ this shows $Q \neq (Q : I)$, hence $Q \subsetneq (Q : I)$. On the other hand, $1 \notin (Q : I)$ because $I \not\subseteq Q$; hence $(Q : I) \subsetneq R$. ∎

NOW we show that the Noetherian property for a ring R has consequences for the structure of R-modules.

(11.17) Theorem. *Let R be a commutative Noetherian ring and let A be an R-module. If A is finitely generated, then so is every submodule of A. In other words, every finitely generated R-module is a Noetherian R-module.*

Proof. We will prove by induction on n that if A can be generated by n elements then every submodule of A is finitely generated. Let $A = R \star x_1 + \cdots + R \star x_n$ and let $U \subseteq A$ be a submodule. Then

$$I := \{r \in R \mid \text{there are } r_2, \ldots, r_n \in R \text{ with } r \star x_1 + r_2 \star x_2 + \cdots + r_n \star x_n \in U\}$$

is an ideal of R, and

$$U_0 := U \cap (R \star x_2 + \cdots + R \star x_n)$$

is a submodule of $R \star x_2 + \cdots + R \star x_n$. Since R is Noetherian, the ideal I is finitely generated, say by $s_1, \ldots, s_M$. Then there are elements $u_1, \ldots, u_M \in U$ with $u_i \in s_i \star x_1 + R \star x_2 + \cdots + R \star x_n$. Let $u \in U$, say $u = r_1 \star x_1 + \cdots + r_n \star x_n$. Then $r_1 \in I$; hence there are elements $t_1, \ldots, t_M \in R$ with $r_1 = t_1 s_1 + \cdots + t_M s_M$. Consequently, $u - t_1 \star u_1 - \cdots - t_M \star u_M \in U_0$. Since $u \in U$ was an arbitrary element of U, we have shown that

$(\star)$ $$U = R \star u_1 + \cdots + R \star u_M + U_0 .$$

If $n = 1$ then $U_0 = \{0\}$, and $(\star)$ is already the claim. If $n \geq 2$, then U_0 is finitely generated by induction hypothesis, say $U_0 = \sum_{j=1}^{N} R \star v_j$, and then $(\star)$ implies that $U = \sum_{i=1}^{M} R \star u_i + \sum_{j=1}^{N} R \star v_j$ which is again the claim. ∎

(11.18) Corollary. *Let $R \subseteq S$ be commutative rings. If R is a Noetherian ring and S is finitely generated as an R-module, then S is a Noetherian ring.*

Proof. This follows immediately from (11.17), because every ideal of S is in particular an R-module. ∎

(11.19) Example. Let $n \in \mathbb{Z} \setminus \{0,1\}$ be a square-free number. Then $\mathbb{Z}[\sqrt{n}]$ is a Noetherian ring. This was already established in (11.4), but can now be obtained as a simple consequence of (11.18).

THE next two results deal with finitely generated algebras over Noetherian rings.

(11.20) Proposition. *If R is a commutative Noetherian ring with identity and if S is a finitely generated R-algebra, then S is also Noetherian.*

Proof. Let $S = R[s_1, \ldots, s_n]$ with elements $s_i \in S$. By (11.5), the polynomial ring $R[x_1, \ldots, x_n]$ is Noetherian; but then S is Noetherian by (11.3)(b), since S is obviously a homomorphic image of $R[x_1, \ldots, x_n]$. ∎

(11.21) Proposition. *Let $R \subseteq S \subseteq T$ be commutative rings. If R is Noetherian and T is finitely generated as an R-algebra and as an S-module, then S is finitely generated as an R-algebra.*

Proof. By assumption, there are elements $x_i, y_j \in T$ with $T = R[x_1, \ldots, x_m]$ and $T = Sy_1 + \cdots + Sy_n$. Hence there are elements $s_{ij}, s_{ijk} \in S$ with

$$(1) \quad x_i = \sum_j s_{ij} y_j \quad \text{and} \quad (2) \quad y_i y_j = \sum_k s_{ijk} y_k$$

for all indices i, j, k. Let $S_0 := R[s_{ij}, s_{ijk}] \subseteq S$; then S_0 is Noetherian by (11.20). Substituting (1) and applying (2) repeatedly, we see that

$$T = R[x_1, \ldots, x_m] = S_0 y_1 + \cdots + S_0 y_n$$

so that T is finitely generated as an S_0-module. Since S_0 is Noetherian, theorem (11.17) implies that every S_0-submodule of T is finitely generated. In particular, S is finitely generated as an S_0-module, say

$$S = S_0 s_1 + \cdots + S_0 s_r = R[s_{ij}, s_{ijk}] s_1 + \cdots + R[s_{ij}, s_{ijk}] s_r .$$

This shows that S is generated as an R-algebra by the elements s_i, s_{ij}, s_{ijk}. ∎

WE will close this section by introducing the class of Artinian rings (in formal analogy to the class of Artinian modules) and show that – despite the formal symmetry between the definitions of Noetherian and Artinian rings – in the case of commutative rings with identity the Artinian rings form only a very special class of Noetherian rings.†

(11.22) Definition. *A ring is called* **Artinian** *if every descending sequence of ideals becomes stationary, or, equivalently, if every non-empty set of ideals possesses a minimal element.*

(11.23) Examples. (a) All simple rings (in particular all fields and skew-fields) are Artinian.
(b) Every finite ring is Artinian.
(c) The ring $\mathbb{Z}$ is not Artinian because if $a \neq 0$ then $\langle a \rangle \supsetneq \langle a^2 \rangle \supsetneq \langle a^3 \rangle \supsetneq \cdots$ is a strictly descending chain of ideals.
(d) The polynomial ring $K[x]$ is not Artinian because $\langle x \rangle \supsetneq \langle x^2 \rangle \supsetneq \langle x^3 \rangle \supsetneq \cdots$ is a strictly descending chain of ideals.
(e) Every homomorphic image of an Artinian ring is Artinian.

WE are now going to prove that a commutative ring with identity is Artinian if and only if it is Noetherian and has the property that every prime ideal is maximal. Before we can prove this theorem we need to prove four lemmas.

(11.24) Lemma. *Let R be a commutative ring with identity. Suppose there are (not necessarily distinct) maximal ideals $M_1, \ldots, M_n$ with $M_1 \cdots M_n = \{0\}$. For $1 \leq k \leq n$ let $R_k := M_1 \cdots M_k$; formally, let $R_0 := R$.*
(a) R is Noetherian if and only if all the quotient modules R_k/R_{k+1} are Noetherian.
(b) R is Artinian if and only if all the quotient modules R_k/R_{k+1} are Artinian.
(c) R is Noetherian if and only if R is Artinian.

Proof. The parts (a) and (b) can be proved simultaneously. We have the short exact sequences

$$
\begin{array}{ccccccccc}
\{0\} & \longrightarrow & R_1 & \longrightarrow & R & \longrightarrow & R/R_1 & \longrightarrow & \{0\} \\
\{0\} & \longrightarrow & R_2 & \longrightarrow & R_1 & \longrightarrow & R_1/R_2 & \longrightarrow & \{0\} \\
& & & & \vdots & & & & \\
\{0\} & \longrightarrow & R_{n-1} & \longrightarrow & R_{n-2} & \longrightarrow & R_{n-2}/R_{n-1} & \longrightarrow & \{0\}
\end{array}
$$

If R is Noetherian (Artinian), we read these sequences from top to bottom and see that all the submodules R_k and all the quotients R_k/R_{k+1} are Noetherian (Artinian), due to (10.16). Conversely, if all the quotients R_k/R_{k+1} and hence in particular $R_{n-1} \cong R_{n-1}/\{0\} = R_{n-1}/R_n$ are Noetherian (Artinian), we read these sequences from bottom to top and see that all modules R_k (and hence R) are Noetherian (Artinian), again due to (10.16).

---

† All that follows in this section will not be referred to again in this text and hence can be skipped without any further loss.

(c) Each quotient R_k/R_{k+1} is a vector space over the field R/M_{k+1}, hence is Noetherian (as an R/M_{k+1}-module or as an R-module, there is no difference) if and only if it is Artinian, due to example (10.12)(d). Hence the claim follows immediately from (a) and (b). ∎

(11.25) Lemma. *Let R be a commutative ring with identity. If R is Artinian, then every prime ideal of R is maximal.*

Proof. Let $P \trianglelefteq R$ be a prime ideal. Then R/P is an integral domain which is again Artinian (as a homomorphic image of an Artinian ring). Let $\xi \neq 0$ in R/P. Since the chain $\langle \xi \rangle \supseteq \langle \xi^2 \rangle \supseteq \langle \xi^3 \rangle \supseteq \cdots$ becomes stationary there is an index n with $\xi^n \in \langle \xi^{n+1} \rangle$, say $\xi^n = \xi^{n+1}\eta$ with $\eta \in R/P$. Since the cancellation rule is valid in R/P, this equation implies $1 = \xi\eta$ so that ξ is invertible. Hence every nonzero element in R/P is invertible so that R/P is a field. But then P is a maximal ideal. ∎

(11.26) Lemma. *Let R be a commutative Artinian ring with $R^2 = R$. Then R possesses only a finite number of maximal ideals.*

Proof. Let $\mathcal{M}$ be the set of all finite intersections of maximal ideals of R. Since R is Artinian, the set $\mathcal{M}$ possesses a minimal element, say $M_1 \cap \cdots \cap M_n$. Let M be an arbitrary maximal ideal of R. Then the minimality of $M_1 \cap \cdots \cap M_n$ implies that $M \cap M_1 \cap \cdots \cap M_n = M_1 \cap \cdots \cap M_n$, i.e., $M_1 \cap \cdots \cap M_n \subseteq M$ and consequently $M_1 \cdots M_n \subseteq M$. Since all maximal ideals of R are prime due to (6.4)(b), this implies $M_i \subseteq M$ for some index i and hence $M_i = M$ because M_i is maximal. This shows that $M_1, \ldots, M_n$ are the only maximal ideals of R. ∎

(11.27) Lemma. *Let R be a commutative Artinian ring with nilradical N. Then $N^k = \{0\}$ for some $k \in \mathbb{N}$.*

Proof. The sequence $N \supseteq N^2 \supseteq N^3 \supseteq \cdots$ becomes stationary, say $N^k = N^{k+1} = N^{k+2} = \cdots =: I$. Suppose $I \neq \{0\}$. Then the set

$$\mathcal{M} := \{ J \trianglelefteq R \mid JI \neq \{0\} \}$$

is not empty because $I, N, R \in \mathcal{M}$, hence possesses a minimal element J_0. Since $J_0 I \neq \{0\}$ there is an element $x \in J_0$ with $xI \neq \{0\}$. Then $\langle x \rangle \in \mathcal{M}$ and $\langle x \rangle \subseteq J_0$, hence $\langle x \rangle = J_0$ due to the minimality of J_0. Also, $(xI)I = xI^2 = xI \neq \{0\}$ so that $xI \in \mathcal{M}$ and $xI \subseteq \langle x \rangle$; hence $xI = \langle x \rangle = J_0$, again due to the minimality of J_0. Then $x \in xI$, say $x = xy$ with $y \in I$. But this equation implies $x = xy = xy^2 = xy^3 = \cdots$. Now $y \in I = N^k \subseteq N$ is nilpotent so that $y^n = 0$ for some n. But then $x = xy^n = 0$ which contradicts the fact $xI \neq \{0\}$. This is the desired contradiction. ∎

AFTER this sequence of lemmas the main work is done, and we can prove the promised theorem.

(11.28) Theorem. *Let R be a commutative ring with identity. Then the following statements are equivalent:*

(a) *R is Artinian;*

(b) *R is Noetherian, and every prime ideal of R is maximal.*

Proof. Let us show first that (a) implies (b). By lemma (11.25) it is enough to prove that R is Noetherian. By lemma (11.26), there are only finitely many maximal ideals $M_1, \ldots, M_n$ in R which are at the same time the only prime ideals of R. Let N be the nilradical of R; then $N = M_1 \cap \cdots \cap M_n$. By lemma (11.27) there is a number k with $\{0\} = N^k = (\bigcap_{i=1}^{n} M_i)^k \supseteq (\prod_{i=1}^{n} M_i)^k = \prod_{i=1}^{n} M_i^k$. Thus R satisfies the hypothesis of lemma (11.24) and hence is Artinian.

Let us now show that (b) implies (a). Since R is Noetherian the zero ideal $\{0\}$ possesses a primary decomposition, say $\{0\} = Q_1 \cap \cdots \cap Q_n$. Taking the radical on both sides, we obtain $N = \sqrt{\{0\}} = P_1 \cap \cdots \cap P_n$ where N is the nilradical of R and where $P_1, \ldots, P_n$ are prime and hence maximal by hypothesis. By (11.15), there is an index k with $N^k = \{0\}$ so that $\{0\} = N^k = (\bigcap_{i=1}^{n} P_i)^k \supseteq (\prod_{i=1}^{n} P_i)^k = \prod_{i=1}^{n} P_i^k$. Hence R satisfies the hypothesis of lemma (11.24) and thus is Noetherian. ■

Exercises

Problem 1. Let R be the ring of all sequences $(a_1, a_2, \ldots)$ with $a_i \in \mathbb{Z}_2$ which eventually become stationary; addition and multiplication on R are defined componentwise. Show that R is not Noetherian by:
   (a) exhibiting an strictly ascending chain of ideals,
   (b) exhibiting a non-empty set of ideals without a maximal element, and
   (c) finding an ideal which is not finitely generated.

Problem 2. (a) Let K be a field. Show that the polynomial ring $K[x_1, x_2, x_3, \ldots]$ in a countably infinite number of variables is not Noetherian.
   (b) Show that a subring of a Noetherian ring need not be Noetherian.
   Hint. Consider the ring R from part (a) and its quotient field Q.

Problem 3. Show that the ring $C[0,1]$ of all continous functions $f : [0,1] \to \mathbb{R}$ is not Noetherian.

Problem 4. Suppose that R is a Noetherian (Artinian) ring and A is a finitely generated R-module. Show that A is a Noetherian (Artinian) R-module.
   Hint. A is a quotient of $R \oplus \cdots \oplus R$ for a finite number of summands.

Problem 5. Let R be a commutative ring.
   (a) Let $Y \subseteq R$ be a finite set such that for each $y \in Y$ there is an element $e_y \in R$ with $e_y y = y$. Show that there is an element $e \in R$ with $ey = y$ for all $y \in Y$.
   Hint. If $e_1 y_1 = y_1$ and $e_2 y_2 = y_2$, then $e := e_1 + e_2 - e_1 e_2$ satisfies $ey_1 = y_1$ and $ey_2 = y_2$.
   (b) Show that if $R[x]$ is Noetherian, then R has an identity element.
   Hint. R is a homomorphic image of $R[x]$ and hence Noetherian; thus there is a finite set $Y \subseteq R$ with $R = \langle Y \rangle$. For each $r \in Y$, the ideal chain $\langle r \rangle \subsetneq \langle r, rx \rangle \subsetneq \langle r, rx, rx^2 \rangle \subsetneq \cdots$ becomes stationary so that $rx^{n+1} \in \langle r, rx, \ldots, rx^n \rangle$, say $rx^{n+1} = \sum_{i=0}^{n} f_i(x) rx^i + \sum_{i=0}^{n} n_i \cdot rx^i$ with $f_i \in R[x]$ and $n_i \in \mathbb{Z}$. Compare coefficients of x^{n+1} to conclude that $r = \alpha r$ for some $\alpha \in R$. Now apply part (a).

Problem 6. Let R be a Noetherian integral domain and let $I \neq \{0\}$ be a nonzero ideal of R. Show that I contains a product of nonzero prime ideals.
   Hint. Proceed by "Noetherian induction", i.e., suppose the set of counterexamples is non-empty and pick a maximal element in this set.

Problem 7. Let I be an ideal of a commutative ring R. A decomposition $I = I_1 \cap \cdots \cap I_m$ of I into irreducible ideals is called **reduced** if no I_k is contained in the intersection of the remaining ideals I_j, i.e., if no I_k is superfluous. Show that the number of components in such a decomposition is uniquely determined.
   Hint. If $I = I_1 \cap \cdots \cap I_m = J_1 \cap \cdots \cap J_n$ are two reduced decompositions, show that for each index k there is an index l with $I = I_1 \cap \cdots \cap I_{k-1} \cap J_l \cap I_{k+1} \cap \cdots \cap I_m$.

Problem 8. Let I and J be ideals of a commutative Noetherian ring R. Show that if n is large enough then

$$(I + J^n) \cap (I : J^n) = I + ((I : J^n) \cap J^n) = I .$$

Hint. The first equation follows from problem 7(c) in section 5. To obtain the second equation, take a reduced primary decomposition $I = Q_1 \cap \cdots \cap Q_r$ where Q_i is P_i-primary so that $(I : J^n) \cap J^n = (Q_1 : J^n) \cap \cdots \cap (Q_r : J^n) \cap J^n$. Now if $J \not\subseteq P_i$ then $(Q_i : J^n) = Q_i$ for all n, and if $J \subseteq P_i$ then $J^n \subseteq Q_i$ if n is sufficiently large.

Problem 9. Let R be a commutative Noetherian ring. Let $\{0\} = Q_1 \cap \cdots \cap Q_n$ be a reduced primary decomposition of $\{0\}$ where Q_i is P_i-primary. Show that $\bigcup_{i=1}^{n} P_i$ is exactly the set of all zero-divisors of R.
Remark. Compare with problem 12 in section 6.

Problem 10. Let R be a commutative Noetherian ring and let P be a maximal ideal. Show that an ideal $I \neq R$ of R is P-primary if and only if it contains a power of P.

Problem 11. Let R be a commutative Noetherian ring and let $I \neq R$ be an ideal of R with a reduced primary decomposition $I = Q_1 \cap \cdots \cap Q_n$ where Q_i is P_i-primary. Show that an ideal $J \trianglelefteq R$ satisfies $(I : J) = I$ if and only if $I \not\subseteq P_k$ for $1 \leq k \leq n$.

Problem 12. Let R be a principal ideal domain and let A be a finitely generated free R-module. Show that every submodule of R is free. Can you generalize your proof to arbitrary (not necessarily finitely generated) free R-modules?
Hint. Modify the proof of theorem (11.17).

Problem 13. Let R be a commutative ring with identity. Show that if every maximal ideal of R has the form $\langle x \rangle$ where $x^2 = x$ then R is Noetherian.

Problem 14. Show that every Artinian integral domain is a field.
Hint. Consider the chain $\langle x \rangle \supseteq \langle x^2 \rangle \supseteq \langle x^3 \rangle \supseteq \cdots$ to find an inverse of $x \neq 0$.

Problem 15. Let R be a commutative Noetherian ring. Show that a power series $\sum_{n=0}^{\infty} a_n x^n$ is nilpotent in $R[[x]]$ if and only if all the coefficients a_n are nilpotent in R.

12. Field extensions

IN this section we will begin a systematic study of field extensions. The theory of field extensions is, of course, contained in the theory of ring extensions, but we will see that it has a very special flavor, due to the special properties of fields. Moreover, as was pointed out at the end of section 10, a good understanding of field extensions will also help us to study integral ring extensions.

(12.1) Definition. *If L is a field containing a smaller field K, then we call K a **subfield** of L and L an **extension** of K. We shall write $(L : K)$ to express that L is an extension of K. Any field Z with $K \subseteq Z \subseteq L$ is called an **intermediate field** of the extension $(L : K)$.*

(12.2) Examples. (a) $\mathbb{R}$ is an intermediate field of the field extension $(\mathbb{C} : \mathbb{Q})$.
(b) $\mathbb{Q} + \mathbb{Q}\sqrt{2}$ is an intermediate field of the field extension $(\mathbb{R} : \mathbb{Q})$.
(c) If L is a field of characteristic p then $L^p := \{x^p \mid x \in L\}$ is a subfield of L.
(d) Let $(L : K)$ be a field extension. Then an arbitrary intersection of intermediate fields of $(L : K)$ is again an intermediate field of $(L : K)$.
(e) If L is a field then the intersection of all subfields of K is again obviously the smallest subfield of L; it is called the **prime field** of L. It is easy to see that the prime field of L is isomorphic to $\mathbb{Q}$ if $\operatorname{char} K = 0$ and isomorphic to $\mathbb{Z}_p$ if $\operatorname{char} K = p \neq 0$. (See problem 1 below.)

MOST often an intermediate field of a given extension $(L : K)$ is obtained by the process of *adjoining* elements of L to K which is described in the following definition.

(12.3) Definition. *Let $(L : K)$ be a field extension.*
(a) *For any subset $A \subseteq L$ we denote by $K(A)$ the intersection of all intermediate fields $K \subseteq Z \subseteq L$ such that $A \subseteq Z$. This is again an intermediate field of $(L : K)$, obviously the smallest one containing all of A. We say that $K(A)$ is obtained from K by **adjoining** the elements of A to the base-field K. It is easy to verify that*

$$K(A) = \{\frac{p(\alpha)}{q(\alpha)} \mid \alpha \in A, \; p, q \in K[x], \; q(\alpha) \neq 0\} \; .$$

Indeed, since $K(A)$ is a field containing K and A, it must contain the right-hand side. But the right-hand side is a field already.
(b) *We write for short $K(\alpha_1, \ldots, \alpha_n)$ instead of $K(\{\alpha_1, \ldots, \alpha_n\})$ to denote an intermediate field obtained by adjoining a finite number of elements. A field extension $(L : K)$ is called **simple** if there is an element $\alpha \in L$ such that $L = K(\alpha)$; each such element is called a **primitive element** of L over K.*
(c) *If $Z_1, \ldots, Z_n$ are intermediate fields of $(L : K)$ then $Z_1 \cdots Z_n := K(Z_1 \cup \cdots \cup Z_n)$ is called the **compositum** of these intermediate fields. This is obviously the smallest intermediate field of $(L : K)$ containing all the fields Z_i.*

(12.4) Examples. (a) We have $\mathbb{Q}(\sqrt{2}) = \mathbb{Q} + \mathbb{Q}\sqrt{2}$.

(b) If $Z_1 = K(A)$ and $Z_2 = K(B)$ then clearly $Z_1 Z_2 = K(A \cup B) = K(A)(B) = K(B)(A)$ for all subsets $A, B \subseteq L$; it does not matter in which order the elements are adjoined. For example, $\mathbb{Q}(\sqrt{2}, \sqrt{3}) = \mathbb{Q}(\sqrt{2})(\sqrt{3}) = \mathbb{Q}(\sqrt{3})(\sqrt{2})$.

(c) Let us show that

$$\mathbb{Q}(\sqrt{2}, \sqrt{3}) = \mathbb{Q}(\sqrt{2} + \sqrt{3}).$$

The inclusion $\mathbb{Q}(\sqrt{2} + \sqrt{3}) \subseteq \mathbb{Q}(\sqrt{2}, \sqrt{3})$ is trivial because $\mathbb{Q}(\sqrt{2}, \sqrt{3})$ is a field containing $\sqrt{2}$ and $\sqrt{3}$, hence also contains $\sqrt{2} + \sqrt{3}$. Conversely, let $x := \sqrt{2} + \sqrt{3}$. Then $\sqrt{3} = x - \sqrt{2}$; so $3 = (x - \sqrt{2})^2 = x^2 - 2\sqrt{2}x + 2$ and consequently $\sqrt{2} = \frac{x^2 - 1}{2x}$. But the right-hand side is an element of $\mathbb{Q}(x) = \mathbb{Q}(\sqrt{2} + \sqrt{3})$. This shows that $\sqrt{2}$ lies in $\mathbb{Q}(\sqrt{2} + \sqrt{3})$; but then so does $\sqrt{3}$ because $\sqrt{3} = x - \sqrt{2}$, and the inclusion $\mathbb{Q}(\sqrt{2}, \sqrt{3}) \subseteq \mathbb{Q}(\sqrt{2} + \sqrt{3})$ is also proved. This shows that $(\mathbb{Q}(\sqrt{2}, \sqrt{3}) : \mathbb{Q})$ is a simple extension (even though this is not clear at first glance).

NOTE the difference between the ring $K[A]$, obtained from K by *ring-adjunction* of the elements of A to K, and the field $K(A)$, obtained from K by *field-adjunction* of A; namely, $K[A]$ consists of all polynomial expressions in the elements of A, whereas $K(A)$ consists of all rational expressions in the elements of A. Clearly, $K(A)$ is the quotient field of $K[A]$. If A is a finite set, there is an easy criterion for the coincidence of $K(A)$ and $K[A]$.

(12.5) Theorem. *Let $(L : K)$ be a field extension and let $\alpha_1, \ldots, \alpha_n \in L$. Then the following statements are equivalent:*
(1) $K(\alpha_1, \ldots, \alpha_n) = K[\alpha_1, \ldots, \alpha_n]$;
(2) $K[\alpha_1, \ldots, \alpha_n]$ *is a field;*
(3) $I := \{f \in K[x_1, \ldots, x_n] \mid f(\alpha_1, \ldots, \alpha_n) = 0\}$ *is a maximal ideal of the polynomial ring $K[x_1, \ldots, x_n]$;*
(4) *the elements $\alpha_1, \ldots, \alpha_n$ are all algebraic over K.*

Proof. The equivalence of (1) and (2) is clear, and the equivalence of (2) and (3) follows from the fact that $K[\alpha_1, \ldots, \alpha_n] \cong K[x_1, \ldots, x_n]/I$. The implication (2)$\Longrightarrow$(4) (known as **Zariski's lemma**) is nontrivial and will be proved in (12.35) below after we will have developed some of the theory of field extensions. Finally, if (4) holds then every element of $K[\alpha_1, \ldots, \alpha_n]$ is algebraic over K, due to (10.24)(b) and (10.24)(a). Thus each nonzero element β of $K[\alpha_1, \ldots, \alpha_n]$ satisfies an equation $\beta^m + a_{m-1}\beta^{m-1} + \cdots + a_1\beta + a_0 = 0$ with coefficients $a_i \in K$ and $a_0 \neq 0$; then $\beta^{-1} = -a_0^{-1}(\beta^{m-1} + a_{m-1}\beta^{m-2} + \cdots + a_1) \in K[\alpha_1, \ldots, \alpha_n]$ which shows that every nonzero element in $K[\alpha_1, \ldots, \alpha_n]$ has an inverse in $K[\alpha_1, \ldots, \alpha_n]$. Thus $K[\alpha_1, \ldots, \alpha_n]$ is a field. ∎

(12.6) Examples. (a) $\mathbb{Q}[e]$ is not a field (for example, $\frac{1}{e} \notin \mathbb{Q}[e]$). Also, $\mathbb{Q}[e, \frac{1}{e}]$ is not a field (for example, $\frac{1}{e+1} \notin \mathbb{Q}[e, \frac{1}{e}]$).

(b) $I := \{f \in \mathbb{R}[x, y] \mid f(i, i) = 0\}$ is a maximal ideal of $\mathbb{R}[x, y]$, because i is algebraic over $\mathbb{R}$.

AN element α is algebraic over a field K if and only if the family of all nonzero polynomials $p \in K[x]$ with $p(\alpha) = 0$ is not empty. We will now show that this set contains a distinguished element which is called the *minimal polynomial* of α over K.

(12.7) Proposition. *Suppose that α is algebraic over a field K so that the ideal*

$$I_\alpha := \{p \in K[x] \mid p(\alpha) = 0\} \trianglelefteq K[x]$$

does not consist of the zero polynomial alone. Then there is a polynomial $p \in I_\alpha \setminus \{0\}$ which satisfies the following equivalent conditions:
(1) p has minimal degree in $I_\alpha \setminus \{0\}$;
(2) every polynomial $q \in I_\alpha$ is divisible by p;
(3) p is irreducible in $K[x]$.
If in addition p is monic, then p is uniquely determined and is called the **minimal polynomial** *of α.*

Proof. Let us show first that the conditions (1)-(3) are indeed equivalent. Remark (7.13) shows that (1) implies (2). Let us prove that (2) implies (3). To get a contradiction, suppose that p satisfies (2) but is reducible. Then $p = p_1 p_2$ where $p_1, p_2 \in K[x]$ have a degree ≥ 1. Since $0 = p(\alpha) = p_1(\alpha)p_2(\alpha)$, we must have $p_1(\alpha) = 0$ or $p_2(\alpha) = 0$, say $p_1(\alpha) = 0$ so that $p_1 \in I_\alpha$. By (2) we have $p_1 = rp$ for some $r \in K[x]$. Consequently, $p = p_1 p_2 = rpp_2$, hence $rp_2 = 1$ which is impossible because the degree of p_2 is ≥ 1. This is the desired contradiction. Finally, let us prove that (3) implies (1). To get a contradiction, suppose that p is irreducible, but does not satisfy (1). Choose a polynomial $q \in I_\alpha \setminus \{0\}$ with minimal degree; then $\deg q < \deg p$. By (7.13) there is an element $r \in K[x]$ with $p = rq$. But then $\deg r = \deg p - \deg q > 0$ so that $p = rq$ is a proper factorization of p which is impossible because p is irreducible.

Obviously, there is a polynomial $p \in I_\alpha \setminus \{0\}$ which satisfies (1) and hence also (2) and (3). We want to show that p is uniquely determined up to a scalar factor. Suppose that $p_1, p_2 \in I_\alpha \setminus \{0\}$ satisfy the conditions (1)-(3). Then by (2), p_1 and p_2 are divisible by each other; i.e., there are elements $f, g \in K[x]$ such that $p_2 = fp_1$ and $p_1 = gp_2$. But then $p_2 = fgp_2$, so $fg = 1$. This shows that f and g are units in $K[x]$, hence constants. ∎

NOTE that in the proof of (12.7) we used the fact that $K[x]$ is a Euclidean domain which depends on the fact that K is a field. The analogue of proposition (12.7) for arbitrary commutative rings does not hold true.

(12.8) Examples. (a) Let p be a prime number and $n \in \mathbb{N}$. Then $\sqrt[n]{p}$ is a root of the polynomial $f(x) = x^n - p$, and since this polynomial is irreducible in $\mathbb{Q}[x]$ by Eisenstein's criterion, it is the minimal polynomial of $\sqrt[n]{p}$ over $\mathbb{Q}$.

(b) Let $x = \sqrt{1 + \sqrt{3}}$. Then $x^2 = 1 + \sqrt{3}$, whence $\sqrt{3} = x^2 - 1$ and consequently $3 = (x^2 - 1)^2 x^4 - 2x^2 + 1$. So x is a root of the polynomial $p(x) = x^4 - 2x^2 - 2 \in \mathbb{Q}[x]$. This polynomial has leading coefficient 1 and is irreducible by Eisenstein's criterion, so it is the minimal polynomial of x.

(c) Let $x := \sqrt{2} + \sqrt{3}$. Then $\sqrt{3} = x - \sqrt{2}$, so $3 = (x - \sqrt{2})^2 = x^2 - 2x\sqrt{2} + 2$ and consequently $2x\sqrt{2}x = x^2 - 1$. Squaring yields $8x^2 = x^4 - 2x^2 + 1$. Hence $\sqrt{2} + \sqrt{3}$ is a root of the polynomial $f(x) = x^4 - 10x^2 + 1$. One readily checks that f is irreducible over $\mathbb{Q}$ and hence is the minimal polynomial of $\sqrt{2} + \sqrt{3}$ over $\mathbb{Q}$.

(d) Let $x := i\sqrt[3]{2}$. Then $x^3 = -2i$, whence $x^6 = -4$. This shows that x is a root of the polynomial $p(x) = x^6 + 4 \in \mathbb{Q}[x]$. A relatively tedious calculation shows that p is irreducible over $\mathbb{Q}$ and hence is the minimal polynomial of $i\sqrt[3]{2}$ over $\mathbb{Q}$.

(e) As a final example, consider $x := \sqrt{2} + \sqrt[3]{3}$. We have $\sqrt[3]{3} = x - \sqrt{2}$, so $3 = (x - \sqrt{2})^3 = x^3 + 6x - \sqrt{2}(3x^2 + 2)$ and consequently $\sqrt{2}(3x^2 + 2) = x^3 + 6x - 3$. Square this last equation to see that x is a root of the polynomial $p(x) = x^6 - 6x^4 - 6x^3 + 12x^2 - 36x + 1$. Once it is established that p is irreducible over $\mathbb{Q}$, we know that p is the minimal polynomial of $\sqrt{2} + \sqrt[3]{3}$ over $\mathbb{Q}$. To verify that p is irreducible in $\mathbb{Z}[x]$ (and hence in $\mathbb{Q}[x]$), we could use the "brute force method" of trying all possible types of factorizations; this is straightforward, but somewhat tedious and hence omitted here. (After we will have developed some elementary theory of field extensions, we will be able to conclude that p is irreducible without such a tedious inspection.)

WE now observe that there is a major advantage in the study of field extensions as opposed to the study of general ring extensions; namely, if $(L : K)$ is a field extension then L can be considered as a vector space over K (thereby ignoring some of the structure of L). This observation – as simple as it may be – makes available to us the power of linear algebra to study field extensions.† As a first example of using linear algebra to study field extensions, we define the *degree* of a field extension.

(12.9) Definition. *Let $(L : K)$ be a field extension and consider L as a vector space over K. Then the dimension $\dim_K L$ of this vector space is denoted by $[L : K]$ and is called the **degree** of the field extension $(L : K)$. A field extension is called **finite** if its degree is finite.*

TO be able to utilize the concept of degree, we need ways to determine the degree of a given field extension. The next two theorems provide such ways.

(12.10) Theorem. *Let $L = K(\alpha)$ be a simple extension of K.*

(a) *If α is algebraic over K and if n is the degree of the minimal polynomial of α over K, then $(1, \alpha, \alpha^2, \ldots, \alpha^{n-1})$ is a vector space basis of L over K; in particular, $[L : K] = n$, i.e.,*

$$\boxed{[K(\alpha) : K] \;=\; degree \ of \ the \ minimal \ polynomial \ of \ \alpha \ over \ K.}$$

(b) *If α is transcendental over K, then the degree $[L : K]$ is infinite.*

---

† For any ring extension $(S : R)$ we can treat S as an R-module. However, the theory of vector spaces is much richer than the theory of arbitrary modules; this is why we can say more about field extensions than about arbitrary ring extensions.

Proof. (a) Let p be the minimal polynomial of α; then $n = \deg p$. From the relation $p(\alpha) = 0$ one can easily deduce that all elements $f(\alpha)/g(\alpha)$ of $K(\alpha)$ can be expressed as a linear combination of the powers $1, \alpha, \ldots, \alpha^{n-1}$, so that these span L as a vector space over K. On the other hand, they are linearly independent over K, due to the minimality of p. Hence these elements form a vector space basis of L over K.

(b) Since there is no polynomial $p \neq 0$ in $K[x]$ with $p(\alpha) = 0$, the infinitely many elements $1, \alpha, \alpha^2, \ldots \in K(\alpha)$ are linearly independent over K, whence $[K(\alpha) : K] = \dim_K K(\alpha) = \infty$. ∎

(12.11) Theorem. *Consider three fields* $K \subseteq Z \subseteq L$.
(a) *If* $(x_i)_{i \in I}$ *is a basis of* L *over* Z *and if* $(y_j)_{j \in J}$ *is a basis of* Z *over* K, *then* $(x_i y_j)_{i \in I, j \in J}$ *is a basis of* L *over* K.
(b) *The following* **degree formula** *holds:*

$$\boxed{[L : K] = [L : Z][Z : K] \,.}$$

In particular, $[L : K]$ *is finite if and only if both* $[L : Z]$ *and* $[Z : L]$ *are finite.*

Proof. It is enough to prove part (a) because (b) is then a simple consequence of (a). First of all, let $x \in L$ be arbitrary. Then we can find elements $z_i \in Z$ (only finitely many different from zero) such that $x = \sum_{i \in I} z_i x_i$. For each i, we can find elements $k_j^{(i)} \in K$ (only finitely many different from zero) such that $z_i = \sum_{j \in J} k_j^{(i)} y_j$. Altogether, we obtain

$$x = \sum_{i \in I,\, j \in J} k_j^{(i)} x_i y_j \,.$$

This shows that the elements $x_i y_j$ span all of L over K.
Now suppose that

$$0 = \sum_{i \in I,\, j \in J} k_j^{(i)} x_i y_j = \sum_{i \in I} \underbrace{\left(\sum_{j \in J} k_j^{(i)} y_j \right)}_{\in Z} x_i \,.$$

Since the elements $(x_i)_{i \in I}$ are linearly independent over Z, this implies that $\sum_{j \in J} k_j^{(i)} y_j = 0$ for all i. Consequently, since the elements $(y_j)_{j \in J}$ are linearly independent over K, we have $k_j^{(i)} = 0$ for all i, j. This shows that the elements $x_i y_j$ are linearly independent over K. ∎

(12.12) Examples. (a) The minimal polynomial of $\sqrt{2}$ over $\mathbb{Q}$ is $x^2 - 2$. A basis of $\mathbb{Q}(\sqrt{2})$ over $\mathbb{Q}$ is given by $(1, \sqrt{2})$ so that $\mathbb{Q}(\sqrt{2}) = \mathbb{Q} + \mathbb{Q}\sqrt{2}$ where the right-hand side in this equation denotes the set of all elements $a + b\sqrt{2}$ with $a, b \in \mathbb{Q}$.

(b) The minimal polynomial of $\sqrt[3]{2}$ over $\mathbb{Q}$ is $x^3 - 2$. A basis of $\mathbb{Q}(\sqrt[3]{2})$ over $\mathbb{Q}$ is given by $(1, \sqrt[3]{2}, \sqrt[3]{4})$ so that $\mathbb{Q}(\sqrt[3]{2}) = \mathbb{Q} + \mathbb{Q}\sqrt[3]{2} + \mathbb{Q}\sqrt[3]{4}$ where the right-hand side in this equation denotes the set of all elements $a + b\sqrt[3]{2} + c\sqrt[3]{4}$ with $a, b, c \in \mathbb{Q}$.

(c) Let us find a basis of $\mathbb{Q}(\sqrt{2}, \sqrt{3})$ over $\mathbb{Q}$. Since $x^2 - 2 \in \mathbb{Q}[x]$ is the minimal polynomial of $\sqrt{2}$ over $\mathbb{Q}$, we know that $(1, \sqrt{2})$ is a basis of $\mathbb{Q}(\sqrt{2})$ over $\mathbb{Q}$. Now the quadratic polynomial $x^2 - 3 \in \mathbb{Q}(\sqrt{2})$ has no root in $\mathbb{Q}(\sqrt{2})$, hence is irreducible over this field. But then we know that $(1, \sqrt{3})$ is a basis of $\mathbb{Q}(\sqrt{2}, \sqrt{3})$ over $\mathbb{Q}(\sqrt{2})$. By (12.11)(a), a basis of $\mathbb{Q}(\sqrt{2}, \sqrt{3})$ over $\mathbb{Q}$ is given by $(1, \sqrt{2}, \sqrt{3}, \sqrt{6})$ so that

$$\mathbb{Q}(\sqrt{2}, \sqrt{3}) \;=\; \mathbb{Q} \oplus \mathbb{Q}\sqrt{2} \oplus \mathbb{Q}\sqrt{3} \oplus \mathbb{Q}\sqrt{6}$$

as a direct sum of $\mathbb{Q}$-vector spaces. A second basis of $\mathbb{Q}(\sqrt{2}, \sqrt{3})$ over $\mathbb{Q}$ can be obtained by invoking (12.4)(c). In fact, we can represent $\mathbb{Q}(\sqrt{2}, \sqrt{3}) = \mathbb{Q}(\sqrt{2} + \sqrt{3})$ as a simple extension and then apply (12.10)(a) to see that $(1, \sqrt{2} + \sqrt{3}, (\sqrt{2} + \sqrt{3})^2, (\sqrt{2} + \sqrt{3})^3)$ is a basis of $\mathbb{Q}(\sqrt{2}, \sqrt{3})$ over $\mathbb{Q}$.

(d) Let K be an arbitrary field and let $K(x)$ be the field of of rational functions over K, i.e., the quotient field of the polynomial ring $K[x]$. We can consider K as a subfield of $K(x)$ by identifying the elements of K with constant polynomials. Then $\big(K(x) : K\big)$ is a field extension of infinite degree.

(e) Let p be a prime number and let $n \in \mathbb{N}$. Then the polynomial $x^n - p \in \mathbb{Q}[x]$ is irreducible by Eisenstein's criterion so that $[\mathbb{Q}(\sqrt[n]{p}) : \mathbb{Q}] = n$ by (12.10)(a). Hence if p, q are primes and m, n are natural numbers such that m does not divide n, then $\sqrt[m]{q} \notin \mathbb{Q}(\sqrt[n]{p})$; this is an immediate consequence of the degree formula.

(f) Let $(L : K)$ be a finite field extension. Then an element $\alpha \in L$ is a primitive element over K if and only if the degree of its minimal polynomial equals $[L : K]$.

SUPPOSE that α is an element which is algebraic over a field K. Once the minimal polynomial of α is known, we also know the degree $[K(\alpha) : K]$, due to (12.10)(a). On the other hand, it is sometimes possible to determine the degree $[K(\alpha) : K]$ in a different way (for example, using the degree formula (12.11)(b)) and then to deduce the minimal polynomial of α over K. This is illustrated in the following two examples.

(12.13) Examples. (a) It is easily checked that the number $\alpha := i\sqrt[3]{2}$ is a root of the polynomial $p(x) = x^6 + 4 \in \mathbb{Q}[x]$. To exhibit p as the minimal polynomial of x over $\mathbb{Q}$, we could try to prove the irreducibility of p. This is tedious, however, because none of the easy criteria in (8.18) applies. But we can argue as follows. We have

$$\mathbb{Q}(i, \sqrt[3]{2}) \;=\; \mathbb{Q}(i\sqrt[3]{2}) \ .$$

Here the inclusion $\supseteq$ is trivial, and $\subseteq$ follows from $i = -\frac{1}{2}(i\sqrt[3]{2})^3 \in \mathbb{Q}(i\sqrt[3]{2})$. Using (8.18)(a), we see that the minimal polynomial of $\mathbb{Q}(\sqrt[3]{2})$ over $\mathbb{Q}$ is $x^3 - 2$ and that the minimal polynomial of $\mathbb{Q}(i, \sqrt[3]{2})$ over $\mathbb{Q}\sqrt[3]{2}$ is $x^2 + 1$. By (12.10)(a), this shows that $[\mathbb{Q}(i, \sqrt[3]{2}) : \mathbb{Q}(\sqrt[3]{2})] = 2$ and $[\mathbb{Q}(\sqrt[3]{2}) : \mathbb{Q}] = 3$. But then $[\mathbb{Q}(i, \sqrt[3]{2}) : \mathbb{Q}] = 2 \cdot 3 = 6$, by the degree formula (12.11)(b). We conclude that the minimal polynomial of $i\sqrt[3]{2}$ has degree 6, hence must be p. It follows as a corollary that the polynomial $p(x) = x^6 + 4$ is irreducible over $\mathbb{Q}$.

(b) Let $x := \sqrt{2} + \sqrt[3]{3}$. We have $\sqrt[3]{3} = x - \sqrt{2}$, so $3 = (x - \sqrt{2})^3 = x^3 + 6x - \sqrt{2}(3x^2 + 2)$ and consequently $\sqrt{2}(3x^2 + 2) = x^3 + 6x - 3$. Square this last equation to see that x is a root of the polynomial $p(X) = X^6 - 6X^4 - 6X^3 + 12X^2 - 36X + 1$. Instead of proving directly that p is irreducible over $\mathbb{Q}$ (which is tedious) we invoke the degree formula again. First of all,

$$\mathbb{Q}(\sqrt{2}, \sqrt[3]{3}) \;=\; \mathbb{Q}(\sqrt{2} + \sqrt[3]{3}) \ .$$

Here $\supseteq$ is trivial, whereas $\subseteq$ follows from the equation $\sqrt{2} = \frac{x^3+6x-3}{3x^2+2}$ proved above because the right-hand side of this equation is obviously an element of $\mathbb{Q}(x) = \mathbb{Q}(\sqrt{2}+\sqrt[3]{3})$. Now the minimal polynomial of $\sqrt{2}$ over $\mathbb{Q}(\sqrt[3]{3})$ is x^2-2, whence $[\mathbb{Q}(\sqrt{2}, \sqrt[3]{3}) : \mathbb{Q}(\sqrt[3]{3})] = 2$. Also, the minimal polynomial of $\sqrt[3]{3}$ over $\mathbb{Q}$ is $x^3 - 3$, whence $[\mathbb{Q}(\sqrt[3]{3}) : \mathbb{Q}] = 3$. The degree formula yields $[\mathbb{Q}(\sqrt{2}, \sqrt[3]{3}) : \mathbb{Q}] = 2 \cdot 3 = 6$, and we conclude that p must be the minimal polynomial of x over $\mathbb{Q}$.

THE next theorem shows how (12.10) and (12.11) can be used to determine all subfields of rational function fields in one variable. To wit, whenever $\theta = \theta(x)$ is a rational function, we can form the subfield $K(\theta)$ of $K(x)$, consisting of all rational functions of the form

$$\frac{a_0 + a_1\theta(x) + \cdots + a_m\theta(x)^m}{b_0 + b_1\theta(x) + \cdots + b_n\theta(x)^n} .$$

We will now prove that all subfields of $K(x)$ are of this type.

(12.14) Theorem. *Let K be a field and let $K(x)$ be the rational function field over K.*

(a) If $\theta \in K(x) \setminus K$, then $\big(K(\theta) : K\big)$ is a transcendental extension, and $\big(K(x) : K(\theta)\big)$ is a finite algebraic extension. More precisely, if $\theta = f/g$ where $f, g \in K[x]$ are relatively prime, then $[K(x) : K(\theta)] = \max(\deg f, \deg g)$.

*(b) (**Lüroth's Theorem.**) Let $K \subsetneqq Z \subseteq K(x)$ be an intermediate field. Then there is a rational function $\theta \in K(x)$ with $Z = K(\theta)$.*

Proof. (a) Consider the polynomial $p(t) := f(t) - \theta(x)g(t) \in K(\theta)[t]$. This polynomial is irreducible in $K[\theta][t] = K[\theta, t]$ because it has degree 1 in θ and because f, g are relatively prime; hence it is also irreducible in $K(\theta)[t]$, due to (8.2)(d). On the other hand, x is clearly a root of p so that p is the minimal polynomial of x over $K(\theta)$. Then $[K(x) : K(\theta)] = \deg p = \max(\deg f, \deg g)$, due to (12.10)(a). Since $[K(x) : K] = \infty$, the degree formula implies that also $[K(\theta) : K] = \infty$ so that $\big(K(\theta) : K\big)$ is a transcendental extension.

(b) Pick any element $\varphi \in Z \setminus K$; then $[Z(x) : Z] = [K(x) : Z] \le [K(x) : K(\varphi)]$ by the degree formula, and since $[K(x) : K(\varphi)] < \infty$ by part (a) this implies that x is algebraic over Z. Let $\alpha(t) = t^n + \frac{f_{n-1}(x)}{g_{n-1}(x)}t^{n-1} + \cdots + \frac{f_1(x)}{g_1(x)}t + \frac{f_0(x)}{g_0(x)} \in Z[t]$ be the minimal polynomial of x over Z where $f_i, g_i \in K[x]$ are coprime for each index i. If we denote the least common multiple of $g_0, \ldots, g_{n-1}$ in $K[x]$ by g, then $A(x, t) := g(x)\alpha(t) = g(x)t^n + \tilde{f}_{n-1}(x)t^{n-1} + \cdots + \tilde{f}_1(x)t + \tilde{f}_0(x) \in K[x, t]$ is primitive as a polynomial in t, and $\deg_t A = \deg \alpha = [K(x) : Z]$ by (12.10)(a). Since x is transcendental over K, there is at least one index k such that $\frac{f_k(x)}{g_k(x)} =: \theta(x)$ lies in $K(x) \setminus K$. We will show that $Z = K(\theta)$ for any such θ. To do so, let $\beta(t) := f_k(t) - \theta(x)g_k(t)$ and $B(x, t) := g_k(x)\beta(t) = g_k(x)f_k(t) - f_k(x)g_k(t) \in K[x, t]$; then $\deg_x B = \deg_t B = \max(\deg f_k, \deg g_k) = [K(x) : K(\theta)]$ by part (a). Since α is the minimal polynomial of x over Z and β is the minimal polynomial of x over $K(\theta) \subseteq Z$ due to part (a), we know that α divides β in $Z[t]$, the more so in $K(x)[t]$. Hence A divides B in $K(x)[t]$; since A is primitive, this implies that A divides B in $K[x, t]$ due to (8.2)(c), say $B(x, t) = A(x, t)\rho(x, t)$. Consequently, $\deg_x A \le \deg_x B$. On the other hand, $\deg_x A \ge \deg g \ge \deg g_k$ and $\deg_x A \ge \deg \tilde{f}_k \ge \deg f_k$, hence $\deg_x A \ge \max(\deg f_k, \deg g_k) = \deg_x B$ so that $\deg_x A = \deg_x B$. This shows that $\deg_x \rho = 0$ so that $\rho(x,t) = \overline{\rho}(t) \in K[t]$. Writing $f_k(x) = \sum_{r=0}^{N} c_r x^r$, $g_k(x) = \sum_{r=0}^{N} d_r x^r$ and $A(x,t) = \sum_{r=0}^{N} a_r(t) x^r$, the equation $B(x,t) = A(x,t) \overline{\rho}(t)$ reads $\sum_{r=1}^{N} \big(d_r f_k(t) - c_r g_k(t)\big) x^r = \sum_{r=1}^{N} a_r(t) \overline{\rho}(t) x^r$. Hence $\overline{\rho}$ divides all the polynomials $d_r f_k - c_r g_k$ in $K[t]$. Since g_k is coprime with f_k, in particular not a scalar multiple of f_k, the vectors $(c_0, \ldots, c_N)$ and $(d_0, \ldots, d_N)$ in K^{N+1} are linearly independent. Therefore, we can find indices i,j with $\det\big(\begin{smallmatrix} c_i & d_i \\ c_j & d_j \end{smallmatrix}\big) \neq 0$. Consequently, there are elements $u, v, \overline{u}, \overline{v} \in K$ such that

$$f_k(t) = u\big(d_i f_k(t) - c_i g_k(t)\big) + v\big(d_j f_k(t) - c_j g_k(t)\big) \quad \text{and}$$
$$g_k(t) = \overline{u}\big(d_i f_k(t) - c_i g_k(t)\big) + \overline{v}\big(d_j f_k(t) - c_j g_k(t)\big);$$

hence $\overline{\rho}$ divides f_k and g_k. Since these two polynomials are coprime, $\overline{\rho}$ must be a constant polynomial $\overline{\rho}(t) = \rho_0 \in K$ so that the equation $B(x,t) = A(x,t) \overline{\rho}(t)$ becomes $B(x,t) = \rho_0 \cdot A(x,t)$. This implies that $[K(x) : K(\theta)] = \deg_t B = \deg_t A = [K(x) : Z]$. Consequently, $[Z : K(\theta)] = 1$, i.e., $Z = K(\theta)$. ∎

THE concept of degree can be used to derive a fundamental fact about finite fields. We know that for each prime number p the residue-class ring $\mathbb{Z}_p$ is a field, and we have also encountered a field with 4 elements (see (5.12)(b)). We will now prove that the cardinality of a finite field must be a prime power; for example, there is no field with 6 elements. Later on, we will also see that every prime power occurs as the cardinality of some finite field.

(12.15) Theorem. *If K is a finite field, then $\operatorname{char} K$ is a prime number p, and K can be considered as a finite-dimensional vector space over $\mathbb{Z}_p$. If $n := [K : \mathbb{Z}_p]$, then $|K| = p^n$.*

Proof. If $\operatorname{char} K$ were zero, then the elements $m \cdot 1$ $(m \in \mathbb{Z})$ would be pairwise different so that K could not be finite; hence $\operatorname{char} K$ is a prime number p. It is easy to see that the prime field of K is isomorphic to $\mathbb{Z}_p$. (See problem 1 below). Hence K can be considered as a vector space over $\mathbb{Z}_p$; then $n := [K : \mathbb{Z}_p] = \dim_{\mathbb{Z}_p} K$ is finite because K is a finite set. If $(k_1, \ldots, k_n)$ is a basis of K over $\mathbb{Z}_p$, then K consists exactly of the p^n different elements $a_1 k_1 + \cdots + a_n k_n$ where $a_i \in \mathbb{Z}_p$. ∎

BY introducing the notion of the degree of a field extension and deriving its basic properties we exploited for the first time methods from linear algebra to study field extensions.† As a second example for the use of concepts from linear algebra in field theory, we define the *trace* and the *norm* of a finite field extension.

---

† Even the modest information on field extensions collected up to this point suffices to solve some famous geometric problems which date back to antiquity and had to wait for almost 2000 years before being solved. In order to not interrupt our systematic study of field extensions, the discussion of these geometric problems has been deferred to the end of this section even though no additional theory is needed for their understanding.

(12.16) Definitions. *Let $(L : K)$ be a finite field extension and let $\alpha \in L$. Then the left-multiplication*

$$m_\alpha : \begin{array}{ccc} L & \to & L \\ \xi & \mapsto & \alpha\xi \end{array}$$

is obviously K-linear, i.e., an endomorphism of the K-vector space L so that the following definitions (applying concepts from linear algebra) make sense.

(a) *The polynomial $f_\alpha(x) := \det(x\mathbf{1} - m_\alpha) \in K[x]$ is called the* **characteristic polynomial** *of α with respect to L and K.*

(b) *The* **trace** *and the* **norm** *of α with respect to $(L : K)$ are defined by $\mathrm{tr}_K^L(\alpha) := \mathrm{tr}\, m_\alpha$ and $N_K^L(\alpha) := \det m_\alpha$. Note that the trace and the norm of any element $\alpha \in L$ are, by definition, elements of the base-field K.*

(12.17) Example. Let K be a field and let $L = K(\theta)$ where $\theta^2 = k \in K$ but $\theta \notin K$; we simply write $\theta =: \sqrt{k}$. (Examples are $K = \mathbb{R}$ and $L = \mathbb{C} = \mathbb{R}(i)$ or $K = \mathbb{Q}$ and $L = \mathbb{Q}(\sqrt{n})$ where $n \in \mathbb{Z} \setminus \{0, 1\}$ is a square-free number.) Then $(1, \sqrt{k})$ is a basis of L over K. For any element $\alpha = a + b\sqrt{k}$ in L we have $m_\alpha(x + y\sqrt{k}) = (a + b\sqrt{k})(x + y\sqrt{k}) = (ax + kby) + (ay + bx)\sqrt{k}$; this shows that the matrix representation of $m_\alpha : L \to L$ with respect to the basis $(1, \sqrt{k})$ of L over K is

$$m_\alpha = \begin{pmatrix} a & kb \\ b & a \end{pmatrix}.$$

Therefore, we have $f_\alpha(x) = x^2 - 2ax + a^2 - kb^2$, $\mathrm{tr}_K^L(\alpha) = 2a$ and $N_K^L(\alpha) = a^2 - kb^2$. Note that if $K = \mathbb{Q}$ and $L = \mathbb{Q}(\sqrt{n})$ then the norm coincides with the concept of norm defined in (3.6)(c).

(12.18) Example. Let us find $\mathrm{tr}_{\mathbb{Q}}^{\mathbb{Q}(\alpha)}(2\alpha + 1)$ and $N_{\mathbb{Q}}^{\mathbb{Q}(\alpha)}(2\alpha + 1)$ where $\alpha \in \mathbb{C}$ satisfies the equation $\alpha^3 + \alpha^2 - 2$. Since the polynomial $f(x) = x^3 + x^2 - 2$ is irreducible in $\mathbb{Q}[x]$ a basis of $\mathbb{Q}(\alpha)$ over $\mathbb{Q}$ is given by $(1, \alpha, \alpha^2)$, and since

$$
\begin{aligned}
(2\alpha + 1) \cdot 1 &= 1 + 2\alpha, \\
(2\alpha + 1) \cdot \alpha &= \alpha + 2\alpha^2, \\
(2\alpha + 1) \cdot \alpha^2 &= 2\alpha^3 + \alpha^2 = 2(-\alpha^2 + 2) + \alpha^2 = 4 - \alpha^2
\end{aligned}
$$

we have

$$m_{2\alpha+1}(a_0 \cdot 1 + a_1 \cdot \alpha + a_2 \cdot \alpha^2) = \begin{pmatrix} 1 & 0 & 4 \\ 2 & 1 & 0 \\ 0 & 2 & -1 \end{pmatrix} \begin{pmatrix} a_0 \\ a_1 \\ a_2 \end{pmatrix}.$$

Thus $\mathrm{tr}_{\mathbb{Q}}^{\mathbb{Q}(\alpha)}(2\alpha + 1) = \mathrm{tr}\, m_{2\alpha+1} = 1$ and $N_{\mathbb{Q}}^{\mathbb{Q}(\alpha)}(2\alpha + 1) = \det m_{2\alpha+1} = 15$.

LET us collect some properties of the trace and the norm of a finite field extension.

(12.19) Proposition. *Let $(L : K)$ be a finite field extension*

(a) *For all elements $\alpha, \beta \in L$ and $k \in K$ we have $\mathrm{tr}_K^L(\alpha + \beta) = \mathrm{tr}_K^L(\alpha) + \mathrm{tr}_K^L(\beta)$, $\mathrm{tr}_K^L(k\alpha) = k\,\mathrm{tr}_K^L(\alpha)$ and $N_K^L(\alpha\beta) = N_K^L(\alpha)N_K^L(\beta)$; i.e., the trace is K-linear, whereas the norm is multiplicative.*

(b) *If $\alpha \in K$ then $\mathrm{tr}_K^L(\alpha) = [L : K] \cdot \alpha$ and $N_K^L(\alpha) = \alpha^{[L:K]}$.*

Proof. This is an immediate consequence of the identities $m_{\alpha+\beta} = m_\alpha + m_\beta$, $m_{k\alpha} = k \cdot m_\alpha$ and $m_{\alpha\beta} = m_\alpha \circ m_\beta$ and the properties of trace and determinant of vector space endomorphisms. ∎

TO gain some insight into the structure of field extension we will now derive some useful results on algebraic extensions. Note that some results in the following theorem can be obtained as simple consequences of theorem (10.24) because the notions of being integral and being algebraic coincide in the context of fields; however, we prefer giving a self-contained proof.

(12.20) Theorem. (a) *Every finite extension is algebraic.*
(b) *A simple extension is finite if and only if it is algebraic.*
(c) *If $[K(\alpha) : K] < \infty$ then $[K(\alpha,\beta) : K(\beta)] \leq [K(\alpha) : K]$.*
(d) *The extension $(L : K)$ is finite if and only if $L = K(\alpha_1,\ldots,\alpha_n)$ with finitely many elements $\alpha_i \in L$ which are algebraic over K.*
(e) *Let Z be an intermediate field of $(L : K)$. Then $(L : K)$ is algebraic if and only if both $(L : Z)$ and $(Z : K)$ are algebraic.*
(f) *Let $A \subseteq L$. If all elements of A are algebraic over K then $K(A) = K[A]$, and $(K(A) : K)$ is an algebraic field extension.*†
(g) *Let $(L : K)$ be an arbitrary field extension. Then the elements of L which are algebraic over K form an intermediate field of $(L : K)$.*

Proof. (a) Let $[L : K] = n$ and $\alpha \in L$. Then the $n+1$ elements $1, \alpha, \ldots, \alpha^n$ are linearly dependent over K, so that $p(\alpha) = 0$ for some nonzero polynomial $p \in K[x]$. This shows that α is algebraic over K. Since $\alpha \in L$ was arbitrary, the claim is proved.
(b) This was already proved in (12.10).
(c) If $[K(\alpha) : K] < \infty$ then α is algebraic over K; let $p \in K[x] \subseteq K(\beta)[x]$ be the minimal polynomial of α over K. Then the minimal polynomial q of α over $K(\beta)$ divides p; hence

$$[K(\alpha,\beta) : K(\beta)] = \deg q \leq \deg p = [K(\alpha) : K] .$$

(d) If $[L : K] = n$, pick any vector space basis $(\alpha_1,\ldots,\alpha_n)$ of L over K; then $L = K(\alpha_1,\ldots,\alpha_n)$. Conversely, let $L = K(\alpha_1,\ldots,\alpha_n)$ where $[K(\alpha_i) : K] < \infty$. Then by the degree formula and part (c), we obtain

$$[L : K] = \prod_{i=1}^{n}[K(\alpha_1,\ldots,\alpha_i) : K(\alpha_1,\ldots,\alpha_{i-1})] \leq \prod_{i=1}^{n}[K(\alpha_i) : K] < \infty .$$

(e) If $(L : K)$, then trivially $(L : Z)$ and $(Z : K)$ are algebraic. Suppose conversely that $(L : Z)$ and $(Z : K)$ are algebraic and let $\alpha \in L$. Then there are elements $z_i \in Z$ such that
$$\alpha^n + z_{n-1}\alpha^{n-1} + \cdots + z_1\alpha + z_0 = 0 ,$$

---

† On the other hand, the equation $K(A) = K[A]$ does *not* imply that all elements of A are algebraic over K; simply take $K = \mathbb{Q}$ and $A := \{p(e)/q(e) \mid p, q \in \mathbb{Q}[x] \setminus \{0\}\}$.

because α is algebraic over Z. Then, trivially, α is algebraic over $M := K(z_0, \ldots, z_{n-1})$ which is a finite extension of K, by part (d). But also $[M(\alpha) : M] < \infty$, by part (b). Consequently,

$$[M(\alpha) : K] = [M(\alpha) : M] \cdot [M : K] < \infty ,$$

so that the extension $(M(\alpha) : K)$ is algebraic, by (a). In particular, the element α is algebraic over K. Since $\alpha \in L$ was arbitrary, the claim is proven.

(f) If A is a finite set the claim is just the implication (4) $\Longrightarrow$ (2) in (12.5). In the general case let $\xi \in K(A)$. Then there is a finite subset $\{\alpha_1, \ldots, \alpha_n\}$ of A with $\xi \in K(\alpha_1, \ldots, \alpha_n) = K[\alpha_1, \ldots, \alpha_n] \subseteq K[A]$. This implies that ξ is algebraic over K, due to parts (d) and (a). Since $\xi \in K(A)$ was arbitrarily chosen, this gives the claim.

(g) Simply apply part (f) with $A := \{\alpha \in L \mid \alpha \text{ is algebraic over } K\}$. ∎

WE can now give a characterization of simple extensions.

(12.21) Theorem. *Let $(L : K)$ be a finite extension. Then $L = K(\alpha)$ for some $\alpha \in L$ if and only if $(L : K)$ only has a finite number of intermediate fields.*

Proof. If K is finite, then so is L; hence $(L : K)$ has trivially only a finite number of intermediate fields. Also, if α is a generator of the cyclic group $L^\times$, then $L = K(\alpha)$. So the claim holds if K is a finite field. Therefore, let K be infinite in the remainder of the proof.

Suppose first that $(L : K)$ has only a finite number of intermediate fields. Pick $\alpha \in L$ such that $[K(\alpha) : K]$ is maximal. We claim that $K(\alpha) = L$. To get a contradiction, assume that $\beta \notin K(\alpha)$ for some $\beta \in L$. Now two of the fields $K(\alpha + k\beta)$ with $k \in K$ must coincide because $(L : K)$ has only finitely many intermediate fields; say $K(\alpha + k_1\beta) = K(\alpha + k_2\beta) =: Z$. Then Z contains the elements $\alpha + k_1\beta$ and $\alpha + k_2\beta$, hence also their difference $(k_1 - k_2)\beta$, hence β, hence also α. But then $K(\alpha + k_1\beta) = Z \supseteq K(\alpha, \beta) \supsetneq K(\alpha)$ contradicting the choice of α.

Suppose conversely that $L = K(\alpha)$. Let $f \in K[x]$ be the minimal polynomial of α over K. Consider an arbitrary intermediate field Z. Let $g \in Z[x]$ be the minimal polynomial of α over Z, say $g(x) = x^r + a_{r-1}x^{r-1} + \cdots + a_1 x + a_0$; this is a divisor of f. We claim that $Z = K(a_0, \ldots, a_{r-1})$. Here the inclusion $\supseteq$ is trivial so that $[L : K(a_0, \ldots, a_{r-1})] \geq [L : Z] = r$. On the other hand, we also have $[L : K(a_0, \ldots, a_{r-1})] \leq r$, because α satisfies an equation of degree r over $K(a_0, \ldots, a_{r-1})$. This shows that $[L : Z] = [L : K(a_0, \ldots, a_{r-1})] = r$ which (together with the fact $K(a_0, \ldots, a_{r-1}) \subseteq Z$) implies that $K(a_0, \ldots, a_{r-1}) = Z$. We have therefore shown that every intermediate field of $(K(\alpha) : K)$ is of the form $Z = K(a_0, \ldots, a_{r-1})$ where $x^r + a_{r-1}x^{r-1} + \cdots + a_1 x + a_0$ is a monic divisor of f. Since f has only finitely many monic divisors, the claim holds. ∎

WE will now show that an arbitrary field extension can be split into an algebraic extension and a purely transcendental extension, a notion yet to be defined. This is quite satisfactory because it allows one to prove certain statements about general field extension by reducing them to the two extreme cases of purely transcendental extensions (whose structure is easily described) and algebraic extensions. Let us start with the following definitions which are spelled out for arbitrary extensions of commutative rings even though they will be used for field extensions only.

(12.22) Definitions. *Let $R \subseteq S$ be commutative rings and let $X \subseteq S$ be a subset of S.*

(a) *The set X is called* **algebraically generating** *for S over R if S is algebraic over $R[X]$, i.e., if every element $\alpha \in S$ satisfies an equation of the form*

$$(\star) \qquad \sum_{k=0}^{n} f_k(x_1, \ldots, x_m)\alpha^k = 0$$

where $x_1, \ldots, x_m \in X$ and $f_0, \ldots, f_n \in R[X_1, \ldots, X_m]$ with $f_n \neq 0$.

(b) *The set X is called* **algebraically independent** *over R if there is no algebraic relation over R between the elements of X, i.e., if $f(x_1, \ldots, x_n) = 0$ with $x_1, \ldots, x_n \in X$ and $f \in R[X_1, \ldots, X_n]$ implies that f is the zero polynomial $f = 0$.*

CLEARLY, a finite number of elements $x_1, \ldots, x_n$ is algebraically independent over R if and only if the surjective homomorphism $\Phi : R[X_1, \ldots, X_n] \to R[x_1, \ldots, x_n]$ given by $f \mapsto f(x_1, \ldots, x_n)$ has zero kernel, i.e., is an isomorphism.

(12.23) Examples. (a) Let K be a field. Then the elements X^3 and Y are algebraically independent over the polynomial ring $K[X, Y]$; hence $K[X^3, Y] \cong K[X, Y]$ even though $K[X^3, Y] \subsetneqq K[X, Y]$.

(b) If R is an integral domain, then the elementary symmetric polynomials $s_1, \ldots, s_n \in R[X_1, \ldots, X_n]$ are algebraically independent over R, due to the uniqueness statement in (4.23)(a).

NOTE that if $(L : K)$ is a field extension and if $X \subseteq L$ then saying that L is algebraic over $K[X]$ is tantamount to saying that L is algebraic over $K(X)$.

(12.24) Proposition. *Let $(L : K)$ be a field extension. For a subset $X \subseteq L$, the following conditions are equivalent:*

(1) *X is algebraically generating for L and algebraically independent over K;*

(2) *X is a maximal algebraically independent set over K;*

(3) *X is a minimal algebraically generating set for L.*

Proof. (1)$\Longrightarrow$(2). Let $\alpha \in L \setminus X$; then there is an equation of the form

$$\Phi(x_1, \ldots, x_m, \alpha) := \sum_{k=0}^{n} f_k(x_1, \ldots, x_m)\alpha^k = 0$$

where Φ is not the zero polynomial; hence $X \cup \{\alpha\}$ is not algebraically independent.

(2)$\Longrightarrow$(1). Let $\alpha \in L$. Then α satisfies an equation of the form $\sum_{k=0}^{n} f_k(x_1, \ldots, x_m)\alpha^k = 0$ with $x_i \in X$ since otherwise $X \cup \{\alpha\}$ would be algebraically independent over K, contradicting the maximality of X.

$(1) \Longrightarrow (3)$. Suppose that there is an element $\alpha \in X$ such that $X \setminus \{\alpha\}$ is algebraically generating. Then there is a nontrivial equation of the form $\sum_{k=0}^{n} f_k(x_1, \ldots, x_m) \alpha^k = 0$ with $x_1, \ldots, x_m \in X \setminus \{\alpha\}$ which contradicts the hypothesis that X is algebraically independent; hence our assumption that $X \setminus \{\alpha\}$ is algebraically generating was wrong.

$(3) \Longrightarrow (1)$. Suppose there is a nontrivial relation $f(x_1, \ldots, x_n) = 0$ between elements $x_1, \ldots, x_n \in X$ over K which involves x_n, say

$$f_m(x_1, \ldots, x_{n-1}) x_n^m + \cdots + f_1(x_1, \ldots, x_{n-1}) x_n + f_0(x_1, \ldots, x_{n-1}) = 0 .$$

This equation has the form $\varphi(x_n) = 0$ where φ is a polynomial with coefficients in $K(X \setminus \{x_n\})$. We claim that in this case each element $\alpha \in L$ satisfies an algebraic equation over $X \setminus \{x_n\}$. Indeed, since X is algebraically generating, each element $\alpha \in L$ satisfies an equation of the form $\sum_{k=0}^{m} f_k(x_1, \ldots, x_n) \alpha^k = 0$ with elements $x_i \in X$. If this equation involves x_n, it can be rewritten as $\psi(x_n) = 0$ where ψ has coefficients in $K((X \setminus \{x_n\}) \cup \{\alpha\})$. But then $\operatorname{Res}(\varphi, \psi) = 0$, which is a nontrivial algebraic relation for α over $K(X \setminus \{x_n\})$. Since $\alpha \in L$ was arbitrary, this shows that $X \setminus \{x_n\}$ is algebraically generating for L, contradicting the minimality of X. ∎

(12.25) Definition. *Let $(L : K)$ be a field extension. A subset $X \subseteq L$ satisfying the equivalent conditions of proposition (12.24) is called a* **transcendence basis** *of L over K.*

NOTE the similarity between the concepts of algebraically generating and independent sets and transcendence bases in the context of field extensions and the concepts of linearly generating and independent sets and bases in the context of vector spaces. (See problem 52 below for a simultaneous derivation of the basis properties in both situations.) In the same way we established the existence of vector space bases, we will now prove a proposition which guarantees the existence of transcendence bases (simply take $A = \emptyset$ and $B = L$ for a mere existence statement).

(12.26) Proposition. *Let $(L : K)$ be a field extension and let $A \subseteq B$ be subsets of L such that A is algebraically independent over K and B is algebraically generating for L. Then there is a transcendence basis X of L over K such that $A \subseteq X \subseteq B$.*

Proof. The family of all algebraically independent sets Y with $A \subseteq Y \subseteq B$ is inductively ordered by set-theoretical inclusion, hence contains a maximal element X by Zorn's lemma. Since B is algebraically generating for L, this implies that X is a maximal algebraically independent set and hence a transcendence basis of L over K. ∎

(12.27) Examples. (a) The set $\{\pi\}$ is a transcendence basis of $\mathbb{Q}(\pi, \sqrt{2})$ over $\mathbb{Q}$.

(b) Let K be a field and let $L := K((x_i)_{i \in I})$ be the field of all rational functions over K in the variables x_i. Then $\{x_i \mid i \in I\}$ is a transcendence basis of L over K. Indeed, the variables x_i are, of course, algebraically independent over K, and an arbitrary element $\alpha \in L$, say $\alpha = p(x_{i_1}, \ldots, x_{i_n})/q(x_{i_1}, \ldots, x_{i_n})$, satisfies an algebraic relation over K, namely $q(x_{i_1}, \ldots, x_{i_n}) \alpha - p(x_{i_1}, \ldots, x_{i_n}) = 0$.

(c) A field extension $(L : K)$ is algebraic if and only if the empty set $\emptyset$ is a transcendence basis of L over K.

(12.28) Exchange Lemma. *Let $(L : K)$ be a field extension and let X and Y be transcendence bases of L over K. Then for any element $x \in X$, there is an element $y \in Y$ such that the set*

$$X_{xy} := (X \setminus \{x\}) \cup \{y\}$$

is again a transcendence basis of L over K. In other words, any element of X can be replaced be a suitably chosen element of Y without violating the properties of a transcendence basis.

Proof. It is impossible that every element of Y is algebraic over $K(X \setminus \{x\})$. Otherwise $K(Y)$ would be algebraic over $K(X \setminus \{x\})$; since L is algebraic over $K(Y)$ and since being algebraic is a transitive property, this would imply that L is algebraic over $K(X \setminus \{x\})$, contradicting the fact that X is a minimal algebraically generating set. Thus there is an element $y \in Y$ which is transcendental over $K(X \setminus \{x\})$. Then the set X_{xy} formed with such an element y is algebraically independent over K.

Now if x were transcendental over $K(X_{xy})$, then $X_{xy} \cup \{x\} = X \cup \{y\}$ would be algebraically independent, contradicting the fact that X is a maximal algebraically independent set. Hence x is algebraic over $K(X_{xy})$, which implies that $K(X_{xy} \cup \{x\}) = K(X \cup \{y\})$ is algebraic over $K(X_{xy})$. Since L is algebraic over $K(X)$ – and hence the more so over $K(X \cup \{y\})$ – and since being algebraic is a transitive property, this implies that L is algebraic over $K(X_{xy})$.

We have shown that X_{xy} is algebraically independent and that L is algebraic over $K(X_{xy})$; this shows that X_{xy} is a transcendence basis of L over K. ∎

(12.29) Theorem. *Any two transcendence bases of a field extension have the same cardinality.*

Proof. Let X and Y be transcendence bases of a given field extension with $|X| \leq |Y|$. Pick index sets $I \subseteq J$ with $X = \{x_i \mid i \in I\}$ and $Y = \{y_j \mid j \in J\}$ and choose a well-ordering of J. Then if i' denotes the immediate successor of $i \in I$, we define a family $(X_i)_{i \in I}$ of transcendence bases, starting with X, by obtaining $X_{i'_0}$ from X_{i_0} by replacing x_{i_0} by a suitable element $y_{j(i_0)}$; this is possible by the exchange lemma and by the principle of transfinite induction. † Here $y_{j(i_0)}$ necessarily differs from all the elements $y_{j(i)}$ with $i < i_0$, since otherwise $X_{i'_0}$ could not be a transcendence basis; this shows that the mapping $j : I \to J$ is injective.

It is easy to verify that $\{y_{j(i)} \mid i \in I\}$ is a transcendence basis contained in Y. Since Y itself is a transcendence basis, i.e., a minimal algebraically generating set, this

---

† To understand the idea of the proof, one should consider the special case that X is finite, say $X = \{x_1, x_2, \ldots, x_m\}$; the set Y may or may not be infinite. Applying the exchange lemma for the first time, we replace x_1 by an element $y_1 \in Y$ to obtain a transcendence basis $\{y_1, x_2, \ldots, x_n\}$. In the next step, x_2 is replaced, and after n steps, we obtain a transcendence basis $\{y_1, y_2, \ldots, y_n\}$.

forces $\{y_{j(i)} \mid i \in I\} = Y$; hence $\{j(i) \mid i \in I\} = J$ so that the mapping $j : I \to J$ is also surjective and thus a bijection. We conclude that $|I| = |J|$, i.e., $|X| = |Y|$. ■

(12.30) Definition. *The* **transcendence degree** *of a field extension* $(L : K)$ *is the cardinality of a transcendence basis of L over K; it is denoted by* $\mathrm{trdeg}(L : K)$.

THIS means that $\mathrm{trdeg}(L : K)$ is the maximal number of elements one can pick in L which are algebraically independent over K. The next theorem proves a formula for the transcendence degree which is analogous to the degree formula (12.11)(b).

(12.31) Theorem. *Let* $K \subseteq Z \subseteq L$ *be fields.*
(a) *If X is a transcendence basis of Z over K and Y is a transcendence basis of L over Z, then $X \cap Y = \emptyset$, and $X \cup Y$ is a transcendence basis of L over K.*
(b) $\mathrm{trdeg}(L : K) = \mathrm{trdeg}(L : Z) + \mathrm{trdeg}(Z : K)$.

Proof. Since (b) is an immediate consequence of (a), it is enough to prove part (a). First, we have $X \subseteq Z$ and $Y \subseteq L \setminus Z$ so that $X \cap Y = \emptyset$. Second, we have to prove that L is algebraic over $K(X \cup Y)$ and that $X \cup Y$ is algebraically independent over K.

By hypothesis, Z is algebraic over $K(X)$ and hence over $K(X \cup Y)$. Therefore, $K(X \cup Y)(Z)$ is algebraic over $K(X \cup Y)$, and hence the field $Z(Y)$ (which is contained in $K(X \cup Y)(Z)$) is algebraic over $K(X \cup Y)$. Since also L is algebraic over $Z(Y)$ and since being algebraic is a transitive property, we conclude that L is algebraic over $K(X \cup Y)$.

To prove the algebraic independence of $X \cup Y$ over K, suppose that $f(x_1, \ldots, x_m, y_1, \ldots, y_n) = 0$ with $x_i \in X$ and $y_j \in Y$ where f has coefficients in K. Then the polynomial $g(Y_1, \ldots, Y_n) := f(x_1, \ldots, x_m, Y_1, \ldots, Y_n) \in Z[Y_1, \ldots Y_m]$ satisfies $g(y_1, \ldots, y_n) = 0$; then the algebraic independence of Y over Z forces $g = 0$. Now write f in the form

$$f(X_1, \ldots, X_m, Y_1, \ldots, Y_n) = \sum_i p_i(X_1, \ldots, X_m) q_i(Y_1, \ldots, Y_n) ;$$

then $g = 0$ implies $p_i(x_1, \ldots, x_m) = 0$ for all i. The algebraic independence of X over K then forces $p_i = 0$ for all i and hence $f = 0$. ■

(12.32) Definition. *A field extension* $(L : K)$ *is called* **purely transcendental** *if there is a transcendence basis X of L over K such that $L = K(X)$.*

(12.33) Examples. (a) Let $L = K\big((x_i)_{i \in I}\big)$ be the rational function field over K in the variables $(x_i)_{i \in I}$; then $(L : K)$ is a purely transcendental field extension.
(b) Let $(L : K)$ be a purely transcendental field extension, say $L = K(S)$ where $S = \{s_i\}_{i \in I}$ is a transcendence basis of L over K. Introduce variables $(x_i)_{i \in I}$ and consider the rational function field $K\big((x_i)_{i \in I}\big)$. Then the evaluation mapping

$$\begin{array}{ccc} K\big((x_i)_{i \in I}\big) & \to & L \\ f(x_{i_1}, \ldots, x_{i_n}) & \mapsto & f(s_{i_1}, \ldots, s_{i_n}) \end{array}$$

is well-defined an injective (because S is algebraically independent over K), but also surjective (because S is algebraically generating for L), hence is an isomorphism so that $L \cong K\big((x_i)_{i \in I}\big)$. This shows that the example in (a) is typical; rational function fields are the only examples for purely transcendental field extensions, up to isomorphism.

EXAMPLES (12.33)(a) and (12.33)(b) completely characterize purely transcendental field extensions; on the other hand, we have gathered a lot of information on algebraic field extension. Therefore, the next theorem is very satisfying since it reduces the study of arbitrary field extensions to the study of purely transcendental extensions and the study of algebraic extensions.

(12.34) Theorem. *Let $(L : K)$ be an arbitrary field extension. Then there is an intermediate field Z such that $(L : Z)$ is algebraic and $(Z : K)$ is purely transcendental.*

Proof. By (12.26) we can find a transcendence basis X of L over K. Then $\big(L : K(X)\big)$ is algebraic (because X is algebraically generating for L), and $\big(K(X) : K\big)$ is purely transcendental (by the very definition of a purely transcendental extension).∎

WE can now prove the converse of (12.20)(f) in the case that A is a finite set. Note that this converse was used in the proof of (12.5) already.

(12.35) Zariski's lemma. *Let $(L : K)$ be a field extension and let $\alpha_1, \ldots, \alpha_n \in L$ be such that $K(\alpha_1, \ldots, \alpha_n) = K[\alpha_1, \ldots, \alpha_n]$. Then the elements $\alpha_1, \ldots, \alpha_n$ are all algebraic over K.*

Proof. The case $n = 1$ is trivial. In the general case, suppose that not all the elements α_i are algebraic over K. After renumbering we may assume that $\alpha_1, \ldots, \alpha_r$ are algebraically independent ($r \geq 1$) and that each of the elements $\alpha_{r+1}, \ldots, \alpha_n$ is algebraic over the field $B := K(\alpha_1, \ldots, \alpha_r)$. Then $K[\alpha_1, \ldots, \alpha_n]$ is a finite algebraic extension of B and hence is a finitely generated B-module. Applying (11.21) to $K \subseteq B \subseteq K[\alpha_1, \ldots, \alpha_n]$ we conclude that B is finitely generated as a K-algebra, say $B = K[\beta_1, \ldots, \beta_s]$. Note that each β_i has the form $\beta_i = p_i(\alpha_1, \ldots, \alpha_r)/q_i(\alpha_1, \ldots, \alpha_r)$ with polynomials $p_i, q_i \in K[X_1, \ldots, X_r]$. Now since $\alpha_1, \ldots, \alpha_r$ are algebraically independent we have $K[\alpha_1, \ldots, \alpha_r] \cong K[X_1, \ldots, X_r]$ which is a unique factorization domain by (8.2). Let q be a prime factor of $q_1 \cdots q_s + 1$. Then the element $q(\alpha_1, \ldots, \alpha_r)^{-1}$ in the field B cannot be a polynomial in $\beta_1, \ldots, \beta_s$ with coefficients in K because q is prime to each of the q_j. † This is the desired contradiction. ∎

---

† If q^{-1} were a polynomial in the $\beta_i = p_i/q_i$ we could clear denominators to obtain an equation $1/q = \Phi(p_1, \ldots, p_s, q_1, \ldots, q_s)/q_1^{\alpha_1} q_2^{\alpha_2} \cdots q_s^{\alpha_s}$. This would imply $q_1^{\alpha_1} q_2^{\alpha_2} \cdots q_s^{\alpha_s} = q \Phi(p_1, \ldots, p_s, q_1, \ldots, q_s)$ so that q would have to divide one of the q_i. But this contradicts our choice of q.

WE want to conclude this section with a nice geometric application of field extensions. It is a striking feature of this application that only the most elementary properties of field extensions are needed to solve problems that remained unsolved for 2000 years; a clear indication for the enormous power of algebraic methods in solving geometrical problems.

IN early Greek mathematics two numbers were called *commensurable* if a multiple of one of them equals some multiple of the other; for example, given two sticks of different lengths, one can ask for natural numbers m and n such that m sticks of the first sort have the same length as n sticks of the second sort.† Proportions played an important role in the Greek theory of harmony, and the Greeks believed that any two numbers whatsoever are commenurable until it was found by the Pythagoreans, in the 5th or 4th century B.C., that the side and the diagonal of a square are not commensurable. †† This fact came as a shock to the Greeks and had to be kept as a secret inside the Pythagorean brotherhood. It is said that Hippasus, a member of the Pythagorean school, was thrown overboard during a voyage because he dared to betray the secret. How deeply the existence of incommensurable lengths (in modern terms: the existence of irrational numbers) moved the Greeks is also underlined by its repeated mentioning in Plato's dialogues.

From then on, the Greeks formulated all their mathematics in geometrical, not in arithmetical terms; the "numbers" of Greek mathematics were exactly those quantities that arise as measurements from geometrical constructions. This leads to the following question: Given some points in a plane, which other points can be constructed from these given points by only using a ruler and a compass? More precisely, the following steps are allowed to construct new points. We may draw a line through two existing points, and we may draw a circle whose center is an existing point and whose radius is the distance between two existing points; all points of intersection between two figures (two lines, two circles, or a circle and a line) drawn in this way can be constructed. Of course, we can iterate the process: In each step, we can use points that have been constructed in previous steps.

For example, we can
(1) bisect a line segment,
(2) drop the perpendicular to a given line from a given point,
(3) draw the parallel to a given line through a given point,
(4) double a square,
(5) bisect an arbitrary angle,
(6) trisect a right angle,
(7) construct a regular hexagon (i.e., divide a circle into six equal parts).

---

† In modern terms, if the sticks have the lengths l_1 and l_2, respectively, then $ml_1 = nl_2$ means $l_1/l_2 = n/m$; hence saying that l_1 and l_2 are commensurable is tantamount to saying that the quotient l_1/l_2 is a rational number. However, what does it mean to say that a stick has length l? Of course, we can arbitrarily define a unit length 1 and then attach the length l to a stick which is l times as long as the unit length – but what does this mean if l is not commensurable with the unit length?

†† In modern terminology this amounts to the fact that $\sqrt{2}$ is an irrational number, because the diagonal of a square with side a is $a\sqrt{2}$.

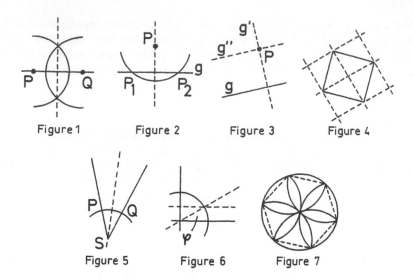

Figure 1　　　Figure 2　　　Figure 3　　　Figure 4

Figure 5　　　Figure 6　　　Figure 7

Let us quickly show that these constructions can indeed be performed.

(1) Given the segment $\overline{PQ}$, draw circles with a radius $r > \frac{1}{2}\overline{PQ}$ around P and around Q and connect the two points of intersection of these circles by a straight line. This line hits the segment $\overline{PQ}$ at its midpoint.

(2) Given a line g and a point P, draw a circle around P which intersects g in two points P_1 and P_2. Then use construction (1) to bisect the segment $\overline{P_1 P_2}$.

(3) Given P and g, first drop the perpendicular g' of g through P and then the perpendicular g'' of g' through P, using construction (2) both times. Then g'' is parallel to g.

(4) Draw the parallels to the diagonals through the endpoints of the square; these four parallels form the sides of a square twice as big as the original one.

(5) Draw a circle around the vertex S of the angle and let P and Q be its points of intersection with the two sides. Then draw circles around P and Q with a radius $r > \frac{1}{2}\overline{SP} = \frac{1}{2}\overline{SQ}$; these intersect in a point T. Then the line $\overline{ST}$ bisects the given angle.

(6) Draw a circle around the vertex, find the midpoint of one of the sides as in (1) and draw the parallel to the second side through this midpoint as in (3). The line from the vertex to the point of intersection of this parallel with the circle creates an angle of 30^0; indeed, the sine of this angle is easily seen to be $\frac{1}{2}$.

(7) Given a circle of radius r, draw a circle of the same radius r around any point on the original circle. Around the two points of intersection of the two circles, draw two more circles of radius r. Repeat this process, until six points on the original circle are constructed (see the rosette in Figure 7). These six points are the vertices of a regular hexagon.

On the other hand, there were geometric problems that the Greeks found themselves unable to solve; for example, they could not trisect arbitrary angles, duplicate a cube, square a circle, or find certain regular polygons, by only using a ruler and a compass. It turns out that all these problems are not solvable in principle. But how is it at all possible to prove that a certain construction cannot be performed?

We will proceed by translating the geometric problem of executing ruler and compass constructions into an arithmetical problem by introducing a system of cartesian coordinates as follows. Pick a line g and construct a line g' perpendicular to g; this is possible by (2) above. Denote the point of intersection of g and g' by O and pick any point $P \neq O$ on g. If we define the distance between O and P to be the unit length 1, then this amounts to creating a system of cartesian coordinates, and the location of any constructible point can be indicated as a pair (a, b) of coordinates with respect to this system.

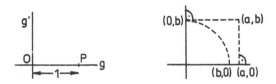

Let us note first that a point (a, b) is constructible if and only if the points $(a, 0)$ and $(0, b)$ are constructible (see the picture on the right). So it makes no difference whether we talk about the constructability of points in the plane $\mathbb{R}^2$ or about the constructability of real numbers. (Here a number x is called constructible if the point $(x, 0)$ is constructible.) Then we can give the following algebraic characterization of constructible numbers.

(12.36) Theorem. *Let A be a subset of $\mathbb{R}$ containing 0 and 1 and let $\mathrm{Const}(A)$ be the set of all numbers which can be constructed only using a ruler and a compass, starting from the given set A.*

(a) *If $a, b \in \mathbb{R}$ belong to $\mathrm{Const}(A)$, then so do $-a$, $a + b$, $1/a$ (if $a \neq 0$), ab and $\sqrt{|a|}$. In particular, $\mathrm{Const}(A)$ is a subfield of $\mathbb{R}$ containing $\mathbb{Q}(A)$.*

(b) *A number $x \in \mathbb{R}$ belongs to $\mathrm{Const}(A)$ if and only if there is an increasing chain of fields*

$$K_0 := \mathbb{Q}(A) \subseteq K_1 \subseteq \cdots \subseteq K_m \quad \text{with} \quad x \in K_m \quad \text{and}$$
$$K_{i+1} = K_i(\sqrt{r_i}) \text{ for some element } r_i > 0 \text{ in } K_i.$$

In particular, if x is constructible, then x is contained in an extension L of K_0 such that the degree $[L : K_0]$ is a power of 2.

(c) *If x is transcendental over $\mathbb{Q}(A)$, then x cannot belong to $\mathrm{Const}(A)$.*

(d) *Let $p \in \mathbb{Q}(A)[x]$ be an irreducible polynomial whose degree n is not a power of 2. Let $\alpha \in \mathbb{C}$ be a root of p. Then α does not belong to $\mathrm{Const}(A)$.*

Proof. (a) To construct $-a$ from a, draw a circle with radius $|a|$ around $(0, 0)$ (see Figure 1 below). To construct $a + b$ from a and b, draw a circle with radius $|b|$ around $(a, 0)$ (see Figure 2 below). To construct $1/a$ from $a > 0$, draw the parallel to $\overline{(1, 0)(0, a)}$ through $(0, 1)$; with the notation of Figure 3, the intercept theorem yields $x/1 = 1/a$, i.e., $x = 1/a$. To construct ab from a, $b > 0$, draw the parallel to $\overline{(a, 0)(0, 1)}$ through $(0, b)$; with the notation of Figure 4, the intercept theorem yields $a/1 = x/b$, i.e., $x = ab$. Finally, to construct $\sqrt{a}$ from $a > 0$, bisect the segment $\overline{(0, 0)(a + 1, 0)}$ and draw a circle with radius $\frac{1}{2}(a+1)$ around the midpoint. Dropping the perpendicular from the point $(1,0)$ as in Figure 5, the altitude theorem yields $h^2 = 1 \cdot a$, i.e., $h = \sqrt{a}$.

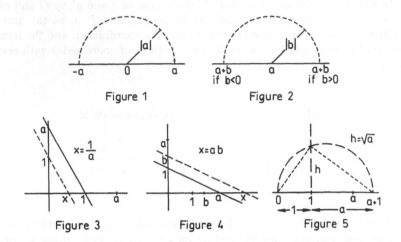

Figure 1   Figure 2

Figure 3   Figure 4   Figure 5

(b) A number x is constructible if and only if x occurs as one of the coordinates of a point of intersection between two lines, two circles, or a line and a circle that are obtained from already existing points. Some elementary analytic geometry shows that x then satisfies a linear equation if we intersect two straight lines, a quadratic equation if we intersect two circles or a circle and a straight line; in each case, the coefficients of the equation are rational expressions in points that are already constructed. So x itself can be written as a rational expression in already constructed numbers and square roots of those numbers. But this says exactly that the constructability of a number is equivalent to the existence of a chain of fields of the indicated type. We have $[K_{i+1} : K_i] = 1$ or 2, depending on whether $\sqrt{r_i}$ lies in K_i already or not. So if $L := K_m$, then $x \in L$ and

$$[L : K_0] = [K_m : K_{m-1}] \cdots [K_1 : K_0] ,$$

where the right-hand side is clearly a power of 2.

(c) This is an immediate consequence of part (b) since $[\mathbb{Q}(A)(x) : \mathbb{Q}(A)]$ is infinite if x is transcendental over $\mathbb{Q}(A)$.

(d) Since p is irreducible over $K := \mathbb{Q}(A)$, we have $[K(\alpha) : K] = \deg p = n$. Suppose that α is constructible from A. Then α is contained in some extension L of K whose degree is a power of 2, say 2^m. But then

$$2^m = [L : K] = [L : K(\alpha)] \cdot [K(\alpha) : K] = n \cdot [L : K(\alpha)] .$$

But this is clearly impossible, since the left-hand side is a power of 2, whereas the right-hand side is not (by assumption, n possesses a prime divisor different from 2). ■

(12.37) Examples. (a) (**Squaring a circle.**) There is no ruler and compass construction that would allow one to transform a circle of radius $r = 1$ into a square with the same area. Indeed, the square would have to have the area $\pi r^2 = \pi$, hence the edge $\sqrt{\pi}$. But $\sqrt{\pi}$ is transcendental over $\mathbb{Q}$ (because π is) and hence not constructible from $\mathbb{Q}$, due to (12.36)(c).

(b) (**Rectifying a circle arc.**) It is not possible with a ruler and a compass alone to transform the circumference of a circle with radius $r = 1$ into a line segment with the same length. Indeed, the line segment would have to have the length $2\pi r = 2\pi$ which is transcendental over $\mathbb{Q}$ and hence not constructible by (12.36)(c).

(c) Let $p \in K[x]$ be an irreducible polynomial whose degree n is not a power of 2. Let α be a root of p (in some extension of K). Then α is not constructible from K. In fact, since p is irreducible over K, we have

$$[K(\alpha) : K] = \text{degree of } p = n .$$

Suppose that α is constructible from K. Then α is contained in some extension L of K whose degree is a power of 2. But then

$$2^m = [L : K] = [L : K(\alpha)] \cdot [K(\alpha) : K] = n \cdot [L : K(\alpha)] .$$

But this is clearly impossible, since the left-hand side is a power of 2, whereas the right-hand side is not (by assumption, n possesses a prime divisor different from 2).

(c) (**Trisection of an angle.**) There is no method which allows one to trisect an arbitrary angle by only using a ruler and a compass.† Indeed, $\frac{\varphi}{3}$ can be constructed from φ if and only if $\cos \frac{\varphi}{3}$ can be constructed from $\cos \varphi$. Now by a well-known trigonometric formula, $\cos \frac{\varphi}{3}$ is a root of the polynomial

$$p(x) := 4x^3 - 3x - \cos \varphi \in \mathbb{Q}(\cos \varphi)[x] .$$

If p is irreducible over $\mathbb{Q}(\cos \varphi)$ then $\cos \frac{\varphi}{3}$ cannot be constructed from $\mathbb{Q}(\cos \varphi)$ by (12.36)(d) because the degree of p is 3 and not a power of 2. Choose for example $\varphi = 60^0$; then $p(x) = 4x^3 - 3x - \frac{1}{2}$ is irreducible over $\mathbb{Q}(\cos 60^0) = \mathbb{Q}$ by (8.23).

(d) (**Delian problem: Doubling a cube.**) There is a legend that the people of Athens were troubled by a plague in 430 B.C. and that they were told by the oracle at Delos that, to rid the city of this plague, they had to double the size of Apollo's cube-shaped altar, preserving the form of a cube. This amounts to the following mathematical problem: Given a cube with a unit edge 1, is it possible to construct a cube of twice the volume by only using a ruler and a compass? If x is the edge of the desired cube, its volume ought to be $x^3 = 2$ so that $x = \sqrt[3]{2}$. But $x = \sqrt[3]{2}$ is not constructible from $\mathbb{Q}$ by (12.36)(d) because the polynomial $x^3 - 2$ is irreducible over $\mathbb{Q}$. Thus there is no ruler and compass construction by which one can double a cube. (We saw before that the corresponding two-dimensional problem – transforming a given square into a square of the double area – is easily solved.)

---

† This is in contrast to the fact that an arbitrary angle can be bisected only using a ruler and a compass. Note, however, that there are special angles (for example 90^0) which can be trisected.

Exercises

Problem 1. (a) Show that the prime field of a field K consists of all quotients $\frac{m \cdot 1}{n \cdot 1}$ with $m, n \in \mathbb{Z}$ and $n \cdot 1 \neq 0$.

(b) Show that the prime field of a field K is isomorphic to $\mathbb{Q}$ if $\operatorname{char} K = 0$ and isomorphic to $\mathbb{Z}_p$ if $\operatorname{char} K = p$.

(c) Show that if $(L : K)$ is a field extension then L and K have the same prime field (and hence the same characteristic).

Problem 2. Let R be an integral domain and let $K \subseteq R$ be a subfield of R. Show that if R is finite-dimensional as a vector space over K then R is a field.

Hint. We have to show that every nonzero element $a \neq 0$ in R is invertible. Now if $a \neq 0$, then the left-multiplication $l_a : r \mapsto ar$ is an injective vector space endomorphism, hence also surjective since $\dim_K R < \infty$.

Problem 3. Let $(L : K)$ be a field extension.

(a) Show that an element $\alpha \in L$ is algebraic over K if and only if $\alpha^{-1} \in K[\alpha]$.

(b) Show that $(L : K)$ is algebraic if and only if every ring R with $K \subseteq R \subseteq L$ is a field.

(c) Let Z_1 and Z_2 be intermediate fields of $(L : K)$. Show that $K[Z_1 \cup Z_2]$ is an integral domain with quotient field $K(Z_1 \cup Z_2)$. Show that if Z_1 and Z_2 are algebraic over K then $K[Z_1 \cup Z_2] = K(Z_1 \cup Z_2)$.

Problem 4. Let K be an infinite field and let $(L : K)$ be a field extension with $K \subsetneqq L$. Show that the quotient group $L^\times / K^\times$ is infinite.†

Hint. Show that if $L^\times / K^\times$ were finite, then the K-vector space L would be a finite union of one-dimensional subspaces.

Problem 5. Are there numbers $x, y \in \mathbb{R} \setminus \mathbb{Q}$ such that $\mathbb{Q}(x) \cong \mathbb{Q}(ix)$ and $\mathbb{Q}(y) \not\cong \mathbb{Q}(iy)$?

Problem 6. Let K be a field and let k be an element of K which is not a square in K. Show that $\left\{ \begin{pmatrix} a & kb \\ b & a \end{pmatrix} \mid a, b \in K \right\}$ is a field which is isomorphic with $K(\sqrt{k})$.

Problem 7. (a) Show that if $\operatorname{char} K \neq 2$ then $K(\sqrt{a}, \sqrt{b}) = K(\sqrt{a} + \sqrt{b})$; for example, $\mathbb{Q}(i, \sqrt{2}) = \mathbb{Q}(i + \sqrt{2})$ and $\mathbb{Q}(\sqrt{3}, \sqrt{5}) = \mathbb{Q}(\sqrt{3} + \sqrt{5})$. Use this result to find the minimal polynomials of $i + \sqrt{2}$ and of $\sqrt{3} + \sqrt{5}$ over $\mathbb{Q}$.

(b) Let $a > 0$ and $b \neq 0$ be rational numbers. Show that

$$\mathbb{Q}(\sqrt{a}, \sqrt[3]{b}) = \mathbb{Q}(\sqrt{a} + \sqrt[3]{b}) = \mathbb{Q}(\sqrt{a} \cdot \sqrt[3]{b}) .$$

If $a, b \in \mathbb{N}$ are prime numbers, find the minimal polynomial of $\sqrt{a} + \sqrt[3]{b}$ and $\sqrt{a} \cdot \sqrt[3]{b}$ over $\mathbb{Q}$.

---

† It is even true that $L^\times / K^\times$ is not finitely generated. See A. Brandis, *Über die multiplikative Struktur von Körpererweiterungen*, Math. Zeitschrift **87** (1985).

(c) Let $L = K(\alpha, \beta)$ and suppose that there are coprime natural numbers $m, n \in \mathbb{N}$ such that α^m and β^n lie in K. Show that $K(\alpha, \beta) = K(\alpha\beta)$. For example, $\mathbb{Q}(\sqrt[3]{2}, \sqrt[4]{5}) = \mathbb{Q}(\sqrt[3]{2} \cdot \sqrt[4]{5})$. (**Hint.** There are integers $x, y \in \mathbb{Z}$ with $xm + yn = 1$.)

Problem 8. (a) Suppose that $\alpha \in \mathbb{C}$ satisfies the equation $\alpha^2 - 4\alpha + 13 = 0$. Show that $\mathbb{Q}(\alpha) = \mathbb{Q}(i)$.
(b) Suppose that $\alpha \in \mathbb{C}$ satisfies the equation $\alpha^3 + \alpha + 1 = 0$. Show that $\mathbb{Q}(\alpha, \sqrt{2}) = \mathbb{Q}(\alpha\sqrt{2})$.

Problem 9. Find the minimal polynomials over $\mathbb{Q}$ of the following complex numbers.

$$\sqrt{2} + \sqrt{5}, \quad \sqrt{2} + \sqrt{3} + \sqrt{5}, \quad \sqrt{2} + \sqrt[5]{5}, \quad \sqrt{2} + \sqrt[5]{3}, \quad \sqrt{3} + \sqrt[5]{3}, \quad \sqrt[3]{3} + \sqrt[4]{2},$$

$$\sqrt[4]{2} + \sqrt[4]{3}, \quad \sqrt[3]{2} - \sqrt[3]{7}, \quad (1 + \sqrt[3]{3})\sqrt[4]{5}, \quad \sqrt[5]{2}\sqrt[4]{3}, \quad \sqrt[7]{5} + 2, \quad \sqrt{2 + \sqrt[3]{2}}, \quad \sqrt[3]{2} + i\sqrt[5]{2}$$

Problem 10. Find the minimal polynomials of $2 + \sqrt{2}$, $\sqrt{2 + \sqrt{2}}$ and $\sqrt{2 + \sqrt{2 + \sqrt{2}}}$ over $\mathbb{Q}$.

Problem 11. (a) Let n be a natural number which is not a square. Find the minimal polynomial of $\sqrt{n} + i$ over $\mathbb{Z}$, over $\mathbb{Q}$, over $\mathbb{R}$ and over $\mathbb{C}$.
(b) For any $\alpha \in \mathbb{Q}$ find the minimal polynomial of $\sqrt[3]{\alpha} + \sqrt[3]{\alpha^2}$ over $\mathbb{Q}$.
(c) Suppose that $\alpha \in \mathbb{C}$ and $\beta \in \mathbb{C}$ satisfy the equations $\alpha^3 - 2\alpha + 1 = 0$ and $2\beta^3 + \beta^2 - 1 = 0$. Find the minimal polynomial of $\alpha + \beta$ over $\mathbb{Q}$.

Problem 12. (a) Suppose $f \in K[x]$ and $g \in K[x]$ are the minimal polynomials of two elements α and β, respectively, in some extension of K. Let $m := \deg f$ and $n := \deg g$. Then the minimal polynomials of $\alpha + \beta$ and $\alpha\beta$ have a degree $\leq mn$, respectively. If m and n are relatively prime, then the minimal polynomial of $\alpha\beta$ has degree exactly mn.
(b) Let $\alpha, \beta \in \mathbb{C}$ such that $\alpha^2 + \alpha - 1 = 0$ and $2\beta^3 + \beta^2 - 1 = 0$. Find the minimal polynomials of $1 + \beta$ and of $\alpha + \beta$.
(c) Suppose that $\alpha, \beta \in \mathbb{C}$ satisfy the equations $\alpha^3 + \alpha + 1 = 0$ and $\beta^2 + \beta - 3 = 0$. Find the minimal polynomials of $\alpha + \beta$ and $\alpha\beta$ over $\mathbb{Q}$.

Problem 13. In each of the following cases determine the minimal polynomial p of α over $\mathbb{Q}$ and a basis of $\mathbb{Q}(\alpha)$ over $\mathbb{Q}$. Also, find out which complex roots of p lie in $\mathbb{Q}(\alpha)$.

$$\text{(a) } \alpha = (1 + i)\sqrt[4]{5} \qquad \text{(b) } \alpha = \sqrt{2} + i \qquad \text{(c) } \alpha = \sqrt{2 + \sqrt[3]{5}}$$

Problem 14. (a) Suppose that α satisfies the equation $\alpha^6 + \alpha + 1$ over $\mathbb{Z}_2$. Find the minimal polynomial of α^{21} over $\mathbb{Z}_2$.
(b) Let $K := \mathbb{Z}_2$ and $L := \mathbb{Z}_2[x]/\langle x^3 + x + 1\rangle$. Find the minimal polynomial over K of each of the elements of L.

Problem 15. (a) Prove directly that if α is algebraic then so is α^2.

(b) Show that α is transcendental over a field K if and only if $f(\alpha)$ is transcendental for all nonconstant polynomials $f \in K[x]$.

(c) Suppose that $\alpha \in \mathbb{C} \setminus \{0\}$ is algebraic over $\mathbb{Q}$. Show that the minimal polynomials of α and $1/\alpha$ over $\mathbb{Q}$ have the same degree.

Problem 16. Let $(L : K)$ be a field extension and let α and β be elements of L.

(a) Show that $K(\alpha) \cong K(\beta)$ with an isomorphism leaving K pointwise fixed if and only if either α and β are both transcendental or else α and β are both algebraic over K and have the same minimal polynomial over K.

(b) Suppose that char $K \neq 2$ and that $a := \alpha^2$ and $b := \beta^2$ lie in $K \setminus \{0\}$. Show that there is an isomorphism $\sigma : K(\alpha) \to K(\beta)$ leaving K pointwise fixed if and only if $K(\alpha) = K(\beta)$ which, in turn, is the case if and only if a/b is a square in K.

Problem 17. For a polynomial $p(x) = a_0 + a_1 x + \cdots + a_n x^n \in \mathbb{Z}[x]$ with $a_n \neq 0$, let $h(p) := n + |a_0| + |a_1| + \cdots + |a_n|$. Prove the following statements!

(a) For any $N \in \mathbb{N}$ the set A_N of all $x \in \mathbb{R}$ with $p(x) = 0$ for some $p \in \mathbb{Z}[x] \setminus \{0\}$ with $h(p) \leq N$ is finite.

(b) The set of all real numbers which are algebraic over $\mathbb{Q}$ is exactly $\bigcup_{N \in \mathbb{N}} A_N$; this set is uncountably infinite.

(c) There are denumerably many real numbers which are transcendental over $\mathbb{Q}$. (Even though this result guarantees the existence of an abundance of transcendental numbers, it is usually very hard to prove that a given number (such as e or π) is transcendental.)

Problem 18. Let $(L : K)$ be an algebraic field extension. Prove the following statements!

(a) If K is infinite then L and K have the same cardinality.

(b) If K is finite then L is either finite or denumberable.

Hint. Associate with each irreducible monic polynomial $f \in K[x]$ its roots in L; this covers all elements of L with no duplication.

Problem 19. Find the four roots $\alpha_i \in \mathbb{C}$ of $f(x) = x^4 - 2x^2 + 2$. Find all the fields $\mathbb{Q}(\alpha_i, \alpha_j)$ which are obtained from $\mathbb{Q}$ by adjoining two of the roots of f and determine their degrees over $\mathbb{Q}$.

Problem 20. In each of the following cases let $\alpha \in \mathbb{C}$ be a root of f. Show that f is irreducible in $\mathbb{Q}[x]$ and find polynomials $p, q \in \mathbb{Q}[x]$ with $pf + qg = 1$. Moreover, represent $g(\alpha)^{-1}$ as a linear combination of powers of α.

(a) $f(x) = x^5 + x^3 + 1$, $g(x) = x^2 + 1$

(b) $f(x) = x^3 + 3x^2 + 6x + 30$, $g(x) = x^2 + 6$

Problem 21. (a) For arbitrary rational number $a, b, c, d \in \mathbb{Q}$, express the inverse of $a + b\sqrt{2} + c\sqrt{3} + d\sqrt{6}$ in $\mathbb{Q}(\sqrt{2}, \sqrt{3})$ with respect to the basis $(1, \sqrt{2}, \sqrt{3}, \sqrt{6})$ of $\mathbb{Q}(\sqrt{2}, \sqrt{3})$ over $\mathbb{Q}$.

(b) Let $\alpha := \sqrt[5]{3}$. Argue that $(1, \alpha, \alpha^2, \alpha^3, \alpha^4)$ is a basis of $\mathbb{Q}(\alpha)$ over $\mathbb{Q}$ and express $(1 + 6\alpha^3 + 4\alpha^6)(4 + 2\alpha - \alpha^2)^{-1}$ with respect to this basis.

(c) Suppose that $\alpha \in \mathbb{C}$ satisfies the equation $\alpha^3 - \alpha + 1 = 0$. Find the inverse of $2\alpha^2 - 2\alpha + 1$ in the form $a_0 + a_1\alpha + a_2\alpha^2$ and find the minimal polynomial of $2\alpha^2 - \alpha + 1$ over $\mathbb{Q}$.

(d) Suppose that $\beta \in \mathbb{C}$ satisfies the equation $\beta^3 - 6\beta^2 + 9\beta + 3 = 0$. Argue that $(1, \beta, \beta^2)$ is a basis of $\mathbb{Q}(\beta)$ over $\mathbb{Q}$ and express $\beta^5, 3\beta^4 - 2\beta + 1$ and $(\beta + 2)^{-1}$ with respect to this basis.

Problem 22. Let $(L : K)$ be a field extension and let V be a vector space over L. By restricting the scalar multiplication to K, we can consider V as a vector space over K. Show that $\dim_K V = [L : K] \cdot \dim_L V$.

Remark. This result is clearly a generalization of the degree formula (12.11)(b).

Problem 23. (a) Find the degrees of $\sqrt{2} + \sqrt{3}$ and of $\sqrt{2} \cdot \sqrt{3}$ over $\mathbb{Q}$.

(b) Find $[\mathbb{Q}(i, \sqrt{2}) : \mathbb{Q}]$.

(c) Let $\alpha := \sqrt{7} \cdot \sqrt[3]{2}$. Show that $\mathbb{Q}(\alpha) = \mathbb{Q}(\sqrt{7}, \sqrt[3]{2})$, determine the degree of $\mathbb{Q}(\alpha)$ over $\mathbb{Q}$ and find the minimal polynomial of α over $\mathbb{Q}$.

(d) Find an element α with $\mathbb{Q}(\sqrt{2}, \sqrt[3]{5}) = \mathbb{Q}(\alpha)$ and characterize those elements $\alpha \in \mathbb{Q}(\sqrt{2}, \sqrt[3]{5})$ with $\mathbb{Q}(\sqrt{2}, \sqrt[3]{5}) \neq \mathbb{Q}(\alpha)$.

(e) Determine the degree of $\mathbb{Q}(\sqrt[4]{2}, \sqrt[4]{3})$ over $\mathbb{Q}$. Is $\mathbb{Q}(\sqrt[4]{2}, \sqrt[4]{3}) = \mathbb{Q}(\sqrt[4]{2} + \sqrt[4]{3})$?

(f) Let $L := \mathbb{Q}(i, \sqrt{3}, e^{2\pi i/3})$. Determine the degree and a basis of L over $\mathbb{Q}$.

(g) Determine the numbers $[\mathbb{Q}(\sqrt{3}, \sqrt[4]{2}) : \mathbb{Q}]$ and $[\mathbb{Q}(\sqrt{3}\sqrt[4]{2}) : \mathbb{Q}]$. Does the equality $\mathbb{Q}(\sqrt{3}, \sqrt[4]{2}) = \mathbb{Q}(\sqrt{3}\sqrt[4]{2})$ hold? Find an element α such that $\mathbb{Q}(\sqrt{3}, \sqrt[4]{2}) = \mathbb{Q}(\alpha)$.

Problem 24. Let $K = \mathbb{Q}(\varepsilon)$ where $\varepsilon := e^{\pi i/4}$. Determine $[K(\sqrt{2}) : K]$ and $[K(\sqrt{3}) : K]$.

Problem 25. Suppose that $\alpha, \beta \in \mathbb{C}$ satisfy the equations $\alpha^3 - \alpha + 1 = 0$ and $\beta^2 - \beta - 1 = 0$. Find an element $\gamma \in \mathbb{C}$ with $\mathbb{Q}(\alpha, \beta) = \mathbb{Q}(\gamma)$ and determine the degree $[\mathbb{Q}(\alpha, \beta) : \mathbb{Q}]$. Do the same if $\alpha^3 - 2\alpha + 3 = 0$ and $\beta^2 + \beta + 2 = 0$.

Problem 26. Suppose that α and β are algebraic of degrees m and n. Show that if m and n are coprime then $\alpha\beta$ is algebraic of degree mn.

Problem 27. Let a be an element of a field K and let $m, n \in \mathbb{N}$ be coprime.

(a) Show that α is a root of $x^{mn} - a$ if and only if α^m is a root of $x^n - a$.

(b) Let α be a root of $x^{mn} - a$ in some extension of K. Show that if $x^m - a$ and $x^n - a$ are irreducible then $[K(\alpha^n) : K] = m$, $[K(\alpha^m) : K] = n$ and $[K(\alpha^{mn}) : K] = mn$.

(c) Show that $x^{mn} - a$ is irreducible if and only if $x^m - a$ and $x^n - a$ are irreducible.

Problem 28. In each of the following cases show that $L = \mathbb{Q}(\alpha)$, determine $[L : \mathbb{Q}]$ and find a basis of L over $\mathbb{Q}$ as well as the minimal polynomial of α over $\mathbb{Q}$.
    (a) $L = \mathbb{Q}(i, \sqrt[3]{3})$, $\alpha = i + \sqrt[3]{3}$
    (b) $L = \mathbb{Q}(i, \sqrt{6})$, $\alpha = i + \sqrt{6}$
    (c) $L = \mathbb{Q}(\sqrt{3}, \sqrt[3]{5})$, $\alpha = \sqrt{3}\sqrt[3]{5}$
    (d) $L = \mathbb{Q}(\sqrt{2}, \sqrt{3}, \sqrt{5})$, $\alpha = \sqrt{2} + \sqrt{3} + \sqrt{5}$
    (e) $L = \mathbb{Q}(\sqrt{3}, \sqrt[4]{2})$, $\alpha = \sqrt{3} + \sqrt[4]{2}$

Problem 29. Suppose $\alpha \in \mathbb{C}$ satisfies $\alpha^2 = 1 + i$. Show that each field in the chain $\mathbb{Q} \subseteq \mathbb{Q}(i\sqrt{2}) \subseteq \mathbb{Q}(i, \sqrt{2}) \subseteq \mathbb{Q}(\sqrt{2}, \alpha)$ is an extension of degree 2 of its predecessor.

Problem 30. (a) Find a subfield L of $\mathbb{C}$ such that $[L : (L \cap \mathbb{R})] > 2$.
    (a) Let $(L : K)$ be a field extension and let $\alpha \in L$. Show that if $[K(\alpha) : K] = 2$ and $K \cap L = K$ then $[L(\alpha) : L] = 2$.

Problem 31. Show that $\mathbb{C}$ is (up to isomorphism) the only finite extension of $\mathbb{R}$, using the classification of the irreducible elements of $\mathbb{R}[x]$ given in (8.24)(b).

Problem 32. (a) Show that if f $[L : K]$ is a prime number, then the extension $(L : K)$ has no proper intermediate fields. Find all elements $\alpha \in L$ such that $L = K(\alpha)$.
    (b) Suppose $[K(\alpha) : K]$ is odd. Show that $K(\alpha) = K(\alpha^2)$.
    Hint. Consider the chain $K \subseteq K(\alpha^2) \subseteq K(\alpha)$.

Problem 33. (a) Let L be the field of all complex numbers which are algebraic over $\mathbb{Q}$. Show that if p is any prime, then $\sqrt[n]{p} \in L$ for any $n \in \mathbb{N}$. Conclude that $(L : \mathbb{Q})$ is an infinite algebraic extension.
    (b) Show that $(\mathbb{Q}(\sqrt[2]{2}, \sqrt[3]{2}, \sqrt[4]{2}, \sqrt[5]{2}, \ldots) : \mathbb{Q})$ is an algebraic extension of infinite degree.
    (c) Let $p_1, p_2, p_3, \ldots$ be the sequence of prime numbers and let $L_n := \mathbb{Q}(\sqrt{p_1}, \sqrt{p_2}, \ldots, \sqrt{p_n}) \subseteq \mathbb{R}$. Show by induction on n that $\sqrt{p_{n+1}} \notin L_n$ and that $[L_{n+1} : L_n] = 2$ for all n. Conclude that $[L_n : \mathbb{Q}] = 2^n$ for all n.

Problem 34. (a) Let $(L : K)$ be a field extension. Show that if there is a natural number n such that $[Z : K] \leq n$ for all intermediate fields $K \subseteq Z \subsetneq L$ then $[L : K] < \infty$.
    (b) Show that if $(L : K)$ is an infinite algebraic extension then for any $n \in \mathbb{N}$ there is an intermediate field Z such that $(Z : K)$ is a finite extension with $[Z : K] \geq n$.

Problem 35. Let K_0 be a field of prime characteristic p, let $L := K_0(x, y)$ be the rational function field in two variables x, y over K_0 and let $K := K_0(x^p, y^p)$ be the subfield of all rational functions in x^p and y^p. Show that the intermediate fields $K(x + ky)$ with $k \in K$ are pairwise different and conclude that $(L : K)$ is a field extension of finite degree p^2 which has an infinite number of intermediate fields.

Problem 36. Let α be transcendental over a field K.

(a) Show that every element in $K(\alpha) \setminus K$ is also transcendental over K.

(b) Show that $K(\alpha^2) \cong K(\alpha)$ even though $K(\alpha^2) \subsetneq K(\alpha)$.

(c) Show that if $K \subsetneq Z \subseteq K(\alpha)$ then α is algebraic over Z.

(d) Show that if $n > 1$ then α^n is transcendental over K with $K(\alpha^n) \neq K(\alpha)$.

(e) Show that if $n > 1$ then $K(\alpha) \supset K(\alpha^n) \supset K(\alpha^{n^2}) \supset K(\alpha^{n^3}) \supset \cdots$ is a strictly decreasing series of fields which does not reach K. In particular, the extension $(K(\alpha) : K)$ has infinitely many intermediate fields.

(f) Show that if $\beta^n = \alpha$ then β is also transcendental over K.

Hint. Consider the chain $K \subseteq K(\alpha) \subseteq K(\beta)$.

(g) Show that if α is algebraic over $K(\beta)$ then β is algebraic over $K(\alpha)$.

Problem 37. Let $K(x)$ be the rational function field over a field K; then $Z := K(x^3 + x + 1, x^{-2} + x^2)$ is a subfield of $K(x)$. Find a rational function θ such that $Z = K(\theta)$.

Problem 38. Show that a rational function $\theta = \theta(x)$ satisfies $K(\theta) = K(x)$ if and only if it has the form $\theta(x) = \frac{ax+b}{cx+d}$ with $ad - bc \neq 0$.

Problem 39. Show that if $(L : K)$ is a finite field extension then $[L(x) : K(x)] = [L : K]$.

Problem 40. Suppose that $\alpha \in \mathbb{C}$ satisfies the equation $\alpha^3 + \alpha + 2 = 0$. Find $\mathrm{tr}_{\mathbb{Q}}^{\mathbb{Q}(\alpha)}(\alpha^2 - \alpha + 1)$ and $N_{\mathbb{Q}}^{\mathbb{Q}(\alpha)}(\alpha^2 - \alpha + 1)$.

Problem 41. Let tr and N be the trace and the norm of the field extension $(\mathbb{Q}(\sqrt{2}, \sqrt{3}) : \mathbb{Q})$. Find $\mathrm{tr}(x)$ and $N(x)$ for all $x \in \{\sqrt{2}, \sqrt{3}, \sqrt{2} + \sqrt{3}, \sqrt{2}, \sqrt{3}\}$.

Problem 42. Let Z_1 and Z_2 be intermediate fields of the field extension $(L : K)$ and let $Z_1 Z_2 = Z_1(Z_2) = Z_2(Z_1)$ be their compositum. Prove the following statements!

(a) If $[Z_1 : K] < \infty$ then $[Z_1 Z_2 : Z_2] \leq [Z_1 : K]$.

Hint. Let $R := K[Z_1, Z_2]$ be the K-subalgebra of L generated by $Z_1 \cup Z_2$. Show that every vector space basis of Z_1 over K generates R as a vector space over Z_2. Conclude that $[R : Z_2] \leq [Z_1 : K] < \infty$ so that R is a field; hence $R = Z_1 Z_2$.

(b) If $(Z_1 : K)$ is algebraic then so is $(Z_1 Z_2 : Z_2)$.

(c) If $(Z_1 : K)$ and $(Z_2 : K)$ are algebraic, then so is $(Z_1 Z_2 : K)$.

(d) $n := [Z_1 Z_2 : K]$ is finite if and only if $n_1 := [Z_1 : K]$ and $n_2 := [Z_2 : K]$ are finite; in this case $n \leq n_1 n_2$, and n is a common multiple of n_1 and n_2. If n_1 and n_2 are coprime, then $[Z_1 Z_2 : K] = n_1 n_2$.

Hint. Consider the chains $K \subseteq Z_i \subseteq Z_1 Z_2$ $(i = 1, 2)$.

(e) Show that $n = n_1 n_2 < \infty$ implies that $Z_1 \cap Z_2 = K$.

Problem 43. Let $\alpha := \sqrt[3]{2}$ and $\beta := \varepsilon\sqrt[3]{2}$ where $\varepsilon := \frac{1}{2}(-1+i\sqrt{3}) = e^{2\pi i/3}$. Prove that

$$\mathbb{Q}(\alpha) \cap \mathbb{Q}(\beta) = \mathbb{Q}, \quad [\mathbb{Q}(\alpha):\mathbb{Q}] = [\mathbb{Q}(\beta):\mathbb{Q}] = 3,$$
$$[\mathbb{Q}(\alpha,\beta):\mathbb{Q}(\beta)] = 2, \quad [\mathbb{Q}(\alpha,\beta):\mathbb{Q}] = 6.$$

(**Remark.** This shows that $[Z_1 Z_2 : Z_2]$ need not be a divisor of $[Z_1 : \mathbb{Q}]$ in problem 42(a) and that the implication converse to that in problem 42(e) does not hold.)

Problem 44. Let α and β be arbitrary elements in some extension L of the field K. Let $[K(\alpha):K] = m$ and $[K(\beta):K] = n$. Show that

$$[K(\alpha,\beta):K(\beta)] = m \quad \text{if and only if} \quad [K(\alpha,\beta):K(\alpha)] = n$$

and that these equivalent statements hold if m and n are relatively prime. Also, if m and n are relatively prime, then $[K(\alpha,\beta):K] = mn$.

Problem 45. Let α,β be algebraic elements over K with minimal polynomials $f,g \in K[x]$, respectively. Show that f is irreducible over $K(\beta)$ if and only if g is irreducible over $K(\alpha)$.

Problem 46. Suppose that $p := [K(\alpha) : K]$ and $q := [K(\beta) : K]$ are prime numbers with $p > q$ and $\operatorname{char} K \neq p$. Show that $K(\alpha,\beta) = K(\alpha + \beta)$. For example, $\mathbb{Q}(\sqrt[3]{2}, \sqrt[4]{5}) = \mathbb{Q}(\sqrt[3]{2} + \sqrt[4]{5})$.

Hint. Let h be the minimal polynomial of $\alpha + \beta$ over K. If the claim were false, then the minimal polynomial of α over K would be $f(x) := h(x + \beta)$.

Problem 47. Let $m,n \in \mathbb{N}$ be relatively prime and let p be a prime number.
(a) Show that the two polynomials $f(x) = x^n \pm p^m$ are irreducible in $\mathbb{Q}[x]$.
Hint. Since $g(x) = x^n \pm p$ is irreducible by Eisenstein's criterion, we can pick a root of $\beta \in \mathbb{C}$ with $[\mathbb{Q}(\beta) : \mathbb{Q}] = n$. Then one of the two numbers $\alpha = \pm\beta^m$ is a root of f, and $\mathbb{Q}(\alpha) = \mathbb{Q}(\beta)$.
(b) Let $f(x) = x^n + up^m$ where p where $u \in \mathbb{Z}$ is not divisible by p. Show that f is irreducible.

Problem 48. Suppose that $x,y \in \mathbb{R}$ are algebraically independent over $\mathbb{Q}$ and let $K := \mathbb{Q}(x + iy) \subseteq L := \mathbb{Q}(i,x,y) \subseteq \mathbb{C}$. Show that $(L : K)$ is not an algebraic extension.
Hint. $\operatorname{trdeg}(L : \mathbb{Q}) = \operatorname{trdeg}(L : K) + \operatorname{trdeg}(K : \mathbb{Q})$.

Problem 49. Let $(L : K)$ be a field extension and let $\alpha_1,\ldots,\alpha_n \in L$. Show that $\alpha_1,\ldots,\alpha_n$ are algebraically independent over K if and only if α_{r+1} is transcendental over $K(\alpha_1,\ldots,\alpha_r)$ for $1 \leq r \leq n-1$.

Problem 50. Let $K \subseteq Z \subseteq L$ be fields such that $(Z : K)$ is algebraic. Show that if $X \subseteq L$ is algebraically independent over K then also over Z.

Problem 51. Let $(L : K)$ be a field extension and let $\overline{K}$ be the algebraic closure of K in L. Is $(L : \overline{K})$ necessarily a purely transcendental extension?

Problem 52. (Abstract dependence relations.) Suppose that X is a set and that $s : \mathfrak{P}(X) \to \mathfrak{P}(X)$ is an assignment with the following properties.

(1) If $A \subseteq B$ then $s(A) \subseteq s(B)$.

(2) If $x \in s(A)$ then $x \in s(A_0)$ for some finite subset $A_0 \subseteq A$.

(3) $A \subseteq s(A)$ for all subsets $A \subseteq X$.

(4) $s\big(s(A)\big) = s(A)$ for all $A \subseteq X$.

(5) Whenever $y \in s(A \cup \{x\})$ but $x \notin s(A)$ then $x \in s(A \cup \{y\})$.

We call $s(A)$ the **span** of A, and if $x \in s(A)$ we say that x **depends** on A.

(a) Show that if V is a vector space and if $s(A)$ denotes the subspace of V spanned by $A \subseteq V$ then s has the above properties.

(b) Show that if $(L : K)$ is a field extension and if $s(A)$ denotes the algebraic closure of $K(A)$ in L then s has the above properties.

(c) In the general situation of a set X and a mapping $s : \mathfrak{P}(X) \to \mathfrak{P}(X)$ with the above properties, let $A \subseteq X$ be **free** in the sense that $a \notin s(A \setminus \{a\})$ for all $a \in A$ and let B be generating in the sense that $s(B) = X$. Show that there is a subset $B_0 \subseteq B$ such that $A \cap B_0 = \emptyset$ and $A \cup B_0$ is both free and generating.

Hint. Pick a maximal element in the family of all sets $S \subseteq B$ such that $A \cap S = \emptyset$ and $A \cup S$ is free.

(d) A subset $A \subseteq X$ is called a **basis** if it is both free and generating. Show that X has a basis and that all bases have the same cardinality.

(e) What does part (d) tell you about bases of a vector space? What about transcendence bases of a field extension?

Problem 53. Let $(L : K)$ be a field extension.

(a) Show that $(L : K)$ is transcendental if and only if $L = K(S)$ for any transcendence basis of L over K.

(b) Show that if $(L : K)$ is purely transcendental then every element $\alpha \in L \setminus K$ is transcendental over K.

(c) Show that it is possible that $(L : K)$ is *not* purely transcendental even though every element $\alpha \in L \setminus K$ is transcendental over K.

Problem 54. Let $(L : K)$ be a field extension of finite transcendence degree. Show that the following statements are equivalent.

(1) $[L : K(X)] < \infty$ for every transcendence basis X of $(L : K)$;

(2) there is a transcendence basis X of $(L : K)$ with $[L : K(X)] < \infty$;

(3) $(L : K)$ is **finitely generated**; i.e., there are finitely many elements $\alpha_1, \ldots, \alpha_n$ in L with $L = K(\alpha_1, \ldots, \alpha_n)$.

Problem 55. Let Z be an intermediate field of the field extension $(L : K)$. Show that $(L : K)$ is finitely generated if and only if $(L : Z)$ and $(Z : K)$ are.

Problem 56. Let Z_1 and Z_2 be intermediate fields of an extension $(L : K)$. Show that $\mathrm{trdeg}(Z_1 Z_2 : Z_2) \le \mathrm{trdeg}(Z_1 : K)$ and that

$$\max_{i=1,2} \mathrm{trdeg}(Z_i : K) \ \le \ \mathrm{trdeg}(Z_1 Z_2 : K) \ \le \ \mathrm{trdeg}(Z_1 : K) + \mathrm{trdeg}(Z_2 : K) \,.$$

Problem 57. (a) Show that $\operatorname{trdeg}(\mathbb{C} : \mathbb{Q}) = \operatorname{trdeg}(\mathbb{R} : \mathbb{Q}) = |\mathbb{R}|$.

(b) Show that the field $\mathbb{C}$ possesses a nondenumerable number of automorphisms.

Hint. Choose a transcendence basis X of $\mathbb{C}$ over $\mathbb{Q}$. Then any bijection $\varphi : X \to X$ can be extended to an automorphism of $\mathbb{Q}(X)$ and then to an automorphism of $\mathbb{C}$.

(c) Let X be a transcendence basis of $(\mathbb{C} : \mathbb{Q})$ and let A be the algebraic closure of $\mathbb{Q}$ in $\mathbb{C}$. Show that $[\mathbb{C} : A(X)]$ is infinite. Is $(\mathbb{C} : A)$ purely transcendental?

(d) Show that there is an nondenumerable number of subfields of $\mathbb{R}$ which are isomorphic to $\mathbb{R}$.

Hint. Let A be the algebraic closure of $\mathbb{Q}$ in $\mathbb{R}$ and choose a transcendence basis of $\mathbb{R}$ over A. Pick $x \in X$. Then X and $X \setminus \{x\}$ have the same cardinality; hence $\mathbb{R} = A(X) \cong A(X \setminus \{x\})$.

(e) Show that there are subfields $K \neq \mathbb{R}$ of $\mathbb{C}$ with $[\mathbb{C} : K] = 2$.

Problem 58. Let $p_1, \ldots, p_n \in K[x_1, \ldots, x_n]$. Show that the following conditions are equivalent.

(1) $p_1, \ldots, p_n$ are algebraically independent;

(2) $p_1, \ldots, p_n$ form a transcendence basis of $K(x_1, \ldots, x_n)$ over K;

(3) the determinant of the Jacobian matrix $(\partial p_i / \partial x_j)$ is a nonzero polynomial.

Problem 59. Is it possible to construct an isosceles triangle with area 1 whose vertices lie on a circle with radius 1 by using a ruler and a compass only?

Problem 60. We can identify the coordinate plane $\mathbb{R}^2$ with the set $\mathbb{C}$ of complex numbers; then a complex number $a + ib$ is constructible if and only if $a, b \in \mathbb{R}$ are constructible (with a ruler and a compass only). Show that if a cubic polynomial in $\mathbb{Q}[x]$ has a root which is constructible then it has a rational root.

Problem 61. (a) Show that a complex number $\omega \neq 0$ is constructible if and only if $\alpha := \omega + \omega^{-1}$ is contructible.

(b) Show that a regular $2n$-gon can be constructed if and only if a regular n-gon can be constructed.

(c) Clearly, a regular n-gon is constructible if and only if the number $\omega := e^{2\pi i/n}$ is constructible. By part (b), we can assume that n is odd. Show that $\alpha := \omega + \omega^{-1} = 2 \cos \frac{\pi}{n}$ satisfies an equation of degree $\frac{n-1}{2}$ over $\mathbb{Q}$.

Hint. Note that ω is a root of

$$p(z) := z^{n-1} + z^{n-2} + \cdots + z^2 + z + 1 = \frac{z^n - 1}{z - 1}$$
$$= (z - \omega)(z - \omega^2) \cdots (z - \omega^{n-2})(z - \omega^{n-1}).$$

(d) Prove that $\cos \frac{2\pi}{5} = \frac{1}{4}(\sqrt{5} - 1)$ and deduce that a regular pentagon can be constructed by only using a ruler and a compass. Can you indicate a construction?

(e) Show that $2 \cos \frac{2\pi}{7} = e^{2\pi i/7} + e^{-2\pi i/7}$ satisfies the equation $x^3 + x^2 - 2x - 1 = 0$ and deduce that there is no ruler and compass construction for a regular heptagon.

Problem 62. Show that it is possible to trisect any angle $\alpha = 360^0/n$ where 3 does not divide n by only using a ruler and a compass. Can you indicate a way to trisect 72^0?

Problem 63. Decide for each of the following numbers n whether or not a regular n-gon can be constructed by using a ruler and a compass alone.

$$1, \ 2, \ 3, \ 6, \ 8, \ 9, \ 10, \ 11, \ 12, \ 14, \ 18, \ 24, \ 40, \ 72, \ 140$$

Problem 64. Suppose that m and n are coprime natural numbers. Show that if a regular m-gon and a regular n-gon are constructible with a ruler and a compass then so is a regular mn-gon.

Problem 65. Show that $\sin x$ is contructible if and only if $\cos x$ is constructible.

Problem 66. (a) Show that if $\alpha \in \mathbb{C}$ satisfies the equation $8\alpha^3 + 4\alpha^2 - 4\alpha = 1$ then α is not constructible.
(b) Show that all four roots of the polynomial $x^4 - 6x^2 + 2$ are constructible.

Problem 67. (a) Show that the polynomial $f(x) = x^4 + x + 1$ is irreducible over $\mathbb{Q}$ and has four different, pairwise conjugate roots α, $\bar{\alpha}$, β and $\bar{\beta}$ in $\mathbb{C}$.
(b) Argue that $\alpha\bar{\alpha}\beta\bar{\beta} = 1$ and $\alpha + \bar{\alpha} + \beta + \bar{\beta} = 0$ and us these facts to show that $\theta := \alpha\bar{\alpha} + \beta\bar{\beta} = |\alpha|^2 + |\beta|^2 \in \mathbb{R}$ satisfies the equation $\theta^4 - 4\theta^2 - \theta = 0$ and hence $\theta^3 - 4\theta - 1 = 0$. Conclude that θ is not a constructible number.
(c) Show that if $\alpha \in \mathbb{C}$ satisfies the equation $\alpha^4 + \alpha + 1 = 0$ then α is not constructible.
Hint. Use part (b) or problem 21(c) in section 8.

13. Splitting fields and normal extensions

ONE of the mathematical problems that incited the development of a theory of field extensions was that of finding roots of polynomials. Given a polynomial $f \in K[x]$, one tries to find an element α with $f(\alpha) = 0$. It may well happen that no such element exists in the given field K, but there is still the chance that a root of f exists in some larger field $L \supseteq K$; for example, the polynomial $f(x) = x^2 + 1 \in \mathbb{R}[x]$ has no root in $\mathbb{R}$, but has the roots $\pm i$ in $\mathbb{C} \supseteq \mathbb{R}$. In fact, it turns out that *every* polynomial $f \in K[x]$ (where K is an arbitrary field) possesses a root in some extension of K. With the right point of view, this is not even hard to prove.

As a matter of fact, the existence of the roots of a polynomial was not considered as a problem before the 19th century, but simply taken for granted. "Numbers" like $\sqrt{-1}$ and $\sqrt{-121}$ were already used by Renaissance mathematicians (long before the system of complex numbers was defined), and mathematicians like Girard, Newton, Vandermonde, Lagrange or Galois assumed that every polynomial has roots "somewhere" without specifying where. They unscrupulously talked about "the roots" of a polynomial without feeling the need to explain what these roots were and without justifying the assumption of their existence; the way they probably thought about the existence of roots is outlined in problem 1 below. It was only Gauss who objected to using these "numbers" of doubtful existence in calculations; he called them "true shadows of shadows". Finally, it was Kronecker who rigorously proved, for any polynomial f over a field K, the existence of a field $L \supseteq K$ containing a root α of f. Moreover, he showed that if f is irreducible then the field $K(\alpha) \subseteq L$ is unique up to isomorphism and hence does not depend on the choice of L.

To motivate Kronecker's proof, let us put the cart before the horse and then work backwards; i.e., let us assume that a polynomial $f \in K[x]$ has a root α in some extension L of K and then try to get an idea how such an extension may be found if it is not given in advance. Obviously, it is no loss of generality to assume that f is irreducible (otherwise decompose the polynomial f into its irreducible factors). Then the evaluation map $K[x] \to L$ given by $p \mapsto p(\alpha)$ is a ring homomorphism with image $K[\alpha] = K(\alpha)$ and kernel $\langle f \rangle$, so that $K(\alpha) \cong K[x]/\langle f \rangle$ by the isomorphism theorem. Here the left-hand side is a field extension of K containing α, whereas the right-hand side can be formed without being given α and L in advance (and hence is independent of the particular choice of α and L). So why don't we simply take this right-hand side as the desired field extension? This is indeed exactly the content of the following proof; let us now give precise formulations and statements.

(13.1) Proposition. *Let K be an arbitrary field and $f \in K[x]$ an irreducible polynomial.*

(a) *The quotient ring $K[x]_f = K[x]/\langle f \rangle$ is a field extension of K and contains a root of f.*

(b) *If α is a root of f in some extension L of K, then the mapping $K[x]_f \to K(\alpha)$ given by $[p] \mapsto p(\alpha)$ is an isomorphism.*

(c) *Let $\sigma : K \to K'$ be a field isomorphism. If α is a root of f (in some extension of K) and if α' a root of $\sigma \star f$ (in some extension of K'), then there is a unique isomorphism $\Sigma : K(\alpha) \to K'(\alpha')$ with $\Sigma(\alpha) = \alpha'$ and $\Sigma|_K = \sigma$.*

Proof (Kronecker). (a) First of all $\langle f \rangle$ is a maximal ideal of $K[x]$ due to (7.4)(e) and due to the fact that $K[x]$ is a principal ideal domain; hence $K[x]_f$ is a field due to (6.14)(d). Second, the ring homomorphism $i : K \to K[x]_f$ given by $k \mapsto [k]$ (where k is considered as a constant polynomial on the right-hand side) is one-to-one; so we can identify K with the subfield $i(K)$ of $K[x]_f$. Now we claim that $[x]$ is a root of f. Indeed, if $f(x) = \sum_{k=0}^{n} a_k x^k$ then

$$f([x]) \;=\; \sum_{k=0}^{n} a_k [x]^k \;=\; \left[\sum_{k=0}^{n} a_k x^k\right] \;=\; [f] \;=\; [0] \,.$$

(b) The homomorphism
$$\begin{aligned} K[x] &\;\to\; K(\alpha) \\ p &\;\mapsto\; p(\alpha) \end{aligned}$$
is surjective because $K(\alpha) = K[\alpha]$ by (12.5) and has $\langle f \rangle = K[x] \cdot f$ as its kernel because f is the minimal polynomial of α. Then the claim follows from the homomorphism theorem.

(c) It is easy to check that the mapping $\Sigma_0 : K[x]_f \to K'[x]_{\sigma \star f}$ given by $[h] \mapsto [\sigma \star h]$ is an isomorphism. By part (b), we have the isomorphisms

$$i : \begin{aligned} K[x]_f &\;\to\; K(\alpha) \\ [p] &\;\mapsto\; p(\alpha) \end{aligned} \quad \text{and} \quad i' : \begin{aligned} K'[x]_{\sigma \star f} &\;\to\; K'(\alpha') \\ [q] &\;\mapsto\; q(\alpha') \end{aligned} \,.$$

Now simply define $\Sigma := i' \circ \Sigma_0 \circ i^{-1}$. Thus the following diagram commutes.

$$
\begin{array}{ccc}
K(\alpha) & \xrightarrow{\;\;\;\;\;\;\Sigma\;\;\;\;\;\;} & K'(\alpha') \\[4pt]
{\scriptstyle i}\big\uparrow & & \big\uparrow{\scriptstyle i'} \\[4pt]
K[x]_f & \xrightarrow{\;\;\;\;\;\;\Sigma_0\;\;\;\;\;\;} & K'[x]_{\sigma \star f}
\end{array}
$$

$\blacksquare$

EVEN though Kronecker's result does not give us much insight into the "nature" of the roots – whatever that means –, it eases our mind: The roots of a polynomial are no longer mysterious objects of doubtful existence, but simply elements of a clearly defined larger field.

A repeated application of Kronecker's result will show that for every polynomial $f \in K[x]$ there is a field $L \supseteq K$ over which f completely splits into linear factors. Indeed, by applying (13.1) for the first time, we can find a root α in some extension L_1 of K and then write $f(x) = (x - \alpha) g(x)$ with $g \in L_1[x]$. Repeating the process, we can find a field $L_2 \supseteq L_1$ containing a root β of g and hence write $f(x) = (x - \alpha)(x - \beta) h(x)$ with $h \in L_2[x]$. Continuing in this way, we can find a field over which f completely splits into linear factors. It turns out that such a field is unique up to isomorphism, if we choose it to be as small as possible; so we can unambiguously talk about the "splitting field" of f. More generally, the notion of the splitting field of a whole family of polynomials is defined as follows.

(13.2) Definition. *Let K be a field.*

(a) *Let $\mathfrak{F} \subseteq K[x]$ be a family of polynomials. A **splitting field** of $\mathfrak{F}$ over K is a field $L \supseteq K$ such that*

(1) *each $f \in \mathfrak{F}$ splits over L†;*

(2) *L is minimal with respect to property (1); i.e., if each $f \in \mathfrak{F}$ splits over an intermediate field Z of $(L : K)$ then $Z = L$.*

(b) *A field extension $(L : K)$ is called **normal** if L is a splitting field of some family $\mathfrak{F} \subseteq K[x]$.*

(13.3) Examples. (a) The fields $\mathbb{C}$ and $\mathbb{R}[x]_{x^2+1}$ are both splitting fields of $x^2 + 1 \in \mathbb{R}[x]$.

(b) The field $\mathbb{Q}(\sqrt[4]{2}, i)$ is a splitting field of $x^4 - 2 \in \mathbb{Q}[x]$.

(c) A splitting field of $x^3 - 2$ over $\mathbb{Q}$ is $\mathbb{Q}(\sqrt[3]{2}, i\sqrt{3})$.

(d) If $n \in \mathbb{N}$ then $\mathbb{Q}(e^{2\pi i/n})$ is a splitting field of $x^n - 1 \in \mathbb{Q}[x]$.

(e) If L is a splitting field of a single polynomial $f \in K[x]$ and if $\alpha_1, \ldots, \alpha_n \in L$ are the roots of f, then clearly $L = K(\alpha_1, \ldots, \alpha_n)$.

(f) Generalizing on (e), let L be a splitting field of a family $\mathfrak{F} \subseteq K[x]$. For each $f \in \mathfrak{F}$, let $X_f \subseteq L$ be the set of all roots of f. Then clearly $L = K(\bigcup_{f \in \mathfrak{F}} X_f) = K[\bigcup_{f \in \mathfrak{F}} X_f]$. This shows that every splitting field over K is algebraic over K.

THE next theorem establishes the existence and uniqueness (up to isomorphism) of a splitting field of an arbitrary family of polynomials. Moreover, it is shown that every field isomorphism $\sigma : K \to K'$ can be extended to an isomorphism between associated normal extensions of K and K'. This is very important because it is not true for arbitrary field extensions $(L : K)$ and $(L' : K')$ that every field isomorphism $\sigma : K \to K'$ can be extended to an isomorphism $\Sigma : L \to L'$. (See problem 13(a) below.)

In general, to compare two extensions L and L' of a field K, it makes sense to study those homomorphisms of L into L' which leave K pointwise fixed. Now nonzero homomorphisms between fields (in contrast to homomorphism between rings) are automatically injective, i.e., are embeddings. Thus we give the following definition.

(13.4) Definition. *Let L and L' be extensions of a field K. Then a nonzero homomorphism $\sigma : L \to L'$ which leaves K pointwise fixed is called a **K-embedding** of L into L'. The set of all K-embeddings of L into L' is denoted by $\mathrm{Emb}_K(L, L')$. An isomorphism $\sigma : L \to L'$ which leaves K pointwise fixed is called a **K-isomorphism**.*

---

† Recall that "to split" means "to split completely into linear factors".

LET us now establish the existence and essential uniqueness of splitting fields.

(13.5) Theorem. *Let K be a field.*
(a) *Every family $\mathfrak{F} \subseteq K[x]$ of polynomials possesses a splitting field.*
(b) *Let $\sigma : K \to K'$ be a field isomorphism and let*

$$
\begin{array}{ccc}
K[x] & \to & K'[x] \\
f & \mapsto & \sigma \star f
\end{array}
$$

be the ring isomorphism induced by σ. If $L \supseteq K$ is a splitting field of $\mathfrak{F} \subseteq K[x]$ and if $L' \supseteq K'$ is a splitting field of $\mathfrak{F}' := \{\sigma \star f \mid f \in \mathfrak{F}\} \subseteq K'[x]$, then σ can be extended to an isomorphism $\Sigma : L \to L'$.
(c) *Any two splitting fields of a family $\mathfrak{F} \subseteq K[x]$ are isomorphic with an isomorphism leaving K pointwise fixed.*
(d) *If L is a splitting field of $f \in K[x]$ with $\deg f = n$, then $[L : K]$ divides $n!$; in particular, $[L : K] \leq n!$.*

Proof. (a) Let us first assume that $\mathfrak{F} = \{f\}$ consists of one single polynomial; in this case we proceed by induction on the degree of f. If $f \in K[x]$ has degree 1, i.e., is linear, then K itself is a splitting field. So let $\deg f > 1$. If f splits over K, then take $L := K$. Otherwise, f possesses an irreducible factor g of degree > 1. By (13.1)(a), g has a root α in some extension of K. Then $f(x) = (x - \alpha)h(x)$ with a polynomial $h \in K(\alpha)[x]$ with smaller degree than f. Let L be a splitting field of h over $K(\alpha)$. Then L is a splitting field of f over K. Thus for one single polynomial, the construction of a splitting field is simply an iteration of the process of adjoining a root as described in (13.1).

In the general case, we have to adjoin not only the roots of a given polynomial, but of all polynomials belonging to a given family $\mathfrak{F} \subseteq K[x]$; so we have to iterate the process in (13.1) over all elements of $\mathfrak{F}$. This requires no additional algebraic considerations, but only some set-theoretic manœuvres.

Let $(f_i)_{i \in I}$ be a well-ordering of $\mathfrak{F}$. Since we can form splitting fields of single polynomials, we can define L inductively by

$$
K_i := \text{a splitting field of } f_i \text{ over } \bigcup_{j < i} K_j , \quad L := \bigcup_{i \in I} K_i ;
$$

note that $K_j \subseteq K_i$ whenever $j < i$. (If i_0 is the smallest element of I then K_{i_0} is to be understood as a splitting field of f_{i_0} over K.) Then obviously every polynomial $f \in \mathfrak{F}$ splits over L, and no proper subfield of L has this property.

(b) Again, let us first treat the case that $\mathfrak{F} = \{f\}$ consists of one single polynomial f. In this case, the proof is by induction on $n := \deg f$. If $n = 1$ then $L = K$ and $L' = K'$, so that the statement is trivially true. Hence let $n > 1$. We write $f = qr$ with an irreducible factor q (if f itself is irreducible then $q = f$ and $r = 1$). Choose a root $\alpha \in L$ of q and a root $\alpha' \in L'$ of $\sigma \star q$. Then σ extends to an isomorphism $\Sigma_0 : K(\alpha) \to K'(\alpha')$ by (13.1)(c). Now we split apart the linear factor $x - \alpha$ from f and write $f(x) = (x - \alpha)\hat{q}(x)r(x)$; then L is the splitting field of $\hat{q}r$ over $K(\alpha)$ and L' is the splitting field of $\sigma \star (\hat{q}r)$ over $K'(\alpha')$. By induction hypothesis, there is an isomorphism $\Sigma : L \to L'$ with $\Sigma|_{K(\alpha)} = \Sigma_0$ and consequently $\Sigma|_K = \Sigma_0|_K = \sigma$. So we obtain the following commutative diagram.

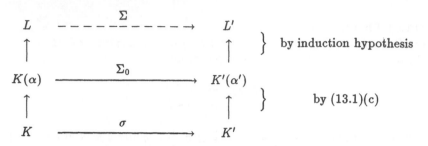

In the general case, let Γ be the set of all triplets (Z, Z', φ) where $K \subseteq Z \subseteq L$ and $K' \subseteq Z' \subseteq L'$ are intermediate fields and where $\varphi : Z \to Z'$ is an isomorphism extending σ. Then

$$(Z_1, Z_1', \varphi_1) \leq (Z_2, Z_2', \varphi_2) :\Longleftrightarrow Z_1 \subseteq Z_2, \; Z_1' \subseteq Z_2', \; \varphi_1 = \varphi_2|_{Z_1}$$

defines an inductive order on Γ. Indeed, every chain $(Z_i, Z_i', \varphi_i)_{i \in I}$ in Γ has an upper bound, namely (Z, Z', φ) where $Z := \bigcup_{i \in I} Z_i$, $Z' := \bigcup_{i \in I} Z_i'$ and where $\varphi : Z \to Z'$ is defined by $\varphi|_{Z_i} = \varphi_i$. So by Zorn's lemma, Γ has a maximal element (Z_0, Z_0', φ_0). We claim that $Z_0 = L$ and $Z_0' = L'$; then φ_0 is the desired extension of σ, and we are done.

Suppose $Z_0 \subsetneq L$; then some $f \in \mathfrak{F}$ does not split over Z_0. But f splits over L; hence L contains a splitting field Z of f. Also, L' is a splitting field of $\mathfrak{F}'$, hence contains a splitting field Z' of $\sigma \star f = \varphi_0 \star f$. Since we already treated the case of a single polynomial, we know that $\varphi_0 : Z_0 \to Z_0'$ can be extended to an isomorphism $\Phi : Z \to Z'$. But this clearly contradicts the maximality of (Z_0, Z_0', φ_0). Similarly (using φ_0^{-1}), one shows that the assumption $Z_0' \subsetneq L'$ leads to a contradiction. Hence $Z_0 = L$ and $Z_0' = L'$, as claimed.

(c) Take $K' = K$ and $\sigma = \mathrm{id}$ in part (b).

(d) We proceed by induction on n. If $n = 1$, then $L = K$, and the claim is trivial. Suppose the claim is proved for all degrees less than n and let $\deg f = n$.

If f is reducible over K, then we can write $f = f_1 f_2$ where $d := \deg f_1 < n$ and $n - d = \deg f_2 < n$. Let $Z \subseteq L$ be the splitting field of f_1 over K; then $[Z : K]$ divides $d!$, by induction hypothesis. Also, L is a splitting field of f_2 over Z, so that $[L : Z]$ divides $(n - d)!$, again by induction hypothesis. But then $[L : K] = [L : Z] \cdot [Z : K]$ divides $(n - d)! \, d!$, hence also $\displaystyle \binom{n}{d} (n - d)! \, d! = n!$.

If f is irreducible over K, we choose a root $\alpha \in L$ of f and have $[K(\alpha) : K] = n$. Then L is the splitting field of $g(x) := f(x)/(x - \alpha)$ over $K(\alpha)$, and $\deg g = n - 1$. Hence $[L : K(\alpha)]$ divides $(n - 1)!$, by induction hypothesis. But then $[L : K] = [L : K(\alpha)] \cdot [K(\alpha) : K] = [L : K(\alpha)] \cdot n$ divides $(n - 1)! \, n = n!$. $\blacksquare$

WHEN we study a polynomial with rational coefficients, we are in the pleasant situation that we can view $\mathbb{Q}$ as a subfield of $\mathbb{C}$. Since $\mathbb{C}$ is algebraically closed, we know *a priori* that the polynomial in question splits completely into linear factors over the complex numbers. Our next goal will be to show that an arbitrary field K is contained in a larger field which is algebraically closed and also that such a larger field is uniquely determined (up to isomorphism) if it is chosen as small as possible. Let us start by characterizing algebraically closed fields.

(13.6) Proposition. *For a field K the following conditions are equivalent:*

(1) K is algebraically closed, i.e., every nonconstant polynomial $f \in K[x]$ splits into linear factors over K;

(2) every nonconstant polynomial $f \in K[x]$ has a root in K;

(3) if $(L : K)$ is an algebraic extension then $L = K$.

Proof. Clearly, (1) implies (2). Conversely, if (2) holds then one proves easily by induction on $\deg f$ that every nonconstant polynomial $f \in K[x]$ splits into linear factors over K. Suppose that (1) holds and let α be an element which is algebraic over K, say with minimal polynomial p. Since p splits over K we have $\alpha \in K$; hence every element which is algebraic over K is already contained in K. This shows that (1) implies (3). Finally, suppose that (3) holds and let $f \in K[x]$. If α be a root of f (in some splitting field of f) then $K(\alpha)$ is an algebraic extension of K, hence must be K so that $\alpha \in K$. Thus (3) implies (2). ∎

WE now define an algebraic closure of a field K as an algebraically closed extension of K which is as small as possible.

(13.7) Definition. *An **algebraic closure** of a field K is a field $L \supseteq K$ such that*

(1) L is algebraically closed,

(2) L is minimal with property (1); i.e., if Z is an algebraically closed intermediate field of $(L : K)$ then $Z = L$.

NEXT, we give various characterizations of algebraic closures; this will not only clarify the concept of an algebraic closure, but will also facilitate the proof of the existence and uniqueness of algebraic closures.

(13.8) Theorem. *Let $(\overline{K} : K)$ be a field extension. Then the following conditions are equivalent:*

(1) $\overline{K}$ is an algebraic closure of K;

(2) $(\overline{K} : K)$ is algebraic, and $\overline{K}$ is algebraically closed;

(3) $(\overline{K} : K)$ is algebraic, and every polynomial $f \in K[x]$ splits over $\overline{K}$;

(4) $\overline{K}$ is a splitting field over K of the family $\mathfrak{F} := K[x]$;

(5) $(\overline{K} : K)$ is algebraic, and for every chain $K \subseteq Z_0 \subseteq L_0$ of algebraic extensions and every homomorphism $\varphi_0 : Z_0 \to \overline{K}$ there is an extension $\Phi : L_0 \to \overline{K}$;

(6) $(\overline{K} : K)$ is algebraic, and for every algebraic extension $(L : K)$ of K there is a K-embedding $\sigma : L \to \overline{K}$ (loosely speaking, this means that $\overline{K}$ is the largest possible algebraic extension of K).

Proof. (1)$\Longrightarrow$(2). Let Z be the field of all elements in L which are algebraic over K; then $(Z:K)$ is an algebraic field extension. We claim that Z is algebraically closed; this will imply $Z = L$ by condition (13.7)(2) so that indeed $(L:K)$ is algebraic. Let $p \in Z[x]$. Then p splits over L, since L is algebraically closed; say $p(x) = c \prod_i (x - \alpha_i)$ with $\alpha_i \in L$. But then the elements α_i are algebraic over Z, hence also over K, by (12.20)(e), hence lie in Z, by the very definition of Z. So p splits over Z.

(2)$\Longrightarrow$(5). If $\varphi = 0$ we can simply take $\Phi = 0$; hence we may assume that φ_0 is nonzero and thus is an embedding. Let Γ be the set of all pairs (Z, φ) where $Z_0 \subseteq Z \subseteq L_0$ is an intermediate field and where $\varphi|_{Z_0} = \varphi_0$. Then

$$(Z_1, \varphi_1) \le (Z_2, \varphi_2) :\Longleftrightarrow Z_1 \subseteq Z_2, \ \varphi_1 = \varphi_2|_{Z_1}$$

defines an inductive order on Γ. Indeed, every chain $(Z_i, \varphi_i)_{i \in I}$ in Γ has an upper bound, namely (Z, φ) where $Z := \bigcup_{i \in I} Z_i$ and where $\varphi : Z \to \overline{K}$ is defined by $\varphi|_{Z_i} = \varphi_i$. So by Zorn's lemma, Γ has a maximal element (Z, Φ). We claim that $Z = L_0$. Suppose there is an element $\alpha \in L_0 \setminus Z$ and let $p \in K[x]$ be the minimal polynomial of α over K. Then $\Phi \star p \in \overline{K}[x]$ has a root $\alpha' \in \overline{K}$, because $\overline{K}$ is algebraically closed. By (13.1)(c), the embedding $\Phi : Z \to \overline{K}$ extends to an embedding $\Phi : Z(\alpha) \to \overline{K}$. But then the maximality of (Z, Φ) forces $Z(\alpha) = Z$ and hence $\alpha \in Z$; this is the desired contradiction.

(5)$\Longrightarrow$(6). Let $(L:K)$ be an algebraic extension. By applying (5) with $Z_0 := K$, $L_0 := L$ and $\varphi_0 :=$ inclusion map of K into $\overline{K}$, we find an extension $\Phi : L \to \overline{K}$. Since Φ is not the zero map, it is automatically an embedding.

(6)$\Longrightarrow$(3). Let $f \in K[x]$ and let Z be a splitting field of K. By (6), there is an embedding $\sigma : Z \to \overline{K}$. Then $\sigma(Z) \subseteq \overline{K}$ is a splitting field for $\sigma \star f$; hence $\sigma \star f$ splits over $\overline{K}$. But since f lies in $K[x]$ and σ is the identity on K, we have $\sigma \star f = f$; hence f splits over $\overline{K}$.

(3)$\Longrightarrow$(4). Suppose every polynomial $f \in K[x]$ splits over an intermediate field $Z \subseteq \overline{K}$ and let $\alpha \in \overline{K}$. Then α is algebraic over K by condition (3); let $p \in K[x]$ be the minimal polynomial of α. Since p splits over Z by hypothesis, we have $\alpha \in Z$. Since $\alpha \in \overline{K}$ was an arbitrary element of $\overline{K}$, this shows $Z = \overline{K}$.

(4)$\Longrightarrow$(3). Let Z be the set of all elements of L which are algebraic over K; then $(Z:K)$ is an algebraic field extension. We claim that every polynomial $f \in K[x]$ splits over Z; this will imply $Z = \overline{K}$ by condition (2) in (13.2)(a), so that indeed $(\overline{K}:K)$ is algebraic. Let $f \in K[x]$; then f splits over $\overline{K}$ by hypothesis; say $f(x) = c \prod_i (x - \alpha_i)$ with $\alpha_i \in \overline{K}$. But then the elements α_i are algebraic over K and hence lie in Z by the very definition of Z. So f splits over Z.

(3)$\Longrightarrow$(2). We have to prove that an arbitrary polynomial $q \in \overline{K}[x]$ splits over $\overline{K}$. Let $\alpha_1, \ldots, \alpha_n$ be the roots of q, counted with multiplicities, in some splitting field of q. Each α_k is algebraic over $\overline{K}$ and hence also over K; let $p_k \in K[x]$ be the minimal polynomial of α_k over K. Then q divides $p := p_1 \cdots p_n \in K[x]$ in $\overline{K}[x]$. Since p splits over $\overline{K}$ by hypothesis, q must also split over $\overline{K}$.

(2)$\Longrightarrow$(1). Let $Z \subseteq \overline{K}$ be algebraically closed and let $\alpha \in \overline{K}$. Since $(\overline{K}:K)$ is algebraic, we have $p(\alpha) = 0$ for some $p \in K[x] \subseteq Z[x]$ so that $\alpha \in \overline{K}$. This shows that $\overline{K} \subseteq Z$, hence $Z = \overline{K}$. ∎

CHARACTERIZATION (4) of an algebraic closure is particularly remarkable; in fact, this characterization tells us that if $\overline{K}$ is minimal with the property that all polynomials in $K[x]$ split over $\overline{K}$, then actually all polynomials in $\overline{K}[x]$ split over $\overline{K}$. Also, since we established existence and uniqueness of arbitrary splitting fields in (13.5), this characterization immediately yields the existence and uniqueness of algebraic closures.

(13.9) Theorem. *Every field has an algebraic closure, and any two algebraic closures of a field K are isomorphic with an isomorphism leaving K pointwise fixed.*

Proof. Since an algebraic closure of a field K is the same as a splitting field of $\mathfrak{F} = K[x]$, this theorem follows immediately from (13.5)(a) and (13.5)(c). ∎

HAVING available the notion of an algebraic closure, we can give the following characterization of normal extensions.

(13.10) Theorem. *Let $(L : K)$ be an algebraic field extension and let $\overline{L}$ be an algebraic closure of L (and hence of K). Then the following conditions are equivalent:*
(1) $(L : K)$ is normal, i.e., L is the splitting field of some family $\mathfrak{F} \subseteq K[x]$ of polynomials over K;
(2) every irreducible polynomial $f \in K[x]$ which has a root in L splits over L;
(3) every K-embedding $\sigma : L \to \overline{L}$ maps L to L;
(4) every K-automorphism $\sigma : \overline{L} \to \overline{L}$ maps L to L.
If $[L : K] < \infty$, then there is another equivalent condition:
(1') L is the splitting field of some polynomial $f \in K[x]$.

Proof. (1)$\Longrightarrow$(4). Suppose L is the splitting field of $(f_i)_{i \in I} \subseteq K[x]$. This means $L = K(\bigcup_{i \in I} X_i)$ where $X_i \subseteq \overline{L}$ is the set of roots of f_i. Now if $\sigma : \overline{L} \to \overline{L}$ leaves K pointwise fixed then $\sigma \star f = f$ for all $f \in K[x]$; hence if $f(x) = c(x - \alpha_1) \cdots (x - \alpha_N)$ then $f(x) = (\sigma \star f)(x) = c(x - \sigma(\alpha_1)) \cdots (x - \sigma(\alpha_N))$ so that σ permutes the roots of f. Therefore, σ permutes the elements of each set X_i, which implies that $\sigma(L) \subseteq L$.

(4)$\Longrightarrow$(3). Given a K-homomorphism $\sigma : L \to \overline{L}$, we can find an extension $\Sigma : \overline{L} \to \overline{L}$, due to (13.5)(b). Then $\Sigma(L) \subseteq L$ by hypothesis, hence $\sigma(L) \subseteq L$.

(3)$\Longrightarrow$(2). Let $f \in K[x]$ be irreducible and let $\alpha \in L$ be a root of f. We have to show that if $\beta \in \overline{L}$ is a root of f, then $\beta \in L$. By (13.1)(c), there is an isomorphism $\sigma_0 : K(\alpha) \to K(\beta)$ with $\sigma_0(\alpha) = \beta$ and $\sigma_0|_K = \text{id}_K$. By (13.5)(b), we can extend σ_0 to an isomorphism $\Sigma : \overline{L} \to \overline{L}$. Applying condition (3) to $\sigma := \Sigma|_L : L \to \overline{L}$, we see that $\sigma(L) \subseteq L$; in particular, $\beta = \sigma(\alpha) \in L$.

(2)$\Longrightarrow$(1). For each element $\alpha \in L$, let $p_\alpha \in K[x]$ be the minimal polynomial of α over K. It is easy to see that (2) implies that L is the splitting field of $\mathfrak{F} := \{p_\alpha \mid \alpha \in L\}$.

Trivially, (1') implies (1) (simply take $\mathfrak{F} := \{f\}$). We claim that if $[L : K] < \infty$, then (1) implies (1'). If L is the splitting field of a family $\mathfrak{F}$ and if $[L : K] < \infty$, then it is a simple consequence of the degree formula that L is the splitting field of some finite subset $\{f_1, \ldots, f_r\} \subseteq \mathfrak{F}$. But this is tantamount to saying that L is the splitting field of $f := f_1 \cdots f_r$. ∎

(13.11) Examples. (a) Every field extension $(L : K)$ of degree 2 is normal. Indeed, take any element $\alpha \in L \setminus K$ and let $p \in K[x]$ be its minimal polynomial. Then p is a quadratic polynomial which has a root in L (namely α) and hence must split over L; on the other hand, $L = K(\alpha)$ by a simple application of the degree formula. Hence L is the splitting field of p.

(b) The field extensions $\big(\mathbb{Q}(\sqrt[4]{2}) : \mathbb{Q}(\sqrt{2})\big)$ and $\big(\mathbb{Q}(\sqrt{2}) : \mathbb{Q}\big)$ are normal, due to part (a). On the other hand, the extension $\big(\mathbb{Q}(\sqrt[4]{2}) : \mathbb{Q}\big)$ is not normal, because the irreducible polynomial $f(x) = x^4 - 2 \in \mathbb{Q}[x]$ has a root in $\mathbb{Q}(\sqrt[4]{2})$, but does not split over $\mathbb{Q}(\sqrt[4]{2})$. This shows that being normal is not a transitive property.

(c) If $(L : K)$ is normal and Z is an intermediate field, then $(L : Z)$ is normal.

(d) If $\overline{K}$ is an algebraic closure of K, then $(\overline{K} : K)$ is a normal extension.

THE next proposition shows that normality is preserved under the formation of composita.

(13.12) Proposition. *Let $K \subseteq Z_1, Z_2 \subseteq L$ be fields.*
(a) *If $(Z_1 : K)$ is normal then so is $(Z_1 Z_2 : Z_2)$.*
(b) *If $(Z_1 : K)$ and $(Z_2 : K)$ are normal then so are $(Z_1 Z_2 : K)$ and $(Z_1 \cap Z_2 : K)$.*

Proof. (a) If Z_1 is the splitting field of $\mathfrak{F} \subseteq K[x]$ then $Z_1 Z_2$ is the splitting field of $\mathfrak{F} \subseteq Z_2[x]$.

(b) Let $\sigma : Z_1 Z_2 \to \overline{Z_1 Z_2}$ be a K-homomorphism. Then $\sigma(Z_1) \subseteq Z_1$ and $\sigma(Z_2) \subseteq Z_2$ by (13.10)(3) because $(Z_1 : K)$ and $(Z_2 : K)$ are normal. But these two inclusions clearly imply that $\sigma(Z_1 Z_2) \subseteq Z_1 Z_2$ and $\sigma(Z_1 \cap Z_2) \subseteq Z_1 \cap Z_2$; applying (13.10)(3) again we conclude that $(Z_1 Z_2 : K)$ is normal. ∎

SINCE normal field extensions have very convenient properties we now try to embed an arbitrary algebraic extension L of a field K into a normal extension N of K. To see what possibilities there are to find such an embedding we consider first the case of a simple extension $L = K(\alpha)$.

(13.13) Proposition. *Suppose that the element α is algebraic over the field K with minimal polynomial $f \in K[x]$ and let N be any normal extension of K containing α. Then the following conditions on an element $\beta \in N$ are equivalent:*
(1) *β is a root of f;*
(2) *β has the same minimal polynomial as α, namely f;*
(3) *there is a K-isomorphism $\sigma : K(\alpha) \to K(\beta)$ with $\sigma(\alpha) = \beta$;*
(4) *there is a unique K-isomorphism $\sigma : K(\alpha) \to K(\beta)$ with $\sigma(\alpha) = \beta$;*
(5) *there is a K-automorphism $\varphi : N \to N$ with $\varphi(\alpha) = \beta$.*
(If these conditions are satisfied then β is called a **conjugate** *of α over K.) Consequently, the number of K-embeddings of $K(\alpha)$ into N equals the number of roots of f; i.e.,*

$$|\mathrm{Emb}_K\big(K(\alpha) : N\big)| \;=\; \text{number of roots of } f \;\leq\; \deg f \;=\; [K(\alpha) : K].$$

Proof. $(1)\Longrightarrow(2)$. The polynomial f is monic, irreducible and satisfies $f(\beta) = 0$, hence is the minimal polynomial of β.

$(2)\Longrightarrow(3)$. This is a special case of $(13.1)(c)$.

$(3)\Longrightarrow(4)$. A K-isomorphism $\sigma : K(\alpha) \to K(\beta)$ is automatically determined by its image of α.

$(4)\Longrightarrow(5)$. The K-isomorphism $\sigma : K(\alpha) \to K(\beta)$ can be extended to an automorphism of N, due to $(13.5)(b)$.

$(5)\Longrightarrow(1)$. Let $f(x) = \sum_k a_k x^k \in K[x]$; then $f(\beta) = f(\varphi(\alpha)) = \sum_k a_k \sigma(\alpha)^k = \sigma(\sum_k a_k \alpha^k) = \sigma(f(\alpha)) = \sigma(0) = 0$.

Thus we have shown that each embedding $\sigma : L \to N$ over K maps α to a root of f and that conversely for any root β of f there is such a σ with $\beta = \sigma(\alpha)$. Since each K-embedding σ of $K(\alpha)$ is uniquely determined by the value $\sigma(\alpha)$ this gives the last claim. ∎

(13.14) Lemma. *Let $K \subseteq Z \subseteq L \subseteq N$ be fields such that N is normal over K. For each element $\sigma \in \mathrm{Emb}_K(Z, N)$, we pick an automorphism $\Sigma : N \to N$ extending σ. Then the mapping*

$$
\begin{array}{ccc}
\mathrm{Emb}_K(Z, N) \times \mathrm{Emb}_Z(L, N) & \to & \mathrm{Emb}_K(L, N) \\
(\sigma, \tau) & \mapsto & \Sigma \circ \tau
\end{array}
$$

is a bijection. Consequently, $|\mathrm{Emb}_K(L, N)| = |\mathrm{Emb}_K(Z, N)| \cdot |\mathrm{Emb}_Z(L, N)|$.

Proof. Clearly, if $\tau : L \to N$ over Z and $\sigma : Z \to N$ over K, then $\Sigma \circ \tau : L \to N$ over K, so that the indicated mapping is well-defined.

Injectivity. Suppose that $\Sigma_1 \circ \tau_1 = \Sigma_2 \circ \tau_2$. Then $\Sigma_2^{-1} \circ \Sigma_1 \circ \tau_1 = \tau_2$ so that $\Sigma_2^{-1} \circ \Sigma_1|_Z = \mathrm{id}_Z$, i.e., $\sigma_1 = \Sigma_1|_Z = \Sigma_2|_Z = \sigma_2$ and hence $\Sigma_1 = \Sigma_2$ (since we chose one extension for each σ). Then $\Sigma_1 \circ \tau_1 = \Sigma_2 \circ \tau_2 = \Sigma_1 \circ \tau_2$; multiplication by Σ_1^{-1} from the left yields $\tau_1 = \tau_2$.

Surjectivity. If $\theta : L \to N$ over K, then $\theta|_Z : Z \to N$ over K, so that $\sigma := \theta|_Z$ belongs to $\mathrm{Emb}_K(Z, N)$. If $\Sigma : N \to N$ is the chosen extension of σ, then $\Sigma^{-1} \circ \theta : L \to N$ over Z, so that $\tau := \Sigma^{-1} \circ \theta$ belongs to $\mathrm{Emb}_Z(L, N)$. Clearly, $\theta = \Sigma \circ \tau$. ∎

WE now define the normal closure of a field extension $(L : K)$ as the smallest extension of L which is normal over K.

(13.15) Definition. *Let $(L : K)$ be an algebraic field extension. A field $L' \supseteq L$ is called a **normal closure** of $(L : K)$ if the following conditions hold:*

(1) *$(L' : K)$ is normal;*

(2) *L' is minimal with property (1); i.e., whenever $K \subseteq L \subseteq Z \subseteq L'$ and $(Z : K)$ is normal then $Z = L'$.*

(13.16) Examples. (a) Suppose that a field extension $(L : K)$ is given by $L = K(A) = K[A]$ where each element $\alpha \in A$ is algebraic over K. Let $\bar{L}$ be an algebraic closure of L (and hence also of K). For each $\alpha \in A$, let $p_\alpha \in K[x]$ be the minimal polynomial of α and let X_α be the set of roots of p_α in $\overline{L}$. Then $K(\bigcup_{\alpha \in A} X_\alpha)$ is clearly a normal closure of $(L : K)$.

(b) The normal closure of $(\mathbb{Q}(\sqrt[3]{2}) : \mathbb{Q})$ is given by $\mathbb{Q}(\sqrt[3]{2}, \sqrt[3]{2}e^{2\pi i/3})$, i.e., by the splitting field of $x^3 - 2$.

(13.17) Theorem. *Let $(L : K)$ be an algebraic field extension.*

(a) *For each $\alpha \in L$, let $p_\alpha \in K[x]$ be the minimal polynomial of α over K. Then N is a normal closure of $(L : K)$ if and only if N is a splitting field of $\mathfrak{F} := \{p_\alpha \mid \alpha \in L\} \subseteq K[x]$.*

(b) *There is a normal closure of $(L : K)$. Moreover, any two normal closures of $(L : K)$ are isomorphic with an isomorphism which leaves K pointwise fixed.*

(c) *If N is a normal closure of $(L : K)$, then $[N : K] < \infty$ if and only if $[L : K] < \infty$.*

Proof. (a) This is merely a restatement of (13.16)(a) (with $A := L$).

(b) Using part (a), this is an immediate consequence of (13.5)(a) and (13.5)(c).

(c) If $[L : K] < \infty$, we can write $L = K(A)$ with a *finite* set $A \subseteq L$, say $A = \{\alpha_1, \ldots, \alpha_n\}$. Let p_k be the minimal polynomial of α_k over K and let X_k be the set of roots of p_k (in some algebraic closure of K). Then $N := K(X_1 \cup \cdots \cup X_n)$ is a normal closure of $(L : K)$, and since $X_1 \cup \cdots \cup X_n$ is a finite set, we have $[N : K] < \infty$. Conversely, if $[N : K] < \infty$ then $[L : K] < \infty$ by the degree formula. ∎

IN the next proposition we show that the normal closure of a finite extension $(L : K)$ is a compositum of fields which are all K-isomorphic to L.

(13.18) Proposition. *Let $(L : K)$ be a finite field extension.*

(a) *If $\sigma_1, \ldots, \sigma_n : L \to \overline{K}$ are the different K-embeddings of L into some fixed algebraic closure $\overline{K}$ of K† then the compositum $N := (\sigma_1 L) \cdots (\sigma_n L)$ is a normal closure of $(L : K)$.*

(b) *If N is any normal closure of $(L : K)$ then there are subfields $L_1, \ldots, L_r$, all K-isomorphic with L, such that $N = L_1 \cdots L_r$.*

Proof. (a) Let us show first that N is normal over K. If $\tau : N \to \overline{K}$ is a K-embedding then $\tau\sigma_i : K \to \overline{K}$ is also a K-embedding, hence must be one of the σ_j. This shows that the elements $\tau\sigma_1, \ldots, \tau\sigma_n$ are up to order the same as $\sigma_1, \ldots, \sigma_n$; this clearly shows that τ maps N into itself. Second, every normal extension of K containing L must also contain all $\sigma_i(L)$ and hence must contain N.

(b) This follows immediately from part (a) because all normal closures of $(L : K)$ are K-isomorphic, due to (13.17)(b). To give a different proof, let $L = K(\alpha_1, \ldots, \alpha_n)$ and let $N = K(\beta_1, \ldots, \beta_r)$ where the β_i are the conjugates of the α_i. Thus for each index i there is an index j such that β_i is a conjugate of α_j; hence there is a K-isomorphism $\sigma : K(\alpha_j) \to K(\beta_i)$ with $\sigma(\alpha_j) = \beta_i$, due to (13.13). Then $L_i := \sigma(L)$ is clearly a subfield of N which is K-isomorphic with L and which contains $\sigma(\alpha_j) = \beta_i$; thus $N = K(\beta_1, \ldots, \beta_r) \subseteq L_1 \cdots L_r \subseteq N$. ∎

---

† The argument used in the proof of (13.13) shows immediately that there is only a finite number of these embeddings.

AS an application of the concept of a splitting field, let us now show how this concept can be used to determine the structure of finite fields. Recall that the characteristic of a finite field K must be a prime number p and that its cardinality $|K|$ must be a power of p, due to (12.15). In the following theorem, we will obtain more precise information.

(13.19) Theorem. (a) *K is a field with $|K| = p^n =: q$ if and only if K is a splitting field of*

$$f(x) := x^q - x \in \mathbb{Z}_p[x] ;$$

in this case, every element of K is a root of f. In particular, for every prime power q, there is – up to isomorphism – a unique field K with $|K| = q$.

(b) *Let K be a field with $|K| = p^n =: q$. Then there is an irreducible polynomial $f \in \mathbb{Z}_p[x]$ such that $K \cong \mathbb{Z}_p[x]_f$. Each such polynomial f has degree n.*

(c) *Let L be a field with $|L| = p^n$. If K is a subfield of L, then $|K| = p^m$ where m divides n. Conversely, for each divisor m of n, there is a unique subfield K of L with $|K| = p^m$, namely $K = \{x \in L \mid x^{p^m} = x\}$.*

Proof. (a) Suppose that $|K| = q$. Then $|K^\times| = q - 1$, so that $x^{q-1} = 1$ for all $x \in K^\times$, by Lagrange's theorem (I.21.18). Consequently, $x^q = x$ for all $x \in K = K^\times \cup \{0\}$; i.e., every element of K is a root of f. Hence f cannot split over any proper subfield of K. But f splits over K, because f has degree q and so cannot have more than q roots in a field. This shows that K is a splitting field of f.

Conversely, let K be a splitting field of f. Then f has q roots in K, counted with multiplicity. But since $f'(x) = qx^{q-1} - 1 = p^n x^{p^n-1} - 1 = -1 \neq 0$, these roots are pairwise distinct. As a splitting field, K is generated by the q roots. But these roots themselves form a field, because the mapping $x \mapsto x^q$ preserves addition and multiplication, by (2.16)(j).

(b) Let α be a generator of $K^\times$ and let f be the minimal polynomial of α over $\mathbb{Z}_p$. Then the mapping

$$\begin{array}{ccc} \mathbb{Z}_p[x]_f & \to & K \\ [\varphi] & \mapsto & \varphi(\alpha) \end{array}$$

is well-defined (since $f(\alpha) = 0$), surjective (since α generates $K^\times$) and one-to-one (since the domain of this ring homomorphism is a field), hence is an isomorphism. Furthermore, if f is any polynomial such that $K \cong \mathbb{Z}_p[x]_f$, then $n = [K : \mathbb{Z}_p] = [\mathbb{Z}_p[x]_f : \mathbb{Z}_p] = \deg f$.

(c) Let $K \subseteq L$. Since char $K = $ char $L = p$, we have $|K| = p^m$ where $m = [K : \mathbb{Z}_p]$. The degree formula yields $n = [L : \mathbb{Z}_p] = [L : K] \cdot [K : \mathbb{Z}_p] = [L : K] \cdot m$ so that m divides n. Suppose conversely that m divides n. Then $p^m - 1$ divides $p^n - 1$ so that $g(x) := x^{p^m-1} - 1$ divides $f(x) := x^{p^n-1} - 1$. Now L is a splitting field of f, by part (a), hence contains a splitting field K of g which is a field with p^m elements, again by part (a). Invoking part (a) for the third time by using the fact that *every* element of K is a root of g, we see that

$$K = \{x \in L \mid x^{p^m} = x\} ;$$

this gives the uniqueness claim. ∎

(13.20) Proposition. *Let K be a finite field with q elements.*

(a) *For each $d \in \mathbb{N}$ there is an irreducible polynomial $f \in K[x]$ with $\deg f = d$; each such polynomial divides $x^{q^d} - x$.*

(b) *An irreducible polynomial $f \in K[x]$ divides $x^{q^n} - x$ if and only if $\deg f$ divides n.*

(c) *The number of irreducible monic polynomials of degree n over K is*

$$\frac{1}{n} \sum_{d \mid n} \mu(d) q^{n/d}$$

where μ is the Möbius function. If n is a prime number this number is $(q^n - q)/n$.

Proof. (a) Given a natural number d, we know from (13.19) that there is a field $L \supseteq K$ with $[L : K] = d$. Let α be a generator of $L^\times$ and let f be the minimal polynomial of α over K; then f is irreducible with $\deg f = [K(\alpha) : K] = [L : K] = d$.

Conversely, given an irreducible polynomial $f \in K[x]$ of degree d, let α be any root of f in some extension of K. Then $K(\alpha)$ is a field which has degree d over K and hence has q^d elements. This entails that $K(\alpha)$ is a splitting field for $x^{q^d} - x$, due to (13.19)(a). Since $x^{q^d} - x$ has q^d different roots and since $K(\alpha)$ has q^d elements this implies that α is a root of $x^{q^d} - x$. We have shown that each root of f is also a root of $x^{q^d} - x$; this implies that f divides $x^{q^d} - x$.

(b) Let $\deg f = d$. If d divides n then $q^d - 1$ divides $q^n - 1$ in $\mathbb{N}$, hence $x^{q^d - 1}$ divides $x^{q^n - 1}$. Multiplying by x we see that $x^{q^d} - x$ divides $x^{q^n} - x$. Since f divides $x^{q^d} - x$ by part (a) we conclude that f divides $x^{q^n} - x$.

Assume conversely that f divides $x^{q^n} - x$. Let L be a splitting field of $x^{q^n} - x$; then $|L| = q^n = |K|^n$ so that $n = [L : K]$. Now let $\alpha \in L$ be a root of f; then $d = [K(\alpha) : K]$ divides $[L : K] = n$.

(c) For each $d \in \mathbb{N}$ we denote by S_d be the set of all irreducible monic polynomials of degree d in $K[x]$ and by $\alpha(d)$ the number of irreducible monic polynomials of degree d in $K[x]$. By part (a) we have $\alpha(d) = |S_d| \neq 0$, and part (b) shows that

$$x^{q^n} - x = \prod_{d \mid n} \prod_{f \in S_d} f(x) .$$

Taking degrees on both sides of this equation we obtain

$$q^n = \sum_{d \mid n} d \cdot \alpha(d) .$$

Then the Möbius inversion formula yields

$$\alpha(n) = \frac{1}{n} \sum_{d \mid n} \mu(d) q^{n/d}$$

which is the claim. If n is a prime number this formula becomes $\alpha(n) = \frac{1}{n}\big(\mu(1)q^n + \mu(n)q^1\big) = (q^n - q)/n$. ∎

THE fact that if $|K| = q$ and if $f \in K[x]$ is irreducible of degree d then $f \mid x^{q^d} - x$ can sometimes be applied to prove that a polynomial over a finite field is irreducible.

(13.21) Example. Let us show that $f(x) = x^5 - x - 1$ is irreducible in $\mathbb{Z}_3[x]$. If f had a linear factor g then g would divide $x^3 - x$; if f had a quadratic factor h then h would divide $x^9 - x$. Hence every proper factor of f would have a common factor with $x^9 - x$, hence also with $x(x^9 - x) = x^{10} - x^2 = (x^5 - x)(x^5 + x)$ which is obviously impossible.

NOTE that the availability of the concept of a splitting field enabled us to completely determine the structure of finite fields. In particular, we obtained the remarkable fact that for each prime power q there is, up to isomorphism, a unique field with q elements. In honor of Evariste Galois the finite fields are also called *Galois fields*.

(13.22) Definition. *Let q be a prime power. Then the unique field with q elements is called the* **Galois field** *of order q and is denoted by* GF(q).

Exercises

Problem 1. Let $f \in K[x]$ be an irreducible polynomial of degree n. Let L be the set of all formal expressions $k_0 + k_1\alpha + \cdots + k_{n-1}\alpha^{n-1}$ with coefficients $k_i \in K$.† Define addition and multiplication on L in the obvious way with the understanding that all powers α^k with $k \geq n$ emerging from a product expansion are reduced to polynomial expressions in $\alpha^0 = 1, \alpha^2, \ldots, \alpha^{n-1}$ by using the formal equation $f(\alpha) = 0$. (In other words: If $p, q \in K[x]$ are polynomials of degree $\leq n-1$, define $p(\alpha)q(\alpha) := r(\alpha)$ where r is the remainder of pq when divided by f in $K[x]$.)

(a) Show that $(L, +, \cdot)$ is a field.

Hint. Let $q \in K[x]$ be a nonzero polynomial of degree $\leq n-1$. To show that $q(\alpha)$ is invertible in L, use the fact that there are polynomials $p_1, p_2 \in K[x]$ with $qp_1 + fp_2 = 1$.

(b) Argue that L is really "the same" as the field $K[x]/\langle f \rangle$.

Problem 2. Let $K = \mathbb{Z}_2$ and $f(x) = x^3 + x + 1 \in K[x]$. Consider K as a subfield of $L := K[x]_f$.

(a) Verify that f is irreducible over K so that L is in fact a field. Find the addition table and the multiplication table of L.

(b) Let $\alpha := [x] \in L$; then $\alpha^3 + \alpha + 1 = 0$ and $L = K(\alpha)$. For each element $\varepsilon \in \{\alpha^2 + 1, \alpha^2 + \alpha, \alpha^2 + \alpha + 1\}$, calculate some powers of ε and find the minimal polynomial of ε over K.

(c) Find a generator of the cyclic group $L^\times$.

(d) Express $[x+1]^{-1} \in L$ and $[x^2 + x + 1]^{99} \in L$ as polynomial expressions in α.

(e) Determine $[M : K]$ where M is a splitting field of p over K.

Problem 3. (a) Determine in each of the following cases whether or not there is a K-isomorphism $K(\alpha) \to K(\beta)$.

$$K = \mathbb{R}, \ \alpha = i\sqrt{2}, \ \beta = i\sqrt{3}, \qquad K = \mathbb{Q}, \ \alpha = \sqrt{2}, \ \beta = i$$

(b) Let $f_1(x) = x^2 - a$ and $g_2(x) = x^2 - x + b$ be irreducible polynomials over a field K of characteristic 2 with splitting fields L_1 and L_2. Is there a K-isomorphism between L_1 and L_2?

Problem 4. In each of the following examples verify that f is irreducible over $\mathbb{Q}$ and find the prime factorization of f over $\mathbb{Q}(\alpha)$ where α is a root of f.

(a) $f(x) = x^4 + 1$

(b) $f(x) = x^3 - 5x^2 + 6x - 1$

(c) $f(x) = x^3 - 6x^2 + 7x - 1$

---

† To avoid talking about "formal expressions" $\sum_i k_i \alpha^i$ (with the symbol α being undefined), one might also define L simply as the set of all n-tuples $(k_0, \ldots, k_{n-1})$; the symbol α is introduced to ease the definition of the multiplication on L. Compare with the introduction to polynomial rings in (2.6) and (2.7).

Problem 5. Consider the polynomial $f(x) = x^3 - 3x + 1$ over an arbitrary field K. Show that if α is a root of f (in some extension of K) then so is $\alpha^2 - 2$. Conclude that f splits over $K(\alpha)$. Deduce that f is either irreducible over K or else splits over K.

Problem 6. For each of the following polynomials over $\mathbb{Q}$, find the splitting field $L \subseteq \mathbb{C}$ and determine $[L : \mathbb{Q}]$.
(a) $x^4 + x^2 + 1$, $x^4 - 2x^2 + 2$, $x^4 - 8x^2 + 15$, $x^4 + x^3 - x^2 - 2x - 2$, $x^4 - 2x^2 + 8x + 6$
(b) $x^4 + 1$, $x^4 - 2$, $x^5 - 1$, $x^6 + 1$, $x^4 + x + 1$, $x^9 + x^3 + 1$
In each case, try to find an element $\alpha \in \mathbb{C}$ such that $L = \mathbb{Q}(\alpha)$.

Problem 7. Find splitting fields of the following polynomials.
(a) $x^2 + x + 1$ over $\mathbb{Z}_2$
(b) $x^3 + x^2 + 1$ over $\mathbb{Z}_3$
(c) $(x^2 + 2)(x - 1)(x^2 + x - 1)$ over $\mathbb{Z}_3$
(d) $x^4 + x^3 + x^2 + x + 1$ over $\mathbb{Z}_5$
(e) $x^5 + x + 1$ over $\mathbb{Z}_2$.
(f) $x^6 + 1$ over $\mathbb{Z}_2$
In each case determine the degree of the splitting field over the base field.

Problem 8. Let $p(x) = x^4 - x \in \mathbb{Z}_2[x]$ and let $L \supseteq \mathbb{Z}_2$ be a splitting field of p. Let $\alpha \in L$ be different from $0, 1$ such that $p(\alpha) = 0$. Show that $p(\alpha + 1) = 0$ and conclude that $L = \{0, 1, \alpha, \alpha + 1\}$. Find the addition table and the multiplication table of L.

Problem 9. (a) Let α, β, γ be the complex root of $f(x) = x^3 + x + 1 \in \mathbb{Q}[x]$. Show that $\mathbb{Q}(\beta/\alpha)$ is a splitting field of f and determine its degree over $\mathbb{Q}$.
(b) Let a be a rational number which is not a cube in $\mathbb{Q}$ and let K be the splitting field of $f(x) = x^3 - a$. What is $[K : \mathbb{Q}]$?

Problem 10. Let $K \subseteq Z \subseteq L$ be fields and let $\mathfrak{F} \subseteq K[x]$ be a family of polynomials. Show that if L is a splitting field of $\mathfrak{F}$ over K, then also over Z. Conversely, show that if L is a splitting field of L over Z and Z has the special form $Z = K(A)$ where each $\alpha \in A$ is a root of some $f \in \mathfrak{F}$, then L is a splitting field of $\mathfrak{F}$ over K.

Problem 11. (a) Let $L = K(A)$ where each element $\alpha \in A$ is a root of a quadratic polynomial over K. Show that $(L : K)$ is normal.
(b) Let $\alpha := (1 + i)\sqrt[4]{5}$. Show that $(\mathbb{Q}(\alpha) : \mathbb{Q}(i\sqrt{5}))$ and $(\mathbb{Q}(i\sqrt{5}) : \mathbb{Q})$ are normal extensions but that $(\mathbb{Q}(\alpha) : \mathbb{Q})$ is not.

Problem 12. (a) Give an example for a field extension of degree 3 which is normal.
(b) Give an example for a field extension of degree 3 which is not normal.
(c) Let $(L : K)$ be a field extension of degree 3 which is not normal and let N be a normal closure of $(L : K)$. Show that $[N : K] = 6$ and that there is a unique intermediate field Z of $(N : K)$ such that $[Z : K] = 2$.

Problem 13. Let $K = \mathbb{Q}(\sqrt{2})$ and $L = \mathbb{Q}(\sqrt[4]{2})$.

(a) Show that the automorphism of K given by $\sigma(a + b\sqrt{2}) = a - b\sqrt{2}$ cannot be extended to an automorphism of L. **Hint.** $\sqrt{2}$ is a square in L, whereas $-\sqrt{2}$ is not.

(b) Find the normal closure L' of $(L : K)$ (as a subfield of $\mathbb{C}$) and find an automorphism $\Sigma : L' \to L'$ extending σ.

Problem 14. Is there a number $x \in \mathbb{R} \setminus \mathbb{Q}$ such that the fields $\mathbb{Q}(x)$ and $\mathbb{Q}(ix)$ are isomorphic? Is there a number $y \in \mathbb{R} \setminus \mathbb{Q}$ such that the fields $\mathbb{Q}(y)$ and $\mathbb{Q}(iy)$ are non-isomorphic?

Problem 15. Let $f \in \mathbb{Q}[x]$ be an irreducible polynomial. Show that if one of the roots of f is constructible with a ruler and a compass then all the roots of f are constructible.

Problem 16. Show that a finite field $K = \{a_1, \ldots, a_n\}$ cannot be algebraically closed. (**Hint.** Consider the polynomial $f(x) = (x - a_1) \cdots (x - a_n) + 1$.)

Problem 17.† (a) Let $p \in \mathbb{C}[x]$ be a nonconstant polynomial such that $|p(z)| \geq |p(0)|$ for all $z \in \mathbb{C}$. Show that $p(0) = 0$.

Hint. Let $p(z) = a_0 + a_k z^k + O(z^{k+1})$ where $a_k \neq 0$. Write $-a_0/a_k = re^{i\varphi}$ and let $\omega := \sqrt[k]{r}e^{i\varphi/k}$ so that $a_k \omega^k = -a_0$. Then $p(t\omega) = a_0 - t^k(a_0 - O(t))$ and hence $|p(t\omega)| \leq | \, |a_0| - t^k|a_0 - O(t)| \, |$. Show that if we had $|a_0| \neq 0$ then this expression could be made smaller than $|a_0|$ by choosing $t > 0$ sufficiently close to 0, contradicting the hypothesis that $|p(z)| \geq |a_0|$ for all $z \in \mathbb{C}$. Hence we must have $a_0 = 0$.

(b) Show that any nonconstant polynomial $f \in \mathbb{C}[x]$ has a root in $\mathbb{C}$.

Hint. Show first that $|f(z)| \to \infty$ as $|z| \to \infty$ and conclude (by restricting f to a suitable compact subset of $\mathbb{C}$) that f attains a minimum, say at $z = z_0$. Then apply part (a) to $p(z) := f(z + z_0)$.

Problem 18. Let L be an algebraic closure of K. Show that if K is finite then L is denumerable and if K is infinite then $|K| = |L|$.

Problem 19. (a) Suppose that $K \subseteq L$ are fields such that L is algebraically closed. Show that $Z := \{\alpha \in L \mid \alpha$ is algebraic over $K\}$ is an algebraic closure of K.

(b) Suppose that the field K is algebraically closed in all of its extensions L; i.e., whenever $L \supseteq K$ is an extension of K then $\{\alpha \in L \mid \alpha$ is algebraic over $K\} = K$. Show that K is algebraically closed.

---

† This problem indicates a proof of the fact that the field $\mathbb{C}$ of complex numbers is algebraically closed. This fact can be easily derived by function-theoretical means (for example Liouville's theorem), but it might seem desirable to give a more elementary proof. Note that since $\mathbb{C}$ is not defined in a purely algebraic way but involving a limiting process, some arguments from calculus are always required to prove this "fundamental theorem of algebra".

Problem 20. In this problem we want to construct an algebraic closure of the field $\mathbb{Z}_p$ where p is a prime. Let R be the integral closure of $\mathbb{Z}$ in $\mathbb{C}$.

(a) Show that there is a maximal ideal P of R with $p \in P$. (Then R/P is a field.)

(b) Show that $P \cap \mathbb{Z} = p\mathbb{Z}$ so that $\mathbb{Z}_p = \mathbb{Z}/p\mathbb{Z} = \mathbb{Z}/(P \cap \mathbb{Z}) \cong (\mathbb{Z} + P)/P$ can be considered as a subfield of R/P.

(c) Show that R/P is algebraically closed.

(d) Show that if R_0 is a subring of R containing P such that R_0/P is algebraically closed then $R = R_0$.

(e) Conclude that R/P is an algebraic closure of $\mathbb{Z}_p$.

Problem 21. Let C be an algebraic closure of $\mathbb{Z}_p$. Then C contains for each $n \in \mathbb{N}$ a unique subfield of order p^n which we denote by $\mathrm{GF}(p^n)$. Show that the subfields of C are exactly the sets $\bigcup_{n \in X} \mathrm{GF}(p^n)$ where $X \subseteq \mathbb{N}$ is a set with the following properties:

(1) $1 \in X$;

(2) if $n \in X$ and $d \mid n$ then $d \in X$;

(3) if $m, n \in X$ then $\mathrm{lcm}(m, n) \in X$.

Conclude that C has a nondenumerable number of subfields.

Problem 22. Let $K \subseteq L \subseteq M$ such that $[L : K] = n$. Show that there are at most n different K-embeddings $\varphi : L \to M$.

Problem 23. For each of the following extensions find the normal closure in $\mathbb{C}$.

(a) $(\mathbb{Q}(\sqrt[3]{2}, \sqrt[4]{5}, i) : \mathbb{Q})$

(b) $(\mathbb{Q}(\sqrt[3]{2}, \sqrt[4]{5}, i) : \mathbb{Q}(\sqrt[3]{2}))$

(c) $(\mathbb{Q}(\sqrt[8]{2}, i) : \mathbb{Q})$

(d) $(\mathbb{Q}(\sqrt[8]{3}, i) : \mathbb{Q})$

(e) $(\mathbb{Q}(\sqrt[8]{3}, i) : \mathbb{Q}(i))$

(f) $(\mathbb{Q}(\sqrt[5]{3}, i) : \mathbb{Q}(i))$

(g) $(\mathbb{Q}(e^{2\pi i/5}) : \mathbb{Q})$

Problem 24. (a) Let $\alpha \in \mathbb{C}$ be a complex number with $\alpha^2 = 1 + i$. Show that $\mathbb{Q}(\alpha, \sqrt{2})$ is a normal closure of $(\mathbb{Q}(\alpha) : \mathbb{Q})$.

(b) Let $n \in \mathbb{N}$ and let p be a prime number. Show that $\mathbb{Q}(\sqrt[n]{p}, e^{2\pi i/n})$ is a normal closure of $(\mathbb{Q}(\sqrt[n]{p}) : \mathbb{Q})$. Conclude that $(\mathbb{Q}(\sqrt[n]{p}) : \mathbb{Q})$ is normal if and only if $n \leq 2$.

Problem 25. Let $\alpha_1, \ldots, \alpha_n$ be algebraic elements over a field K with minimal polynomials $f_1, \ldots, f_n \in K[x]$, respectively. Show that $(K(\alpha_1, \ldots, \alpha_n) : K)$ is a normal extension if and only if $K(\alpha_1, \ldots, \alpha_n)$ is the splitting field of $f_1 \cdots f_n$.

Problem 26. Let $K \subseteq Z \subseteq L$ be fields and let N be a normal closure of $(L : K)$. Is N necessarily a normal closure of $(L : Z)$?

Problem 27. (a) Show that an algebraic extension $(L : K)$ is normal if and only if each irreducible polynomial $f \in K[x]$ has a prime decomposition in $L[x]$ such that all factors have the same degree.

(b) Let $(L : K)$ be a finite normal extension. Show that if $f \in K[x]$ is irreducible over K and if g, h are irreducible monic factors of f over L, then there is a K-homomorphism $\sigma : L \to L$ with $h = \sigma \star g$.

(c) Use the example $K = \mathbb{Q}$ and $L = \mathbb{Q}(\sqrt[3]{2})$ to show that the claim in (b) need not be true if $(L : K)$ is not normal.

Problem 28. Let $(L_1 : K)$ and $(L_2 : K)$ be finite normal extensions. Show that the following two conditions are equivalent:

(1) there is a K-homomorphism $\sigma : L_1 \to L_2$;

(2) there are polynomials $f, g \in K[x]$ with $g \mid f$ such that L_1 is a splitting field of g and L_2 is a splitting field of f over K.

Problem 29. (a) Let $K := \mathbb{Z}_2[x]/\langle x^3 + x + 1 \rangle$ and let $\alpha := [x]$. Show that K is a field with 8 elements and that α is a generator of $(K^\times, \cdot)$.

(b) Let $K := \mathbb{Z}_3[x]/\langle x^2 + 1 \rangle$ and let $\alpha := [x]$. Show that K is a field with 9 elements and that $\alpha + 1$ generates $(K^\times, \cdot)$.

Problem 30. Explain how to construct fields with 25 and 49 elements.

Problem 31. Let K be a field with 8 elements. Is $x^3 + x^2 + 1 \in K[x]$ irreducible?

Problem 32. (a) Find all irreducible quadratic polynomials in $\mathbb{Z}_5[x]$.

(b) Determine all irreducible polynomials of degree ≤ 3 over $\mathbb{Z}_3$.

(c) How many irreducible monic polynomials of degree 2 are there in $\mathbb{Z}_{29}[x]$.

(d) How many monic irreducible polynomials of degree 9 are there in $\mathbb{Z}_3[x]$?

(e) How many irreducible monic polynomials of degree 8 are there over $\mathrm{GF}(256)$?

Problem 33. (a) How many subfields are there in $\mathrm{GF}(p^n)$?

(b) How many elements α with $\mathrm{GF}(p^n) = \mathbb{Z}_p(\alpha)$ are there in $\mathrm{GF}(p^n)$?

Problem 34. Let $L := \mathbb{Z}_2[x]_{x^6 + x + 1}$ and $\alpha := [x] \in L$.

(a) Show that α generates $L^\times$.

(b) Find the minimal polynomial of α^{21} over $\mathbb{Z}_2$.

Problem 35. Let K be a finite field with q elements. Show that if α is algebraic over K then $\alpha^{q^m} = \alpha$ for some $m \in \mathbb{N}$.

Problem 36. Let $f \in K[x]$ be a polynomial with splitting field L. Prove that if *all* elements of L are roots of f, then char K is a prime p, and $f(x) = x^{p^n} - x$ for some n.

Problem 37. Let K be a finite field with $q = p^n$ elements where char $K = p$.

(a) Using the Frobenius homomorphism $x \mapsto x^2$, show that if $p = 2$ then every element of K is a square.

(b) Show that if $p > 2$, then an element $x \neq 0$ in K is a square if and only if $x^{(q-1)/2} = 1$ and is a non-square if and only if $x^{(q-1)/2} = -1$. Conclude that the squares form a subgroup of $(K^\times, \cdot)$ of index 2.

Hint. Fix $x \neq 0$ and let L be an extension of K which contains an element y with $y^2 = x$. Using the fact that $K = \{\alpha \in L \mid \alpha^q = \alpha\}$, show that $y \in K$ if and only if $y^{q-1} \neq 1$.

(c) Show that every element of a finite field can be written as the sum of two squares.

Hint. If $|K| = 2^n$ then every element of K is a square. If $|K| = q$ is odd, argue that for any $a \in K$ the sets $\{a - x^2 \mid x \in K\}$ and $\{y^2 \mid y \in K\}$ have $\frac{1}{2}(q+1)$ elements and hence must have an element in common.

Problem 38. Let K be a field with q elements. Using the fact that $K^\times$ is cyclic of order $q - 1$, show that the element $\sum_{x \in K} x^n \in K$ is given by

$$\sum_{x \in K} x^n = \begin{cases} -1, & \text{if } n \in \mathbb{N} \text{ is a multiple of } q - 1; \\ 0, & \text{if } n \in \mathbb{N} \text{ is not a multiple of } q - 1. \end{cases}$$

Hint. If $s := \sum_{x \in K} x^n$ and $y \neq 0$ then $s = y^n s$.

Problem 39. Let K be a finite field with algebraic closure $\overline{K}$. Let $\alpha \in \overline{K}$. Show that $q := |K(\alpha)|$ is a finite number and that $\alpha^{q-1} = 1$. (In particular, every nonzero element of $\overline{K}$ is a root of unity.)

Problem 40. Let p and q be distinct primes. Show that if p divides none of the numbers $q - 1$, $q^2 - 1$, $q^3 - 1$, ..., $q^{p-2} - 1$ then $(x^p - 1)/(x - 1)$ is irreducible over $\mathbb{Z}_q$.

Hint. $x^p - 1$ splits over the field with q^d elements if and only if p divides $q^d - 1$.

Problem 41. Let $\alpha_1, \ldots, \alpha_n$ be the roots of a polynomial $f \in K[x]$ (in some extension of K). Show that every element of $K(\alpha_1, \ldots, \alpha_n)$ can be written as a sum of terms of the form $c \cdot \alpha_1^{i_1} \alpha_2^{i_2} \cdots \alpha_{n-1}^{i_{n-1}}$ where $c \in K$ and where $i_1 \leq n - 1$, $i_2 \leq n - 1$, ..., $i_{n-1} \leq 1$. (In particular, any of the n roots can be expressed by the $n - 1$ other roots over K.) **Hint.** Use problem 42 in section 4.

Problem 42. (a) Show that a polynomial $f \in K[x]$ is irreducible if and only if it has a root α (in some extension of K) such that $[K(\alpha) : K] = \deg f$.

(b) Let K be a field and let $a \in K$. Suppose that $m, n \in \mathbb{N}$ are relatively prime. Use part (a) to show that $x^{mn} - a \in K[x]$ is irreducible if and only if $x^m - a$ and $x^n - a$ are irreducible.

(c) Use part (b) to show that $x^6 + 4 \in \mathbb{Q}[x]$ is irreducible. (Compare with (12.13)(a).)

14. Separability of field extensions

WE have seen that for every polynomial $f \in K[x]$ over a field K there is a field $L \supseteq K$ over which f splits, say $f(x) = c(x - \alpha_1) \cdots (x - \alpha_n)$ where $n = \deg f$. Now it is possible that the roots $\alpha_1, \ldots, \alpha_n$ are not pairwise distinct even if f is irreducible.

(14.1) Example. Let K_0 be a field of characteristic $p \neq 0$ (for example, we might take $K_0 = \mathbb{Z}_p$) and let $K = K_0(t)$ be the rational function field in the indeterminate t over K_0. Then the polynomial $f(x) := x^p - t \in K[x]$ is irreducible by Eisenstein's criterion (clearly, t is a prime element in $K_0[t]$), but has only a single root in its splitting field: If α is a root of f, then $\alpha^p = t$ and hence $f(x) = x^p - t = x^p - \alpha^p = (x - \alpha)^p$, due to the arithmetic rules in characteristic p.

TO investigate this peculiar phenomenon, we introduce the following terminology.

(14.2) Definitions. *Let K be a field.*
(a) *An irreducible polynomial $f \in K[x]$ is called* **separable** *if it has no multiple roots (in its splitting field).*†
(b) *An element α which is algebraic over K is called* **separable** *over K if its minimal polynomial over K is separable.*
(c) *An algebraic field extension $(L : K)$ is called* **separable** *if every element $\alpha \in L$ is separable over K.*

LET us give a characterization of separable polynomials and then look at some examples.

(14.3) Proposition. *Let K be a field and let $f \in K[x]$ be an irreducible polynomial. Then the following conditions are equivalent:*
(1) *f divides f';*
(2) *$f' = 0$;*
(3) *$\operatorname{char} K$ is a prime number p, and f has the form $f(x) = g(x^p)$ for some polynomial $g \in K[x]$;*
(4) *f is not separable.*

Proof. (1)$\Longrightarrow$(2). This implication is clear because f' has smaller degree than f.
(2)$\Longrightarrow$(3). If $f(x) = \sum_{k=0}^{n} a_k x^k$ then $f'(x) = \sum_{k=1}^{n} k a_k x^{k-1}$; hence $f' = 0$ if and only if $k \cdot 1 = 0$ in K whenever $a_k \neq 0$. This gives the claim.
(3)$\Longrightarrow$(4). Let α be a root of f in some splitting field of f; then α^p is a root of g which means that $x - \alpha^p$ divides $g(x)$. But then $(x - \alpha)^p = x^p - \alpha^p$ divides $g(x^p) = f(x)$ which shows that α is at least a p-fold root of f.
(4)$\Longrightarrow$(2). If $f(x) = g(x^p) = \sum_{k=0}^{n} a_k x^{pk}$ then $f'(x) = \sum_{k=1}^{n} pk a_k x^{pk-1} = 0$.
(2)$\Longrightarrow$(1). This implication is trivial. ∎

---

† This terminology (due to van der Waerden) expresses the idea that the roots of f "lie separately".

(14.4) Examples. (a) If char $K = 0$, then every algebraic field extension $(L : K)$ is separable; this is an immediate consequence of proposition (14.3).

(b) Let $K = \mathbb{Z}_p(t)$ and let α be a root of $f(x) = x^p - t \in K[x]$. Then the field extension $(K(\alpha) : K)$ is not separable, as was shown in example (14.1).

(c) If $(L : K)$ is a separable field extension and if Z is an intermediate field, then $(L : Z)$ and $(Z : K)$ are also separable extensions.

Here the separability of $(Z : K)$ is trivial. To prove that $(L : Z)$ is separable, let $\alpha \in L$ and let f and g be the minimal polynomials of α over K and Z, respectively. Then f is separable by hypothesis, and since g divides f in $Z[x]$, the polynomial g must also be separable. This shows that α is separable over Z.

(14.5) Definition. *A field is called* **perfect** *if all of its algebraic extensions are separable (or equivalently, if every irreducible polynomial with coefficients in this field is separable).*

IT is quite easy to give a complete characterization of perfect fields. This characterization will show that most fields we have to do with are, in fact, perfect. Thus the property of being perfect is not as exclusive as the word may suggest.

(14.6) Theorem. *Let K be a field.*

(a) *If char $K = 0$ then K is perfect.*

(b) *If char $K = p \neq 0$, then K is perfect if and only if every element of K has a p-th root in K; i.e., if and only if the Frobenius homomorphism $\sigma : K \to K$ given by $\sigma : x \mapsto x^p$ is surjective (and hence an automorphism of K).*

(c) *If K is finite then K is perfect.*

(d) *If K is algebraically closed then K is perfect.*

Proof. (a) This follows immediately from (14.4)(a).

(b) Suppose that K is perfect. Let $\alpha \in K$ and let L be a splitting field of $f(x) := x^p - \alpha$. We claim that f has only a single root $\beta \in L$; indeed, if β' is a second root, then $(\beta' - \beta)^p = \beta'^p - \beta^p = f(\beta') - f(\beta) = 0$. Therefore, $f(x) = (x - \beta)^p$. Now the minimal polynomial g of β over K is separable by hypothesis and divides f; this implies $g(x) = x - \beta$, whence $\beta \in K$.

Suppose conversely that K is not perfect. Then there is an irreducible polynomial $p \in K[x]$ which is not separable, i.e., satisfies $p' = 0$ and hence has the form $p(x) = \sum_i a_i x^{pi}$. By assumption, we can find for each of the coefficients $a_i \in K$ an element $b_i \in K$ with $a_i = b_i^p$. But then $p(x) = \sum_i b_i^p x^{pi} = (\sum_i b_i x^i)^p$, contradicting the irreducibility of p.

(c) The Frobenius homomorphism $\varphi : K \to K$, given by $\varphi(x) = x^p$, is injective, hence also surjective because K is finite. Hence every element of K possesses a p-th root in K, so that part (b) gives the claim.

(d) Distinguishing the cases char $K = 0$ and char $K \neq 0$, this is an immediate consequence of parts (a) and (b). ∎

DESPITE the terminology "perfect" (which suggests a rather special property), almost all the fields of practical interest *are* perfect, as theorem (14.6) shows. Let us present at least one example of a field which is not perfect.

(14.7) Example. Let p be a prime number and let $K = \mathbb{Z}_p(x)$ be the rational function field over $\mathbb{Z}_p$. Then K is not perfect by example (14.1).

BEFORE we enter into a systematic discussion of separability, let us prove a fact which is quite important for theoretical as well as for practical purposes, namely, that a finite separable extension is automatically simple, i.e., can be obtained by adjoining one single element to the basefield. We have already seen special cases of this phenomenon in section 12 where we derived the equations

$$\mathbb{Q}(\sqrt{2}, \sqrt{3}) = \mathbb{Q}(\sqrt{2} + \sqrt{3}), \quad \mathbb{Q}(i, \sqrt[3]{2}) = \mathbb{Q}(i\sqrt[3]{2}) \quad \text{and} \quad \mathbb{Q}(\sqrt{2}, \sqrt[3]{3}) = \mathbb{Q}(\sqrt{2}, \sqrt[3]{3}) .$$

(See also the problems in section 12.) However, there *are* finite field extensions which are not simple, as is shown by the following example.

(14.8) Example. Let $K := \mathbb{Z}_2(t)$ be the field of all rational functions over $\mathbb{Z}_2$, i.e., the field of all formal expressions $\frac{r(t)}{s(t)}$ with $r, s \in \mathbb{Z}_2[t]$ and $s \neq 0$. Consider the quadratic polynomials $F(x) := x^2 - t$ and $G(x) := x^2 - (t + t^3)$ in $K[x]$ and let α and β be roots of F and G, respectively, in a suitable extension of K, so that

$$\alpha^2 = t \quad \text{and} \quad \beta^2 = t + t^3 .$$

We leave it as an exercise to show that F is irreducible over $K(\beta)$ and that G is irreducible over $K(\alpha)$. (See problem 1 below.) Once this is established, we can conclude that

$$(1) \qquad \underbrace{[K(\alpha,\beta) : K]}_{} = \underbrace{[K(\alpha,\beta) : K(\alpha)]}_{=\,\deg G\,=\,2} \cdot \underbrace{[K(\alpha) : K]}_{=\,\deg F\,=\,2} = 2 \cdot 2 = 4 .$$

On the other hand, every element $\gamma \in K(\alpha,\beta)$ can be written in the form $\gamma = f(t) + g(t)\alpha + h(t)\beta$ with $f, g, h \in K$, due to the fact that α^2 and β^2 lie in K. Taking squares on both sides, we see that $\gamma^2 = f(t)^2 + g(t)^2\alpha^2 + h(t)^2\beta^2 = f(t)^2 + t\,g(t)^2 + (t+t^3)\,h(t)^2 \in K$; hence

$$(2) \qquad\qquad [K(\gamma) : K] \leq 2 \quad \text{for all } \gamma \in K(\alpha,\beta) .$$

A comparison of (1) and (2) shows immediately that there is no element γ with $K(\alpha,\beta) = K(\gamma)$; thus $\big(K(\alpha,\beta) : K\big)$ is not a simple field extension.

HAVING seen this example, we will now show that, roughly speaking, separability ensures simplicity.

(14.9) Theorem on primitive elements. (a) *Let α, β be algebraic elements over a field K such that α is separable. Then there is an element $\gamma \in K(\alpha, \beta)$ such that $K(\alpha, \beta) = K(\gamma)$. If K is infinite, the element γ can be chosen as a K-linear combination of α and β.*

(b) *Every finite separable field extension $(L : K)$ is simple. Moreover, if the field K is infinite and if $L = K(\alpha_1, \ldots, \alpha_n)$, then a primitive element γ can be chosen in the form $\gamma = \sum_i k_i \alpha_i$ with $k_i \in K$.*

Proof. (a) Let $L := K(\alpha, \beta)$. If K is finite, then so is L. By (4.14) the group $L^\times = L \setminus \{0\}$ is cyclic; in this case, every generator γ of $L^\times$ has the desired property. So we can assume that K is infinite.

Let $\alpha = \alpha_1, \alpha_2, \ldots, \alpha_r$ be the roots of f (which are pairwise different by assumption) and $\beta = \beta_1, \ldots, \beta_s$ the roots of g (which need not be pairwise different, because g need not be separable). We can suppose that $\alpha \notin K$; otherwise the claim is trivial. For any $x \in K$ let

$$W(x) := \{\alpha_i x + \beta_j \mid 2 \leq i \leq r, \ 1 \leq j \leq s\} \ .$$

Now the set $\{\frac{\beta_j - \beta}{\alpha - \alpha_i} \mid 2 \leq i \leq r, 1 \leq j \leq s\}$ is finite. So since $|K| = \infty$, there is an element $y \in K$ which is not contained in this set. Consequently,

$$\alpha y + \beta \notin W(y) \ ;$$

otherwise there were indices i, j such that $\alpha y + \beta = \alpha_i y + \beta_j$, i.e., $y = \frac{\beta_j - \beta}{\alpha - \alpha_i}$. Now we claim that

$$\gamma := \alpha y + \beta$$

is a primitive element of $K(\alpha, \beta)$. Obviously, $K(\gamma) \subseteq K(\alpha, \beta)$. To prove the converse inclusion, consider the polynomials

$$\widehat{g}(x) := g(\gamma - yx) \in K(\gamma)[x] \quad \text{and} \quad h := \gcd(f, \widehat{g}) \in K(\gamma)[x] \ .$$

We have $f(\alpha) = 0$ and $\widehat{g}(\alpha) = g(\gamma - y\alpha) = g(\beta) = 0$. So $x - \alpha$ is a common divisor of f and $\widehat{g}$, hence also of h (in a suitable extension, namely in $K(\alpha, \gamma)[x]$). Now we claim that α is the only root of h. Indeed, every other root would have to be a root of f, hence one of the elements α_i ($2 \leq i \leq r$), but this is impossible because $\widehat{g}(\alpha_i) = g(\gamma - y\alpha_i) \neq 0$ (note that $\gamma - y\alpha_i \neq \beta_j$ for all j because $\gamma \notin W(y)$).

Having only α as a root, h is of the form $h(x) = (x - \alpha)^N$ for some $N \in \mathbb{N}$ which means that α is an N-fold root of h, the more so of f. But f has only simple roots! Hence $N = 1$ so that $h(x) = x - \alpha$. Now by definition $h \in K(\gamma)[x]$, whence $\alpha \in K(\gamma)$ and consequently $\beta = \gamma - y\alpha \in K(\gamma)$. This shows the reverse inclusion $K(\alpha, \beta) \subseteq K(\gamma)$.

(b) Writing $L = K(\alpha_1, \ldots, \alpha_n)$, the claim follows immediately from (a) by induction on n. ∎

(14.10) Example. If $\alpha, \beta \in \mathbb{C}$ are algebraic over $\mathbb{Q}$ then there is a rational number $r \in \mathbb{Q}$ such that $\mathbb{Q}(\alpha, \beta) = \mathbb{Q}(\alpha + r\beta)$.

FOR most purposes, this is all one needs to know about separable field extensions. Thus the reader may wish to skip the following systematic study.

LET $(L : K)$ be an algebraic field extension. Whether or not an element $\alpha \in L$ is separable over K depends on the number of roots of its minimal polynomial $f \in K[x]$ (in some splitting field Z). Now, as we saw in (13.13), this number equals the number of possibilities of embedding $K(\alpha)$ into Z (or into any normal extension of K containing α). Therefore, it is not too surprising that we can obtain a characterization of separable field extensions via embeddings. We will do this first for finite extensions and afterwards treat the general case.

(14.11) Lemma. *Let $(L : K)$ be a finite field extension.*
(a) *Let N be a normal extension of K containing L. Then $|\mathrm{Emb}_K(L, N)| \leq [L : K]$ with equality if and only if $(L : K)$ is a separable extension.*
(b) *If Z is an intermediate field of $(L : K)$ such that $(L : Z)$ and $(Z : K)$ are separable, then $(L : K)$ is a separable extension.*

Proof. (a) We proceed by induction on $n := [L : K]$, the case $n = 1$ (i.e., $L = K$) being trivial. Let $n > 1$ and pick $\alpha \in L \setminus K$; then $[L : K(\alpha)] < n$. By (13.14) we have

(1) $\qquad |\mathrm{Emb}_K(L, N)| = |\mathrm{Emb}_K(K(\alpha), N)| \cdot |\mathrm{Emb}_{K(\alpha)}(L, N)|$,

and (13.13) shows that

(2) $\qquad |\mathrm{Emb}_K(K(\alpha), N)| = \text{number of roots of } f \leq \deg f = [K(\alpha) : K];$

clearly, we have equality in (2) if and only if α is separable over K. Moreover, by induction hypothesis, we have

(3) $\qquad |\mathrm{Emb}_{K(\alpha)}(L, N)| \leq [L : K(\alpha)]$

with equality if and only if $(L : K(\alpha))$ is separable. Conditions (1), (2) and (3) imply that

(4) $\qquad |\mathrm{Emb}_K(L, N)| \leq [K(\alpha) : K] \cdot [L : K(\alpha)] = [L : K]$.

It remains to determine when equality holds.

Suppose that $(L : K)$ is separable and let $\alpha \in L \setminus K$. Then $(L : K(\alpha))$ and $(K(\alpha) : K)$ are separable by (14.4)(c) so that equality must hold in (2) and (3), hence also in (4). Suppose conversely that equality holds in (4). Then we must have equality in (2) and (3); in particular, α must be separable over K. But this is true for an arbitrary element $\alpha \in L \setminus K$; hence $(L : K)$ is a separable field extension.
(b) This is an immediate consequence of part (a) and (13.14). ∎

THE characterization just obtained allows us to draw useful conclusions on separable field extensions. For example, it is not immediately clear from the definition that the separability of an element α implies the separability of elements such as $\alpha^3 + \alpha^2$ or $\alpha^{17} - \alpha + 1$. However, using (14.11), this will be shown in the next theorem. Once this result is available, we can generalize lemma (14.11) to possibly infinite algebraic extensions.

(14.12) Theorem. *Let* $(L : K)$ *be an algebraic field extension.*

(a) *If* $\alpha \in L$ *is separable over* K, *then so are all elements* $p(\alpha)$ *where* $p \in K[x]$; *this means that the field extension* $(K(\alpha) : K)$ *is separable.*

(b) *If* $L = K(A)$, *then* $(L : K)$ *is separable if and only if every element* $\alpha \in A$ *is separable over* K.

(c) *If* $(L : K)$ *is separable and if* N *is a normal closure of* $(L : K)$ *then* $(N : K)$ *is separable.*

(d) **(Behavior under the formation of composita.)** *Let* Z_1, Z_2 *be intermediate fields of* $(L : K)$. *If* $(Z_1 : K)$ *is separable then so is* $(Z_1 Z_2 : Z_2)$. *If* $(Z_1 : K)$ *and* $(Z_2 : K)$ *are separable then so is* $(Z_1 Z_2 : K)$.

(e) **(Transitivity of separability.)** *Let* Z *be an intermediate field of* $(L : K)$. *Then the extension* $(L : K)$ *is separable if and only if* $(L : Z)$ *and* $(Z : K)$ *are separable extensions.*

Proof. (a) Let f be the minimal polynomial of α over K and let N be any normal extension of K containing α. In (13.13) we saw that $|\mathrm{Emb}_K(K(\alpha), N)|$ is the number of roots of f; hence (14.11) implies that $(K(\alpha) : K)$ is a separable extension if and only if f is separable, i.e., if and only if α is separable.

(b) Trivially, if $(K(A) : K)$ is a separable extension, then every element of A is separable over K. Suppose conversely that every element of A is separable over K; we want to conclude that $(K(A) : K)$ is a separable extension. Assume first that A is finite so that we can proceed by induction on $|A|$. The case $|A| = 1$ was settled in part (a). Suppose the claim has been proved for $|A| = n$ and consider an extension $(K(\alpha_1, \ldots, \alpha_n, \alpha_{n+1}) : K)$. Then $(K(\alpha_1, \ldots, \alpha_n, \alpha_{n+1}) : K(\alpha_{n+1}))$ is separable by induction hypothesis, and $(K(\alpha_{n+1}) : K)$ is separable by part (a). Consequently, $(K(\alpha_1, \ldots, \alpha_n, \alpha_{n+1}) : K)$ is separable by (14.11)(b).

In the general case, let $\beta \in K(A)$. Then there is a finite subset A_0 of A with $\beta \in K(A_0)$. Since $(K(A_0) : K)$ is separable by what we just proved, the element β is separable. Since $\beta \in K(A)$ was an arbitrary element of L, we conclude that $(L : K)$ is a separable extension.

(c) Write $L = K(A)$; then $N = K(A')$ where A' is obtained from A by taking with each element $\alpha \in A$ also all conjugates of α. Now all elements of A are separable over K by part (b); but since (trivially) an element is separable if and only if all of its conjugates are, this implies that even all elements of A' are separable over K. Applying part (b) in the other direction, we conclude that $N = K(A')$ is separable over K.

(d) If $(Z_1 : K)$ is separable we can write $Z_1 = K(A)$ where each $\alpha \in A$ is separable over K. Then each such α is the more so separable over Z_2, and since $Z_1 Z_2 = Z_2(A)$ we conclude from part (b) that $(Z_1 Z_2 : Z_2)$ is separable. If also $(Z_2 : K)$ is separable we can write $Z_2 = K(B)$ where each $\beta \in B$ is separable over K; then $Z_1 Z_2 = K(A \cup B)$, and part (b) shows that $(Z_1 Z_2 : K)$ is separable in this case.

(e) One half of the statement was already proved in example (14.4)(c). Hence it is enough to show that if $(L : Z)$ and $(Z : K)$ are separable then so is $(L : K)$. This claim was established in (14.11)(b) for a finite extension. In the general case, let $\alpha \in L$ and let $f(x) = x^n + z_{n-1}x^{n-1} + \cdots + z_1 x + z_0 \in Z[x]$ be the minimal polynomial of α over Z; since $(L : Z)$ is separable by assumption, the polynomial f has no multiple roots. Now f is also the minimal polynomial of α over $Z' := K(z_0, \ldots, z_{n-1})$, and this field is separable over K by part (b). Since f has no multiple roots, the element α is separable over Z'; hence $(Z'(\alpha) : Z')$ is separable by part (a). Since also $(Z' : K)$ is separable and since the transitivity is clear for finite extensions by (14.11)(b), we conclude that $(Z'(\alpha) : K)$ is separable; in particular, α is separable over K. Since $\alpha \in L$ was arbitrarily chosen, this shows that $(L : K)$ is separable. ∎

(14.13) Theorem. Let $(L : K)$ be an algebraic field extension and let N be a normal extension of K containing L. Then

$$|\mathrm{Emb}_K(L, N)| \leq [L : K]$$

with equality if and only if $(L : K)$ is a separable extension.

Proof. (a) Write $L = K(A)$ with some set A of algebraic elements and let $(\alpha_i)_{i \in I}$ be a well-ordering of A; for the sake of convenience, we arrange that there is a maximal index $i_{\max}$. We now form the intermediate fields $Z_i := K(\alpha_j \mid j \leq i)$, obtained from K by successively adjoining the elements of A; then $K_{i_{\max}} = L$. We are done if we can show that for all $i \in I$ the following statement holds:

$(\star)$ $|\mathrm{Emb}_K(Z_i, N)| \leq [Z_i : K]$ with equality if and only if $(Z_i : K)$ is separable.

By the principle of transfinite induction, it is enough to show that the validity of $(\star)$ for all indices less than a given index j implies the validity of $(\star)$ for the index j. Let i be the immediate predecessor of j so that $Z_j = Z_i(\alpha_j)$. Considering the chain $K \subseteq Z_i \subseteq Z_j$ and applying (13.14), we have

(1) $$|\mathrm{Emb}_K(Z_j, N)| = |\mathrm{Emb}_K(Z_i, N)| \cdot |\mathrm{Emb}_{Z_i}(Z_j, N)| .$$

Also, if f is the minimal polynomial of α_j over Z_i then

(2)
$$\begin{aligned}
|\mathrm{Emb}_{Z_i}(Z_j, N)| &= |\mathrm{Emb}_{Z_i}(Z_i(\alpha_j) : N)| = \text{number of roots of } f \\
&\leq \deg f = [Z_i(\alpha_j) : Z_i] = [Z_j : Z_i]
\end{aligned}$$

with equality if and only if α_j is separable over Z_i. Moreover, by induction hypothesis, we have

(3) $$|\mathrm{Emb}_K(Z_i, N)| \leq [Z_i : K]$$

with equality if and only if $(Z_i : K)$ is separable. Conditions (1), (2) and (3) imply that

(4) $$|\mathrm{Emb}_K(Z_j, N)| \leq [Z_i : K] \cdot [Z_j : Z_i] = [Z_j : K] .$$

It remains to determine when equality holds.

Suppose $(Z_j : K)$ is separable. Then $(Z_j : Z_i)$ and $(Z_i : K)$ are separable by (14.4)(c) so that equality must hold in (2) and (3), hence also in (4). Suppose, conversely, that equality holds in (4). Then we must have equality in (2) and (3) so that the extensions $(Z_j : Z_i)$ and $(Z_i : K)$ are separable. Now (14.12)(d) implies that $(Z_j : K)$ is separable † as desired. ∎

TO ease a further study of separable extensions, we now introduce a concept which is, in some sense, the complete opposite of separability.

(14.14) Definition. (a) *An element which is algebraic over a field K is called* **purely inseparable** *over K if its minimal polynomial has only a single root (in its splitting field), i.e., can be written as $(x - \alpha)^n$ for some $n \in \mathbb{N}$.*

(b) *An algebraic field extension $(L : K)$ is called* **purely inseparable** *if every element $\alpha \in L$ is purely inseparable over K.*

OUR next goal is to prove that every algebraic field extension can be split into a separable extension and a purely inseparable extension. Before we can do so, we have to obtain some preliminary results.

(14.15) Proposition. *Let $(L : K)$ be an algebraic field extension.*

(a) *An element $\alpha \in L$ is both separable and purely inseparable over K if and only if it belongs to K.*

(b) *Suppose $\operatorname{char} K = p \neq 0$. For any element $\alpha \in L$ there is an integer $n \geq 0$ such that α^{p^n} is separable over K.*

(c) *Suppose $\operatorname{char} K = p \neq 0$. An element $\alpha \in L$ is purely inseparable over K if and only if there is an integer $n \geq 0$ with $\alpha^{p^n} \in K$. If such a number n exists and is chosen as small as possible then $x^{p^n} - \alpha^{p^n}$ is the minimal polynomial of α over K which implies that $[K(\alpha) : K] = p^n$.*

(d) *The extension $(L : K)$ is purely inseparable if and only if $L = K(A)$ for some set $A \subseteq L$ consisting of purely inseparable elements over K.*

(e) *Let Z be an intermediate field of $(L : K)$. If $\alpha \in L$ is purely inseparable over K, then it is also purely inseparable over Z.*

(f) **(Transitivity of pure inseparability.)** *Let Z be an intermediate field of $(L : K)$. Then $(L : K)$ is purely inseparable if and only if $(L : Z)$ and $(Z : K)$ are purely inseparable.*

(g) **(Behavior under the formation of composita.)** *Let Z_1, Z_2 be intermediate field of $(L : K)$. If $(Z_1 : K)$ is purely inseparable then so is $(Z_1 Z_2 : Z_2)$. If $(Z_1 : K)$ and $(Z_2 : K)$ are purely inseparable then so is $(Z_1 Z_2 : K)$.*

Proof. (a) If α is purely inseparable, its minimal polynomial over K is $f(x) = (x - \alpha)^n$ for some $n \in \mathbb{N}$. If α is also separable, then f has no multiple roots, which

---

† This argument was not yet available in the proof of (14.11)(b); therefore, we had to prove the finite case and the general case of the claim in question separately even though the proofs are very similar.

forces $n = 1$. This shows that α is both separable and purely inseparable if and only if its minimal polynomial over K is $f(x) = x - \alpha$, which is the case if and only if $x - \alpha \in K[x]$, i.e., if $\alpha \in K$.

(b) Let f be the minimal polynomial of α and choose n such that $f \in K[x^{p^n}]$ but $f \notin K[x^{p^{n+1}}]$; then there is a polynomial g with $f(x) = g(x^{p^n})$ and $g \notin K[x^p]$ (so that g is separable). We claim that g is irreducible; in fact, if $g = g_1 g_2$ then $f(x) = g(x^{p^n}) = g_1(x^{p^n})g_2(x^{p^n})$ which implies that g_1 or g_2 is a constant polynomial, due to the irreducibility of f. Thus g is irreducible and satisfies $g(\alpha^{p^n}) = f(\alpha) = 0$, hence is the minimal polynomial of α^{p^n}. Then α^{p^n} is separable because its minimal polynomial g is.

(c) Suppose $\alpha \in L$ is purely inseparable over K. Then the minimal polynomial f of α has a splitting of the form $f(x) = (x - \alpha)^m$. Write $m = p^n q$ where $p \nmid q$. Then

$$f(x) = (x - \alpha)^{p^n q} = (x^{p^n} - \alpha^{p^n})^q = (x^{p^n})^q - q\alpha^{p^n}(x^{p^n})^{q-1} + \text{lower powers of } x .$$

Since $f \in K[x]$, we have $q\alpha^{p^n} \in K$, which implies $\alpha^{p^n} \in K$ because $p \nmid q$. Suppose conversely that $\alpha^{p^n} \in K$ for some $n \in \mathbb{N}_0$; we can assume that n is chosen as small as possible. Then α is a root of the polynomial $f(x) = (x - \alpha)^{p^n} = x^{p^n} - \alpha^{p^n} \in K[x]$. We claim that f is the minimal polynomial of α (which clearly implies that α is purely inseparable). Indeed, every nonconstant factor g of f in $K[x]$ is also a factor of f in $L[x]$ and hence must have the form $(x - \alpha)^{p^m} = x^{p^m} - \alpha^{p^m}$ for some $m \leq n$ (up to a constant factor); but if $m < n$ then $\alpha^{p^m} \notin K$.

(d) If $(L : K)$ is purely inseparable, then all elements of $A := L$ are purely inseparable over K, and $L = K(A)$. Suppose conversely that $L = K(A)$ where every element of A is inseparable over K; we want to show that L is separable over K. Since in characteristic 0 every field extension is separable, we may assume that $\operatorname{char} K$ is a prime number p. Let $A = \{\alpha_i \mid i \in I\}$. By part (c), there is for each index i a number $n_i \in \mathbb{N}_0$ with $\alpha_i^{p^{n_i}} \in K$. Since every element of L is a polynomial expression in the elements α_i ($i \in I$), we conclude that for every element $\alpha \in L$ there is a number $n \geq 0$ with $\alpha^{p^n} \in K$. Considering part (c) again, this shows that every element α is purely inseparable over K.

(e) The minimal polynomial of α over Z divides the minimal polynomial of α over K; hence since the latter possesses only a single root then so does the former, which is what we wanted to show.

(f) If $(L : K)$ is purely inseparable, then $(Z : K)$ is trivially purely inseparable, and $(L : Z)$ is also purely inseparable by part (e). Suppose conversely that $(L : Z)$ and $(Z : K)$ are purely inseparable. To show that $(L : K)$ is purely inseparable, we can clearly assume that $\operatorname{char} K$ is a prime number p (otherwise $K = Z = L$). Let $\alpha \in L$. Since $(L : K)$ is purely inseparable, there is an m with $\alpha^{p^m} \in Z$, and since $(Z : K)$ is purely inseparable, there is an n with $(\alpha^{p^m})^{p^n} \in K$; here we used part (c) each time. Hence $\alpha^{p^m p^n} = \alpha^{p^{m+n}} \in K$; applying part (c) in the opposite direction, we see that α is purely inseparable over K. Since $\alpha \in L$ was an arbitrary element, this shows that $(L : K)$ is purely inseparable.

(g) If $(Z_1 : K)$ is purely inseparable we can write $Z_1 = K(A)$ where each $\alpha \in A$ is purely inseparable over K. But then each $\alpha \in A$ is also purely inseparable over Z_2 by part (e), and since $Z_1 Z_2 = Z_2(A)$ part (d) shows that $(Z_1 Z_2 : Z_2)$ is purely inseparable. If also $(Z_2 : K)$ is separable, say $Z_2 = K(B)$ where each $\beta \in B$ is purely inseparable over K, then we can write $Z_1 Z_2 = K(A \cup B)$, and part (d) shows that $(Z_1 Z_2 : K)$ is purely inseparable in this case. ∎

IN our next theorem we will show that an arbitrary algebraic field extension can be split into a separable extension and a purely inseparable extension.

(14.16) Theorem. *Let $(L : K)$ be an algebraic field extension.*
(a) *The sets*

$$S := \{x \in L \mid x \text{ is separable over } K\} \quad \text{and}$$
$$P := \{x \in L \mid x \text{ is purely inseparable over } K\}$$

*(called the **separable closure** and the **purely inseparable closure** of K in L, respectively) are field extensions of K with $S \cap P = K$.*
(b) *If $(L : K)$ is normal then so is $(S : K)$.*
(c) *$(S : K)$ is a separable extension whereas $(P : K)$ is a purely inseparable extension. Moreover, $(L : S)$ is a purely inseparable extension.*
(d) *$(L : P)$ is separable if and only if $L = SP$.*

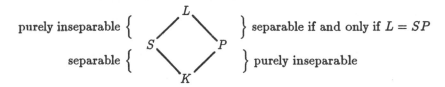

purely inseparable $\Big\{$ $\Big\}$ separable if and only if $L = SP$

separable $\Big\{$ $\Big\}$ purely inseparable

Proof. If char $K = 0$ then $S = L$ and $P = K$ so that the claim is trivial; hence we may assume that char $K = p \neq 0$.

(a) Let $x, y \in S$. Then $K(x, y)$ is a separable field extension of K by (14.12)(b); hence $x \pm y$, xy and $1/x$ (if $x \neq 0$) are separable and belong to S. This shows that S is a field. A completely analogous argument using (14.15)(d) shows that P is a field. The last claim, namely that $S \cap P = K$, is merely a restatement of (14.15)(a).

(b) Suppose that an irreducible polynomial $f \in K[x]$ has a root α in S. Then f splits over L by (13.10). Since f (being the minimal polynomial of the separable element α) is separable, all the roots of f actually lie in S. Hence f splits over S. By (13.10) this means that $(S : K)$ is normal.

(c) Since we know by now that S and P are fields, the first two claims are immediate consequences of (14.12)(b) and (14.15)(d). Finally, for every element $\alpha \in L$ there is an integer $n \geq 0$ with $\alpha^{p^n} \in S$, due to (14.15)(b); by (14.15)(c) this means that S is a purely inseparable extension.

(d) If L is separable over P then the more so over SP. On the other hand L is purely inseparable over S, hence the more so over SP. Consequently, L is both separable and purely inseparable over SP which implies by (14.15)(a) that $L = SP$. Suppose conversely that $L = SP$. Write $S = K(A)$ where each element $\alpha \in A$ is separable over K. Then each $\alpha \in A$ is also separable over P, and $L = P(A)$. Due to (14.12)(b) this shows that $(L : P)$ is separable. ∎

THE theorem just obtained allows us to define a measure of "how separable or inseparable" an algebraic extension is.

(14.17) Definition. *Let $(L : K)$ be an algebraic field extension and let S be the separable closure of K in L. Then $[L : K]_s := [S : K]$ is called the **separable degree** of $(L : K)$, and $[L : K]_i := [L : S]$ is called the **inseparable degree** of $(L : K)$.*

OBVIOUSLY, $[L : K]_i = 1$ if and only if $(L : K)$ is separable (which is always the case if char $K = 0$). Also, the degree formula shows that $[L : K]_s[L : K]_i = [L : K]$.

WE will now show that the separable degree of a finite field extension $(L : K)$ equals the number of possible K-embeddings of L into the normal closure of $(L : K)$. In view of theorem (14.13), this result is not too surprising and, in fact, makes this theorem more precise. As a consequence, we will be able to prove a formula for the separable degree which is analogous to the degree formula and for the formula of the transcendence degree.

(14.18) Theorem. *Let $(L : K)$ be an algebraic field extension, let S be the separable closure of K in L and let N be a normal extension of K such that $K \subseteq S \subseteq L \subseteq N$.*

(a) *The mapping* $\begin{array}{ccc} \mathrm{Emb}_K(L, N) & \to & \mathrm{Emb}_K(S, N) \\ \sigma & \mapsto & \sigma|_S \end{array}$ *is a bijection.*

(b) *The separable degree of $(L : K)$ equals the number of K-embeddings of L into N, i.e.,*

$$[L : K]_s = |\mathrm{Emb}_K(L, N)| .$$

(c) *If α is an element of L with minimal polynomial f, then the separable degree $[K(\alpha) : K]_s$ of $K(\alpha)$ over K equals the number of roots of f.*

(d) *If Z is any intermediate field of $(L : K)$, then*

$$[L : K]_s = [L : Z]_s[Z : K]_s \quad and \quad [L : K]_i = [L : Z]_i[Z : K]_i .$$

Proof. (a) If char $K = 0$ then $S = L$, and the claim is trivial. So suppose that char $K = p \neq 0$.

Injectivity. Suppose $\sigma|_S = \tau|_S$ and let $\alpha \in L$. Since L is purely inseparable over S, there is a natural number n such that $\alpha^{p^n} \in S$. Since σ and τ coincide on S, this implies $\sigma(\alpha)^{p^n} = \sigma(\alpha^{p^n}) = \tau(\alpha^{p^n}) = \tau(\alpha)^{p^n}$ and hence $\left(\sigma(\alpha) - \tau(\alpha)\right)^{p^n} = 0$. Consequently, $\sigma(\alpha) - \tau(\alpha) = 0$. This shows that $\sigma(\alpha) = \tau(\alpha)$ for all $\alpha \in L$, i.e., that $\sigma = \tau$.

Surjectivitiy. Each embedding $\sigma_0 : S \to N$ over K extends to an automorphism $\Sigma : N \to N$; restricting Σ, we obtain an embedding $\sigma := \Sigma|_L : L \to N$ over K with $\sigma|_S = \sigma_0$.

(b) We have $|\mathrm{Emb}_K(L, N)| = |\mathrm{Emb}_K(S, N)|$ by part (a), $|\mathrm{Emb}_K(S, N)| = [S : K]$ by (14.12)(4) and $[S : K] = [L : K]_s$ by definition.

(c) Apply part (b) in the special case $L = K(\alpha)$.

(d) The first equation is an immediate consequence of part (b) and lemma (13.14). Then the second equation follows from the first by using the equation $[L : K]_i[L : K]_s = [L : K]$ and the degree formula. ∎

(14.19) Proposition. *Let $(L : K)$ be a finite extension in characteristic $p \neq 0$. Then the inseparable degree $[L : K]_i$ is a power of p.*

Proof. We can write $L = K(\alpha_1, \ldots, \alpha_n)$ where each element α_i is purely inseparable over K, hence also over $K(\alpha_1, \ldots, \alpha_{i-1})$. Since $[K(\alpha_1, \ldots, \alpha_i) : K(\alpha_1, \ldots, \alpha_{i-1})]$ is a power of p, due to (13.15)(c), the degree formula shows that $[L : K] = [K(\alpha_1, \ldots, \alpha_n) : K]$ is also a power of p. ∎

FOR finite extensions we can obtain a further characterization of separability by applying properties of trace and norm. The key result is the following proposition which expresses trace and norm in terms of embeddings.

(14.20) Proposition. *Let $(L : K)$ be a finite field extension and let N be a normal extension of K containing L. Furthermore, let $\alpha \in L$ and let*

$$ r := [K(\alpha) : K]_s ; \quad m := [K(\alpha) : K]_i ; \quad k := [L : K(\alpha)] ; \quad d := [K(\alpha) : K] . $$

Note that r equals the number of K-embeddings of $K(\alpha)$ into N; say $\mathrm{Emb}_K(K(\alpha), N) = \{\sigma_1, \ldots, \sigma_k\}$.

(a) If $p_\alpha(x) = x^d + a_{d-1}x^{d-1} + \cdots + a_1 x + a_0 \in K[x]$ is the minimal polynomial of α and if f_α is the characteristic polynomial of α over K, then the following equations hold:

(1)
$$ p_\alpha(x) = \big(x - \sigma_1(\alpha)\big)^m \cdots \big(x - \sigma_r(\alpha)\big)^m ; $$

(2)
$$ f_\alpha(x) = p_\alpha(x)^k = \big(x - \sigma_1(\alpha)\big)^{mk} \cdots \big(x - \sigma_r(\alpha)\big)^{mk} ; $$

(3)
$$ \mathrm{tr}_K^L(\alpha) = -ka_{d-1} = [L : K(\alpha)]_s \cdot [L : K]_i \big(\sigma_1(\alpha) + \cdots + \sigma_r(\alpha)\big) ; $$

(4)
$$ N_K^L(\alpha) = (-1)^{dk} a_0^k = \big(\sigma_1(\alpha) \cdots \sigma_r(\alpha)\big)^{[L:K(\alpha)] \cdot [L:K]_i} . $$

(b) If $\Phi_1, \ldots, \Phi_R$ are the different K-embeddings of L into N, then for all $\alpha \in L$ we have
$$ \mathrm{tr}_K^L(\alpha) = [L : K]_i \big(\Phi_1(\alpha) + \cdots + \Phi_R(\alpha)\big) \quad \text{and} $$
$$ N_K^L(\alpha) = \big(\Phi_1(\alpha) \cdots \Phi_R(\alpha)\big)^{[L:K]_i} . $$

Moreover, writing $\alpha_i := \Phi_i(\alpha)$ for $1 \leq i \leq R$, then for every polynomial $q \in K[x]$ we have

($\star$)
$$ \mathrm{tr}_K^L(q(\alpha)) = [L : K]_i \big(q(\alpha_1) + \cdots + q(\alpha_R)\big) \quad \text{and} $$
$$ N_K^L(q(\alpha)) = \big(q(\alpha_1) \cdots q(\alpha_R)\big)^{[L:K]_i} . $$

Proof. (a) We know that $\alpha_1 := \sigma_1(\alpha), \ldots, \alpha_r := \sigma_r(\alpha)$ are the different roots of p_α; let m_i be the multiplicity of α_i as a root of p_α. Then $p_\alpha(x) = (x - \alpha_1)^{m_1} \cdots (x - \alpha_r)^{m_r}$. For any fixed index i let σ be the unique element of $\{\sigma_1, \ldots, \sigma_r\}$ with $\sigma(\alpha) = \alpha_i$; then $(\sigma \star p_\alpha)(x) = \big(x - \sigma(\alpha_1)\big)^{m_1} \cdots \big(x - \sigma(\alpha_r)\big)^{m_r}$. Now since p_α lies in $K[x]$ and since σ leaves K pointwise fixed, we have $\sigma \star p_\alpha = p_\alpha$, i.e.,

$$(x - \alpha_1)^{m_1} \cdots \underbrace{(x - \alpha_i)^{m_i}}_{(x - \alpha_i)^{m_1}} \cdots (x - \alpha_r)^{m_r} = \big(x - \sigma(\alpha_1)\big)^{m_1} \cdots \big(x - \sigma(\alpha_r)\big)^{m_r}.$$

Since σ is one-to-one, the elements $\sigma(\alpha_1), \ldots, \sigma(\alpha_r)$ are pairwise different; hence the fact that $N[x]$ is a unique factorization domain implies that $(x - \alpha_i)^{m_1} = (x - \alpha_i)^{m_i}$ and consequently that $m_i = m_1$. This shows that all the roots of p_α have the same multiplicity $m_1 = \cdots = m_k =: m'$. Hence the degree of f is $\deg f = m' \cdot r = m' \cdot [K(\alpha) : K]_s$; on the other hand, we have $\deg f = [K(\alpha) : K] = [K(\alpha) : K]_i[K(\alpha) : K]_s$. A comparison shows that $m' = [K(\alpha) : K]_i = m$. This proves equation (1).

To prove the second equation, we observe that $(1, \alpha, \ldots, \alpha^{d-1})$ is a basis of $K(\alpha)$ over K. Also, let $\theta_1, \ldots, \theta_k$ be a basis of L over $K(\alpha)$. Then

$$B := (\theta_1, \alpha\theta_1, \alpha^2\theta_1, \ldots, \alpha^{d-1}\theta_1; \ldots; \theta_k, \alpha\theta_k, \alpha^2\theta_k, \ldots, \alpha^{d-1}\theta_k)$$

is a basis of L over K, and the matrix representation of m_α with respect to this basis is

$$\underbrace{\begin{pmatrix} A & & & \\ & A & & \\ & & \ddots & \\ & & & A \end{pmatrix}}_{k \text{ blocks on the diagonal}} \quad \text{where} \quad A = \begin{pmatrix} 0 & 0 & 0 & \cdots & -a_0 \\ 1 & 0 & 0 & \cdots & -a_1 \\ & 1 & 0 & \cdots & -a_2 \\ & & 1 & & -a_3 \\ & & & \ddots & \vdots \\ & & & & -a_{d-1} \end{pmatrix}.$$

Since the characteristic polynomial of A is p_α, the second claim follows.

To prove the equations (3) and (4), we note that if $N := \deg f_\alpha$ then $f_\alpha(x) = x^N - \text{tr}_K^L(\alpha)x^{N-1} + \cdots + (-1)^N N_K^L(\alpha)$ by (I.9.13). On the other hand, we have just shown that

$$f_\alpha(x) = p_\alpha(x)^k = (x^d + a_{d-1}x^{d-1} + \cdots + a_1x + a_0)^k = x^{dk} + ka_{d-1}x^{dk-1} + \cdots + a_0^k;$$

comparing coefficients, we see first of all that $N = dk$ and then (using equation (1)) that

$$\text{tr}_K^L(\alpha) = -ka_{d-1} = km\big(\sigma_1(\alpha) + \cdots + \sigma_r(\alpha)\big)$$

and

$$N_K^L(\alpha) = (-1)^{dk}a_0^k = (-1)^{dk}(-1)^{rmk}\big(\sigma_1(\alpha) \cdots \sigma_r(\alpha)\big)^{mk}$$
$$= (-1)^{dk+rmk}\big(\sigma_1(\alpha) \cdots \sigma_r(\alpha)\big)^{mk}.$$

Since $mk = [L : K(\alpha)] \cdot [K(\alpha) : K]_i = [L : K(\alpha)]_s[L : K]_i$ and $dk + rmk = 2[L : K]$ by (14.18)(d), this gives the claim.

(b) Due to (13.14), each element $\Phi \in \mathrm{Emb}_K(L, N)$ has the form $\Phi = \Sigma \circ \tau$ where $\tau \in \mathrm{Emb}_{K(\alpha)}(L, N)$ and Σ is an extension of an element $\sigma \in \mathrm{Emb}_K(K(\alpha), N)$; then $\Phi(\alpha) = \Sigma(\tau(\alpha)) = \Sigma(\alpha) = \sigma(\alpha)$. This means that the sequence $(\Phi_1(\alpha), \ldots, \Phi_R(\alpha))$ contains the same elements as the sequence $(\sigma_1(\alpha), \ldots, \sigma_r(\alpha))$, except that each element is listed $|\mathrm{Emb}_{K(\alpha)}(L, N)|$ times, i.e., $[L : K(\alpha)]_s$ times. Therefore,

$$[L : K]_i(\Phi_1(\alpha) + \cdots + \Phi_R(\alpha)) = [L : K]_i[L : K(\alpha)]_s(\sigma_1(\alpha) + \cdots + \sigma_r(\alpha)) = \mathrm{tr}_K^L(\alpha)$$

and

$$\left(\Phi_1(\alpha) \cdots \Phi_R(\alpha)\right)^{[L:K]_i} = \left(\sigma_1(\alpha) \cdots \sigma_r(\alpha)\right)^{[L:K]_i[L:K(\alpha)]_s} = N_K^L(\alpha) \ ,$$

where we used equations (3) and (4) in part (a). To prove the second claim, it suffices (considering the result just obtained) to show that $\Phi_i(q(\alpha)) = q(\Phi_i(\alpha))$ for all i. Let $q(x) = \sum_r k_r x^r$ with coefficients $k_r \in K$. Since Φ_i leaves K pointwise fixed, we have

$$\Phi_i(q(\alpha)) = \Phi_i\left(\sum_r k_r \alpha^r\right) = \sum_r k_r \Phi_i(\alpha)^r = q(\Phi_i(\alpha)) \ .$$

∎

THIS last result enables us to determine the behavior of trace and norm under composing field extensions.

(14.21) Proposition. *If Z is an intermediate field of a finite field extension $(L : K)$ then*

$$\mathrm{tr}_K^L = \mathrm{tr}_K^Z \circ \mathrm{tr}_Z^L \qquad and \qquad N_K^L = N_K^Z \circ N_Z^L \ .$$

Proof. Let $F := \mathrm{Emb}_Z(L, N)$ and $G := \mathrm{Emb}_K(Z, N)$. We choose for any element $g \in G$ an extension $\widehat{g} : N \to N$; then $H := \mathrm{Emb}_K(L, N) = \{\widehat{g} \circ f \mid f \in F, g \in G\}$, due to (13.14). Using (14.20)(b) and the formula $[L : K]_i = [L : Z]_i[Z : K]_i$, we obtain

$$\mathrm{tr}_K^L(\alpha) = [L : K]_i \sum_{\substack{f \in F, \\ g \in G}} \widehat{g}(f(\alpha)) = [Z : K]_i \sum_{g \in G} \widehat{g}\big(\underbrace{[L : K]_i \sum_{f \in F} f(\alpha)}_{= \mathrm{tr}_Z^L(\alpha) \in Z}\big)$$

$$= [Z : K]_i \sum_{g \in G} g\big(\mathrm{tr}_Z^L(\alpha)\big) = \mathrm{tr}_K^Z\big(\mathrm{tr}_Z^L(\alpha)\big)$$

and

$$N_K^L(\alpha) = \left[\prod_{\substack{f \in F, \\ g \in G}} \widehat{g}(f(\alpha))\right]^{[L:K]_i} = \left[\prod_{g \in G} \widehat{g}\big(\underbrace{[\prod_{f \in F} f(\alpha)]^{[L:Z]_i}}_{= N_Z^L(\alpha) \in Z}\big)\right]^{[Z:K]_i}$$

$$= \left[\prod_{g \in G} g\big(N_Z^L(\alpha)\big)\right]^{[Z:K]_i} = N_K^Z\big(N_Z^L(\alpha)\big) \ .$$

∎

THE formulas obtained in (14.20)(b) are particularly simple for separable extensions. In fact, if $(L : K)$ is a finite separable extension and if $\alpha \in L$ then $\mathrm{tr}_K^L = \sum_i \sigma_i(\alpha)$ and $N_K^L(\alpha) = \prod_i \sigma_i(\alpha)$ where the mappings σ_i are the different embeddings of L into a normal closure of $(L : K)$. Let us give two examples in which these formulas are applied.

(14.22) **Example.** Let us find $\mathrm{tr}_{\mathbb{Q}}^{\mathbb{Q}(\alpha)}(2\alpha + 1)$ and $N_{\mathbb{Q}}^{\mathbb{Q}(\alpha)}(2\alpha + 1)$ where $\alpha \in \mathbb{C}$ satisfies the equation $\alpha^3 + \alpha^2 - 2$. (This was done in (12.18) already, using only the definitions of trace and norm.) If α, β, γ are the three roots of $f(x) := x^3 + x^2 - 2$ then the conjugates of $2\alpha + 1$ are $2\alpha + 1$, $2\beta + 1$ and $2\gamma + 1$. Hence

$$\mathrm{tr}_{\mathbb{Q}}^{\mathbb{Q}(\alpha)}(2\alpha + 1) = (2\alpha + 1) + (2\beta + 1) + (2\gamma + 1) = 2\underbrace{(\alpha + \beta + \gamma)}_{= -1} + 3 = 1$$

and

$$N_{\mathbb{Q}}^{\mathbb{Q}(\alpha)}(2\alpha + 1) = (2\alpha + 1)(2\beta + 1)(2\gamma + 1)$$
$$= 1 + 2\underbrace{(\alpha + \beta + \gamma)}_{= -1} + 4\underbrace{(\alpha\beta + \beta\gamma + \gamma\alpha)}_{= 0} + 8\underbrace{(\alpha\beta\gamma)}_{= 2} = 15 .$$

(14.23) **Example.** Let $\varepsilon = e^{2\pi i/p}$ where p is an odd prime and let $\mathrm{tr} = \mathrm{tr}_{\mathbb{Q}}^{\mathbb{Q}(\varepsilon)}$ and $N = N_{\mathbb{Q}}^{\mathbb{Q}(\varepsilon)}$ be the norm and the trace of $\mathbb{Q}(\varepsilon)$ over $\mathbb{Q}$. Note that the minimal polynomial of ε over $\mathbb{Q}$ is

$$f(x) = \frac{x^p - 1}{x - 1} = x^{p-1} + \cdots + x + 1 = (x - \varepsilon)(x - \varepsilon^2) \cdots (x - \varepsilon^{p-1}) ,$$

this polynomial being irreducible by (8.19)(c). Hence a basis of $\mathbb{Q}(\varepsilon)$ over $\mathbb{Q}$ is given by $(1, \varepsilon, \ldots, \varepsilon^{p-2})$, and there are $p - 1$ embeddings of $\mathbb{Q}(\varepsilon)$ into $\mathbb{C}$, given by $\sigma_k(\varepsilon) = \varepsilon^k$. Thus if $x = a_0 + a_1\varepsilon + \cdots + a_{p-2}\varepsilon^{p-2}$ is an arbitrary element of $\mathbb{Q}(\varepsilon)$ then

$$\mathrm{tr}(x) = \sum_{k=1}^{p-1} \sigma_k(x) = \sum_{k=1}^{p-1} (a_0 + a_1\varepsilon^k + a_2\varepsilon^{2k} + \cdots + a_{p-2}\varepsilon^{(p-1)k})$$

$$= (p-1)a_0 + a_1 \sum_{k=1}^{p-1} \varepsilon^k + a_2 \sum_{k=1}^{p-1} \varepsilon^{2k} + \cdots + a_{p-2} \sum_{k=1}^{p-1} \varepsilon^{(p-1)k}$$

$$= (p-1)a_0 + (a_1 + a_2 + \cdots + a_{p-2}) \sum_{\kappa=1}^{p-1} \varepsilon^{\kappa} = (p-1)a_0 - (a_1 + \cdots + a_{p-2})$$

and

$$N(x) = \prod_{k=1}^{p-1} \sigma_k(x) = \prod_{k=1}^{p-1} (a_0 + a_1\varepsilon^k + a_2\varepsilon^{2k} + \cdots + a_{p-2}\varepsilon^{(p-1)k})$$

$$= \sum_{i_0, \ldots, i_{p-2}} a_{i_0}\varepsilon^{i_0} a_{i_1} \varepsilon^{2i_1} a_{i_2} \varepsilon^{3i_2} \cdots a_{i_{p-2}} \varepsilon^{(p-1)i_{p-2}}$$

$$= \sum_{i_0, \ldots, i_{p-2}} a_{i_0} a_{i_1} a_{i_2} \cdots a_{i_{p-2}} \varepsilon^{i_0 + 2i_1 + 3i_2 + \cdots + (p-1)i_{p-2}}$$

where the sum ranges over all possible choices of indices. In particular we find that

$$N_{\mathbb{Q}}^{\mathbb{Q}(\varepsilon)}(\varepsilon) = \varepsilon^{1+2+3+\cdots+(p-1)} = \varepsilon^{p(p-1)/2} = 1.$$

Moreover,

$$N_{\mathbb{Q}}^{\mathbb{Q}(\varepsilon)}(1-\varepsilon^r) = \prod_{k=1}^{p-1}\sigma_k(1-\varepsilon^r) = \prod_{k=1}^{p-1}(1-\varepsilon^{kr}) = \prod_{\kappa=1}^{p-1}(1-\varepsilon^\kappa) = f(1) = p \,.$$

WE observe next that if $(L:K)$ is a finite field extension then the trace function tr_K^L which is a K-linear mapping from L into K also induces a K-bilinear mapping from $L \times L$ into K, namely $(\alpha,\beta) \mapsto \mathrm{tr}_K^L(\alpha\beta)$. If we want to find the matrix representation of this bilinear mapping with respect to a basis of L we are led to the notion of the *discriminant* of the field extension $(L:K)$.

(14.24) **Definition.** *The discriminant* Dis_K^L *of a finite field extension* $[L:K]$ *is the mapping which assigns to any basis* $(\alpha_1,\ldots,\alpha_n)$ *of* L *the element*

$$\mathrm{Dis}_K^L(\alpha_1,\ldots,\alpha_n) := \det\big(\mathrm{tr}_K^L(\alpha_i\alpha_j)\big)_{i,j} \,.$$

(14.25) **Proposition.** *If*

$$\begin{pmatrix} \beta_1 \\ \vdots \\ \beta_n \end{pmatrix} = \begin{pmatrix} k_{11} & \cdots & k_{1n} \\ \vdots & & \vdots \\ k_{n1} & \cdots & k_{nn} \end{pmatrix} \begin{pmatrix} \alpha_1 \\ \vdots \\ \alpha_n \end{pmatrix}$$

where the matrix $P := (k_{ij})$ *has entries in* K *then*

$$\mathrm{Dis}_K^L(\beta_1,\ldots,\beta_n) = (\det P)^2 \, \mathrm{Dis}_K^L(\alpha_1,\ldots,\alpha_n) \,.$$

Proof. This follows immediately from the theory of bilinear forms. (See (I.15.6) and the accompanying remarks.) Let us give an independent proof, however. Introducing the matrices $A := \big(\mathrm{tr}_K^L(\alpha_i\alpha_j)\big)_{i,j}$ and $B := \big(\mathrm{tr}_K^L(\beta_i\beta_j)\big)_{i,j}$ we have

$$B_{ij} = \mathrm{tr}_K^L(\beta_i\beta_j) = \mathrm{tr}_K^L\big((\sum_r k_{ir}\alpha_r)(\sum_s K_{js}\alpha_s)\big) = \sum_{r,s} k_{ir}k_{js}\,\mathrm{tr}_K^L(\alpha_r\alpha_s)$$

$$= \sum_{r,s} P_{ir}P_{js}A_{rs} = \sum_s \underbrace{\sum_r P_{ir}A_{rs}}_{=(PA)_{is}} \underbrace{P_{js}}_{=(P^T)_{sj}} = (PAP^T)_{ij}$$

for all indices i,j. Hence $B = PAP^T$ and consequently $\det B = \det(PAP^T) = (\det P)^2 \det A$ which is what we wanted to show. ∎

SINCE the trace of a finite field extensions can be expressed in terms of embeddings, due to (14.20), it is clear that the discriminant can also be expressed in terms of embeddings. We will do so (for separable extensions) in the following proposition.

(14.26) Proposition. *Let $(L : K)$ be a finite separable field extension of degree* $n = [L : K]$.

(a) *If $\sigma_1, \ldots, \sigma_n$ are the different K-embeddings of L into an algebraic closure $\overline{K}$ of K then*

$$\operatorname{Dis}_K^L(\alpha_1, \cdots, \alpha_n) = \det(\sigma_i(\alpha_j))^2 .$$

(b) *If α is a primitive element of $(L : K)$ and if $f \in K[x]$ is the minimal polynomial of α over K then*

$$\operatorname{Dis}_K^L(1, \alpha, \ldots, \alpha^{n-1}) = (-1)^{n(n-1)/2} N_K^L(f'(\alpha)) = \prod_{j>i}(\sigma_i(\alpha) - \sigma_j(\alpha))^2 = D(f) .$$

Proof. (a) Let $A := (\sigma_i(\alpha_j))_{i,j}$; then

$$(A^T A)_{ij} = \sum_{r=1}^{n}(A^T)_{ir} A_{rj} = \sum_{r=1}^{n} A_{ri} A_{rj}$$

$$= \sum_{r=1}^{n} \sigma_r(\alpha_i)\sigma_r(\alpha_j) = \sum_{r=1}^{n} \sigma_r(\alpha_i\alpha_j) = \operatorname{tr}_K^L(\alpha_i\alpha_j) .$$

Hence $\operatorname{Dis}_K^L(\alpha_1, \ldots, \alpha_n) = \det(\operatorname{tr}(\alpha_i\alpha_j))_{i,j} = \det(A^T A) = (\det A)^2$.

(b) Let $\alpha = \alpha_1, \alpha_2, \ldots, \alpha_n$ be the roots of f so that $f(x) = \prod_{j=1}^{n}(x - \alpha_j)$. Then $f'(\alpha) = \prod_{j=2}^{n}(\alpha - \alpha_j)$ and hence

$$N_K^L(f'(\alpha)) = \prod_{i=1}^{n}\prod_{j=2}^{n} \sigma_i(\alpha - \alpha_j) = \prod_{i=1}^{n}\prod_{j=2}^{n}(\sigma_i(\alpha) - (\sigma_i\sigma_j)(\alpha))$$

(1)

$$= \prod_{i=1}^{n}\prod_{k\neq i}(\sigma_i(\alpha) - \sigma_k(\alpha)) = (-1)^{n(n-1)/2} \prod_{i>k}(\sigma_i(\alpha) - \sigma_k(\alpha))^2 ;$$

here the last equation is obtained by combining any two terms $\sigma_i(\alpha) - \sigma_k(\alpha)$ and $\sigma_k(\alpha) - \sigma_i(\alpha)$ to the single term $-(\sigma_i(\alpha) - \sigma_k(\alpha))$ where $i > k$. On the other hand, combining part (a) with the formula for the Vandermonde determinant obtained in (I.7.16) we see that

(2)

$$\operatorname{Dis}_K^L(1, \alpha, \ldots, \alpha^{n-1}) = \det \begin{pmatrix} 1 & \sigma_1(\alpha) & \cdots & \sigma_1(\alpha)^{n-1} \\ 1 & \sigma_2(\alpha) & \cdots & \sigma_2(\alpha)^{n-1} \\ \vdots & \vdots & & \vdots \\ 1 & \sigma_n(\alpha) & \cdots & \sigma_n(\alpha)^{n-1} \end{pmatrix}^2$$

$$= \prod_{i>k}(\sigma_i(\alpha) - \sigma_k(\alpha))^2 = D(f)$$

where the last equation stems from the fact that the elements $\sigma_i(\alpha)$ are exactly the roots of f. The claim now follows by comparing (1) and (2). ∎

(14.27) Example. Let $K = \mathbb{Q}$ and $L = \mathbb{Q}(\sqrt{n})$ where $n \in \mathbb{Z} \setminus \{0, 1\}$ is square-free. Then there are two K-embeddings of L into $\mathbb{C}$, namely the inclusion map σ_1 and the conjugation map $\sigma_2 : a + b\sqrt{n} \mapsto a - b\sqrt{n}$. Consequently,

$$\mathrm{Dis}_K^L(a + b\sqrt{n}, c + d\sqrt{n}) = \det \begin{pmatrix} a + b\sqrt{n} & c + d\sqrt{n} \\ a - b\sqrt{n} & c - d\sqrt{n} \end{pmatrix}^2 = 4n(ad - bc)^2 .$$

(14.28) Example. Let $\varepsilon = e^{2\pi i/p}$ where p is an odd prime and let $N = N_{\mathbb{Q}}^{\mathbb{Q}(\varepsilon)}$. The minimal polynomial f of ε over $\mathbb{Q}$ is $f(x) = (x^p - 1)/(x - 1)$; hence $f(x)(x - 1) = x^p - 1$ and therefore, by the product rule, $f'(x)(x - 1) + f(x) = px^{p-1}$. Plugging in $x = \varepsilon$ this gives $f'(\varepsilon)(\varepsilon - 1) = p\varepsilon^{p-1} = p\varepsilon^{-1}$ so that $f'(\varepsilon) = p/(\varepsilon(\varepsilon - 1))$. Therefore, (14.26)(b) yields

$$\mathrm{Dis}_{\mathbb{Q}}^{\mathbb{Q}(\varepsilon)}(1, \varepsilon, \dots, \varepsilon^{p-2}) = (-1)^{(p-1)(p-2)/2} N\left(\frac{p}{\varepsilon(\varepsilon - 1)}\right)$$

$$= (-1)^{(p-1)/2} \frac{N(p)}{N(\varepsilon)N(\varepsilon - 1)} = (-1)^{(p-1)/2} p^{p-2}$$

because $N(p) = p^{p-1}$ due to (12.19)(b), $N(\varepsilon) = 1$ due to (14.23) and $N(\varepsilon - 1) = N(-1)N(\varepsilon - 1) = (-1)^{p-1}p = p$ due to (14.23).

WE can now give a new characterization of separability.

(14.29) Theorem. *For a finite field extension $(L : K)$, the following statements are equivalent:*
(1) $\mathrm{tr}_K^L : L \to K$ *is surjective;*
(2) $\mathrm{tr}_K^L \neq 0$;
(3) *the bilinear form* $B : \begin{array}{ccc} L \times L & \to & K \\ (\alpha, \beta) & \mapsto & \mathrm{tr}_K^L(\alpha\beta) \end{array}$ *is nondegenerate;*
(4) $\mathrm{Dis}_K^L(\alpha_1, \dots, \alpha_n) \neq 0$ *for every basis $(\alpha_1, \dots, \alpha_n)$ of L over K;*
(5) *there is a basis $(\alpha_1, \dots, \alpha_n)$ of L over K such that $\mathrm{Dis}_K^L(\alpha_1, \dots, \alpha_n) \neq 0$;*
(6) $(L : K)$ *is a separable extension.*

Proof. Let us first prove the equivalence of the first four conditions. Trivially, (1) implies (2). Conversely, if (2) holds, there is an element $\alpha \in L$ with $\mathrm{tr}^L_K(\alpha) = x \neq 0$. Then for any element $k \in K$ we have $\mathrm{tr}^L_K(kx^{-1}\alpha) = kx^{-1}\,\mathrm{tr}^L_K(\alpha) = kx^{-1}x = k$; this shows that tr^L_K is surjective so that (1) holds. Moreover, if (1) holds, there is an element $x \neq 0$ with $\mathrm{tr}^L_K(x) \neq 0$. Let $\beta \in L \setminus \{0\}$; then $B(x\beta^{-1}, \beta) = \mathrm{tr}^L_K(x\beta^{-1}\beta) = \mathrm{tr}^L_K(x) \neq 0$ so that $\beta \notin \mathrm{Rad}\,B$. This shows that B is nondegenerate; hence (2) implies (3). Conversely, (3) trivially implies (2). The equivalence of (3), (4) and (5) simply expresses the fact that a bilinear form is nondegenerate if and only if it is represented by an invertible matrix.

Let us show that (2) implies (6). Suppose that (6) does not hold so that $(L : K)$ is not separable. Then $\mathrm{char}\,K$ is a prime number p, and $[L : K]_i > 1$. Proposition (14.19) shows that $[L : K]_i$ is proper power of p, in particular a multiple of p. But then (12.19)(b) implies immediately that $\mathrm{tr}^L_K(\alpha) = 0$ for all $\alpha \in L$.

To show that (6) implies (5) we assume that $(L : K)$ is separable; then there is an element $\alpha \in L$ with $L = K(\alpha)$, due to (14.9). Then $(1, \alpha, \dots, \alpha^{n-1})$ is basis of L over K, and if $\alpha_2, \dots, \alpha_n$ are the (pairwise different) conjugates of α then

$$\mathrm{Dis}^L_K(1, \alpha, \cdots, \alpha^{n-1}) = \prod_{i>j}(\alpha_i - \alpha_j)^2 \neq 0$$

by (14.26)(a). ∎

Exercises

Problem 1. Let $K = \mathbb{Z}_2(t)$ be the rational function field over $\mathbb{Z}_2$ and consider the polynomials $F(x) := x^2 - t$ and $G(x) := x^2 - (t + t^3)$ in $K[x]$. Show that if α and β are roots of F and G, respectively, then F is irreducible over $K(\beta)$ and G is irreducible over $K(\alpha)$. (This result was used in (14.8).)

Problem 2. Show that for any polynomial $f \in \mathbb{Z}[x]$ there are only finitely many primes p such that f is irreducible and inseparable over $\mathbb{Z}_p$. Is it still true that there are only finitely many primes p such that at least one of the irreducible factors of f over $\mathbb{Z}_p$ is inseparable?

Problem 3. Let K be a field of prime characteristic p and let α be algebraic over K. Show that $(K(\alpha) : K)$ is separable if and only if $K(\alpha) = K(\alpha^{p^n})$ for all $n \in \mathbb{N}$.

Problem 4. Let $(L : K)$ be a finite extension in characteristic $p \neq 0$. Show that if $[L : K]$ is not divisible by p then $(L : K)$ is a separable extension.

Problem 5. Let K be a field of characteristic p and let $n \in \mathbb{N}_0$. Show that the polynomial $f(x) = x^{p^n} - a \in K[x]$ is reducible if and only if there is an element $b \in K$ with $a = b^p$, i.e., if a has a p-th root in K.

Problem 6. Let K be a field of characteristic $p \neq 0$ and let α, β be algebraic elements over K such that $\alpha^p \in K$, $\beta^p \in K$ and $[K(\alpha, \beta) : K] = p^2$. Show that $(K(\alpha, \beta) : K)$ is not a simple extension and conclude that both α and β are not separable. Also, exhibit an infinite number of intermediate fields.
Remark. This generalizes example (14.8).

Problem 7. Let $(L : K)$ be an algebraic field extension.
(a) Show that if K is perfect then L is perfect.
(b) Show that if L is perfect and $(L : K)$ is separable then K is perfect.
(c) Show that if L is perfect and $(L : K)$ is finite then K is perfect.

Problem 8. Show that a field K is perfect if and only if every algebraic closure of K is separable over K.

Problem 9. Show that the rational function field $K(x)$ over a field K is perfect if and only if char $K = 0$.

Problem 10. Let K be a field of characteristic p and let $\overline{K}$ be an algebraic closure of K.

(a) Show that $\sqrt[p]{K} := \{x \in \overline{K} \mid x^p \in K\}$ is an intermediate field of $(\overline{K} : K)$ and that $K = \{x^p \mid x \in \sqrt[p]{K}\} = (\sqrt[p]{L})^p$.

(b) Show that

$$K \subseteq \sqrt[p]{K} \subseteq \sqrt[p]{\sqrt[p]{K}} \subseteq \cdots$$

is an ascending chain of fields whose union is the smallest perfect field containing K. (It is therefore called the **perfect closure** of K.)

Problem 11. Let L be a subfield of $\mathbb{C}$. Show that if $[L : \mathbb{Q}] = n$ then there are exactly n embeddings of L into $\mathbb{C}$.

(b) Suppose that $\alpha \in \mathbb{C}$ satisfies the equation $\alpha^4 + \alpha^2 + 1$. How many embeddings of $\mathbb{Q}(\alpha)$ into $\mathbb{C}$ are there?

Problem 12. Let K be a field of prime characteristic p and let $f(x) = x^p - x - a$ with $a \in K$. Show that f either completely splits into linear factors over K or else is irreducible. Show that if α is a root of f then $(K(\alpha) : K)$ is a separable extension.

Problem 13. Let $(L : K)$ be an algebraic field extension where K has prime characteristic p. For any $q \in \mathbb{N}$ we let $L^q := \{x^q \mid x \in L\}$. Show that $(L : K)$ is separable if and only if $L^p K = K$ if and only if $L^{p^n} K = L$ for all $n \geq 1$.

Problem 14. Show that an algebraic extension $(L : K)$ is purely inseparable if and only if the only embedding of K into some extension of L is the inclusion map.

Problem 15. Let $(L : K)$ be a field extension in characteristic $p \neq 0$ and let $\alpha \in L$. Show that $K(\alpha^p) = K(\alpha)$ if and only if α is algebraic and separable over K. Moreover, show that if $\alpha^p \in K$ but $\alpha \notin K$ then the extension $(K(\alpha) : K)$ is purely inseparable of degree p.

Problem 16. Let $(L : K)$ be a field extension in characteristic $p \neq 0$ and let $Z = KL^p$ the compositum of K and of $L^p = \{x^p \mid x \in L\}$.

(a) Show that $(L : Z)$ is purely inseparable.

(b) Show that if $(L : K)$ is separable then $L = Z$.

(c) Show that if $[L : K] < \infty$ and if $L = Z$ then $(L : K)$ is separable.

Problem 17. Let $(L : K)$ be a field extension and let $\alpha, \beta \in L \setminus \{0\}$. Show that if α is separable and β purely inseparable over K then $K(\alpha, \beta) = K(\alpha + \beta) = K(\alpha\beta)$.

Problem 18. Suppose that $(L : K)$ is an algebraic field extension such that every polynomial $f \in K[x]$ has a root in L. Show that L is an algebraic closure of K.

Hint. Prove the statement first for separable extensions (using the primitive element theorem) and for purely inseparable extensions. Then use the fact that an algebraic extension can be split into a purely inseparable and a separable extension.

Problem 19. Let $(L : K)$ be an algebraic extension and let S and P be the separable and the purely inseparable closure of K in L, respectively. Moreover, let Z be an intermediate field of $(L : Z)$. Prove the following statements!

(a) $(L : Z)$ is purely inseparable if and only if $S \subseteq Z$.

(b) If $(L : Z)$ is separable then $P \subseteq Z$.

(c) If $Z \cap S = K$ then $Z \subseteq P$.

Problem 20. Let $\overline{K}$ be an algebraic closure of K and let S be the separable closure of K in $\overline{K}$. Show that S is the splitting field of the family $\mathcal{F}$ of all separable polynomials in $K[x]$.

Problem 21. Let $K = \mathbb{Z}_2(t)$ and let $L = K(\alpha)$ where α is a root of the irreducible polynomial $x^4 + t \in K[x]$. Let S and P be the separable and the purely inseparable closure of K in L, respectively. Show that $L \neq SP$.

Problem 22. Let K_0 be a field of characteristic $p \neq 0$ and let $K := K_0(x, y)$ be the rational function field in two variables over K_0. Moreover, let $L = K(\alpha)$ where $\alpha^{2p} + x\alpha^p + y = 0$.

(a) Show that $(L : K)$ is an extension of degree $2p$ which is neither separable nor purely inseparable.

(b) Show that the extension $(L : K)$ has a single proper intermediate field and that this field is not purely inseparable over $Z = K$.

Problem 23. Let $K \subseteq L$ be finite fields of characteristic p, say $|K| = p^m$ and $|L| = p^{mn}$ where $n = [L : K]$.

(a) Let $\sigma_0 : K \to K$ be the Frobenius automorphism of K, given by $x \mapsto x^p$. Show that the automorphism group $\text{Aut } K$ has exactly m elements, namely $\sigma_0, \sigma_0^2, \ldots, \sigma_0^m = \text{id}$.

(b) Show that there are exactly n automorphisms of L leaving K pointwise fixed, namely $\sigma^m, \sigma^{2m}, \ldots, \sigma^{nm} = \text{id}_L$ where σ is the Frobenius automorphism of L.

(c) For $\alpha \in L$, let f_α be the characteristic polynomial of α with respect to $(L : K)$. Show that

$$f_\alpha(x) = (x - \alpha)(x - \alpha^q)(x - \alpha^{q^2}) \cdots (x - \alpha^{q^{n-1}})$$

where $q := p^m$ and conclude that

$$\text{tr}_K^L(\alpha) = \alpha + \alpha^q + \cdots + \alpha^{q^{n-1}} \quad \text{and} \quad N_K^L = \alpha^{1 + q + q^2 + \cdots + q^{n-1}}.$$

Also, show that the norm $N_K^L : L \to K$ is surjective.

(d) Let $\alpha \in L$. Show that the polynomial $f(x) = x^q - x - \alpha$ (where $q = p^m$) has a root in L if and only if $\text{tr}_K^L(\alpha) = 0$.

Problem 24. Let $(L : K)$ be a finite field extension and let Z_1, Z_2 be intermediate fields of $(L : K)$. Suppose that $(Z_1 : K)$ is separable and that $(Z_2 : K)$ is purely inseparable. Show that $[Z_1 Z_2 : Z_2] = [Z_1 : K] = [Z_1 Z_2 : K]_s$ and $[Z_1 Z_2 : Z_1] = [Z_2 : K] = [Z_1 Z_2 : K]_i$.

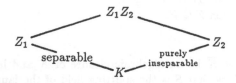

Problem 25. Let $(L : K)$ be a separable field extension of degree $[L : K] = n$ and let $\sigma_1, \ldots, \sigma_n$ be the different K-embeddings of L into a normal closure N of $(L : K)$.

(a) Let $(\alpha_1, \ldots, \alpha_n)$ be a basis of L over K. Show that the vectors $v_i \in L^n$ defined by $v_i := \big(\sigma_i(\alpha_1), \ldots, \sigma_i(\alpha_n)\big)^T$, are linearly independent.

(b) Let α be a primitive element of $(L : K)$ so that $(\beta_1, \ldots, \beta_n) := (1, \alpha, \alpha^2, \ldots, \alpha^{n-1})$ is a basis of L over K. Show that $\det\big(\sigma_i(\beta_j)\big)_{i,j} \neq 0$.

(c) Conclude from part (a) that $\sigma_1, \ldots, \sigma_n$ are linearly independent.

Problem 26. Let $(L : K)$ be a finite separable field extension, say $L = K(\alpha)$. Moreover, let f be the minimal polynomial of α, say $f(x) = (x - \alpha_1)(x - \alpha_2) \cdots (x - \alpha_n)$.

(a) Using the Lagrange interpolation formula, show that

$$\sum_{k=1}^{n} \frac{f(x)}{x - \alpha_k} \frac{\alpha_k^i}{f'(\alpha_k)} = x^i \quad \text{for } 0 \leq i \leq n .$$

(b) Let $f(x) = (x - \alpha)(b_{n-1} x^{n-1} + \cdots + b_1 x + b_0)$ in $L[x]$. Show that the dual basis of $(1, \alpha, \ldots, \alpha^{n-1})$ with respect to the nondegenerate bilinear form tr_K^L is

$$\Big(\frac{b_0}{f'(\alpha)}, \frac{b_1}{f'(\alpha)}, \ldots, \frac{b_{n-1}}{f'(\alpha)} \Big) .$$

Problem 27. Let $f(x) = x^n + px + q \in K[x]$ where K is an arbitrary field. Show that the discriminant of f is $D(f) = (-1)^{n(n-1)/2}\big(n^n q^{n-1} + (1 - n)^{n-1} p^n\big)$.

Hint. Since the discriminant of a polynomial f can be expressed as a polynomial with integer coefficients in the coefficients of f (independently of the particular base-field), we can assume without loss of generality that $\mathrm{char}\, K = 0$ and that f is irreducible over K. Show that $\alpha \cdot \big(f'(\alpha) + np - p\big) = -nq$ and use the formula $D(f) = (-1)^{n(n-1)/2} N_K^{K(\alpha)}\big(f'(\alpha)\big)$ which was established in (14.26).

15. Field theory and integral ring extensions

IN this section we want to exploit the theory of field extensions to study integral ring extensions. The situation will usually be as follows: R is an integral domain with quotient field K, L is an field extension of K and S is the integral closure of R in L. Then knowledge on $(L : K)$ will be used to obtain information on $(S : R)$.

Let us begin by show that in this situation R coincides with K if and only if S is an intermediate field of $(L : K)$.

(15.1) Lemma. *Let $R \subseteq S$ be integral domains such that S is integral over R. Then S is a field if and only if R is.*

Proof. Suppose that R is a field and let $s \in S \setminus \{0\}$. Let

$$s^n + r_{n-1}s^{n-1} + \cdots + r_1 s + r_0 = 0 \quad (r_i \in R)$$

be an equation of integral dependence for s of smallest possible degree. Then $r_0 \neq 0$ (otherwise $s^{n-1} + r_{n-1}s^{n-2} + \cdots + r_1 = 0$ which is an equation of smaller degree); hence r_0^{-1} exists because R is a field. But then we can write

$$s(s^{n-1} + r_{n-1}s^{n-2} + \cdots + r_1)(-r_0^{-1}) = 1$$

which shows that s is invertible. We have shown that every nonzero element of S is invertible which means that S is a field.

Suppose, conversely, that S is a field. Let $r \in R \setminus \{0\}$; then r^{-1} exists in S, and since S is integral over R there is an equation

$$r^{-m} + r_{m-1}r^{-(m-1)} + \cdots + r_1 r^{-1} + r_0 = 0 \quad (r_i \in R).$$

Multiplying by r^{m-1}, we see that

$$r^{-1} = -(r_{m-1} + \cdots + r_1 r^{m-2} + r_0 r^{m-1}) \in R.$$

Hence every nonzero element in R has in inverse in R, so that R is a field. ∎

AS a corollary we can relate properties of ideals in S and in R.

(15.2) Proposition. *Let $R \subseteq S$ be integral domains such that S is integral over R. Then an ideal $Q \trianglelefteq S$ is maximal if and only if $P := Q \cap R \trianglelefteq R$ is maximal in R.*

Proof. We know from (6.14)(d) that Q is maximal in S if and only if S/Q is a field and that P is maximal in R if and only if $R/P = R/(Q \cap R) \cong (Q + R)/R$ is a field. Since S/Q is integral over $(Q + R)/R$ these two conditions are equivalent, due to (15.1). ∎

WE will now restrict our attention to the case that R is integrally closed in its quotient field K. Note that this condition is satisfied whenever R is a unique factorization domain, due to (10.25); in particular, all the results to be derived are valid for the ring $R = \mathbb{Z}$. If in this case L is a finite extension of K and if $\alpha \in R$ then the coefficients of the minimal polynomial, the norm and the trace of α (which are a priori known to lie in K) are actually elements of R; this is shown in the next two propositions.

(15.3) Proposition. *Let R be an integral domain which is integrally closed in its quotient field K. Moreover, let $(L : K)$ be a finite field extension and let S be the integral closure of R in L.*

(a) *If $\alpha \in S$ then the conjugates of α (in a normal closure of L over K) are again integral over R.*

(b) *If $\alpha \in S$ then $\mathrm{tr}_K^L(\alpha)$ and $N_K^L(\alpha)$ are elements of R.*

(c) *An element $\alpha \in S$ is a unit in S if and only if $N_K^L(\alpha)$ is a unit in R.*

(d) *If $\alpha \in R$ is such that $N(\alpha)$ is irreducible in R then α is irreducible in S.*

(e) *Every element of L can be written as a quotient s/r with $s \in S$ and $r \in R$; in particular, L is the quotient field of S.*

(f) *Suppose that the extension $(L : K)$ is normal. Let P be a maximal ideal of R and let Q and Q' be maximal ideals of S with $P = Q \cap R = Q' \cap R$. Then there is a K-automorphism σ of L with $Q' = \sigma Q$.*

Proof. Let $\Phi_1 = \mathrm{id}$, $\Phi_2, \ldots, \Phi_n$ be the different K-embeddings of L into a normal closure N of $(L : K)$.

(a) By assumption the element α satisfies an equation of the form $\alpha^n + r_{n-1}\alpha^{n-1} + \cdots + \cdots + r_1\alpha + r_0 = 0$ with coefficients $r_i \in R$. Just apply Φ_i to this equation and observe that $\Phi_i|_R \equiv \mathrm{id}$ to see that all the elements $\Phi_i(\alpha)$ are again integral over R.

(b) By the very definition of the trace and the norm we have $\mathrm{tr}_K^L(\alpha) \in K$ and $N_K^L(\alpha) \in K$. On the other hand, part (a) together with (14.20) (b) shows that these two elements are integral over R; hence they belong to the integral closure of R in K which, by hypothesis, is R itself.

(c) Let α be a unit in S, say $\alpha\beta = 1$. Then $1 = N(1) = N(\alpha)N(\beta)$ in R which shows that $N(\alpha)$ is a unit in R. Conversely, if $N(\alpha)$ is a unit in R with inverse r then $1 = N(\alpha)r = \left(\Phi_1(\alpha) \cdots \Phi_n(\alpha)\right)^{[L:K]_i} r$ which shows that α is a unit in S, the inverse being $\alpha^{[L:K]_i - 1}\left(\Phi_2(\alpha) \cdots \Phi_n(\alpha)\right)^{[L:K]_i} r$ where we used (14.20)(b) again.

(d) If $\alpha = \alpha_1\alpha_2$ in S then $N(\alpha) = N(\alpha_1)N(\alpha_2)$ in R which, by hypothesis, implies that $N(\alpha_1)$ or $N(\alpha_2)$ must be a unit in R. By part (c) this implies that α_1 or α_2 must be a unit in S.

(e) Let $\alpha \in L$. Since α is algebraic over K we have an equation of the form

$$(\star) \qquad \alpha^n + \frac{r_{n-1}}{r'_{n-1}}\alpha^{n-1} + \cdots + \frac{r_1}{r'_1}\alpha + \frac{r_0}{r'_0} = 0 \,.$$

Let $\rho := r_0' r_1' \cdots r_{n-1}' \in R$ and multiply $(\star)$ by ρ^n to get an equation of the form

$$(\rho\alpha)^n + r_{n-1}''(\rho\alpha)^{n-1} + \cdots + r_1''(\rho\alpha) + r_0'' = 0$$

which shows that $\rho\alpha$ is integral over R, say $\rho\alpha = s \in S$. But then $\alpha = s/\rho$ as desired.

(f) Suppose that the claim is not true. Then the ideals Q and $\sigma^{-1}Q'$ where σ ranges over the K-automorphisms of L are pairwise distinct. By the Chinese remainder theorem there is an element $s \in S$ with $s \equiv 0 \bmod Q$ and $s \equiv 1 \bmod \sigma^{-1}Q'$ for all σ. Then $r := N_K^L(s) = \prod_\sigma \sigma(s)$ lies in R by part (b) and satisfies $r \equiv 0 \bmod Q$ and $r \equiv 1 \bmod Q'$ so that $r \in R \cap Q = P$ but $r \notin Q' \cap R = P$ which is clearly a contradiction. ■

(15.4) Proposition. *Let R be an integral domain which is integrally closed in its quotient field K and let α be an algebraic element in some field extension of K. Then the following conditions are equivalent:*

(1) α is algebraic over K, and the minimal polynomial $p \in K[x]$ of α over K has coefficients in R;

(2) α is integral over R.

Proof. The implication (1)$\Longrightarrow$(2) is trivial; let us prove conversely that (2) implies (1). Hence assume that α is integral over R and let $p \in K[x]$ be the minimal polynomial of α. The roots of p, i.e., the conjugates of α, are integral over R by (15.3)(a); hence the coefficients of p (which are the elementary symmetric polynomials in the roots of p) are also integral over R. On the other hand, these coefficients lie in K. Since R is integrally closed in K this implies that the coefficients of p lie in R. ■

NOTE that this proposition immediately yields the claims in examples (10.20)(c) and (d). To see another application of (15.4), let us determine the integral closure of $\mathbb{Z}$ in $\mathbb{Q}(\sqrt{n})$.

(15.5) Proposition. *Let $n \in \mathbb{Z} \setminus \{0, 1\}$ be a square-free number. Then the integral closure of $\mathbb{Z}$ in $\mathbb{Q} + \mathbb{Q}\sqrt{n}$ is*

$$\mathcal{O}_n := \begin{cases} \mathbb{Z} + \mathbb{Z}\sqrt{n}, & \text{if } n \not\equiv 1 \text{ modulo } 4; \\ \mathbb{Z} + \mathbb{Z}(\frac{1+\sqrt{n}}{2}), & \text{if } n \equiv 1 \text{ modulo } 4. \end{cases}$$

Proof. Let $a + b\sqrt{n} \in \mathbb{Q} + \mathbb{Q}\sqrt{n}$. Since $p(x) := \left(x - (a + b\sqrt{n})\right)\left(x - (a - b\sqrt{n})\right) = x^2 - 2ax + (a^2 - nb^2) \in \mathbb{Q}[x]$, the minimal polynomial of $a + b\sqrt{n}$ over $\mathbb{Q}$ is $x - (a + b\sqrt{n})$ if $b = 0$ and p if $b \neq 0$. Suppose $b \neq 0$. Then proposition (15.4) shows that $a + b\sqrt{n}$ is integral over $\mathbb{Z}$ if and only if

$$2a \in \mathbb{Z} \quad \text{and} \quad a^2 - nb^2 \in \mathbb{Z}.$$

For these equations to hold, there are two possibilities.

The first possibility is that $a \in \mathbb{Z}$ and hence $m := nb^2 \in \mathbb{Z} \setminus \{0\}$. Write $b = r/s$ with $r, s \in \mathbb{Z}$ coprime. Then $nr^2 = ms^2$; this shows that if p is a prime factor of s then $p^2 \mid n$ which is impossible since n is square-free. Consequently, we have $s = \pm 1$ and hence $b \in \mathbb{Z}$.

The second possibility is that a is half of an odd integer, say $a = \frac{2m+1}{2}$ with $m \in \mathbb{Z}$. Then $nb^2 \in \mathbb{Z} + a^2 = \mathbb{Z} + \frac{4m^2 + 4m + 1}{4} = \mathbb{Z} + \frac{1}{4}$, say $nb^2 = k + \frac{1}{4}$. Writing $b = \frac{r}{s}$ again, this reads $4nr^2 = (4k+1)s^2$. Since n is square-free and $r, s \in \mathbb{Z}$ are relatively prime, this implies that $s^2 = 4$, hence $s = \pm 2$, and that r is odd. Thus $b = \frac{2M+1}{2}$ for some $M \in \mathbb{Z}$. Then

$$\mathbb{Z} \ni a^2 - nb^2 = \frac{4m^2 + 4m + 1}{4} - n \cdot \frac{4M^2 + 4M + 1}{4} \in \mathbb{Z} + \frac{1 - n}{4}$$

which implies $\frac{1-n}{4} \in \mathbb{Z}$, i.e., $n \equiv 1$ modulo 4; only in this case can the second possibility occur. This proves that the integral closure of $\mathbb{Z}$ in $\mathbb{Z} + \mathbb{Z}\sqrt{n}$ is contained in $\mathcal{O}_n$. The converse inclusion is easy to see since $\sqrt{n}$ and $\frac{1}{2}(1 + \sqrt{n})$, respectively, satisfy monic equations over $\mathbb{Z}$. ∎

SPECIALIZING to the case $R = \mathbb{Z}$ we will now prove the important result that in the situation

we can always choose a basis of L over $\mathbb{Q}$ which is at the same time a basis of S over $\mathbb{Z}$. Part of this statement is true in the general case, as the following proposition shows.

(15.6) Proposition. *Suppose that the integral domain R is integrally closed in its quotient field K. Let $(L : K)$ be a finite separable field extension and let S be the integral closure of R in L. Then there is a basis $(\beta_1, \ldots, \beta_n)$ of L over K such that $S \subseteq R\beta_1 + \cdots + R\beta_n$.*

Proof. Each element $v \in L$ is algebraic over K and hence satisfies an equation $r_m v^m + r_{m-1} v^{m-1} + \cdots + r_1 v + r_0 = 0$ with coefficients $r_i \in R$; to see this, just clear denominators in an algebraic equation of v over K. Multiplying by $r_m^{m-1} \in R$, this gives a monic equation for $\alpha := r_m v$ over R which implies that $\alpha \in S$. This argument shows that if we start with an arbitrary basis $(v_1, \ldots, v_n)$ of L over K, we can multiply its members by suitable elements of R to obtain a new basis $(\alpha_1, \ldots, \alpha_n)$ whose members α_i lie in S. Let us take such a basis $(\alpha_1, \ldots, \alpha_n)$.

Since the bilinear form $(\alpha, \beta) \mapsto \operatorname{tr}_K^L(\alpha\beta)$ is nondegenerate due to the separability of $(L : K)$, we can find a dual basis $(\beta_1, \ldots, \beta_n)$ of $(\alpha_1, \ldots, \alpha_n)$ with respect to this bilinear form so that $\operatorname{tr}_K^L(\alpha_i \beta_j) = \delta_{ij}$. We claim that $(\beta_1, \ldots, \beta_n)$ has the desired property. Let $x \in S$, say $x = \sum_j x_j \beta_j$ with coefficients $x_j \in K$. Since both x and α_i lie in S, we have $x\alpha_i \in S$ for all i. Then by (15.4) the minimal polynomial of $x\alpha_i$ over K has coefficients in R; since $\operatorname{tr}_K^L(x\alpha_i)$ is (up to sign) one of the coefficients of this polynomial, we have $\operatorname{tr}_K^L(x\alpha_i) \in R$. But then

$$R \ni \operatorname{tr}_K^L(x\alpha_i) = \operatorname{tr}_K^L(\sum_j x_j \alpha_i \beta_j) = \sum_j x_j \operatorname{tr}_K^L(\alpha_i \beta_j) = \sum_j x_j \delta_{ij} = x_i \, ;$$

hence $x_i \in R$ for all i so that $x \in R\beta_1 + \cdots + R\beta_n$. Since x was an arbitrary element of S this shows that $S \subseteq R\beta_1 + \cdots + R\beta_n$. ∎

(15.7) Theorem. *Let $(L : \mathbb{Q})$ be a finite field extension and let S be the integral closure of $\mathbb{Z}$ in L. Then there is a basis $(\alpha_1, \ldots, \alpha_n)$ of L over $\mathbb{Q}$ which is at the same time a basis of S over $\mathbb{Z}$ so that $S = \mathbb{Z}\alpha_1 + \cdots + \mathbb{Z}\alpha_n$. Consequently, S is a Noetherian ring.*

Proof. By (15.6) we know that S is contained in a finitely generated $\mathbb{Z}$-module and hence is a finitely generated $\mathbb{Z}$-module itself, due to (11.17). Moreover, S (considered as an abelian group) is trivially torsion-free and hence is free, due to (I.25.19)(b).†
Thus there are elements $\alpha_1, \ldots, \alpha_m \in S$ such that $S = \mathbb{Z}\alpha_1 \oplus \cdots \oplus \mathbb{Z}\alpha_m$. This implies that $(\alpha_1, \ldots, \alpha_m)$ is linearly independent over $\mathbb{Q}$ and, due to (15.3)(e), also generating for L over $\mathbb{Q}$. Hence $(\alpha_1, \ldots, \alpha_m)$ is a basis of L over $\mathbb{Q}$ (which implies in particular that $m = [L : K]$). Finally, the fact that S is a finitely generated $\mathbb{Z}$-module implies by (11.17) that S is a Noetherian $\mathbb{Z}$-module and hence a Noetherian ring. ∎

IT is convenient to introduce some terminology at this point.

(15.8) Definitions. *A finite-dimensional field extension L of $\mathbb{Q}$ is called an **algebraic number field**. If L is an algebraic number field and if S is the integral closure of $\mathbb{Z}$ in L then an **integral basis** of L is a basis of L over $\mathbb{Q}$ which is at the same time a basis of S over $\mathbb{Z}$.*

(15.9) Example. Let $n \in \mathbb{Z} \setminus \{0, 1\}$ be a square-free number. Then an integral basis of $\mathbb{Q}(\sqrt{n})$ is given by $(1, \sqrt{n})$ if $n \not\equiv 1$ modulo 4 and by $(1, \frac{1+\sqrt{n}}{2})$ if $n \equiv 1$ modulo 4. This is an immediate consequence of (15.5).

THERE are many different ways to choose an integral basis $(\alpha_1, \ldots, \alpha_n)$ for an algebraic number field K. However, it will turn out that the discriminant $\mathrm{Dis}_K^L(\alpha_1, \ldots, \alpha_n)$ is the same for all these bases, as we show in the next proposition.

(15.10) Proposition. *Let L be an algebraic number field with an integral basis $(\alpha_1, \ldots, \alpha_n)$ and an arbitrary basis $(\beta_1, \ldots, \beta_n)$. Then the quotient*

$$\mathrm{Dis}_{\mathbb{Q}}^L(\beta_1, \ldots, \beta_n) / \mathrm{Dis}_{\mathbb{Q}}^L(\beta_1, \ldots, \beta_n)$$

is the square of an integer. If both $(\alpha_1, \ldots, \alpha_n)$ and $(\beta_1, \ldots, \beta_n)$ are integral bases then

$$\mathrm{Dis}_{\mathbb{Q}}^L(\beta_1, \ldots, \beta_n) = \mathrm{Dis}_{\mathbb{Q}}^L(\beta_1, \ldots, \beta_n) .$$

---

† This argument goes through if $\mathbb{Z}$ is replaced by an arbitrary principal ideal domain R because the theory of modules over principal ideal domains is practically identical with the theory of $\mathbb{Z}$-modules, i.e., of abelian groups.

Proof. Let P be the invertible matrix which provides the transition from the basis $(\alpha_1, \ldots, \alpha_n)$ to the basis $(\beta_1, \ldots, \beta_n)$ so that

$$\begin{pmatrix} \beta_1 \\ \vdots \\ \beta_n \end{pmatrix} = \begin{pmatrix} k_{11} & \cdots & k_{1n} \\ \vdots & & \vdots \\ k_{n1} & \cdots & k_{nn} \end{pmatrix} \begin{pmatrix} \alpha_1 \\ \vdots \\ \alpha_n \end{pmatrix}.$$

Since $(\alpha_1, \ldots, \alpha_n)$ is an integral bases we have $P \in \mathbb{Z}^{n \times n}$ and hence $\det P \in \mathbb{Z}$; since $\mathrm{Dis}_K^L(\beta_1, \ldots, \beta_n) = (\det P)^2 \, \mathrm{Dis}_K^L(\alpha_1, \ldots, \alpha_n)$ by (14.25), the first claim follows. If $(\beta_1, \ldots, \beta_n)$ is also an integral basis we can exchange the roles of both bases. Since 1 is the only square of an integer whose reciprocal is again the square of an integer, the second claim follows. ∎

THIS gives rise to the following definition.

(15.11) Definition. *The **discriminant** of an algebraic number field L is the integer $\mathrm{Dis}_{\mathbb{Q}}^L(\alpha_1, \ldots, \alpha_n)$ where $(\alpha_1, \ldots, \alpha_n)$ is any integral basis of L.*

(15.12) Example. If $n \in \mathbb{Z} \setminus \{0, 1\}$ is square-free then the discriminant of $\mathbb{Q}(\sqrt{n})$ is

$$\begin{cases} 4n, & \text{if } n \not\equiv 1 \text{ modulo } 4; \\ n, & \text{if } n \equiv 1 \text{ modulo } 4. \end{cases}$$

This follows immediately from (15.9) and (14.27).

THE determination of the discriminant of an algebraic number field is an important problem in algebraic number theory. Sometimes this determination can be eased by choosing a primitive element α and then applying (15.10) with $(\beta_1, \ldots, \beta_n) = (1, \alpha, \ldots, \alpha^{n-1})$ because the discriminant of a basis of this form can be computed by (14.26). Another criterion which restricts the possible values for the discriminant is the following result.

(15.13) Stickelberger's theorem. *Let $(\alpha_1, \ldots, \alpha_n)$ be an integral basis of an algebraic number field L. Then $\mathrm{Dis}_{\mathbb{Q}}^L(\alpha_1, \ldots, \alpha_n) \equiv 0$ or 1 modulo 4.*

Proof. If $\sigma_1, \ldots, \sigma_n$ are the different embeddings of $\mathbb{Q}$ into L then

$$\det\big(\sigma_i(\alpha_j)\big) = \sum_{\tau \in \mathrm{Sym}_n} \mathrm{sign}(\tau) \prod_{i=1}^{n} \sigma_i(\alpha_{\tau(i)})$$

$$= \underbrace{\sum_{\tau \in \mathrm{Alt}_n} \prod_{i=1}^{n} \sigma_i(\alpha_{\tau(i)})}_{=:\, P} - \underbrace{\sum_{\tau \in \mathrm{Sym}_n \setminus \mathrm{Alt}_n} \prod_{i=1}^{n} \sigma_i(\alpha_{\tau(i)})}_{=:\, Q}$$

so that $\text{Dis}_{\mathbb{Q}}^L(\alpha_1, \ldots, \alpha_n) = (P - Q)^2 = (P + Q)^2 - 4PQ$. Now $P + Q$ and PQ are symmetric expressions in the elements α_i and their conjugates and hence lie in $\mathbb{Q}$; on the other hand these numbers are integral over $\mathbb{Z}$ and hence are integers themselves. Since every square in $\mathbb{Z}$ is congruent with 0 or 1 modulo 4 this gives the claim. ∎

WE will now turn to fields that are obtained from $\mathbb{Q}$ by adjoining a root of unity, i.e., an element $\varepsilon \in \mathbb{C}$ such that $\varepsilon^n = 1$ for some $n \in \mathbb{N}$. Let us first give a characterization of roots of unity in terms of the absolute values of their conjugates.

(15.14) Lemma. (a) *Let* $n \in \mathbb{N}$ *and* $\varepsilon > 0$ *be fixed numbers. Let* $S(n, \varepsilon)$ *be the set of those numbers* $\alpha \in \mathbb{C}$ *which are integral over* $\mathbb{Z}$, *which satisfy* $[\mathbb{Q}(\alpha) : \mathbb{Q}] \leq n$ *and whose conjugates all have an absolute value* $\leq \varepsilon$. *Then* $S(n, \varepsilon)$ *is a finite set.*
(b) *Let* $\alpha \in \mathbb{C}$ *be integral over* $\mathbb{C}$ *such that all conjugates of* α *have absolute value* 1. *Then* α *is a root of unity.*

Proof. (a) Let $\alpha \in S(n, \varepsilon)$ and let $f \in \mathbb{Q}[x]$ be the minimal polynomial of α over $\mathbb{Q}$. Then f has coefficients in $\mathbb{Z}$ by (15.4), and $\deg f = [\mathbb{Q}(\alpha) : \mathbb{Q}] \leq n$. Moreover, the coefficients of f are just the elementary symmetric polynomials in the conjugates of α and hence are bounded by a constant C which depends only on n and ε. Since the set of all polynomials $f \in \mathbb{Z}[x]$ of degree $\leq n$ with all coefficients $\leq C$ is clearly finite this implies that $S(n, \varepsilon)$ is finite.
(b) Let $n := [\mathbb{Q}(\alpha) : \mathbb{Q}]$. Then all powers of α are contained in $S(n, 1)$ which is a finite set by part (a). Hence the numbers α^k with $k \in \mathbb{Z}$ cannot all be different. Consequently, there is a number $m \in \mathbb{N}$ with $\alpha^m = 1$. ∎

WE will now determine the integral closure of $\mathbb{Z}$ in $\mathbb{Q}(\varepsilon)$ where ε is a root of unity of prime order.

(15.15) Theorem. *Let* $\varepsilon = e^{2\pi i/p}$ *where* p *is a prime number. Then* $\mathbb{Z}[\varepsilon]$ *is the integral closure of* $\mathbb{Z}$ *in* $\mathbb{Q}(\varepsilon)$.

Proof. Throughout the proof we will denote by $\text{tr} = \text{tr}_{\mathbb{Q}}^{\mathbb{Q}(\varepsilon)}$ and by $N = N_{\mathbb{Q}}^{\mathbb{Q}(\varepsilon)}$ the trace and the norm of the field extension $(\mathbb{Q}(\varepsilon) : \mathbb{Q})$. Moreover, let $\sigma_1, \ldots, \sigma_{p-1}$ be the different embeddings of $\mathbb{Q}$ into $\mathbb{Q}(\varepsilon)$ where σ_k is determined by the condition $\sigma_k(\varepsilon) = \varepsilon^k$.
Let S be the integral closure of $\mathbb{Z}$ in $\mathbb{Q}(\varepsilon)$; then clearly $\mathbb{Z}[\varepsilon] \subseteq S$. To prove the converse inclusion we show first that

$(\star)$
$$S(1 - \varepsilon) \cap \mathbb{Z} = \mathbb{Z} \cdot p.$$

Here the inclusion "$\supseteq$" holds because $p = N(1 - \varepsilon) = \prod_{k=1}^{p-1}(1 - \varepsilon^k)$ by (14.23) which shows that p is divisible by $1 - \varepsilon$ in $\mathbb{Z}[\varepsilon] \subseteq S$. Conversely, if $m \in \mathbb{Z}$ is divisible by $1 - \varepsilon$ in S, say $m = \alpha(1 - \varepsilon)$, then $m^{p-1} = N(m) = N(\alpha)N(1 - \varepsilon) = N(\alpha)p$ which shows that p divides m^{p-1} and hence m so that $m \in \mathbb{Z}p$. Thus $(\star)$ is established.

Now let $x \in S$. Then on the one hand $x(1-\varepsilon)$ lies in S so that $\mathrm{tr}\big(x(1-\varepsilon)\big) \in \mathbb{Z}$ by (15.3)(b). On the other hand, if we write $x_k = \sigma_k(x)$ for $1 \le k \le p-1$, we have

$$
\begin{aligned}
\mathrm{tr}\big(x(1-\varepsilon)\big) &= \sigma_1\big(x(1-\varepsilon)\big) + \sigma_2\big(x(1-\varepsilon)\big) + \cdots + \sigma_{p-1}\big(x(1-\varepsilon)\big) \\
&= x_1(1-\varepsilon) + x_2(1-\varepsilon^2) + \cdots + x_{p-1}(1-\varepsilon^{p-1}) \\
&= (1-\varepsilon)\underbrace{\big(x_1 + (1+\varepsilon)x_2 + \cdots + x_{p-1}(1+\varepsilon + \cdots + \varepsilon^{p-2})\big)}_{\in\, S} \\
&\in (1-\varepsilon)S .
\end{aligned}
$$

Thus $\mathrm{tr}\big(x(1-\varepsilon)\big) \in S(1-\varepsilon) \cap \mathbb{Z} = \mathbb{Z}p$. Now we write $x = a_0 + a_1\varepsilon + \cdots + a_{p-2}\varepsilon^{p-2}$ with coefficients $a_i \in \mathbb{Q}$. Then

$$
\begin{aligned}
\mathrm{tr}\big(x(1-\varepsilon)\big) &= \mathrm{tr}(x - x\varepsilon) \\
&= \mathrm{tr}\big(a_0(1-\varepsilon) + a_1(\varepsilon - \varepsilon^2) + \cdots a_{p-2}(\varepsilon^{p-2} - \varepsilon^{p-1})\big) \\
&= a_0\big(\mathrm{tr}(1) - \mathrm{tr}(\varepsilon)\big) + a_1\big(\mathrm{tr}(\varepsilon) - \mathrm{tr}(\varepsilon^2)\big) + \cdots a_{p-2}\big(\mathrm{tr}(\varepsilon^{p-2}) - \mathrm{tr}(\varepsilon^{p-1})\big) \\
&= a_0 p
\end{aligned}
$$

where we used in the last equation that $\mathrm{tr}(1) = p-1$ and $\mathrm{tr}(\varepsilon^k) = -1$ for $1 \le k \le p-1$, due to (14.23). Thus $a_0 p \in \mathbb{Z}p$ which means that $a_0 \in \mathbb{Z}$. We have shown that if $x = a_0 + a_1\varepsilon + \cdots + a_{p-2}\varepsilon^{p-2} \in S$ then $a_0 \in \mathbb{Z}$. Applying the same argument to

$$
x' := (x - a_0)\varepsilon^{p-1} = \frac{x - a_0}{\varepsilon} = a_1 + a_2\varepsilon + \cdots + a_{p-2}\varepsilon^{p-3}
$$

(which is integral again) we see that $a_1 \in \mathbb{Z}$. Continuing in this way we find that all coefficients lie in $\mathbb{Z}$ so that $x \in \mathbb{Z}[\varepsilon]$. ∎

WITH this last result in mind, we can present another application of (15.3).

(15.16) Example. Let $\varepsilon = e^{2\pi i/p}$ where p is an odd prime. Then the elements $1 - \varepsilon^k$ with $1 \le p-1$ are pairwise associate irreducible elements of the ring $\mathbb{Z}[\varepsilon]$. Each such element satisfies $N_{\mathbb{Q}}^{\mathbb{Q}(\varepsilon)}(1 - \varepsilon^k) = p$ by (14.23) and hence is irreducible in $\mathbb{Z}[\varepsilon]$, due to (15.3)(d). Moreover, we have the factorization $1 - \varepsilon^k = (1-\varepsilon)(1+\varepsilon+\cdots+\varepsilon^{k-1})$ in $\mathbb{Z}[\varepsilon]$; taking norms we find that $p = pN(1+\varepsilon+\cdots+\varepsilon^{k-1})$ which, by (15.3)(c), implies that $1 + \varepsilon + \cdots + \varepsilon^{k-1}$ is a unit in $\mathbb{Z}[\varepsilon]$. Hence $1 - \varepsilon^k$ is an associate of $1 - \varepsilon$, for any $1 \le k \le p-1$.

SINCE (15.15) implies that $(1, \varepsilon, \ldots, \varepsilon^{p-2})$ is an integral basis of $\mathbb{Q}(\varepsilon)$ we have another example for the discriminant of a number field.

(15.17) Example. Let $\varepsilon = e^{2\pi i/p}$ where p is an odd prime number. Then the discriminant of $\mathbb{Q}(\varepsilon)$ is $(-1)^{(p-1)/2}p^{p-2}$; this is an immediate consequence of (14.28).

IN a later section we will systematically study roots of unity. We will then use the above results to derive Kummer's lemma which played a key role in the proof of theorem (9.23) but was used without proof in section 9.

Exercises

Problem 1. Let A be the algebraic closure of $\mathbb{Z}$ in $\mathbb{C}$.
(a) Show that if $\alpha \in A \setminus \{0\}$ has the minimal polynomial $f \in \mathbb{Q}[x]$ then $\alpha^{-1} \in A$ if and only if $f(0) = \pm 1$.
(b) Show that $\alpha \in A$ if and only if $f(\alpha) \in A$ for some monic polynomial $f \in \mathbb{Z}[x]$ if and only if $f(\alpha) \in A$ for all monic polynomials $f \in \mathbb{Z}[x]$.

Problem 2. Let $(L : \mathbb{Q})$ be a finite field extension and let S be the integral closure of $\mathbb{Z}$ in L. Show that each nonzero ideal of S contains a natural number.

Problem 3. (a) Let $\alpha \in \mathbb{C}$ be such that $\alpha^3 + \alpha + 1 = 0$. Show that $\mathbb{Z}[\alpha]$ is the integral closure of $\mathbb{Z}$ in $\mathbb{Q}(\alpha)$.
(b) Let $\alpha \in \mathbb{C}$ be such that $\alpha^3 + \alpha^2 - 2\alpha + 8 = 0$. Show that $\mathbb{Z}[\alpha]$ is not the integral closure of $\mathbb{Z}$ in $\mathbb{Q}(\alpha)$. **Hint.** Show that $4/\alpha$ is integral over $\mathbb{Z}$.

Problem 4. Find the integral closure of $\mathbb{Z}$ in $\mathbb{Q}(i, \sqrt{2})$, in $\mathbb{Q}(\sqrt{2}, \sqrt{3})$ and in $\mathbb{Q}(\sqrt[4]{2})$.

Problem 5. Let $\varepsilon = e^{2\pi i/3}$. Find the algebraic closure S of $\mathbb{Z}$ in $\mathbb{Q}(\varepsilon)$. Moreover, find all units and all prime elements in R.

Problem 6. Find all prime elements in $\mathcal{O}_2$ and in $\mathcal{O}_5$.

Problem 7. Let $n \in \mathbb{Z} \setminus \{0, 1\}$ be a square-free number and let $K := \mathbb{Q} + \mathbb{Q}\sqrt{n}$.
(a) Show that if $\alpha \in K$ is integral over $\mathbb{Z}$ then $N(\alpha) \in \mathbb{Z}$.
(b) Show that there is always an element $\alpha \in K$ which is not integral over $\mathbb{Z}$ but nevertheless satisfies $N(\alpha) \in \mathbb{Z}$.
Hint. If $|n - 1| > 4$, one can take $\alpha = \frac{n+1}{n-1} + \frac{2}{n-1}\sqrt{n}$. Treat those values for n for which $|n - 1| \leq 4$ separately.

Problem 8. Let $n \in \mathbb{Z} \setminus \{0, 1\}$ be a square-free number and let $\mathcal{O}_n$ be the integral closure of $\mathbb{Z}$ in $\mathbb{Q} + \mathbb{Q}\sqrt{n}$. Show that for a number $\alpha \in \mathbb{Q} + \mathbb{Q}\sqrt{n}$ the following conditions are equivalent:
(1) $\alpha \in \mathcal{O}_n$,
(2) there is a monic quadratic polynomial $q \in \mathbb{Z}[x]$ with $q(\alpha) = 0$,
(3) there is a monic polynomial $p \in \mathbb{Z}[x]$ with $p(\alpha) = 0$,
(4) there is a monic polynomial $f \in \mathcal{O}_n[x]$ with $f(\alpha) = 0$,
(5) there is an element $\beta \in \mathbb{Q} + \mathbb{Q}\sqrt{n}$ with $\alpha^m \in \mathcal{O}_n\beta$ for all $m \in \mathbb{N}$.
Hint. If (5) is satisfied, let $\alpha^m = x_m\beta$ with $x_m \in \mathcal{O}_n$. Then $I_m := \langle x_1, \ldots, x_m \rangle$ defines an ascending chain $I_1 \subseteq I_2 \subseteq I_3 \subseteq \cdots$ of ideals in $\mathcal{O}_n$.

Problem 9. Let $(L : \mathbb{Q})$ be a normal extension of degree n and let S be the integral closure of $\mathbb{Z}$ in S. Show that if $\sigma_1, \ldots, \sigma_n$ are the different embeddings of L into $\mathbb{C}$ then $\sum_{i=1}^{n} \sigma_i(\alpha)\overline{\sigma_i(\alpha)} \geq n$ for all $\alpha \in S \setminus \{0\}$.

Proof. The geometric mean of n positive numbers cannot exceed their arithmetic mean.

Problem 10. (a) Show by induction on n that if $a_1, \ldots, a_n \in \mathbb{Z} \setminus \{0\}$ have 1 as a greatest common divisor then there is a matrix $A \in \mathbb{Z}_{n \times n}$ with determinant ± 1 which has $(a_1, \ldots, a_n)$ as its first row.

(b) Show that every algebraic number field possesses an integral basis which has 1 as one of its elements.

Problem 11. Let $L \subseteq L'$ be algebraic number fields and let $S \subseteq S'$ be the integral closures of $\mathbb{Z}$ and in L and in L', respectively.

(a) Show that $mS' \cap S = mS$ for all $m \in \mathbb{Z}$.

(b) Show that every integral basis of L can be extended to an integral basis of L'.

Problem 12. Let S be the integral closure of $\mathbb{Z}$ in an algebraic number field L and let D be the discriminant of L. Let $\alpha_1, \ldots, \alpha_n$ be elements of S which form a basis of L over $\mathbb{Q}$. Show that the following conditions are equivalent:

(1) $(\alpha_1, \ldots, \alpha_n)$ is an integral basis of L;

(2) $|\mathrm{Dis}_{\mathbb{Q}}^{L}(\alpha_1, \ldots, \alpha_n)| = |D|$;

(3) $|\mathrm{Dis}_{\mathbb{Q}}^{L}(\alpha_1, \ldots, \alpha_n)| \leq |\mathrm{Dis}_{\mathbb{Q}}^{L}(\beta_1, \ldots, \beta_n)|$ for all bases $(\beta_1, \ldots, \beta_n)$ of L over $\mathbb{Q}$ which consist entirely of elements of S.

Problem 13. Find the discriminants of the algebraic number fields $\mathbb{Q}(\sqrt[3]{2})$, $\mathbb{Q}(\sqrt[4]{2})$ and $\mathbb{Q}(\sqrt[5]{2})$.

Problem 14. Find the discriminants of the algebraic number fields $\mathbb{Q}(i, \sqrt{2})$ and $\mathbb{Q}(\sqrt{2}, \sqrt{3})$.

16. Affine algebras

AT the end of section 10 we started investigating extensions of commutative rings $R \subseteq S$ where S is obtained from R by adjoining a finite number of elements $s_1, \ldots, s_n$ so that S is a finitely generated R-algebra. Since the surjective homomorphism $\Phi : R[X_1, \ldots, X_n] \to R[s_1, \ldots, s_n]$ given by $p \mapsto p(s_1, \ldots, s_n)$ induces an isomorphism $R[s_1, \ldots, s_n] \cong R[X_1, \ldots, X_n]/I$ where $I := \ker \Phi = \{p \in R[X_1, \ldots, X_n] \mid p(s_1, \ldots, s_n) = 0\}$, the class of finitely generated R-algebras consists exactly of all homomorphic images of the polynomial rings $R[X_1, \ldots, X_n]$. Therefore, it is very natural in the context of polynomial rings to also study finitely generated algebras.

We will now restrict our investigation to the case that $R = K$ is a field. The interest in rings of the form $K[s_1, \ldots, s_n]$ stems from affine geometry where one is interested in solutions sets of polynomial equations (for example conic sections); therefore, rings of this form are called *affine algebras*.

(16.1) Definition. *An* **affine algebra** *over a field K is a finitely generated K-algebra, i.e., a ring of the form $K[\alpha_1, \ldots, \alpha_n]$.*

IT is an important fact that every affine K-algebra $K[x_1, \ldots, x_n]$ is integral over some subalgebra $K[y_1, \ldots, y_d]$ such that $y_1, \ldots, y_d$ are algebraically independent so that $K[y_1, \ldots, y_d] \cong K[Y_1, \ldots, Y_d]$. This fact is known as *Noether's normalization theorem*. Thus if we completely understand polynomial rings over K as well as integral ring extensions, then we also completely understand finitely generated K-algebras in general. To prove Noether's theorem, we need the following lemma.

(16.2) Lemma. *Let K be a field and let $f \in K[x_1, \ldots, x_n]$ be a nonconstant polynomial.*

(a) *There are natural numbers $k_2, \ldots, k_n$ such that the substitution*

$$y_2 := x_2 - x_1^{k_2}, \quad y_3 := x_3 - x_1^{k_3}, \quad \ldots, \quad y_n := x_n - x_1^{k_n}$$

leads to an equation $f(x_1, \ldots, x_n) = cx_1^N + r(x_1, y_2, \ldots, y_n)$ where $c \in K^\times$ and where the degree of r in x_1 is less than N.

(b) *If K is infinite, then a substitution of the form*

$$y_2 := x_2 - a_2 x_1, \quad y_3 := x_3 - a_3 x_1, \quad \ldots, \quad y_n := x_n - a_n x_1$$

with $a_i \in K$ leads to the same form.

Proof. (a) Using the indicated substitution, we obtain

$$(1) \quad f(x_1, \ldots, x_n) = f(x_1, y_2 + x_1^{k_2}, \ldots, y_n + x_1^{k_n}) = g(x_1, \ldots, x_n, y_2, \ldots, y_n)$$

where the polynomial $g \in K[X_1, \ldots, X_n, Y_2, \ldots, Y_n]$ is defined by

$$(2) \quad \begin{aligned} g(X_1, \ldots, X_n, Y_2, \ldots, Y_n) &:= f(X_1, Y_2 + X_1^{k_1}, \ldots, Y_n + X_1^{k_n}) \\ &= \sum_{i_1, \ldots, i_n} c_{i_1, \ldots, i_n} X_1^{i_1 + i_2 k_2 + \cdots + i_n k_n} + q(X_1, Y_2, \ldots, Y_n) . \end{aligned}$$

Now we specify how $k_2, \ldots, k_n$ should be chosen, namely such that $i_1 + i_2 k_2 + \cdots + i_n k_n \neq j_1 + j_2 k_2 + \cdots + j_n k_n$ whenever $(i_1, \ldots, i_n)$ and $(j_1, \ldots, j_n)$ are different multi-indices occurring in the polynomial f.† Any such choice for $k_2, \ldots, k_n$ ensures that no terms in the sum in (2) cancel so that the sum has higher degree in X_1 than the polynomial q. Thus (2) can be rewritten in the form

$$(2') \qquad g(X_1, \ldots, X_n, Y_2, \ldots, Y_n) = cX_1^N + r(X_1, Y_2, \ldots, Y_n)$$

where r is a polynomial whose degree in X_1 is less than N.

(b) Decompose f into its homogeneous components, say $f = f_0 + f_1 + \cdots + f_m$ where $f_m \neq 0$ and $m \geq 1$. Reindexing the variables in the proof of (8.5)(a), we see that $f_m(x_1, \ldots, x_n) = f_m(1, -a_2, \ldots, -a_n)x_1^m + \widehat{q}(x_1, y_2, \ldots, y_n)$ where the degree of $\widehat{q}$ in x_1 is less than m; then clearly f itself has the same form. Since f_m is homogeneous and $\neq 0$, the polynomial $f_m(1, X_2, \ldots, X_n)$ is $\neq 0$. By theorem (8.19) we can choose $a_2, \ldots, a_n \in K$ such that $c := f_m(1, -a_2, \ldots, -a_n) \neq 0$; whence the claim. ∎

NOW we are ready to prove the first version of Noether's normalization theorem.

(16.3) Noether's Normalization Theorem (weak version). *Let K be a field and let R be a finitely generated K-algebra, say $R = K[x_1, \ldots, x_n]$ with elements $x_i \in R$. Then there is a number $0 \leq m \leq n$ and there are elements $u_1, \ldots, u_m \in R$ such that the following statements hold:*

(1) $u_1, \ldots, u_m$ are algebraically independent over K;

(2) R is integral over $K[u_1, \ldots, u_m]$.

Whenever these conditions are satisfied, we call the algebra $K[u_1, \ldots, u_m]$ a **Noetherian normalization** *of R.*

Proof. We proceed by induction on n. The case $n = 1$ (i.e., $R = K[x]$) is rather trivial; if x is algebraic (hence integral) over K, we choose $m = 0$, otherwise x is algebraically independent over K, and we choose $m = 1$ and $u = x$. Let us suppose that the statement is established for some number $n - 1$. Then if $x_1, \ldots, x_n$ are algebraically independent, we are done. Otherwise there is a nonzero polynomial $f \in K[X_1, \ldots, X_n]$ with $f(x_1, \ldots, x_n) = 0$. Using one of the substitutions indicated in (16.2), this equation can be rewritten as

$$0 = f(x_1, \ldots, x_n) = cx_1^N + r(x_1, y_2, \ldots, y_n)$$

where the degree of the polynomial r in x_1 is less than N. Dividing by c, we see that this equation implies that x_1 is integral over $K[y_2, \ldots, y_n]$. For $2 \leq i \leq n$, we have $x_i = y_i + x_1^{k_i}$ so that x_i is also integral over $K[y_2, \ldots, y_n]$. This shows that $R = K[x_1, \ldots, x_n]$ is integral over $K[y_2, \ldots, y_n]$.

---

† One possible choice is as follows. Pick a natural number $p > \deg f$ and let $k_i := p^{i-1}$. To see that this choice yields the desired property, just observe that if $(i_1, \ldots, i_n) \neq (j_1, \ldots, j_n)$ are multi-indices occurring in f, then $i_k, j_k < p$ for all k since $p > \deg f$ and hence $i_1 + i_2 p + i_3 p^3 + \cdots + i_n p^{n-1} \neq j_1 + j_2 p + j_3 p^3 + \cdots + j_n p^{n-1}$, due to the uniqueness of the "p-adic" expansion of a natural number.

By induction hypothesis, there are elements $u_1, \ldots, u_m \in K[y_2, \ldots, y_n]$ which are algebraically independent over K such that $K[y_2, \ldots, y_n]$ is integral over $K[u_1, \ldots, u_m]$. Since being integral is a transitive property, we see that R is also integral over $K[u_1, \ldots, u_m]$. ∎

LET us repeat the content of Noether's normalization theorem. Any finitely generated algebra over a field K can be obtained by first adjoining a maximal number of algebraically independent elements and then performing an integral ring extension.

(16.4) Example. Consider the quotient ring $R := \mathbb{R}[X, Y]/I$ where $I := \langle XY - 1 \rangle$. Letting $x := X + I$ and $y := Y + I$, we have $R = K[x, y]$ where x and y satisfy the relation $xy - 1 = 0$. If we denote by V the hyperbola $V := \{(x, y) \in \mathbb{R}^2 \mid xy = 1\}$, we can identify R with the ring of all functions $f : V \to \mathbb{R}$ which are restrictions of polynomial functions $p : \mathbb{R}^2 \to \mathbb{R}$. Now the element x is not integral over $\mathbb{R}[y] \cong \mathbb{R}[Y]$. In fact, if there were polynomials $p_i \in \mathbb{R}[Y]$ with $x^n + p_{n-1}(y)x^{n-1} + \cdots + p_1(y)x + p_0(y) = 0$ for all $(x, y) \in V$, we could divide by x^n to obtain

$$1 + \frac{p_{n-1}(y)}{x} + \cdots + \frac{p_1(y)}{x^{n-1}} + \frac{p_0(y)}{x^n} = 0$$

for all $(x, y) \in V$. Letting $y \to 0$ and observing that then $x = 1/y \to \pm\infty$, we obtain the contradiction $1 = 0$.

Geometrically, the fact that x is not integral over $\mathbb{R}[y]$ stems from the fact that there is no point $(x, y) \in V$ with $y = 0$. Repeating the proof of Noether's theorem in this special case, we apply the substitution $x' = x$, $y' = y - ax$ with $a \neq 0$, according to (16.2)(b). Then the equation $xy - 1 = 0$ becomes $0 = x'(y' + ax') - 1 = ax'^2 + x'y' - 1$ so that $x'^2 + \frac{1}{a}y'x' - \frac{1}{a} = 0$ so that x' is integral over $\mathbb{R}[y']$. This shows that $\mathbb{R}[y']$ is a Noetherian normalization of R. Geometrically, the indicated substitution means a tilting of the x-axis, which guarantees that every line parallel to the x'-axis now intersects V in exactly two points.

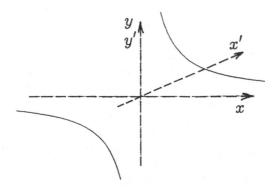

(16.5) Example. In the situation that R is an affine K-algebra, N is a Noetherian normalization of R and N' is a Noetherian normalization of N, then N' is also a Noetherian normalization of R; this follows easily from the fact that being integral is a transitive property.

WE can use Noether's Normalization Theorem to obtain a second proof of Zariski's lemma (12.35).

(16.6) Zariski's Lemma. *Let $(L : K)$ be a field extension such that $L = K[x_1, \ldots, x_n]$ is an affine K-algebra. Then $(L : K)$ is an algebraic field extension.*

Proof. By Noether's theorem (16.3) we can find a Noetherian normalization $K[u_1, \ldots, u_m]$ of L. Since L is a field, then so is $K[u_1, \ldots, u_m]$, due to (15.1). But $K[u_1, \ldots, u_m]^\times = K^\times$ because $u_1, \ldots, u_m$ are algebraically independent; hence $m = 0$. This means that L is integral (i.e., algebraic) over K. ∎

ZARISKI'S lemma yields some information on ideals of polynomials rings.

(16.7) Proposition. *Let M be a maximal ideal of the polynomial ring $K[x_1, \ldots, x_n]$ where K is a field.*
(a) The intersection $M \cap K[x_1, \ldots, x_{n-1}]$ is a maximal ideal of $K[x_1, \ldots, x_{n-1}]$.
(b) The ideal M is generated by n elements.

Proof. (a) We have $K[x_1, \ldots, x_n]/M = K[\xi_1, \ldots, \xi_n]$ where $\xi_i = x_i + M$ is the residue class of x_i modulo M. Since M is maximal, the ring $K[\xi_1, \ldots, \xi_n]$ is a field, hence an algebraic field extension, due to Zariski's lemma (16.6). Consequently, $K[\xi_1, \ldots, \xi_{n-1}]$ is a subfield of $K[\xi_1, \ldots, \xi_n]$ (and not just a subring). Now restricting the quotient map $\Phi : K[x_1, \ldots, x_n] \to K[\xi_1, \ldots, \xi_n]$ to $K[x_1, \ldots, x_{n-1}]$, we obtain a surjective homomorphism $\varphi : K[x_1, \ldots, x_{n-1}] \to K[\xi_1, \ldots, \xi_{n-1}]$ with kernel $\ker \varphi = M \cap K[x_1, \ldots, x_{n-1}]$. Hence $K[x_1, \ldots, x_{n-1}]/(M \cap K[x_1, \ldots, x_{n-1}])$ is isomorphic to $K[\xi_1, \ldots, \xi_{n-1}]$ which is a field. But this means that $M \cap K[x_1, \ldots, x_{n-1}]$ is a maximal ideal of $K[x_1, \ldots, x_{n-1}]$.

(b) We proceed by induction on n. The case $n = 1$ is clear, because $K[x_1]$ is Euclidean and hence a principal ideal domain. Let $n \geq 2$. If M is a maximal ideal of $K[x_1, \ldots, x_n]$, then $M \cap K[x_1, \ldots, x_{n-1}]$ is a maximal ideal of $K[x_1, \ldots, x_{n-1}]$ by part (a) and hence by induction hypothesis generated by $n - 1$ elements $f_1, \ldots, f_{n-1}$. Then $L := K[x_1, \ldots, x_{n-1}]/\langle f_1, \ldots, f_{n-1} \rangle$ is a field. Consider the surjective homomorphism $\varphi : K[x_1, \ldots, x_n] \to L[x_n]$ which is obtained by writing each element $f \in K[x_1, \ldots, x_n]$ as $f(x_1, \ldots, x_n) = \sum_{i=0}^{d} \varphi_i(x_1, \ldots, x_{n-1})x_n^i$ and taking the coefficients φ_i modulo $\langle f_1, \ldots, f_{n-1} \rangle$. The kernel of φ is the ideal $\langle\langle f_1, \ldots, f_{n-1} \rangle\rangle$ generated by $f_1, \ldots, f_{n-1}$ in $K[x_1, \ldots, x_n]$ (sic!). Hence $K[x_1, \ldots, x_n]/\langle\langle f_1, \ldots, f_{n-1} \rangle\rangle \cong L[x_n]$. Thus the ideal $M/\langle\langle f_1, \ldots, f_{n-1} \rangle\rangle$ is isomorphic to an ideal of $L[x_n]$. Since $L[x_n]$ is a principal ideal domain, there is a polynomial $F_n \in L[x_1, \ldots, x_n]$ which generates this ideal. It is now easy to see that if F_n is represented by $f_n \in K[x_1, \ldots, x_n]$ then $M = \langle\langle f_1, \ldots, f_n \rangle\rangle$. ∎

(16.8) Example. The ideal $I := \{f \in \mathbb{R}[x,y] \mid f(i,i) = 0\}$ is a maximal ideal of $\mathbb{R}[x,y]$, due to (12.6)(b). Hence (16.7)(b) shows that I can be generated by two elements. In fact, $I = \langle x^2 + 1, x - y \rangle$; for example, the element $y^2 + 1$ of I can be written as $y^2 + 1 = (x^2 + 1) - (x + y)(x - y)$.

GIVEN a finitely generated K-algebra R and an ideal $I \subsetneq R$, it is often desirable to choose not just any old Noetherian normalization of R, but a Noetherian normalization $K[y_1, \ldots, y_d]$ which is adapted to the ideal I in the sense that $I \cap K[y_1, \ldots, y_d]$ is generated by some of the elements y_i. As we will show in the next lemma, such a choice of a Noetherian normalization automatically yields a Noetherian normalization for the quotient algebra R/I and hence lends itself to be used in induction proofs where the induction step is carried out by a transition from R to a quotient ring R/I.

(16.9) Lemma. *Let R be a finitely generated K-algebra and let $I \subsetneq R$ be an ideal of R. For any element $r \in R$, we denote the residue-class of r modulo I by $\bar{r} := r + I$. If N is a Noetherian normalization of R such that $N = K[y_1, \ldots, y_d]$ and $I \cap N = \langle y_{\delta+1}, \ldots, y_d \rangle$ for some $\delta \le d$, then $\overline{N} = K[\overline{y_1}, \ldots, \overline{y_d}] = K[\overline{y_1}, \ldots, \overline{y_\delta}]$ is a Noetherian normalization of $\overline{R} = R/I$.*

Proof. Consider the quotient map $\Phi : R \to R/I = \overline{R}$ and its restriction $\varphi : N \to (N + I)/I = \overline{N}$. The kernel of φ is $N \cap \ker \Phi = N \cap I = \langle y_{\delta_1}, \ldots, y_d \rangle$; hence $K[\overline{y_1}, \ldots, \overline{y_d}] = K[\overline{y_1}, \ldots, \overline{y_\delta}]$. Also, since φ induces an isomorphism $K[y_1, \ldots, y_\delta] \cong K[\overline{y_1}, \ldots, \overline{y_\delta}]$, the elements $\overline{y_1}, \ldots, \overline{y_\delta}$ are algebraically independent over K. Moreover, $\overline{R}$ is integral over $\overline{N}$ since R is integral over N. This shows that $\overline{N}$ is a Noetherian normalization of $\overline{R}$. ∎

WE are now going to show that whenever I is a proper ideal of a finitely generated K-algebra $K[x_1, \ldots, x_n]$, then we can find a Noetherian normalization which is adapted to I; we will call this fact the strong version of Noether's normalization theorem. (The weak version just guarantees the existence of some Noetherian normalization without any special properties). We first treat the case that $x_1, \ldots, x_n$ are algebraically independent and then turn to the general case.

(16.10) Proposition. *Let $R = K[x_1, \ldots, x_n]$ be a finitely generated algebra where $x_1, \ldots, x_n$ are algebraically independent over K. Moreover, let $I \subsetneq R$ be an ideal of R. Then there is a number $0 \le \delta \le n$ and there is a Noetherian normalization $N = K[y_1, \ldots, y_n]$ of R such that $I \cap N = \langle y_{\delta+1}, \ldots, y_n \rangle$. If K is infinite, we can choose the y_i's as linear combinations of the x_k's, say $y_i = \sum_k a_{ik} x_k$ with $a_{ik} \in K$.*

Proof. Let us first treat the special case that I is a principal ideal, say $I = \langle f(x_1, \ldots, x_n) \rangle$ where $f \in K[X_1, \ldots, X_n]$ is a nonconstant polynomial. Let $y_1 := f(x_1, \ldots, x_n)$ and $y_2, \ldots, y_n$ as in lemma (16.2) and $N := K[y_1, \ldots, y_n]$. Then $y_1, \ldots, y_n$ are algebraically independent over K since $x_1, \ldots, x_n$ are. Next, we observe that $R = K[y_1, \ldots, y_n, x_1]$. Hence to show that R is integral over N, it is enough to show that x_1 is integral over N; but this follows immediately by dividing the equation

$0 = f(x_1, \ldots, x_n) - y_1 = cx_1^N + r(x_1, y_2, \ldots, y_n) - y_1$ occurring in (16.2) by c. Finally, let us show that $I \cap N = \langle y_1 \rangle$. Here the inclusion $\supseteq$ is trivial. Conversely, let $\xi \in I \cap K[y_1, \ldots, y_n]$; then $\xi \in I = Ry_1$, say $\xi = ry_1$. Since R is integral over N, we can write

$$r^d + p_{d-1}(y_1, \ldots, y_n)r^{d-1} + \cdots + p_1(y_1, \ldots, y_n)r + p_0(y_1, \ldots, y_n) = 0 .$$

Multiplying this equation by y_1^d, we obtain

$$\xi^d + y_1\, p_{d-1}(y_1, \ldots, y_n)\xi^{d-1} + \cdots + y_1\, p_1(y_1, \ldots, y_n)\xi + y_1^d\, p_0(y_1, \ldots, y_n) = 0 .$$

This shows that y_1 divides ξ^d in N. Since $N = K[y_1, \ldots, y_n] \cong K[Y_1, \ldots, Y_n]$ is a unique factorization domain, we see that the element y_1 is prime. Hence $y_1 \mid \xi^d$ implies $y_1 \mid \xi$ so that $\xi \in Ny_1 \subseteq Ry_1$. Thus the claim is established in the special case that I is a principal ideal.

For the general case, we proceed by induction on n. The case $n = 1$ is already contained in the special case treated above, because every ideal in $K[x_1] \cong K[X_1]$ is a principal ideal; so let $n \geq 2$. Pick a nonzero element $y_1 \in I$, say $y_1 = f(x_1, \ldots, x_n)$ with a nonconstant polynomial f, and let $y_2, \ldots, y_n$ as in (16.2). By induction hypothesis, there is a Noetherian normalization $K[z_1, \ldots, z_{n-1}]$ of $K[y_2, \ldots, y_n]$ with $I \cap K[z_1, \ldots, z_{n-1}] = \langle z_{\delta+1}, \ldots, z_{n-1} \rangle$ for some $0 \leq \delta \leq n - 1$. If K is infinite, then the z_i's can be chosen as linear combinations of the y_i's (by induction hypothesis) and the y_i's as linear combinations of the x_i's (by lemma (16.2)(b)) so that the z_i's are also linear combinations of the x_i's in this case.

Now let $z_n := y_1$; then clearly $z_1, \ldots, z_n$ are algebraically independent over K. Also, $R = K[x_1, \ldots, x_n]$ is integral over $K[y_1, \ldots, y_n]$ (by the choice of the y_i's), and $K[y_1, \ldots, y_n]$ is integral over $K[z_1, \ldots, z_n]$; by transitivity, R is integral over $K[z_1, \ldots, z_n]$. Finally, let us show that $I \cap K[z_1, \ldots, z_n] = \langle z_{\delta+1}, \ldots, z_n \rangle$. Here the inclusion $\supseteq$ is trivial. To prove $\subseteq$ we observe that an arbitrary element of $K[z_1, \ldots, z_n]$ has the form $\xi = p(z_1, \ldots, z_{n-1}) + z_n \cdot q(z_1, \ldots, z_{n-1})$. Since $z_n = y_1 \in I$, we see that $\xi \in K[z_1, \ldots, z_n] \cap I$ implies that $p(z_1, \ldots, z_{n-1}) \in I$ and hence $p(z_1, \ldots, z_{n-1}) \in I \cap K[z_1, \ldots, z_{n-1}] = \langle z_{\delta+1}, \ldots, z_{n-1} \rangle$. This shows that $\xi \in \langle z_{\delta+1}, \ldots, z_{n-1} \rangle + z_n \cdot K[z_1, \ldots, z_{n-1}] = \langle z_{\delta+1}, \ldots, z_n \rangle$. ∎

(16.11) Noether Normalization Theorem (strong version). *Let $R = K[x_1, \ldots, x_n]$ be a finitely generated K-algebra and let $I \subsetneq R$ be an ideal of R. Then there are numbers $0 \leq \delta \leq d \leq n$ and elements $z_1, \ldots, z_d \in R$ such that the following conditions are satisfied:*

(1) $z_1, \ldots, z_d$ are algebraically independent;
(2) R is integral over $K[z_1, \ldots, z_d]$;
(3) $I \cap K[z_1, \ldots, z_d] = \langle z_{\delta+1}, \ldots, z_d \rangle$.

Proof. If $x_1, \ldots, x_n$ are algebraically independent, then the statement holds (with $d = n$) according to (16.10). In the general case, we can write $R \cong K[X_1, \ldots, X_n]/J$. By proposition (16.10) there is a Noetherian normalization $K[Y_1, \ldots, Y_n]$ of $K[X_1, \ldots, X_n]$ such that $J \cap K[Y_1, \ldots, Y_n] = \langle Y_{\delta+1}, \ldots, Y_n \rangle$ (where the Y_i's can be chosen as linear combinations of the X_k's if K is infinite). Let $x_i = X_i + J$ and

$y_i = Y_i + J$. Then $K[y_1, \ldots, y_n]$ is a Noetherian normalization of $R = K[x_1, \ldots, x_n]$, due to lemma (16.9). Applying (16.10) to $R' := K[y_1, \ldots, y_d]$ and $I' := I \cap R'$, we find a number $0 \le \delta \le d$ and elements $z_1, \ldots, z_d \in R'$ such that $K[z_1, \ldots, z_d]$ is a Noetherian normalization of R' with $I' \cap K[z_1, \ldots, z_d] = \langle z_{\delta+1}, \ldots, z_d \rangle$. If K is infinite, the elements z_i can be chosen as linear combinations of the y_i's. Then $K[z_1, \ldots, z_d]$ is a Noetherian normalization of R by (16.5), and since $K[z_1, \ldots, z_d] \subseteq R'$, we have $I \cap K[z_1, \ldots, z_d] = I' \cap K[z_1, \ldots, z_d] = \langle z_{\delta+1}, \ldots, z_d \rangle$. ∎

FINALLY we will prove a more refined version of Noether's normalization theorem (16.3) in the case that the field K is perfect.

(16.12) Theorem. *Let K be a perfect field and let $R = K[x_1, \ldots, x_n]$ be a finitely generated K-algebra which is an integral domain.*

(a) There is a Noetherian normalization $N := K[y_1, \ldots, y_d]$ of R such that $L := Q(R) = K(x_1, \ldots, x_n)$ is a separable field extension of $L_0 := Q(N) = K(y_1, \ldots, y_d)$.

(b) There is a further element $y_{d+1} \in L$ such that $L = L_0(y_{d+1})$. If K is infinite, this element can be chosen as a K-linear combination of $y_1, \ldots, y_d$ so that $y_{d+1} = \sum_i \lambda_i y_i \in N \subseteq R$.†

Proof. (a) If $\operatorname{char} K = 0$ then every algebraic field extension of K is separable, and the claim is already contained in (16.3). Suppose $\operatorname{char} K = p$; in this case, the hypothesis that K be perfect means exactly that every element of K has a p-th root in K. We have $R \cong K[X_1, \ldots, X_n]/I$ where I is the kernel of the evaluation map

$$\Phi : \begin{array}{ccc} K[X_1, \ldots, X_n] & \to & R \\ f & \mapsto & f(x_1, \ldots, x_n) \end{array}.$$

The fact that R is an integral domain is tantamount to saying that I is a prime ideal, and since $K[X_1, \ldots, X_n]$ is a unique factorization domain this means that either $I = \{0\}$ or else I contains an irreducible polynomial f. Since the first possibility is trivial (if $I = \{0\}$, then $N = R$ and $L = L_0$), assume the latter. Suppose that f is inseparable in each variable X_i; then f has the form

$$f(X_1, \ldots, X_n) = \sum a_{i_1, \ldots, i_n} X_1^{p i_1} \cdots X_n^{p i_n}.$$

Picking a p-th root $b_{i_1, \ldots, i_n} \in K$ for each coefficient $a_{i_1, \ldots, i_n}$, this reads

$$f(X_1, \ldots, X_n) = \sum b_{i_1, \ldots, i_n}^p X_1^{p i_1} \cdots X_n^{p i_n} = \left(\sum b_{i_1, \ldots, i_n} X_1^{i_1} \cdots X_n^{i_n} \right)^p$$

contradicting the irreducibility of f. Hence f is separable in at least one of the variables, without loss of generality in X_1. Then as in the proof of (16.13) we have a *separable* monic equation

$$0 = \frac{1}{c} f(x_1, \ldots, x_n) = x_1^N + \frac{1}{c} r(x_1, y_2, \ldots, y_n)$$

---

† This means that the field extension $\big(K(x_1, \ldots, x_n) : K \big)$ can be split into a purely transcendental extension $(L_0 : K)$ and a simple algebraic extension $(L : L_0)$. In other words, we can write $L = K(y_1, \ldots, y_d, y_{d+1})$ where there is only one algebraic dependence relation between the generators y_i. The geometric significance of this result will be exhibited in the next section.

which shows that x_1 is integral over $K[y_2, \ldots, y_n]$ and separable over $K(y_2, \ldots, y_n)$. The separability of x_1 over $K(y_2, \ldots, y_n)$ implies that $K(x_1, y_2, \ldots, y_n)$ is separable over $K(y_2, \ldots, y_n)$. By induction hypothesis, there are elements $u_1, \ldots, u_m$ such that $K[y_2, \ldots, y_n]$ is integral over $K[u_1, \ldots, u_m]$ and $K(y_2, \ldots, y_n)$ is separable over $K(u_1, \ldots, u_m)$. Since integrality and separability are transitive properties, the claim follows.

(b) The existence of y_{d+1} is an immediate consequence of the Primitive Element Theorem of separable field extensions. Suppose that K is infinite. Then since $L = L_0(x_1, \ldots, x_n)$ this theorem implies that we can find a primitive element c of $(L : L_0)$ which is an L_0-linear combination of $x_1, \ldots, x_n$, i.e., $c = \sum_i \lambda_i x_i$ with $\lambda_i \in L_0$, say $\lambda_i = \frac{p_i(y_1, \ldots, y_d)}{q_i(y_1, \ldots, y_d)}$ so that $c = \frac{p_1}{q_1} x_1 + \cdots + \frac{p_n}{q_n} x_n = \frac{1}{q_1 \cdots q_n}(P_1 x_1 + \cdots + P_n x_n)$ (where the coefficients $P_i := p_i \prod_{j \neq i} q_j$ lie in $K[y_1, \ldots, y_d]$). Then $y_{d+1} := q_1 \cdots q_n c = P_1 x_1 + \cdots + P_n x_n$ lies in R, and $L = L_0(c) = L_0(q_1 \cdots q_n c) = L_0(y_{d+1})$. ∎

LET us point out what we will do in the remainder of this section before we actually formulate and prove our results in detail. We will define the *dimension* of a commutative ring R as the maximal possible length of a chain $\{0\} \subsetneq P_1 \subsetneq \cdots \subsetneq P_d$ of prime ideals in R. (This way of measuring the "size" of a commutative ring will be filled with some geometric intuition in the next two sections when we discuss the correlation between ring theory and algebraic geometry.) Then we will show that if $(S : R)$ is an integral ring extension then R and S have the same dimension so that, roughly speaking, an integral extension of a ring is not much larger than the original ring; this will require the comparison of chains of prime ideals in R and S. Finally, we will be able to explicitly determine the dimension of affine algebras. This is the outline of our remaining program for this section; let us go ahead to its execution (which is, unfortunately, quite technical).

(16.13) **Definitions.** *Let R be a commutative ring. The* **height** *$\mathrm{ht}(P)$ of a prime ideal $P \trianglelefteq R$ is the supremum of all numbers $d \geq 0$ such that there is a chain*

$$P_0 \subsetneq P_1 \subsetneq \cdots \subsetneq P_d = P$$

of prime ideals of R. The **Krull** *dimension or simply* **dimension** *$\dim R$ of R is the supremum of all occurring heights of prime ideals, i.e., the supremum of all lengths of prime ideal chains in R.*

(16.14) **Examples.** (a) If K is a field then the only prime ideal is $\{0\}$; hence $\dim K = 0$.

(b) If R is a principal domain but not a field then $\dim R = 1$. In fact, the nonzero prime ideals of R are exactly the ideals $\langle p \rangle$ where p is a prime element. But then p is automatically irreducible, and hence $\langle p \rangle$ is a maximal ideal due to (7.4)(e). Thus the maximal ideals of prime ideals in R are the chains $\{0\} \subsetneq \langle p \rangle$ where p is a prime element; this gives the claim. In particular, $\dim \mathbb{Z} = 1$ and $\dim K[X] = 1$ for any field K.

(c) Let K be a field and let $R = K[x_1, \ldots, x_n]$ be the polynomial ring over K in n variables $x_1, \ldots, x_n$. Then $\{0\} \subsetneq \langle x_1 \rangle \subsetneq \langle x_1, x_2 \rangle \subsetneq \cdots \subsetneq \langle x_1, \ldots, x_n \rangle$ is a chain of prime ideals in $K[x_1, \ldots, x_n]$; hence $\dim K[x_1, \ldots, x_n] \geq n$. We will show later that, in fact, $\dim K[x_1, \ldots, x_n] = n$. (See (16.21) below.)

(d) The ring $R := K[x_1, x_2, x_3, \ldots]$ in an countably infinite number of variables satisfies $\dim R = \infty$ because $\{0\} \subsetneqq \langle x_1 \rangle \subsetneqq \langle x_1, x_2 \rangle \subsetneqq \langle x_1, x_2, x_3 \rangle \subsetneqq \cdots$ is an infinite chain of prime ideals.

(e) Let R be a commutative ring with identity. Then R is Artinian if and only if R is Noetherian and $\dim R = 0$; this is merely a restatement of theorem (11.28).

WE are going to show that if S is an integral dimension of R then $\dim S = \dim R$. To do so, we have to compare chains of prime ideals in R with those is S; this is done in the next two theorems. Let us start with a definition.

(16.15) Definition. *Let $R \subseteq S$ be rings. We say that an ideal J of S **lies over** an ideal I of R if $I = J \cap R$.*

(16.16) Lying-over Theorem. *Suppose $R \subseteq S$ are commutative rings such that S is integral over R. Let P be a prime ideal of R.*
(a) *There is a prime ideal Q of S lying over P.*
(b) *If $Q_1, Q_2 \trianglelefteq S$ are prime ideals lying over P with $Q_1 \subseteq Q_2$, then $Q_1 = Q_2$.*

Proof. (a) The set
$$\mathcal{M} := \{J \trianglelefteq S \mid J \cap R \subseteq P\}$$
is not empty (since $\{0\} \in \mathcal{M}$) and possesses a maximal element Q, due to Zorn's lemma. We claim that Q is a prime ideal lying over P.

Let us show first that Q lies over P, i.e., that $Q \cap R = P$. By definition of $\mathcal{M}$, we have $Q \cap R \subseteq P$. Suppose $Q \cap R \subsetneqq P$ and let $a \in P \setminus Q$. Then $Q \subsetneqq Q + Sa$ and consequently $(Q + Sa) \cap R \not\subseteq P$, due to the maximality of Q. Hence we can find elements $q \in Q$ and $s \in S$ with $q + sa \in R \setminus P$. Using the fact that S is integral over R, we find elements $a_0, \ldots, a_{n-1} \in R$ with $s^n + a_{n-1}s^{n-1} + \cdots + a_1 s + a_0 = 0$. Then

$$R \ni (q + sa)^n + \underbrace{a_{n-1}a(q + sa)^{n-1} + \cdots + a_1 a^{n-1}(q + sa) + a_0 a^n}_{\in P \text{ because } a \in P \text{ and } a_0, \ldots, a_{n-1}, q + sa \in R}$$

$$= a^n \underbrace{\left(s^n + a_{n-1}s^{n-1} + \cdots + a_1 s + a_0\right)}_{= 0} + (\text{a multiple of } q) \in Q$$

which is an element in $R \cap Q \subseteq P$. This implies $(q + sa)^n \in P$, hence $q + sa \in P$ because P is prime. But this contradicts the fact that $q + sa \in R \setminus P$. Hence our assumption $Q \cap R \subsetneqq P$ was false.

Let us now show that Q is a prime ideal. If $J \trianglelefteq S$ is any ideal with $Q \subsetneqq J$ then $Q \cap R \not\subseteq P$ due to the maximality of Q in $\mathcal{M}$, hence $Q \cap (R \setminus P) \neq \emptyset$; this shows that Q is maximal in the set of all ideals of S disjoint with the multiplicatively closed set $R \setminus P$. But then (6.4)(a) implies that Q is a prime ideal.

(b) Suppose $Q_1 \subsetneqq Q_2$ and pick an element $q \in Q_2 \setminus Q_1$. Since q is integral over R, there is a minimal number n such that there are elements $a_0, \ldots, a_{n-1} \in R$ with

($\star$)
$$\underbrace{q^n + a_{n-1}q^{n-1} + \cdots + a_1 q + a_0}_{\in Q_2} \in Q_1 .$$

Then $a_0 \in Q_1 - Q_2 = Q_2$, hence $a_0 \in Q_2 \cap R = P = Q_1 \cap R$. Consequently, $(\star)$ implies $q^n + a_{n-1}q^{n-1} + \cdots + a_1 q \in Q_1$, i.e.,

$$(\star\star) \qquad\qquad q(q^{n-1} + a_{n-1}q^{n-2} + \cdots + a_1) \in Q_1 .$$

Since $q \notin Q_1$ and Q_1 is prime, this implies $q^{n-1} + a_{n-1}q^{n-2} + \cdots + a_1 \in Q_1$, contradicting our choice of n. This is the desired contradiction. ∎

THE lying-over theorem can be extended to a theorem which guarantees the existence of a chain of prime ideals lying over a given chain.

(16.17) **Going-up Theorem.** *Suppose $R \subseteq S$ are commutative rings such that S is integral over R. Let $P_0 \subsetneq P_1 \subsetneq \cdots \subsetneq P_n$ be a chain of prime ideals in R.*

(a) If $Q_0 \subsetneq Q_1 \subsetneq \cdots \subsetneq Q_m$ is a chain of prime ideals in S with $m < n$ such that Q_i lies over P_i for $0 \le i \le m$ then there are prime ideals $Q_m \subsetneq Q_{m+1} \subsetneq \cdots \subsetneq Q_n$ in S such that Q_i lies over P_i for $m+1 \le i \le n$; i.e., the "upper chain" can be extended to a chain lying over the given chain in R.

In S: $\qquad\qquad Q_0 \subsetneq Q_1 \subsetneq \cdots \subsetneq Q_m \subsetneq \mathcal{ONE\ CAN\ GO\ UP}!$

In R: $\qquad\qquad P_0 \subsetneq P_1 \subsetneq \cdots \subsetneq P_m \subsetneq P_{m+1} \subsetneq \cdots \subsetneq P_n$

(b) If, for any fixed index i, there is no prime ideal of R strictly between P_i and P_{i+1}, then there is no prime ideal of S strictly between Q_i and Q_{i+1}.

Proof. (a) To prove the extendability of the chain $Q_0 \subsetneq \cdots \subsetneq Q_m$, we can clearly assume $m = 0$ and $n = 1$; the general case then follows easily by induction.

Hence we have prime ideals $P_0 \subsetneq P_1$ in R and a prime ideal Q_0 in S with $Q_0 \cap R = P_0$. Since S is integral over R, the quotient ring S/Q_0 is integral over $(R + Q_0)/Q_0$ (which is isomorphic to $R/(Q_0 \cap R) = R/P_0$). Indeed, for any $s \in S$ one can simply read an equation of the form $s^n + a_{n-1}s^{n-1} + \cdots + a_1 s + a_0 = 0$ modulo Q_0. Now $(P_1 + Q_0)/Q_0 \cong P_1/P_0$ is a prime ideal of $(R + Q_0)/Q_0 \cong R/P_0$, due to (6.15). Hence there is a prime ideal of S/Q_0 lying over $(P_1 + Q_0)/Q_0$ due to (16.16)(a); this prime ideal has necessarily the form Q_1/Q_0 where Q_1 is a prime ideal of S with $Q_0 \subsetneq Q_1 \subseteq S$. Let us prove that Q_1 lies over P_1.

The fact that Q_1/Q_0 lies over $(P_1 + Q_0)/Q_0$ means that $(Q_0 + P_1)/Q_0 = (Q_1/Q_0) \cap ((Q_0 + R)/Q_0) = (Q_1 \cap (Q_0 + R))/Q_0$ or, equivalently,

$$(\star) \qquad\qquad Q_0 + P_1 = Q_1 \cap (Q_0 + R) .$$

We claim that $(\star)$ implies the desired equality $Q_1 \cap R = P_1$. First of all, let $p_1 \in P_1$. Then by $(\star)$ there are elements $q_1 \in Q_1$, $q_0 \in Q_0$ and $r \in R$ with $p_1 = q_1 = q_0 + r$. Then $q_0 = p_1 - r \in R$ so that $p_1 \in R$, i.e., $p_1 \in Q_1 \cap R$. This shows $P_1 \subseteq Q_1 \cap R$. Conversely, let $x \in Q_1 \cap R$. Then by $(\star)$ there are elements $q_0 \in Q_0$ and $p_1 \in P_1$ with $x = q_0 + p_1$. Then $q_0 = x - p_1 \in R$ so that $q_0 \in Q_0 \cap R = P_0$ and hence $x \in P_0 + P_1 = P_1$. This shows $Q_1 \cap R \subseteq P_1$.

(b) Suppose Q is a prime ideal strictly between Q_i and Q_{i+1}. By (16.16)(b), Q cannot lie over either P_i nor P_{i+1}. Hence the prime ideal $Q \cap R$ of R lies strictly between P_i and P_{i+1}. ∎

NOW we can prove the announced theorem that integral extensions are dimension-preserving.

(16.18) Theorem. *Let $R \subseteq S$ be commutative rings with S integral over R.*
(a) *If Q is a prime ideal of S and $P := Q \cap R$, then $\mathrm{ht}(Q) \leq \mathrm{ht}(P)$; moreover, Q is maximal in S if and only if P is maximal in R.*
(b) *$\dim R = \dim S$.*

Proof. (a) The first statement is an immediate consequence of the going-up theorem. Suppose P is maximal and let $Q' \trianglelefteq S$ be a maximal ideal with $Q \subseteq Q'$. Then $P' := Q' \cap R$ is a prime ideal of R with $P \subseteq P'$, hence $P = P'$ by the maximality of P. But then (16.16)(b) implies that $Q = Q'$. This shows that Q is maximal. Conversely, suppose that P is not maximal, say $P \subsetneq P'$ with a prime ideal P'. By (16.16) there are prime ideals $Q \subsetneq Q'$ lying over P and P', respectively. Hence Q is not maximal in S.

(b) Let $P_0 \subsetneq P_1 \subsetneq \cdots \subsetneq P_d$ be a chain of prime ideals of R of length d. Then there is a chain $Q_0 \subsetneq Q_1 \subsetneq \cdots \subsetneq Q_d$ of prime ideals in S of length d, due to (16.17)(a). This shows that $\dim R \leq \dim S$. Conversely, let $Q_0 \subsetneq Q_1 \subsetneq \cdots \subsetneq Q_d$ be a chain of prime ideals in S and let $P_i := Q_i \cap R$. Then the prime ideals P_i of R are pairwise different, due to (16.16)(b); hence $P_0 \subsetneq P_1 \subsetneq \cdots \subsetneq P_d$ is a chain of prime ideals in R of length d. This shows that $\dim S \leq \dim R$. ∎

UNDER more restrictive hypotheses, we can also prove a "going-down theorem" which allows one to extend a given chain of overlying ideals in the "downward" direction. We need a preparatory lemma.

(16.19) Lemma. *Suppose the integral domain R is integrally closed in its quotient field K and let $I \trianglelefteq R$ be a radical ideal of R.*
(a) *I is integrally closed in K.*
(b) *If $(L : K)$ is a field extension and if $\alpha \in L$ is integral over I, then the minimal polynomial of α over K has coefficients in I.*

Proof. (a) If $\alpha \in K$ is integral over I, then α is also integral over R and hence lies in R because R is integrally closed in K. Let $\alpha^m + y_{m-1}\alpha^{m-1} + \cdots + y_1\alpha + y_0 = 0$ be an equation of integral dependence for α over I; then $\alpha^m = -(y_0 + y_1\alpha + \cdots + y_{m-1}\alpha^{m-1}) \in I$, hence $\alpha \in \sqrt{I} = I$.

(b) This follows immediately from part (a) and (15.4). ∎

(16.20) Going-down Theorem. *Suppose $R \subseteq S$ are integral domains such that S is integral over R and such that R is integrally closed in its quotient field K.*

(a) *Let $P_0 \subsetneqq P_1 \subsetneqq \cdots \subsetneqq P_n$ be a chain of prime ideals in R. Then if $Q_0 \subsetneqq Q_{m+1} \subsetneqq \cdots \subsetneqq Q_n$ is a chain of prime ideals in S with $m < n$ such that Q_i lies over P_i for $m + 1 \leq i \leq n$, then there are prime ideals $Q_0 \subsetneqq Q_1 \subsetneqq \cdots \subsetneqq Q_m \subsetneqq Q_{m+1}$ in S such that Q_i lies over P_i for $0 \leq i \leq m$; i.e., the "upper chain" can be extended to a chain lying over the given chain in R.*

In S: $\mathcal{ONE\ CAN\ GO\ DOWN}$! $\underset{\neq}{\subseteq} Q_{m+1} \underset{\neq}{\subseteq} \cdots \underset{\neq}{\subseteq} Q_n$

In R: $P_0 \underset{\neq}{\subseteq} P_1 \underset{\neq}{\subseteq} \cdots \underset{\neq}{\subseteq} P_m \underset{\neq}{\subseteq} P_{m+1} \underset{\neq}{\subseteq} \cdots \underset{\neq}{\subseteq} P_n$

(b) *If the prime ideal $Q \trianglelefteq S$ lies over $P \trianglelefteq R$, then $\mathrm{ht}(Q) = \mathrm{ht}(P)$.*

Proof. (a) We can clearly assume $m = 0$ and $n = 1$; the general case then follows easily by induction. Hence we are given prime ideals $P_0 \subsetneqq P_1$ in R and a prime ideal $Q_1 \trianglelefteq S$ lying over P_1. Then the sets $\Sigma_0 := R \setminus P_0$ and $\sigma_1 := S \setminus Q_1$ are multiplicatively closed; hence so is $\Sigma := \Sigma_0 \Sigma_1 = \{rs \mid r \in \Sigma_0, s \in \Sigma_1\}$. We denote by $P_0 S$ the ideal of S generated by P_0, i.e., $P_0 S := \{\sum_i s_i p_i \mid p_i \in P_0, s_i \in S\}$. We will prove that $P_0 S \cap \Sigma = \emptyset$. Once this is clear, proposition (6.4)(a) guarantees the existence of a prime ideal Q_0 of S with $P_0 S \subseteq Q_0$ and $Q_0 \cap \Sigma = \emptyset$. To see that this prime ideal has the desired properties, observe that $Q_0 \cap \Sigma_1 \subseteq Q_0 \cap \Sigma = \emptyset$ so that $Q_0 \subseteq S \setminus \Sigma_1 = Q_1$; also, $P_0 \subseteq P_0 S \subseteq Q_0$ and hence

$$Q_0 \cap R = Q_0 \cap (P_0 \cup \Sigma_0) = \underbrace{(Q_0 \cap P_0)}_{= P_0} \cap \underbrace{(Q_0 \cap \Sigma_0)}_{\subseteq Q_0 \cap \Sigma = \emptyset} = P_0$$

so that Q_0 lies over P_0.

So let us prove that $P_0 S \cap \Sigma = \emptyset$. To obtain a contradiction, assume that there is an element $x \in P_0 S \cap \Sigma$. We can write $x = \sum_i p_i s_i$ with $p_i \in P_0$ and $s_i \in S$. We claim that each of the elements $p_i s_i$ (and hence x itself) is integral over P_0. Indeed, since $s_i \in S$ is integral over R, there is an equation $s_i^n + r_{n-1} s_i^{n-1} + \cdots + r_1 s_i + r_0 = 0$ with coefficients in R. Multiplying by p_i^n, we obtain

$$(p_i s_i)^n + \underbrace{r_{n-1} p_i}(p_i s_i)^{n-1} + \cdots + \underbrace{r_1 p_i^{n-1}}(p_i s_i) + \underbrace{r_0 p_i^n} = 0$$
$$\qquad\qquad\quad \in P_0 \qquad\qquad\qquad\quad \in P_0 \qquad\quad \in P_0$$

which shows that $p_i s_i$ is integral over P_0. Then (16.19)(b) implies that the minimal polynomial $f(X) = X^N + a_{N-1} X^{N-1} + \cdots + a_1 X + a_0$ of x over K has coefficients in P_0. Until now, we only used the fact that $x \in P_0 S$; since also $x \in \Sigma = \Sigma_0 \Sigma_1$, we can write $x = rs$ with $r \in \Sigma_0$ and $s \in \Sigma_1$. Then the minimal polynomial of $s = x/r$ over K is

$$g(X) = X^N + \frac{a_{N-1}}{r} X^{N-1} + \cdots + \frac{a_1}{r^{N-1}} X + \frac{a_0}{r^N}.$$

Since $s \in S$ is integral over R, we have $g \in R[X]$ by (15.4) so that the elements $\rho_i := a_{N-i}/r^i \in K$ actually lie in R. Then $r^i \rho_i = a_{N-i} \in P_0$; since $r \notin P_0$ and P_0 is a prime ideal, this implies $\rho_i \in P_0$ so that $g \in P_0[X]$. Since $g(s) = 0$, we have $s^N = -(\rho_{N-1} s^{N-1} + \cdots + \rho_1 s + \rho_0) \in P_0 S \subseteq P_1 S \subseteq Q_1 S \subseteq Q_1$; since A_1 is a prime ideal, this implies $s \in Q_1$, contradicting the fact that $s \in \Sigma_1 = S \setminus Q_1$.

(b) By (16.18)(a) we have $\mathrm{ht}(Q) \leq \mathrm{ht}(P)$. The converse inequality is an immediate consequence of part (a), because for any prime ideal chain in R ending with P we can find a prime ideal chain of the same length in S ending with Q. ■

WE have now gathered enough information to prove results on the dimension of finitely generated algebras over a field.

(16.21) Theorem. *Let K be a field and let $R = K[x_1, \ldots, x_n]$ be a finitely generated K-algebra. If $N = K[y_1, \ldots, y_d]$ is a Noetherian normalization of R, then the Krull dimension of R is $\dim R = d$.*

Proof. By (16.18)(b) and example (16.14)(c) we have $\dim R = \dim N \geq d$. Thus it remains to show that whenever $Q_0 \subsetneq Q_1 \subsetneq \cdots \subsetneq Q_m$ is a chain of prime ideals in R then $m \leq d$. If $d = 0$ we have $N = K$ (a field) so that the claim is trivial in this case. Also, if $d = 1$, then $N \cong K[x_1] \cong K[X_1]$ is a unique factorization domain, and the claim was proved in (16.14)(b) for this special case. Hence assume $d \geq 2$. Letting $P_i := Q_i \cap N$, we obtain a chain $P_0 \subsetneq P_1 \subsetneq \cdots \subsetneq P_m$ where the inclusions are strict by (16.16)(b). By (16.10) there is a Noetherian normalization $N' = K[z_1, \ldots, z_d]$ of N (and hence of R) such that $P_1 \cap N' = \langle z_{\delta+1}, \ldots, z_d \rangle$. Then $P_1 \neq \{0\}$ implies $P_1 \cap N' \neq \{0\}$ (by (16.16)(b)) so that $\delta < d$, i.e., $\delta \leq d - 1$. Taking residue-classes modulo P_1, we see by lemma (16.9) that $K[\overline{z_1}, \ldots, \overline{z_\delta}]$ is a Noetherian normalization of $K[\overline{y_1}, \ldots, \overline{y_d}] = \overline{N}$ and hence of $\overline{R}$. Now $\overline{P_1} \subsetneq \overline{P_2} \subsetneq \cdots \subsetneq \overline{P_m}$ is a chain of prime ideals of length $m - 1$ in $\overline{R}$; hence $m - 1 \leq \delta$ by induction hypothesis. Thus $m \leq \delta + 1 \leq d$, which is what we wanted to show. ■

IF we assume in addition that $R = K[x_1, \ldots, x_n]$ is an integral domain we can show more, namely, that all saturated prime ideal chains in R have the same length (which is $\dim R = d$, of course). †

(16.22) Theorem. *Let $R = K[x_1, \ldots, x_n]$ be a finitely generated K-algebra which has no zero-divisors (and hence is an integral domain) and let $N = K[y_1, \ldots, y_d]$ be a Noetherian normalization of R. Suppose that $Q_0 \subsetneq Q_1 \subsetneq \cdots \subsetneq Q_m$ is a chain of prime ideals in R and let $P_0 \subsetneq P_1 \subsetneq \cdots \subsetneq P_m$ be the corresponding chain in N, given by $P_i = Q_i \cap N$.*
*(a) If $Q_0 \subsetneq Q_1 \subsetneq \cdots \subsetneq Q_m$ is **saturated** in the sense that no prime ideal can properly be inserted into the chain, then $P_0 \subsetneq P_1 \subsetneq \cdots \subsetneq P_m$ is also saturated.*
(b) If $Q_0 \subsetneq Q_1 \subsetneq \cdots \subsetneq Q_m$ is a saturated chain, then $m = d$.

Proof. (a) Suppose we can insert a prime ideal P_* between P_i and P_{i+1}. Choose a Noetherian normalization $N' = K[z_1, \ldots, z_d]$ of N such that $P_i \cap K[z_1, \ldots, z_d] =$

---

† This property is sometimes expressed by saying that $K[x_1, \ldots, x_n]$ is *equidimensional*. It is not too hard to find examples of commutative rings which are not equidimensional. (See problem 3(b) below.)

$\langle z_{\delta+1}, \ldots, z_d \rangle$. Taking residue classes modulo P_i, we see by (16.9) that $\overline{N'} = K[\overline{z_1}, \ldots, \overline{z_\delta}]$ is a Noetherian normalization of $\overline{N} = (N + P_i)/P_i = K[\overline{y_1}, \ldots, \overline{y_d}]$ (and hence of $\overline{R} = R/P_i$). Then $\{\overline{0}\} \subsetneq \overline{P_*} \subsetneq \overline{P_{i+1}}$ implies $\{\overline{0}\} \subsetneq \overline{P_*} \cap \overline{N'} \subsetneq \overline{P_{i+1}} \cap \overline{N'}$, due to part (b) of the lying-over theorem. In particular, there is a prime ideal strictly between $\{\overline{0}\} = P_i/P_i$ and $\overline{P_{i+1}} = P_{i+1}/P_i$. Using the isomorphism $N/P_i \cong (N + Q_i)/Q_i$, we see that there is a prime ideal strictly between $\{\overline{0}\} = Q_i/Q_i$ and $(P_{i+1} + Q_i)/Q_i$. Also, Q_{i+1}/Q_i lies over $(P_{i+1} + Q_i)/Q_i$ because

$$\frac{Q_{i+1}}{Q_i} \cap \frac{N + Q_i}{Q_i} = \frac{Q_{i+1} \cap (N + Q_i)}{Q_i} = \frac{(Q_{i+1} \cap N) + Q_i}{Q_i} = \frac{P_i + Q_i}{Q_i}.$$

Hence we are in the following situation.

$$\text{In } R/Q_i: \qquad\qquad\qquad\qquad Q_{i+1}/Q_i$$
$$\text{In } (N + Q_i)/Q_i: \quad \{0\} \subsetneq \star \subsetneq (P_{i+1} + Q_i)/Q_i$$

Now the going-down theorem guarantees the existence of a prime ideal of R/Q_i strictly between $\{0\} = Q_i/Q_i$ and Q_{i+1}/Q_i. (Why is this theorem applicable?) Hence there is a prime ideal of R strictly between Q_i and Q_{i+1} which is the desired contradiction.

(b) We proceed by induction on d, the case $d = 0$ being trivial. Let $d \geq 1$ and choose a Noetherian normalization $K[z_1, \ldots, z_d]$ of N with $P_1 \cap K[z_1, \ldots, z_d] = \langle z_{\delta+1}, \ldots, z_d \rangle$. This ideal has height 1 by (16.20)(b) which forces $\delta + 1 = d$ (otherwise we would have strict inclusions $\{0\} \subsetneq \langle z_{\delta+1} \rangle \subsetneq \langle z_{\delta+1}, \ldots, z_d \rangle$ of prime ideals). Now

$$\{0\} = \frac{P_1}{P_1} \subsetneq \frac{P_2}{P_1} \subsetneq \cdots \subsetneq \frac{P_m}{P_1}$$

is a saturated chain of prime ideals in $(K[z_1, \ldots, z_d] + P_1)/P_1 = K[\overline{z_1}, \ldots, \overline{z_d}] = K[\overline{z_1}, \ldots, \overline{z_\delta}] = K[\overline{z_1}, \ldots, \overline{z_{d-1}}]$ of length $m - 1$. By induction hypothesis, we have $m - 1 = d - 1$. Hence $m = d$, which is what we wanted to show. ∎

IN the next section we will study the geometrical background from which the interest in affine algebras originated.

Exercises

Problem 1. Let $R \subseteq S$ be finitely generated K-algebras where K is a field. Show that if M is a maximal ideal of S then $M \cap R$ is a maximal ideal of R.

Problem 2. Suppose the integral domain is integrally closed in its quotient field K. Show that if $f, g \in K[x]$ are monic polynomials with $fg \in R[x]$ then $f, g \in R[x]$.

Hint. Argue that the roots of f and g (in some splitting field of $\{f, g\}$) and hence the coefficients of f and g must be integral over R.

Problem 3. (a) Let $p \in \mathbb{N}$ be a prime number. Show that $\{0\} \subsetneqq \langle p \rangle \subsetneqq \langle p, x \rangle$ and $\{0\} \subsetneqq \langle x \rangle \subsetneqq \langle p, x \rangle$ are saturated chains of prime ideals in $\mathbb{Z}[x]$. What can you conclude about the Krull dimension of $\mathbb{Z}[x]$?

(b) Let $R = \{a/b \mid a, b \in \mathbb{Z}, b \text{ odd}\}$. Show that $\{0\} \subsetneqq \langle 2 \rangle \subsetneqq \langle 2, x \rangle$ and $\{0\} \subsetneqq \langle 2x - 1 \rangle$ are both saturated chains of prime ideals in $R[x]$. What can you conclude about the Krull dimension of $R[x]$?

Problem 4. The ring $S := \mathbb{Z}[\sqrt[3]{2}]$ is integral over $\mathbb{Z}$. Find the prime ideals of S lying over $\langle 7 \rangle$, $\langle 5 \rangle$ and $\langle 31 \rangle$.

Problem 5. Let $R \subseteq S$ be commutative rings and let P be a prime ideal of R. Let $\langle P \rangle$ be the ideal of S generated by P. Show that there is a prime ideal Q of S with $Q \cap S = P$ if and only if $\langle P \rangle \cap R = P$.

Problem 6. Suppose that $R \subseteq S$ where S is integral over R and let $Q \trianglelefteq S$ be a prime ideal lying over $P \trianglelefteq R$.

(a) Use the methods of this section to show that Q is maximal in S if and only if P is maximal in R. (This provides a new proof of (15.2).)

(b) Show that there is no ideal $J \trianglelefteq S$ (prime or not) with $J \cap R = P$ and $Q \subsetneqq J$.

Problem 7. Given commutative rings $R \subseteq S$, let $P \trianglelefteq R$ be a prime ideal of R and let $\langle P \rangle$ be the ideal of S generated by P. Show that there is a prime ideal $Q \trianglelefteq S$ with $Q \cap R = P$ if and only if $\langle P \rangle \cap R = P$.

Problem 8. Let x be a variable over the field K.

(a) Show that $S := K[x]$ is integral over $R := [x^2 - 1]$.

(b) Show that $Q := \langle x - 1 \rangle$ is a prime ideal of S and determine $P := Q \cap R$.

(c) Show that S/Q is not integral over R/P. (**Hint.** Consider $1/(x + 1)$.)

Problem 9. Let $(S : R)$ be an integral extension of commutative rings and let $I \trianglelefteq R$ be an ideal of R. Moreover, let $J \trianglelefteq S$ be the ideal of S generated by I.

(a) Show that if $I \subsetneq R$ then $J \subsetneq S$.

(b) Show that $\sqrt{J}$ consists of all elements $\alpha \in S$ which satisfy an equation $\alpha^n + b_{n-1}\alpha^{n-1} + \cdots + b_1\alpha + b_0 = 0$ with coefficients $b_0, \ldots, b_{n-1} \in I$.

(c) Let $R \subseteq S$ be integral domains such that R is integrally closed in its quotient field K and assume that I is a prime ideal of R. Show that the minimal polynomial of any element $\alpha \in J$ over K has coefficients in I (apart from the leading coefficient which is 1).

Problem 10. Show that if $(S : R)$ is an integral extension of integral domains and if $\{0\} \neq J \trianglelefteq S$ is an ideal of S then $J \cap R \neq \{0\}$.

Problem 11. In the following two examples an integral domain R and a prime ideal P are specified. Show in both cases that R is integrally closed in its quotient field but that R/P is not.

(a) $R = \mathbb{Z}[x]$, $P = \langle x^2 - 5 \rangle$

(b) $R := K[X, Y]$ (where K is a field), $P = \langle Y^2 - X^3 \rangle$

In part (b) find the integral closure of R/P in its quotient field and show that it is isomorphic to a polynomial ring over K.

Problem 12. Let K be a field and let $R = K[X, Y]/\langle X^2Y - 1 \rangle$ so that $R = K[x, y]$ where x and y satisfy the relation $x^2 y = 1$. Find a Noetherian normalization N of R and integral equations for x and y over N.

Problem 13. Let R be an integral domain with quotient field K and let $(R_i)_{i \in I}$ be subrings of R which all contain the identity element of R.

(a) Show that each R_i is integrally closed in its quotient field, then so is $\bigcap_{i \in I} R_i$.

(b) Show that R is integrally closed in K if and only if $R[x]$ is integrally closed in $R(x) = K(x)$.

Hint. If $f \in R(x)$ is integral over $R[x]$, then the more so over $K[x]$. But $K[x]$ is a unique factorization domain. For the converse implication, use part (a) along with the fact that $R[x] \cap K = R$.

Problem 14. (This exercise provides an alternative proof of Zariski's lemma.) Let $L = K[x_1, \ldots, x_n]$ be a finitely generated K-algebra where K is a field. Show by induction on n that if L is a field then L is algebraic over K.

Hint. Assume by induction hypothesis that $x_2, \ldots, x_n$ are algebraic over $K(x_1)$. If x_1 is algebraic over K we are done, otherwise prove the following statements.

(1) There is a nonzero polynomial $h \in K[x_1]$ such that $h(x_1)x_2, \ldots, h(x_1)x_n$ are integral over $K[x_1]$.

(2) For any polynomial $f \in K[X_1, \ldots, X_n]$ there is a natural number $m \in \mathbb{N}$ such that $h^m f$ is integral over $K[x_1]$.

(3) Derive a contradiction by making a suitable choice for f.

Problem 15. (This exercise provides an alternative proof of the Going-down theorem.) Let $R \subseteq S$ be an integral extension of integral domains and assume that R is integrally closed in its quotient field. Let $P_1 \subsetneq P_2$ be prime ideals in R and let Q_2 be a prime ideal of S lying over P_2.

(a) Show that $\Sigma := \{rs \mid r \in R \setminus P_1,\, s \in S \setminus Q_2\}$ is multiplicatively closed with $\Sigma \cap P_1 S = \emptyset$.

(b) By Zorn's lemma we can find a maximal element Q_1 in

$$\mathfrak{M} := \{Q \trianglelefteq S \mid P_1 S \subseteq Q,\, \Sigma \cap Q = \emptyset\}.$$

Show that Q_1 is a prime ideal of S with $Q_1 \subsetneq Q_2$ lying over P_1.

Problem 16. Show that if R is an integral domain with $\dim R = 0$ then R is a field.

Problem 17. Let R be an integral domain with quotient field K and let $(L : K)$ be an algebraic field extension. Finally, let S be the integral closure of R in L.

(a) Show that the integral closure of R in K is $S \cap K$.

(b) Show that L is the quotient field of S.

(c) Show that an element of L lies in S, i.e., is integral over R, if and only if its minimal polynomial over K has coefficients in $S \cap K$.

Problem 18. Let $R \subseteq S$ be an integral ring extension of integral domains. Show that every homomorphism $\varphi : R \to C$ into an algebraically closed field C can be extended to a homomorphism $\Phi : S \to C$.

17. Ring theory and algebraic geometry

JUST as systems of linear equations are the subject of linear algebra, systems of arbitrary polynomial equations are studied in algebraic geometry. Geometrically, to solve a system of linear equations means to find the points of intersection of lines, planes, and higher-dimensional affine spaces. Similarly, to solve a system of polynomial equations means to study intersections of circles, parabolas and other curves, but also higher-dimensional objects which we will call "affine varieties". Let us give the general definition of an affine variety as the solution set of a system of polynomial equations.

(17.1) Definition. *Let $K \subseteq L$ be fields. A subset $V \subseteq L^n$ is called an **affine K-variety** if there are polynomials $f_1, \ldots, f_m \in K[X_1, \ldots, X_n]$ such that*

$$V = \{(x_1, \ldots, x_n) \in L^n \mid f_i(x_1, \ldots, x_n) = 0 \text{ for } 1 \leq i \leq m\} ;$$

i.e., if V is the set of common zeros of $f_1, \ldots, f_m$.

(17.2) Examples. (a) Varieties of the form $V = \{(x_1, \ldots, x_n) \in L^n \mid \sum a_{ij}x_ix_j + \sum b_ix_i + c = 0\}$ are called **quadrics**; they were studied in section 15 of Volume I already, using methods of linear algebra.

(b) A variety of the form $V = \{(x, y) \in L^2 \mid f(x, y) = 0\}$ is called a **plane curve**. Several examples (with $K = L = \mathbb{R}$) are given in the pictures below.

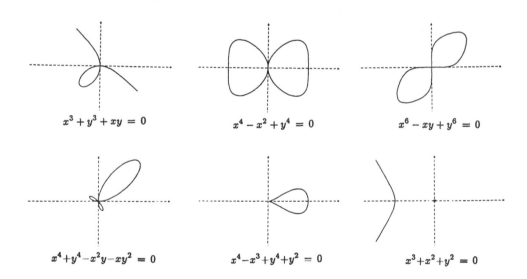

$x^3 + y^3 + xy = 0$

$x^4 - x^2 + y^4 = 0$

$x^6 - xy + y^6 = 0$

$x^4 + y^4 - x^2y - xy^2 = 0$

$x^4 - x^3 + y^4 + y^2 = 0$

$x^3 + x^2 + y^2 = 0$

On the next page we present a family of plane curves depending on a parameter a; observe the change in shape as a varies.

$$y^2 = x(x-1)^2 + ax = x((x-1)^2 + a)$$

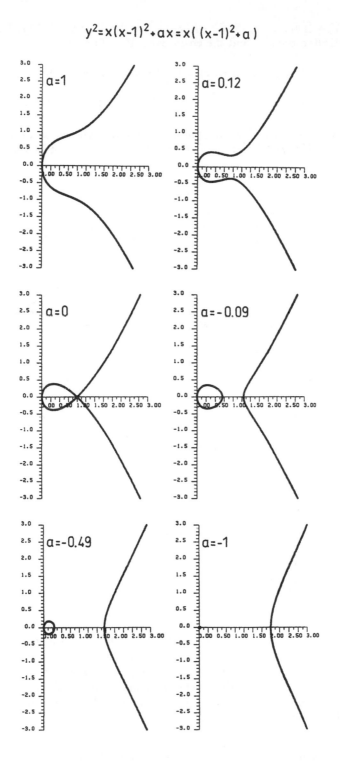

(c) Let $V = \{(x_1, \ldots, x_n) \in L^n \mid f_i(x_1, \ldots, x_n) = 0 \text{ for } 1 \leq i \leq m\}$ be a K-variety. Then the **cylinder over V** and the **cone over V**, defined as

$$\mathrm{Cyl}(V) := V \times L = \{(x_1, \ldots, x_n, t) \in L^{n+1} \mid f_i(x_1, \ldots, x_n) = 0 \text{ for } 1 \leq i \leq m\} \text{ and}$$

$$\mathrm{Cone}(V) := \{(\xi_1, \ldots, \xi_n, t) \in L^{n+1} \mid t^{\deg f_i} f_i(\frac{\xi_1}{t}, \ldots, \frac{\xi_n}{t}) = 0 \text{ for } 1 \leq i \leq m\},$$

are again K-varieties. The following pictures show the cylinder and the cone over the plane curve $V = \{(x, y) \in \mathbb{R}^2 \mid x^4 - x^2 + y^2 = 0\}$ which was sketched above.

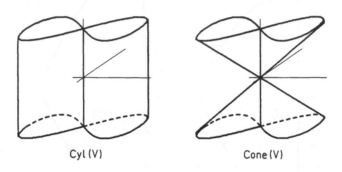

Cyl (V)   Cone (V)

(17.3) Remark. Note that we allowed arbitrary fields in our definition. Thus we can consider a "curve" in L^2 where $L = \mathbb{Z}_3$ is the finite field with three elements. Here the "plane" L^2 has only 9 points, and the "curve" $y^2 = x^3$ consists of the three points $(0, 0)$, $(1, 1)$ and $(1, 2)$ as indicated in the first picture below.

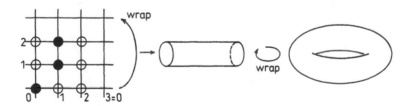

To visualize the vector space L^2 it is sometimes helpful to consider the lattice $\mathbb{Z}^2$ and then identify points whose coordinates are the same modulo 3 to obtain a "geometric model" for L^2. This is tantamount to gluing together the line $x = 3$ and $x = 0$ and the lines $y = 3$ and $y = 0$ in the Euclidean plane. The result is that we represent the "plane" L^2 as nine points on the torus, as indicated in the second picture above. (Compare with the discussion in (I.5.5).).

THUS by "algebraic geometry", we mean the study of solution sets of systems of algebraic equations over any field. One might well argue whether the investigation of varieties over fields other than $\mathbb{R}$ or $\mathbb{C}$ should still be called "geometry", but since much of the elementary theory of varieties can be formulated for arbitrary fields, there is no need for restrictive assumptions concerning the base-field at this point. Also, it is quite useful to distinguish between the base-field L of the vector space in which a variety V sits and the coefficient field K of the polynomial equations by which V is defined. Let us illustrate this last remark by an example in which a number-theoretical problem is solved by a geometric technique.

(17.4) Example. Let us try to find all integer solutions of the equation $a^2 + b^2 = c^2$.† Clearly, if $c = 0$, then $a = b = 0$; so suppose $c \neq 0$. Letting $x := a/c$ and $y := b/c$ our equation becomes $x^2 + y^2 = 1$; hence we are asked to find all rational points $(x, y) \in \mathbb{Q}^2$ with $x^2 + y^2 = 1$. (Clearly, any two rational numbers can be written as fractions with the same denominator.) To be able to utilize geometric techniques, we consider the circle $\{(x, y) \in \mathbb{R}^2 \mid x^2 + y^2 = 1\}$ which is a $\mathbb{Q}$-variety in $\mathbb{R}^2$. A "parameterization" of this circle can be obtained as follows. We "sweep out" the plane $\mathbb{R}^2$ by all straight lines passing through $(0, 1)$, i.e., by all lines of the form $y = tx + 1$, and determine for each of these lines the points of intersection with V.

If $y = tx + 1$, the equation $x^2 + y^2 = 1$ becomes $1 = x^2 + (tx + 1)^2 = (1 + t^2)x^2 + 2tx + 1$, i.e., $x\big((1 + t^2)x + 2t\big) = 0$. The solution $x = 0$ corresponds to the point $(0, 1)$. If $x \neq 0$, then we can conclude $x = -\frac{2t}{1 + t^2}$ and hence $y = tx + 1 = \frac{1 - t^2}{1 + t^2}$. Thus we obtain the parameterization

$$x(t) = \frac{-2t}{1 + t^2}, \qquad y(t) = \frac{1 - t^2}{1 + t^2}.$$

If t varies, then this parameterization covers all points of V except $(0, -1)$ (which can be obtained as the limiting case $t \to \pm\infty$, i.e., as the point of intersection of V with the vertical line $x = 0$). Now clearly, if we plug in a rational value for t, then we obtain a rational point (x, y) on V. On the other hand, if $x(t)$ and $y(t)$ are both rational numbers, then so is $\frac{y(t) - 1}{x(t)} = t$. Thus the rational solutions $(x, y) \in \mathbb{Q}^2$ are exactly the points $(\frac{-2t}{1 + t^2}, \frac{1 - t^2}{1 + t^2})$ with $t \in \mathbb{Q}$ and the point $(0, -1)$.

Returning to our initial problem of finding all triples $(a, b, c) \in \mathbb{Z}^3$ with $a^2 + b^2 = c^2$, we obtain exactly the solutions $(0, b, -b)$ (corresponding to $(0, -1)$) and (a, b, c) where $\frac{a}{c} = \frac{-2t}{1 + t^2}$, $\frac{b}{c} = \frac{1 - t^2}{1 + t^2}$. Writing $t = \frac{r}{s}$ and $\frac{c}{r^2 + s^2} = \frac{u}{v}$, we obtain

$$(a, b, c) = \left(-\frac{2rs}{r^2 + s^2}c, \frac{s^2 - r^2}{s^2 + r^2}c, c\right) = \left(-2rs\frac{u}{v}, (s^2 - r^2)\frac{u}{v}, (s^2 + r^2)\frac{u}{v}\right).$$

---

† This question was already discussed in problem 24 of section 9.

We see that a number-theoretical problem which consists of finding elements in K^n lying on a K-variety $V \subseteq L^n$ can be (at least partially) solved if we succeed in finding a "parametrization" of V by polynomials (or rational functions) whose coefficients lie in K.

IN our study of linear algebra, we already observed the power of Descartes' idea of describing geometric objects by algebraic equations and then solving geometric problems by translating them into algebraic problems. In the setting of affine varieties, which are described by polynomial equations, this interplay between geometry and algebra will be realized as a correspondence between properties of varieties and properties of polynomial rings. Let us first set up a correspondence between subsets of L^n and subsets of $K[X_1, \ldots, X_n]$ where $(L : K)$ is a field extension.

(17.5) Definition. *Let $K \subseteq L$ be fields.*
(a) *For any nonempty set $F \subseteq K[X_1, \ldots, X_n]$ (possibly infinite) of polynomials we let*

$$\mathcal{V}(F) := \{(x_1, \ldots, x_n) \in L^n \mid f(x_1, \ldots, x_n) = 0 \text{ for all } f \in F\} .$$

(b) *For any subset $V \subseteq L^n$ we let*

$$\mathcal{I}(V) := \{f \in K[X_1, \ldots, X_n] \mid f(x_1, \ldots, x_n) = 0 \text{ for all } (x_1, \ldots, x_n) \in V\} .$$

ALL the sets $\mathcal{V}(F)$ with a *finite* set F of polynomials are, by definition, varieties. It turns out, however, that even for an infinite family F the set $\mathcal{V}(F)$ is a variety. This and other useful information is contained in the next proposition which lists the basic rules for the assignments $\mathcal{V}$ and $\mathcal{I}$.

(17.6) Proposition. *Let $K \subseteq L$ be fields. Moreover, let V, V_1, V_2 be subsets of L^n and F, F_1, F_2 be subsets of $K[X_1, \ldots, X_n]$.*
(a) *If $F_1 \subseteq F_2$ then $\mathcal{V}(F_1) \supseteq \mathcal{V}(F_2)$.*
(b) *$\mathcal{V}(F_1 \cup F_2) = \mathcal{V}(F_1) \cap \mathcal{V}(F_2)$.*
(c) *$\mathcal{V}(F) = \mathcal{V}(\langle F \rangle)$ where $\langle F \rangle$ is the ideal of $K[X_1, \ldots, X_n]$ generated by F.*
(d) *$\mathcal{V}(F)$ is a K-variety in L^n.*
(e) *$\mathcal{I}(V)$ is a radical ideal of $K[X_1, \ldots, X_n]$.*
(f) *If $V_1 \subseteq V_2$ then $\mathcal{I}(V_1) \supseteq \mathcal{I}(V_2)$.*
(g) *$V \subseteq \mathcal{V}(\mathcal{I}(V))$ and $F \subseteq \mathcal{I}(\mathcal{V}(F))$.*
(h) *$(\mathcal{V} \circ \mathcal{I} \circ \mathcal{V})(F) = F$ and $(\mathcal{I} \circ \mathcal{V} \circ \mathcal{I})(V) = V$.*

Proof. (a) Let $F_1 \subseteq F_2$. If x is a common root of all polynomials $f \in F_1$, then the more so a root of all polynomials $f \in F_2$.
(b) A point $x \in L^n$ is a common root of all $f \in F_1 \cup F_2$ if and only if it is a common root of all $f \in F_1$ and a common root of all $f \in F_2$.
(c) We have $F \subseteq \langle F \rangle$ and hence $\mathcal{V}(F) \supseteq \mathcal{V}(\langle F \rangle)$ by part (a). Conversely, let $x \in \mathcal{V}(F)$ and $g \in \langle F \rangle$. Then there are polynomials $f_1, \ldots, f_r \in F$ and $p_1, \ldots, p_r \in$

$K[X_1,\ldots,X_n]$ such that $g = p_1 f_1 + \cdots + p_n f_n$. Since $x \in \mathcal{V}(F)$, we have $f_1(x) = \cdots = f_r(x) = 0$ and hence $g(x) = 0$. Since $x \in \mathcal{V}(F)$ and $g \in \langle F \rangle$ were chosen arbitrarily, this shows that $\mathcal{V}(F) \subseteq \mathcal{V}(\langle F \rangle)$.

(d) Since $K[X_1,\ldots,X_n]$ is Noetherian, the ideal $\langle F \rangle$ is finitely generated, say $\langle F \rangle = \langle f_1,\ldots,f_r \rangle$. Then $\mathcal{V}(F) = \mathcal{V}(\langle F \rangle) = \mathcal{V}(f_1,\ldots,f_r)$ is a variety.

(e) If f and g vanish on V, then so do $f + g$ and all products pf with $p \in K[X_1,\ldots,X_n]$; this shows that $\mathcal{I}(V)$ is an ideal of $K[X_1,\ldots,X_n]$. Moreover, if $f^m \in I$ for some m then $f(x)^m = 0$ for all $x \in V$, which implies that $f(x) = 0$ for all $x \in V$; therefore, $\mathcal{I}(V)$ is a radical ideal of $K[X_1,\ldots,X_n]$.

(f) Let $V_1 \subseteq V_2$. If f vanishes on V_2, then the more so on V_1.

(g) Both inclusions are tautological statements. Indeed, if $x \in V$ then every $f \in \mathcal{I}(V)$ vanishes on x so that x belongs to the zero set of $\mathcal{I}(V)$; this proves $V \subseteq \mathcal{V}(\mathcal{I}(V))$. On the other hand, by the very definition of $\mathcal{V}(F)$, each $f \in F$ vanishes on $\mathcal{V}(F)$, hence belongs to $\mathcal{I}(\mathcal{V}(F))$; this proves the inclusion $F \subseteq \mathcal{I}(\mathcal{V}(F))$.

(h) Let $V \subseteq L^n$. Then $V \subseteq \mathcal{V}(\mathcal{I}(V))$ by (g), hence $\mathcal{I}(V) \supseteq (\mathcal{I} \circ \mathcal{V} \circ \mathcal{I})(V)$ by part (f). Conversely, $\mathcal{I}(V) \subseteq (\mathcal{I} \circ \mathcal{V})(\mathcal{I}(V))$ by part (g) (with $F := \mathcal{I}(V)$).

Let $F \subseteq K[X_1,\ldots,X_n]$. Then $F \subseteq \mathcal{I}(\mathcal{V}(F))$ by (g), hence $\mathcal{V}(F) \supseteq (\mathcal{V} \circ \mathcal{I} \circ \mathcal{V})(F)$ by part (a). Conversely, $\mathcal{V}(F) \subseteq (\mathcal{V} \circ \mathcal{I})(\mathcal{V}(F))$ by part (g) (with $V := \mathcal{V}(F)$). ∎

(17.7) Proposition. *Let I and J be ideals of $K[X_1,\ldots,X_n]$.*
(a) $\mathcal{V}(IJ) = \mathcal{V}(I \cap J) = \mathcal{V}(I) \cup \mathcal{V}(J)$.
(b) $\mathcal{V}(I + J) = \mathcal{V}(I) \cap \mathcal{V}(J)$.

Proof. (a) Since $IJ \subseteq I \cap J$ is contained in both I and J we have $\mathcal{V}(IJ) \supseteq \mathcal{V}(I \cap J) \supseteq \mathcal{V}(I) \cup \mathcal{V}(J)$ by (17.5)(a). It remains to show that $\mathcal{V}(IJ) \subseteq \mathcal{V}(I) \cup \mathcal{V}(J)$. Suppose $x \notin \mathcal{V}(I) \cup \mathcal{V}(J)$. Then there are polynomials $f \in I$ and $g \in J$ with $f(x) \neq 0$ and $g(x) \neq 0$. Consequently, $fg \in IJ$ with $(fg)(x) = f(x)g(x) \neq 0$ so that $x \notin \mathcal{V}(IJ)$.

(b) Using (17.6)(b) and (17.6)(c), we obtain $\mathcal{V}(I) \cap \mathcal{V}(J) = \mathcal{V}(I \cup J) = \mathcal{V}(\langle I \cup J \rangle) = \mathcal{V}(I + J)$. ∎

(17.8) Proposition. *Let V, V_1, V_2 be K-varieties in L^n.*
(a) $V = \mathcal{V}(\mathcal{I}(V))$.
(b) *If $V_1 \subsetneq V_2$ then $\mathcal{I}(V_1) \supsetneq \mathcal{I}(V_2)$.*

Proof. (a) We have $V = \mathcal{V}(F)$ for some subset $F \subseteq K[X_1,\ldots,X_n]$. Then (17.6)(h) implies $\mathcal{V}(\mathcal{I}(V)) = (\mathcal{V} \circ \mathcal{I} \circ \mathcal{V})(F) = \mathcal{V}(F) = V$.

(b) We know from (17.6)(f) that $\mathcal{I}(V_1) \supseteq \mathcal{I}(V_2)$. Suppose $\mathcal{I}(V_2) = \mathcal{I}(V_1)$. Then $V_2 = \mathcal{V}(\mathcal{I}(V_2)) = \mathcal{V}(\mathcal{I}(V_1)) = V_1$. ∎

IN general this is about all one can say about the correspondence between radical ideals and varieties. In the case that the field L is algebraically closed, however, it will turn out that this is actually a 1-1-correspondence. This is the content of Hilbert's famous *Nullstellensatz*.

(17.9) Theorem. (Hilbert's Nullstellensatz.) *Let $K \subseteq L$ be fields where L is algebraically closed.*

(a) *If $I \neq K[X_1, \ldots, X_n]$ is a proper ideal of $K[X_1, \ldots, X_n]$ then $\mathcal{V}(I) \neq \emptyset$.*

(b) *If $I \neq K[X_1, \ldots, X_n]$ is a proper ideal of $K[X_1, \ldots, X_n]$ then $\mathcal{I}(\mathcal{V}(I)) = \sqrt{I}$.*

(c) *The mappings $\mathcal{I}$ and $\mathcal{V}$ set up a 1-1-correspondence between K-varieties in L^n and radical ideals of $K[X_1, \ldots, X_n]$.*

Proof. (a) Let $M \trianglelefteq K[X_1, \ldots, X_n]$ be a maximal ideal of $K[X_1, \ldots, X_n]$ containing I. Then $K' := K[X_1, \ldots, X_n]/M$ is a field which is finitely generated as a K-algebra by the n elements $\xi_i := X_i + M$. Then Zariski's lemma (12.35) implies that $(K' : K)$ is an algebraic field extension. By theorem (13.8) the embedding homomorphism $i : K \to L$ extends to a homomorphism $\varphi : K' \to L$. We claim that $\big(\varphi(\xi_1), \ldots, \varphi(\xi_n)\big) \in L^n$ is an element of $\mathcal{V}(M) \subseteq \mathcal{V}(I)$. Indeed, let $f \in M$, say $f(X_1, \ldots, X_n) = \sum_{i_1, \ldots, i_n} a_{i_1, \ldots, i_n} X_1^{i_1} \cdots X_n^{i_n}$ with coefficients $a_{i_1, \ldots, i_n} \in K$. Since φ is a homomorphism of K-algebras, this implies

$$f\big(\varphi(\xi_1), \ldots, \varphi(\xi_n)\big) = \sum_{i_1, \ldots, i_n} a_{i_1, \ldots, i_n} \varphi(\xi_1)^{i_1} \cdots \varphi(\xi_n)^{i_n} = \varphi\Big(\sum_{i_1, \ldots, i_n} a_{i_1, \ldots, i_n} \xi_1^{i_1} \cdots \xi_n^{i_n}\Big)$$

$$= \varphi\Big(\sum_{i_1, \ldots, i_n} a_{i_1, \ldots, i_n} X_1^{i_1} \cdots X_n^{i_n} + M\Big) = \varphi(f + M) = \varphi(M) = \varphi([0]) = 0.$$

(b) The inclusion $\supseteq$ is trivial. Indeed, if $f \in \sqrt{I}$ then $f^n \in I \subseteq \mathcal{I}(\mathcal{V}(I))$ for some $n \in \mathbb{N}$ and therefore $f(x)^n = 0$ for all $x \in \mathcal{V}(I)$. Clearly, this implies $f(x) = 0$ for all $x \in \mathcal{V}(I)$ which means $f \in \mathcal{I}(\mathcal{V}(I))$.

To prove the converse inclusion $\subseteq$ we take an element $f \neq 0$ in $\mathcal{I}(\mathcal{V}(I))$. We show that $f \in \sqrt{I}$ by a cunning application of (a) called *Rabinowitsch's trick*. Introduce a new variable T and consider in the polynomial ring $K[X_1, \ldots, X_n, T]$ the ideal J generated by I and the polynomial $F(X_1, \ldots, X_n, T) := f(X_1, \ldots, X_n)T - 1$. † Suppose $J \neq K[X_1, \ldots, X_n, T]$. Then $\mathcal{V}(J) \neq \emptyset$ by part (a), say $(x_1, \ldots, x_n, t) \in \mathcal{V}(J) \subseteq \mathcal{V}(I)$. Then $(x_1, \ldots, x_n) \in \mathcal{V}(I)$ which implies $f(x_1, \ldots, x_n) = 0$ and hence $F(x_1, \ldots, x_n, t) = 0 \cdot t - 1 = -1 \neq 0$, contradicting the fact that $(x_1, \ldots, x_n, t)$ is a common root of J and hence a root of F. Thus our assumption $J \neq K[X_1, \ldots, X_n, T]$ was false; we have $J = K[X_1, \ldots, X_n, T]$, in particular $1 \in J$. Thus there are polynomials $f_i \in I$ and $r, r_i \in K[X_1, \ldots, X_n, T]$ such that

$$(\star) \qquad\qquad 1 = \sum_{i=1}^{s} r_i f_i + r F.$$

Define a K-algebra homomorphism $\Theta : K[X_1, \ldots, X_n, T] \to K(X_1, \ldots, X_n)$ by $\Theta(X_i) := X_i$ and $\Theta(T) := 1/f$; note that the definition of Θ implies $\Theta(F) = 0$. Applying Θ to both sides of the equation $(\star)$, we obtain

$$1 = \sum_{i=1}^{s} \Theta(r_i) f_i.$$

---

† The elements of I are treated as polynomials in the variables $X_1, \ldots, X_n, T$ via $\bar{f}(X_1, \ldots, X_n, T) := f(X_1, \ldots, X_n)$.

Now each $\Theta(r_i)$ can be written in the form $\Theta(r_i) = p_i/f^{n_i}$ with $p_i \in K[X_1, \ldots, X_n]$ and $n_i \in \mathbb{N}_0$. Let $n := \max(n_1, \ldots, n_s)$ and multiply the last equation by f^n to obtain

$$f^n = \sum_{i=1}^{s} p_i f^{n-n_i} \underbrace{f_i}_{\in I} \in I$$

which clearly implies $f \in \sqrt{I}$.

(c) Proposition (17.8)(a) states that $\mathcal{V} \circ \mathcal{I}$ is the identity map on the set of all K-varieties in L^n, and part (b) of Hilbert's Nullstellensatz implies that $\mathcal{I} \circ \mathcal{V}$ is the identity on the set of all radical ideals of $K[X_1, \ldots, X_n]$; whence the claim. ∎

THE name "Nullstellensatz" (which is German for "zero theorem") refers to part (a), of course, which states that every proper ideal of $K[X_1, \ldots, X_n]$ has a zero. Clearly, part (b) implies part (a), but since (a) was used to prove (b), both statements have been spelled out as parts of the theorem. Let us now derive a result which is closely related to the Nullstellensatz.

(17.10) Theorem. *Let K be an algebraically closed field. Then the maximal ideals of the polynomial ring $K[X_1, \ldots, X_n]$ are exactly the sets $\langle X_1 - a_1, \ldots, X_n - a_n \rangle = \{f \in K[X_1, \ldots, X_n] \mid f(a_1, \ldots, a_n) = 0\}$ with $a_1, \ldots, a_n \in K$.*

Thus the 1-1-correspondence between radical ideals of $K[X_1, \ldots, X_n]$ and K-varieties in K^n can be restricted to a 1-1-correspondence between maximal ideals of $K[X_1, \ldots, X_n]$ and points in K^n. †

Proof. As in the proof of (17.9)(a), let $M \trianglelefteq K[X_1, \ldots, X_n]$ be a maximal ideal so that $K' = K[X_1, \ldots, X_n]/M = K[\xi_1, \ldots, \xi_n]$ is an algebraic field extension of K; let $i : K \to K'$ be the embedding homomorphism. Since K is algebraically closed and hence does not possess any proper algebraic extension, we have $K' = i(K)$ so that i is an isomorphism. Letting $a_k := i^{-1}(\xi_k)$, we have $X_k - a_k \in M$ for all k and hence $\langle X_1 - a_1, \ldots, X_n - a_n \rangle \subseteq M$. Since the left-hand side in this inclusion is already a maximal ideal, equality must hold. ∎

(17.11) Proposition. *Let $K \subseteq L$ be fields.*

(a) If V and W are two K-varieties in L^n then $V \cup W$ is also a K-variety in L^n. Hence finite unions of varieties are again varieties.

(b) If $(V_i)_{i \in I}$ is a family of K-varieties in L^n then $\bigcap_{i \in I}$ is also a K-variety in L^n. Hence arbitrary intersections of varieties are again varieties.

(c) If $V \subseteq L^m$ and $W \subseteq L^n$ are two K-varieties then $V \times W$ is a K-variety in L^{m+n}. Hence finite Cartesian products of varieties are again varieties.

Proof. (a) If V is the zero set of $f_1, \ldots, f_r$ and W the zero set of $g_1, \ldots, g_s$ then $V \cup W$ is the zero set of the polynomials $f_i g_j$ $(1 \leq i \leq r, 1 \leq j \leq s)$.

---

† Compare with problem 30 in section 5.

(b) If V_i is the zero set of a family F_i of polynomials, then $\bigcap_{i \in I} V_i$ is the zero set of $\bigcup_{i \in I} F_i$.

(c) Suppose V is the zero set of $f_1, \ldots, f_r \in K[X_1, \ldots, X_m]$ and W the zero set of $g_1, \ldots, g_s \in K[Y_1, \ldots, Y_n]$. For $1 \leq i \leq r$ and $1 \leq j \leq s$ we define $\overline{f}_i(X_1, \ldots, X_m, Y_1, \ldots, Y_n) := f_i(X_1, \ldots, X_m)$ and $\overline{g}_j(X_1, \ldots, X_m, Y_1, \ldots, Y_n) := g_j(Y_1, \ldots, Y_n)$; i.e., we treat f_i and g_j as polynomials in the $m+n$ variables $X_1, \ldots, X_m$, $Y_1, \ldots, Y_n$. Then $V \times W$ is the zero set of $\overline{f}_1, \ldots, \overline{f}_r, \overline{g}_1, \ldots, \overline{g}_s$. ∎

WE just observed that arbitrary intersections and finite unions of K-varieties are again K-varieties. Moreover, the empty set and all of L^n are trivially also K-varieties. Hence there is a unique topology on L^n which has exactly the K-varieties as its closed sets. This topology is called the *Zariski topology* on L^n. If $V \subseteq L^n$ is a K-variety in L^n then the subspace topology on V induced from L^n is called the *Zariski topology* on V; the formal definition is as follows.

(17.12) Definition. *Let V be a K-variety in L^n. Then there is a unique topology on V whose closed subsets are the sets $\{x \in V \mid f(x) = 0 \text{ for all } f \in F\}$ for a (finite) family $F \subseteq K[X_1, \ldots, X_n]$. This topology is called the* **Zariski topology** *on V.*

IT is clear that a basis of the Zariski topology on V is given by the sets $D_f := \{x \in V \mid f(x) \neq 0\}$ where $f \in K[X_1, \ldots, X_n]$. We should observe that if L is a subfield of $\mathbb{C}$ then V also has a natural topology inherited from the Euclidean topology on $\mathbb{C}^n$. This "natural" topology is finer than the Zariski topology, because all polynomial mappings are continuous with respect to the natural topology. In general, the Zariski topology is much coarser than the natural topology (we will see illustrations of this statement), and as far as topological statements on varieties are concerned, we should avoid using our intuition which is based on properties of the natural topology. The next result shows how the fact that the polynomial ring is Noetherian (which is a finiteness property) is reflected in a topological finiteness property, namely quasi-compactness.

(17.13) Proposition. *Let V be a K-variety in L^n. Then every subset A of V is quasi-compact.*

Proof. Let $(U_i)_{i \in I}$ be an open cover of A; we have to find a finite subcover. Since every open set in V is a union of sets of the form D_g, the inclusion $A \subseteq \bigcup_{i \in I} U_i$ can be rewritten in the form $A \subseteq \bigcup_{i \in I, g \in G_i} D_g$ where G_i is a finite set of polynomials, for each $i \in I$. Then

$$(\star) \qquad V \setminus A \supseteq V \setminus \bigcup_g D_g = \bigcap_g (V \setminus D_g) = \bigcap_g (V \cap \mathcal{V}(g)) .$$

Now the ideal generated by all the occurring functions g is already generated by a finite number of g's, say by $g_1, \ldots, g_n$ so that $\bigcap_g \mathcal{V}(g) = \bigcap_{r=1}^n \mathcal{V}(g_r)$. Plugging this into $(\star)$ and reading backwards, we obtain $V \setminus \bigcup_{r=1}^m D_{g_r} \subseteq V \setminus A$ which means $A \subseteq \bigcup_{r=1}^m D_{g_r}$. Now $g_1, \ldots, g_r$ belong to a finite set $I_0 \subseteq I$ of indices; then $(U_i)_{i \in I_0}$ is the desired finite subcover. ∎

WE now turn to the problem of decomposing a variety into simpler ones which can be more easily investigated.

(17.14) Definition. *A K-variety is called* **irreducible** *if $V = V_1 \cup V_2$ with K-varieties V_1, V_2 implies that $V = V_1$ or $V = V_2$; otherwise V is called* **reducible.**

(17.15) Examples. (a) The variety $V = \{(x,y,z) \in L^3 \mid xz = yz = 0\}$ is reducible because it can be written as $\{(x,y,z) \in K^3 \mid z = 0\} \cup \{(x,y,z) \in K^3 \mid x = y = 0\}$.
   (b) The variety $V = \{(x,y) \in \mathbb{R}^2 \mid x^4y^2 + x^4 - x^2y^2 + y^4 - 2x^2 - 7y^2 + 9 = 0\}$ is reducible, because it can be written as

$$V = \{(x,y) \in \mathbb{R}^2 \mid ((x-1)^2 + y^2 - 4)^2 + ((x+1)^2 + y^2 - 4)^2 = 0\}$$
$$= \{(x,y) \in \mathbb{R}^2 \mid (x-1)^2 + y^2 = 4\} \cup \{(x,y) \in \mathbb{R}^2 \mid (x+1)^2 + y^2 = 4\} .$$

Geometrically, V is the union of two circles.
   (c) The variety $\{(x,y,z) \in L^3 \mid xyz = 0\}$ is reducible because it is the union of the three coordinate axes which are clearly varieties.

NOTE that all the above examples were examples of reducible varieties. This was not by accident because we do not yet know a simple criterion which would allow us to check for irreducibility. Such a criterion will be provided in the next proposition; namely, a variety V is irreducible if and only if its ideal $\mathcal{I}(V)$ is a prime ideal. Thus a geometric condition is shown to be equivalent to a purely algebraic condition.

(17.16) Proposition. *Let V be a K-variety in L^n. Then the following conditions are equivalent.*
   (a) *V is irreducible.*
   (b) *V is Zariski-connected.*
   (c) *If $U_1, U_2 \subseteq V$ are nonempty open sets then $U_1 \cap U_2 \neq \emptyset$.*
   (d) *Any nonempty open subset of V is dense.*
   (e) *The ideal $\mathcal{I}(V)$ is prime.*

Proof. By the definition of the Zariski topology, (a) and (b) both mean that V cannot be written as a union of two proper closed subsets. Then (c) is only a restatement in terms of complements because $U_1 \cap U_2 = \emptyset$ if and only if $V = (V \setminus U_1) \cup (V \setminus U_2)$. The equivalence of (c) and (d) follows from the fact that a subset of a topological space is dense if and only if it intersects every nonempty open set. Thus the only nontrivial part of the proof is to show the equivalence of (a) and (e). Suppose V is irreducible and let f_1, f_2 be polynomials with $f_1 f_2 \in \mathcal{I}(V)$. Let $V_1 := \mathcal{V}(f_1)$ and $V_2 := \mathcal{V}(f_2)$. Then $V = (V \cap V_1) \cup (V \cap V_2)$; since V is irreducible, this implies $V = V \cap V_1$ or $V = V \cap V_2$, i.e., $V \subseteq V_1$ or $V \subseteq V_2$. But $V \subseteq V_1$ implies $f_1 \in \mathcal{I}(V)$, and $V \subseteq V_2$ implies $f_2 \in \mathcal{I}(V)$. Hence $f_1 f_2 \in \mathcal{I}(V)$ implies $f_1 \in \mathcal{I}(V)$ or $f_2 \in \mathcal{I}(V)$ so that $\mathcal{I}(V)$ is prime.

Suppose conversely that V is reducible; we have to show that $\mathcal{I}(V)$ is not prime in this case. Since V is reducible, there are varieties $V_1 \subsetneqq V$ and $V_2 \subsetneqq V$ such that $V = V_1 \cup V_2$. By (17.8)(b), we have $\mathcal{I}(V_1) \supsetneqq \mathcal{I}(V)$ and $\mathcal{I}(V_2) \supsetneqq \mathcal{I}(V)$. Pick $f_1 \in \mathcal{I}(V_1) \setminus \mathcal{I}(V)$ and $f_2 \in \mathcal{I}(V_2) \setminus \mathcal{I}(V)$. Then $f_1 f_2 \in \mathcal{I}(V_1)\mathcal{I}(V_2) \subseteq \mathcal{I}(V_1) \cap \mathcal{I}(V_2) = \mathcal{I}(V_1 \cup V_2) = \mathcal{I}(V)$ even though none of the factors f_1, f_2 belongs to $\mathcal{I}(V)$. This shows that $\mathcal{I}(V)$ is not prime. ∎

(17.17) Examples. (a) Let $K \subseteq L$ be fields and let $f \in K[X_1, \ldots, X_n]$ be a polynomial. Then the variety $V = \{x \in L^n \mid f(x) = 0\}$ is irreducible if and only if the ideal $\mathcal{I}(V) = \langle f \rangle$ is a prime ideal, which is the case if and only if the polynomial f is irreducible.

(b) The polynomial $f(x, y) = x^3 + x^2 + y^2 \in \mathbb{R}[x, y]$ is irreducible; hence the variety $V := \mathcal{V}(f)$ is Zariski-connected by part (a). On the other hand, the picture of V given in (17.2)(a) shows immediately that V is not connected with respect to the Euclidean topology on $\mathbb{R}^2$. This is due to the fact that the Zariski topology is much coarser than the Euclidean topology.

(c) If L is algebraically closed, then the 1-1-correspondence between radical ideals and varieties can be restricted to a 1-1-correspondence between prime ideals and irreducible varieties.

(17.18) Proposition. *Let $V \subseteq L^n$ be a K-variety.*

(a) *There are irreducible varieties $V_1, \ldots, V_n$ such that $V = V_1 \cup \cdots \cup V_n$.*

(b) *If $V = V_1 \cup \cdots \cup V_r$ is a decomposition of V into irreducible varieties which is* **reduced** *in the sense that no V_i can be omitted* †, *then $V_1, \ldots, V_r$ are uniquely determined.*

Proof. (a) Suppose that V is a variety which cannot be written as a union of irreducible varieties. In particular, V itself is not irreducible, hence can be written as a union $V = V_1 \cup V_1'$ with $V_1 \subsetneqq V$ and $V_1' \subsetneqq V$. But then at least one of V_1 and V_1', say V_1, cannot be written as a union of irreducible varieties, because V cannot. Hence we can repeat the argument and obtain a representation $V_1 = V_2 \cup V_2'$ with $V_2 \subsetneqq V_1$ and $V_2' \subsetneqq V_1$. Continuing in this way, we obtain a strictly decreasing sequence $V \supsetneqq V_1 \supsetneqq V_2 \supsetneqq V_3 \supsetneqq \cdots$ of varieties, hence by (17.8)(b) a strictly increasing sequence $\mathcal{I}(V) \subsetneqq \mathcal{I}(V_1) \subsetneqq \mathcal{I}(V_2) \subsetneqq \mathcal{I}(V_3) \subsetneqq \cdots$ of ideals in $K[X_1, \ldots, X_n]$. But this is impossible because the ring $K[X_1, \ldots, X_n]$ is Noetherian.

(b) Suppose that we have two reduced decompositions $V = \bigcup_i V_i = \bigcup_j W_j$. Then for any index i we have $V_i = V_i \cap V = V_i \cap \left(\bigcup_j W_j \right) = \bigcup_j (V_i \cap W_j)$. Since V_i is irreducible, there is an index j with $V_i = V_i \cap W_j$, i.e., $V_i \subseteq W_j$. Exchanging the roles of the two decompositions, there is an index k with $W_j \subseteq V_k$. But then $V_i \subseteq W_j \subseteq V_k$. Since we started with a reduced decomposition we conclude $i = k$ and hence $V_i = W_j$. ∎

(17.19) Remark. This decomposition of a variety into irreducible subvarieties is the geometric analogue to the primary decomposition of the corresponding ideal. More

---

† This is tantamount to saying that $V_i \not\subseteq V_j$ for $i \neq j$.

precisely, suppose $V = \mathcal{V}(I)$ for some ideal $I \trianglelefteq K[X_1, \ldots, X_n]$ and let $I = Q_1 \cap \cdots \cap Q_n$ be a reduced primary decomposition of I where Q_i is P_i-primary. Then the irreducible components of V are exactly the sets $\mathcal{V}(P)$ where P is a minimal element of the set $\{P_1, \ldots, P_n\}$. If $P \in \{P_1, \ldots, P_n\}$ is not minimal then the variety $\mathcal{V}(P)$ is contained in one of the irreducible components of V.

(17.20) Proposition. *Let $V = V_1 \cup \cdots \cup V_n$ be the reduced decomposition of a variety into its irreducible components.*

(a) *An open set $U \subseteq V$ is dense in V if and only if $U \cap V_i \neq \emptyset$ for $1 \leq i \leq n$.*

(b) *Every dense open subset $U \subseteq V$ contains a subset of the form D_g which is still dense in V.*

Proof. (a) If $U \cap V_{i_0} = \emptyset$ for some index i_0, then $U \subseteq \bigcup_{i \neq i_0} V_i$ and hence $\overline{U} \subseteq \bigcup_{i \neq i_0} V_i \subsetneqq V$; hence if U is dense in V, then $U \cap V_i \neq \emptyset$ for all i. Suppose conversely that this condition is satisfied; then $U \cap V_i$ is dense in V_i for any i, due to (17.16)(d). We show that U is dense in V by proving that U intersects any nonempty open subset $\Omega \subseteq V$. If $\Omega \neq \emptyset$, then $\Omega \cap V_{i_0} \neq \emptyset$ for some index i_0. Since $U \cap V_{i_0}$ is dense in V_{i_0}, this implies $(U \cap V_{i_0}) \cap (\Omega \cap V_{i_0}) \neq \emptyset$, hence $U \cap \Omega \neq \emptyset$.

(b) Let $A := V \setminus U$. By part (a), we have $V_i \not\subseteq A$ and hence $\mathcal{I}(V_i) \not\supseteq \mathcal{I}(A)$ for all i. Then $\mathcal{I}(A) \not\supseteq \mathcal{I}(V_1) \cup \cdots \cup \mathcal{I}(V_n)$ due to (6.25), so that there is an element $g \in \mathcal{I}(A)$ with $g \notin \mathcal{I}(V_1) \cup \cdots \cup \mathcal{I}(V_n)$. Then $D_g \subseteq U$ and $D_g \cap V_i \neq \emptyset$ for all i, so that D_g is dense in V due to part (a). ∎

WE now define the dimension of a variety V by studying the irreducible subvarieties of V.

(17.21) Definition. *The (geometric) dimension $\dim_{\text{geo}} V$ of a variety V is defined as the supremum of all numbers d such that there is a chain $V_0 \subsetneqq V_1 \subsetneqq \ldots \subsetneqq V_d$ of irreducible subvarieties of V.*

IT is clear that if $V = V_1 \cup \cdots \cup V_n$ is the decomposition of V into irreducible components, then $\dim V = \max_{1 \leq i \leq n} \dim V_i$ so that it is enough to discuss the concept of dimension for irreducible varieties.[†] Let us discuss two examples in an intuitive way. We assume, of course, that the "plane" L^2 is two-dimensional; a chain of irreducible subvarieties is given by $\{(0,0)\} \subsetneqq L \times \{0\} \subsetneqq L^2$. Also, we expect the paraboloid $V = \{(x, y, z) \in L^3 \mid z = x^2 + y^2\}$ to be a "surface", i.e., a two-dimensional variety; here suitable chains of subvarieties (which can easily be sketched) are given by

$$\{(0,0)\} \subsetneqq \{(x, 0, z) \mid z = x^2\} \subsetneqq V \quad \text{or} \quad \{(1,0)\} \subsetneqq \{(x, y, 1) \mid x^2 + y^2 = 1\} \subsetneqq V \ .$$

However, due to the abundance of possible chains of irreducible subvarieties of a given variety V, it can become rather difficult to prove that V has a certain dimension. Note

---

[†] One might even argue whether a dimension should be assigned to a reducible variety, for example, whether the variety $V = \{(x, y, z) \in \mathbb{R}^3 \mid z(x^2 + y^2) = 0\}$ should be called two-dimensional.

that it is not even immediately clear from definition (17.21) whether or not the geometric dimension of a variety is always finite. We defer a study of the notion of dimension to the next section.

IN order to learn something about a variety it is useful to study functions on this variety.

(17.22) Definitions. *Let $V \subseteq L^n$ be a K-variety. A function $f : V \to L$ is called a* **polynomial function** *(more precisely, a K-polynomial function) if there is a polynomial $p \in K[X_1, \ldots, X_n]$ such that $f(x_1, \ldots, x_n) = p(x_1, \ldots, x_n)$ for all $(x_1, \ldots, x_n) \in V$.*

The set of all K-polynomial functions on V is a ring if we define addition and multiplication pointwise; it is called the **coordinate ring** *of V and denoted by $K[V]$.*

(17.23) Remark. There are in general many different polynomials which represent the same polynomial function on a variety V. For example, let $V \subseteq L^2$ be the parabola $y = x^2$. Then the functions $f(x, y) = y^2$ and $g(x, y) = yx^2$ coincide on all points $(x, y) \in V$, hence determine the same polynomial function on V. More generally, two polynomials $f, g \in K[X_1, \ldots, X_n]$ will represent the same polynomial function on V if and only if $f - g$ vanishes on V, i.e., if $f - g \in \mathcal{I}(V)$. To make this point more precise, we can consider the mapping

$$
\begin{array}{ccc}
K[X_1, \ldots, X_n] & \to & K[V] \\
f & \mapsto & f|_V
\end{array}
$$

where $f|_V$ denotes the restriction to V of the polynomial function associated with f. This mapping is a surjective homomorphism with kernel $\mathcal{I}(V)$; hence

$(\star)$
$$
K[V] \cong \frac{K[X_1, \ldots, X_n]}{\mathcal{I}(V)}
$$

by the homomorphism theorem. The name "coordinate ring" comes from the fact that $K[V]$ is obviously the smallest ring of functions on V which contains all the coordinate functions $(x_1, \ldots, x_n) \mapsto x_i$. As an immediate consequence of $(\star)$, we obtain the following result.

(17.24) Proposition. *Let $V \subseteq L^n$ be a K-variety. Then V is irreducible if and only if $K[V]$ is an integral domain.*

Proof. Clearly, $K[V]$ is a commutative ring with the constant function 1 as its identity element. This ring is an integral domain if and only if it is free of zero divisors. Since $K[V] \cong K[X_1, \ldots, X_n]/\mathcal{I}(V)$ this is the case if and only if $\mathcal{I}(V)$ is a prime ideal, due to (6.14)(a). Thus (17.16) gives the claim. ∎

(17.25) Examples. (a) Let $V = \{(x, y) \in K^2 \mid xy = 1\}$. Then

$$
K[V] = \{\varphi : V \to K \mid \varphi(x, y) = \frac{g(x)}{x^n} \text{ with } g \in K[X] \text{ and } n \in \mathbb{N}_0\} \ .
$$

Every function on V of the form $\varphi(x,y) = g(x)/x^n = g(x)y^n$ is indeed polynomial so that the inclusion $\supseteq$ is clear. To prove $\subseteq$ we just note that the most general polynomial function on V takes the form

$$\sum_{k=0}^{n} a_k(x)y^k = \sum_{k=0}^{n} \frac{a_k(x)}{x^k} = \frac{\sum_{k=0}^{n} a_k(x)x^{n-k}}{x^n} \quad \text{where } a_0, \ldots, a_n \in K[X] \ .$$

(b) Let $V = \{(x,y) \in L^2 \mid y^2 = x^3\}$. Then

$$K[V] = \{\varphi : V \to K \mid \varphi(x,y) = P(x) + yQ(x) \text{ with } P, Q \in K[X]\} \ .$$

Here the inclusion $\supseteq$ is trivial, and $\subseteq$ follows from the fact that the most general polynomial function on V has the form

$$a_0(x) + a_1(x)y + a_2(x)y^2 + a_3(x)y^3 + \cdots$$
$$= a_0(x) + a_1(x)y + a_2(x)x^3 + a_3(x)x^3y + \cdots$$
$$= \big(a_0(x) + x^3 a_2(x) + \cdots\big) + y\big(a_1(x) + x^3 a_3(x) + \cdots\big) \ .$$

(17.26) Definitions. *Let $V \subseteq L^m$ and $W \subseteq L^n$ be K-varieties. A mapping $\varphi : V \to W$ is called a* **polynomial map** *or a* **homomorphism of varieties** *if all the component functions $\varphi_1, \ldots, \varphi_n : V \to L$ are polynomial functions on V.*
If φ is bijective such that φ^{-1} is also a polynomial map, then φ is called a **polynomial isomorphism** *or an* **isomorphism of varieties**.

(17.27) Examples. (a) Define $f, g : L^2 \to L^2$ by $f(x,y) = (x, x^2 + y)$ and $g(u,v) = (u, -u^2 + v)$. Then f and g are isomorphisms which are inverses of each other.
(b) Consider the parabola $V = \{(x,y) \in L^2 \mid y = x^k\}$ and the straight line $W = L^1 = L$. Then inverse isomorphisms $\varphi : V \to W$ and $\psi : W \to V$ are given by $\varphi(x,y) = x$ and $\psi(t) = (t, t^k)$.
(c) Let $V = L^1 = L$ and $W = \{(x,y) \in L^2 \mid y^2 = x^3\}$. The mapping $\varphi : V \to W$ given by $\varphi(t) = (t^2, t^3)$ is a bijective polynomial mapping; the inverse function $\psi : W \to V$ is given by $\psi(0,0) = 0$ and $\psi(x,y) = \frac{y}{x}$ if $(x,y) \neq (0,0)$. However, φ is not an isomorphism of varieties because ψ is not a polynomial map. This seems obvious, but how can we prove it exactly? One way is as follows: If ψ were a polynomial function, there would be polynomials $P, Q \in K[X]$ with $\psi(x,y) = P(x) + yQ(x)$, due to (17.23)(b). This would imply $t = \psi(\varphi(t)) = \psi(t^2, t^3) = P(t^2) + t^3 Q(t^2)$ for all $t \in L$ which is clearly impossible.

THE next proposition answers the question which (abstract) rings can occur as coordinate rings of varieties.

(17.28) Proposition. *Let K be a field. For a ring R, the following conditions are equivalent:*

(1) $R \cong K[V]$ *for some K-variety;*

(2) R *is a finitely generated K-algebra without nonzero nilpotent elements.*

Here V is irreducible if and only if R has no zero-divisors.

Proof. $(1)\Longrightarrow(2)$. Given a K-variety $V \subseteq L^n$, we have $K[V] \cong K[\xi_1,\ldots,\xi_n]$ where $\xi_k := X_k + \mathcal{I}(V)$; hence $K[V]$ is (isomorphic to) a finitely generated K-algebra. Clearly, $K[V]$ has no nonzero nilpotent elements, because $f^n = 0$ for a function $f : V \to L$ implies $f = 0$.

$(2)\Longrightarrow(1)$. Write $R = K[\xi_1,\ldots,\xi_n]$ and define a surjective homomorphism $\theta : K[X_1,\ldots,X_n] \to R$ of K-algebras by $\theta(X_i) := \xi_i$; then $I := \ker\theta$ is a radical ideal of $K[X_1,\ldots,X_n]$ because R has no nonzero nilpotent elements. Let L be an algebraically closed field containing K and let $V := \mathcal{V}(I) \subseteq L^n$; then Hilbert's Nullstellensatz implies that $\mathcal{I}(V) = \mathcal{I}\big(\mathcal{V}(I)\big) = \sqrt{I} = I$. Therefore,

$$K[V] \cong \frac{K[X_1,\ldots,X_n]}{\mathcal{I}(V)} = \frac{K[X_1,\ldots,X_n]}{I} \cong R .$$

The last statement follows immediately from the observation that two functions $f, g \in K[V]$ satisfy $fg = 0$ if and only if $V = \mathcal{V}(fg) = \mathcal{V}(f) \cup \mathcal{V}(g)$. ∎

WE now prove that a polynomial mapping between varieties induces an algebra homomorphism between the corresponding coordinate rings. This will allow us to reformulate certain geometric problems as algebraic problems.

(17.29) Theorem. *Let $V \subseteq L^m$ and $W \subseteq L^n$ be K-varieties.*

(a) *A map $\varphi : V \to W$ is polynomial if and only if $g\circ\varphi \in K[V]$ whenever $g \in K[W]$.*

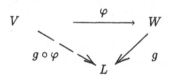

(b) *Every polynomial map $\varphi : V \to W$ induces a homomorphism $\varphi^\star : K[W] \to K[V]$ of K-algebras via*

$$\varphi^\star : \begin{array}{ccc} K[W] & \to & K[V] \\ g & \mapsto & g\circ\varphi . \end{array}$$

(c) *The assignment $\varphi \mapsto \varphi^\star$ sets up a 1-1-correspondence between polynomial maps $V \to W$ and K-algebra homomorphisms $K[W] \to K[V]$. This assignment has the properties*

$$(\mathrm{id}_V)^\star = \mathrm{id}_{K[V]} \qquad and \qquad (\varphi \circ \psi)^\star = \psi^\star \circ \varphi^\star$$

if $\psi : U \to V$ and $\varphi : V \to W$ are polynomial maps.

(d) *A polynomial map $\varphi : V \to W$ is an isomorphism of varieties if and only if $\varphi^\star : K[W] \to K[V]$ is an isomorphism of K-varieties. Moreover, $\varphi^\star$ is one-to-one if and only if $\overline{\varphi(V)} = W$, i.e., if the image of φ is Zariski-dense in W.*

Proof. (a) Let $\varphi = (\varphi_1, \ldots, \varphi_n)$; then $g \circ \varphi$ is given by $x \mapsto g(\varphi_1(x), \ldots, \varphi_n(x))$ where $x \in V$. Clearly, if g and $\varphi_1, \ldots, \varphi_n$ are polynomial functions then so is $g \circ \varphi$. Suppose conversely that $g \circ \varphi$ is polynomial for all $g \in K[W]$; then we can take in particular the coordinate function $g(y_1, \ldots, y_n) = y_i$ for which $g \circ \varphi = \varphi_i$. Hence $\varphi_1, \ldots, \varphi_n$ are polynomial.

(b) The mapping φ^* is well-defined by part (a); it remains to show that φ^* is an algebra homomorphism. We have $\varphi^*(c) = c$ for every constant function; moreover,

$$\varphi^*(g_1 + g_2) = (g_1 + g_2) \circ \varphi = g_1 \circ \varphi + g_2 \circ \varphi = \varphi^*(g_1) + \varphi^*(g_2) \quad \text{and}$$
$$\varphi^*(g_1 g_2) = (g_1 g_2) \circ \varphi = (g_1 \circ \varphi)(g_2 \circ \varphi) = \varphi^*(g_1)\varphi^*(g_2) .$$

(c) Let us show first that the assignment $\varphi \mapsto \varphi^*$ is one-to-one. We have $\varphi^* = \psi^*$ if and only if $g \circ \varphi = g \circ \psi$ for all $g \in K[W]$. Choosing the coordinate functions $g(y_1, \ldots, y_n) = y_i$, this implies $\varphi_i = \psi_i$ for all i and hence $\varphi = \psi$.

Let us show now that the assignment $\varphi \mapsto \varphi^*$ is also onto, i.e., that every K-algebra homomorphism $\Phi : K[W] \to K[V]$ is of the form $\Phi = \varphi^*$ for some polynomial map $\varphi : V \to W$. Given Φ, we let $\varphi_i := \Phi(g_i)$ $(1 \le i \le n)$ where $g_i(y_1, \ldots, y_n) = y_i$ is the i-th coordinate function. We then define a polynomial map $\varphi : V \to L^n$ by $\varphi(x) := (\varphi_1(x), \ldots, \varphi_n(x))$. Once we have shown that $\varphi(V) \subseteq W$, i.e., that φ is in fact a polynomial mapping from V into W, we are done, because then we have $\varphi^*(g_i) = g_i \circ \varphi = \varphi_i = \Phi(g_i)$ for all coordinate functions g_i and hence $\varphi^* = \Phi$ because $K[W]$ is generated by $g_1, \ldots g_n$ as a K-algebra.

To prove $\varphi(V) \subseteq W = \mathcal{V}(\mathcal{I}(W))$, it is enough to show that $p(\varphi(x)) = 0$ whenever $x \in V$ and $p \in \mathcal{I}(W)$. Now if $p \in \mathcal{I}(W) \subseteq K[X_1, \ldots, X_n]$ then $p(g_1, \ldots, g_n) = 0$ as an element of $K[W]$. Consequently, using the fact that Φ is an algebra homomorphism, we obtain

$$0 = \Phi(0) = \Phi(p(g_1, \ldots, g_n)) = p(\Phi(g_1), \ldots, \Phi(g_n)) = p(\varphi_1, \ldots, \varphi_n) .$$

Hence $p(\varphi(x)) = p(\varphi_1(x), \ldots, \varphi_n(x)) = 0$ for all $x \in V$.

The remaining statements are clear; we have $\mathrm{id}^*(g) = g \circ \mathrm{id} = g$ and $(\varphi \circ \psi)^*(g) = g \circ (\varphi \circ \psi) = (g \circ \varphi) \circ \psi = \psi^*(g \circ \varphi) = \psi^*(\varphi^*(g))$ for all g.

(d) We have $\psi \circ \varphi = \mathrm{id}_V$ and $\varphi \circ \psi = \mathrm{id}_W$ if and only if $\varphi^* \circ \psi^* = \mathrm{id}_{K[V]}$ and $\psi^* \circ \varphi^* = \mathrm{id}_{K[W]}$; since the assignment $\varphi \mapsto \varphi^*$ is a 1-1-correspondence this gives the first claim.

To prove the second claim we observe that $g \in \ker \varphi^*$ if and only if $g \circ \varphi = 0$, i.e., if g vanishes on $\varphi(V)$ which is tantamount to saying that g vanishes on $\overline{\varphi(V)}$ because g is Zariski-continuous. Clearly, if $\overline{\varphi(V)} = W$ this means $g = 0$ so that $\ker \varphi^* = \{0\}$. Conversely, if $\overline{\varphi(V)} \subsetneq W$ then there is an element $g \in \mathcal{I}(\varphi(V)) \backslash \mathcal{I}(W)$, due to (17.8)(b). Then clearly $g \in \ker \varphi^*$ but $g \ne 0$. ∎

(17.30) Example. Consider the varieties $V = \mathbb{R}^1$ and $W = \{(x, y) \in L^2 \mid y^2 = x^3 + \alpha x^2\}$ where $\alpha \in \mathbb{R}$ is a given constant and consider the polynomial mapping $\varphi : V \to W$ given by $\varphi(t) = (t^2 - \alpha, t^3 - \alpha t)$.[†] In the pictures below, the mapping φ is sketched in the cases $\alpha = 1$, $\alpha = 0$ and $\alpha = -1$.

[†] This parameterization of W is obtained by intersecting W with the straight lines $y = tx$; compare with example (17.4) above.

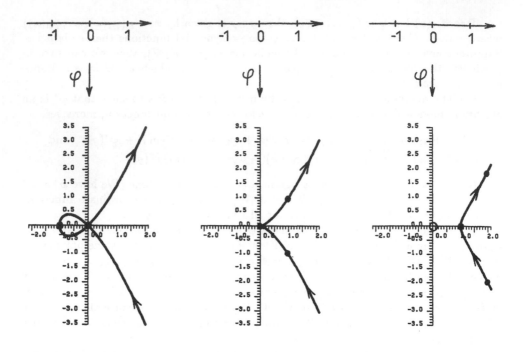

If $\alpha > 0$, then $\varphi(\sqrt{\alpha}) = \varphi(-\sqrt{\alpha}) = (0,0)$ so that φ is not injective; if $\alpha < 0$ then $\varphi(t) \neq (0,0)$ for all $t \in \mathbb{R}$ so that φ is not surjective. If $\alpha = 0$, then φ is a bijection, but not a polynomial isomorphism, due to (17.27)(c). To understand this fact from a different perspective, let us determine the homomorphism $\varphi^* : K[W] \to K[V]$. Using the identifications $K[V] \cong K[T]$ and $K[W] \cong K[X,Y]/\langle Y^2 - X^3 \rangle$ and writing $x = X + \langle Y^2 - X^3 \rangle$, $y = Y + \langle Y^2 - X^3 \rangle$, the mapping φ^* is given by $x \mapsto T^2$ and $y \mapsto T^3$. Since the coordinate functions x and y generated $K[W]$ over K, we have $\operatorname{im} \varphi^* = K[T^2, T^3] \subsetneq K[T]$ so that φ^* is not surjective. This gives a new proof that φ is not an isomorphism of varieties. Since $K[T^2, T^3] = \{f \in K[T] \mid f'(0) = 0\}$ the fact that $\operatorname{im} \varphi^* = K[T^2, T^3]$ can be interpreted geometrically: The mapping φ "squeezes together" the rays $t \geq 0$ and $t \leq 0$ to form a cusp at $(0,0) \in W$; thus W allows only tangent vectors at $(0,0)$ in one direction.

(17.31) **Definition.** *Let $V \subseteq L^m$ and $W \subseteq L^n$ be K-varieties. A polynomial map $\varphi : V \to W$ is called of **finite type** if $\overline{\varphi(V)} = W$ and if $K[V]$ is integral over $\varphi^*(K[W])$.*

THE next theorem shows that the notion of a mapping of finite type is very close to that of an isomorphism, at least if the base-field is algebraically closed.

(17.32) Theorem. *Let $\varphi : V \to W$ be a mapping of finite type between K-varieties.*

(a) *For every point $w \in W$, the pre-image $\varphi^{-1}(w)$ cannot be infinite. (Hence φ is "almost" injective.)*

(b) *If K is algebraically closed, then φ is surjective.*

Proof. (a) Let x_i be the i-th coordinate function on V. Since $K[V]$ is integral over $\varphi^*(K[W])$, there are polynomial functions $q_k : W \to L$ such that

$$x_i^n + (q_{n-1} \circ \varphi)x_i^{n-1} + \cdots + (q_1 \circ \varphi)x_i + (q_0 \circ \varphi) = 0 .$$

Hence if $v \in \varphi^{-1}(w)$, then

$(\star)$ $\qquad x_i(v)^n + q_{n-1}(w)x_i(v)^{n-1} + \cdots + q_1(w)x_i(v) + q_0(w) = 0 .$

This equation has at most a finite number of solutions; hence there is only a finite number of possible x_i-coordinates of points in $\varphi^{-1}(w)$. Since this holds for all indices i, the claim follows.

(b) The idea of the proof is that if w varies over all of W, none of the roots of equation $(\star)$ – which exist because K is algebraically closed – can tend to infinity, since the coefficient of the highest power of $x_i(v)$ is 1. Hence if w varies, the elements of $\varphi^{-1}(w)$ can coincide, but they cannot vanish.

To give a formal proof, let $y_1, \ldots, y_n$ be the coordinate functions on W. Then

$$v \in \varphi^{-1}(w) \iff \varphi(v) = w \iff y_i\big(\varphi(v)\big) = w_i \text{ for all } i \iff (\varphi^*y_i)(v) = w_i \text{ for all } i .$$

Hence $\varphi^{-1}(w)$ is empty if and only if the ideal $\langle \varphi^*y_i - w_i \mid 1 \le i \le n \rangle \trianglelefteq K[V]$ has no zero. By Hilbert's Nullstellensatz, this is the case if and only if $\langle \varphi^*y_1 - w_1, \ldots, \varphi^*y_n - w_n \rangle = K[V]$. Letting $J := \langle y_1 - w_1, \ldots, y_n - w_n \rangle = \{\psi \in K[W] \mid \psi(w) = 0\} \trianglelefteq K[W]$, this means that $K[V](\varphi^*J) = K[V]$. By (10.23)(b), applied to $R := \varphi^*(K[W])$, $S := K[V]$ and $I := \varphi^*J$, this is possible if and only if $\varphi^*J = \varphi^*(K[W])$. Since φ^* is one-to-one by (17.29)(d), this means $J = K[W]$, which is clearly impossible. ∎

WE now want to define the notion of a rational function on a variety. Intuitively, if V is a K-variety in L^n, then a rational function on V should be defined as a function of the form $x \mapsto p(x)/q(x)$ where $p, q \in K[V]$ are polynomial functions with $q \ne 0$. An inherent difficulty of the notion of rational functions is that such functions will in general not be defined on all of V, but that their domain will be a proper subset of V, due to the zeros of the denominator. This can lead to serious problems if we want to define sums and products of rational functions. For example, let $V = \{(x,y) \in L^2 \mid xy = 0\}$ and consider the functions $f(x,y) = (x^2 + 1)/x$ and $g(x,y) = (x + 1)/y$. Then f is defined on $\{(x,y) \in V \mid x \ne 0\} = \{(x,0) \mid x \ne 0\}$, whereas g is defined on $\{(x,y) \in V \mid y \ne 0\}$. Hence $f + g$ and fg cannot be defined, because there is no point $x \in V$ at which f and g could be simultaneously evaluated. We will avoid this problem by allowing only rational functions which are defined on a dense open subset of V; in this context, proposition (17.20) allows us to easily verify that a given subset of a variety is dense.[†]

[†] All topological statements refer to the Zariski topology on V.

Since in any topological space the intersection of two dense open subsets is again dense and open, there will be no problem in defining sums and products of rational functions in this more restricted sense.

There is a second feature which is not really a problem, but which we must be aware of. This feature is best explained by an example. Let $V = \{(x, y, z, w) \in L^4 \mid xy = zw\}$ and consider the two functions $f(x, y, z, w) = x/z$ and $g(x, y, z, w) = w/y$. We observe that f is defined on $U_1 = \{(x, y, z, w) \in V \mid z \neq 0\}$ and g on $\{(x, y, z, w) \in V \mid y \neq 0\}$. Now the definition of V implies that f and g coincide on their common domain $U_1 \cap U_2$. Thus instead of treating f and g as two different functions, we consider them as one and the same rational function with domain $U_1 \cup U_2$, namely as the function

$$F(x, y, z, w) := \begin{cases} x/z, & \text{if } z \neq 0; \\ w/y, & \text{if } y \neq 0. \end{cases}$$

Since f and g coincide on $U_1 \cap U_2$, there is no ambiguity in this definition.

AFTER these preparatory remarks, it should be clear why the definition of a rational function is chosen as follows.

(17.33) Definition. *Let V be a K-variety in L^n. A **rational function** on V (more precisely, a K-rational function on V) is a mapping $f : U \to L$, defined on a dense open subset $U \subseteq V$, such that there are families $(p_i)_{i \in I}$ and $(q_i)_{i \in I}$ in $K[V]$ with $U = \bigcup_{i \in I} D_{q_i}$ and $f = p_i/q_i$ on D_{q_i}.*

*If the set U is chosen maximal with respect to this condition, then U is called the **domain** of f and is denoted by $\operatorname{dom} f$; the **image** of f is defined as $\operatorname{im} f := f(\operatorname{dom} f)$.*

Finally, the set of all rational functions on V is denoted by $K(V)$. Due to the fact that the intersection of two dense open subsets is again open and dense, two rational functions can be added, subtracted and multiplied on the intersection of their domains of definition; thus it is clear that $K(V)$ is a K-algebra with $K[V] \subseteq K(V)$.

NOTE that the definition of a rational function implies that $p_i q_j = p_j q_i$ on $D_{q_i} \cap D_{q_j} = D_{q_i q_j}$ which is tantamount to saying that $q_i q_j (p_i q_j - p_j q_i) = 0$ in $K[V]$. To justify the above definition of the domain of a rational function, we have to make sure that there is always an unambiguously defined maximal set on which a rational function is defined. This is done in the following theorem.

(17.34) Identity and Extension Theorem. *Let V be a K-variety in L^n.*

(a) If $\varphi_1 : U_1 \to L$ and $\varphi_2 : U_2 \to L$ are rational functions which coincide on some open subset $U \subseteq U_1 \cap U_2$ which is dense in V, then φ_1 and φ_2 coincide on their common domain $U_1 \cap U_2$.

(b) Suppose $\varphi : U_0 \to L$ is a rational function where U_0 is open and dense in V. Then there is a unique extension of φ onto a uniquely determined maximal domain.

Proof. (a) Let $\varphi(x) := \varphi_1(x) - \varphi_2(x)$ and let $A := \{x \in U_1 \cap U_2 \mid \varphi(x) = 0\}$. To prove that A is a closed subset of $U_1 \cap U_2$ it is enough to verify that $(U_1 \cap U_2) \setminus A$ is open. Let $x \in (U_1 \cap U_2) \setminus A$. Then there is an open neighborhood U_0 of x such that $\varphi(y) = p(y)/q(y)$ on U_0. Then $U_0 \cap D_p$ is an open neighborhood of x which is contained in $(U_1 \cap U_2) \setminus A$. Hence if we take the closure with respect to $U_1 \cap U_2$ on both sides of the inclusion $U \subseteq A$ and use the fact that U is dense in V, we obtain $U_1 \cap U_2 \subseteq \overline{A} = A \subseteq U_1 \cap U_2$ and hence $A = U_1 \cap U_2$, as claimed.

(b) Consider all pairs (U, φ_U) such that U is open and dense in V and $\varphi_U : U \to L$ is a rational function with $\varphi_U|_{U \cap U_0} \equiv \varphi|_{U \cap U_0}$. Let $U_1 := \bigcup_{(U,\varphi_U)} U$ and define $\Phi : U_1 \to L$ by $\Phi(x) := \varphi_U(x)$ if $x \in U$. Then Φ is unambiguously defined due to part (a), and is clearly the unique maximal extension of φ. $\blacksquare$

(17.35) Examples. (a) Let $V = \{(x, y) \in L^2 \mid y^2 = x^2 + x^3\}$. Then $f(x, y) = x/y$ is a rational function on V with $\operatorname{dom} f = V \setminus \{(0, 0)\}$.

(b) Let $V = \{(x, y) \in L^2 \mid x^2 + y^2 = 1\}$. Then $f(x, y) = (1 - y)/x$ is a rational function on V with $\operatorname{dom} f = V \setminus \{(0, -1)\}$. It seems first that f is also not defined at the point $(0, 1)$; note however that $(1 - y)x = x(1 + y)$ on V, and $x/(1 + y)$ is well-defined at $(0, 1)$.

IT is perfectly possible that a rational function $\varphi \in K(V)$ has all of V as its domain without being a polynomial function; take for example $\varphi(x) = 1/(x^2 + 1)$ on $V := \mathbb{R}$. However, the situation is different if L is algebraically closed; in this case $K(V) = K[V]$. This is proved in the next proposition.

(17.36) Proposition. *Let $K \subseteq L$ be fields where L is algebraically closed and let $V \subseteq L^n$ be a K-variety. Moreover, let $\varphi : U \to L$ be a rational function defined on a dense and open subset U of V.*

(a) If $D_g \subseteq U$ for some $g \neq 0$ in $K[V]$, then there is a number $N \in \mathbb{N}$ and there is a polynomial function $f \in K[V]$ such that $\varphi = f/g^N$ on D_g.

(b) If $U = V$, i.e., if φ is defined everywhere, then $\varphi \in K[V]$.

Proof. (a) Let $U = \bigcup_{i \in I} D_{g_i}$ and $\varphi = f_i/g_i$ on D_{g_i}. By (17.13), we can assume that I is finite, say $I = \{1, \ldots, n\}$. Then $g_i g_j (f_i g_j - f_j g_i) = 0$ by the remark following (17.33). Letting $p_i := f_i g_i$ and $q_i := g_i^2$, we have $\varphi = p_i/q_i$ on $D_{q_i} = D_{g_i}$ and $p_i q_j = p_j q_i$ for all i, j. Now $D_g \subseteq U = \bigcup_{i=1}^n D_{q_i}$; taking complements, we see that $\mathcal{V}(g) \supseteq \mathcal{V}(q_1, \ldots, q_n)$ and hence $\mathcal{I}(\mathcal{V}(g)) \subseteq \mathcal{I}(\mathcal{V}(q_1, \ldots, q_n))$. By Hilbert's Nullstellensatz, this means $\sqrt{\langle g \rangle} \subseteq \sqrt{\langle q_1, \ldots, q_n \rangle}$; in particular, there is a natural number $N \in \mathbb{N}$ with $g^N \in \langle q_1, \ldots, q_n \rangle$. This means that there are functions $h_i \in K[V]$ such that $g^N = \sum_{i=1}^n h_i q_i$. Letting $f := \sum_{i=1}^n h_i p_i$, we obtain $g^N p_j = \sum_{i=1}^n h_i q_i p_j = \sum_{i=1}^n h_i p_i q_j = f q_j$ so that $f/q^N = p_j/q_j = \varphi$ on D_{q_j}. Since this holds for all j, we have $\varphi = f/g^N$ on D_g.

(b) Simply take $g := 1$ (constant function with the value 1) in part (a); then $D_g = V$, and the claim follows. $\blacksquare$

IN the next two propositions we determine the algebraic structure of the ring $K(V)$.

(17.37) Proposition. *Let V be a K-variety in L^n.*

(a) *If V is irreducible, then $K(V)$ is the quotient field of $K[V]$.*

(b) *If $V = V_1 \cup \cdots \cup V_r$ is the decomposition of V into its irreducible components then*

$$\Phi: \begin{array}{ccc} K(V) & \to & K(V_1) \times \cdots \times K(V_r) \\ f & \mapsto & (f|_{V_1}, \ldots, f|_{V_r}) \end{array}$$

is an isomorphism of K-algebras.

Proof. (a) Let $Q(K[V])$ be the quotient field of $K[V]$; its elements are formal quotients f/g with $f, g \in K[V]$ such that $g \neq 0$. Each such fraction gives rise to a rational function f/g which is at least defined on D_g; note that D_g is open and nonempty, hence dense in V. Since $f/g = F/G$ as formal quotients clearly implies $f/g = F/G$ as rational functions, this yields a well-defined K-algebra homomorphism $\Phi : Q(K[V]) \to K(V)$. If $f/g \in \ker \Phi$, then $f \equiv 0$ on the dense subset D_g and hence everywhere so that $f/g = 0$; this shows that Φ has zero kernel, i.e., is injective. Now let $\varphi : U \to L$ be any rational function, say $\varphi = f/g$ on $D_g \subseteq U$. Since D_g is open and nonempty, hence dense in V, this implies $\varphi = f/g$ as a rational function. This argument shows that Φ is surjective.

(b) Let $U = \operatorname{dom} f$. Then $V_i \cap U$ is dense in V_i for all i; see (17.19). Thus $f|_{V_i}$ is at least defined on $V_i \cap U$ and is a rational function there. Hence Φ is well-defined and obviously a K-algebra homomorphism. We show that Φ is an isomorphism by exhibiting the inverse isomorphism $\Psi : K(V_1) \times \cdots \times K(V_n) \to K(V)$. Let f_i be a rational function on V_i with domain $\operatorname{dom} f_i$. Then $U_i := (\operatorname{dom} f_i) \setminus \bigcup_{j \neq i} V_j$ is not empty and open, hence dense in V_i. Moreover, $U_i \cap U_j = \emptyset$ for $i \neq j$. Then $U := U_1 \cup \cdots \cup U_r$ is an open dense subset of V. Since this is a disjoint union, we can unambiguously define a rational function $f : U \to L$ by $f(u) := f_i(u)$ if $u \in U_i$ which uniquely extends to a rational function f on V. Call this function $\Psi(f_1, \ldots, f_n)$. By the definitions of Φ and Ψ it is clear that $\Psi \circ \Phi$ and $\Phi \circ \Psi$ are the identity maps on $K(V)$ and $K(V_1) \times \cdots \times K(V_r)$, respectively. ∎

(17.38) Proposition. *Let $K \subseteq K'$ be fields. Then the following conditions are equivalent:*

(1) $K' \cong K(V)$ *where V is an irreducible K-variety;*

(2) K' *is obtained from K by adjoining a finite number of elements.*

Proof. (1)$\Longrightarrow$(2). Write $K[V] = K[\xi_1, \ldots, \xi_n]$ as in (17.28). This is an integral domain with quotient field $K(V) = K(\xi_1, \ldots, \xi_n)$ which is an extension of K of the desired form.

(2)$\Longrightarrow$(1). If $K' = K(\alpha_1, \ldots, \alpha_n)$ then $R := K[\alpha_1, \ldots, \alpha_n]$ is a finitely generated K-algebra without zero-divisors; hence $R \cong K[V]$ for some K-variety V, due to (17.28). But if the rings R and $K[V]$ are isomorphic, then so are their quotient fields K' and $K(V)$. ∎

(17.39) Example. Let us determine $K(V)$ where $V = \{(x,y) \in L^2 \mid y^2 = x^3\}$. We claim that $K(V)$ consists of all functions

$$f(x,y) = u(x) + v(x)y \quad \text{where} \quad u,v \in K(X) \, .$$

It is clear that every function of this form is a rational function on V. To prove the converse, we observe that $K(V)$ is the quotient field of $K[V]$ since V is irreducible. Now due to (17.25)(b), every element of $K[V]$ has the form $p(x) + q(x)y$ where $p, q \in K[X]$. Hence $K(V)$ consists of all quotients

$$\frac{p_1 + q_1 y}{p_2 + q_2 y} = \frac{p_1 + q_1 y}{p_2 + q_2 y} \cdot \frac{p_2 - q_2 y}{p_2 - q_2 y} = \frac{p_1 p_2 - q_1 q_2 y^2 + q_1 p_2 y - q_2 p_1 y}{p_2^2 - q_2^2 y^2}$$

$$= \frac{p_1 p_2 - q_1 q_2 x^3}{p_2^2 - q_2^2 x^3} + \frac{q_1 p_2 - q_2 p_1}{p_2^2 - q_2^2 x^3} \, y \, ;$$

this clearly implies the claim.

$\mathrm{W}\mathrm{E}$ now introduce the notion of a rational mapping between varieties.

(17.40) Definition. *Let $V \subseteq L^m$ and $W \subseteq L^n$ be K-varieties. A partial map $\varphi : V \to W$ is called a K-**rational map** if all the component functions $\varphi_1, \ldots, \varphi_n$ are rational functions on V.*

Note that φ is defined on $\operatorname{dom}\varphi_1 \cap \cdots \cap \operatorname{dom}\varphi_n$ which is a finite intersection of dense and open subsets and hence dense and open itself.

$\mathrm{T}\mathrm{H}\mathrm{E}$ difficulty that rational functions are not defined everywhere makes itself felt, of course, when we try to compose two rational functions $f : U \to V$ and $g : V \to W$. The composition $g \circ f$ is (a priori!) defined on $(\operatorname{dom} f) \cap f^{-1}(\operatorname{dom} g)$, and this set need not be dense in U; it might even be empty. On the other hand, if this set is dense in U so that $g \circ f$ is defined, it might be that $\operatorname{dom}(g \circ f)$ is larger than this set because we always think of a rational function as being extended to its largest possible domain of definition. We now define a class of functions which always can be composed with other rational functions.

(17.41) Definition. *A rational map $\varphi : V \to W$ is called **dominant** if $\varphi(\operatorname{dom}\varphi)$ is dense in W.*

$\mathrm{C}\mathrm{L}\mathrm{E}\mathrm{A}\mathrm{R}\mathrm{L}\mathrm{Y}$, if $\varphi : U \to V$ and $\psi : V \to W$ are rational maps such that φ is dominant, then $\psi \circ \varphi$ is a well-defined rational map (which is at least defined on $\varphi^{-1}(\operatorname{dom}\psi) \cap \operatorname{dom}\varphi$ which is open and dense).

To characterize dominant maps algebraically, we observe that any rational map $\varphi : V \to W$ induces a K-algebra homomorphism $\varphi^* : K[W] \to K(V)$ via $\varphi^*(g) = g \circ \varphi$. This mapping can be extended to a field homomorphism $\varphi^* : K(W) \to K(V)$ if and only if it is injective (we have to define $\varphi^*(f/g) := (\varphi^* f)/(\varphi^* g)$), and this definition works if and only if $\varphi^* g \neq 0$ whenever $g \neq 0$, i.e., if φ^* is injective). It turns out that this condition characterizes dominant maps; this is shown in the next proposition which is an analogue of (17.29).

(17.42) Proposition. *Let $\varphi : V \to W$ be a rational map between K-varieties.*
(a) φ is dominant if and only if $\varphi^* : K[W] \to K(V)$ is injective and hence can be extended to a field homomorphism $\varphi^* : K(W) \to K(V)$.
(b) Every K-homomorphism $\Phi : K(W) \to K(V)$ is of the form $\Phi = \varphi^*$ where $\varphi : V \to W$ is a dominant map.

Proof. To prove (a), we observe that $g \in \ker \varphi^*$, i.e., $g \circ \varphi = 0$, if and only if $\varphi(\operatorname{dom}\varphi) \subseteq \mathcal{V}(g)$; hence φ^* has a nonzero kernel if and only if $\varphi(\operatorname{dom}\varphi)$ is contained in a proper subvariety of V. This means exactly that φ^* is injective if and only if $\varphi(\operatorname{dom}\varphi)$ is dense in W, i.e., if φ is dominant. The proof of part (b) is completely analogous to the corresponding statement in (17.29) and is left as an exercise. ∎

WE now define the notion of a birational isomorphism which is the analogue for rational functions to the notion of a polynomial isomorphism for polynomial mappings.

(17.43) Definition. *Let $V \subseteq L^m$ and $W \subseteq L^n$ be K-varieties. A rational function $\varphi : V \to W$ is called a* **birational isomorphism** *(more precisely, a K-birational isomorphism), if there is a K-rational mapping $\psi : W \to V$ such that $\varphi \circ \psi$ and $\psi \circ \varphi$ are both defined and are (or rather extend to) the identity maps on V and W, respectively. The map ψ (which is uniquely determined) is denoted by φ^{-1} and is called the inverse of φ.*

(17.44) Examples. (a) Let $V = L^1$ and $W = \{(x,y) \in L^2 \mid y^2 = x^3\}$. Define rational functions $\varphi : V \to W$ and $\psi : W \to V$ by $\varphi(t) = (t^2, t^3)$ and $\psi(x,y) = y/x$; then $\operatorname{dom}\varphi = V$ and $\operatorname{dom}\psi = W \setminus \{(0,0)\}$. It is easy to see that φ and ψ are inverses of each other. Comparing this observation with (17.27)(c) and (17.30) we can say that the varieties V and W are not isomorphic, but birationally isomorphic.
(b) Let $V = \{(x,y) \in L^2 \mid x^3 + y^3 = 1\}$ and $W = \{(u,v) \in L^2 \mid v^2 = 4u^3 - 1\}$. Define $\varphi : V \to W$ by $u(x,y) = x/(1-y)$, $v(x,y) = (3-y)/(y-1)$ and $\psi : W \to V$ by $x(u,v) = -2u/(v+1)$, $y(u,v) = (v+3)/(v+1)$. One easily verifies that φ and ψ are inverses of each other. Hence V and W are birationally isomorphic.

(17.45) Proposition. *A rational map $\varphi : V \to W$ is a birational isomorphism if and only if it induces a field isomorphism $\varphi^* : K(W) \to K(V)$.*

Proof. Noting that φ is a birational isomorphism with inverse ψ if and only if φ and ψ are both dominant and satisfy $\varphi \circ \psi = \operatorname{id}_W$ and $\psi \circ \varphi = \operatorname{id}_V$, this is an immediate consequence of (17.42)(a) and the fact that $(\varphi \circ \psi)^* = \psi^* \circ \varphi^*$ and $(\psi \circ \varphi)^* = \varphi^* \circ \psi^*$. ∎

WE now have two equivalence relations between varieties; being (polynomially) isomorphic and being birationally isomorphic. It is clear that the latter is the coarser of the two equivalences; any two isomorphic varieties are the more so birationally isomorphic, whereas two varieties may be birationally isomorphic without being isomorphic. The next theorem gives an idea of how coarse the equivalence relation between varieties defined by birational isomorphism actually is.

(17.46) Theorem. *Let K be a perfect field. Every irreducible K-variety is birationally isomorphic to a hypersurface. If $|K| = \infty$, then the birational isomorphism in question can be chosen as a linear projection.*

Proof. Let $V \subseteq L^n$ be an irreducible K-variety. Due to (16.12) there are elements $y_1, \ldots, y_{d+1} \in K(V)$ such that $K(V) = K(y_1, \ldots, y_d)$, $y_1, \ldots, y_d$ are algebraically independent and y_{d+1} is algebraic over $K(y_1, \ldots, y_d)$; if $|K| = \infty$, then each y_i can be chosen as a K-linear combination of the coordinate functions $x_1, \ldots, x_n$. Let $\Phi(y_1, \ldots, y_d, y_{d+1}) = \sum_{k=0}^{n} a_k(y_1, \ldots, y_d) y_{d+1}^k$ be the minimal polynomial of y_{d+1} over $K(y_1, \ldots, y_d)$. Then

$$V_0 := \{(z_1, \ldots, z_{d+1}) \in L^{d+1} \mid \Phi(z_1, \ldots, z_{d+1}) = 0\}$$

is clearly a K-hypersurface in L^{d+1}, and we claim that the mapping $\varphi : V \to V_0$ given by

$$x \mapsto \begin{pmatrix} y_1(x) \\ \vdots \\ y_{d+1}(x) \end{pmatrix}$$

is a birational isomorphism (which is clearly a linear projection in the case $|K| = \infty$ if the functions y_i are chosen as linear combinations of $x_1, \ldots, x_n$). We prove that φ is a birational isomorphism by verifying that $\varphi^* : \begin{array}{ccc} K(V_0) & \to & K(V) \\ g & \mapsto & g \circ \varphi \end{array}$ is an isomorphism of fields. The surjectivity of φ^* is clear because $K(V) = K(y_1, \ldots, y_d, y_{d+1})$. To check the injectivity, we observe that if $g \in \ker \varphi^*$, i.e., if $g \circ \varphi = 0$, then $g(y_1(x), \ldots, y_{d+1}(x)) = 0$ for all $x \in V$ which means that $g(y_1, \ldots, y_d, y_{d+1})$ is the zero function. Hence g, considered as a polynomial in y_{d+1}, must divide Φ because Φ is the minimal polynomial of y_{d+1}, say $g(y_1, \ldots, y_{d+1}) = h(y_1, \ldots, y_{d+1}) \Phi(y_1, \ldots, y_{d+1})$. But this implies that g is the zero function on V_0, which is what we wanted to show. ∎

Exercises

Problem 1. Given a real number $a > 0$, let C be the set of all points in $\mathbb{R}^2$ for which the product of the disctances to $(1,0)$ and $(-1,0)$ equals a. (Then C is called a **Cassini curve**.) Derive the equation for such a curve and draw several examples, covering the three cases $a < 1$, $a = 1$ and $a > 1$.

Problem 2. Let $\ell_0 \subseteq \mathbb{R}^2$ be the line with the equation $x = 1$ and let $\mathcal{L}$ be the set of all lines through $(0,0)$ which are not parallel to ℓ_0. Fix a number $a > 0$. Then for each line $\ell \in \mathcal{L}$ let $P(\ell)$ be the point of intersection between ℓ and ℓ_0 and let $P_1(\ell)$ and $P_2(\ell)$ be the two points on ℓ whose distance to $P(\ell)$ is a. Then the set $\{P_1(\ell), P_2(\ell) \mid \ell \in \mathcal{L}\}$ is called a **conchoid**.
  (a) Draw a figure to visualize the construction of a conchoid.
  (b) Show that the conchoid is given by the equation $(x^2 + y^2)(x - 1)^2 = a^2 x^2$.
  (c) Sketch conchoids for various choices of a, covering the cases $a < 1$, $a = 1$ and $a > 1$.

Problem 3. Let K be a field.
  (a) Show that if $V \subseteq K^1$ is a variety, then $V = K^1$ or V is a finite set.
  (b) Show that if $V \subseteq K^2$ is a variety, then $V = K^2$ or V is a finite union of points and plane curves.
  Hint. If $V = \mathcal{V}(f_1, \ldots, f_n)$, let $f = \gcd(f_1, \ldots, f_n)$. Then $\mathcal{V}(\frac{f_1}{f}, \ldots, \frac{f_n}{f})$ is finite.
  (c) Show that if $f, g \in K[x, y]$ are coprime then $\mathcal{V}(\langle f, g \rangle) \subseteq K^2$ is a finite set.

Problem 4. Let $K \subseteq L$ be fields and let $V \subseteq L^n$ be the zero set of a polynomial $f \in K[x_1, \ldots, x_n]$. Show that if $n \geq 2$ and if L is algebraically closed then V is an infinite set.

Problem 5. For an algebraically closed field K and coprime natural numbers m and n, let $V := \{(x, y) \in K^2 \mid x^n = y^m\}$. Moreover, for any element $k \in K$ let $f_k(T) := T^m - k$ and $g_k(T) := T^n - k$. Show that

$$V = \{(t^m, t^n) \mid t \in K\} = \{(x, y) \in K^2 \mid \mathrm{Res}(f_x, g_y) = 0\}$$

where $\mathrm{Res}(f, g)$ denotes the resultant of f and g.

Problem 6. For each of the following polynomials $p \in \mathbb{Q}[x, y]$ decide whether $\{(x, y) \in \mathbb{Q}^2 \mid p(x, y) = 0\}$ is dense in $\{(x, y) \in \mathbb{R}^2 \mid p(x, y) = 0\}$.
  (a) $p(x, y) = y - f(x)$ where $f \in K[x]$.
  (b) $p(x, y) = y^2 - x^3$.
  (c) $p(x, y) = x^2 + y^2 - 1$.
  (d) $p(x, y) = x^2 - y^2 - 1$.
  (e) $p(x, y) = x^4 + y^4 - 1$.

Problem 7. Let K be a finite field of characteristic p.

(a) Show that if the variety $V \subseteq K^n$ is given by $V = \mathcal{V}(f_1, \ldots, f_m)$ where each of the polynomials $f_i \in K[x_1, \ldots, X_n]$ has a degree $< n$ then $|V|$ is divisible by p.

Hint. Using problem 13 in section 4, show that the characteristic function $\chi :$ $K^n \to K$ of V, given by $\chi(x) := 1$ if $x \in V$ and $\chi(x) := 0$ if $x \notin V$, can be written as a polynomial function $\chi(x_1, \ldots, x_n) = \sum_{i_1, \ldots, i_n} a_{i_1, \ldots, i_n} x_1^{i_1} \cdots x_n^{i_n}$ of degree less than $n(q-1)$, i.e., $i_1 + \cdots + i_n < n(q-1)$ for all occurring multi-indices. Then in K we have

$$|V| = \sum_{x \in K^n} \chi(x) = \sum_{\substack{x \in K^n, \\ i_1, \ldots, i_n}} a_{i_1, \ldots, i_n} x_1^{i_1} \cdots x_n^{i_n} = \sum_{i_1, \ldots, i_n} a_{i_1, \ldots, i_n} \Big(\sum_{x_1 \in K} x_1^{i_1} \Big) \cdots \Big(\sum_{x_1 \in K} x_n^{i_n} \Big).$$

Now at least one of the indices i_k must be less than $q - 1$. Show that if i is such an index then $\sum_{x \in K} x^i = 0$ to conclude that $|V| = 0$ as an element of K.

(b) Let $\mathcal{F}$ be the family of all polynomials $f \in K[X_1, \ldots, X_n]$ of degree less than n which have no constant term. Show that there are elements $a_1, \ldots, a_n$, not all zero, such that $f(a_1, \ldots, a_n) = 0$ for all $f \in \mathcal{F}$.

Hint. If $V = \mathcal{V}(\mathcal{F})$, then $(0, \ldots, 0) \in V$; hence $V \neq \emptyset$. (Compare with problem 14 in section 4.)

Problem 8. Show that if V and W are K-varieties in L^n, then $\mathcal{I}(V \cup W) = \mathcal{I}(V) \cap \mathcal{I}(W)$ and $V \cup W = \mathcal{V}(\mathcal{I}(V) \cdot \mathcal{I}(W))$. Also, show that if $(V_i)_{i \in I}$ is a family of K-varieties in L^n, then $\bigcap_{i \in I} V_i = \mathcal{V}(\sum_{i \in I} \mathcal{I}(V_i))$. Which parts of these statements remain true if V, W, V_i are arbitrary subsets of L^n?

Problem 9. Suppose the field K is not algebraically closed.

(a) Show that for any $m \in \mathbb{N}$ there is a polynomial $\phi_m \in K[X_1, \ldots, X_m]$ whose only zero in K^m is $(0, \ldots, 0)$. (**Hint.** Use induction on m. To get an idea, you might consider $\phi_m(X_1, \ldots, X_m) = X_1^2 + \cdots + X_m^2$ in the case $K = \mathbb{R}$.)

(b) Let $n \in \mathbb{N}$. Show that if $V \subseteq K^n$ is a K-variety then there is a polynomial $f \in K[X_1, \ldots, X_n]$ with $V = \mathcal{V}(f)$; i.e., every K-variety in K^n can be written as the zero set of a single polynomial. (**Hint.** If $V = \mathcal{V}(f_1, \ldots, f_m)$ consider $f := \phi_m(f_1, \ldots, f_m)$.)

Problem 10. (a) Let $I = \langle x^2 + y^2 - 1, y - 1 \rangle \trianglelefteq K[x, y]$. Determine $V := \mathcal{V}(I)$ and $\mathcal{I}(V)$ and show that $I \subsetneq \mathcal{I}(V)$.

(b) Let $I = \langle xy, xz, yz \rangle \trianglelefteq K[x, y, z]$. Determine $V := \mathcal{V}(I)$ and $\mathcal{I} = (V)$.

(c) Let $I = \langle x^2 + y^2 + z^2, xy + yz + zx \rangle \trianglelefteq K[x, y, z]$. Determine $V := \mathcal{V}(I)$ and $\mathcal{I}(V)$.

Problem 11. (a) Let $V \subseteq L^2$ be a finite set. Show that $\mathcal{I}(V)$ can be generated by two elements.

Hint. Suppose first that the points of V have pairwise different x-coordinates $x_1, \ldots, x_n$. Find a polynomial p such that $V = \{(x, y) \mid \prod_i (x - x_i) = 0, \, y = p(x)\}$.

(b) Let $V = \{(x, y, z) \in L^3 \mid xyz = 0\}$. Show that $\mathcal{I}(V)$ cannot be generated by two elements.

(c) Let K be an infinite field and let $V \subseteq K^n$ be a finite set. Show that $\mathcal{I}(V) \trianglelefteq K[X_1, \ldots, X_n]$ can be generated by n elements.

Hint. Use interpolation polynomials as in problem 19 of section 4.

Problem 12. Let K be a field and let $f, g \in K[x]$ be irreducible polynomials with $\mathcal{V}(f) = \mathcal{V}(g)$. Show that it is possible in this situation that f and g are linearly independent over K. However, show that if K is algebraically closed then there is a constant $c \in K^\times$ with $g = cf$. (**Hint.** Hilbert's Nullstellensatz yields $\sqrt{\langle f \rangle} = \sqrt{\langle g \rangle}$.)

Problem 13. Let $K \subseteq L$ be fields where L is algebraically closed and let $I \trianglelefteq K[X_1, \ldots, X_n]$. Show that there is a natural number N such that if $f \in \mathcal{I}(\mathcal{V}(I))$ then $f^N \in I$. (**Hint.** Use Hilbert's Nullstellensatz and Hilbert's basis theorem.)

Problem 14. Let R be an affine algebra over a field K, say $R = K[\xi_1, \ldots, \xi_n]$. Show that the Jacobson radical and the nilradical of R coincide.

Hint. Let L be an algebraic closure of K and apply Hilbert's Nullstellensatz with the fields K and L.

Problem 15. Let K be an algebraically closed field. Show that the nonzero prime ideals of $K[x, y]$ are exactly the following ones:
 (a) the principal ideals $\langle f \rangle$ where $f \in K[x, y]$ is irreducible,
 (b) the maximal ideals $\langle x - \xi, y - \eta \rangle$ where $(\xi, \eta) \in K^2$.
Hint. If I is a prime ideal then $I = \sqrt{I} = \mathcal{I}(\mathcal{V}(I))$. Show that $\mathcal{V}(I)$ is a finite set.

Problem 16. Let V be a K-variety in L^n. Show that the mappings $\mathcal{I}$ and $\mathcal{V}$ induce an assignment between the subvarieties of V and the radical ideals of $K[V]$. Also show that if L is algebraically closed, then this is a 1-1-correspondence.

Problem 17. (a) Show that the Zariski topology is not a Hausdorff topology, in general.
 (b) Show that the Zariski topology on $V \times W$ is, in general, not the product of the Zariski topologies on V and W.

Problem 18. (a) Show that the set $V_0 := \{(t^2, t^4) \mid t \in \mathbb{R}\}$ is not a variety. Determine its Zariski closure $V := \overline{V_0} = \mathcal{V}(\mathcal{I}(V_0))$ (which is clearly the smallest variety containing V_0) and show that V is irreducible.
 (b) In general, let $K \subseteq L$ be fields where L is infinite and let $f_1, \ldots, f_n \in K[X_1, \ldots, X_m]$ be polynomials. Show that the Zariski closure of

$$V_0 := \left\{ \begin{pmatrix} f_1(t_1, \ldots, t_m) \\ \vdots \\ f_n(t_1, \ldots, t_m) \end{pmatrix} \middle| (t_1, \ldots, t_m) \in L^m \right\} \subseteq L^n$$

is an irreducible K-variety in L^n. (This fact is somewhat inexactly expressed by saying that parameterizable varieties are irreducible. Compare with the general fact that continuous images of connected spaces are connected.)

Problem 19. For an infinite field K, let $I := \langle xy - z^2, x^3 - yz \rangle \trianglelefteq K[x, y, z]$ and $V := \mathcal{V}(I) \subseteq K^3$. Show that the decomposition of V into irreducible components is given by $V = V_1 \cup V_2$ where

$$V_1 := \{(0,0,t) \mid t \in K\} = \{(x,y,z) \mid z = 0\} \quad \text{and}$$
$$V_2 := \{(t^3, t^4, t^5) \mid t \in K\} = \{(x,y,z) \mid xz = y^2, \, x^3 = yz, \, z^2 = x^2 y\} \ .$$

Hint. Compare with problem 33 in section 6.

Problem 20. Show that if V and W are irreducible varieties then the cartesian product $V \times W$ is again irreducible.

Problem 21. Let X and Y be nonempty open subsets of an irreducible variety V. Show that $X \cap Y \neq \emptyset$.

Problem 22. Consider a quadratic polynomial in n variables $x_1, \ldots, x_n$ over an infinite field K, say $q(x) = x^T A x + b^T x + c$ with $A^T = A$. Show that if the variety $V := \{x \in K^n \mid q(x) = 0\}$ is reducible then rank $A \leq 2$.

Problem 23. For each of the following ideals $I \trianglelefteq K[X, Y]$, find $\mathcal{V}(I)$, decompose $\mathcal{V}(I)$ into its irreducible components and determine $\mathcal{I}(\mathcal{V}(I))$.
(a) $I_1 = \langle X^2 - Y^2, \, X^3 + XY^2 - X^2 Y - Y^3 - X + Y \rangle$
(b) $I_2 = \langle X^2 + Y^2 - 1, \, Y - 1 \rangle$

Problem 24. For each of the following ideals $I \trianglelefteq K[X, Y, Z]$, find $\mathcal{V}(I)$, decompose $\mathcal{V}(I)$ into its irreducible components and determine $\mathcal{I}(\mathcal{V}(I))$.
(a) $I_1 = \langle XY, \, XZ, \, YZ \rangle$
(b) $I_2 = \langle XY, \, (X - Y)Z \rangle$
(c) $I_3 = \langle X^2 + Y^2 + Z^2, \, XY + XZ + YZ \rangle$

Problem 25. Decompose the variety $V = \{(x,y,z) \in L^3 \mid y^2 = xz, \, z^2 = y^3\}$ into its irreducible components. Show that all of these are birationally isomorphic to L^1.

Problem 26. (a) Define $\varphi : L^2 \to L^2$ by $\varphi(x,y) = (x, xy)$. Determine the image of φ! Decide whether this image is open, closed, or dense in L^2.
(b) Do the same as in (a) for the mapping $\varphi : L^3 \to L^3$ given by $\varphi(x,y,z) = (x, xy, xyz)$.

Problem 27. Suppose that $\varphi : V \to W$ is a polynomial map. Show that the graph of φ, defined as $\Gamma := \{(x, \varphi(x)) \mid x \in V\} \subseteq V \times W$, is a K-variety which is isomorphic to V.

Problem 28. Let V be a K-variety in L^n. The set of all isomorphisms $\varphi : V \to V$ forms a group under composition which is called the automorphism group of V and is denoted by $\mathrm{Aut}\, V$.
(a) Show that $\mathrm{Aut}\, L^1$ consists exactly of all mappings $f(x) = ax + b$ where $a, b \in K$, $a \neq 0$.
(b) Show that all mappings of the form $(x, y) \mapsto (x, y + p(x))$ with $p \in K[X]$ form a subgroup of $\mathrm{Aut}\, L^2$.

Problem 29. Show that if a K-variety consists of a finite number n of points then $K[V] \cong K \oplus \cdots \oplus K$ with n summands.

Problem 30. Let $V = \{(x,y) \in L^2 \mid y^2 = x^2 + x^3\}$.
(a) Show that a polynomial map $\varphi : L^1 \to V$ is given by $\varphi(t) = (t^2 - 1, t(t^2 - 1))$.
(b) Show that $\varphi^* : K[V] \to K[L^1]$ is injective.
(c) Show that $\operatorname{im} \varphi^* = \{g \in K[L^1] \mid g(1) = g(-1)\}$.
(d) Is φ an isomorphism of varieties? Is φ a birational isomorphism?

Problem 31. (a) Let $V := \{(x,y,z) \in L^3 \mid y^2 = xz, x^3 = yz, z^2 = x^2 y\}$. Show that a polynomial parameterization $\varphi : L \to V$ is given by $\varphi(t) = (t^3, t^4, t^5)$. Is φ an isomorphism of varieties? Is φ a birational isomorphism?
(b) Decompose $X := \{(x,y,z) \in L^3 \mid y^2 = xz, x^3 = yz\}$ into its irreducible components. (**Hint.** The ideal $I = \langle Y^2 - XZ, X^3 - YZ \rangle$ is not prime because $X \notin I$ and $Z^2 - X^2 Y \notin I$ but $X(Z^2 - X^2 Y) \in I$.)

Problem 32. Let L be a field and define a mapping $\varphi : L^1 \to L^3$ by $\varphi(t) := (t, t^2, t^3)$. Show that $V := \operatorname{im} \varphi$ is a variety. Is $\varphi : L^1 \to V$ a birational isomorphism or even a polynomial isomorphism?

Problem 33. Show that for each constant $a \neq 0$ the **lemniscate** $(x^2 + y^2)^2 = a^2(x^2 - y^2)$ possesses a rational parameterization. To do so, consider the points of intersection of the lemniscate with the circles $x^2 + y^2 = t(x - y)$.

Problem 34. Let $n > 2$ and let L be a field such that n is not a multiple of char K. (This is always the case if char $K = 0$.) Show that the curve $x^n + y^n = 1$ does not possess a rational parameterization. (**Hint.** Use problem 33 in section 4.)

Problem 35. Find parameterizations for the following curves!

$$y^2 = x^2 \frac{a + x}{a - x}, \quad y^2 = \frac{x^3}{a - x}, \quad x^3 + y^3 + 3axy = 0, \quad y^2 = \frac{x^2}{(x + 1)(x + 2)}, \quad x^3 = y^3 - y^4$$

Problem 36. Let $\mathbb{Q} \subseteq L$ and $V = \{(x,y,z) \in L^3 \mid 2x^2 + y^2 = 5\}$. Does V possess a rational parameterization? Find all rational numbers a, b, c with $2a^2 + b^2 = 5c^2$.

Problem 37. Let K be a field with char $K \neq 2, 3$. Show that the curve $y^2 = x^3 + ax + b$ possesses a rational parameterization if and only if the polynomial $X^3 + aX + b$ has a multiple root.

Problem 38. Let $V = \{(x,y) \in L^2 \mid y^3 = x^4 + x^3\}$. Define $\varphi : V \to L^1$ by $\varphi(x,y) = x/y$. Show that φ is a birational isomorphism and find φ^{-1}.

Problem 39. Let $V := \{(x,y) \in L^2 \mid xy = 1\}$ and $W = L^1$. Show that the mapping $\varphi : V \to W$ given by $\varphi(x,y) = x$ has dense image, but is not of finite type.

18. Localization

IN order to study an affine variety, one can consider the variety as a whole and hence ask for global properties (like connectedness or homeomorphism type), but one can also concentrate on a specific point on that variety and study only a neighborhood of this point. To give an example, consider the following pictures of algebraic curves in $\mathbb{R}^2$.

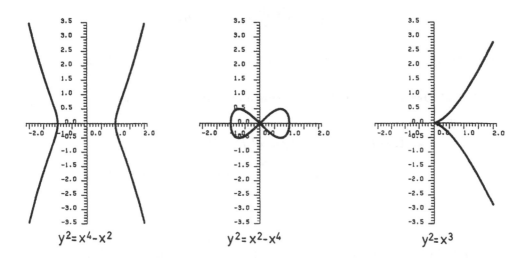

$$y^2 = x^4 - x^2 \qquad\qquad y^2 = x^2 - x^4 \qquad\qquad y^2 = x^3$$

Being asked which of the points on these curves are most "interesting", one undoubtedly will choose the origin in each case. What is it that makes this point special and distinguishes it from the other points on the curve? Our task will be to find a distinction between "regular" and "singular" points on a variety.

THE key notion we will need is that of the tangent space of a K-variety $V \subseteq L^n$ at a point $x \in V$. For any vector $a \neq 0$ in L^n, let l_a be the line through x in the direction a, i.e.,

$$l_a := x + La = \{x + ta \mid t \in L\}\,.$$

We want to define what it means that this line intersects V at x with a certain multiplicity m. To do so, we investigate the behavior of each polynomial $f \in \mathcal{I}(V)$ along the line l_a; i.e., we study the polynomial $F(t) := F_{x,a}(t) = f(x + ta)$ in one variable t. Expanding by powers of t and observing that $f(x) = 0$ since $f \in \mathcal{I}(V)$, this yields an equation

$$f(x + ta) = f_1(x, a)t + f_2(x, a)t^2 + \cdots + f_N(x, a)t^N$$

where the f_i's are polynomials in x and a. Now we can give the following definition.

(18.1) Definition. *In the situation described above, we say that the line l_a intersects V at x with multiplicity m if m is the smallest number with the property that $f_m(x, a) \neq 0$ for some $f \in \mathcal{I}(V)$. We say that l_a **touches** V at x if l_a intersects V at x with multiplicity $m \geq 2$, i.e., if $f_1(x, a) = 0$ for all $f \in \mathcal{I}(V)$.*

347

WE are now going to define the tangent space of the variety V at a point $x \in V$ as the set of all "direction vectors" $a \neq 0$ such that l_a touches V at x. Before we do so, let us observe that the polynomials f_i can be related to the derivatives of f by using Taylor's formula; namely,

$$k! \cdot f_k(x,a) = \sum_{i_1,\ldots,i_k=1}^{n} \frac{\partial^k f}{\partial x_{i_1} \cdots \partial x_{i_k}}(x) a_{i_1} \cdots a_{i_k}$$

for all k; in particular,

$$f_1(x,a) = \sum_{i=1}^{n} \frac{\partial f}{\partial x_i}(x) a_i = f'(x)a$$

where $f'(x) = \left(\frac{\partial f}{\partial x_1}(x), \ldots, \frac{\partial f}{\partial x_n}(x) \right)$ is sometimes called the gradient of f at x.

(18.2) Definition. *The* tangent space *of V at x is defined as*

$$
\begin{aligned}
T_x V &= \{ a \in L^n \mid l_a \text{ touches } V \text{ at } x \} \\
&= \{ a \in L^n \mid f'(x)a = 0 \text{ for all } f \in \mathcal{I}(V) \} \\
&= \{ (a_1, \ldots, a_n) \in L^n \mid \sum_{i=1}^{n} \frac{\partial f}{\partial x_i}(x) \cdot a_i = 0 \ \text{ for all } f \in \mathcal{I}(V) \} .
\end{aligned}
$$

TO fill definition (18.2) with some geometric intuition, let us consider the special case that V is a curve in $\mathbb{R}^2$. The tangent of V at a point x is the line l characterized by the property that in some sufficiently small neighborhood of x the only point of intersection between l and V is x itself, whereas any slight rotation of l around x produces a new point of intersection very close to x; i.e, by choosing a' sufficiently close to a we can arrange that the line $l' = x + La'$ has a point of intersection $x' = x + ta'$ with V such that $t = t(a') \neq 0$ is arbitrarily close to 0.

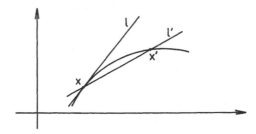

But this implies that $f(x + ta') = f(x') = 0$ for all $f \in \mathcal{I}(V)$ and hence

$$0 = \frac{1}{t} f(x + ta') = f_1(x,a') + f_2(x,a')t + \cdots + f_N(x,a')t^{N-1} ;$$

as $a' \to a$ and $t \to 0$, this equation becomes $f_1(x,a) = 0$, i.e., $f'(x)a = 0$ as in definition (18.2).

WE associated with each variety V and each point $x \in V$ the tangent space $T_x V$. We will show that this assignment is *functorial* in the sense that every polynomial mapping $\varphi : V \to W$ (and even every rational mapping having $x \in V$ in its domain) induces a linear map $\varphi'(x) : T_x V \to T_{\varphi(x)} W$. Intuitively, we think of $\varphi'(x)$ as the "Jacobian matrix" consisting of the partial derivatives of the component functions of φ. However, we should notice that there are some problems in defining the "partial derivatives" of a function on a variety, as is shown in the following example.

(18.3) Example. Let $V = \{(x,y) \in L^2 \mid y = x^2\}$ and define $\varphi : V \to L$ by either $\varphi(x,y) := x^2$ or $\varphi(x,y) := y$; clearly, both definitions yield the same polynomial function on V. Using the first definition and taking the partial derivatives in a naive way, we would obtain $\left(\frac{\partial \varphi}{\partial x}(x,y), \frac{\partial \varphi}{\partial y}(x,y)\right) = (2x, 0)$, using the second definition, we would obtain $\left(\frac{\partial \varphi}{\partial x}(x,y), \frac{\partial \varphi}{\partial y}(x,y)\right) = (0, 1)$; clearly, both results do not coincide. Thus the "gradient" $\varphi'(x,y) = \left(\frac{\partial \varphi}{\partial x}(x,y), \frac{\partial \varphi}{\partial y}(x,y)\right)$ of φ at a point $(x,y) \in V$ does not seem to make sense, because it depends on the particular representation of the function φ.

It is a key observation, however, that if we apply this "gradient" to a tangent vector $(a,b) \in T_{(x,y)} V$, then we always obtain the same result, no matter which representation of φ we chose. To see this for the two "possibilities" for the gradient given above, we note that $T_{(x,y)} V = \{(a,b) \in L^2 \mid -2xa + b = 0\}$ so that

$$(2x, 0) \binom{a}{b} = 2xa = b = (0,1) \binom{a}{b}$$ for all tangent vectors (a,b) at (x,y). As we will show in the following lemma, this is not an accident, but a general fact.

(18.4) Lemma. *Consider a K-variety $V \subseteq L^n$, a rational function $f : V \to L$ and a point $x \in \operatorname{dom} f$ on V. Then a map $f'(x) : T_x V \to L$ can unambiguously be defined as follows: If f has a representation $f = p/q$ of f in a neighborhood of x and if $a \in T_x V$ is a tangent vector at x, then*

$$f'(x)a := \left(\frac{\partial f}{\partial x_1}(x), \ldots, \frac{\partial f}{\partial x_n}(x)\right) \begin{pmatrix} a_1 \\ \vdots \\ a_n \end{pmatrix} := \sum_{i=1}^{n} \frac{\partial f}{\partial x_i}(x) \, a_i$$

$$= \sum_{i=1}^{n} \frac{\frac{\partial p}{\partial x_i}(x) q(x) - p(x) \frac{\partial q}{\partial x_i}(x)}{q(x)^2} \, a_i \, .$$

Proof. The definition is what we expect if we think of the familiar quotient rule of calculus; however, we have to show that this definition does not depend on the special choice of p and q in the representation $f = p/q$. Since all we have at hand is the notion of partial derivatives of formal polynomials, we have to reduce this problem to a problem concerning formal polynomials (and not functions on V). So suppose we have two different representations $f = p/q = P/Q$ of f in a neighborhood U of x; then $p(y)Q(y) - P(y)q(y) = 0$ for all $y \in U$. By the definition of the Zariski topology, this neighborhood may be assumed to have the form $U = D_h = \{y \in V \mid h(y) \neq 0\}$ for some polynomial function $h \in K[V]$. Hence $p(y)Q(y) - P(y)q(y) = 0$ whenever $h(y) \neq 0$, which is tantamount to saying that $(pQ - Pq)h = 0$ on all of V, i.e., $(pQ - Pq)h \in \mathcal{I}(V)$. Since $a \in T_x V$, we have $g'(x)a = 0$ for all $g \in \mathcal{I}(V)$; applying this with $g := (pQ - Pq)h$, we obtain

$$0 = \sum_{i=1}^{n} \frac{\partial h}{\partial x_i}(x) \underbrace{((pQ - Pq)(x))}_{= 0} a_i + \underbrace{h(x)}_{\neq 0} \sum_{i=1}^{n} ((\frac{\partial p}{\partial x_i}Q + p\frac{\partial Q}{\partial x_i} - \frac{\partial P}{\partial x_i}q - P\frac{\partial q}{\partial x_i})(x)) \, a_i \, .$$

Dividing by $h(x)q(x)Q(x) \neq 0$ and observing that $p(x)/q(x) = P(x)/Q(x)$, this becomes

$$0 = \sum_{i=1}^{n} (\frac{\frac{\partial p}{\partial x_i}(x)}{q(x)} + \frac{P(x)\frac{\partial Q}{\partial x_i}(x)}{Q(x)^2} - \frac{\frac{\partial P}{\partial x_i}(x)}{Q(x)} - \frac{p(x)\frac{\partial q}{\partial x_i}(x)}{q(x)^2}) \cdot a_i$$

which is exactly the desired relation

$$\sum_{i=1}^{n} \frac{\frac{\partial p}{\partial x_i}(x)q(x) - p(x)\frac{\partial q}{\partial x_i}(x)}{q(x)^2} \cdot a_i = \sum_{i=1}^{n} \frac{\frac{\partial P}{\partial x_i}(x)Q(x) - P(x)\frac{\partial Q}{\partial x_i}(x)}{Q(x)^2} \cdot a_i \, .$$

∎

(18.5) Proposition. (a) *Let $V \subseteq L^n$ and $W \subseteq L^m$ be K-varieties. Each rational map $\varphi : V \to W$ with $x \in \mathrm{dom}\,\varphi$ induces a linear mapping $\varphi'(x) : T_x V \to T_{\varphi(x)}W$ where*

$$\varphi'(x)a := \begin{pmatrix} \frac{\partial \varphi_1}{\partial x_1}(x) & \cdots & \frac{\partial \varphi_1}{\partial x_n}(x) \\ \vdots & & \vdots \\ \frac{\partial \varphi_m}{\partial x_1}(x) & \cdots & \frac{\partial \varphi_m}{\partial x_n}(x) \end{pmatrix} \begin{pmatrix} a_1 \\ \vdots \\ a_n \end{pmatrix} \, .$$

(b) *If $\varphi : U \to V$ and $\psi : V \to W$ are rational maps such that $x \in \mathrm{dom}\,\varphi$ and $\varphi(x) \in \mathrm{dom}\,\psi$, then*

$$(\psi \circ \varphi)'(x) = \psi'(\varphi(x)) \circ \varphi'(x) : T_x U \to T_{\psi(\varphi(x))}W \quad \text{(chain rule)} \, .$$

(c) *If $\varphi : V \to W$ is a birational isomorphism such that $x \in \mathrm{dom}\,\varphi$ and $\varphi(x) \in \mathrm{dom}\,\varphi^{-1}$, then $\varphi(x) : T_x V \to T_{\varphi(x)}W$ is a vector space isomorphism.*

Proof. (a) Due to lemma (18.4) (applied to the component functions $\varphi_1, \ldots, \varphi_m$ of φ), we have a well-defined mapping $\varphi'(x) : T_x V \to L^m$; it remains to show that $\varphi'(x)a \in T_{\varphi(x)}W$ whenever $a \in T_x V$; i.e., that $g'(\varphi(x))\varphi'(x)a = 0$ for all $g \in \mathcal{I}(W)$. Now if $g \in \mathcal{I}(W)$ then $g \circ \varphi$ is the zero-function on V; so once the chain rule in part (b) is established, we can conclude that $0 = (g \circ \varphi)'(x) = g'(\varphi(x))\varphi'(x)$ (as a linear map defined on $T_x V$), as desired.

(b) This follows immediately from the fact that the definition of the "derivative" of a rational map in (18.4) and (18.5)(a) reduces everything to taking formal derivatives of rational functions, where the chain rule is known to hold.

(c) Simply apply part (b) with $\psi := \varphi^{-1}$ along with the obvious fact that $\mathrm{id}'(x) = \mathrm{id}$ for all x. ∎

(18.6) Proposition. *For any $r \in \mathbb{N}_0$ the set $S_r := \{p \in V \mid \dim T_p V \geq r\}$ is Zariski-closed. (This is sometimes expressed by saying that the mapping $x \mapsto \dim T_x V$ from V into $\mathbb{N}_0$ is upper semi-continuous.)*

Proof. Let $\mathcal{I}(V) = \langle f_1, \ldots, f_m \rangle$; then

$$T_p V = \{(a_1, \ldots, a_n) \in L^n \mid \begin{pmatrix} \frac{\partial f_1}{\partial x_1}(p) & \cdots & \frac{\partial f_1}{\partial x_n}(p) \\ \vdots & & \vdots \\ \frac{\partial f_m}{\partial x_1}(p) & \cdots & \frac{\partial f_m}{\partial x_n}(p) \end{pmatrix} \begin{pmatrix} a_1 \\ \vdots \\ a_n \end{pmatrix} = \begin{pmatrix} 0 \\ \vdots \\ 0 \end{pmatrix} \}$$

$$= \ker \begin{pmatrix} \frac{\partial f_1}{\partial x_1}(p) & \cdots & \frac{\partial f_1}{\partial x_n}(p) \\ \vdots & & \vdots \\ \frac{\partial f_m}{\partial x_1}(p) & \cdots & \frac{\partial f_m}{\partial x_n}(p) \end{pmatrix}$$

so that $\dim T_p V = \dim \ker \left(\frac{\partial f_i}{\partial x_j}(p) \right)_{i,j} = n - \operatorname{rank} \left(\frac{\partial f_i}{\partial x_j}(p) \right)_{i,j}$. Therefore, $p \in S_r$ if and only if $\operatorname{rank} \left(\frac{\partial f_i}{\partial x_j}(p) \right)_{i,j} \leq n - r$ which is the case if and only if every $(n-r+1) \times (n-r+1)$-minor of $\left(\frac{\partial f_i}{\partial x_j}(p) \right)_{i,j}$ vanishes. This last condition shows that S_r is given by a system of polynomial equations over K and hence is a K-variety, i.e., a Zariski-closed set. $\blacksquare$

(18.7) Definition. *Let V be an irreducible K-variety in L^n. Then the* **algebraic dimension** *of V is defined as*

$$\dim_{\mathrm{alg}} V = \min \{ \dim T_x V \mid x \in V \} .$$

A point $x \in V$ is called **regular** *if $\dim T_x V = \dim_{\mathrm{alg}} V$ and* **singular** *if $\dim T_x V > \dim_{\mathrm{alg}} V$.*

(18.8) Proposition. *Let V be a K-variety in L^n. Then the set V_{reg} of regular points is a nonempty open subset of V. If V is irreducible, then V_{reg} is dense in V.*

Proof. Let $d = \dim_{\mathrm{alg}} V$. By the definition of dimension, this means in the notation of (18.6) that $S_d = V$ and $S_{d+1} \subsetneq V$. Then $V_{\mathrm{reg}} = S_d \setminus S_{d+1} = V \setminus S_{d+1}$, which is non-empty and open since S_{d+1} is closed by (18.6). The last statement is clear because an open nonempty subset of an irreducible variety is automatically dense. $\blacksquare$

(18.9) Examples. (a) Let $V = \{(x, y) \in \mathbb{R}^2 \mid x^2 + y^2 = 0\}$. Then V consists of the single point $p = (0, 0)$, and we have $T_p V = \mathbb{R}^2$ which implies that $\dim_{\mathrm{alg}} V = 2$. This is an example in which the "algebraic dimension" of a variety differs in a striking way from the "geometric dimension" of this variety as defined in (17.21); we have $\dim_{\mathrm{geo}} V = 0$, of course.

(b) Let $V = \{(x, y) \in \mathbb{C}^2 \mid x^2 + y^2 = 0\}$. Then V consists of all points (x, ix) and $(x, -ix)$ with $x \in \mathbb{C}$. As in part (a), we have $T_{(0,0)} V = \mathbb{C}^2$; however, if $x \neq 0$, then $T_{(x, \pm ix)} V = \{(a, b) \in \mathbb{C}^2 \mid 2xa \pm 2ixb = 0\} = \{(a, b) \in \mathbb{C}^2 \mid a \pm ib = 0\}$ is a one-dimensional subspace of $\mathbb{C}^2$. Hence $\dim_{\mathrm{alg}} V = 1$; the point $(0, 0)$ is singular, all other points of V are regular.

(18.10) Proposition. *Let $K \subseteq L$ be fields such that K is perfect and L is algebraically closed. If $V = \mathcal{V}(f) \subseteq L^n$ where $f \in K[X_1, \ldots, X_n]$ is irreducible, then $\dim_{\mathrm{alg}} V = n - 1$.*

Proof. We have $T_p V = \{a \in L^n \mid \sum_{i=1}^n \frac{\partial f}{\partial x_i}(p) a_i = 0\}$. If $\frac{\partial f}{\partial x_i}(p) = 0$ for all i, then $T_p V = L^n$; otherwise $T_p V$ is a hyperplane in L^n and hence has dimension $n - 1$. Thus the claim is proved if we can show that there are indeed points p with $\frac{\partial f}{\partial x_i}(p) \neq 0$ for at least one index i. Suppose that this is not the case to obtain a contradiction. Then the polynomials $\frac{\partial f}{\partial x_i}$ all belong to $\mathcal{I}(V) = (\mathcal{I} \circ \mathcal{V})(\langle f \rangle) = \sqrt{\langle f \rangle} = \langle f \rangle$; here we used that L is algebraically closed (by invoking Hilbert's Nullstellensatz) and that f is irreducible (by utilizing the fact that $\langle f \rangle$ is a prime and hence a radical ideal). This means that f divides all the partial derivatives $\frac{\partial f}{\partial x_i}$ in $K[X_1, \ldots, X_n]$. Since $\deg \frac{\partial f}{\partial x_i} < \deg f$, this is only possible if $\frac{\partial f}{\partial x_i} = 0$ for $1 \leq i \leq n$ (as polynomials, not just as functions on V!). Hence f is inseparable in each variable X_i. If $\operatorname{char} K = 0$ this is impossible; if $\operatorname{char} K = p \neq 0$, this implies that f has the form $f(X_1, \ldots, X_n) = \sum_{i_1, \ldots, i_n} a_{i_1, \ldots, i_n} X_1^{p i_1} \cdots X_n^{p i_n}$. Using the hypothesis that K is perfect, we can pick elements $b_{i_1, \ldots, i_n} \in K$ with $b_{i_1, \ldots, i_n}^p = a_{i_1, \ldots, i_n}$. Then

$$f(X_1, \ldots, X_n) = \sum_{i_1, \ldots, i_n} b_{i_1, \ldots, i_n}^p X_1^{p i_1} \cdots X_n^{p i_n} = \left(\sum_{i_1, \ldots, i_n} b_{i_1, \ldots, i_n} X_1^{i_1} \cdots X_n^{i_n} \right)^p$$

which contradicts the irreducibility of f. ∎

(18.11) Theorem. *If V and W are birationally isomorphic irreducible varieties, then $\dim_{\mathrm{alg}} V = \dim_{\mathrm{alg}} W$.*

Proof. Let $\varphi : V \to W$ be a birational isomorphism so that both φ and φ^{-1} are dominant maps. Since the sets V_{reg} and W_{reg} of regular elements in V and W, respectively, are open and dense by (18.8) and since the intersection of two open and dense sets is again open and dense, the set $V_{\mathrm{reg}} \cap \operatorname{dom} \varphi \cap \varphi^{-1}((\operatorname{dom} \varphi^{-1}) \cap W_{\mathrm{reg}})$ is not empty. This means that there is an element $x \in V_{\mathrm{reg}}$ such that $\varphi(x) \in W_{\mathrm{reg}}$. Since then $T_x V$ and $T_{\varphi(x)} W$ are isomorphic by (18.5)(c), we have $\dim_{\mathrm{alg}} V = \dim T_x V = \dim T_{\varphi(x)} W = \dim_{\mathrm{alg}} W$. ∎

(18.12) Theorem. *Let V be an irreducible K-variety in L^n where K is perfect and where L is algebraically closed. Then*

$$\dim_{\mathrm{alg}} V = \operatorname{trdeg}(K(V) : K) = \dim K[V] = \dim_{\mathrm{geo}} V .$$

Proof. Let us derive the first equality for the special case that V is a hypersurface $V = \mathcal{V}(f)$ where $f \in K[X_1, \ldots, X_n]$ is irreducible. Then $\dim_{\mathrm{alg}} V = n - 1$ by (18.10). On the other hand, $K[V] \cong K[X_1, \ldots, X_n]/\langle f \rangle$. If f involves the variable X_n, say, this implies $K(V) \cong K(X_1, \ldots, X_{n-1})[X_n]/\langle f \rangle$; this is a field emerging from K by adjoining the algebraically independent elements $X_1, \ldots, X_{n-1}$ and then performing a simple algebraic extension, hence has transcendence degree $n - 1$ over K. This establishes the first equality for the special case of a hypersurface. Next we use the fact that an arbitrary K-variety is birationally isomorphic to a hypersurface; this fact was proved in (17.46). Now if V is birationally isomorphic to the hypersurface V_0, then $\dim_{\text{alg}} V = \dim_{\text{alg}} V_0$ by (18.11); on the other hand, $K(V) \cong K(V_0)$ as fields by (17.45) and hence $\text{trdeg}\big(K(V) : K\big) = \text{trdeg}(K(V_0) : K)$.

The second equality is simply Noether's normalization theorem (16.3).

The third equality comes from the 1-1-correspondence between prime ideals of $K[V]$ and irreducible subvarieties of V given by Hilbert's Nullstellensatz. ∎

INTUITIVELY, one should be able to determine the tangent space at a point x by only knowing a neighborhood of x in V (and not necessarily all of V). This is indeed the case; to prove this statement we introduce the local ring at a point.

(18.13) Definition. *Let V be a K-variety in L^n and let $x \in V$. The **local ring** $\mathcal{O}_x = \mathcal{O}_{x,V}$ of x is the set of all functions which are defined on some neighborhood of x and can, on this neighborhood, be written as $f(y) = p(y)/q(y)$ with $p, q \in K[V]$, $q(x) \neq 0$. Addition and multiplication in $\mathcal{O}_x$ are defined on the common domain of any two functions in $\mathcal{O}_x$ which is again a neighborhood of x.*

THE ring $\mathcal{O}_x$ has a peculiar algebraic property.

(18.14) Proposition. *Let $V \subseteq L^n$ be a K-variety and let $x \in V$ be a point in V. Then the local ring $\mathcal{O}_x$ at x has a unique maximal ideal, namely $\mathcal{M}_x := \{f \in \mathcal{O}_x \mid f(x) = 0\}$.*

Proof. It is obvious that $\mathcal{M}_x$ is an ideal of $\mathcal{O}_x$. On the other hand, every element $f \notin \mathcal{O}_x \setminus \mathcal{M}_x$ is a unit because $f(x) \neq 0$ implies $1/f \in \mathcal{O}_x$. This implies that every proper ideal of $\mathcal{O}_x$ is contained in $\mathcal{M}_x$. ∎

WE now prove that the tangent space $T_x V$ can be determined by only knowing the local ring $\mathcal{O}_x$.

(18.15) Proposition. *Let V be a K-variety in L^n and let $x \in V$. Let $\mathcal{O}_x$ be the local ring at x with its maximal ideal $\mathcal{M}_x = \{f \in \mathcal{O}_x \mid f(x) = 0\}$. Then*

$$\theta : \begin{array}{ccc} \mathcal{M}_x & \to & (T_x V)^\star \\ f & \mapsto & f'(x) \end{array}$$

is a surjective L-linear mapping with $\ker \theta = \mathcal{M}_x^2$. Consequently, the homomorphism theorem for vector spaces implies that

$$(T_x V)^\star \cong \frac{\mathcal{M}_x}{\mathcal{M}_x^2} \, .$$

Proof. Lemma (18.4) shows that θ is a well-defined mapping, and the L-linearity follows immediately from the fact that taking derivatives of formal polynomials is linear and that $f'(x)$ is defined in terms of the partial derivatives of any formal polynomial underlying f.

Let us show that θ is onto. Given $\lambda \in (T_x V)^*$, we can extend λ to an element of $(L^n)^*$, say $\lambda(x_1, \ldots, x_n) = \lambda_1 x_1 + \cdots + \lambda_n x_n$. Define $f \in \mathcal{O}_x$ by $f(y) = \lambda(y) - \lambda(x)$; then $f'(y) = \lambda$ for all y, in particular $f'(x) = \lambda$.

Let us now show that $\ker \theta = \mathcal{M}_x^2$. Suppose that $f \in \ker \theta$, i.e., $f'(x) = 0$. Writing $f = F/G$, we have $F(x) = f(x)G(x) = 0$ and $F'(x) = G(x)f'(x) + f(x)G'(x) = 0$ in the sense that $F'(x)a = 0$ for all $a \in T_x V$. Since $T_x V = \{a \in L^n \mid g'(x)a = 0$ for all $g \in \mathcal{I}(V)\} = \{g'(x) \mid g \in \mathcal{I}(V)\}^\perp$, this means $F'(x) \in (T_x V)^\perp = \{g'(x) \mid g \in \mathcal{I}(V)\}^{\perp\perp} = \{g'(x) \mid g \in \mathcal{I}(V)\}$, say $F'(x) = \gamma'(x)$ with $\gamma \in \mathcal{I}(V)$. Since also $F(x) = \gamma(x) = 0$, this shows that

$$F(X_1, \ldots, X_n) = \gamma(X_1, \ldots, X_n) + \underbrace{\sum_{i,j=1}^n a_{ij}(X_i - x_i)(X_j - x_j) + \text{higher-order terms}}_{=:\ H(X_1, \ldots, X_n)}$$

as formal polynomials. Clearly, the polynomial function $h : V \to L$ induced by H belongs to $\mathcal{M}_x^2$, and since $\gamma \in \mathcal{I}(V)$, this implies $f = \frac{F}{G} = \frac{1}{G}F = \frac{1}{G}h \in \frac{1}{G}\mathcal{M}_x^2 \subseteq \mathcal{M}_x^2$. ∎

THIS last result allows one to derive (18.5)(a) in a more algebraic way. In fact, a rational map $\varphi : V \to W$ with $x \in \text{dom}\,\varphi$ clearly induces a mapping

$$\varphi^* : \begin{array}{ccc} \mathcal{O}_{\varphi(x)} & \to & \mathcal{O}_x \\ g & \mapsto & g \circ \varphi \end{array}$$

which necessarily maps $\mathcal{M}_{\varphi(x)}$ into $\mathcal{M}_x$ and hence gives rise to a linear mapping from $\mathcal{M}_{\varphi(x)}/\mathcal{M}_{\varphi(x)}^2 \cong (T_{\varphi(x)}W)^*$ into $\mathcal{M}_x/\mathcal{M}_x^2 \cong (T_x V)^*$.

THERE is a more algebraic method of determining the local ring at a point x, indeed a method that lends itself to be used in different contexts.

Suppose that a function has two representations $p_1(y)/q_1(y) = p_2(y)/q_2(y)$ on some neighborhood of x. By the definition of the Zariski topology, this neighborhood may be taken in the form $D_h = \{x \in V \mid h(x) \neq 0\}$ for some polynomial function $h \in K[V]$. But $p_1/q_1 = p_2/q_2$ on D_h means $p_1(y)q_2(y) = p_2(y)q_1(y)$ whenever $h(y) \neq 0$ which is tantamount to saying $(p_1q_2 - p_2q_1)h = 0$. Hence we have the following (algebraically nice) way of defining the local ring at x.

(18.16) Definition. *Let $V \subseteq L^n$ be a K-variety and let $x \in V$ be a point in V. Consider the set X of all pairs (p, q) with $p, q \in K[V]$ and $q(x) \neq 0$. Define an equivalence relation $\sim$ on X via*

$$(p_1, q_1) \sim (p_2, q_2) \ :\Longleftrightarrow\ (p_1q_2 - p_2q_1)h = 0 \text{ for some } h \in K[V] \text{ with } h(x) \neq 0$$

*and denote the equivalence class of (p, q) by p/q. Then the **local ring** of V at x is defined as the set $X/\sim$ of equivalence classes with the operations*

$$\frac{p_1}{q_1} + \frac{p_2}{q_2} := \frac{p_1q_2 + p_2q_1}{q_1q_2} \qquad \text{and} \qquad \frac{p_1}{q_1} \cdot \frac{p_2}{q_2} := \frac{p_1p_2}{q_1q_2}\ .$$

IT turns out that this procedure can be vastly generalized.

(18.17) Theorem. *Let R be a commutative ring and let $S \neq \emptyset$ be a multiplicatively closed subset of R with $0 \notin S$.*

(a) *The relation*

$$(r, s) \sim (r', s') \quad :\Longleftrightarrow \quad (rs' - r's)x = 0 \text{ for some } x \in S$$

is an equivalence relation on $R \times S$; we denote the equivalence class of (r, s) by r/s and the set of all equivalence classes by $S^{-1}R$.

(b) *$S^{-1}R$ becomes a commutative ring with identity if we define addition and multiplication by*

$$\frac{r}{s} + \frac{r'}{s'} := \frac{rs' + r's}{ss'} \qquad and \qquad \frac{r}{s} \cdot \frac{r'}{s'} := \frac{rr'}{ss'} \; .$$

(c) *If R has no zero-divisors then $S^{-1}R$ is an integral domain.*

(d) *For each $r \in R$ the element $(rs)/s$ does not depend on the particular choice of $s \in S$. The mapping $f : R \to S^{-1}R$ given by $r \mapsto (rs)/s$ is a well-defined ring homomorphism with $\ker f = \{r \in R \mid rs = 0 \text{ for some } s \in S\}$. If $s \in S$ then $f(s)$ is a unit. If R is an integral domain then f is an embedding so that we can consider R as a subring of $S^{-1}R$ and $S^{-1}R$ as a subring of the quotient field of R.*

Proof. Even though the statement of this theorem is relatively long, the proof is straightforward and is left as an exercise. (See problem 3 below.) ∎

(18.18) Examples. (a) Let V be an affine variety. Then the local ring $\mathcal{O}_x$ at a point $x \in V$ can be written as $\mathcal{O}_x = S^{-1}R$ where $R = K[V]$ and $S = \{q \in K[V] \mid q(x) \neq 0\}$.

(b) If R is a commutative ring and if P is a prime ideal of R, then $S := R \backslash P$ is multiplicatively closed with $0 \notin S$. In this case we write R_P instead of $S^{-1}R$ and call R_P the **localization** of the ring R at the prime ideal P. Example (a) is a special case of this construction because if x is a point on a K-variety V then $P := \{q \in K[V] \mid q(x) = 0\}$ is a prime ideal of $K[V]$.

(c) If R is an integral domain and $S = R \backslash \{0\}$, then $S^{-1}R$ is the quotient field of R.

(d) Let R be a commutative ring and let S be the set of all elements of R which are not zero-divisors. Then S is multiplicatively closed, and $S^{-1}R$ is called the **full ring of fractions** of R. If R is an integral domain, then $S = R \backslash \{0\}$, and we are in the situation of example (a).

IN the next theorem we relate the ideal structure of the ring $S^{-1}R$ to that of R.

(18.19) Theorem. *Let R be a commutative ring and $S \neq \emptyset$ a multiplicatively closed subset of R with $0 \notin S$.*

(a) There is a correspondence between the ideals of R and the ideals of $S^{-1}R$; namely, we associate with each ideal $I \trianglelefteq R$ an ideal $S^{-1}I \trianglelefteq S^{-1}R$ and with each ideal $J \trianglelefteq S^{-1}R$ and ideal $J_R \trianglelefteq R$ as follows:

$$S^{-1}I := \{\tfrac{x}{s} \mid x \in I, s \in S\} \trianglelefteq S^{-1}R ;$$

$$J_R := \{r \in R \mid \tfrac{rs}{s} \in J \text{ for some (hence all) } s \in S\} \trianglelefteq R .$$

Note that if R has no zero-divisors then we can embed R into $S^{-1}R$ via f and thus consider R as a subring of S; if we do so, then $J_R = J \cap R$.

ideals of R	$I \longmapsto S^{-1}I$	ideals of $S^{-1}R$
	$J_R \longmapsfrom J$	

(b) The composition of the two assignments introduced in part (a) are given by

$$S^{-1}(J_R) = J \quad \text{and}$$
$$(S^{-1}I)_R = \{r \in R \mid rs \in I \text{ for some } s \in S\} \supseteq I$$

for all ideals $J \trianglelefteq S^{-1}R$ and $I \trianglelefteq R$. In particular, every ideal of $S^{-1}R$ is of the form $S^{-1}I$ for some ideal $I \trianglelefteq R$.

(c) If $I \trianglelefteq R$ then $S^{-1}I = S^{-1}R$ if and only if $S \cap I \neq \emptyset$.

(d) If $(I_\alpha)_{\alpha \in A}$ is a family of ideals of R, then

$$S^{-1}(\sum_{\alpha \in A} I_\alpha) = \sum_{\alpha \in A}(S^{-1}I_\alpha),$$

$$S^{-1}(\prod_{\alpha \in A} I_\alpha) = \prod_{\alpha \in A}(S^{-1}I_\alpha),$$

$$S^{-1}(\bigcap_{\alpha \in A} I_\alpha) \subseteq \bigcap_{\alpha \in A}(S^{-1}I_\alpha).$$

Moreover, if $I \trianglelefteq R$ then $S^{-1}(\sqrt{I}) = \sqrt{S^{-1}I}$.

(e) If $(A_i)_{i \in I}$ is a family of ideals of $S^{-1}R$, then

$$\sum_{i \in I}(A_i)_R \subseteq (\sum_{i \in I} A_i)_R,$$

$$\prod_{i \in I}(A_i)_R \subseteq (\prod_{i \in I} A_i)_R,$$

$$(\bigcap_{i \in I} A_i)_R = \bigcap_{i \in I}(A_i)_R.$$

Moreover, if $A \trianglelefteq S^{-1}R$ then $(\sqrt{A})_R = \sqrt{A_R}$.

(f) Let Q be a P-primary ideal of R. If $P \cap S \neq \emptyset$ then $S^{-1}Q = S^{-1}R$. If $P \cap S = \emptyset$ then $S^{-1}Q$ is $S^{-1}P$-primary, and $(S^{-1}Q)_R = Q$.

(g) The correspondence given in part (a) can be restricted to a one-to-one correspondence between the prime (primary) ideals of R disjoint with S and the prime (primary) ideals of $S^{-1}R$.

Proof. (a) If I is an ideal of R, then $S^{-1}R$ is an ideal of $S^{-1}R$; in fact, if $x_1, x_2 \in I$, $s_1, s_2 \in S$ and $r \in R$ then $(x_1/s_1) + (x_2/s_2) = (s_2 x_1 + s_1 x_2)/(s_1 s_2)$ and $(x_1/s_1) \cdot (x_2/s_2) = (x_1 x_2)/(s_1 s_2)$ both lie in $S^{-1}R$, because $s_2 x_1 + s_1 x_2 \in Rx_1 + Rx_2 \subseteq I$, $x_1 r \in Rx_1 \subseteq I$ and $s_1 s_2 \in SS \subseteq S$. On the other hand, if J is an ideal of $S^{-1}R$ then J_R is an ideal of R; this follows immediately from (6.5) because we can write $J_R = f^{-1}(J)$ with the homomorphism $f : R \to S^{-1}R$ defined in (18.17)(d).

(b) Let us show first that $S^{-1}(J_R) = J$ for each ideal J of $S^{-1}R$. If $r \in J_R$ and $s \in S$, then $(rs)/s \in J$ (since $r \in J_R$) and $r/s = ((rs)/s) \cdot (s/s^2) \in J \cdot (S^{-1}R) \subseteq J$ since J is an ideal of $S^{-1}R$; this proves the inclusion $\subseteq$. Conversely, let $r/s \in J$. Since J is an ideal, we have $(r/s) \cdot (s^2/s) \in J$, i.e., $(rs^2)/s^2 \in J$. Since $s^2 \in S$, this shows that $r \in J_R$ and hence $r/s \in S^{-1}(J_R)$.

Let us show now that $r \in (S^{-1}R)_R$ if and only if $rs \in I$ for some $s \in S$. If $r \in (S^{-1}I)_R$ then $(rs')/s' \in S^{-1}I$ for some $s' \in S$, say $(rs')/s' = x/s''$ with $x \in I$ and $s'' \in S$. But this means $(rs's'' - xs')s''' = 0$ for some $s''' = 0$. Letting $s := s's''s'''$, we obtain $rs = xs's''' \in Rx \subseteq I$. Conversely, if $rs \in I$ for some $s \in S$, then $(rs)/s \in S^{-1}I$ which implies $r \in (S^{-1}I)_R$.

(c) If there is an element $s \in S \cap I$ then $s/s \in S^{-1}I$ so that the ideal $S^{-1}I$ contains the identity element of $S^{-1}R$ and hence must be all of $S^{-1}R$. Conversely, suppose $S^{-1}R = S^{-1}I$. Take any element $s \in S$. Then $s/s \in S^{-1}I$; hence there are elements $x \in I$ and $s' \in S$ with $s/s = x/s'$, say $(ss' - xs)s'' = 0$ with $s'' \in S$. Consequently, $ss's'' = xss'' \in S \cap I$.

(d) Let A_0 be a finite subset of A, say $|A_0| = n$, and let $s \in S$. Then if $x_\alpha \in I_\alpha$, we have

$$\frac{\sum_{\alpha \in A_0} x_\alpha}{s} = \sum_{\alpha \in A_0} \frac{x_\alpha}{s} \in \sum_{\alpha \in A_0} (S^{-1}I_\alpha) \subseteq \sum_{\alpha \in A} (S^{-1}I_\alpha) \quad \text{and}$$

$$\frac{\prod_{\alpha \in A_0} x_\alpha}{s} = \frac{s^n}{s} \prod_{\alpha \in A_0} \frac{x_\alpha}{s} \in \frac{s^n}{s} \prod_{\alpha \in A_0} (S^{-1}I_\alpha) \subseteq \prod_{\alpha \in A} (S^{-1}I_\alpha).$$

This proves the inclusions $S^{-1}(\sum_\alpha I_\alpha) \subseteq \sum_\alpha (S^{-1}I_\alpha)$ and $S^{-1}(\prod_\alpha I_\alpha) \subseteq \prod_\alpha (S^{-1}I_\alpha)$. Conversely, given an element $\sum_{\alpha \in A_0}(x_\alpha/s_\alpha)$ in $\sum_\alpha (S^{-1}I_\alpha)$, let $\sigma := \prod_{\alpha \in A_0} s_\alpha$ and $\sigma_\alpha := \prod_{\beta \neq \alpha} s_\beta$; then $\sum_{\alpha \in A_0}(x_\alpha/s_\alpha) = (\sum_{\alpha \in A_0} \sigma_\alpha x_\alpha)/s \in S^{-1}(\sum_{\alpha \in A_0} I_\alpha) \subseteq S^{-1}(\sum_{\alpha \in A} I_\alpha)$. Similarly, $\prod_{\alpha \in A_0}(x_\alpha/s_\alpha) = (\prod_{\alpha \in A_0} x_\alpha)/s \in S^{-1}(\prod_{\alpha \in A_0} I_\alpha) \subseteq S^{-1}(\prod_{\alpha \in A} I_\alpha)$. The last inclusion $S^{-1}(\bigcap_\alpha I_\alpha) \subseteq \bigcap_\alpha (S^{-1}I_\alpha)$ is trivial.

Now let I be an ideal of R. Let $x \in \sqrt{I}$ (say $x^n \in I$) and $s \in S$; then $(x/s)^n = x^n/s^n \in S^{-1}I$ so that $x/s \in \sqrt{S^{-1}I}$. This proves the inclusion $S^{-1}\sqrt{I} \subseteq \sqrt{S^{-1}I}$. To prove the converse inclusion, let $r/s \in \sqrt{S^{-1}I}$, say $r^n/s^n \in S^{-1}I$. Then there are elements $x \in I$ and $s' \in S$ with $r^n/s^n = x/s'$, i.e., $(r^n s' - xs^n)s'' = 0$ for some $s'' \in S$. Consequently, $r^n s's'' = xs^n s'' \in I$ which implies $(rs's'')^n \in I$ (because I is an ideal) and hence $rs's'' \in \sqrt{I}$. But then $r/s = (rs's'')/(ss's'') \in S^{-1}\sqrt{I}$.

(e) Let I_0 be a finite subset of I, say $|I_0| = n$, and let $s \in S$. If $a_i \in (A_i)_R$ then $(a_i s)/s \in A_i$ and hence $(\sum_{i \in I_0} a_i)s/s = \sum_{i \in I_0}(a_i s)/s \in \sum_{i \in I_0} A_i \subseteq \sum_{i \in I} A_i$ so that $\sum_{i \in I_0} a_i \in (\sum_{i \in I} A_i)_R$. Similarly, $(\prod_{i \in I_0} a_i)s^n/s^n = \prod_{i \in I_0}(a_i s)/s \in \prod_{i \in I_0} A_i \subseteq \prod_{i \in I} A_i$ so that $\prod_{i \in I_0} a_i \in (\prod_{i \in I} A_i)_R$. To prove the equation $(\bigcap_i A_i)_R = \bigcap_i (A_i)_R$, we simply observe that

$$r \in \left(\bigcap_{i \in I} A_i \right)_R \iff \frac{rs}{s} \in \bigcap_{i \in I} A_i \iff \frac{rs}{s} \in A_i \text{ for all } i$$

$$\iff r \in (A_i)_R \text{ for all } i \iff r \in \bigcap_{i \in I} (A_i)_R.$$

Finally, let us show that $(\sqrt{A})_R = \sqrt{A_R}$ for each ideal $A \trianglelefteq S^{-1}R$. Let $s \in S$. Then $r \in (\sqrt{A})_R$ if and only if $(rs)/s \in \sqrt{A}$, i.e., if $(r^n s^n)/s^n \in A$ for some $n \in \mathbb{N}$. But this means exactly that $r^n \in A_R$ for some $n \in \mathbb{N}$, i.e., that $r \in \sqrt{A_R}$.

(f) Suppose that Q is P-primary. Let us show first that $(S^{-1}Q)_R = Q$ in this case. By part (b) we know already that $Q \subseteq (S^{-1}Q)_R$. Conversely, let $r \in (S^{-1}Q)_R$; then $rs \in Q$ for some $s \in S$, due to part (b). If we had $r \notin Q$, we would obtain $s^n \in Q$ for some $n \in \mathbb{N}$ and hence the contradiction $s^n \in Q \cap S = \emptyset$; this shows that $r \in Q$.

If $P \cap S \neq \emptyset$, then part (c) shows that $S^{-1}R = S^{-1}P = S^{-1}\sqrt{Q} = \sqrt{S^{-1}Q}$. But then (6.9)(g) (which can be applied because $S^{-1}R$ is a ring with identity) implies that $S^{-1}R = S^{-1}Q$. Let us consider the second case, namely, that $P \cap S = \emptyset$. To show that $S^{-1}Q$ is $S^{-1}P$-primary, it is enough to show that $S^{-1}Q$ is primary, because $\sqrt{S^{-1}Q} = S^{-1}(\sqrt{Q}) = S^{-1}P$. Suppose that $(x_1/s_1)(x_2/s_2) \in S^{-1}Q$ but $(x_1/s_1) \notin S^{-1}Q$. Then $x_1 \notin Q$ and $(x_1 x_2)/(s_1 s_2) = q/s$ for some $q \in Q$, $s \in S$ so that $(sx_1 x_2 - qs_1 s_2)t = 0$ for some element $t \in S$ which implies $sx_1 x_2 t = qs_1 s_2 t \in Q$. Suppose $x_2^n \notin Q$ for all $n \in \mathbb{N}$; then $sx_1 t \in Q$, and since $x_1 \notin Q$ this implies $(st)^m \in Q$ for some m and hence $st \in \sqrt{Q} = P$, contradicting the fact that $S \cap P = \emptyset$. This shows that $x_2^n \in Q$ for some n. Consequently, $(x_2/s_2)^n = x_2^n/s_2^n \in S^{-1}Q$.

(g) If $I \trianglelefteq R$ is a prime (primary) ideal disjoint with S, then $S^{-1}I \trianglelefteq S^{-1}R$ is also a prime ideal, due to part (f). Conversely, if $J \trianglelefteq S^{-1}R$ is a prime (primary) ideal of $S^{-1}R$, then $J_R = f^{-1}(J)$ is a prime ideal of R due to (6.5) which is necessarily disjoint with S by part (c). Since $(S^{-1}I)_R = I$ by part (f) and $S^{-1}(J_R) = J$ by part (b), this gives the claim. ∎

LET P be a prime ideal of a commutative ring R and let $S := R \setminus P$. Then (18.19)(g) shows that the prime ideals of the localization $R_P = S^{-1}R$ are in 1-1-correspondence with the prime ideals of R containing P. This fact allows one quite often to reduce problems concerning the ring R to problems concerning its localizations R_P. An example is given in proposition (18.22) below; to prove this proposition we need two preparatory results.

(18.20) Proposition. *Let R be an integral domain with quotient field K. Then all the localizations R_P of R can be considered as subrings of K, and $\bigcap_{P \text{ prime}} R_P = R$.*

Proof. The inclusion $\supseteq$ is clear by (18.17)(d). Suppose conversely that an element $k \in K \setminus R$ is given; we want to find a prime ideal P such that $k \notin R_P$. We can write $k = a/b$ where $a, b \in R$ are coprime. Since $k \notin R$, the denominator b cannot be a unit in R. Let p be a prime factor of b and let $P := Rp$; this is a prime ideal due to (7.4)(f). We claim that $k \notin R_P$. Suppose $k \in R_P$. Then $k = r/s$ with $s \notin P$, i.e., $r/s = a/b$ so that $as = rb \in Rb \subseteq Rp = P$; on the other hand $s \notin P$ so that we must have $a \in P$ and hence $p \mid a$. But this contradicts the fact that a and b are coprime. ∎

(18.21) Proposition. *Suppose $A \subseteq B$ is a ring extension. If C is the integral closure of A in B then $S^{-1}C$ is the integral closure of $S^{-1}A$ in $S^{-1}B$.*

Proof. Let $x \in C$ and $s \in S$. Since x is integral over A, there are elements $a_0, \ldots, a_{n-1} \in A$ with $x^n + a_{n-1}x^{n-1} + \cdots + a_1 x + a_0 = 0$. Then for every $s \in S$ we have

$$\left(\frac{x}{s}\right)^n + \frac{a_{n-1}}{s}\left(\frac{x}{s}\right)^{n-1} + \cdots + \frac{a_1}{s^{n-1}}\left(\frac{x}{s}\right) + \frac{a_0}{s^n} = \frac{x^n + a_{n-1}x^{n-1} + \cdots + a_1 x + a_0}{s^n} = \frac{0}{s^n} = 0;$$

since all the coefficients a_{n-i}/s^i belong to $S^{-1}A$ this shows that x/s is integral over $S^{-1}A$.

Conversely, suppose that $b/s \in S^{-1}B$ is integral over $S^{-1}A$. Then there are elements $a_i \in A$, $s_i \in S$ such that

$$\left(\frac{b}{s}\right)^n + \frac{a_{n-1}}{s_{n-1}}\left(\frac{b}{s}\right)^{n-1} + \cdots + \frac{a_1}{s_1}\left(\frac{b}{s}\right) + \frac{a_0}{s_0} = 0 \,.$$

Let $t := s_1 \cdots s_n \in S$. Multiplying the above equation by $s^n t^n$, one obtains an equation which shows that bt is integral over A which means $bt \in C$. Hence $b/s = (bt)/(st) \in S^{-1}C$. ∎

(18.22) Proposition. *Let R be an integral domain. Then the following statements are equivalent.*
(1) R is integrally closed in its quotient field.
(2) $S^{-1}R$ is integrally closed in its quotient field for each multiplicatively closed subset $S \subseteq R$ with $0 \notin S$.
(3) Each localization R_P of R is integrally closed in its quotient field.

Proof. (1)$\Longrightarrow$(2). Apply (18.21) with $A := R$, $B := K$ (the quotient field of R) and $C = R$.

(2)$\Longrightarrow$(3). Take $S := R \setminus P$ where P is a prime ideal of R.

(3)$\Longrightarrow$(1). Suppose $k \in K$ is integral over R. Write $k = a/b$ with coprime elements $a, b \in R$ and let p be a prime factor of b so that $P := Rp$ is a prime ideal of R. Now k is integral over R, the more so over R_P, and since R_P is integrally closed we have $k \in R_P$. Thus there are elements $r \in R, s \in R \setminus P$ with $r/s = k = a/b$. This yields $as = rb \in Rp$. Since $s \notin P$ and P is a prime ideal, this implies $a \in Rp$. Hence every prime factor of b also divides a; hence a is divisible by b so that $k \in R$. ∎

THE procedure of studying a ring R via its localizations R_P is analogous to studying a variety V "locally", i.e., by investigating a small neighborhood of each point $x \in V$. For example, proposition (18.22) tells us that a ring R can be exhibited as being integrally closed by showing that all of its localizations are integrally closed; one expresses this fact sometimes by saying that being integrally closed is a "local property". The reason why the study of the rings R_P is often easier than the study of R itself is that these localizations have a very special property, as is shown in the following proposition.

(18.23) Proposition. *R_P has a unique maximal ideal, namely $S^{-1}P$ where $S = R \setminus P$.*

Proof. Let $M \trianglelefteq R_P$ be a maximal ideal of R_P. Then P is prime, hence takes the form $M = S^{-1}Q$ for a prime ideal Q of R with $Q \cap S = \emptyset$, i.e., $Q \subseteq P$. Hence $S^{-1}Q \subseteq S^{-1}P \subsetneq S^{-1}R$. Then the maximality of M implies $M = S^{-1}Q = S^{-1}P$. ∎

THIS result gives rise to the following definition.

(18.24) Definition. *A **local ring** is a commutative ring with identity which has a unique maximal ideal.*

(18.25) Examples. (a) Every field is a local ring (with maximal ideal $\{0\}$).

(b) Let V be an affine variety and $x \in V$ a point in V. Then $\mathcal{O}_x$ is a local ring, due to (18.14). This example is prototypical and is the reason for the terminology "local ring".

(c) Let K be a field. Then the power series ring $K[[x_1, \ldots, x_n]]$ is a local ring with unique maximal ideal $\langle x_1, \ldots, x_n \rangle$.

(d) If p is prime and $n \in \mathbb{N}$ then $\mathbb{Z}_{p^n}$ is a local ring with unique maximal ideal $\langle p \rangle$.

(e) Let $p \in \mathbb{N}$ be a prime number. Then $\mathbb{Z}_{(p)} := \{m/n \in \mathbb{Q} \mid m, n \in \mathbb{Z}, p \nmid n\}$ is a local ring with unique maximal ideal $\{(pm)/n \mid m, n \in \mathbb{Z}, p \nmid n\}$.

(f) In general, let R be a commutative ring with identity and let P be a prime ideal of R. Letting $S := R \setminus P$, the ring $R_P := S^{-1}P$ is a local ring with unique maximal ideal $S^{-1}P$, due to (18.23).

LET us conclude this section by determining the unique maximal ideal of a local ring.

(18.26) Proposition. *Let R be a commutative ring with identity.*

(a) *If R is a local ring with maximal ideal M then M is the set of all non-units.*

(b) *If the set S of all non-units of R is contained in an ideal $M \neq R$ then R is local with maximal ideal M.*

Proof. (a) Let a be a non-unit. Then Ra is a proper ideal of R, hence contained in a maximal ideal by (5.18), hence contained in M. This shows that M contains all non-units. But M cannot contain any unit because $M \neq R$. Hence M consists exactly of all non-units.

(b) We observe first that $S = M$ because otherwise M would contain a unit, contradicting the hypothesis $M \neq R$. Let $I \neq R$ be any ideal. Then I contains only non-units; hence $I \subseteq S \subseteq M$. Thus M is the only maximal ideal of R. ∎

Exercises

Problem 1. Let V be a curve in $\mathbb{R}^2$, say $V = \{(x,y) \in \mathbb{R}^2 \mid f(x,y) = 0\}$ with $f \in \mathbb{R}[X,Y]$, and let (x_0,y_0) be a singular point of V so that $(f_x(x_0,y_0), f_y(x_0,y_0)) = (0,0)$. Let

$$\Delta(x_0,y_0) := \det \begin{pmatrix} f_{xx}(x_0,y_0) & f_{xy}(x_0,y_0) \\ f_{yx}(x_0,y_0) & f_{yy}(x_0,y_0) \end{pmatrix}.$$

(a) Using the Taylor expansion of f around (x_0,y_0), show that if $\Delta(x_0,y_0) < 0$ then (x_0,y_0) is a double-point of V, i.e., a point in which V intersects itself and has two tangent lines. The slopes of these tangent lines are the solutions k of the quadratic equation $f_{yy}(x_0,y_0)k^2 + 2f_{xy}(x_0,y_0)k + f_{xx}(x_0,y_0) = 0$.

(b) Using the Taylor expansion of f around (x_0,y_0), show that if $\Delta(x_0,y_0) > 0$ then (x_0,y_0) is an isolated point of V in the sense that there is a disk around (x_0,y_0) in which no other point of V is contained.

(c) What can you say if $\Delta(x_0,y_0) = 0$?

(d) Find the critical points of the following curves and determine their nature!

$$f(x,y) = (x^2 + y^2)^2 - 2a^2(x^2 - y^2)$$
$$g(x,y) = x^3 + y^3 - x^2 - y^2$$
$$h(x,y) = (y - x^2)^2 - x^5$$

Problem 2. Let V be an irreducible K-variety in L^n where L is algebraically closed. Show that if $\dim V = n - 1$ if and only if $V = \mathcal{V}(f)$ for some irreducible polynomial $f \in K[X_1,\ldots,X_n]$.

Problem 3. Prove theorem (18.17).

Problem 4. Find all multiplicatively closed subsets S of $R = \mathbb{Z}_6$ and determine the corresponding rings $S^{-1}R$.

Problem 5. Let R be the ring of all entire functions $f : \mathbb{C} \to \mathbb{C}$ and let $z_0 \in \mathbb{C}$. Show that $S := \{f \in R \mid f(z_0) \neq 0\}$ is a multiplicatively closed subset of R and that the $S^{-1}R$ can be identified with the ring of germs of holomorphic functions at z_0.

Problem 6. Let $R = \mathbb{Z} + i\mathbb{Z}$.

(a) Show that $S = \{x \in \mathbb{Z} \mid 5\nmid x\}$ is a multiplicative subset of R.

(b) Show that $T = \{x \in R \mid 5\nmid x\}$ is not a multiplicative subset of R.

(c) Show that $S^{-1}R$ has exactly three prime ideals.

(d) Find the structure of $S^{-1}R/\langle 5\rangle$.

Problem 7. (a) Show that every ring lying between $\mathbb{Z}$ and $\mathbb{Q}$ can be written as $R = S^{-1}\mathbb{Z}$ where S is a multiplicative subset of $\mathbb{Z}$.

(b) Find examples for multiplicatively closed subsets S of $\mathbb{Z}$ such that $S^{-1}\mathbb{Z} \cong \mathbb{Q}$.

Problem 8. Let S be a multiplicatively closed subset of a ring R and let $I \trianglelefteq R$ be an ideal with $I \cap S = \emptyset$. Show that $\overline{S} := \{s + I \mid s \in S\}$ is a multiplicatively closed subset of the quotient ring $\overline{R} := R/I$ and that $(S^{-1}R)/(S^{-1}I) \cong \overline{S}^{-1}\overline{R}$.

Problem 9. Let R be a commutative ring. A multiplicatively closed subset $S \subseteq R$ is called **saturated** if $xy \in S$ if and only if $x \in S$ and $y \in S$.
(a) Show that the set S of all non-zero-divisors is saturated.
(b) Show that S is saturated if and only if $R \setminus S$ is a union of prime ideals.
(c) Show that for any multiplicatively closed subset $S \subseteq R$ there is a unique smallest saturated set $\overline{S}$ containing S and that $R \setminus \overline{S}$ is the union of all prime ideals which do not intersect S.
(d) Let $S = 1 + I$ where I is an ideal of R. Find $\overline{S}$.

Problem 10. Let R be a commutative ring and S be a multiplicatively closed subset of R with $0 \notin S$.
(a) Suppose $g : R \to R'$ is a ring homomorphism such that $g(s)$ is a unit in R' whenever $s \in S$. Show that there is a unique homomorphism $h : S^{-1}R \to R'$ with $g = h \circ f$.

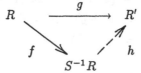

(b) Suppose that g has the additional properties that $g(r) = 0$ implies $rs = 0$ for some $s \in S$ and that every element of R' has the form $g(a)g(s)^{-1}$ with $a \in R$ and $s \in S$. Show that in this case h is an isomorphism.

Problem 11. Let $R \neq \{0\}$ be a commutative ring and let $\mathfrak{S}$ be the set of all multiplicatively closed subsets S of R with $0 \notin S$.
(a) Show that $\mathfrak{S}$ has maximal elements.
(b) Show that S is maximal in $\mathfrak{S}$ if and only if $R \setminus S$ is a minimal prime ideal.
(c) Show that every minimal prime ideal of R consists entirely of zero-divisors.

Problem 12. Show that if R is a principal ideal ring and $S \subseteq R$ is multiplicatively closed, then $S^{-1}R$ is again a principal ideal ring.

Problem 13. Show that if R is a unique factorization domain and $S \subseteq R$ is multiplicatively closed, then $S^{-1}R$ is again a unique factorization domain. Moreover, the prime elements of $S^{-1}R$ are exactly the elements $(ps)/s$ where p is a prime element of R with $\langle p \rangle \cap S = \emptyset$.

Problem 14. Let R be a commutative Noetherian ring and let $S \subseteq R$ be a multiplicatively closed set. Show that $S^{-1}R$ is Noetherian.

Problem 15. Suppose the finitely generated K-algebra R is an integral domain with quotient field $Q(R)$ and $S \subseteq R$ is multiplicatively closed with $0 \notin S$; then clearly $Q(S^{-1}R) = Q(R)$ and hence $\mathrm{trdeg}(Q(S^{-1}R) : K) = \mathrm{trdeg}(Q(R) : K)$. Use this fact to formulate a theorem stating that under suitable hypotheses localization preserves dimension.

Problem 16. Let $A \subseteq B$ be commutative rings with the same identity element. Suppose B is integral over A.
(a) Show that if $S \subseteq A$ is multiplicatively closed then $S^{-1}B$ is integral over $S^{-1}A$.
(b) Suppose that $Q \lhd B$ is a prime ideal of B and let $P := Q \cap R$. Show that Q is maximal in B if and only if P is maximal in R.
Hint. Use part (a) and (15.1).

Problem 17. Let K be a field and let $R := K[x^2 - 1]$ and $S := K[x]$. Prove the following statements!
(a) S is integral over R.
(b) $Q := \langle x - 1 \rangle$ is a maximal ideal of S, and $P := Q \cap R$ is a maximal ideal of R.
(c) S_Q is not integral over R_P.
Hint. Show that the element $1/(x + 1)$ is not integral.

Problem 18. Show that $S^{-1}(I : J) = (S^{-1}I) : (S^{-1}J)$ under any of the following hypotheses:
(a) J is finitely generated;
(b) $I = A_R$ and $J = B_R$ with ideals $A, B \lhd S^{-1}R$.

Problem 19. Let $P \subseteq Q$ be prime ideals of a commutative ring R and let $S := R \setminus P$. (Recall that the prime ideals of $R_P = S^{-1}R$ are in 1-1-correspondence with the prime ideals of R contained in P and that the prime ideals of R/Q are in 1-1-correspondence with the prime ideals of R containing Q.)
(a) Show that $S^{-1}R/S^{-1}P \cong \overline{S}(R/P)$ where $\overline{S} := \{s + P \mid s \in S\}$. Hence the operations of localizing at P and taking the quotient modulo Q commute.
(b) Show that the prime ideals of $S^{-1}R/S^{-1}Q$ are in 1-1-correspondence with the prime ideals of R lying between P and Q.

Problem 20. Let R be a commutative ring with identity and let P and Q be ideals of R. Suppose that $S \subseteq R$ is a multiplicatively closed set which does not contain a zero divisor of R, but contains the identity element. Show that if $S^{-1}Q$ is $S^{-1}P$-primary in $S^{-1}R$, then Q is P-primary in R.
Remark. This is a partial converse of (18.19)(e).

Problem 21. Let Q be P-primary. Show that the n-th symbolic power of Q is given by $Q^{(n)} = (Q^n R_P)_R$.

Problem 22. Let R be an integral domain. Show that $R = \bigcap R_M$ where the intersection runs over all maximal ideals M of R.

Problem 23. Let R be a commutative ring.

(a) Show that R has no nonzero nilpotent elements if and only if all localizations R_P of P have no nonzero nilpotent elements. (Hence the property of having no nonzero nilpotent elements is a local property.)

(b) Find an example such that all localizations R_P of R are integral domains, but that R is not. (Hence the property of being an integral domain is *not* a local property.)

Problem 24. Let R be a local ring. Show that the only elements $x \in R$ with $x^2 = x$ are 0 and 1.

Hint. If $x^2 = x$ with $x \notin \{0,1\}$ then x and $1 - x$ are both non-units.

Problem 25. Let $f : A \to B$ be a surjective ring homomorphism. Show that if A is a local ring and $B \neq \{0\}$, then B is a local ring.

19. Factorization of ideals

IN this chapter we want to investigate the phenomenon that decomposition into irreducible elements need not be unique in a factorization domain R. The mathematician Ernst Eduard Kummer (1810-1893) had the brilliant idea that in such a case uniqueness might be enforced by enlarging the ring R by certain "ideal elements" in a way analogous to the geometric procedure of adjoining "points at infinity" to an affine plane to obtain a projective plane. Vaguely speaking, Kummer's idea was as follows: If we enlarge the ring, we have more elements available and hence have more possibilities of factoring elements; hence we might be able to further decompose elements which are irreducible in R with the aid of the "ideal elements" adjoined from outside the ring which now might serve as last multiplicative building blocks. This might have the consequence that two essentially different factorizations of a ring element in R can be refined to obtain the same finer factorization into "ideal elements" in both cases. Let us take a typical example for non-unique factorization, say $(1 + \sqrt{-5})(1 - \sqrt{-5}) = 2 \cdot 3$ in $\mathbb{Z}[\sqrt{-5}]$. Following Kummer, we want to adjoin "ideal elements" p_1, p_2, p_3, p_4 to $\mathbb{Z}[\sqrt{-5}]$ such that

$$(\star) \qquad \underbrace{(1 + \sqrt{-5})}_{= p_1 p_2} \underbrace{(1 - \sqrt{-5})}_{= p_3 p_4} = \underbrace{2}_{= p_1 p_3} \cdot \underbrace{3}_{= p_2 p_4} .$$

But where should we find these "ideal elements"? Anticipating the result of Kummer's work, we consider the four ideals

$$P_1 := \langle 2, 1 + \sqrt{-5} \rangle ,$$
$$P_2 := \langle 3, 1 + \sqrt{-5} \rangle ,$$
$$P_3 := \langle 2, 1 - \sqrt{-5} \rangle \qquad \text{and}$$
$$P_4 := \langle 3, 1 - \sqrt{-5} \rangle$$

of $\mathbb{Z}[\sqrt{-5}]$. It is not hard to verify that these are prime ideals of $\mathbb{Z}[\sqrt{-5}]$ and that

$$(\star\star) \quad P_1 P_2 = \langle 1 + \sqrt{-5} \rangle, \quad P_3 P_4 = \langle 1 - \sqrt{-5} \rangle, \quad P_1 P_3 = \langle 2 \rangle, \quad P_2 P_4 = \langle 3 \rangle .$$

The striking similarity of $(\star)$ and $(\star\star)$ suggests that we should look at the ideals of a factorization domain R to find Kummer's "ideal elements"; in fact, Kummer introduced the notion of an ideal along these lines, and the name "ideal" was coined following Kummer's idea of "ideal elements".

The same problem of nonunique factorization occurs in the setting of algebraic geometry (and is intimately connected with the problem of a suitable definition of rational functions). To see an example, let R be the coordinate ring of the variety $V = \{(x, y) \in L^2 \mid x^2 + y^2 = 2\}$. In $R = K[V]$, we have the following example for non-unique factorization:

$$(x - 1)(x + 1) = (1 + y)(1 - y) ;$$

note that the four functions $x - 1$, $x + 1$, $1 + y$ and $1 - y$ are all irreducible in $K[V]$. This problem can be solved in a similar way as above: The ideals

$$P_1 := \langle x - 1, 1 - y \rangle, \ P_2 := \langle x - 1, 1 + y \rangle, \ P_3 := \langle x + 1, 1 - y \rangle, \ P_4 := \langle x + 1, 1 + y \rangle$$

are prime ideals in $K[V]$, and we have

$$\langle x - 1 \rangle \;=\; P_1 P_2 \;, \quad \langle x + 1 \rangle \;=\; P_3 P_4 \;, \quad \langle 1 - y \rangle \;=\; P_1 P_3 \;, \quad \langle 1 + y \rangle \;=\; P_2 P_4 \;.$$

In this example the decomposition can even be visualized by considering, with each ideal I involved, the subvariety $\mathcal{V}(I)$ of V; note that

$$\mathcal{V}(P_1) = \{(1,1)\}, \; \mathcal{V}(P_2) = \{(1,-1)\}, \; \mathcal{V}(P_3) = \{(-1,1)\}, \; \mathcal{V}(P_4) = \{(-1,-1)\}.$$

Using the fact that $\mathcal{V}(IJ) = \mathcal{V}(I) \cup \mathcal{V}(J)$ for all ideals I and J, the ideal factorization can be visualized as follows.

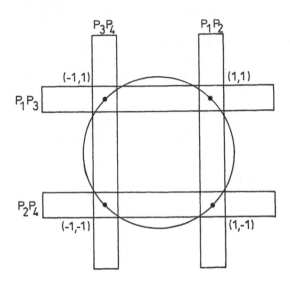

ONCE we suspect that the ideals of R could be used to learn something about factorization in R, we are then reminded that we have already developed a certain calculus of ideals and were able to express the basic notions of divisibility in terms of ideals. More specifically, let R be an integral domain and let $\mathcal{I}(\mathcal{R})$ be the set of all ideals in R. Then the mapping

$$\begin{aligned} R &\;\rightarrow\; \mathcal{I}(\mathcal{R}) \\ a &\;\mapsto\; \langle a \rangle \end{aligned}$$

which associates with each element $a \in R$ the principal ideal generated by a preserves multiplication, maps prime elements to prime ideals, prime powers to primary ideals and irreducible elements to maximal elements in the set of all principal ideals of R. Hence as far as multiplicative properties are concerned, we can operate in $\mathcal{I}(\mathcal{R})$ as well as in R. Even better, the troublesome fact that prime factors are uniquely determined only up to multiplication by units disappears because $\langle a \rangle = \langle b \rangle$ if and only if $b = ua$ for some unit u. For principal ideal domains, the transition from R to $\mathcal{I}(\mathcal{R})$ is particularly easy; for such a domain, we can formulate the unique factorization property immediately in terms of ideals.

(19.1) Proposition. *Let I be an ideal of a principal ideal domain R such that $\{0\} \subsetneq I \subsetneq R$.*

(a) There are prime ideals $P_1, \ldots, P_m$, unique up to order, such that $I = P_1 \cdots P_m$.

(b) There are primary ideals $Q_1, \ldots, Q_n$, unique up to order, such that $I = Q_1 \cdots Q_n = Q_1 \cap \cdots \cap Q_n$.

Proof. Let $I = \langle x \rangle$. Then x has a prime factorization $x = p_1 \cdots p_m$ where the prime factors (which are not necessarily pairwise different) are unique up to order and up to multiplication by units. This elementwise factorization translates into the ideal factorization $\langle x \rangle = \langle p_1 \rangle \cdots \langle p_m \rangle$ where each $P_i = \langle x_i \rangle$ is a prime ideal. Also, collecting like prime factors, we can write $x = p_1^{k_1} \cdots p_n^{k_n}$ where the primes $p_1, \ldots, p_n$ are pairwise different. This elementwise factorization becomes $\langle x \rangle = \langle p_1^{k_1} \rangle \cdots \langle p_n^{k_n} \rangle$ where each of the ideals $Q_i = \langle p_i^{k_i} \rangle$ is a primary ideal. The fact that the product $Q_1 \cdots Q_n$ coincides with the intersection $Q_1 \cap \cdots \cap Q_n$ follows from (5.23); see also the remarks preceding (6.20). ∎

WE generalize these statements by the following definition.

(19.2) Definition. *An integral domain R is called a **Dedekind domain** if every proper ideal of R can be written as a product of prime ideals.†*

SINCE our focus has shifted from multiplicative decompositions of ring elements to multiplicative decompositions of ideals, we can also try to transfer multiplicative notions like identity elements and invertibility to the level of ideals. Now $IR = RI = I$ for all ideals I of an integral domain R, so that R is an identity element for the multiplication of ideals. On the other hand, it is clear that we cannot have $IJ = R$ unless $I = J = R$; hence the notion of invertibility does not make sense for ideals. This can be circumvented by generalizing the notion of an ideal.

(19.3) Definitions. *Let R be an integral domain with quotient field K.*

*(a) An R-submodule A of K is called a **fractional ideal** of R if there is an element $r \neq 0$ in R such that $rA \subseteq R$ (and hence $rA \subseteq R \cap A$).*

*(b) A fractional ideal A is called **invertible** if there is a fractional ideal B with $AB = R$.*

IT is easy to see that the fractional ideals of R are exactly the sets kI where $k \neq 0$ is an element of K and where I is an ideal of R. Let us give some examples of fractional ideals.

(19.4) Examples. Let R be an integral domain with quotient field K.

(a) Every "true" ideal of R is also a fractional ideal.

(b) Every finitely generated R-submodule of K is a fractional ideal of R. Indeed,

---

† It will turn out that such a decomposition is automatically unique up to the order of the factors; this will be proved in (19.9).

if $A = Rk_1 + \cdots + Rk_n$ with $k_i = r_i/s_i \in K$ then let $s := s_1 \cdots s_n$; this element satisfies $sA \subseteq R$.

(c) Every principal ideal $A \neq \{0\}$ of R is invertible as a fractional ideal. Indeed, if $A = rR$ then $B := (1/r)R$ is inverse to A.

LET us now observe that the calculus of ideals can be generalized to fractional ideals.

(19.5) Proposition. *Let R be an integral domain with quotient field K and let A, B be fractional ideals of R.*

(a) *The intersection $A \cap B$, the sum $A + B := \{a + b \mid a \in A, b \in B\}$ and the product $AB := \{\sum_{i=1}^{n} a_i b_i \mid n \in \mathbb{N}, a_i \in A, b_i \in B\}$ are also fractional ideals of R.*

(b) *If $B \neq \{0\}$ then $[A : B] := \{x \in K \mid xB \subseteq A\}$ is also a fractional ideal of R.*

(c) *$AR = RA = A$ and $A[R : A] = [R : A]A \subseteq R$.*

Proof. (a) It is clear that $A \cap B$, $A + B$ and AB are R-submodules of K, since A and B are. Moreover, if $rA \subseteq R$ and $sB \subseteq R$ then $r(A \cap B) \subseteq R$ and $s(A \cap B) \subseteq R$, $rs(A + B) \subseteq R$ and $rs(AB) \subseteq R$.

(b) It is easily checked that $[A : B]$ is an R-submodule of K (because A is). Thus it remains for us to find an element $\rho \in R \setminus \{0\}$ with $\rho[A : B] \subseteq R$. Since A and B are fractional ideals, there are elements $r, s \in R \setminus \{0\}$ with $rA \subseteq R$ and $sB \subseteq B$. Pick any element $b_0 \in B \setminus \{0\}$; then $b := sb_0 \neq 0$ lies in $B \cap R$. Taking $\rho := rb$, we obtain $\rho[A : B] = rb[A : B] \subseteq rA \subseteq R$, as desired.

(c) Both statements are trivial. ∎

NOTE that if I and J are true ideals of R then their ideal quotient is $(I : J) = [I : J] \cap R$.

(19.6) Proposition. *Let R be an integral domain with quotient field K.*

(a) *If A, B are fractional ideals with $AB = R$ then B is uniquely determined, namely, $B = [R : A]$. We write $B = A^{-1}$.*

(b) *If $AC = BC$ where C is invertible then $A = B$ (**cancellation rule**).*

(c) *A product $A_1 \cdots A_n$ of fractional ideals is invertible if and only if each factor A_i is invertible, and $(A_1 \cdots A_n)^{-1} = A_1^{-1} \cdots A_n^{-1}$.*

(d) *Suppose $P_1 \cdots P_m = Q_1 \cdots Q_n$ with prime ideals $P_k, Q_l \trianglelefteq R$. If each P_k is invertible (as a fractional ideal), then $m = n$ and $P_k = Q_k$ for all k (after reindexing if necessary).*

(e) *If A is invertible then A is finitely generated as an R-module.*

Proof. (a) If $AB = R$ then trivially $B \subseteq [R : A]$. On the other hand, $[R : A] = R \cdot [R : A] = BA[R : A] \subseteq BR = RB \subseteq B$; hence $B = [R : A]$.

(b) Simply multiply both sides of the equation $AC = BC$ by C^{-1} to obtain $AR = BR$, i.e., $A = B$.

(c) If $A_1 \cdots A_n$ is invertible (with inverse B), then the equation $A_i(\prod_{j \neq i} A_j B) = A_1 \cdots A_n B = R$ shows that each factor A_i is invertible (with inverse $\prod_{j \neq i} A_j B$).

Conversely, if each factor A_i is invertible, then obviously $A_1^{-1} \cdots A_n^{-1}$ is an inverse of $A_1 \cdots A_n$.

(d) We proceed by induction on m. If $m = 1$, we have $Q_1 \cdots Q_n = P_1$; since P_1 is prime, this implies that $Q_i \subseteq P_1$ for some index i, due to (6.3)(5). On the other hand $P_1 \subseteq Q_1 \cdots Q_n \subseteq Q_i$; this shows $P_1 = Q_i$. Let $Q_0 := \prod_{j \neq i}$; then $P_1 = Q_i Q_0$. Multiplying both sides by $P_1^{-1} = Q_i^{-1}$, we obtain $R = Q_0$ which is clearly a contradiction to the fact that $Q_j \subsetneq R$ for all j and hence $Q_0 \subsetneq R$.

Now let $m > 1$. Pick one of the elements $P_1, \ldots, P_m$ which is minimal with respect to inclusion (i.e., does not strictly contain any of the other elements P_j); without loss of generality, we may assume that this element is P_1. Since $Q_1 \cdots Q_n = P_1 \cdots P_m \subseteq P_1$ and since P_1 is prime, we have $Q_i \subseteq P_1$ for some index i; after reindexing if necessary we may assume that $i = 1$ so that $Q_1 \subseteq P_1$. On the other hand, we have $P_1 \cdots P_m \subseteq Q_1 \cdots Q_n \subseteq Q_1$, and since Q_1 is prime, there is an index k such that $P_k \subseteq Q_1$ so that $P_k \subseteq Q_1 \subseteq P_1$. Then the minimality of P_1 implies that $P_k = P_1$ and hence that $P_1 = Q_1$. Multiplying the equation $P_1 \cdots P_m = Q_1 \cdots Q_n$ by $P_1^{-1} = Q_1^{-1}$, we obtain $P_2 \cdots P_m = Q_2 \cdots Q_n$. Now an application of the induction hypothesis to this last equation clearly yields the claim.

(e) Since $1 \in R = A^{-1}A$, there are elements $x_i \in A^{-1} = [R : A]$ and $y_i \in A$ such that $1 = \sum_{i=1}^{n} x_i y_i$. Now take an arbitrary element $a \in A$. Then

$$a = a \cdot 1 = \sum_{i=1}^{n} \underbrace{(a x_i)}_{\in R} y_i \in Ry_1 + \cdots + Ry_n .$$

This shows $A = Ry_1 + \cdots + Ry_n$, i.e., the elements $y_1, \ldots y_n$ generate A as an R-module. ∎

THE next two results show that the use of fractional ideals helps to obtain important information on Dedekind domains.

(19.7) Lemma. *Let $P \neq \{0\}$ be a prime ideal in a Dedekind domain R.*
(a) *If $P \subsetneq P + Ra \subsetneq R$ then $P = P^2 + Pa$.*
(b) *If P is invertible as a fractional ideal then P is a maximal ideal of R.*

Proof. (a) Since R is a Dedekind domain, there are prime ideals $P_1, \ldots, P_m$ and $Q_1, \ldots, Q_n$ such that

$$P + Ra = P_1 \cdots P_m \qquad \text{and} \qquad P + Ra^2 = Q_1 \cdots Q_n .$$

Now $\overline{R} := R/P$ is an integral domain due to (6.14)(a). Consider the images of these ideals under the quotient map $\pi : R \to R/P = \overline{R}$ to obtain

$$\langle \overline{a} \rangle = \overline{P_1} \cdots \overline{P_m} \qquad \text{and} \qquad \langle \overline{a}^2 \rangle = \overline{Q_1} \cdots \overline{Q_n} .$$

These two ideals are invertible as fractional ideals of $\overline{R}$ by example (19.4)(c). Hence $\overline{P_1}, \cdots, \overline{P_m}, \overline{Q_1}, \cdots, \overline{Q_n}$ are all invertible, due to (19.6)(c). On the other hand, all these ideals are prime ideals of $\overline{P}$ due to (6.15) because $\ker \pi = P$ is contained in each P_i and each Q_j. Since

$$\overline{Q_1} \cdots \overline{Q_n} = \langle \overline{a^2} \rangle = \langle \overline{a} \rangle^2 = (\overline{P_1} \cdots \overline{P_m})^2 = \overline{P_1}^2 \cdots \overline{P_m}^2$$

we conclude that $n = 2m$ and (after reordering, if necessary) $\overline{P_i} = \overline{Q_{2i-1}} = \overline{Q_{2i}}$ for $1 \le i \le m$, due to (19.6)(d). Hence $P_i = \pi^{-1}(\overline{P_i}) = \pi^{-1}(\overline{Q_{2i-1}}) = Q_{2i-1}$ and similarly $P_i = Q_{2i}$. Consequently,

$$P \subseteq P + Ra^2 = Q_1 \cdots Q_n = (P_1 \cdots P_m)^2 = (P + Ra)^2 \subseteq P^2 + Ra.$$

Now let $p \in P$ be an arbitrary element. Then we can write $p = x + ra$ with $x \in P^2$ and $r \in R$. Hence $ra = p - x \in P$, but $a \notin P$; therefore $r \in P$ because P is prime. This shows that $P \subseteq P^2 + Pa$. The converse inclusion is trivial.

(b) We have to show that $P + Ra = R$ whenever $a \in R \setminus P$. Take such an element $a \in R \setminus P$. By (a) we have $P = P^2 + Pa$. Hence $R = P^{-1}P = P^{-1}(P^2 + Pa) = P + Ra$. ∎

(19.8) **Theorem.** *Let R be a Dedekind domain. Then every prime ideal $P \ne \{0\}$ of R is a maximal ideal and is invertible (as a fractional ideal).*

Proof. Let $p \in P \setminus \{0\}$. Since R is a Dedekind domain, there are prime ideals P_i such that $\langle p \rangle = P_1 \cdots P_n$. Then $P_1 \cdots P_n = \langle p \rangle \subseteq P$; since P is prime, this implies $P_i \subseteq P$ for some i. On the other hand, the ideal $\langle p \rangle$ is invertible by example (19.4)(c), hence so are $P_1, \ldots, P_n$ due to (19.6)(c). Consequently, (19.7)(b) shows that each P_i is a maximal ideal of R. Hence $P_i = P$ so that P is maximal and invertible. ∎

OUR definition of Dedekind domains does not allow us to effectively check whether or not a given integral domain is Dedekind. Therefore, we want to characterize Dedekind domains by handier conditions which are relatively easy to verify. The next theorem is the first step towards this goal; namely, we collect conditions which are necessarily satisfied in every Dedekind domain. After some additional work, we will see that these conditions are also sufficient and hence characterize Dedekind domains.

(19.9) **Theorem.** *Let R be an integral domain with quotient field K. Then each of the following conditions implies the next:*
   (1) *R is a Dedekind domain;*
   (2) *every proper ideal can be uniquely written as a product of prime ideals,*
   (3) *every nonzero ideal of R is invertible as a fractional ideal,*
   (4) *every nonzero fractional ideal of R is invertible;*
   (5) *R is Noetherian and integrally closed in K, and every nonzero prime ideal is maximal.*

Proof. (1)$\Longrightarrow$(2). Let $I \ne \{0\}$. Then there are prime ideals $P_1, \ldots, P_n$ with $I = P_1 \cdots P_n$. Due to (19.8), the ideal $P_1, \ldots, P_n$ are in fact maximal ideals and hence pairwise comaximal; this implies $I = P_1 \cdots P_n = P_1 \cap \cdots \cap P_n$, due to (5.23). But this is a primary decomposition of I into prime ideals; hence the uniqueness claim follows from the uniqueness theorem (6.22) for primary decompositions.

(2)$\Longrightarrow$(3). Every prime of R is invertible as a fractional ideal, due to (19.8). In combination with (19.6)(c), this yields the claim.

(3)$\Longrightarrow$(4). This is clear since every fractional ideal of R has the form kI with $k \in K^\times$ and $I \trianglelefteq R$.

(4)$\Longrightarrow$(5). Let I be an arbitrary ideal of R. Then I is invertible, hence finitely generated by (19.6)(e). This shows that R is Noetherian.

Let $\alpha \in K$ be integral over R; we want to show that $\alpha \in R$. The ring $S := R[\alpha]$ generated by R and α is a finitely generated submodule by (10.22); therefore, S is a fractional ideal by example (19.4)(b), hence invertible by assumption (4). Thus $S = RS = (S^{-1}S)S = S^{-1}(SS) = S^{-1}S = R$, i.e., $R[\alpha] = R$ so that $\alpha \in R$. This shows that R is integrally closed in K.

Suppose that there is a nonzero prime ideal which is not maximal. Then there is a maximal ideal M with $P \subsetneqq M$. By assumption (4), the ideal M is invertible as a fractional ideal. Then $M^{-1}P \subseteq M^{-1}M = R$ so that $M^{-1}P \trianglelefteq R$. Thus $P = M(M^{-1}P)$ can be written as a product of two ideals of R. Since P is prime and $M \not\subseteq P$, we have $M^{-1}P \subseteq P$ by (6.3)(4). Therefore,

$$R = M^{-1}M \subseteq M^{-1}R = M^{-1}(PP^{-1}) = \underbrace{(M^{-1}P)}_{\subseteq P} P^{-1} \subseteq PP^{-1} = R \, ;$$

whence $R = M^{-1}R = M^{-1}$ so that $R = MM^{-1} = MR = M$. This is the desired contradiction; hence also the third condition is established. ∎

WE have seen that every Dedekind domain is Noetherian and integrally closed in its quotient field. To proceed further, we now prove a result about integral domains with these properties. Recall that a prime ideal P in a commutative ring R is called *minimal* if there is no prime ideal P' with $\{0\} \subsetneqq P' \subsetneqq P$.

(19.10) Proposition. *Let R be a Noetherian integral domain which is integrally closed. If P is a minimal prime ideal and if Q is P-primary, then there is a number $m \in \mathbb{N}$ with $Q = P^{(m)}$.*

Proof. If $Q = P$ then the claim is trivial; so assume that $Q \subsetneqq P$. Pick an element $a \neq 0$ in P. We claim that

$(\star)$ $$Ra \subsetneqq (Ra : P) \, .$$

Indeed, if $Ra = Q_1 \cap \cdots \cap Q_n$ is a reduced primary decomposition where Q_i is P_i-primary with exponent k_i then $P_1^{k_1} \cdots P_n^{k_n} \subseteq Q_1 \cdots Q_n \subseteq Q_1 \cap \cdots \cap Q_n = Ra \subseteq P$. Since P is prime, there is an index i_0 with $P_{i_0} \subseteq P$; now the minimality of P implies that $P_{i_0} = P$. By (11.16) we have $(Q_{i_0} : P) \supsetneqq Q_{i_0}$ whereas $(Q_i : P) = Q_i$ for all $i \neq i_0$, due to (6.19)(c). Consequently,

$$
\begin{aligned}
(Ra : P) &= (Q_1 \cap \cdots \cap Q_n) : P \\
&= \underbrace{(Q_1 : P)}_{= Q_1} \cap \underbrace{(Q_2 : P)}_{= Q_2} \cap \cdots \cap \underbrace{(Q_{i_0} : P)}_{\supsetneqq Q_{i_0}} \cap \cdots \cap \underbrace{(Q_n : P)}_{= Q_n} \\
&\supseteq Q_1 \cap \cdots \cap Q_n = Ra \,.
\end{aligned}
$$

We claim that strict inclusion must hold. Otherwise, we would have two primary decompositions $Ra = Q_1 \cap \cdots \cap Q_{i_0} \cap \cdots \cap Q_n = Q_1 \cap \cdots \cap (Q_{i_0} : P) \cap \cdots \cap Q_n$. Since P is minimal, the set $\{P\}$ is isolated for Ra so that the P-component is uniquely determined; hence $Q_{i_0} = (Q_{i_0} : P)$ which contradicts (11.16). Thus $(\star)$ is established.

Hence there is an element $b \in (Ra : P) \setminus Ra$. This means $bP \subseteq Ra$, but $b \notin Ra$. Consider the element b/a in the quotient field of R. Then $b/a \notin R$, and $I := (b/a)P$ is an ideal of R. Suppose $I \subseteq P$, i.e., $(b/a)P \subseteq P$. Then P is an $R[b/a]$-module which is finitely generated as an R-module (since R is Noetherian) but whose annihilator in $R[b/a]$ is zero. (If $(rb)/a$ annihilates P then $rbP = \{0\}$ which implies $r = 0$ by the cancellation rule.) By (10.22) this means that b/a is integral over R. But R is integrally closed by hypothesis; hence we obtain $b/a \in R$ which is a contradiction. Hence our assumption $I \subseteq P$ was false; we have $I \not\subseteq P$.

Now let k be the exponent of Q. Then $(b/a)^k Q \not\subseteq P$ because otherwise $I^k = ((b/a)P)^k \subseteq (b/a)^k Q \subseteq P$ which would imply $I \subseteq P$ because P is prime. Hence we can choose the largest number $r \in \{0, 1, \ldots, k-1\}$ with $(b/a)^r Q \subseteq P$ and let $J := (b/a)^{r+1} Q$. Then $J = (b/a)(b/a)^r Q \subseteq (b/a)P = I \subseteq R$ so that $J \trianglelefteq R$, but $J \not\subseteq P$. Moreover,

$$
JP^{r+1} = (\frac{b}{a}P)^{r+1} Q = I^{r+1} Q \,.
$$

We claim that this implies $Q = P^{(r+1)}$. To prove this claim, choose any reduced primary decomposition

$$
P^{r+1} = P^{(r+1)} \cap \underbrace{Q_2 \cap \cdots \cap Q_n}_{=: A}
$$

of P^{r+1}. Then $P^{(r+1)} AJ \subseteq (P^{(r+1)} \cap A)J = P^{r+1} J = QI^{r+1} \subseteq Q$ so that $P^{(r+1)} \subseteq Q : (AJ) = (Q : A) : J$. Assume that $A \subseteq P$. Then $P_2 \cdots P_n \subseteq P_2 \cap \cdots \cap P_n = \sqrt{A} \subseteq \sqrt{P^{r+1}} = P$ and hence $P_i \subseteq P$ for some index i because P is prime. Consequently, $P_i = P$ by the minimality of P; but this contradicts the fact that $P, P_2, \ldots, P_n$ – stemming from a reduced decomposition – are pairwise different. This shows that the assumption $A \subseteq P$ was false; so we have $A \not\subseteq P$ and hence $(Q : A) = Q$ due to (6.19)(c). Furthermore, $(Q : A) : J = Q : J = Q$ because $J \not\subseteq P$ also. Consequently, $P^{(r+1)} \subseteq Q$.

Conversely, we have $QI^{r+1} = P^{r+1} J \subseteq P^{r+1} \subseteq P^{(r+1)}$ and therefore $Q \subseteq (P^{(r+1)} : I^{r+1})$ which equals $P^{(r+1)}$ because $I^{r+1} \not\subseteq P$ (otherwise $I \subseteq P$). This shows that $Q \subseteq P^{(r+1)}$. $\blacksquare$

THE above proposition was the missing link which allows us to prove the following theorem which characterizes Dedekind domains.

(19.11) Theorem. *Let R be an integral domain with quotient field K. Then the following conditions are equivalent:*

(1) *R is a Dedekind domain;*

(2) *every proper ideal can be uniquely written as a product of prime ideals,*

(3) *every nonzero ideal of R is invertible as a fractional ideal,*

(4) *every nonzero fractional ideal of R is invertible;*

(5) *R is Noetherian and integrally closed in K, and every nonzero prime ideal is maximal;*

(6) *R is Noetherian, every nonzero prime ideal is maximal, and every primary ideal in R is a prime power.*

Proof. The implications $(1)\Longrightarrow(2)\Longrightarrow(3)\Longrightarrow(4)\Longrightarrow(5)$ have been proved in (19.9) already.

$(5)\Longrightarrow(6)$. The assumption that all nonzero prime ideals are maximal is equivalent to saying that all nonzero prime ideals are minimal. Thus (19.10) shows that every primary ideal of R is a symbolic prime power. But since all prime ideals are maximal, every symbolic power is already a true power. Indeed, suppose $P^m = P^{(m)} \cap Q_2 \cap \cdots \cap Q_n$ with $n \geq 2$ where Q_i is P_i-primary. Then $P = \sqrt{P^m} = \sqrt{P^{(m)} \cap Q_2 \cap \cdots \cap Q_n} = P \cap P_2 \cap \cdots \cap P_n \subseteq P_n$ which implies $P = P_n$ by the maximality of P. But this is a contradiction.

$(6)\Longrightarrow(1)$. Let I be a proper ideal of R. Since R is Noetherian, we have a reduced primary decomposition $I = Q_1 \cap \cdots \cap Q_n$ where, say, Q_i is P_i-primary. By assumption (6), the ideals $P_1, \ldots, P_n$ are maximal; since they stem from a reduced primary decomposition, they are also pairwise different and hence pairwise comaximal. Therefore, $Q_1, \ldots, Q_n$ are pairwise comaximal due to (6.9)(g). But then $I = Q_1 \cap \cdots \cap Q_n = Q_1 \cdots Q_n$ by the Chinese Remainder Theorem. By assumption (6), every primary ideal in R is a prime power; hence there are natural numbers k_i with $Q_i = P_i^{k_i}$. This yields $I = P_1^{k_1} \cdots P_n^{k_n}$. ∎

WHEREAS the definition of a Dedekind domain expressed a desirable property which is hard to verify directly, the condition (19.11)(e) (which characterizes Dedekind domains) is relatively easy to check and allows one to reveal a given integral domain as a Dedekind domain. In fact, we can exhibit a whole class of rings as Dedekind domains; namely, we will show that the integral closure of $\mathbb{Z}$ in an algebraic number field is always a Dedekind domain.

(19.12) Theorem. *Let L be an algebraic number field and let S be the integral closure of $\mathbb{Z}$ in L. Then L is a Dedekind domain.*

Proof. S is Noetherian by (15.7) and integrally closed in its quotient field L by (10.24)(e). (The fact that L is the quotient field of S was established in (15.3)(e).) What remains is to show that every nonzero prime ideal Q is S is maximal. Applying part (b) of the lying-over theorem (16.16) with $Q_1 = \{0\}$ and $Q_2 = Q$ we see that $Q \cap \mathbb{Z} \neq \{0\}$. Hence $Q \cap \mathbb{Z}$ is a nontrivial prime ideal and hence a maximal ideal in $\mathbb{Z}$. But then Q is a maximal ideal of S, due to (15.2) or to the going-up theorem (16.17).∎

(19.13) Example. Let $n \in \mathbb{Z} \setminus \{0,1\}$ be a square-free number and let

$$\mathcal{O}_n = \begin{cases} \mathbb{Z} + \mathbb{Z}\sqrt{n}, & \text{if } n \not\equiv 1 \text{ modulo } 4; \\ \mathbb{Z} + \mathbb{Z}(\frac{1+\sqrt{n}}{2}), & \text{if } n \equiv 1 \text{ modulo } 4. \end{cases}$$

Then $\mathcal{O}_n$ is the integral closure of $\mathbb{Z}$ in $\mathbb{Q}(\sqrt{n})$ due to (15.5) and hence is a Dedekind domain.

(19.14) Example. Let $\varepsilon = e^{2\pi i/p}$ where p is an odd prime. Then $\mathbb{Z}[\varepsilon]$ is the integral closure of $\mathbb{Z}$ in $\mathbb{Q}(\varepsilon)$, due to (15.15), and hence is a Dedekind domain. ∎

THE end of this section will be used to develop a theory of factorization of ideals in a Dedekind domain. Let us start by introducing some notation.

(19.15) Notation. If P is a prime ideal in a Dedekind domain R, we let $P^0 := R$. Then if $(P_i)_{i \in I}$ are the prime ideals of R, every nonzero ideal $A \trianglelefteq R$ possesses a unique multiplicative decomposition $A = \prod_{i \in I} P_i^{e_i}$ with $e_i \geq 0$ for all i such that $I_A := \{i \in I \mid e_i \neq 0\}$ is finite.

(19.16) Proposition. *Let R be a Dedekind domain with prime ideals $(P_i)_{i \in I}$. If $A = \prod_{i \in I} P_i^{e_i}$ and $B = \prod_{i \in I} P_i^{f_i}$ are ideals of R, then the following conditions are equivalent:*
  (1) $A \supseteq B$;
  (2) $e_i \leq f_i$ for all $i \in I$;
  (3) $A \mid B$ in the sense that there is an ideal $C \trianglelefteq R$ with $B = CA$.

Proof. (1)$\Longrightarrow$(2). Multiply both sides of the inclusion $\prod_i P_i^{e_i} \supseteq \prod_i P_i^{f_i}$ by the fractional ideal $(\prod_i P_i^{\min(e_i, f_i)})^{-1}$ to obtain an inclusion $\prod_{i \in I_0} P_i^{a_i} \supseteq \prod_{i \in I_1} P_i^{b_i}$ where $I_0 := \{i \in I \mid a_i > b_i\}$ and $I_1 := \{i \in I \mid a_i < b_i\}$ are disjoint.† If $I_0 \neq \emptyset$, we can pick an element $i_0 \in I_0$. Then $\prod_{i \in I_1} P_i^{b_i} \subseteq P_{i_0} \subsetneqq R$ and hence $P_i \subseteq P_{i_0}$ for some $i \in I_1$ since P_{i_0} is prime. But all prime ideals in R are maximal; hence $P_i = P_{i_0}$ and thus $i = i_0$ which contradicts the fact that I_0 and I_1 are disjoint. Therefore, our assumption that $I_0 \neq \emptyset$ was wrong; so we have $I_0 = \emptyset$ which is the claim.
  (2)$\Longrightarrow$(3). Take $C := \prod_i P_i^{f_i - e_i}$.
  (3)$\Longrightarrow$(1). $B = CA \subseteq A$. ∎

$\dagger$ For example, $P_1^2 P_2^3 P_3^2 \supseteq P_1 P_2^5 P_3^2$ leads to $P_1 \supseteq P_2^2$.

SINCE we are used to factorizations of natural numbers where $a \mid b$ implies that a is smaller than b, there might be a slight problem (psychological rather than mathematical) in the fact that if $A \mid B$ if and only if A is larger than B. This comes from the fact, of course, that multiplying ideals leads to smaller ideals; for example, taking higher and higher powers makes an ideal smaller and smaller. It might be helpful to observe the analogy between the semigroups $\mathcal{I}(R)$ and $[0,1]$ (where the zero ideal $\langle 0 \rangle = \{0\}$ corresponds to 0 and where $\langle 1 \rangle = R$ corresponds to 1).

THE next theorem shows that we find the greatest common divisor and the least common multiple of two ideals in a Dedekind domain (written as products of prime ideals) in exactly the same way as we find the greatest common divisor and the least common multiple of two elements in a unique factorization domain (written as products of prime elements, i.e., irreducible elements).

(19.17) Theorem. *Let R be a Dedekind domain with prime ideals $(P_i)_{i \in I}$. For any two ideals $A = \prod_{i \in I} P_i^{e_i}$ and $B = \prod_{i \in I} P_i^{f_i}$, the following statements hold:*

(a) $A + B = \prod_{i \in I} P_i^{\min(e_i, f_i)}$ *(which is the greatest common divisor of A and B in $\mathcal{I}(R)$);*

(b) $A \cap B = \prod_{i \in I} P_i^{\max(e_i, f_i)}$ *(which is the least common multiple of A and B in $\mathcal{I}(R)$).*

Proof. (a) Let $C := A + B$. Then C has a prime factorization $C = \prod_i P_i^{g_i}$. Since $A \subseteq C$ and $B \subseteq C$, we have $g_i \leq e_i$ and $g_i \leq f_i$ for all i, due to (19.16), hence $g_i \leq \min(e_i, f_i)$. Using (19.16) again, this means $C \supseteq \prod_i P_i^{\min(e_i, f_i)}$. On the other hand,

$$
C = \prod_i P_i^{e_i} + \prod_i P_i^{f_i} = \prod_i P_i^{\min(e_i, f_i)} \left(\prod_i P_i^{e_i - \min(e_i, f_i)} + \prod_i P_i^{f_i - \min(e_i, f_i)} \right)
$$
$$
\subseteq \prod_i P_i^{\min(e_i, f_i)} .
$$

(b) Let $C := A \cap B$ and write $C = \prod_i P_i^{g_i}$. Since $C \subseteq A$ and $C \subseteq B$, we have $g_i \geq e_i$ and $g_i \geq f_i$, hence $g_i \geq \max(e_i, f_i)$ for all i and therefore $C \subseteq \prod_i P_i^{\max(e_i, f_i)}$, due to (19.16). On the other hand, $\prod_i P_i^{\max(e_i, f_i)} \subseteq \prod_i P_i^{e_i} = A$ and $\prod_i P_i^{\max(e_i, f_i)} \subseteq \prod_i P_i^{f_i} = B$, hence $\prod_i P_i^{\max(e_i, f_i)} \subseteq A \cap B$. ∎

(19.18) Theorem. *Let R be a Dedekind domain.*

(a) *If $\{0\} \neq A \subseteq B$ are ideals of R, there is an element $x \in R$ such that $A + \langle x \rangle = B$.*

(b) *Let $I \trianglelefteq R$ and let $x \in I$ be a nonzero element in I. Then there is an element $y \in I$ such that $I = \langle x, y \rangle$. (Hence every ideal in a Dedekind domain is generated by two elements.)*

(c) *If A, B are nonzero ideals of R, there is an ideal $A^* \neq \{0\}$ such that $A A^*$ is a principal ideal and $B + A^* = R$.*

Proof. (a) We have $A \subseteq B$; hence we can write $A = \prod_i P_i^{f_i}$ and $B = \prod_i P_i^{e_i}$ with $e_i \leq f_i$. Let $C_0 := \prod_i P_i^{e_i+1}$ and $C_j := P_j^{e_j} \prod_{i \neq j} P_i^{e_i+1}$ so that $C_j \subseteq P_i^{e_i+1}$ if $i \neq j$. Then $C_j \supsetneq C_0$ so that we can pick elements $x_j \in C_j \setminus C_0$. We claim that $x := \sum_j x_j$ has the desired property.

We have $\langle x \rangle \subseteq \sum_j \langle x_j \rangle \subseteq \sum_j C_j = \gcd_j C_j = B$ and hence $B \mid \langle x \rangle$ so that $\langle x \rangle = BC$ for some ideal C. We claim that none of the ideals P_i divides C. Otherwise we would have $P_i^{e_i+1} \mid BC = \langle x \rangle$, i.e., $\langle x \rangle \subseteq P_i^{e_i+1}$, hence

$$\langle x_i \rangle = \langle x - \sum_{j \neq i} x_j \rangle \subseteq \langle x \rangle + \sum_{j \neq i} \langle x_j \rangle \subseteq \langle x \rangle + \sum_{j \neq i} C_j \subseteq P_i^{e_i+1}$$

and consequently $\langle x_i \rangle \subseteq P_i^{e_i+1} \cap C_i = C_0$ which contradicts the choice of x_i. Thus, in fact, none of the ideals P_i divides C; consequently,

$$A + \langle x \rangle = A + BC = B(\prod_i P_i^{f_i - e_i} + C) = B \cdot R = B.$$

(b) Pick $x \in I \setminus \{0\}$ so that $\{0\} \neq \langle x \rangle \subseteq I$. By part (a), there is an element $y \in R$ with $\langle x \rangle + \langle y \rangle = I$, i.e., $\langle x, y \rangle = I$.

(c) We have $AB \subseteq A$ and hence $AB + \langle x \rangle = A$ for some $x \in R$, due to part (a). Then $\langle x \rangle \subseteq A$, i.e., $A \mid \langle x \rangle$, say $\langle x \rangle = AA^*$. Then $A = AB + AA^* = A(B + A^*)$; multiplying both sides by the fractional ideal A^{-1} yields $R = B + A^*$. ∎

(19.19) Theorem. *Let R be a Dedekind domain. Then the following conditions are equivalent:*
(1) every prime ideal of R is a principal ideal;
(2) R is a principal ideal domain;
(3) R is a unique factorization domain.

Proof. (1)$\Longrightarrow$(2). This implication follows easily from the fact that every nonzero ideal of R is a finite product of prime ideals.

(2)$\Longrightarrow$(3). This implication is true for general integral domains, as was proved in (7.10)(b).

(3)$\Longrightarrow$(1). Let P be a nonzero prime ideal of R. By (19.18)(c), there is an ideal P^* with $PP^* = \langle x \rangle$ for some element $x \in R^\heartsuit$. By hypothesis, there is a factorization $x = x_1 \cdots x_n$ of x into irreducible elements. Then $PP^* = \langle x_1 \rangle \cdots \langle x_n \rangle$ where each $\langle x_i \rangle$ is a prime ideal. Since the prime decomposition of an ideal of R is unique, this implies that P must be one of the ideals $\langle x_i \rangle$. Since P was chosen arbitrarily, we have shown that every prime ideal of R is a principal ideal. ∎

Exercises

Problem 1. Let R be a commutative ring with identity. Show that if every nonzero ideal generated by two elements is invertible (as a fractional ideal), then every finitely generated ideal is invertible.

Hint. To prove that $I = \langle x_1, \ldots, x_n \rangle$ is invertible use induction on n. Letting $A := \langle x_1, \ldots, x_{n-1} \rangle$, $B := \langle x_2, \ldots, x_n \rangle$ and $C := \langle x_1, x_n \rangle$, show that $I(x_1 A^{-1} C^{-1} + x_n B^{-1} C^{-1}) = R$.

Problem 2. Let R be an integral domain. Show that the following three statements are equivalent:

(a) R is a Dedekind domain;

(b) R is Noetherian, and every nonzero ideal of R generated by two elements is invertible as a fractional ideal;

(c) R is Noetherian, and whenever M is a maximal ideal of R there is no ideal I with $M^2 \subsetneq I \subsetneq M$.

Problem 3. Let R be a Dedekind domain and let $M \trianglelefteq R$ be a maximal ideal of R. Prove the following statements.

(a) Every M-primary ideal of R is a power of M.

(b) There is no ideal $I \trianglelefteq R$ with $M^2 \subsetneq I \subsetneq M$.

Hint. To prove (a) use the fact that every nonzero M-primary ideal can be written as a product of prime ideals. To prove (b), show first that any ideal I with $M^2 \subseteq I \subseteq M$ is M-primary and then use (a).

Problem 4. Let $V = \{(x, y) \in L^2 \mid x^2 + y^2 = 1\}$.

(a) Show that the coordinate ring $K[V]$ is a Dedekind domain.

(b) Find the unique factorization of $\langle y^2 - (x-1)(x^2-1) \rangle \trianglelefteq K[V]$ into prime ideals.

Problem 5. Let $V = \{(x, y) \in L^2 \mid x^2 = y^3\}$.

(a) Show that the coordinate ring $K[V]$ is not integrally closed and hence is not a Dedekind domain.

(b) Find a nonzero ideal of $K[V]$ which cannot be factored into prime ideals.

Problem 6. Let $I \neq \{0\}$ be an ideal in a Dedekind domain R.

(a) Show that there are maximal ideals $M_1, \ldots, M_n$ of R with $I = M_1 \cdots M_n = M_1 \cap \cdots \cap M_n$.

(b) Show that R/I is a principal ideal domain.
Hint. Use part (a) and the Chinese remainder theorem.

(c) Use part (b) to give a new proof that I can be generated by at most two elements.

Problem 7. Let $R = \mathbb{Z}[\sqrt{-5}]$. Show that

$$I := \{a + b\sqrt{-5} \mid a, b \in \mathbb{Z}, \ a \equiv b \, (2)\}$$

is an ideal of R which is not a principal ideal. Find an element $y \in I$ such that $I = \langle 4 - 2\sqrt{-5}, \ y \rangle$.

Problem 8. Let n be a square-free integer with $n \equiv 1$ modulo 4.
(a) Show that $\langle 2, \sqrt{n} \rangle$ is not a principal ideal in $\mathbb{Z}[\sqrt{n}]$.
(b) Show that $\langle 2, \sqrt{n} \rangle$ is a principal ideal in $\mathbb{Z}[(1 + \sqrt{n})/2]$.

Problem 9. Let I, J and K be ideals of a Dedekind domain R. Prove the distributive laws
$$I \cap (J + K) \ = \ (I \cap J) + (I \cap K) \quad \text{and}$$
$$I + (J \cap K) \ = \ (I + J) \cap (I + K).$$

Problem 10. Let R be a Dedekind domain. For a polynomial $f(x) = \sum_{k=0}^{n} a_k x^k$ in $R[x]$ we call $C(f) := \langle a_0, \ldots, a_n \rangle \trianglelefteq R$ the **content** of f. Prove that $C(fg) = C(f)C(g)$ for all $f, g \in R[x]$. (Compare with Gauß' lemma (8.2)(a).)

Problem 11. (a) Show that each nonzero element of a Dedekind domain is contained in only finitely many maximal ideals of R.
(b) Show that if a Dedekind has only a finite number of maximal ideals then it is a principal ideal domain.

Problem 12. Let R be a Dedekind domain and let S be a multiplicatively closed subset of R. Show that $S^{-1}R$ is a Dedekind domain again.

Problem 13. (a) Consider the ideals $P_1 = \langle 3, 4 + \sqrt{-5} \rangle$, $P_2 = \langle 3, 4 - \sqrt{-5} \rangle$, $P_3 = \langle 7, 4 + \sqrt{-5} \rangle$ and $P_4 = \langle 7, 4 - \sqrt{-5} \rangle$ in $\mathbb{Z} + \mathbb{Z}\sqrt{-5}$. Determine $P_1 P_2$, $P_3 P_4$, $P_1 P_3$ and $P_2 P_4$.
(b) Find coprime elements α and β in $\mathbb{Z} + \mathbb{Z}\sqrt{-5}$ such that there are no elements $x, y \in \mathbb{Z} + \mathbb{Z}\sqrt{-5}$ with $x\alpha + y\beta = 1$.

20. Introduction to Galois theory: Solving polynomial equations

IN this section we will briefly consider the history of solving polynomial equations. Starting from the well-known quadratic formula which was known to the ancients already, we will present the formulas which were found in the renaissance period to solve equations of degree 3 and 4. With increasing degree, these formulas become more and more complicated; it is not at all important to memorize them, and we will see that their derivation – though elementary – involves arithmetical tricks which somehow "magically" lead to the solution, but it remains obscure why these tricks work. It certainly took a great deal of trial and error before mathematicians hit upon the right "trick", and a solution based on "tricks" does not easily lend itself to generalization.

Nevertheless, there was good hope to also find formulas to solve equations of degree 5 and higher. However, all attempts in this direction failed, and it became important to analyze more precisely "why" there are solutions in degree ≤ 4 to see the obstacles for higher degrees. Completely new ideas (which enormously stimulated the development of modern algebra and which will be the subject of the subsequent sections) entered the field; these ideas constitute what is now known as Galois theory. But before we develop this rather sophisticated part of mathematics, we want to review the starting point from which it spread, namely the question how to solve polynomial equations.

LET us start with the well-known quadratic formula.

(20.1) Quadratic equations. To solve the equation

(1)
$$x^2 + px + q = 0 \,,$$

we add $(p/2)^2$ to both sides to get

$$(x + \frac{p}{2})^2 + q = (\frac{p}{2})^2 \,.$$

Now the linear substitution $y := x + (p/2)$ reduces (1) to the "pure" quadratic equation

(2)
$$y^2 = (\frac{p}{2})^2 - q \,.$$

Hence if $\sqrt{(p/2)^2 - q}$ denotes any one of the two solutions of this equation[†], then the solutions of (1) are given by

(3)
$$x = -\frac{p}{2} \pm \sqrt{(\frac{p}{2})^2 - q} \,.$$

---

† Clearly, if y_1 is one solution of the equation $y^2 = (p/2)^2 - q =: \alpha$, then $y_2 := -y_1$ is the other, and these two roots are in fact different unless $\alpha = 0$ or unless the base field (which was not specified at all) has characteristic 2. However, there is no purely algebraic way of distinguishing between these roots. Only if the base field is contained in $\mathbb{R}$ and if $\alpha > 0$, there is a canonical choice; namely, in this case $\sqrt{\alpha}$ is defined as the *positive* one of the two roots, but this choice involves the order relation on $\mathbb{R}$, not just the field axioms, and hence is not purely algebraic.

WHEREAS already the Babylonians of about 2000 B.C. knew how to solve quadratic equations, it was only in the 16th century that Italian mathematicians found a formula to solve cubic equations.† Let us present this formula in modern notation.

(20.2) Cubic equations. Given the general cubic equation

$$(4) \qquad x^3 + ax^2 + bx + c = 0 ,$$

we apply the linear substitution $y := x + (a/3)$ which reduces (4) to an equation without quadratic term, namely

$$(5) \qquad \boxed{y^3 + py + q = 0} \qquad \text{where} \quad p := b - \frac{a^2}{3} \quad \text{and} \quad q := c - \frac{a}{3}b + 2(\frac{a}{3})^3 .$$

This substitution follows a rather general recipe to transform a given polynomial into a polynomial whose second coefficient vanishes. (See problem 1 below.) The next step, on the other hand, is very tricky, and it becomes clear only from hindsight why it works: Namely, we now try to find a solution of the form $y = \sqrt[3]{A} + \sqrt[3]{B}$. This yields

$$y^3 = A + B + 3(\sqrt[3]{A} + \sqrt[3]{B})\sqrt[3]{AB}$$

so that (5) reads

$$A + B + q + (\sqrt[3]{A} + \sqrt[3]{B})(3\sqrt[3]{AB} + p) = 0 .$$

This last equation clearly holds if we make sure that both the rational and the irrational part of the left-hand side vanish; i.e., if

$$A + B = -q \quad \text{and} \quad AB = -(\frac{p}{3})^3 .$$

This system of two equations has the solution

$$A = -\frac{q}{2} + \sqrt{(\frac{p}{3})^3 + (\frac{q}{2})^2} , \quad B = -\frac{q}{2} - \sqrt{(\frac{p}{3})^3 + (\frac{q}{2})^2} \qquad \text{(or vice versa)}$$

so that a solution of (5) is given by **Cardano's formula**

$$(6) \qquad \boxed{y = \sqrt[3]{-\frac{q}{2} + \sqrt{(\frac{p}{3})^3 + (\frac{q}{2})^2}} + \sqrt[3]{-\frac{q}{2} - \sqrt{(\frac{p}{3})^3 + (\frac{q}{2})^2}} .}$$

---

† Around 1515, Scipione del Ferro († 1526), professor of mathematics at Bologna, obtained the formula first, but he did not publish it. The formula was rediscovered in 1535 by Niccolò Fontana ($\sim$ 1500-1557), nicknamed Tartaglia ("Stammerer"), from Brescia. He told his result to Girolamo Cardano (1501-1576) who published it – contrary to his promise to keep the secret of the formula – in his work "Ars Magna, sive de regulis algebraicis" (*The Great Art, or On the Rules of Algebra*).

Using $p = b - (a^2/3)$, $q = c - (a/3)b + 2(a/3)^3$ and $y = x + (a/3)$, we can easily write down a solution of the initial equation (4).

FOR several reasons, Cardano's formula is far less convenient than the quadratic formula.

(a) The formula gives only one solution of a cubic equation, not all solutions. For instance, Cardano constructed the example

$$x^3 + 16 \;=\; 12x$$

to obtain 2 as a solution, but his formula yields the solution -4. One way out of this difficulty is to recall that we generally expect a field element to have three different cubic roots (in some algebraically closed field extension) which cannot be distinguished algebraically.† Hence if we choose different possibilities for the cubic roots in Cardano's formula††, we might get all different solutions of equation (5); however, this raises the question whether each of the three choices for the first cubic root in Cardano's formula can be combined with each choice for the second cubic root. (See problem 4 below.)

(b) It is sometimes difficult to see whether or not a solution is rational. In his "Ars Magna", Cardano applied his formula to the equation

$$x^3 + 6x \;=\; 20$$

and obtained the real solution

$$\sqrt[3]{10 + \sqrt{108}} \;+\; \sqrt[3]{-10 + \sqrt{108}}\,.$$

Now obviously 2 is a solution of the given equation, and since the function $f(x) = x^3 + 6x$ increases monotonically, this is the only real solution. Hence

$$\sqrt[3]{10 + \sqrt{108}} \;+\; \sqrt[3]{-10 + \sqrt{108}} \;=\; 2$$

which is far from obvious.

(c) It is easy to check that 4 is a solution of the equation

$$x^3 \;=\; 15x + 4\,,$$

but Cardano's formula yields the expression

$$\sqrt[3]{2 + \sqrt{-121}} \;+\; \sqrt[3]{2 - \sqrt{-121}}\,.$$

How should we interpret the occurring square roots of negative numbers? The appearance of such roots in applications of the quadratic formula simply signalizes the absence

---

† When α is a real number, then we denote by $\sqrt[3]{\alpha}$ the unique *real* solution of the equation $y^3 = \alpha$, but if the base field is $\mathbb{Q}$ then this is not a natural choice, because $\mathbb{R}$ is obtained from $\mathbb{Q}$ by an analytical, not an algebraic process.

†† In much the same way we could have written the solution of the quadratic equation $x^2 + px + q = 0$ as $x = -(p/2) + \sqrt{(p/2)^2 - q}$ with the understanding that $\sqrt{(p/2)^2 - q}$ can be any of the two roots of $(p/2)^2 - q$.

of real solutions, hence does not cause any trouble. However, a cubic polynomial with real coefficients has always a real root, and indeed, Cardano and his contemporaries felt compelled to study square roots of negative numbers. Rafaele Bombelli (ca. 1526-1573), in his treatise "Algebra" of 1572, manipulated these mysterious objects like ordinary numbers without speculating too much about what these objects really were. Calculating for example

$$(2 + \sqrt{-1})^3 \;=\; 8 + 12\sqrt{-1} + 6(\sqrt{-1})^2 + (\sqrt{-1})^3 \;=\; 8 + 12\sqrt{-1} - 6 - \sqrt{-1}$$
$$=\; 2 + 11\sqrt{-1} \;=\; 2 + \sqrt{-121} \;,$$

he obtained

$$\sqrt[3]{2 + \sqrt{-121}} \;+\; \sqrt[3]{2 - \sqrt{-121}} \;=\; (2 + \sqrt{-1}) + (2 - \sqrt{-1}) \;=\; 4$$

which is in fact the expected solution.

ONE of Cardano's students, Ludovico Ferrari (1522-1565), found a formula to solve quartic equations which we now want to present.

(20.3) Quartic equations. Given an arbitrary quartic equation

(7)
$$x^4 + ax^3 + bx^2 + cx + d \;=\; 0 \;,$$

The linear substitution $y := x + (a/4)$ leads to an equation without a cubic term, namely

(8)
$$y^4 + py^2 + qy + r \;=\; 0$$

where

$$p \;=\; b - 6\left(\frac{a}{4}\right)^2 \;, \quad q \;=\; c - \left(\frac{a}{2}\right)b + \left(\frac{a}{2}\right)^3 \;, \quad r \;=\; d - \left(\frac{a}{4}\right)c + \left(\frac{a}{4}\right)^2 b - 3\left(\frac{a}{4}\right)^4 \;.$$

If $q = 0$, this equation is biquadratic and can be easily solved. (Do it!) If $q \neq 0$ we rewrite (8) as

(8')
$$\left(y^2 + \frac{p}{2}\right)^2 \;=\; -qy - r + \left(\frac{p}{2}\right)^2$$

and add a term u in the square on the left-hand side;

(8'')
$$\left(y^2 + \frac{p}{2} + u\right)^2 \;=\; -qy - r + \left(\frac{p}{2}\right)^2 + u^2 + 2uy^2 + pu \;.$$

Now again a very tricky argument enters: If we succeed in choosing u such that also the right-hand side of (8'') is a square, then we can take the square root on both sides, thus simplifying (8'') substantially. By looking at the coefficients of y^2 and y, we see that there is only one way that the right-hand side can be a square, namely

$$-qy - r + \left(\frac{p}{2}\right)^2 + u^2 + 2uy^2 + pu \;=\; \left(\sqrt{2u}\,y - \frac{q}{2\sqrt{2u}}\right)^2 \;.$$

Now one easily checks that this last equation is equivalent to the auxiliary equation (called Ferrari's resolvent cubic equation)

$$(9) \quad -r + (\tfrac{p}{2})^2 + pu + u^2 \;=\; \frac{q^2}{8u} \,, \qquad \text{i.e.,} \qquad 8u^3 + 8pu^2 + (2p^2 - 8r)u - q^2 \;=\; 0.$$

Note that $u \neq 0$ since $q \neq 0$. Now if we take a solution u of (9) (which is possible because there is a formula to solve cubic equations!), then (8″) reduces to

$$y^2 + \frac{p}{2} + u \;=\; \pm(\sqrt{2u}\, y - \frac{q}{2\sqrt{2u}})$$

yielding the four solutions

$$y \;=\; \varepsilon \sqrt{\frac{u}{2}} \pm \sqrt{-\frac{u}{2} - \frac{p}{2} - \frac{\varepsilon q}{2\sqrt{2u}}} \,, \qquad \text{where } \varepsilon = \pm 1 \,.$$

ENCOURAGED by the discovery of the formulas to solve cubic and quartic equations, mathematicians tried to find analogous formulas for quintic and even higher-order equations – all in vain. It became clear only in the 19th century that it is indeed impossible to find a general formula to solve equations of order $n \geq 5$, and it took a lot of efforts and new ideas to gain this insight. Before we build up the machinery that will enable us to discuss this question thoroughly, let us point out how our modern approach is related to the problem under consideration. The mathematicians of the 16th century were exclusively interested in equations with rational or even integer coefficients, but they knew that they had to leave the domain of rational numbers to obtain solutions of polynomial equations. But even when they came across really mysterious objects like

$$\sqrt[3]{\sqrt{-121}} \,,$$

they insisted that their solutions could be treated as true numbers – the usual arithmetic laws (commutativity, associativity, distributivity) should be valid for these new numbers as well. We have already prepared the foundations to clarify what kind of objects these new numbers are; namely, if we start with a polynomial f with coefficients in a field K, its roots will be elements of a clearly defined larger field $L \supseteq K$, for example, a splitting field of f. So instead of discussing polynomial equations in a straightforward manner, we will discuss field extensions related to these equations and their properties.

This style of approaching the topic has been called the "Dedekind-Artin linearization of Galois theory" (*Kaplansky*). In the style of classical algebra, the leading theme of Galois theory was the solution of polynomial equations, the search for the roots of a given polynomial. These roots usually did not lie in the base-field of the polynomial; one had to ascend to a larger field. Now in his exposition "Galois Theory" (1942), Artin changed the point of view: He did not start with a smaller field and try to extend it, but he treated the larger field as a given datum. Consequently, he developed a theory of field extensions rather than a theory of polynomials. This approach turns out to be fruitful, because the structure theory of fields becomes available to study polynomial equations.

Exercises

Problem 1. Let $f(x) = x^n + a_{n-1}x^{n-1} + \cdots + a_1 x + a_0$ be a monic polynomial over a field K such that $\mathrm{char}\,K$ does not divide n. Show that the substitution $y := x + (a_{n-1}/n)$ yields a polynomial of the form $f(y) = y^n + b_{n-2}y^{n-2} + \cdots + b_1 y + b_0$, i.e., a polynomial whose second coefficient vanishes.

Problem 2. Solve the quadratic equations $x^2 + 2\sqrt{5}\,x - 12i = 0$ and $x^2 - 2i\sqrt{7}\,x - 24i = 0$ in $\mathbb{C}$.

Problem 3. To solve the equation $x^3 + px + q = 0$, Vieta suggested the substitution $x = y - (p/(3y))$. Use this substitution to obtain a quadratic equation for y^3. Solve for y and recover Cardano's formula.

Problem 4. Given the equation $x^3 + px + q = 0$, let $\varepsilon := e^{2\pi i/3}$.
   (a) Show that one can find numbers a and b with $-3ab = p$ and $a^3 + b^3 = q$ by solving a quadratic equation for a^3.
   (b) Show that $\varepsilon^2 + \varepsilon + 1 = 0$ and use this fact to derive the factorization

$$x^3 - 3abx + (a^3 + b^3) = (x + a + b)(x + \varepsilon a + \varepsilon^2 b)(x + \varepsilon^2 a + \varepsilon b) .$$

   (c) Combine parts (a) and (b) to show that the three solutions of the equation $x^3 + px + q = 0$ are $P + Q$, $\varepsilon P + \varepsilon^2 Q$ and $\varepsilon^2 P + \varepsilon Q$ where

$$D := \left(\tfrac{p}{3}\right)^3 + \left(\tfrac{q}{2}\right)^2 , \quad P := \sqrt[3]{-\tfrac{q}{2} + \sqrt{D}} , \quad Q := \sqrt[3]{-\tfrac{q}{2} - \sqrt{D}} .$$

(This is a more precise version of Cardano's formula.)

Problem 5. Find all solutions of the following cubic equations.
   (a) $x^3 - 3x - 2 = 0$
   (b) $x^3 + 3x - 2 = 0$
   (c) $x^3 + x^2 - 2x - 1 = 0$
   (d) $2x^3 - 5x^2 + 18x - 45 = 0$

Problem 6. (a) Show that $\sqrt{5 + \sqrt{24}} = \sqrt{3} + \sqrt{2}$ and $\sqrt{5 - \sqrt{24}} = \sqrt{3} - \sqrt{2}$.
   (b) Show that

$$\sqrt[3]{1 + \tfrac{2}{3}\sqrt{\tfrac{7}{3}}} + \sqrt[3]{1 - \tfrac{2}{3}\sqrt{\tfrac{7}{3}}} = 1 \quad \text{and} \quad \sqrt[3]{2 + \sqrt{5}} + \sqrt[3]{2 - \sqrt{5}} = 1 .$$

Problem 7. With each quartic polynomial of the form $f(x) = x^4 + px^2 + qx + r$ we associate Ferrari's resolvent cubic $\varphi(u) := 8u^3 + 8pu^2 + (2p^2 - 8r)u - q^2$.

(a) We saw in (20.3) that if u is a root of φ then the equation $f(x) = 0$ is equivalent to

(⋆)
$$x^2 + \frac{p}{2} + u = \pm(\sqrt{2u}\,x - \frac{q}{2\sqrt{2u}})\;;$$

for each possible choic of sign, this is a quadratic equation for x. Let α_1, α_2 be the two solutions of the equation with "+" and α_3, α_4 the ones for "−". Show that $u = -\frac{1}{2}(\alpha_1 + \alpha_2)(\alpha_3 + \alpha_4)$.

Hint. $\alpha_1\alpha_2$ and $\alpha_3\alpha_4$ can be expressed by Vieta's formula for the two quadratic equations (⋆); on the other hand, the coeffient p of the polynomial $f(x) = x^4 + px^2 + qx + r$ is an elementary symmetric polynomial in the roots $\alpha_1, \alpha_2, \alpha_3, \alpha_4$ of f.

(b) Let $\alpha_1, \alpha_2, \alpha_3, \alpha_4$ be the roots of f. Show that the roots of φ are given by

$$u_1 = -\frac{1}{2}(\alpha_1 + \alpha_2)(\alpha_3 + \alpha_4),$$

$$u_2 = -\frac{1}{2}(\alpha_1 + \alpha_3)(\alpha_2 + \alpha_4),$$

$$u_3 = -\frac{1}{2}(\alpha_1 + \alpha_4)(\alpha_2 + \alpha_3).$$

(c) In the notation of part (b), check that

$$(u_1 - u_2)(u_1 - u_3)(u_2 - u_3) = -(\alpha_1 - \alpha_2)(\alpha_1 - \alpha_3)(\alpha_1 - \alpha_4)(\alpha_2 - \alpha_3)(\alpha_2 - \alpha_4)(\alpha_3 - \alpha_4)$$

and conclude that $D(\varphi) = D(f)$, i.e., that f and φ have the same discriminant.

Problem 8. Associate with a quartic polynomial $f(x) = x^4 - s_1x^3 + s_2x^2 - s_3x + s_4$ the cubic polynomials

$$p(x) = x^3 - 2s_2x^2 + (s_2^2 + s_1s_3 - 4s_4)x - (s_1s_2s_3 - s_1^2s_4 - s_3^2)^2 \quad \text{and}$$
$$q(x) = x^3 - s_2x^2 + (s_1s_3 - 4s_4)x - (s_1^2s_4 - 4s_2s_4 + s_3^2)$$

and let $\alpha_1, \alpha_2, \alpha_3$ and α_4 be the (unknown) roots of f.

(a) Show that the roots of p are

$$u_1 = (\alpha_1 + \alpha_2)(\alpha_3 + \alpha_4), \quad u_2 = (\alpha_1 + \alpha_3)(\alpha_2 + \alpha_4), \quad u_3 = (\alpha_1 + \alpha_4)(\alpha_2 + \alpha_3) \;.$$

(b) Show that the roots of q are

$$v_1 = \alpha_1\alpha_2 + \alpha_3\alpha_4, \quad v_2 = \alpha_1\alpha_3 + \alpha_2\alpha_4, \quad v_3 = \alpha_1\alpha_4 + \alpha_2\alpha_3 \;.$$

(c) Show that

$$\prod_{i<j}(\alpha_i - \alpha_j) = -(u_1 - u_2)(u_1 - u_3)(u_2 - u_3) = -(v_1 - v_2)(v_1 - v_3)(v_2 - v_3) \;.$$

Problem 9. Generalizing the notion of a symmetric polynomial, we call a polynomial $f \in K[x_1,\ldots,x_n]$ in n variables invariant under a group $G \leq \mathrm{Sym}_n$ if $f(x_{\sigma(1)},\ldots,x_{\sigma(n)}) = f(x_1,\ldots,x_n)$ for all $\sigma \in G$. Show that a polynomial in four variables $x_1,\ldots,x_4$ is invariant under the group

$$V := \{\mathrm{id},\ (12)(34),\ (13)(24),\ (14)(23)\}$$

if and only if it can be written as a polynomial in $(x_1+x_2)(x_3+x_4)$, $(x_1+x_3)(x_2+x_4)$ and $(x_1+x_4)(x_2+x_3)$. Compare with problem 7 above!

Problem 10. To solve the quartic equation $y^4 + py^2 + qy + r = 0$ with $q \neq 0$, find the conditions for a,b,c,d to satisfy

$$y^4 + py^2 + qy + r = (y^2 + ay + b)(y^2 + cy + d) .$$

Express b, c and d by a and get a cubic equation for a^2. Why does this procedure (which is due to Descartes) indicate a way to solve the given quartic equation?

Problem 11. In order to solve the quartic equation $x^4 + ax^3 + bx^2 + cx + d = 0$ one can try to find a factorization

$(\star)$
$$\begin{aligned} x^4 + ax^3 + bx^2 + cx + d &= (x^2 + px + q)^2 - (rx + s)^2 \\ &= \big((x^2 + px + q) + (rx + s)\big)\big((x^2 + px + q) - (rx + s)\big) . \end{aligned}$$

(a) How can the solutions of the equation $x^4 + ax^3 + bx^2 + cx + d = 0$ be obtained from the factorization $(\star)$?
(b) Derive the exact conditions for p, q, r, s for the factorization $(\star)$ to hold.
(c) Show that if q is a solution of the cubic equation

$$8q^3 - 4bq^2 + (2ac - 8d)q + (4bd - a^2d - c^2) = 0$$

and if

$$p := \frac{a}{2}, \quad s := \sqrt{q^2 - d}, \quad r := \sqrt{\frac{a^2}{4} + 2q - b}$$

then $(\star)$ holds.

Problem 12. Let $f(x) = ax^4 + 4bx^3 + 6cx^2 + 4dx + e$ be a quartic polynomial over a field K with $\mathrm{char}\,K \neq 2, 3$.
(a) Show that to factorize f in the form

$$f(x) = (px^2 + 2qx + r)(p'x^2 + 2q'x + r')$$

one has to choose p,q,r,p',q',r' such that

$$a = pp',$$
$$2b = pq' + p'q,$$
(⋆)
$$6c = pr' + p'r + 4qq',$$
$$2d = qr' + q'r,$$
$$e = rr'.$$

(b) Show that with the substitution $s := qq' - c$ the system (⋆) is equivalent to the matrix equation

(⋆⋆)
$$\begin{pmatrix} p & p' & 0 \\ q & q' & 0 \\ r & r' & 0 \end{pmatrix} \begin{pmatrix} p' & q' & r' \\ p & q & r \\ 0 & 0 & 0 \end{pmatrix} = 2 \begin{pmatrix} a & b & c - 2s \\ b & c + s & d \\ c - 2s & d & e \end{pmatrix}.$$

(c) Take the determinant on both sides of (⋆⋆) to see that s satisfies the cubic equation

$$4s^3 - (ae - 4bd + 3c^2)s - \det \begin{pmatrix} a & b & c \\ b & c & d \\ c & d & e \end{pmatrix} = 0.$$

Show that once s is obtained from this equation then p, q, r, p', q', r' can be easily obtained.

Problem 13. Find the solutions of the following quartic equations.
(a) $x^4 - 13x^2 + 36 = 0$
(b) $x^4 - 4x^3 + 8x + 20 = 0$
(c) $x^4 + 3x^3 - x - 3 = 0$
(d) $x^4 - 2x^3 + 3x^2 - 2x + 2 = 0$
(e) $x^4 + 4x^3 + 4x^2 - 4x - 5 = 0$

Problem 14. A ladder of length ℓ is to be leaned against a wall such that it touches a box whose cross-section is a square of side s (see the picture). At what distance x from the box does one have to put the foot of the ladder? (To get a numerical result, take $\ell = 5$ and $s = 1$; note that numerical methods like Newton's are much better suited to get concrete solutions for practical purposes than our algebraic solution of quartic equations.)

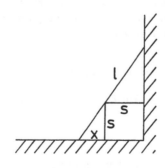

21. The Galois group of a field extension

WE observed that one feature that distinguishes the study of field extensions from that of arbitrary ring extensions is the fact that a field can be considered as a vector space over any of its subfields, an observation which makes the power of linear algebra available for the study of field extensions. A second characteristic feature comes to light when we consider – according to the general approach of studying with any algebraic structure also the mappings which preserve this structure – not just fields but also field homomorphisms. Namely, nonzero homomorphisms between fields (in contrast to homomorphisms between rings) are automatically injective and hence faithfully transfer the structure of the first field to the second field. Since one of the main motivations of the study of field extensions stems from the problem of finding the roots of polynomials, let us point out the role of field homomorphisms in this context.

(21.1) Remark. Any field homomorphism $\sigma : K \to K'$ induces a homomorphism

$$
\begin{array}{ccc}
K[x] & \to & K'[x] \\
f & \mapsto & \sigma \star f
\end{array}
$$

between the associated polynomial rings by letting $(\sigma \star f) := \sum_{k=0}^{n} \sigma(a_k) x^k \in K'[x]$ where $f(x) = \sum_{k=0}^{n} a_k x^k \in K[x]$. Now if σ can be extended to a homomorphism $\sigma : L \to L'$ between extensions L of K and L' of K', then

$$
(\sigma \star f)(\sigma(\alpha)) \;=\; \sum_{k=0}^{n} \sigma(a_k)\sigma(\alpha)^k \;=\; \sigma(\sum_{k=0}^{n} a_k \alpha^k) \;=\; \sigma(f(\alpha))
$$

for all $\alpha \in L$. In particular, if $\alpha \in L$ is a root of f then $\sigma(\alpha)$ is a root of $\sigma \star f$. In the special situation that σ is a K-automorphism of L (which means that σ is the identity on K) we have $\sigma \star f = f$ for all $f \in K[x]$; consequently, σ permutes the roots of f in L.

THUS if f is a polynomial with coefficients in a field K and if L is a field extension of K (for example a splitting field of f) there is a close relation between the roots of f in L and the K-automorphisms of L. It is easily checked that these automorphisms form a group, the group operation being the composition of mappings. Due to its importance this group is given a special name.

(21.2) Definition. *The* **Galois group** *of a field extension* $(L : K)$ *is*

$$
G_K^L \;:=\; \{\sigma \in \operatorname{Aut} L \mid \sigma(k) = k \text{ for all } k \in K\} \,.
$$

LET us give some examples of Galois groups.

(21.3) Example. If K is the prime field of L, then obviously $G_K^L = \operatorname{Aut} L$.

(21.4) Example. Let us determine the Galois group $G_{\mathbb{Q}}^{\mathbb{Q}(\sqrt{2},\sqrt{3})}$ of the field extension $(\mathbb{Q}(\sqrt{2},\sqrt{3}) : \mathbb{Q})$. Every element $\sigma \in G_{\mathbb{Q}}^{\mathbb{Q}(\sqrt{2},\sqrt{3})}$ is uniquely determined by the values $\sigma(\sqrt{2})$ and $\sigma(\sqrt{3})$. To wit: every element of $\mathbb{Q}(\sqrt{2},\sqrt{3})$ has the form $a + b\sqrt{2} + c\sqrt{3} + d\sqrt{6}$ where $a,b,c,d \in \mathbb{Q}$, and

$$\sigma(a + b\sqrt{2} + c\sqrt{3} + d\sqrt{6}) = a + b\sigma(\sqrt{2}) + c\sigma(\sqrt{3}) + d\sigma(\sqrt{2})\sigma(\sqrt{3}) .$$

Now $\sqrt{2}$ and $\sqrt{3}$ satisfy the equations $x^2 = 2$ and $x^2 = 3$, respectively; hence so do $\sigma(\sqrt{2})$ and $\sigma(\sqrt{3})$. Consequently, $\sigma(\sqrt{2}) = \pm\sqrt{2}$ and $\sigma(\sqrt{3}) = \pm\sqrt{3}$. Each possible choice of signs yields in fact an automorphism of $\mathbb{Q}(\sqrt{2},\sqrt{3})$. We omit the somewhat tedious (but straightforward) verification, since this claim will be a simple consequence of the main theorem of Galois theory to be presented later. Thus $G_{\mathbb{Q}}^{\mathbb{Q}(\sqrt{2},\sqrt{3})}$ has the following 4 elements.

$$\sigma_0 : \begin{matrix} \sqrt{2} \mapsto \sqrt{2} \\ \sqrt{3} \mapsto \sqrt{3} \end{matrix} \qquad \sigma_1 : \begin{matrix} \sqrt{2} \mapsto -\sqrt{2} \\ \sqrt{3} \mapsto \sqrt{3} \end{matrix} \qquad \sigma_2 : \begin{matrix} \sqrt{2} \mapsto \sqrt{2} \\ \sqrt{3} \mapsto -\sqrt{3} \end{matrix} \qquad \sigma_3 : \begin{matrix} \sqrt{2} \mapsto -\sqrt{2} \\ \sqrt{3} \mapsto -\sqrt{3} \end{matrix}$$

Since each field automorphism $\sigma \in G_{\mathbb{Q}}^{\mathbb{Q}(\sqrt{2},\sqrt{3})}$ is uniquely determined by the vector space endomorphism $\sigma|_{\mathbb{Q}\sqrt{2}+\mathbb{Q}\sqrt{3}}$, we can use the identifications

$$\sigma_0 \simeq \begin{pmatrix} 1 & 0 \\ 0 & 1 \end{pmatrix}, \quad \sigma_1 \simeq \begin{pmatrix} -1 & 0 \\ 0 & 1 \end{pmatrix}, \quad \sigma_2 \simeq \begin{pmatrix} 1 & 0 \\ 0 & -1 \end{pmatrix}, \quad \sigma_3 \simeq \begin{pmatrix} -1 & 0 \\ 0 & -1 \end{pmatrix} .$$

This shows that $G_{\mathbb{Q}}^{\mathbb{Q}(\sqrt{2},\sqrt{3})}$ is isomorphic to Klein's group $\mathbb{Z}_2 \times \mathbb{Z}_2$:

$$\boxed{G_{\mathbb{Q}}^{\mathbb{Q}(\sqrt{2},\sqrt{3})} \cong \mathbb{Z}_2 \times \mathbb{Z}_2 .}$$

(21.5) Example. An element σ of $G_{\mathbb{Q}}^{\mathbb{Q}(\sqrt{2})}$ is uniquely determined by the value $\sigma(\sqrt{2})$. Now $\sqrt{2}$ satisfies the equation $x^2 = 2$; hence so does $\sigma(\sqrt{2})$, which implies $\sigma(\sqrt{2}) = \pm\sqrt{2}$. This leads to the two automorphisms $\sigma = \mathrm{id}$ and $\sigma = \tau$ where $\tau : \mathbb{Q}(\sqrt{2}) \to \mathbb{Q}(\sqrt{2})$ is defined by $\tau(a + b\sqrt{2}) := a - b\sqrt{2}$. Hence

$$\boxed{G_{\mathbb{Q}}^{\mathbb{Q}(\sqrt{2})} = \{\mathrm{id}, \tau\} \cong \mathbb{Z}_2 .}$$

(21.6) Example. Let us determine $G_{\mathbb{Q}}^{\mathbb{Q}(\sqrt[3]{2})}$. Every element $\sigma \in G_{\mathbb{Q}}^{\mathbb{Q}(\sqrt[3]{2})}$ is uniquely determined by the value $\sigma(\sqrt[3]{2})$. Now $\sqrt[3]{2}$ satisfies the equation $x^3 = 2$, hence so does $\sigma(\sqrt[3]{2})$; but since also $\sigma(\sqrt[3]{2}) \in \mathbb{Q}(\sqrt[3]{2}) \subseteq \mathbb{R}$ we conclude that $\sigma(\sqrt[3]{2}) = \sqrt[3]{2}$. This shows that $G_{\mathbb{Q}}^{\mathbb{Q}(\sqrt[3]{2})}$ is the trivial group whose only element is the identity on $\mathbb{Q}(\sqrt[3]{2})$, i.e.,

$$\boxed{G_{\mathbb{Q}}^{\mathbb{Q}(\sqrt[3]{2})} = \{\mathrm{id}\} .}$$

(21.7) Example. Let us determine $G_{\mathbb{R}}^{\mathbb{C}}$. For all $\sigma \in G_{\mathbb{R}}^{\mathbb{C}}$ and all complex numbers $a + ib$ we have $\sigma(a + ib) = \sigma(a) + \sigma(i)\sigma(b) = a + \sigma(i)b$; so σ is uniquely determined by the value $\sigma(i)$. Now i satisfies the equation $x^2 = -1$; hence so does $\sigma(i)$; whence $\sigma(i) = \pm i$. If $\sigma(i) = i$ then σ is the identity on $\mathbb{C}$; if $\sigma(i) = -i$ then σ is complex conjugation; both maps are obviously automorphisms of $\mathbb{C}$. Hence

$$\boxed{G_{\mathbb{R}}^{\mathbb{C}} = \{\mathrm{id}, \tau\}}$$

where τ is complex conjugation.

(21.8) Example. Let us show that

$$\boxed{G_{\mathbb{Q}}^{\mathbb{R}} = \{\mathrm{id}\} \ .}$$

Let $\sigma \in G_{\mathbb{Q}}^{\mathbb{R}}$. Then $a < b$ implies $\sigma(a) < \sigma(b)$. Indeed, if $a < b$, then $b - a = y^2$ for some $y \neq 0$, whence $\sigma(b) - \sigma(a) = \sigma(b - a) = \sigma(y^2) = \sigma(y)^2 > 0$. Suppose $\sigma \neq \mathrm{id}$ to get a contradiction. Then $\sigma(x) \neq x$ for some $x \in \mathbb{R}$. If $\sigma(x) > x$, pick a rational number r such that $x < r < \sigma(x)$. Apply σ to this last inequality; this yields $\sigma(x) < \sigma(r) < \sigma^2(x)$. But $\sigma(r) = r$ so that $\sigma(x) < r$ contradicting the choice of r. The assumption $\sigma(x) < x$ is falsified in a completely analogous way.

(21.9) Example. We want to determine the Galois group $G_K^{K(x)}$ where K is an arbitrary field and where $K(x)$ is the rational function field over K. Let $\sigma \in G_K^{K(x)}$ and let $\theta \in K(x)$ be the image of the rational function x under σ. (It is clear that σ is uniquely determined by its value on x.) Write $\theta = f/g$ with coprime polynomials $f, g \in K[x]$; since σ is an automorphism of $K(x)$ we have $K(x) = K(\theta)$ and hence $\max(\deg f, \deg g) = 1$, due to (12.14)(a). This shows that $f(x) = ax + b$ and $g(x) = cx + d$ with $a, b, c, d \in K$, and since $\theta \in K(x) \setminus K$ we must have $ad - bc \neq 0$. Therefore, the elements of $G_K^{K(x)}$ are exactly the mappings of the form

$$\varphi(x) \mapsto \varphi(\frac{ax + b}{cx + d}) \quad \text{with } ad - bc \neq 0.$$

Thus $G_K^{K(x)}$ is isomorphic to the group of all Möbius transformations.

A crucial idea behind associating with any field extension $(L : K)$ its Galois group G_K^L is the possibility of translating field-theoretical problems concerning $(L : K)$ into group-theoretical problems concerning G_K^L which might be easier to handle. To realize this idea, we try to set up a correspondence between the intermediate fields of $(L : K)$ and the subgroups of G_K^L.

(21.10) Definitions. *Let $(L : K)$ be a field extension. With every subgroup U of the Galois group G_K^L we associate its* **fixed field**

$$\boxed{F(U) := \{x \in L \mid \sigma(x) = x \text{ for all } \sigma \in U\}}$$

which is clearly an intermediate field of $(L : K)$. Conversely, with each intermediate field Z of the extension $(L : K)$ we associate its own Galois group

$$G(Z) := G_Z^L = \{\sigma \in \operatorname{Aut} L \mid \sigma(z) = z \text{ for all } z \in Z\}$$

which is clearly a subgroup of G_K^L.

THIS gives a correspondence

$$\{\text{ intermediate fields of } (L : K)\} \quad \overset{F}{\underset{G}{\rightleftharpoons}} \quad \{\text{ subgroups of } G_K^L\}$$

(called the **Galois correspondence** of the field extension $(L : K)$) whose properties are explored in the next proposition.

(21.11) Proposition. *Let $(L : K)$ be a field extension and let F and G be the mappings between subgroups of G_K^L and intermediate fields of $(L : K)$ as defined above.*

(a) *$U \subseteq G_{F(U)}^L = (G \circ F)(U)$ for every subgroup $U \le G_K^L$.*

(b) *The mapping F is inclusion-reversing; i.e., if $U_1 \subseteq U_2$ are subgroups of G_K^L then $F(U_2) \subseteq F(U_1)$.*

(c) *The mapping F "changes unions to intersections"; i.e., if $(U_i)_{i \in I}$ is a family of subgroups of G_K^L then $F(\langle \bigcup_{i \in I} U_i \rangle) = \bigcap_{i \in I} F(U_i)$.*

(d) *If $U \le G_K^L$ then $F(\sigma U \sigma^{-1}) = \sigma(F(U))$ for all $\sigma \in G_K^L$.*

(e) *$Z \subseteq F(G_Z^L) = (F \circ G)(Z)$ for every intermediate field Z of $(L : K)$.*

(f) *The mapping G is inclusion-reversing; i.e., if $Z_1 \subseteq Z_2$ are intermediate fields of $(L : K)$ then $G_{Z_2}^L \subseteq G_{Z_1}^L$.*

(g) *The mapping G "changes unions to intersections"; i.e., if $(Z_i)_{i \in I}$ is a family of intermediate fields of $(L : K)$ then $G_{\langle \bigcup_{i \in I} Z_i \rangle}^L = \bigcap_{i \in I} G_{Z_i}^L$.*

(h) *If Z is an intermediate field of $(L : K)$ then $G_{\sigma(Z)}^L = \sigma G_Z^L \sigma^{-1}$ for all $\sigma \in G_K^L$.*

(i) *We have $G = G \circ F \circ G$ and $F = F \circ G \circ F$.*

Proof. (a) If $\sigma \in U$ then $\sigma(x) = x$ for all $x \in F(U)$, whence $\sigma \in G_{F(U)}^L$.

(b) Let $U_1 \subseteq U_2$. If $x \in F(U_2)$ then $\sigma(x) = x$ for all $\sigma \in U_2$, the more so for all $\sigma \in U_1$; hence $x \in F(U_1)$.

(c) An element $x \in L$ lies in $F(\langle \bigcup_i U_i \rangle)$ if and only if x is fixed under all $\sigma_i \in U_i$, for each i, i.e., if x lies in each $F(U_i)$.

(d) An element $x \in L$ lies in $F(\sigma U \sigma^{-1})$ if and only if $\sigma \varphi \sigma^{-1}(x) = x$ for all $\varphi \in U$, i.e., if $\varphi(\sigma^{-1}(x)) = \sigma^{-1}(x)$ for all $\varphi \in U$. But this means exactly that $\sigma^{-1}(x) \in F(U)$, i.e., that $x \in \sigma(F(U))$.

(e) If $z \in Z$ then $\sigma(z) = z$ for all $\sigma \in G_Z^L$, hence $z \in F(G_Z^L)$.

(f) Let $Z_1 \subseteq Z_2$. If $\sigma \in G_{Z_2}^L$ then $\sigma(z) = z$ for all $z \in Z_2$, the more so for all $z \in Z_1$; hence $\sigma \in G_{Z_1}^L$.

(g) An automorphism $\sigma : L \to L$ belongs to $G_{\langle \bigcup_i Z_i \rangle}^L$ if and only if σ is the identity on $\bigcup_i Z_i$, i.e., the identity on each Z_i, which means that $\sigma \in \bigcap_i G_{Z_i}^L$.

(h) An element $\varphi \in G_K^L$ lies in $G_{\sigma(Z)}^L$ if and only if $\varphi(w) = w$ for all $w \in \sigma(Z)$, i.e., if $\varphi(\sigma(z)) = \sigma(z)$ for all $z \in Z$. But this says exactly that $\sigma^{-1}\varphi\sigma(z) = z$ for all $z \in Z$ which means that $\sigma^{-1}\varphi\sigma \in G_Z^L$, i.e., that $\varphi \in \sigma G_Z^L \sigma^{-1}$.

(i) Let U be a subgroup of G_L^L. By (a) and (b), we have $(F \circ G \circ F)(U) \subseteq F(U)$. Applying (e) to $Z := F(U)$ also yields the reverse inclusion; so taken together, $(F \circ G \circ F)(U) = F(U)$. Since U was arbitrary, this shows that $F \circ G \circ F = F(U)$.

Now let Z be an intermediate field of $(L : K)$. By (e) and (f), we have $(G \circ F \circ G)(Z) \subseteq G(Z)$. Applying (a) to $U := G(Z)$ also yields the inverse inclusion; so altogether $(G \circ F \circ G)(Z) = G(Z)$. Since Z was arbitrary, this shows that $G \circ F \circ G = G$. ∎

(21.12) Example. Let us study in some detail the field extension $(\mathbb{Q}(\sqrt{2}, \sqrt{3}) : \mathbb{Q})$. Its Galois group $G_{\mathbb{Q}}^{\mathbb{Q}(\sqrt{2}, \sqrt{3})}$ has exactly four elements which we called $\sigma_0 = \mathrm{id}, \sigma_1, \sigma_2, \sigma_3$ in example (21.4). The subgroup lattice of $G_{\mathbb{Q}}^{\mathbb{Q}(\sqrt{2}, \sqrt{3})}$ is

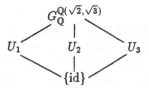

where $U_k := \{\sigma_0, \sigma_k\}$ $(k = 1, 2, 3)$. Let us determine the fixed field of each of these subgroups. Trivially, we have

$$F(\{\mathrm{id}\}) = \mathbb{Q}(\sqrt{2}, \sqrt{3}).$$

To determine $F(U_1)$ we observe that an arbitrary element of $\mathbb{Q}(\sqrt{2}, \sqrt{3})$ has the form $x = a + b\sqrt{2} + c\sqrt{3} + d\sqrt{6}$. Such an element belongs to $F(U_1)$, i.e., satisfies $\sigma_1(x) = x$, if and only if $a - b\sqrt{2} + c\sqrt{3} - d\sqrt{6} = a + b\sqrt{2} + c\sqrt{3} + d\sqrt{6}$ which means $b = d = 0$, i.e., $x = a + c\sqrt{3}$. This shows that the fixed field of U_1 is $\mathbb{Q}(\sqrt{3})$. Completely analogous calculations can be done for U_2 and U_3; as a result, we see that

$$F(U_1) = \mathbb{Q}(\sqrt{3}), \quad F(U_2) = \mathbb{Q}(\sqrt{2}), \quad F(U_3) = \mathbb{Q}(\sqrt{6}).$$

To compute the fixed group of the full Galois group $G_{\mathbb{Q}}^{\mathbb{Q}(\sqrt{2}, \sqrt{3})}$ we could proceed in the same way, but we can also use (21.11)(b). Indeed, $U_k \subseteq G_{\mathbb{Q}}^{\mathbb{Q}(\sqrt{2}, \sqrt{3})}$ implies $F(G_{\mathbb{Q}}^{\mathbb{Q}(\sqrt{2}, \sqrt{3})}) \subseteq F(U_k)$ $(k = 1, 2, 3)$. Consequently,

$$F(G_{\mathbb{Q}}^{\mathbb{Q}(\sqrt{2}, \sqrt{3})}) \subseteq F(U_1) \cap F(U_2) \cap F(U_3) = \mathbb{Q}(\sqrt{3}) \cap \mathbb{Q}(\sqrt{2}) \cap \mathbb{Q}(\sqrt{6}) = \mathbb{Q};$$

so we have

$$F(G_{\mathbb{Q}}^{\mathbb{Q}(\sqrt{2}, \sqrt{3})}) = \mathbb{Q}.$$

THE following terminology allows for a very succinct restatement of some statements concerning Galois correspondences.

(21.13) Definitions. *Let* $(L : K)$ *be a field extension.*

(a) *An intermediate field* Z *of* $(L : K)$ *is called* **closed** *if* $Z = F(U)$ *for some subgroup* $U \leq G_K^L$. *Conversely, a subgroup* $U \leq G_K^L$ *is called* **closed** *if there is an intermediate field* Z *of* $(L : K)$ *with* $U = G(Z) = G_Z^L$.

(b) *An intermediate field* Z *of* $(L : K)$ *is called* **stable** *if* $\sigma(Z) \subseteq Z$ *for all* $\sigma \in G_K^L$; *this is tantamount to saying that* $\sigma(Z) = Z$ *for all* $\sigma \in G_K^L$.†

THE properties of closedness and stability can be characterized in a slightly different way, as the following proposition shows.

(21.14) Proposition. *Let* $(L : K)$ *be a field extension.*

(a) *An intermediate field* Z *of* $(L : K)$ *is closed if and only if* $Z = F(G_Z^L)$.

(b) *A subgroup* U *of* G_K^L *is closed if and only if* $U = G_{F(U)}^L$.

Proof. (a) If Z is closed, say $Z = F(U)$, then $F(G_Z^L) = (F \circ G)(Z) = (F \circ G \circ F)(U) = F(U) = Z$. The converse implication is trivial.

(b) If U is closed, say $U = G_Z^L$, then $G_{F(U)}^L = (G \circ F)(U) = (G \circ F \circ G)(Z) = G(Z) = G_Z^L = U$. The converse implication is trivial. ■

IN the next proposition, we gather some elementary consequences of the definitions given in (21.13).

(21.15) Proposition. *Let* $(L : K)$ *be a field extension.*

(a) *Any intersection of closed intermediate fields of* $(L : K)$ *is closed.*

(b) *Any intersection of closed subgroups of* G_K^L *is closed.*

(c) *The mappings* F *and* G *set up a 1-1-correspondence between the closed intermediate fields of* $(L : K)$ *and the closed subgroups of* G_K^L.

(d) *If an intermediate field is stable then its group* $G(Z) = G_Z^L$ *is normal in* G_K^L *with quotient group*

$$G_K^L / G_Z^L \cong \{\sigma \in G_K^Z \mid \sigma \text{ can be extended to an automorphism of } L\} \ .$$

(e) *If a subgroup* $U \leq G_K^L$ *is normal then its fixed field* $F(U)$ *is stable.*

(f) *If a subgroup* $U \leq G_K^L$ *is normal then so is* $(G \circ F)(U) = G_{F(U)}^L$.

(g) *If an intermediate field* Z *is stable then so is* $(F \circ G)(Z) = F(G_Z^L)$.

Proof. Most of the statements are immediate consequences of (21.11); in fact, part (a) follows from (21.11)(c), part (b) from (21.11)(g), part (c) from (21.11)(i), the

---

† Suppose that $\sigma(Z) \subseteq Z$ for all $\sigma \in G_K^L$. If $z \in Z$ and $\sigma \in G_K^L$ then also $\sigma^{-1} \in G_K^L$ and hence $\sigma^{-1} z \in Z$. But then $z = \sigma(\sigma^{-1} z) \in \sigma(Z)$; since $z \in Z$ was arbitrary, this implies $Z \subseteq \sigma(Z)$ so that $\sigma(Z) = Z$.

first half of part (d) from (21.11)(h) and part (e) from (21.11)(d). To prove the second half of part (d) we note that the restriction map $R : G_K^L \to G_K^Z$ given by $\Sigma \mapsto \Sigma|_Z$ (which is well-defined because Z is stable) is a homomorphism, its kernel being G_Z^L and its image consisting exactly of those elements G_K^Z which can be extended to L. Hence the homomorphism theorem yields the claim concerning the quotient group. Moreover, if U is normal then $F(U)$ is stable by (e) and then $(G \circ F)(U)$ is normal by (d); this proves (f). Finally, if Z is stable then $G(Z)$ is normal by (d) and then $(F \circ G)(Z)$ is stable by (e); this proves (g). ∎

THERE is also a quantitative aspect of the Galois correspondence between intermediate fields of $(L : K)$ and subgroups of G_K^L; namely, we can relate the degrees of intermediate fields to the indices of the corresponding subgroups and vice versa.

(21.16) Theorem. *Let* $(L : K)$ *be a field extension.*
(a) *If* $K \subseteq Z_1 \subseteq Z_2 \subseteq L$ *are intermediate fields with* $[Z_2 : Z_1] = n < \infty$ *then* $[G_{Z_1}^L : G_{Z_2}^L] \leq n$.
(b) *If* $\{\mathrm{id}\} \subseteq U_1 \subseteq U_2 \subseteq G_K^L$ *are subgroups with* $[U_2 : U_1] = n < \infty$ *then* $[F(U_1) : F(U_2)] \leq n$.
(c) *If in* (a) *additionally* $F(G_{Z_1}^L) = Z_1$, *i.e., if* Z_1 *is closed, then* $F(G_{Z_2}^L) = Z_2$ *and* $[G_{Z_1}^L : G_{Z_2}^L] = n$.
(d) *If in* (b) *additionally* $G_{F(U_1)}^L = U_1$, *i.e., if* U_1 *is closed, then* $G_{F(U_2)}^L = U_2$ *and* $[F(U_1) : F(U_2)] = n$.

Proof. (a) We proceed by induction on n, the case $n = 1$ (i.e., $Z_1 = Z_2$) being trivial. Let $n > 1$. If there is a field Z properly between Z_1 and Z_2 we can conclude

$$[G_{Z_1}^L : G_{Z_2}^L] = [G_{Z_1}^L : G_Z^L] \cdot [G_Z^L : G_{Z_2}^L] \leq [Z : Z_1] \cdot [Z_2 : Z] = [Z_2 : Z_1] = n,$$

where the first equation is the index formula (I.21.19)(b), the following inequality is the induction hypothesis and the next equation is the degree formula. Thus in this case the claim holds. If there is *no* proper intermediate field then $Z_2 = Z_1(\alpha)$ for some α. Let $f \in Z_1[x]$ be the minimal polynomial of α over Z_1 so that $\deg f = n$. Two cosets $\sigma G_{Z_2}^L$ and $\tau G_{Z_2}^L$ of $G_{Z_1}^L$ modulo $G_{Z_2}^L$ are equal if and only if $\sigma \tau^{-1} \in G_{Z_2}^L$ which is the case if and only if $(\sigma \tau^{-1})(\alpha) = \alpha$, i.e., $\sigma(\alpha) = \tau(\alpha)$. Now if $\sigma \in G_{Z_1}^L$ then $\sigma \star f = f$ which implies that $\sigma(\alpha)$ is a root of f. Therefore, $[G_{Z_1}^L : G_{Z_2}^L]$ (which is the number of cosets $\sigma G_{Z_2}^L$) does not exceed the number of roots of f which is at most $\deg f = n$.

(b) All elements in a left coset φU_1 of U_1 in U_2 coincide on $F(U_1)$; indeed, if $\sigma \in U_1$ and $x \in F(U_1)$ then $(\varphi \circ \sigma)(x) = \varphi(\sigma(x)) = \varphi(x)$. Hence we can unambiguously apply a coset φU_1 to an element of $F(U_1)$. Suppose that $[F(U_1) : F(U_2)] > n$; then we can find $n + 1$ elements $x_1, \ldots, x_{n+1} \in F(U_1)$ which are linearly independent over $F(U_2)$. Let $C_1, \ldots, C_n$ be the cosets of U_1 in U_2. Then the system

$$a_1(C_1 x_1) + \cdots + a_{n+1}(C_1 x_{n+1}) = 0$$
(⋆)
$$\cdots$$
$$a_1(C_n x_1) + \cdots + a_{n+1}(C_n x_{n+1}) = 0$$

of n linear equations for the $n + 1$ unknowns $a_1, \ldots, a_{n+1}$ has a nontrivial solution $(a_1, \ldots, a_{n+1}) \neq (0, \ldots, 0)$. We pick a nontrivial solution with as many zeros as possible; after reindexing, we may assume that this solution is $(a_1, \ldots, a_r, 0, \ldots, 0)$ with

$a_i \neq 0$ for $1 \leq i \leq r$, and after dividing by a_1 we may also assume that $a_1 = 1$. It is *not* possible that all of $a_2, \ldots, a_r$ lie in $F(U_2)$; namely, one of the cosets, say C_i, is U_1 itself, and then the i-th row of $(\star)$ reads $a_1 x_1 + \cdots + a_{n+1} x_{n+1} = 0$ which would imply $a_1 = \cdots = a_{n+1} = 0$ because $x_1, \ldots, x_{n+1}$ are linearly independent over $F(U_2)$. Without loss of generality assume that $a_2 \notin F(U_2)$. Then there is an element $\sigma \in U_2$ with $\sigma(a_2) \neq a_2$. Applying σ to each of the n equations in $(\star)$ we obtain

$$
(\star\star) \quad
\begin{aligned}
\sigma(a_1)((\sigma C_1) x_1) \;+\; &\cdots \;+\; \sigma(a_{n+1})((\sigma C_1) x_{n+1}) \;=\; 0 \\
&\cdots \\
\sigma(a_1)((\sigma C_n) x_1) \;+\; &\cdots \;+\; \sigma(a_{n+1})((\sigma C_n) x_{n+1}) \;=\; 0.
\end{aligned}
$$

Observing that σ simply permutes the cosets C_i so that $\{\sigma C_1, \ldots, \sigma C_n\} = \{C_1, \ldots, C_n\}$ we see that the equations for $\sigma(a_1), \ldots, \sigma(a_n)$ are just a permutation of the original equations (for $a_1, \ldots, a_n$); hence since $(1, a_2, \ldots, a_r, 0, \ldots, 0)$ is a solution of $(\star)$ then so is $(\sigma(1), \sigma(a_2), \ldots, \sigma(a_r), \sigma(0), \ldots, \sigma(0)) = (1, \sigma(a_2), \ldots, \sigma(a_r), 0, \ldots, 0)$. Subtracting the two solutions, we get a new solution $(0, \sigma(a_2) - a_2, \ldots, \sigma(a_r) - a_r, 0, \ldots, 0)$ which is non-trivial because $\sigma(a_2) \neq a_2$ and which has more zeros than the first solution which clearly contradicts our choice of the a_i.

(c) We have $[G_{Z_1}^L : G_{Z_2}^L] \leq [Z_2 : Z_1] < \infty$ by part (a) and then $[F(G_{Z_1}^L) : F(G_{Z_2}^L)] \leq [G_{Z_1}^L : G_{Z_2}^L]$ by part (b). Since also $F(G_{Z_1}^L) = Z_1$ by hypothesis, this yields

$$
\begin{aligned}
n \cdot [F(G_{Z_2}^L) : Z_2] \;=\; [Z_2 : Z_1] \cdot [F(G_{Z_2}^L) : Z_2] \;&=\; [F(G_{Z_2}^L) : Z_1] \;=\; [F(G_{Z_2}^L) : F(G_{Z_1}^L)] \\
&\leq\; [G_{Z_1}^L : G_{Z_2}^L] \;\leq\; [Z_2 : Z_1] \;=\; n \;.
\end{aligned}
$$

Dividing by n, we see that $[F(G_{Z_2}^L) : Z_2] \leq 1$, hence $F(G_{Z_2}^L) = Z_2$, and the above inequality becomes $n \leq [G_{Z_2}^L : G_{Z_1}^L] \leq n$ which yields $[G_{Z_2}^L : G_{Z_1}^L] = n$.

(d) The proof of part (d) is very similar to that of part (c) and is left as an exercise. ∎

AS a special case of (21.16)(a) we obtain the following result.

(21.17) Proposition. *If $(L : K)$ is a finite field extension then $|G_K^L| \leq [L : K]$.*

Proof. Take $Z_1 := K$ and $Z_2 := L$ in (21.16)(a). ∎

THIS is an important estimate for the size of Galois groups, and we want to offer a second proof for this result, based on the following theorem which is interesting in its own right.

(21.18) Dedekind's Theorem. *Pairwise different automorphisms $\sigma_1, \ldots, \sigma_n$ of a field L are linearly independent in the L-vector space of all mappings from L into L.*

Proof (Artin). The proof is by induction on n. The case $n = 1$ is trivial since $\sigma_1 \neq 0$. Suppose that the claim is proved for $n - 1$ and assume $\sum_{i=1}^{n} \lambda_i \sigma_i = 0$; we have to show that all coefficients $\lambda_i \in L$ vanish. More precisely, our assumption is

$$
(1) \qquad \sum_{i=1}^{n} \lambda_i \sigma_i(x) = 0 \quad \text{for all } x \in L .
$$

Since $\sigma_1 \neq \sigma_n$, there is an element $\alpha \in L$ such that $\sigma_1(\alpha) \neq \sigma_n(\alpha)$. Replace x by αx in (1) and use $\sigma_i(\alpha x) = \sigma_i(\alpha)\sigma_i(x)$ to obtain

$$
(2) \qquad \lambda_1 \sigma_1(\alpha)\sigma_1(x) + \lambda_2 \sigma_2(\alpha)\sigma_2(x) + \cdots + \lambda_n \sigma_n(\alpha)\sigma_n(x) = 0 \quad (x \in L) .
$$

On the other hand, multiply (1) by $\sigma_1(\alpha)$ to obtain

$$
(3) \qquad \lambda_1 \sigma_1(\alpha)\sigma_1(x) + \lambda_2 \sigma_1(\alpha)\sigma_2(x) + \cdots + \lambda_n \sigma_1(\alpha)\sigma_n(x) = 0 \quad (x \in L) .
$$

Now subtract (3) from (2); this gives

$$
\lambda_2 \big(\sigma_2(\alpha) - \sigma_1(\alpha)\big) \sigma_2(x) + \cdots + \lambda_n \big(\sigma_n(\alpha) - \sigma_1(\alpha)\big)\sigma_n(x) = 0 \quad (x \in L) .
$$

By induction hypothesis, this implies $\lambda_i \big(\sigma_i(\alpha) - \sigma_1(\alpha)\big) = 0$ for $2 \leq i \leq n$; in particular, $\lambda_n \big(\sigma_n(\alpha) - \sigma_1(\alpha)\big) = 0$, whence $\lambda_n = 0$ since $\sigma_n(\alpha) - \sigma_1(\alpha) \neq 0$ by the choice of α. So the initial equation reduces to $\sum_{i=1}^{n-1} \lambda_i \sigma_i = 0$; here the induction hypothesis is applicable and shows that also the coefficients $\lambda_1, \ldots, \lambda_{n-1}$ vanish. $\blacksquare$

NOW we are ready to give a second proof of proposition (21.17) above.

(21.19) Proposition. *If $(L : K)$ is a finite field extension, then $|G_K^L| \leq [L : K]$.*

Proof. We denote by $\mathrm{Hom}_K(L, L)$ the set of all K-linear mappings from L into L, i.e., the set of all mappings $\sigma : L \to L$ such that $\sigma(x + y) = \sigma(x) + \sigma(y)$ and $\sigma(kx) = k\sigma(x)$ for all $x, y \in L$ and all $k \in K$. This set can be made into a vector space over L (not just over K!) by defining $(\sigma + \tau)(x) := \sigma(x) + \tau(x)$ and $(\lambda\sigma)(x) := \lambda \cdot \sigma(x)$. Now by Dedekind's theorem, the elements of G_K^L (which is clearly a subset of $\mathrm{Hom}_K(L, L)$) are linearly independent over L as mappings from L to L, hence the more so as elements of $\mathrm{Hom}_K(L, L)$. This shows that $|G_K^L| \leq \dim_L \mathrm{Hom}_K(L, L)$. Hence we are done if we can show that the dimension of $\mathrm{Hom}_K(L, L)$ over L equals $[L : K]$. Let $n := [L : K] = \dim_L K$ and let $(\alpha_1, \ldots, \alpha_n)$ be a basis of L over K. Define K-linear mappings $f_i \in \mathrm{Hom}_K(L, L)$ by $f_i(\alpha_j) := \delta_{ij}$. We are done if we can show that $(f_1, \ldots, f_n)$ is a basis of $\mathrm{Hom}_K(L, L)$.

(1) The f_i are linearly independent over L, because $\sum_{i=1}^{n} \lambda_i f_i = 0$ implies $0 = \sum_{i=1}^{n} \lambda_i f_i(\alpha_j) = \sum_{i=1}^{n} \lambda_i \delta_{ij} = \lambda_j$ for all j.

(2) The f_i span all of $\mathrm{Hom}_K(L, L)$. Indeed, if $f \in \mathrm{Hom}_K(L, L)$ and $x = \sum_{i=1}^{n} k_i \alpha_i \in L$ are arbitrary, we have $f(x) = \sum_{i=1}^{n} k_i f(\alpha_i) = \sum_{i=1}^{n} f_i(x) f(\alpha_i)$ so that $f = \sum_{i=1}^{n} f(\alpha_i) f_i$. $\blacksquare$

IN proposition (21.15)(c) we stated that there is a 1-1-correspondence between the *closed* intermediate fields of an extension $(L : K)$ and the *closed* subgroups of its Galois group G_K^L. This fact is of little value as long as we do not have any criteria which allow us to decide whether or not a given intermediate field or a given subgroup is closed. In the next theorem, however, we will show that every *finite* group of a Galois group is automatically closed. Since the Galois group of a finite field extension is finite by (21.17), this means that in the most important case of field extensions, namely, finite extensions, only closed subgroups do occur!

AS a technical tool to prove that every finite subgroup U of a Galois group G_K^L is closed, we introduce the trace operator of $(L : K)$ with respect to the subgroup $U \le G_K^L$.

(21.20) Definition. *Let $(L : K)$ be a field extension and let U be a finite subgroup of G_K^L. Then the* **U-trace** *of an element $\alpha \in L$ is defined as $\operatorname{tr}_U(\alpha) := \sum_{\sigma \in U} \sigma(\alpha)$.*

THE most important properties of the U-trace just defined are collected in the following proposition.

(21.21) Proposition. *Let $(L : K)$ be a field extension and let U be a finite subgroup of G_K^L.*
(a) tr_U is a K-linear mapping from L into the fixed field $F(U)$ of U; if $\operatorname{char} K$ does not divide $|U|$, then the mapping $\operatorname{tr}_U : L \to F(U)$ is surjective.
(b) tr_U cannot be the zero mapping.

Proof. (a) If $\tau \in U$ then $\tau\big(\operatorname{tr}_U(\alpha)\big) = \tau\big(\sum_{\sigma \in U} \sigma(\alpha)\big) = \sum_{\sigma \in U} \tau\sigma(\alpha) = \sum_{\sigma' \in U} \sigma'(\alpha) = \operatorname{tr}_U(\alpha)$; hence $\operatorname{tr}_U(\alpha)$ lies in $F(U)$ for any $\alpha \in L$. The K-linearity of tr_U is obvious, also the fact that $\operatorname{tr}_U(1) = |U|$. Finally, if $\operatorname{char} K$ does not divide $|U|$, we can divide elements of L by $|U|$. Now if $\alpha \in F(U)$ then $\operatorname{tr}_U(\alpha) = \sum_{\sigma \in U} \sigma(\alpha) = \sum_{\sigma \in U} \alpha = |U| \cdot \alpha$ and hence $\operatorname{tr}_U(\alpha/|U|) = \operatorname{tr}_U(\alpha)/|U| = \alpha$; this shows that tr_U is surjective in this case.
(b) By Dedekind's theorem (21.18) the elements of U are linearly independent as mappings from L to L. Thus no nontrivial linear combination of these elements can be the zero mapping; in particular $\sum_{\sigma \in U} \sigma \not\equiv 0$. ∎

WE can now prove the announced theorem that if $(L : K)$ is any field extension then every finite subgroup of G_K^L is closed.

(21.22) Artin's Theorem. (a) *Let $(L : K)$ be a field extension. If U is a finite subgroup of G_K^L, then $[L : F(U)] = |U|$ and $G\big(F(U)\big) = U$.*
(b) If $(L : K)$ is a finite field extension then $G \circ F = \operatorname{id}$.

Proof. Let $n := |U|$. If we can show that any $n + 1$ elements of L are linearly dependent over the fixed field $F(U)$ of U, then we have $[L : F(U)] = \dim_{F(U)} L \le n = |U|$. So let $\alpha_1, \dots, \alpha_{n+1} \in L$ be arbitrary elements. We denote the different elements of U by $\sigma_1, \ldots, \sigma_n$. Then the linear system

$$(\star) \quad \begin{pmatrix} \sigma_1^{-1}(\alpha_1) & \sigma_1^{-1}(\alpha_2) & \cdots & \sigma_1^{-1}(\alpha_{n+1}) \\ \sigma_2^{-1}(\alpha_1) & \sigma_2^{-1}(\alpha_2) & \cdots & \sigma_2^{-1}(\alpha_{n+1}) \\ \vdots & \vdots & & \vdots \\ \sigma_n^{-1}(\alpha_1) & \sigma_n^{-1}(\alpha_2) & \cdots & \sigma_n^{-1}(\alpha_{n+1}) \end{pmatrix} \begin{pmatrix} x_1 \\ x_2 \\ \vdots \\ x_n \\ x_{n+1} \end{pmatrix} = \begin{pmatrix} 0 \\ 0 \\ \vdots \\ \vdots \\ 0 \end{pmatrix}$$

has a nontrivial solution $(\xi_1, \ldots, \xi_{n+1})^T \neq (0, \ldots, 0)^T$ $(\xi_i \in L)$. To see this, just observe that the system consists of n homogeneous equations for $n + 1$ unknowns. By renumbering the elements of U if necessary, we can assume that $\xi_1 \neq 0$. Still without loss of generality, we may even assume that $\mathrm{tr}_U(\xi_1) \neq 0$; otherwise we could replace ξ_1 by some element of the form $\theta \xi_1$ with $\theta \in L$. To wit: we have $\mathrm{tr}_U \neq 0$ by (21.21)(b) so that $\mathrm{tr}_U \theta \xi_1 \neq 0$ for some $\theta \in L$, and if $(\xi_1, \ldots, \xi_{n+1})^T$ is a solution of $(\star)$ then so is $(\theta \xi_1, \ldots, \theta \xi_{n+1})^T$.

Now ξ satisfies the j-th row of $(\star)$, i.e.,

$$\sigma_j^{-1}(\alpha_1)\xi_1 + \cdots + \sigma_j^{-1}(\alpha_{n+1})\xi_{n+1} = 0 \,.$$

Applying σ_j to this equation, we obtain

$$\alpha_1 \sigma_j(\xi_1) + \cdots + \alpha_{n+1}\sigma_j(\xi_{n+1}) = 0 \,.$$

Adding these n equations for $1 \leq j \leq n$, we get

$$\alpha_1 \underbrace{\mathrm{tr}_U(\xi_1)}_{\neq 0} + \cdots + \alpha_{n+1}\,\mathrm{tr}_U(\xi_{n+1}) = 0 \,.$$

Now the left-hand side is a nontrivial linear combination of the elements α_i with coefficients in $F(U)$. Consequently, $\alpha_1, \ldots, \alpha_{n+1}$ are linearly dependent over $F(U)$.

We have shown that $[L : F(U)] \leq |U| < \infty$ so that in particular $\big(L : F(U)\big)$ is a finite field extension. But then (21.17) is applicable and yields $|G_{F(U)}^L| \leq [L : F(U)]$. On the other hand $|U| \leq |G_{F(U)}^L|$ because trivially $U \subseteq G_{F(U)}^L$. Combining all the inequalities established, we obtain

$$|U| \leq |G_{F(U)}^L| \leq [L : F(U)] \leq |U| \,.$$

Hence $|U| = |G_{F(U)}^L|$ which together with the trivial inclusion $U \subseteq G_{F(U)}^L$ yields $U = G_{F(U)}^L$. But this is the claim.

(b) If $[L : K]$ is finite then G_K^L is a finite group by (21.19) so that every subgroup U of G_K^L is automatically finite; hence part (a) applies. ∎

(21.23) Example. Let $K(x)$ be the rational function field over a field K and define six elements $\sigma_i \in G_K^{K(x)}$ as follows:

$$
\begin{aligned}
(\sigma_1 f)(x) &= f(x), \\
(\sigma_2 f)(x) &= f(1-x), \\
(\sigma_3 f)(x) &= f(\frac{1}{x}), \\
(\sigma_4 f)(x) &= f(\frac{x}{x-1}), \\
(\sigma_5 f)(x) &= f(\frac{1}{1-x}), \\
(\sigma_6 f)(x) &= f(\frac{x-1}{x}).
\end{aligned}
$$

It is easy to see that $U := \{\sigma_1, \ldots, \sigma_6\}$ is a subgroup of $\operatorname{Aut} K(x)$ which is isomorphic to Sym_3; hence Artin's theorem (21.22) shows that $[K(x) : F(U)] = |U| = 6$. To identify $F(U)$ we let

$$
g(x) := \frac{(x^2 - x + 1)^3}{x^2(x-1)^2} \in K(x).
$$

It is easily verified that $g \in F(U)$ and hence $K(g) \leq F(U)$ so that $[K(x) : K(g)] \geq 6$. On the other hand we have $(x^2 - x + 1)^3 - g(x) \cdot x^2(x-1)^2 = 0$ which shows that x satisfies a polynomial equation of degree 6 over $K(g)$ so that $[K(x) : K(g)] \leq 6$. We conclude that $K(g) = F(U)$ and $[K(x) : K(g)] = 6$. Note that this also illustrates Lüroth's theorem (12.14)(b).

LET us pause for a moment to reflect on where we are standing in the course of our considerations. We are trying to set up a 1-1-correspondence between the intermediate fields of a field extension $(L : K)$ and the subgroups of its Galois group G_K^L because we want to be able to translate field-theoretical problems into group-theoretical ones. To this end, we introduced the mappings F (associating with each subgroup $U \leq G_K^L$ its fixed field) and G (associating with each intermediate field its Galois group G_Z^L). The most favorable situation would be if F and G were inverses of each other, because this would allow for a faithful transfer between the field-theoretical structure of $(L : K)$ and the group-theoretical structure of G_K^L. This is desirable because it is in general far easier finding the subgroups of G_K^L than listing all the intermediate fields of $(L : K)$ since it is very easy to overlook some intermediate fields and also counting some of them twice without realizing it. In the case that F and G are inverses of each other the equations $F \circ G \circ F = F$ and $G \circ F \circ G = G$ would simply become $F \circ G = \operatorname{id}$ and $G \circ F = \operatorname{id}$. Thus we ask under which circumstances the equations $G \circ F = \operatorname{id}$ and $F \circ G = \operatorname{id}$ hold; note that the first of these equations has already been established for all *finite* field extensions. A necessary condition for the equation $F \circ G = \operatorname{id}$ is that $K = (F \circ G)(K) = F(G_K^L)$, i.e., that the base-field K itself is closed. It will turn out that in the case of *algebraic* extensions this condition already implies that *all* intermediate fields of $(L : K)$ are closed. Therefore, we give the following definition.

(21.24) Definition. *A field extension $(L : K)$ is called a **Galois extension** or of Galois type, if $F(G_K^L) = K$, i.e., if for any $\alpha \in L \setminus K$ there is an automorphism of L which leaves K pointwise fixed, but actually moves α.*

(21.25) Examples. (a) $\big(\mathbb{Q}(\sqrt{2},\sqrt{3}) : \mathbb{Q}\big)$ is a Galois extension, according to (21.12). On the other hand, $\big(\mathbb{Q}(\sqrt[3]{2}) : \mathbb{Q}\big)$ has a trivial Galois group according to (21.6) and hence is not a Galois extension.

(b) Let $K = \mathbb{Z}_2(t)$ be the rational function field over $\mathbb{Z}_2$ and let L be a splitting field of $f(x) := x^2 - t \in K[x]$. Then $L = K(\alpha)$ where α satisfies $\alpha^2 = t$. Each element $\sigma \in G_K^L$ is uniquely determined by the element $\sigma(\alpha)$, and since $\sigma(\alpha)^2 = \sigma(\alpha^2) = \sigma(t) = t = \alpha^2$ we must have $\sigma(\alpha) = \pm\alpha$. But $-\alpha = \alpha$ because we are in characteristic 2; hence $\sigma(\alpha) = \alpha$ for all $\sigma \in G_K^L$. This shows that $G_K^L = \{\mathrm{id}\}$ and consequently $F(G_K^L) = F(\{\mathrm{id}\}) = L \neq K$. Thus $(L : K)$ is not a Galois extension.

(c) Let $(L : K)$ be any field extension. Then an intermediate field Z of $(L : K)$ is closed if and only if $(L : Z)$ is a Galois extension; this is simply a restatement of (21.14)(a).

(d) Let L be a field and let Γ be a finite subgroup of Aut L. If $K := F(\Gamma)$ is the fixed field of Γ then $(L : K)$ is a Galois extension with Galois group $G_K^L = \Gamma$. Indeed, we have $F(G_K^L) = F(G_{F(\Gamma)}^L) = (F \circ G \circ F)(\Gamma) = F(\Gamma) = K$, and since Γ is finite (hence closed) we have $G_K^L = G_{F(\Gamma)}^L = \Gamma$.

(e) Let K be an arbitrary field, let $L := K(x_1, \ldots, x_n)$ be the rational function field in n variables over K and let S be the subfield of L consisting of all symmetric rational functions. Since S is the fixed field of $\Gamma := \mathrm{Sym}_n$ (which can obviously be considered as a group of automorphisms of L) we see from part (d) that $(L : S)$ is a Galois extension with $G_S^L \cong \mathrm{Sym}_n$.

(f) Let $K(x)$ be the rational function field over a field K. Then $\big(K(x) : K\big)$ is a Galois extension if and only if K is infinite. (See problems 10 and 11 below.)

KNOWING that every finite subgroup of a Galois group is closed, we can easily establish that every finite group occurs as the Galois group of some field extension.

(21.26) Proposition. *Let U be a finite group. Then there is a Galois extension $(L : K)$ with $G_K^L \cong U$.*

Proof. Due to (I.23.20) we can identify U with a subgroup of some symmetric group Sym_n. Let K_0 be any field and let $L := K_0(x_1, \ldots, x_n)$ be the rational function field over K_0 in n variables. For each permutation $(x_1, \ldots, x_n) \mapsto (x_{\sigma(1)}, \ldots, x_{\sigma(n)})$ of the variables there is a unique automorphism $\overline{\sigma} : L \to L$ with $\overline{\sigma}(x_i) = x_{\sigma(i)}$. Let Γ be the set of those automorphisms which correspond to elements of U. Then example (21.25)(d) shows that we can choose $K := F(\Gamma)$.[†] ∎

DEFINITION (21.24) was chosen to express a desirable property; it is not well suited to an easy determination of whether or not a given field extension is of Galois type. However, in the next section we will be able to give a complete characterization of *algebraic* Galois extensions.

[†] If U is the full symmetric group Sym_n then K is the field of all symmetric rational functions, and we are in the situation of example (21.25)(e).

Exercises

Problem 1. Determine the Galois groups $G_{\mathbb{Q}(\sqrt{2})}^{\mathbb{Q}(\sqrt{2}, \sqrt[3]{2})}$ and $G_{\mathbb{Q}(\sqrt{2})}^{\mathbb{Q}(\sqrt{2}, \sqrt[4]{2})}$.

Problem 2. Let $(L : K)$ be a finite field extension and let $\sigma : L \to L$ be a nonzero K-homomorphism. Show that then $\sigma \in G_K^L$.

Problem 3. Show that $\mathbb{Q}(x^2)$ is a closed subfield of $(\mathbb{Q}(x) : \mathbb{Q})$, but that $\mathbb{Q}(x^3)$ is not.

Problem 4. Suppose that Z is a closed subfield of $(L : K)$. Show that the normalizer of G_Z^L in G_K^L is $\{\sigma \in G_K^L \mid \sigma(Z) = Z\}$.

Problem 5. Let $(L : K)$ be a field extension and let $\sigma_1, \ldots, \sigma_n : K \to L$ be pairwise different nonzero field homomorphisms. Show that $Z := \{x \in K \mid \sigma_1(x) = \cdots = \sigma_n(x)\}$ is a subfield of K with $[K : Z] \geq n$.

Problem 6. Let L be a field, let U be a subgroup of $\mathrm{Aut}(L)$ and let K be the fixed field of U. Show that an element $\alpha \in L$ is algebraic over K if and only if the set $\{\sigma(\alpha) \mid \sigma \in U\}$ is finite.

Problem 7. Let G be a monoid (i.e., a semigroup with identity element 1) and let K be a field. A character of G in K is a mapping $\chi : G \to K^\times$ with $\chi(1) = 1$ and $\chi(g_1 g_2) = \chi(g_1)\chi(g_2)$. Show that pairwise different characters of G in K are linearly independent in the K-vector space of all functions $f : G \to K$.

Problem 8. Given a field K, let $\alpha_1, \ldots, \alpha_n \in K^\times$ and $a_1, \ldots, a_n \in K$. Show that if $a_1 \alpha_1^m + \cdots + a_n \alpha_n^m = 0$ for all $m \in \mathbb{Z}$ then $a_1 = \cdots = a_n = 0$.
Hint. Consider the mappings $\chi_i : \mathbb{Z} \to K^\times$ given by $\chi_i(m) := \alpha_i^m$ and apply problem 7.

Problem 9. Let $\sigma_1, \ldots, \sigma_n : K \to L$ different field embeddings. Show that there are elements $a_1, \ldots, a_n \in K$ such that $\det(\sigma_i(\alpha_j)) \neq 0$.

Problem 10. Let K be a finite field with q elements.
(a) Show that $G_K^{K(x)}$ has $q^3 - q$ elements.
(b) Show that $F(G_K^{K(x)}) = K(\varphi)$ where

$$\varphi(x) = \frac{(x^{q^2} - x)^{q+1}}{(x^q - x)^{q^2+1}}$$

and conclude that $(K(x) : K)$ is not a Galois extension.
(c) Show that if U is the subgroup of $G_K^{K(x)}$ which consists of all mappings σ of the form $(\sigma\theta)(x) = \theta(ax + b)$ with $a \neq 0$ then $F(U) = K((x^q - x)^{q-1})$.

(d) Show that if V is the subgroup of $G_K^{K(x)}$ which consists of all mappings σ of the form $(\sigma\theta)(x) = \theta(x + b)$ then $F(V) = K(x^q - x)$.

Problem 11. Let K be an infinite field and let $K(x)$ be the rational function field over K.

(a) Show that $(K(x) : K)$ is a Galois extension.

Hint. If $f/g \in F(G_K^{K(x)})$ then $f(x + a)/g(x + a) = f(x)/g(x)$ for all $a \in K$ because all the mappings $\varphi \mapsto \varphi(\ast + a)$ are automorphisms of $K(x)$. This means that the polynomial $h(x, y) := f(x + y)g(x) - f(x)g(x + y)$ vanishes on all of K^2 and hence must be the zero polynomial because $|K| = \infty$. Deduce that f/g is a constant.

(b) Show that the only closed subgroups of $G_K^{K(x)}$ are the finite subgroups and G itself. (**Hint.** By Lüroth's theorem we have $[K(x) : Z] < \infty$ for any intermediate field $K \subsetneq Z \subseteq K(x)$).

Problem 12. Let $K(x)$ be the rational function field over a field K.

(a) Show that for a subgroup $U \leq G_K^{K(x)}$ the following conditions are equivalent:
(1) $(K(x) : F(U))$ is a Galois extension;
(2) U is finite;
(3) there is a nonconstant rational function with $U = \{\sigma \in G \mid \sigma \star \varphi = \varphi\}$.

(b) Let U be a finite subgroup of $G_K^{K(x)}$ and let $\theta \in K(x)$. Show that $F(U) = K(\theta)$ if and only if $U = \{\sigma \in G_K^{K(x)} \mid \sigma\theta = \theta\}$.

(c) Consider the subgroup $V := \{\mathrm{id}, \alpha, \beta\} \leq G_K^{K(x)}$ where

$$(\alpha f)(x) = f(\frac{1}{1-x}) \quad \text{and} \quad (\beta f)(x) = f(\frac{x-1}{x}).$$

(In the terminology of (21.23) this means that $V = \{\sigma_1, \sigma_5, \sigma_6\} \leq U$.) Find the fixed field $F(V)$ of V and determine the degree $[K(x) : F(V)]$.

Problem 13. Let K be a field of characteristic 0 and let σ be the element of $G_K^{K(x)}$ defined by $(\sigma f)(x) = f(x + 1)$. Show that $U := \langle \sigma \rangle$ is an infinite cyclic group; determine its fixed field $F(U)$ and the degree $[K(x) : F(U)]$.

Problem 14. Let $L = \mathbb{Q}(\sqrt[4]{2})$ and $Z = \mathbb{Q}(\sqrt{2})$.

(a) Show that $G_\mathbb{Q}^L = \{\mathrm{id}, \sigma\}$ where $\sigma(\sqrt[4]{2}) = -\sqrt[4]{2}$.

(b) Show that $(L : Z)$ and $(Z : \mathbb{Q})$ are both Galois extensions but that $(L : \mathbb{Q})$ is not. (Hence the property of being of Galois type is not transitive.)

Problem 15. Suppose that $(L : Z)$ and $(Z : K)$ are Galois extensions and that every element of G_K^Z can be extended to an automorphism of L. Show that $(L : K)$ is a Galois extension.

Problem 16. Let K be an infinite field and let $Z = K(x)$ and $L = K(x, y)$ with variables x, y. Show that $(L : K)$, $(Z : K)$ and $(L : Z)$ are all Galois extensions, but that Z is not a stable intermediate field of $(L : K)$.

SEEN from an abstract point of view, one can consider Galois theory as a way of studying a field extension $(L : K)$ by investigating an associated algebraic structure, namely the Galois group G_K^L. In the following problem, we will consider another example for an algebraic structure which can be associated with a field extension $(L : K)$ and gives rise to a correspondences analogous to the Galois correspondence.

Problem 17 (Jacobson-Bourbaki-correspondence). Let $(L : K)$ be a field extension. The set

$$\operatorname{End} L := \{T : L \to L \mid T(\alpha + \beta) = T(\alpha) + T(\beta) \text{ for all } \alpha, \beta \in L\}$$

is an L-algebra if we define addition, multiplication and scalar multiplication by

$$(S + T)(\alpha) := S(\alpha) + T(\alpha), \quad (ST)(\alpha) := S\big(T(\alpha)\big) \quad \text{and} \quad (\lambda T)(\alpha) := \lambda \cdot T(\alpha) \ .$$

With each intermediate field Z of $(L : K)$ we associate a subalgebra $\operatorname{End}_Z(L)$ of $\operatorname{End}(L)$ by defining

$$\operatorname{End}_Z(L) := \{T \in \operatorname{End}(L) \mid T \text{ is } Z\text{-linear}\}$$

so that $T \in \operatorname{End}_Z(L)$ if and only if $T(z\alpha) = z \cdot T(\alpha)$ for all $z \in Z$ and $\alpha \in L$. On the other hand, we associate with each subalgebra $\mathfrak{A}$ of $\operatorname{End}(L)$ a subfield $\operatorname{Comm}(\mathcal{A})$ of L by defining

$$\operatorname{Comm}(\mathcal{A}) := \{\alpha \in L \mid D \circ m_\alpha = m_\alpha \circ D \text{ for all } D \in \mathcal{A}\}$$

where $m_\alpha : L \to L$ is multiplication by α; this means that $\alpha \in \operatorname{Comm}(\mathcal{A})$ if and only if $D(\alpha\beta) = \alpha \cdot D(\beta)$ for all $D \in \mathcal{A}$ and all $\beta \in L$.

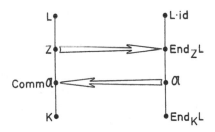

Show that the emerging Jacobson-Bourbaki-correspondence between the intermediate fields of $(L : K)$ and the subalgebras of $\operatorname{End}(L)$ can be restricted to a 1-1-correspondence between those intermediate fields Z of $(L : K)$ with $[L : Z] < \infty$ and those subalgebra $\mathcal{A}$ of $\operatorname{End}(L)$ with $\dim_L(\mathcal{A}) < \infty$. Moreover, show that $\dim_L(\operatorname{End}_Z L) = [L : Z]$ for all these intermediate fields.

22. Algebraic Galois extensions

IN this section we will first derive a complete characterization of *algebraic* Galois extension. Once this is done we can show that for *finite* Galois extension the mappings F and G which constitute the Galois correspondence are inverse bijections and hence faithfully relate the field-theoretical structure of $(L : K)$ with the group-theoretical structure of G_K^L; this is called the *Main Theorem of Galois theory*. Let us begin by characterizing algebraic Galois extensions.

(22.1) Theorem. *Let $(L : K)$ be an algebraic field extension. Then the following statements are equivalent:*

(1) $(L : K)$ *is a Galois extension, i.e.,* $F(G_K^L) = K$;

(2) K *is a closed subfield of* $(L : K)$, *i.e.,* $K = F(U)$ *for some subgroup* $U \le G_K^L$;

(3) $(L : K)$ *is normal and separable;*

(4) L *is a splitting field of a family* $\mathfrak{F} \subseteq K[x]$ *consisting entirely of separable polynomials.*

If $[L : K] < \infty$, we have three more equivalent conditions:

(5) L *is the splitting field of a polynomial $f \in K[x]$ all of whose irreducible factors are separable;*

(6) L *is the splitting field of a separable irreducible polynomial $g \in K[x]$ such that $L = K(\alpha)$ for some root of g;*

(7) $|G_K^L| = [L : K]$.

Proof. (1)$\Longrightarrow$(2). Just take $U := G_K^L$.

(2)$\Longrightarrow$(1). If $K = F(U)$ then $F(G_K^L) = (F \circ G)(K) = (F \circ G \circ F)(U) = F(U) = K$.

(1)$\Longrightarrow$(3). Let $\alpha \in L$ and let $f \in K[x]$ be the minimal polynomial of α over K. Let $\alpha = \alpha_1, \ldots, \alpha_r$ be the different roots of f in L and define $g(x) := (x - \alpha_1) \cdots (x - \alpha_r)$. Since each $\sigma \in G_K^L$ permutes the roots of f, we have $\sigma \star g = g$ for all $\sigma \in G_K^L$; hence the coefficients of g lie in $F(G_K^L) = K$ so that $g \in K[x]$. Since g divides f and since f is the minimal polynomial of α over K, this implies $g = f$; hence $\deg f = \deg g = r$. This shows that the number of different roots of f in L is $\deg f$ (which is the maximal possible number); hence f is separable and splits over L. This observation yields two conclusions: first, the minimal polynomial f of an arbitrary element $\alpha \in L$ is separable (which shows that $(L : K)$ is separable); second, if an irreducible polynomial $f \in K[x]$ has a root $\alpha \in L$ then it splits over L (which by (13.10) means that $(L : K)$ is normal).

(3)$\Longrightarrow$(4). Since $(L : K)$ is normal, L is the splitting field of some family $\mathfrak{G} \subseteq K[x]$. Then L is clearly also the splitting field of

$$\mathfrak{F} := \{f \in K[x] \mid f \text{ is a monic irreducible factor of some } g \in \mathfrak{G}\}.$$

On the other hand, each $f \in \mathfrak{F}$ splits over L and hence is the minimal polynomial of some element $\alpha \in L$, hence must be separable by hypothesis.

(4)$\Longrightarrow$(3). We have $L = K(A)$ where A is the set of roots of $\mathfrak{F} \subseteq K[x]$. For each $\alpha \in A$, the minimal polynomial of α over K lies in $\mathfrak{F}$ and is therefore separable, which means that α is separable. Then (14.12)(b) shows that $(L : K)$ is separable. The normality of $(L : K)$ is clear because L is a splitting field over K.

(4)$\Longrightarrow$(1). Suppose this implication is established in the case that $[L : K] < \infty$;

then we can argue as follows in the general case. Let $\alpha \in L \setminus K$, let $f \in K[x]$ be the minimal polynomial of α over K and let $Z \subseteq L$ be the splitting field of f. Then $(Z : K)$ is a *finite* normal extension which is also separable (because (4) implies (3)); since we assumed the implication (4)$\Longrightarrow$(1) for finite extensions, this shows that $(Z : K)$ is a Galois extension. Hence since $\alpha \in Z \setminus K$, there is an element $\sigma \in G_K^Z$ with $\sigma(\alpha) \neq \alpha$. Since L is normal over K, hence the more so over Z, we can extend $\sigma : Z \to Z$ to an automorphism $\Sigma : L \to L$. Then Σ belongs to G_K^L and satisfies $\Sigma(\alpha) \neq \alpha$. This shows that every element in $L \setminus K$ can be moved by an element of G_K^L, which is what we wanted to show.

From now on we assume that $[L : K] < \infty$. Under this assumption, we will prove the implications (4)$\Longrightarrow$(5)$\Longrightarrow$(6) $\Longrightarrow$(7)$\Longrightarrow$(1), which clearly will finish the proof.

(4)$\Longrightarrow$(5). If L is the splitting field of a family $\mathfrak{F}$ as in (4), then clearly L is the splitting field of some finite subset $\{f_1, \ldots, f_r\} \subseteq \mathfrak{F}$ as can be easily deduced from the degree formula. But this is tantamount to saying that L is the splitting field of $f := f_1 \cdots f_r$.

(5)$\Longrightarrow$(6). Let $\alpha_1, \ldots, \alpha_n$ be the roots of f so that $L = K(\alpha_1, \ldots, \alpha_n)$. Since each α_i is the root of an irreducible factor of f and hence separable, we know by (14.12)(b) that $(L : K)$ is a separable extension and hence possesses a primitive element α. The minimal polynomial $g \in K[x]$ of α splits over L by (13.10) because L is a splitting field and hence normal over K; this shows that L is the splitting field of g.

(6)$\Longrightarrow$(7). We have $[L : K] = [K(\alpha) : K] = \deg g =: d$. Let $\alpha_1 = \alpha, \alpha_2, \ldots, \alpha_d$ be the roots of g (they are pairwise different by hypothesis). For each $1 \leq i \leq d$ there is an element $\sigma_i \in G_K^L$ with $\sigma_i(\alpha) = \alpha_i$; hence $|G_K^L| \geq [L : K]$. On the other hand $|G_K^L| \leq [L : K]$ by (21.17) or (21.19); hence condition (7) follows.

(7)$\Longrightarrow$(1). From (7) we obtain $|G_K^L| = [L : K] = [L : F(G_K^L)] \cdot [F(G_K^L) : K]$. On the other hand, applying Artin's Theorem (21.22)(a) with $U := G_K^L$, we obtain $|G_K^L| = [L : F(G_K^L)]$. Comparing the two equations, we see that $[F(G_K^L) : K] = 1$ which means $F(G_K^L) = K$. ∎

THE characterization of Galois extensions just obtained can be used to derive some general information on the structure of normal extensions. Recall from (14.16) that an arbitrary algebraic extension can be obtained by first performing a separable extension and then a purely inseparable extension. We now show that for a *normal* extension the order of these steps can be reversed. Moreover, we will see that every normal extension can be written as a compositum of a purely inseparable extension and a Galois extension.

(22.2) Theorem. *Let $(L : K)$ be a normal field extension. Let S and P be the separable closure and the purely inseparable closure of K in L, respectively. Then the following statements hold:*
(a) $P = F(G_K^L)$;
(b) *both $(L : P)$ and $(S : K)$ are Galois extensions;*
(c) $L = SP$;
(d) *the restriction map $R : G_P^L \to G_K^S$ is an isomorphism;*
(e) $|G_K^L| = [L : K]_s$, *i.e., the size of the Galois group equals the separable degree of $(L : K)$;*
(f) *an intermediate field Z of $(L : K)$ is stable (i.e., satisfies $F(G_Z^L) = Z$) if and only if $(Z : K)$ is normal.*

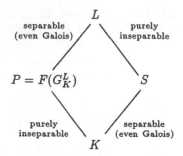

Proof. If char $K = 0$ then $P = K$ and $S = L$ so that the claim is trivial; hence we may assume that char $K = p \neq 0$.

(a) Let $\alpha \in L$ and let f be the minimal polynomial of α. Then

$$\alpha \in P \iff \alpha \text{ is the only root of } f$$
$$\iff \alpha \text{ is the only conjugate of } \alpha \text{ over } K$$
$$\iff \{\sigma(\alpha) \mid \sigma \in G_K^L\} = \{\alpha\}$$
$$\iff \sigma(\alpha) = \alpha \text{ for all } \sigma \in G_K^L$$
$$\iff \alpha \in F(G_K^L).$$

(b) First, $(S : K)$ is separable and normal by (14.16), hence is a Galois extension by (22.1). Second, $(L : P)$ is a Galois extension because

$$F(G_P^L) = (F \circ G)(P) = (\underbrace{F \circ G \circ F}_{= F} \circ G)(K) = (F \circ G)(K) = F(G_K^L) = P .$$

(c) $(L : P)$ is algebraic (since $(L : K)$ is) and hence is an algebraic Galois extension. Then (22.1) yields that $(L : P)$ is separable which by (14.16)(d) implies that $L = SP$.

(d) Every automorphism of L maps separable elements to separable elements; hence S is a stable intermediate field of $(L : K)$ so that the restriction map $R : G_P^L \to G_K^S$ is well-defined. Moreover, $(L : S)$ is normal because $(L : K)$ is; hence every automorphism of S can be extended to an automorphism of L. This shows that R is surjective. Finally, if $\sigma \in \ker R$ then σ is the identity on S and on P, hence on $SP = L$; this shows that R is injective.

(e) We have $G_K^L = G(K) = G \circ F \circ G(K) = G_{F(G_K^L)}^L = G_P^L$ by part (a); hence using part (d) we see that $|G_K^L| = |G_P^L| = |G_K^S| = [S : K] = [L : K]_s$.

(f) Let $\overline{L}$ be an algebraic closure of L (and hence of Z because $(L : Z)$ is algebraic since $(L : K)$ is). Using (13.10) and the fact that every element of G_K^L can be extended to an element of $G_K^{\overline{L}}$, we easily see that the following equivalences hold:

$$(Z : K) \text{ normal} \iff \sigma(Z) \subseteq Z \text{ for all } \sigma \in G_K^{\overline{Z}} = G_K^{\overline{L}}$$
$$\iff \sigma(Z) \subseteq Z \text{ for all } \sigma \in G_K^L$$
$$\iff Z \text{ is stable.}$$

$\blacksquare$

WE are now ready to prove the main theorem of Galois theory.

(22.3) Main Theorem of Galois Theory. *Let $(L : K)$ be an algebraic Galois extension.*

(a) *If Z is any intermediate field of $(L : K)$ then Z is closed, i.e., $(L : Z)$ is a Galois extension again.*

(b) *The Galois correspondence is a one-to-one correspondence between the intermediate fields of $(L : K)$ and the closed subgroups of G_K^L.*

(c) *For an intermediate field $K \leq Z \leq L$, the following statements are equivalent:*
(1) *$(Z : K)$ is a Galois extension;*
(2) *Z is a stable intermediate field of $(L : K)$, i.e., $\sigma(Z) \subseteq Z$ for all $\sigma \in G_K^L$;*
(3) *G_Z^L is a normal subgroup of G_K^L.*
If these statements hold, then $G_K^Z \cong G_K^L / G_Z^L$.

Proof. (a) The properties of normality and separability are inherited from $(L : K)$ to $(L : Z)$; hence the characterization theorem (22.1) gives the claim.

(b) This is clear from part (a) and (21.15)(c).

(c) Since $(Z : K)$ inherits the separability from $(L : K)$, the equivalence of (1) and (2) follows immediately from (22.2)(f). Due to (21.11)(h), the subgroup G_Z^L is normal if and only if $G_{\sigma(Z)}^L = G_Z^L$ for all $\sigma \in G_K^L$ which, by part (a), is the case if and only if $\sigma(Z) = F(G_{\sigma(Z)}^L) = F(G_Z^L) = Z$ for all $\sigma \in G_K^L$, i.e., if Z is stable. This gives the equivalence of (2) and (3). Finally, the isomorphism $G_K^L / G_Z^L \cong G_K^Z$ follows immediately from (21.15)(d) and the observation that *every* element of G_K^Z can be extended to an automorphism of L because $(L : K)$ and hence $(L : Z)$ is normal. ∎

SINCE we know that for a finite extension $(L : K)$ *every* subgroup of G_K^L is closed, we can give a more precise version of the main theorem in this case.

(22.4) Main Theorem of Galois Theory for finite extensions. *Let $(L : K)$ be a finite Galois extension.*

(a) *If $K \leq Z \leq L$ is any intermediate field, then $F(G_Z^L) = Z$ and $|G_Z^L| = [L : Z]$; i.e., the extension $(L : Z)$ is of Galois type again.*

(b) *If $\{\mathrm{id}\} \leq U \leq G_K^L$ is any subgroup, then $G_{F(U)}^L = U$ and $|U| = [L : F(U)]$.*

(c) *For an intermediate field $K \leq Z \leq L$, the following conditions are equivalent:*
(1) *$(Z : K)$ is a Galois extension;*
(2) *Z is stable, i.e., $\sigma(Z) \subseteq Z$ for all $\sigma \in G_K^L$;*
(3) *G_Z^L is a normal subgroup of G_K^L.*
If these conditions hold, then the restriction map $G_K^L \to G_K^Z$ induces an isomorphism from G_K^L / G_Z^L onto G_K^Z.

Proof. The only point which still needs to be proved is the equality of the degree of an intermediate field of $(L : K)$ with the index of the corresponding subgroup of G_K^L. Now if $U \leq G_K^L$ then $[L : F(U)] = |U|$ by Artin's Theorem (21.22)(a). If Z is any intermediate field of $(L : K)$, we can apply this result with $U := G_Z^L$ to obtain $|G_Z^L| = [L : F(G_Z^L)] = [L : Z]$; here we used in the last equation that all intermediate fields of $(L : K)$ are closed. ∎

SUMMARIZING our findings, we see that for a finite Galois extension $(L : K)$ the Galois correspondence constitutes a bijection between the intermediate fields of $(L : K)$ and the subgroups G_K^L, that this correspondence preserves conjugacy and normality relations and that the degree of L over an intermediate field equals the index of the corresponding subgroup in G_K^L.

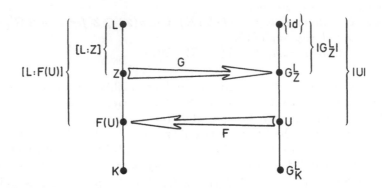

LET us give some examples which illustrate the main theorem of Galois theory.

(22.5) Example. $(\mathbb{Q}(\sqrt{2}, \sqrt{3}) : \mathbb{Q})$ is a Galois extension because $\mathbb{Q}(\sqrt{2}, \sqrt{3})$ is the splitting field of $p(x) = (x^2 - 2)(x^2 - 3)$. Hence $|G_{\mathbb{Q}}^{\mathbb{Q}(\sqrt{2},\sqrt{3})}| = [\mathbb{Q}(\sqrt{2}, \sqrt{3}) : \mathbb{Q}] = 4$ so that the four maps $\sigma_0, \sigma_1, \sigma_2, \sigma_3$ in (21.4) are really automorphisms of $\mathbb{Q}(\sqrt{2}, \sqrt{3})$. (Strictly speaking, we proved in (21.4) only that these maps are the only possible candidates as members of $G_{\mathbb{Q}}^{\mathbb{Q}(\sqrt{2},\sqrt{3})}$, but we did not execute the tedious verification that they are really automorphisms.)

Whereas it is *a priori* not even clear that the extension $(\mathbb{Q}(\sqrt{2},\sqrt{3}) : \mathbb{Q})$ has only a finite number of intermediate fields, the main theorem of Galois theory tells us that there are exactly 3 proper intermediate fields, namely the fixed fields of the 3 proper subgroups U_1, U_2, U_3 of $G_{\mathbb{Q}}^{\mathbb{Q}(\sqrt{2},\sqrt{3})}$.

$$
\begin{array}{c}
\mathbb{Q}(\sqrt{2}, \sqrt{3}) \\
\mathbb{Q}(\sqrt{3}) \quad \mathbb{Q}(\sqrt{2}) \quad \mathbb{Q}(\sqrt{6}) \\
\mathbb{Q}
\end{array}
\qquad
\begin{array}{l}
\mathbb{Q}(\sqrt{3}) = F(U_1) \\
\mathbb{Q}(\sqrt{2}) = F(U_2) \\
\mathbb{Q}(\sqrt{6}) = F(U_3)
\end{array}
$$

(22.6) Example. Let $\varepsilon := e^{2\pi i/3}$; then $\mathbb{Q}(\sqrt[3]{2}, \varepsilon)$ is the splitting field of $p(x) := x^3 - 2$, the roots of p being $\sqrt[3]{2}$, $\varepsilon\sqrt[3]{2}$ and $\varepsilon^2\sqrt[3]{2}$. Now the minimal polynomial of $\sqrt[3]{2}$ over $\mathbb{Q}$ is $x^3 - 2$, whereas the minimal polynomial of ε over $\mathbb{Q}(\sqrt[3]{2})$ is $x^2 + (\sqrt[3]{2})x + \sqrt[3]{4}$. Consequently,

$$
[\mathbb{Q}(\sqrt[3]{2}, \varepsilon) : \mathbb{Q}] = \underbrace{[\mathbb{Q}(\varepsilon, \sqrt[3]{2}) : \mathbb{Q}(\sqrt[3]{2})]}_{= \deg(x^2 + \sqrt[3]{2}x + \sqrt[3]{4}) = 2} \cdot \underbrace{[\mathbb{Q}(\sqrt[3]{2}) : \mathbb{Q}]}_{= \deg(x^3 - 2) = 3} = 2 \cdot 3 = 6 .
$$

Hence $G_{\mathbb{Q}}^{\mathbb{Q}(\sqrt[3]{2},\,\varepsilon)}$ has six elements. Since each element in this Galois group must permute the three roots of $x^3 - 2$ and is uniquely determined by the values on these roots, we conclude that $G_{\mathbb{Q}}^{\mathbb{Q}(\sqrt[3]{2},\,\varepsilon)} \cong \mathrm{Sym}_3$.

THE next example illustrates all aspects of the main theorem of Galois theory.

(22.7) Example. Let L be the splitting field of the polynomial $f(x) = x^4 - 2 \in \mathbb{Q}[x]$. Since the roots of f are $\pm\sqrt[4]{2}$ and $\pm i\sqrt[4]{2}$ we have

$$L := \mathbb{Q}(\pm\sqrt[4]{2}, \pm i\sqrt[4]{2}) = \mathbb{Q}(\sqrt[4]{2}, i\sqrt[4]{2}) = \mathbb{Q}(\sqrt[4]{2}, i) .$$

Now the polynomial f is irreducible by Eisenstein's criterion, so it is the minimal polynomial of $\sqrt[4]{2}$. Consequently,

$$[\mathbb{Q}(\sqrt[4]{2}) : \mathbb{Q}] = \text{degree of } x^4 - 2 = 4 .$$

Also, $x^2 + 1$ is the minimal polynomial of i over $\mathbb{Q}(\sqrt[4]{2})$, because $\pm i \notin \mathbb{Q}(\sqrt[4]{2}) \subseteq \mathbb{R}$. Consequently,

$$[\mathbb{Q}(\sqrt[4]{2}, i) : \mathbb{Q}(\sqrt[4]{2})] = \text{degree of } x^2 + 1 = 2 .$$

By the degree formula,

$$[L : \mathbb{Q}] = [L : \mathbb{Q}(\sqrt[4]{2})] \cdot [\mathbb{Q}(\sqrt[4]{2}) : \mathbb{Q}] = 2 \cdot 4 = 8 .$$

Now let φ be an arbitrary element of $G_{\mathbb{Q}}^L$. The numbers i and $\sqrt[4]{2}$ satisfy the equations $x^2 = -1$ and $x^4 = 2$, respectively; hence so do $\varphi(i)$ and $\varphi(\sqrt[4]{2})$. This shows that $\varphi(i) = \pm i$ (two possibilities) and $\varphi(\sqrt[4]{2}) \in \{\pm\sqrt[4]{2}, \pm i\sqrt[4]{2}\}$ (four possibilities). Since φ is uniquely determined by the values $\varphi(i)$ and $\varphi(\sqrt[4]{2})$, there are at most $2 \cdot 4 = 8$ automorphisms of $\mathbb{Q}(\sqrt[4]{2}, i)$. But now the main theorem of Galois theory yields

$$|G_{\mathbb{Q}}^L| = [L : \mathbb{Q}] = 8$$

so that each possible choice of signs indeed determines an automorphism of $\mathbb{Q}(\sqrt[4]{2}, i)$. Next, we define the elements σ and τ of $G_{\mathbb{Q}}^L$ by

$$\sigma(i) = i, \ \sigma(\sqrt[4]{2}) = i\sqrt[4]{2} \quad \text{and} \quad \tau(i) = -i, \ \tau(\sqrt[4]{2}) = \sqrt[4]{2} .$$

The following list shows that all of the 8 elements of $G_{\mathbb{Q}}^L$ can be expressed by σ and τ.

φ	id	σ	σ^2	σ^3	τ	$\sigma\tau$	$\sigma^2\tau$	$\sigma^3\tau$
$\varphi(\sqrt[4]{2})$	$\sqrt[4]{2}$	$i\sqrt[4]{2}$	$-\sqrt[4]{2}$	$-i\sqrt[4]{2}$	$\sqrt[4]{2}$	$i\sqrt[4]{2}$	$-\sqrt[4]{2}$	$-i\sqrt[4]{2}$
$\varphi(i)$	i	i	i	i	$-i$	$-i$	$-i$	$-i$

(All eight combinations of signs do occur.) Since $\sigma^4(\xi) = \tau^2(\xi) = \xi$ and $\sigma\tau(\xi) = \tau\sigma^3(\xi)$ for $\xi \in \{\sqrt[4]{2}, i\}$, we know that $G_{\mathbb{Q}}^L$ has two generators σ, τ satisfying the relations

$$\sigma^4 = \tau^2 = \text{id} \quad \text{and} \quad \sigma\tau = \tau\sigma^3 .$$

This shows that $G_{\mathbb{Q}}^L$ is the dihedral group D_4; see (I.20.11). The subgroup lattice is

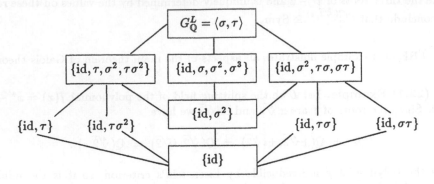

where the framed subgroups are normal, the others are not. It is trivial that $G_{\mathbb{Q}}^L$ and $\{id\}$ are normal. The subgroups with 4 elements (hence of index 2) are normal by (24.5) or by (27.5)(b). Finally, $\{id, \sigma^2\}$ is normal because $\tau^{-1}\sigma^2\tau = \sigma^2$. The equations

$$\sigma\tau\sigma^{-1} = \tau\sigma^2, \quad \sigma^{-1}(\tau\sigma^2)\sigma = \tau, \quad \tau(\tau\sigma)\tau^{-1} = \sigma\tau, \quad \tau^{-1}(\sigma\tau)\tau = \tau\sigma$$

show that the four remaining subgroups are not normal.

The intermediate fields of $(L : \mathbb{Q})$ are exactly the fixed fields $F(U)$ where $U \le G_{\mathbb{Q}}^L$; the subgroup lattice

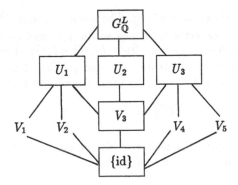

yields immediately the lattice of intermediate fields.

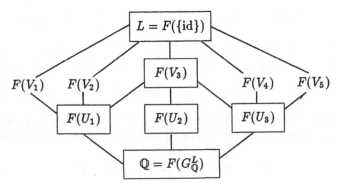

Let us identify the various fixed fields. Trivially, $F(\{\mathrm{id}\}) = L$. Also, $F(G_{\mathbb{Q}}^L) = \mathbb{Q}$ by the main theorem. Furthermore,

$$F(\{\mathrm{id}, \tau, \sigma^2, \tau\sigma^2\}) = \mathbb{Q}(\sqrt{2}) \, ,$$
$$F(\{\mathrm{id}, \sigma, \sigma^2, \sigma^3\}) = \mathbb{Q}(i) \, ,$$
$$F(\{\mathrm{id}, \sigma^2, \tau\sigma, \sigma\tau\}) = \mathbb{Q}(i\sqrt{2}) \, .$$

(Check!) For these three fixed fields, we have $[L : F(U)] = |U| = 4$ by the main theorem; hence the degree formula yields

$$[F(U) : \mathbb{Q}] = \frac{[L : \mathbb{Q}]}{[L : F(U)]} = \frac{8}{4} = 2 \, .$$

Also,

$$F(\{\mathrm{id}, \tau\}) = \mathbb{Q}(\sqrt[4]{2}) \, ,$$
$$F(\{\mathrm{id}, \tau\sigma^2\}) = \mathbb{Q}(i\sqrt[4]{2}) \, ,$$
$$F(\{\mathrm{id}, \sigma^2\}) = \mathbb{Q}(\sqrt{2}, i) \, ,$$
$$F(\{\mathrm{id}, \tau\sigma\}) = \mathbb{Q}((1 - i)\sqrt[4]{2}) \, ,$$
$$F(\{\mathrm{id}, \sigma\tau\}) = \mathbb{Q}((1 + i)\sqrt[4]{2}) \, .$$

Some of these equations are fairly obvious, others need a computation to be verified. Let us check that $F(\{\mathrm{id}, \sigma\tau\}) = \mathbb{Q}((1+i)\sqrt[4]{2})$. An arbitrary element of L has the form

$$x = a + b\xi + c\xi^2 + d\xi^3 + ei + fi\xi + gi\xi^2 + hi\xi^3 \quad \text{where } \xi := \sqrt[4]{2} \, .$$

Since $\sigma\tau(\xi) = i\xi$ and $\sigma\tau(i) = -i$, such an element x lies in $F(\{\mathrm{id}, \sigma\tau\})$, i.e., satisfies $\sigma\tau(x) = x$, if and only if

$$a + bi\xi - c\xi^2 - di\xi^3 - ei + f\xi + gi\xi^2 - h\xi^3$$
$$= a + b\xi + c\xi^2 + d\xi^3 + ei + fi\xi + gi\xi^2 + hi\xi^3$$

which means that $f = b$, $c = 0$, $h = -d$, $e = 0$, $f = b$ and $h = -d$ whereas g is arbitrary, which means that

$$x = a + b(1+i)\xi + d \underbrace{(1-i)\xi^3}_{= -\frac{1}{2}((1+i)\xi)^3} + g \underbrace{i\xi^2}_{= \frac{1}{2}((1+i)\xi)^2} \, .$$

This shows that $x \in F(\{\mathrm{id}, \sigma\tau\})$ if and only if $x \in \mathbb{Q}((1+i)\xi)$.

So the intermediate fields of $(L : \mathbb{Q})$ are given by the following lattice.

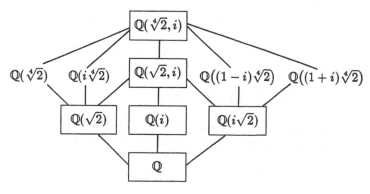

All of the framed fields are Galois extensions of $\mathbb{Q}$, the unframed fields are not. This follows from the fact that $Z = F(U)$ is a Galois extension of K if and only if U is a normal subgroup of G_K^L. In this case the isomorphism $G_K^Z \cong G_K^L/G_Z^L$ holds; i.e., $G_K^{F(U)} \cong G_K^L/G_{F(U)}^L = G_K^L/U$. Let us verify this isomorphism for $U := \{\mathrm{id}, \sigma^2\}$. The Galois group $G_K^{F(U)} = G_{\mathbb{Q}}^{\mathbb{Q}(\sqrt{2}, i)}$ has 4 elements and is generated by α and β where

$$\alpha : \begin{array}{ccc} i & \mapsto & i \\ \sqrt{2} & \mapsto & -\sqrt{2} \end{array}, \qquad \beta : \begin{array}{ccc} i & \mapsto & -i \\ \sqrt{2} & \mapsto & \sqrt{2} \end{array}.$$

We have $\alpha^2 = \beta^2 = \mathrm{id}$ and $\alpha\beta = \beta\alpha$; so

$$G_K^{F(U)} \cong \mathbb{Z}_2 \times \mathbb{Z}_2 .$$

It is easy to check that also

$$G_K^L/U = \langle \sigma, \tau \rangle / \{\mathrm{id}, \sigma^2\} \cong \mathbb{Z}_2 \times \mathbb{Z}_2$$

so that indeed $G_K^{F(U)} \cong G_K^L/U$.

The fact that an intermediate field is a Galois extension of $\mathbb{Q}$ if and only if it is framed in the above diagram also illustrates the characterization of Galois extensions. Indeed, the fields

$$\mathbb{Q}, \quad \mathbb{Q}(\sqrt[4]{2}, i), \quad \mathbb{Q}(\sqrt{2}), \quad \mathbb{Q}(i), \quad \mathbb{Q}(i\sqrt{2}) \quad \text{and} \quad \mathbb{Q}(\sqrt{2}, i)$$

are splitting fields of rational polynomials, namely of

$$x, \quad x^4 - 2, \quad x^2 - 2, \quad x^2 + 1, \quad x^2 + 2, \quad x^4 - x^2 - 2 ;$$

the fields

$$\mathbb{Q}(\sqrt[4]{2}), \quad \mathbb{Q}(i\sqrt[4]{2}), \quad \mathbb{Q}((1-i)\sqrt[4]{2}) \quad \text{and} \quad \mathbb{Q}((1+i)\sqrt[4]{2})$$

are not.

(22.8) Example. Extending our discussion in example (22.6), let us determine the Galois group $G_{\mathbb{Q}}^L$ where L is the splitting field of the polynomial $f(x) = x^3 - 2 \in \mathbb{Q}[x]$. The roots of f are $\sqrt[3]{2}, \varepsilon\sqrt[3]{2}, \varepsilon^2\sqrt[3]{2}$ where $\varepsilon := e^{2\pi i/3} = \frac{1}{2}(-1 + i\sqrt{3})$; thus

$$L = \mathbb{Q}(\sqrt[3]{2}, \varepsilon\sqrt[3]{2}, \varepsilon^2\sqrt[3]{2}) = \mathbb{Q}(\sqrt[3]{2}, \varepsilon) = \mathbb{Q}(\sqrt[3]{2}, i\sqrt{3}) .$$

Being a cubic polynomial without rational roots the polynomial f is irreducible and hence is the minimal polynomial of $\sqrt[3]{2}$. Consequently,

$$[\mathbb{Q}(\sqrt[3]{2}) : \mathbb{Q}] = \text{degree of } x^3 - 2 = 2 .$$

Also, $x^2 + 3$, being a quadratic polynomial without a root in $\mathbb{Q}(\sqrt[3]{2})$, is irreducible over $\mathbb{Q}(\sqrt[3]{2})$ and therefore is the minimal polynomial of $i\sqrt{3}$ over $\mathbb{Q}(\sqrt[3]{2})$. Consequently,

$$[\mathbb{Q}(\sqrt[3]{2}, i\sqrt{3}) : \mathbb{Q}(\sqrt[3]{2})] = \text{degree of } x^2 + 3 = 2 .$$

By the degree formula,

$$[L:\mathbb{Q}] = [L:\mathbb{Q}(\sqrt[3]{2})]\cdot[\mathbb{Q}(\sqrt[3]{2}):\mathbb{Q}] = 2\cdot 3 = 6.$$

Now let φ be an arbitrary element of $G_{\mathbb{Q}}^L$. The numbers $i\sqrt{3}$ and $\sqrt[3]{2}$ satisfy the equations $x^2 = -3$ and $x^3 = 2$, respectively; hence so do $\varphi(i)$ and $\varphi(\sqrt[3]{2})$. This shows that $\varphi(i\sqrt{3}) = \pm i\sqrt{3}$ (two possibilities) and $\varphi(\sqrt[3]{2}) = \varepsilon^k\sqrt[3]{2}$ with $k \in \{0,1,2\}$ (three possibilities). Since φ is uniquely determined by the values $\varphi(i\sqrt{3})$ and $\varphi(\sqrt[3]{2})$, there are at most $2\cdot 3 = 6$ automorphisms of $\mathbb{Q}(\sqrt[3]{2},i\sqrt{3})$. But now the main theorem of Galois theory yields

$$|G_{\mathbb{Q}}^L| = [L:\mathbb{Q}] = 6$$

so that each possible choice for the possible values $\varphi(i\sqrt{3})$ and $\varphi(\sqrt[3]{2})$ indeed determines an automorphism of $\mathbb{Q}(\sqrt[3]{2},i\sqrt{3})$. Next, we define the elements σ and τ of $G_{\mathbb{Q}}^L$ by

$$\sigma(i\sqrt{3}) = -i\sqrt{3},\ \sigma(\sqrt[3]{2}) = \sqrt[3]{2} \quad \text{and} \quad \tau(i\sqrt{3}) = -i\sqrt{3},\ \tau(\sqrt[3]{2}) = \varepsilon\sqrt[3]{2}.$$

The following list shows that all of the 6 elements of $G_{\mathbb{Q}}^L$ can be expressed by σ and τ.

φ	id	σ	τ	$\sigma\tau$	$\tau\sigma$	$\sigma\tau\sigma$
$\varphi(\sqrt[3]{2})$	$\sqrt[3]{2}$	$\varepsilon\sqrt[3]{2}$	$\varepsilon^2\sqrt[3]{2}$	$\varepsilon\sqrt[3]{2}$	$\varepsilon^2\sqrt[3]{2}$	$\sqrt[3]{2}$
$\varphi(i\sqrt{3})$	$i\sqrt{3}$	$-i\sqrt{3}$	$-i\sqrt{3}$	$i\sqrt{3}$	$i\sqrt{3}$	$-i\sqrt{3}$

(All six combinations do occur.) The subgroups and the corresponding fixed fields are

$$U_1 = \{\text{id},\sigma\},\quad U_2 = \{\text{id},\tau\}\quad U_3 = \{\text{id},\sigma\tau\sigma\}\quad U_4 = \{\text{id},\sigma\tau,\tau\sigma\};$$
$$V_1 = \mathbb{Q}(\sqrt[3]{2}),\quad V_2 = \mathbb{Q}(\varepsilon^2\sqrt[3]{2}),\quad V_3 = \mathbb{Q}(\varepsilon\sqrt[3]{2}),\quad V_4 = \mathbb{Q}(i\sqrt{3}).$$

The groups U_1, U_2, U_3 are not normal in $G_{\mathbb{Q}}^L)$ which reflects the fact that $(V_1:\mathbb{Q})$, $(V_2:\mathbb{Q})$ and $(V_3:\mathbb{Q})$ are not Galois extensions. On the other hand U_4 is a normal subgroup of $G_{\mathbb{Q}}^L$; accordingly, V_4 is a Galois extension of $\mathbb{Q}$, being the splitting field of x^2+3.

OUR next objective is to determine the Galois group of an extension of finite fields.

(22.9) Proposition. Let $K \subseteq L$ be finite fields in characteristic p, say $|K| = p^m$ and $|L| = p^n$. Let $\sigma : L \to L$ be the Frobenius automorphism of L, given by $\sigma(x) = x^p$.
(a) $(L:K)$ is a Galois extension.
(b) $\operatorname{Aut} L = G_{\mathbb{Z}_p}^L$ is cyclic of order n, and σ is a generator of $\operatorname{Aut} L$.
(c) G_K^L is cyclic of order $n/m = [L:K]$, and $\sigma^m : x \mapsto x^{p^m}$ is a generator of G_K^L.
(d) The norm mapping $N_K^L : L^\times \to K^\times$ is surjective.

Proof. (a) L is the splitting field of the polynomial $f(x) = x^{p^n} - x \in K[x]$ which is separable because $f'(x) = p^n x^{p^n-1} - 1 = -1 \neq 0$. Hence $(L:K)$ is a Galois extension.
(b) From (a) we know that $(L:\mathbb{Z}_p)$ is a Galois extension; hence $|\operatorname{Aut} L| = |G_{\mathbb{Z}_p}^L| = [L:\mathbb{Z}_p] = n$. Thus to prove (b) it is enough to show that the elements id, σ, ..., σ^{n-1} (which are all automorphism of L) are pairwise different. Let α be a generator of $K^\times$ and suppose that $\sigma^i(\alpha) = \sigma^j(\alpha)$ for $0 \leq j < i \leq n-1$. Then $\alpha^{p^i} = \alpha^{p^j}$, i.e., $\alpha^{p^i-p^j} = 1$

where $0 \leq p^i - p^j \leq p^i - 1 < p^n - 1$. Since the element α has order $|K^\times| = p^n - 1$, we conclude that $p^j - p^i = 0$. This implies $i = j$ so that the elements $\mathrm{id}, \sigma, \ldots, \sigma^{n-1}$ are in fact pairwise different.

(c) We know from (13.19) that K is a splitting field of $x^{p^m} - x$; hence $K = \{x \in L \mid x^{p^m} = x\} = \{x \in L \mid \sigma^m(x) = x\} = F(\langle \sigma^m \rangle)$. Hence $G_K^L = G(K) = (G \circ F)(\langle \sigma^m \rangle) = \langle \sigma^m \rangle$. Since σ^m has clearly order $n/m = [L : \mathbb{Z}_p]/[K : \mathbb{Z}_p] = [L : K]$, this gives the claim.

(d) Let α be a generator of $L^\times$ and let $\varphi := \sigma^m$ so that $G_K^L = \langle \varphi \rangle$. To simplify notation, let $q := p^m = |K|$ and $d := n/m = |G_K^L| = [L : K]$. Then

$$
N_K^L(\alpha) \;=\; \prod_{\theta \in G_K^L} \theta(\alpha) \;=\; \alpha \cdot \varphi(\alpha) \cdots \varphi^{d-1}(\alpha)
$$

$$
= \; \alpha \cdot \alpha^q \cdot \alpha^{q^2} \cdots \alpha^{q^{d-1}} \;=\; \alpha^{1 + q + q^2 + \cdots + q^{d-1}} \; .
$$

Since the order of α is $|L^\times| = q^d - 1 = (q - 1)(1 + q + q^2 + \cdots + q^{d-1})$, this shows that the order of $N_K^L(\alpha)$ is $q - 1 = |K^\times|$ so that $N_K^L(\alpha)$ generates $K^\times$. But this clearly implies that N_K^L is onto. ∎

AFTER spending a lot of effort to establish the main theorem of Galois theory, we now want to start harvesting the fruit of our work. The main application will be a complete translation of the problem of solving polynomial equations into a purely group-theoretical problem; this will be done in the subsequent sections. We will conclude this section by presenting two other applications, the first one being a quick new proof of the Primitive Element Theorem (14.9).

(22.10) Theorem. *Let $(L : K)$ be a finite separable field extension. Then there is an element $\gamma \in L$ with $L = K(\gamma)$. If $L = K(\alpha_1, \ldots, \alpha_n)$ and if K is infinite, then γ can be chosen in the form $\gamma = \sum_i k_i \alpha_i$.*

Proof. Let N be a normal closure of $(L : K)$. Then $(N : K)$ is a finite normal and separable field extension, i.e., a finite Galois extension. Since the finite group G_K^N has only a finite number of subgroups, the Galois correspondence shows that $(N : K)$ and hence $(L : K)$ has only a finite number of intermediate fields. Then (12.21) yields the claim.

Now suppose that K is infinite. Each of the intermediate fields of $(L : K)$ is, of course, a vector space over K. Since $|K| = \infty$, a finite union of proper vector subspaces of L cannot be all of L; hence there is an element $\gamma \in L$ which lies outside the union of all proper intermediate fields of $(L : K)$. But then the intermediate field $K(\gamma)$ cannot be proper, hence must all of L. ∎

AS a second application, let us give a necessary and sufficient condition for the constructibility of numbers with a ruler and a compass only. Observe that the criterion (12.36)(b) was mainly used to show that certain constructions are *impossible*; the following criterion is also convenient to prove off-hand that certain constructions are indeed *possible*. For example, we will use this criterion in a later section to characterize those numbers n for which a regular n-gon is constructible.

(22.11) Theorem. *A number $\alpha \in \mathbb{R}$ is constructible with a ruler and a compass from $a_1, \ldots, a_n \in \mathbb{R}$ if and only if α is contained in a Galois extension L of $K := \mathbb{Q}(a_1, \ldots, a_n)$ whose degree $[L : K]$ is a power of 2. (This is tantamount to saying that if $N \subseteq \mathbb{C}$ is the normal closure of $\big(K(\alpha) : K\big)$ then $[N : K]$ is a power of 2.)*

Proof. Suppose first that α is constructible from $a_1, \ldots, a_n$. Then by (12.36)(b) there is a chain of fields

$$\mathbb{Q}(a_1, \ldots, a_n) \;:=\; K \;:=\; K_0 \,\subseteq\, K_1 \,\subseteq\, \ldots \,\subseteq\, K_m$$

such that $\alpha \in K_m$ and $K_{i+1} = K_i(\sqrt{r_i})$ with $r_i > 0$ for all i. We claim that K_m is contained in a *Galois* extension L of K such that $[L : K]$ is a power of 2. We prove the existence of L by induction on m. The case $m = 0$ is trivial because $(K : K)$ is a Galois extension of degree $1 = 2^0$. Suppose that the claim holds for some m and consider a chain

$$K := K_0 \subseteq K_1 \subseteq \ldots \subseteq K_m \subseteq K_{m+1} \text{ with } K_{m+1} = K_m(\sqrt{r_m}),\, r_m \in K_m,\, r_m > 0.$$

By induction hypothesis there is a Galois extension $(L_0 : K)$ whose degree is a power of 2 such that $K_m \subseteq L_0$. As a Galois extension, L_0 is the splitting field of a polynomial $f \in K[x]$ all of whose irreducible factors are separable. (Since $\operatorname{char} \mathbb{Q} = 0$ we do not have to worry at all about separability, but all the arguments hold for an arbitrary field K.) Now let

$$g(x) \;:=\; \prod_{\sigma \in G_K^{L_0}} \big(x^2 - \sigma(r_m)\big) \,.$$

This polynomial (all of whose irreducible factors are obviously separable) is defined such that all its coefficients are fixed by $G_K^{L_0}$ and therefore lie in $F(G_K^{L_0}) = K$; thus $g \in K[x]$. Now let L be the splitting field of the polynomial $fg \in K[x]$. Then $(L_0 : K)$ is a Galois extension such that $K_m \supseteq L$ and also $\sqrt{r_m} \in L$ (since $\sqrt{r_m}$ is a root of g). Thus L contains $K_{m+1} = K_m(\sqrt{r_m})$. Now L is obtained from L_0 by the adjunction of the roots of g, and these roots are exactly the square roots of the elements $\sigma(r_m)$ $(\sigma \in G_K^{L_0})$ which lie in L_0. Consequently, $[L : L_0]$ is a power of 2. Since also $[L_0 : K]$ is a power of 2, the same holds for

$$[L : K] \;=\; [L : L_0] \cdot [L_0 : K] \,.$$

Let us suppose conversely that $\alpha \in L$ where $(L : K)$ is of Galois type with $[L : K] = 2^m$ for some $m \in \mathbb{N}_0$. By the main theorem of Galois theory we have $|G_K^L| = [L : K] = 2^m$. By Sylow's theorems there is a chain of subgroups

$$G_K^L \;\supseteq\; U_1 \;\supseteq\; U_2 \;\supseteq\; \cdots \;\supseteq\; U_m \;=\; \{\mathrm{id}\}$$

such that $|U_k| = 2^{m-k}$. Consider the corresponding chain of fixed fields, namely

$$F(G_K^L) \;=\; K \;\subseteq\; F(U_1) \;\subseteq\; F(U_2) \;\subseteq\; \cdots \;\subseteq\; F(U_m) \;=\; F(\{\mathrm{id}\}) \;=\; L \,,$$

and let $K_i := F(U_i)$. Then

$$[K_{i+1} : K_i] \overset{(1)}{=} \frac{[L : K_i]}{[L : K_{i+1}]} = \frac{[L : F(U_i)]}{[L : F(U_{i+1})]} \overset{(2)}{=} \frac{|U_i|}{|U_{i+1}|} = \frac{2^{m-i}}{2^{m-i-1}} = 2$$

where we used the degree formula at (1) and the main theorem of Galois theory at (2). This shows that is K_{i+1} is obtained from K_i by adjoining a square root. So we have found a chain of fields as in (12.36)(b). ∎

WE now show that if $(L : K)$ is a finite Galois extension then we can find a basis of L over K of a very special type.

(22.12) Definition. *Let $(L : K)$ be a finite field extension. A basis of L over K which has the special form $\{\sigma(\alpha) \mid \sigma \in G_K^L\}$ for some element $\alpha \in L$ is called a* **normal basis** *of $(L : K)$.*

(22.13) Remarks. (a) If $(L : K)$ possesses a normal basis then $(L : K)$ is necessarily a Galois extension. Indeed, if $\{\sigma(\alpha) \mid \sigma \in G_K^L\}$ is a basis of L over K then $|G_K^L| \geq |\{\sigma(\alpha) \mid \sigma \in G_K^L\}| = [L : K]$; hence $|G_K^L| = [L : K]$ by (21.17) or (21.19).

(b) If $\{\sigma(\alpha) \mid \sigma \in G_K^L\}$ is a normal basis, then α is a primitive element of $(L : K)$. In fact, let p be the minimal polynomial of α over K; then $p(x) = \prod_{\sigma \in G_K^L} (x - \sigma(\alpha))$. This implies that $[K(\alpha) : K] = \deg p = |G_K^L| = [L : K]$; hence $K(\alpha) = L$.

(22.14) Examples. (a) Let $L := \mathbb{Q}(\sqrt{2}, \sqrt{3})$. It is easy to check that $\alpha := \sqrt{2} + \sqrt{3}$ generates a normal basis of L over $\mathbb{Q}$. Indeed, the conjugates of α are the elements $\pm\sqrt{2} \pm \sqrt{3}$, and these elements constitute a basis of L over $\mathbb{Q}$.

(b) Let p be a prime number and let $\varepsilon = e^{2\pi i/p}$. Then $\{\sigma(\varepsilon) \mid \sigma \in G_{\mathbb{Q}}^{\mathbb{Q}(\varepsilon)}\} = \{\varepsilon^k \mid k$ is coprime with $p\} = \{\varepsilon, \varepsilon^2, \ldots, \varepsilon^{p-1}\}$ is a normal basis of $(\mathbb{Q}(\varepsilon) : \mathbb{Q})$. To see this, we only have to observe that the elements $\varepsilon, \varepsilon^2, \ldots, \varepsilon^{p-1}$ are linearly independent over $\mathbb{Q}$, which follows immediately from the fact that the elements $1, \varepsilon, \ldots, \varepsilon^{p-2}$ are.

AFTER these introductory remarks and examples, we can now prove the existence of normal bases.

(22.15) Normal Basis Theorem. *Every finite Galois extension possesses a normal basis.*

Proof. Let $(L : K)$ be a Galois extension with $[L : K] = n$. To prove the existence of a normal basis of $(L : K)$, we distinguish two cases.

(a) First case: K is infinite. Let $\sigma_1, \ldots, \sigma_n$ be the elements of G_K^L. We want to find an element $\alpha \in L$ such that $\sigma_1(\alpha), \ldots, \sigma_n(\alpha)$ are linearly independent over K which means that $\sum_{i=1}^n x_i \sigma_i(\alpha) = 0$ with coefficients $x_i \in K$ implies that $x_1 = \cdots = x_n = 0$. This is tantamount to saying that the linear system

$$\begin{pmatrix} \sigma_1^{-1}\sigma_1(\alpha) & \sigma_1^{-1}\sigma_2(\alpha) & \cdots & \sigma_1^{-1}\sigma_n(\alpha) \\ \vdots & \vdots & & \vdots \\ \sigma_n^{-1}\sigma_1(\alpha) & \sigma_n^{-1}\sigma_2(\alpha) & \cdots & \sigma_n^{-1}\sigma_n(\alpha) \end{pmatrix} \begin{pmatrix} x_1 \\ \vdots \\ x_n \end{pmatrix} = \begin{pmatrix} 0 \\ \vdots \\ 0 \end{pmatrix}$$

has only the trivial solution in K^n or, equivalently, in L^n, which is the case if and only if $\det\bigl(\sigma_i^{-1}\sigma_j(\alpha)\bigr)_{i,j} \neq 0$. Thus we have to find an element $\alpha \in L$ such that the indicated determinant does not vanish.

Let β be a primitive element of $(L : K)$; then the minimal polynomial of β over K is $f(x) := \prod_{\sigma \in G_K^L}\bigl(x - \sigma(\beta)\bigr)$. For $1 \leq i, j \leq n$ let

$$f_{ij}(x) \; := \; \frac{f(x)}{x - \sigma_i^{-1}\sigma_j(\beta)} \; = \; \prod_{\sigma \neq \sigma_i^{-1}\sigma_j} \bigl(x - \sigma(\beta)\bigr) \; ;$$

moreover, let $d(x) := \det\bigl(f_{ij}(x)\bigr)_{i,j} \in L[x]$. Then $f_{ij}(\beta) = 0$ if and only if $\beta = \sigma(\beta)$ for some $\sigma \neq \sigma_i^{-1}\sigma_j$, i.e., if $\mathrm{id} \neq \sigma_i^{-1}\sigma_j$ which means $\sigma_i \neq \sigma_j$, i.e., $i \neq j$. Hence

$$d(\beta) \; = \; \det\begin{pmatrix} f_{11}(\beta) & & 0 \\ & \ddots & \\ 0 & & f_{nn}(\beta) \end{pmatrix} \; = \; f_{11}(\beta)\cdots f_{nn}(\beta) \; \neq \; 0$$

which shows that d is not the zero polynomial. Using the fact that K is an infinite field, we conclude that there is an element γ in K (and not just one in L) such that $d(\gamma) \neq 0$. But then the element $\alpha := f(\gamma)/(\gamma - \beta)$ satisfies

$$\det\bigl(\sigma_i^{-1}\sigma_j(\alpha)\bigr) \; = \; \det\bigl(\sigma_i^{-1}\sigma_j \frac{f(\gamma)}{\gamma - \beta}\bigr) \; = \; \det\bigl(\frac{f(\gamma)}{\gamma - \sigma_i^{-1}\sigma_j(\beta)}\bigr)$$

$$= \; \det\bigl(f_{ij}(\gamma)\bigr) \; = \; d(\gamma) \neq 0 \, ,$$

as desired. Here we used in the second equation that γ and hence $f(\gamma)$ lie in K and are therefore invariant under G_K^L.

(b) Second case: K is finite. In this case L is finite too, and hence G_K^L is cyclic by (22.9), say $G_K^L = \langle \sigma \rangle$. Then $\sigma : L \to L$ (treated as a linear endomorphism of the K-vector space L) satisfies $\sigma^n - \mathrm{id} = 0$, but $p(\sigma) \neq 0$ for all polynomials p of degree less than n (because $\mathrm{id}, \sigma, \sigma^2, \ldots, \sigma^{n-1}$ are linearly independent by Dedekind's theorem). Hence σ has minimal polynomial $\lambda^n - 1$, and since the degree of $\lambda^n - 1$ is $n = [L : K] = \dim_K L$ the minimal polynomial and the characteristic polynomial of σ coincide. This implies that σ has a cyclic vector, i.e., a vector $\alpha \in L$ such that $\alpha, \sigma(\alpha), \ldots, \sigma^{n-1}(\alpha)$ form a basis of L over K. (See problem 25 below.) ∎

IN the next theorem we show how we can construct from a normal basis of a Galois extension a normal basis of an intermediate Galois extension.

(22.16) Theorem. *Let $(L : K)$ be a finite Galois extension and let $\{\sigma(\alpha) \mid \sigma \in G_K^L\}$ be a normal basis of $(L : K)$. Moreover, let $U \leq G_K^L$ and let $\sigma_1 U, \ldots, \sigma_m U$ be the different left cosets of U in G_K^L.*
(a) $\mathrm{tr}_U(\alpha)$ is a primitive element of $F(U)$ over K.
(b) Let $U_i := \sigma_i U \sigma_i^{-1}$. Then $\{\sigma_1^{-1}\bigl(\mathrm{tr}_{U_1}(\alpha)\bigr), \ldots, \sigma_m^{-1}\bigl(\mathrm{tr}_{U_m}(\alpha)\bigr)\}$ is a basis of $F(U)$ over K.
(c) If $U \trianglelefteq G_K^L$ then $\mathrm{tr}_U(\alpha)$ generates a normal basis of $F(U)$ over K.

Proof. (a) If $i \neq j$ then the automorphisms $\sigma_i u, \sigma_j u$ with $u \in U$ are pairwise different and hence linearly independent by Dedekind's theorem (21.18); consequently, the nontrivial linear combination $\sum_{u \in U}(\sigma_i u - \sigma_j u)$ is not the zero mapping on $L = K(\alpha)$ which implies $\sigma_i(\mathrm{tr}_U(\alpha)) = \sum_{u \in U} \sigma_i u(\alpha) \neq \sum_{u \in U} \sigma_j u(\alpha) = \sigma_j(\mathrm{tr}_U(\alpha))$. This shows that the element $\mathrm{tr}_U(\alpha)$ has at least m conjugates in L. Hence $[K(\mathrm{tr}_U(\alpha)) : K]$ (which is the degree of the minimal polynomial of $\mathrm{tr}_U(\alpha)$ over K) is not less than $m = [G_K^L : U] = [F(U) : K]$. Hence, $[K(\mathrm{tr}_U(\alpha)) : K] \geq [F(U) : K] = [F(U) : K(\mathrm{tr}_U(\alpha))] \cdot [K(\mathrm{tr}_U(\alpha)) : K]$ which implies $[F(U) : K(\mathrm{tr}_U(\alpha))] = 1$, i.e., $F(U) = K(\mathrm{tr}_U(\alpha))$.

(b) We have to show that each element $x \in F(U)$ has a unique representation

$$(\star) \qquad x = \sum_{i=1}^{m} a_i \sigma_i^{-1}(\mathrm{tr}_{U_i}(\alpha)) \quad \text{with } a_i \in K .$$

Since x lies in L, there is a unique representation $x = \sum_{\sigma \in G_K^L} a_\sigma \cdot \sigma(\alpha)$ with $a_\sigma \in K$. Since $G_K^L = \bigcup_{i=1}^{m} U\sigma_i^{-1} = \{u\sigma_i^{-1} \mid 1 \leq i \leq m,\, u \in U\}$, this reads

$$(\star\star) \qquad x = \sum_{i=1}^{m} \sum_{u \in U} a_{u\sigma_i^{-1}} u\sigma_i^{-1}(\alpha) = \sum_{i=1}^{m} \sigma_i^{-1} \sum_{u \in U} a_{u\sigma_i^{-1}} \sigma_i u\sigma_i^{-1}(\alpha) .$$

Now if $\tau \in U$ then $\tau x = x$ because $x \in F(U)$; hence

$$\sum_{\sigma \in G_K^L} a_\sigma \cdot \sigma(\alpha) = x = \tau x = \sum_{\sigma \in G_K^L} a_\sigma \tau\sigma(\alpha) = \sum_{\sigma' \in G_K^L} a_{\tau^{-1}\sigma'} \sigma'(\alpha) .$$

Comparing coefficients, we see that $a_{\tau^{-1}\sigma} = a_\sigma$ so that the coefficient a_σ depends only on the right coset $U\sigma$. Hence $(\star\star)$ becomes

$$x = \sum_{i=1}^{m} \sigma_i^{-1} \sum_{u \in U} a_{\sigma_i^{-1}} \sigma_i u\sigma_i^{-1}(\alpha) = \sum_{i=1}^{m} a_{\sigma_i^{-1}} \sigma_i^{-1}\left(\sum_{u \in U} \sigma_i u\sigma_i^{-1}(\alpha)\right)$$

$$= \sum_{i=1}^{m} a_{\sigma_i^{-1}} \sigma_i^{-1}\left(\sum_{\overline{u} \in U_i} \overline{u}(\alpha)\right) = \sum_{i=1}^{m} a_{\sigma_i^{-1}} \sigma_i^{-1}\left(\mathrm{tr}_{U_i}(\alpha)\right)$$

which is the desired representation.

(c) If $U \trianglelefteq G_K^L$ then $U_1 = \cdots = U_m = U$ so that in this case $\{\sigma_1^{-1}(\mathrm{tr}_U(\alpha)), \ldots, \sigma_m^{-1}(\mathrm{tr}_U(\alpha))\}$ is a basis of $F(U)$ over K. ∎

LET us now show how the Galois group of a compositum is related to the Galois groups of its constituents.

(22.17) Theorem. *Let Z_1, Z_2 be intermediate fields of a field extension $(L:K)$.*

(a) *If $(Z_1:K)$ is an algebraic Galois extension then so is $(Z_1Z_2:Z_2)$, and the restriction map $R: G_{Z_2}^{Z_1Z_2} \to G_K^{Z_1}$ is an injective group homomorphism with image $G_{Z_1 \cap Z_2}^{Z_1}$ (so that $G_{Z_2}^{Z_1Z_2} \cong G_{Z_1 \cap Z_2}^{Z_1}$). This implies that every element of $G_{Z_1 \cap Z_2}^{Z_1}$ can be extended to an element of $G_{Z_2}^{Z_1Z_2}$.*

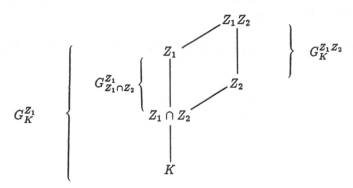

(b) *If both $(Z_1:K)$ and $(Z_2:K)$ are algebraic Galois extensions then so is $(Z_1Z_2:K)$, and the mapping*

$$\theta: \begin{array}{ccc} G_K^{Z_1Z_2} & \to & G_K^{Z_1} \times G_K^{Z_2} \\ \sigma & \mapsto & (\sigma|_{Z_1}, \sigma|_{Z_2}) \end{array}$$

is an injective group homomorphism whose image is

$$U := \{(\sigma_1, \sigma_2) \in G_K^{Z_1} \times G_K^{Z_2} \mid \sigma_1|_{Z_1 \cap Z_2} = \sigma_2|_{Z_1 \cap Z_2}\}.$$

In particular, if $Z_1 \cap Z_2 = K$ then θ is an isomorphism.

Proof. (a) The properties of algebraicity, normality and separability are inherited from $(Z_1:K)$ to $(Z_1Z_2:Z_2)$; hence $(Z_1Z_2:Z_2)$ is an algebraic Galois extension. Next we observe that R is clearly a group homomorphism. The injectivity is also clear, because if $\sigma \in G_{Z_2}^{Z_1Z_2}$ (which implies $\sigma|_{Z_2} = \mathrm{id}_{Z_2}$) lies in $\ker R$ (which means $\sigma|_{Z_1} = \mathrm{id}_{Z_1}$) the clearly $\sigma|_{Z_1Z_2} = \mathrm{id}_{Z_1Z_2}$. Finally, let $U := \mathrm{im}\, R \leq G_K^{Z_1}$. Then obviously $F(U) = Z_1 \cap F(G_{Z_2}^{Z_1Z_2}) = Z_1 \cap Z_2$. Applying G on both sides yields $(G \circ F)(U) = G(Z_1 \cap Z_2) = G_{Z_1 \cap Z_2}^{Z_1}$. But $(G \circ F)(U) = U$! If $[Z_1:K] < \infty$ this is clear by Artin's Theorem (21.22). If $[Z_1:K] = \infty$ we have to anticipate a result which will be proved below; namely, if $(L:K)$ is an algebraic Galois extension then G_K^L can be equipped with a topology such that G_K^L becomes a compact topological group and such that a subgroup of G_K^L is closed if and only if it is topologically closed. This topology is such that the mapping R in our setting is continuous. Consequently, U is the image of the compact group $G_{Z_2}^{Z_1Z_2}$ under the continuous mapping R, hence is compact and thus (topologically) closed in $G_K^{Z_1}$.

(b) The properties of algebraicity, normality and separability are inherited from $(Z_1:K)$ and $(Z_2:K)$ to $(Z_1Z_2:K)$; hence $(Z_1Z_2:K)$ is an algebraic Galois extension. Next we observe that θ is well-defined (and hence clearly an injective group homomorphism) because if $\sigma \in G_K^{Z_1Z_2}$ then $\sigma(Z_1) \subseteq Z_1$ and $\sigma(Z_2) \subseteq Z_2$ due to the fact that Z_1 and Z_2 are stable intermediate fields of $(Z_1Z_2:K)$. Finally, let us show that

$\operatorname{im}\theta = U$. Here the inclusion $\operatorname{im}\theta \subseteq U$ is clear; namely, if $\sigma_1 = \sigma|_{Z_1}$ and $\sigma_2 = \sigma|_{Z_2}$ then $\sigma_1|_{Z_1 \cap Z_2} = \sigma|_{Z_1 \cap Z_2} = \sigma_2|_{Z_1 \cap Z_2}$. To prove the converse inclusion $U \subseteq \operatorname{im}\theta$ let $(\sigma_1, \sigma_2) \in U$. Then $\sigma_1|_{Z_1 \cap Z_2} = \sigma_2|_{Z_1 \cap Z_2}$ can be extended to an element $\sigma \in G_K^{Z_1 Z_2}$ because $Z_1 Z_2$ is normal over K and hence over $Z_1 \cap Z_2$. Now part (a) shows that $\sigma|_{Z_1}^{-1} \circ \sigma_1 \in G_{Z_1 \cap Z_2}^{Z_1}$ can be extended to an element $\varphi_1 \in G_{Z_2}^{Z_1 Z_2}$ and that $\sigma|_{Z_2}^{-1} \circ \sigma_2 \in G_{Z_1 \cap Z_2}^{Z_2}$ can be extended to an element $\varphi_2 \in G_{Z_1}^{Z_1 Z_2}$. Let $\varphi := \sigma \circ \varphi_1 \circ \varphi_2 \in G_K^{Z_1 Z_2}$; we claim that $\theta(\varphi) = (\sigma_1, \sigma_2)$. Indeed,

$$\varphi|_{Z_1} = \sigma|_{Z_1} \circ \underbrace{\varphi_1|_{Z_1} \circ \varphi_2|_{Z_1}}_{= \operatorname{id}_{Z_1}} = \sigma|_{Z_1} \circ \sigma|_{Z_1}^{-1} \circ \sigma_1 = \sigma_1 \qquad \text{and}$$

$$\varphi|_{Z_2} = \sigma|_{Z_2} \circ \underbrace{\varphi_1|_{Z_2} \circ \varphi_2|_{Z_2}}_{= \operatorname{id}_{Z_2}} = \sigma|_{Z_2} \circ \sigma|_{Z_2}^{-1} \circ \sigma_2 = \sigma_2 \,.$$

∎

WE now turn to the study of infinite Galois extensions. † Note that the main theorem of Galois theory, theorem (22.3) above, deals with arbitrary algebraic Galois extensions, finite or infinite; in any case there is a 1-1-correspondence between the intermediate fields of an algebraic Galois extension $(L : K)$ and the *closed* subgroups of G_K^L. What made the situation particularly favorable in the case $[L : K] < \infty$ is the fact that in this case *all* subgroups of G_K^L are closed. This is no longer true in the case of infinite Galois extensions, as the following example shows.

(22.18) Example. Let $K := \mathbb{Z}_p$ and let L be an algebraic closure of K. Then $(L : K)$ is normal by (13.8)(4) and is separable because K is a finite field and hence perfect by (14.6)(c); this shows that $(L : K)$ is an algebraic Galois extension. Let $\varphi : \begin{array}{ccc} L & \to & L \\ x & \mapsto & x^p \end{array}$ be the Frobenius automorphism of L and let $U := \langle \varphi \rangle$. We claim that $G_{F(U)}^L \neq U$ which means that the subgroup U of G_K^L is not closed.

To see this, we first note that $F(U) = \{x \in L \mid x^p = x\} = K$ and hence $G_{F(U)}^L = G_K^L$. So it suffices to show that $U \subsetneq G_K^L$. For each element α in the intermediate field

$$Z := \{x \in L \mid x^{p^{2^m}} = x \text{ for some } m \in \mathbb{N}\}$$

the degree $[K(\alpha) : K]$ is a power of 2. Since for each natural number $n \in \mathbb{N}$ there is an element $\beta \in L$ with $[K(\beta) : K] = n$ by (13.19)(b), this shows that $Z \subsetneq L$. Consequently, there is an element $\sigma \in G_Z^L$ with $\sigma \neq \operatorname{id}$. Then σ belongs to G_K^L, but is not a power of φ. Indeed, if σ were an element of U then we could find a natural number n with $\sigma = \varphi^n$ or $\sigma = \varphi^{-n}$; but this would entail $Z \subseteq F(\sigma) = F(\varphi^n) = \{x \in L \mid x^{p^n} = x\} = \operatorname{GF}(p^n)$. Now $[\operatorname{GF}(p^n) : \mathbb{Z}_p] = n$ whereas $[Z : \mathbb{Z}_p] = \infty$; this is the desired contradiction.

---

† The following material will not be needed elsewhere in the book and can safely be skipped. Note that a result concerning such extensions was anticipated in the proof of (22.17) above.

THUS the main problem in studying infinite Galois extensions is to determine which subgroups of G_K^L are closed. We have already observed in Artin's theorem (21.22)(a) that every *finite* group is closed. In the sequel we want to establish the crucial fact that also every group *of finite index* is closed. The basic result we need is the following theorem.

(22.19) Theorem. *If $(L : K)$ is an algebraic Galois extension and if $U \leq G_K^L$ is a subgroup of finite index, then $[F(U) : K] = [G_K^L : U]$.*

ONCE (22.19) is established, we can argue as follows: Given an algebraic Galois extension $(L : K)$ and a subgroup $U \leq G_K^L$ of finite index, then $G_{F(U)}^L$ has also finite index because $U \subseteq G_{F(U)}^L \subseteq G_K^L$. Applying (22.19) both with U and with $G_{F(U)}^L = (G \circ F)(U)$, we obtain

$$[G_K^L : G_{F(U)}^L] = [F(G_{F(U)}^L) : K] = [(F \circ G \circ F)(U) : K] = [F(U) : K] = [G_K^L : U] ;$$

together with the inclusion $U \subseteq G_{F(U)}^L$ this implies that $U = G_{F(U)}^L$, i.e., that U is closed. To be able to give a proof of (22.19) we need some auxiliary concepts and results.

(22.20) Definition. *Let L be a field, $G \subseteq \mathrm{Aut}\, L$ a subgroup of $\mathrm{Aut}\, L$ and V a vector space over L. An action $(g, v) \mapsto g \star v$ of G on V is called an* **automorphic action** *if $\sigma \star (v + v') = (\sigma \star v) + (\sigma \star v')$ and $\sigma \star (\alpha v) = \sigma(\alpha)(\sigma \star v)$ for all $v, v' \in V$ and all $\alpha \in L$.*

(22.21) Theorem. *Let L be a field. Suppose the group $G \leq \mathrm{Aut}\, L$ acts automorphically on the L-vector space V. Let*

$$\mathrm{Fix}(G) := \{v \in V \mid \sigma \star v \text{ for all } v \in V\} \quad and$$
$$F(G) := \{\alpha \in L \mid \sigma(\alpha) = \alpha \text{ for all } \sigma \in G\}$$

be the fixed point space and the fixed field of G, respectively.

(a) *If G is finite then for each $v \in V$ the element $\bar{v} := \sum_{\sigma \in G} \sigma \star v$ lies in $\mathrm{Fix}(G)$, and if $\overline{\alpha v} = 0$ for all $\alpha \in L$ then $v = 0$.*

(b) *If $v_1, \ldots, v_n \in \mathrm{Fix}(G)$ are linearly independent over $F(G)$, then also over L.*

(c) *Let $W := \mathrm{span}_L \mathrm{Fix}(G)$. If the orbits of G in V are finite, then $W = V$.*

(d) *Suppose the orbits of G in V are finite. Then every basis of $\mathrm{Fix}(G)$ over $F(G)$ is a basis of V over L; in particular, $\dim_{F(G)} \mathrm{Fix}(G) = \dim_L V$.*

Proof. (a) If $\tau \in G$ then

$$\tau \star \bar{v} = \tau \star \sum_{\sigma \in G}(\sigma \star v) = \sum_{\sigma \in G} \tau \star (\sigma \star v) = \sum_{\sigma \in G}(\tau\sigma) \star v = \sum_{\sigma' \in G} \sigma' \star v = \bar{v} .$$

Suppose that $\overline{\alpha v} = 0$ for all $\alpha \in L$. This means $0 = \sum_{\sigma \in G} \sigma \star (\alpha v) = \sum_{\sigma \in G} \sigma(\alpha)(\sigma \star v)$ for all $\alpha \in L$ so that $\sum_{\sigma \in G}(\sigma \star v)\sigma$ is the zero mapping. By Dedekind's theorem (21.18) this means that $\sigma \star v = 0$ for all $\sigma \in G$, in particular $v = \mathrm{id} \star v = 0$.

(b) Suppose the claim is not true. Then we can find $v_1, \ldots, v_n \in \mathrm{Fix}(G)$ which are linearly independent over $F(G)$, but linearly dependent over L; we may assume that

n is chosen minimal with this property. Then there are coefficients $\alpha_1, \ldots, \alpha_n \in L$, not all zero, such that $\sum_{i=1}^n \alpha_i v_i = 0$. By the minimality of n we have $\alpha_i \neq 0$ for all i; dividing all coefficients by α_n we may assume without loss of generality that $\alpha_n = 1$. Applying $\sigma \in G$ to the equation $\sum_{i=1}^n \sigma(\alpha_i) v_i = 0$ because all v_i lie in Fix(G). Since $\sigma(\alpha_n) = \sigma(1) = 1 = \alpha_n$, subtraction of these two equation yields $\sum_{i=1}^{n-1} (\sigma(\alpha_i) - \alpha_i) v_i = 0$. The minimality of n now implies that $\sigma(\alpha_i) = \alpha_i$ for $1 \leq i \leq n - 1$, and $\sigma(\alpha_n) = \alpha_n$ is true anyway because $\alpha_n = 1$. Since this holds for any $\sigma \in G$, we have $\alpha_1, \ldots, \alpha_n \in F(G)$ which contradicts the linear independence of $v_1, \ldots, v_n$ over $F(G)$.

(c) Suppose first that G is finite. Define an automorphic action of G on V/W by $\sigma \bullet (v + W) = (\sigma \star v) + W$. Writing $[v] := v + W$ and using the notation from (a), we have for all $\alpha \in L$ the equation $[\overline{\alpha v}] = \overline{\alpha}(v + W) = \overline{\alpha} \overline{v} + W = [\overline{\alpha v}] = [0]$ because $\overline{\alpha v} \in \text{Fix}(G) \subseteq W$. Then part (a) implies that $[v] = [0]$, i.e., that $v \in W$. Hence $V \subseteq W$ which is what we wanted to show.

The general case is reduced to the special case of a finite group as follows. Let $v \in W = \text{span}_L \text{Fix}(G)$. Since the orbit $G \star v$ is finite, the set

$$ N := \{\sigma \in G \mid \sigma \star w = w \text{ for all } w \in G \star v\} $$

is a normal subgroup of finite index, due to (I.26.7). Consequently, the quotient group $G_0 := G/N$ is a finite group which can clearly be regarded as a group of automorphisms of the field $L_0 := F(N)$ so that $G_0 \leq \text{Aut} \, L_0$. Now the automorphic action of G on V induces an automorphic action of G_0 on $V_0 := \text{span}_{L_0} \text{Fix}(G)$ via $(\sigma N) \star u := \sigma \star u$ (this is unambiguously defined because N acts trivially on V_0). Since G_0 is finite, we know already that $W_0 := \text{span}_{L_0} \text{Fix}(G_0)$ equals V_0. Since v lies in V_0†, this implies $v \in W_0 = \text{span}_{L_0} \text{Fix}(G_0) \subseteq \text{span}_{L_0} \text{Fix}(G) \subseteq \text{span}_L \text{Fix}(G)$.

(d) This follows immediately from (b) and (c). ∎

WE will apply this theorem to obtain the following result.

(22.22) Proposition. *Suppose that $G \leq \text{Aut} \, L$ acts on a finite set X and has finite orbits in L. Denote by $\mathcal{F}$ the L-vector space of all functions $f : X \to L$. Then the dimension of*

$$ \{f \in \mathcal{F} \mid f \circ \sigma = \sigma \circ f \text{ for all } \sigma \in G\} $$
$$ = \{f \in \mathcal{F} \mid f(\sigma \star x) = \sigma(f(x)) \text{ for all } x \in X \text{ and all } \sigma \in G\} $$

over $F(G)$ equals the cardinality $|X|$ of X.

Proof. We define an automorphic action of G on $\mathcal{F}$ by

$$ (\sigma \bullet f)(x) := \sigma(f(\sigma^{-1} \star x)) ; $$

---

† Since $v \in \text{span}_L \text{Fix}(G)$ we can write $v = \sum_i \lambda_i v_i$ where $\lambda_i \in L$ and where the vectors $v_i \in \text{Fix}(G)$ are linearly independent over L; applying any $n \in N$ to this equation, we obtain $v = n \star v = \sum_i n(\lambda_i)(n \star v_i) = \sum_i n(\lambda_i) v_i$; comparing coefficients, we see that $n(\lambda_i) = \lambda_i$ for all i so that all λ_i lie in $F(N) = L_0$.

somewhat sloppily we might write $\sigma \bullet f = \sigma f \sigma^{-1}$. (The proof that this is indeed an automorphic action is straightforward and is left as an exercise. See problem 37 below.) Also, the orbits of G in $\mathcal{F}$ are finite; in fact, if $f \in \mathcal{F}$ then the set

$$\{(\sigma \bullet f)(x) \mid \sigma \in G, \, x \in X\} = \{\sigma(f(\sigma^{-1} \star x)) \mid \sigma \in G, \, x \in X\} \subseteq G \star f(X)$$

is finite because $f(X)$ is finite and G has finite orbits, and this implies that $\{\sigma \bullet f \mid \sigma \in G\} = G \bullet f$ is finite.

Hence (22.21)(d) is applicable and yields $\dim_{F(G)} \mathrm{Fix}(G) = \dim_L \mathcal{F}$. Now on the one hand, we clearly have $\dim_L \mathcal{F} = |X|$; on the other hand, a function $f \in \mathcal{F}$ belongs to $\mathrm{Fix}(G)$ if and only if $\sigma(f(\sigma^{-1} \star x)) = f(x)$ for all $x \in X$ and all $\sigma \in G$, i.e., if $f(\tau \star x) = \tau(f(x))$ for all $x \in X$ and all $\tau \in G$. This gives the claim. ∎

FINALLY, we want to apply (22.22) to prove (22.19). Now in (22.22) we used the hypothesis that a group $G \leq \mathrm{Aut}\, L$ has finite orbits in L, and to apply this proposition we have to find out under what circumstances this hypothesis holds. This is done in the next proposition.

(22.23) Proposition. *Let L be a field and let $G \leq \mathrm{Aut}\, L$. Then the following conditions are equivalent:*

(1) *the orbits of G in L are finite;*

(2) $(L : F(G))$ *is algebraic;*

(3) *the action $G \times L \to L$ is continuous where L is equipped with the discrete topology, G with the Krull topology and $G \times L$ with the product topology.*

Proof. (1)$\Longrightarrow$(2). Suppose that the orbit of $\alpha \in L$ is finite, say $G \star \alpha = \{\alpha_1, \ldots, \alpha_n\}$. Then $f(x) := \prod_{i=1}^n (x - \alpha_i)$ satisfies $\sigma \star f = f$ for all $\sigma \in G$, hence lies in $F(G)[x]$. Since α is a root of f, this shows that α is algebraic over $F(G)$.

(2)$\Longrightarrow$(1). Suppose $\alpha \in L$ is algebraic over $F(G)$ and let f be the minimal polynomial of α over $F(G)$. Then each $\sigma \in G$ satisfies $\sigma \star f = f$, hence maps α to another root of f. Since f has only a finite number of roots, this implies that the orbit $G \star \alpha$ of α is finite.

The equivalence of (1) and (3) was proved in (I.29.33)(h) already. ∎

WE are now ready to prove (22.19); in fact, to prove the following slightly stronger result.

(22.24) Theorem. *Let L be a field and let $G \leq \mathrm{Aut}\, L$ be a group such that $(L : F(G))$ is an algebraic extension. If $U \leq G$ has finite index then $[F(U) : F(G)] = [G : U]$.*

Proof. The set $X := G/U$ is finite, G acts on X via $\sigma \star (\tau U) := (\sigma \tau) U$, and the orbits of G in L are finite due to (22.23). For each $\alpha \in F(U)$ we define a mapping $\widehat{\alpha} : X \to L$ by $\widehat{\alpha}(\tau U) := \tau(\alpha)$. We claim that a mapping $f : X \to L$ satisfies

$f(\sigma \star x) = \sigma(f(x))$ for all $x \in X$ and all $\sigma \in G$ if and only if $f \in \widehat{F(U)}$, i.e., if $f = \widehat{\alpha}$ for some $\alpha \in F(U)$. Once this is established, (22.22) shows that

$$[G:U] = |G/U| = |X| = \dim_{F(G)} \widehat{F(U)} = \dim_{F(G)} F(U) = [F(U):F(G)].$$

To proceed, assume first that $f = \widehat{\alpha}$ for some $\alpha \in F(U)$. Then for all $\sigma \in G$ and all elements $\tau U \in X$ we have $f(\sigma \star (\tau U)) = f(\sigma\tau U) = \widehat{\alpha}(\sigma\tau U) = (\sigma\tau)(\alpha) = \sigma(\tau(\alpha)) = \sigma(\widehat{\alpha}(\tau U)) = \sigma(f(\tau U))$; this shows $f \circ \sigma = \sigma \circ f$.

Suppose conversely that $f \circ \sigma = \sigma \circ f$ for all $\sigma \in G$. Then $\alpha := f(U) = f(\mathrm{id}\, U)$ belongs to $F(U)$ because if $\sigma \in U$ then $\sigma(\alpha) = \sigma(f(U)) = f(\sigma U) = f(U) = \alpha$, and $f = \widehat{\alpha}$ because for all $\tau \in G$ we have $f(\tau U) = f(\tau\,\mathrm{id}\,U) = \tau(f(\mathrm{id}\,U)) = \tau(\alpha) = \widehat{\alpha}(\tau U)$. ∎

UP to this point we have shown that if $(L:K)$ is an algebraic Galois extension then all finite subgroups of G_K^L and all subgroups of finite index are closed. However, we still have not obtained a complete characterization of the closed subgroups of G_K^L. This problem was solved in 1928 by Krull who made the important observation that G_K^L can be equipped with a topology τ such that a subgroup $U \le G_K^L$ is closed in the algebraic sense of definition (21.13) and (21.14) if and only if U is topologically closed with respect to the topology τ. Let us introduce this topology in a more general setting.

(22.25) Definition. *Let X, Y be sets and let Y^X be the set of all mappings $f : X \to Y$. The **finite topology** on Y^X is the topology determined by the subbase consisting of all sets*

$$U(x,y) := \{f : X \to Y \mid f(x) = y\}$$

where $x \in X$ and $y \in Y$.

SOME basic properties of the finite topology on a function space are as follows.

(22.26) Proposition. *Let X, Y be sets and let Y^X be equipped with the finite topology.*
(a) Y^X is a Hausdorff space.
(b) Let $g \in Y^X$. Then the neighborhoods of g are exactly the sets

$$V_{X_0} := \{f : X \to Y \mid f \equiv g \text{ on } X_0\}$$

where X_0 is a finite subset of X.
(c) Let $\mathcal{F} \subseteq Y^X$ and $g \in Y^X$. Then g lies in the closure $\overline{\mathcal{F}}$ of $\mathcal{F}$ if and only if for every finite subset $X_0 \subseteq X$ there is an element $f \in \mathcal{F}$ with $f \equiv g$ on X_0.
(d) A net (generalized sequence) $(f_i)_{i \in I}$ of functions $f_i \in Y^X$ converges to a function $f \in Y^X$ if and only if for every finite subset $X_0 \subseteq X$ there is an index $i_0 \in I$ such that $f_i \equiv f$ on X_0 for all $i \ge i_0$.

Proof. (a) If $f \ne g$ then $f(x_0) \ne g(x_0)$ for some $x_0 \in X$; but then $U(x_0, f(x_0))$ and $U(x_0, g(x_0))$ are disjoint neighborhoods of f and g, respectively.

(b) If $X_0 = \{x_1, \ldots, x_n\}$ then $V_{X_0} = U\big(x_1, g(x_1)\big) \cap \cdots \cap U\big(x_n, g(x_n)\big)$; hence the claim follows immediately from the definition of the finite topology.

(c) Since g lies in $\overline{\mathcal{F}}$ if and only if each neighborhood of g intersects $\mathcal{F}$, the claim follows immediately from part (c).

(d) We have $f_i \to f$ if and only for each neighborhood V of f there is an index $i_0 \in I$ with $f_i \in V$ for all $i \geq i_0$; hence the claim follows immediately from part (b). ∎

THE concept of finite topologies will be used in the next definition to topologize Galois groups.

(22.27) Definition. *Let $(L : K)$ be a field extension and let τ be the finite topology on $L^L = \{f : L \to L\}$. Then the subspace topology induced by τ on $G_K^L \subseteq L^L$ is called the* **finite topology** *on G_K^L.*

(22.28) Proposition. *Let $(L : K)$ be an algebraic Galois extension and let G_K^L be equipped with the finite topology.*

(a) For each $\alpha \in L$ let C_α be the set of conjugates of α. Then

$$\prod_{\alpha \in L} C_\alpha = \{f : L \to L \mid f(\alpha) \in C_\alpha \text{ for all } \alpha \in L\}$$

(regarded as a subspace of L^L) is compact.

(b) If Z is any intermediate field of $(L : K)$ then G_Z^L is a compact topological group.†

(c) G_K^L is discrete if and only if $[L : K] < \infty$.

(d) A subgroup U is closed in the algebraic sense of (21.13)(a) if and only if U is topologically closed as a subspace of G_K^L.

Proof. (a) We first do not treat $\prod_\alpha C_\alpha$ as a subspace of L^L, but equip this set with a different topology. Namely, equip each set C_α with the discrete topology and then $\prod_\alpha C_\alpha$ with the product topology. Since each C_α is finite and hence compact, the product $\prod_\alpha C_\alpha$ is compact by Tychonoff's theorem. To show that $\prod_\alpha C_\alpha$ is also compact if equipped with the subspace topology from L^L, we have to show that the inclusion map $i : \prod_\alpha C_\alpha \to L^L$ is continuous (because then i maps compact sets to compact sets). To establish the continuity of i, we simply have to verify that all sets $i^{-1}\big(U(x,y)\big)$ with $x, y \in L$ are open, and since $i^{-1}\big(U(x,y)\big) = \{(f_\alpha)_{\alpha \in L} \mid f_x = y\}$ this fact follows immediately from the definition of the product topology.

(b) Since an element $\sigma \in G_Z^L$ maps each element $\alpha \in L$ to one of its conjugates, we have $G_Z^L \subseteq \prod_\alpha C_\alpha$. Hence if we can show that G_Z^L is closed in L^L, then G_Z^L is a closed subspace of the compact set $\prod_\alpha C_\alpha$ and hence compact itself. Thus let us show that G_Z^L is closed in L^L.

Let $\theta \in \overline{G_Z^L}$. Moreover, let $\alpha, \beta \in L$ and $z \in Z$. By (22.26)(c) we can find an element $\sigma \in G_Z^L$ which coincides with θ on the finite set $\{\alpha, \beta, z, \alpha + \beta, \alpha\beta\}$. This

---

† We do not have to worry about a distinction between compactness and quasi-compactness since finite topologies always have the Hausdorff property.

implies $\theta(z) = \sigma(z) = z$, $\theta(\alpha + \beta) = \sigma(\alpha + \beta) = \sigma(\alpha) + \sigma(\beta) = \theta(\alpha) + \theta(\beta)$ and $\theta(\alpha\beta) = \sigma(\alpha\beta) = \sigma(\alpha)\sigma(\beta) = \theta(\alpha)\theta(\beta)$. Since α, β and z were arbitrary elements, this shows that $\theta : L \to L$ is a field homomorphism which is the identity on Z. Then θ is automatically injective; to exhibit θ as an element of G_Z^L, it is therefore enough to check that θ is surjective. Given $\alpha \in L$, let $f \in Z[x]$ be the minimal polynomial of α over Z and let $Z' \subseteq L$ be the splitting field of f. Then $(Z' : Z)$ is normal; hence $\theta(Z') \subseteq Z'$. Now we use some linear algebra: $\theta|_{Z'}$ is an injective endomorphism of the finite-dimensional Z-vector space Z', hence automatically surjective so that $\theta(Z') = Z'$. Hence there is an element $\beta \in Z' \subseteq L$ with $\theta(\beta) = \alpha$. Since $\alpha \in L$ was an arbitrary element this shows that θ is surjective.

To verify that G_Z^L is a topological group we have to check that multiplication and inversion are continuous. To prove the continuity of the multiplication, we have to show that $(\sigma_i, \tau_i) \to (\sigma, \tau)$ implies that $\sigma_i \tau_i \to \sigma\tau$. Let $X_0 \subseteq L$ be a finite set; then so is $\tau(X_0)$. Since $\tau_i \to \tau$ there is an index i_1 such that $\tau_i \equiv \tau$ on X_0 for $i \geq i_1$, and since $\sigma_i \to \sigma$ there is an index i_2 such that $\sigma_i \equiv \sigma$ on $\tau(X_0)$ for all $i \leq i_2$. Pick i_3 such that $i_3 \geq i_1$ and $i_3 \geq i_2$. Then for all $i \geq i_3$ and all $x \in X_0$ we have $(\sigma_i \tau_i)(x) = \sigma_i(\tau_i(x)) = \sigma_i(\tau(x)) = \sigma(\tau(x)) = (\sigma\tau)(x)$; hence $\sigma_i \tau_i \equiv \sigma\tau$ on X_0 for all $i \geq i_3$. In view of (22.26)(d) this means that $\sigma_i \tau_i \to \sigma\tau$. The continuity of the inversion map is proved in a completely analogous way.

(c) A compact set is discrete if and only if it is finite. Hence G_K^L is discrete if and only if G_K^L is finite which is easily seen to be the case if and only if $[L : K] < \infty$. (See problem 32 below.)

(d) If U is closed in the sense of (21.13)(a) then $U = G_{F(U)}^L$ which is compact and hence topologically closed by part (b). Suppose conversely that U is topologically closed; we want to show that $G_{F(U)}^L$ is contained in $U = \overline{U}$. By (22.26) this means that we have to find for any finite set $\{\alpha_1, \ldots, \alpha_n\} \subseteq L$ an element $\tau \in U$ with $\tau(\alpha_i) = \sigma(\alpha_i)$ for $1 \leq i \leq n$. Let L_0 be the splitting field of $f_1 \cdots f_n$ over $F(U)$ where f_i is the minimal polynomial of α_i over $F(U)$; then $(L_0 : F(U))$ is a *finite* Galois extension which implies that L_0 is stable. (This is the first time that we used our hypothesis that $(L : K)$ be of Galois type.) Let $U_0 := \{\tau|_{L_0}|\ \tau \in U\}$. Then clearly $F(U_0) = F(U) = (F \circ G \circ F)(U) = F(G_{F(U)}^{L_0})$; since we know that the Galois correspondence is one-to-one for finite Galois extensions, we may apply F^{-1} to obtain $U_0 = G_{F(U)}^{L_0}$. This implies that $\sigma|_{L_0} \in U_0$ which means that σ coincides on L_0 with some $\tau \in U$, which is exactly what we wanted to show. ∎

EVEN though the criterion spelled out in (22.28)(d) is not too explicit (after all, how do we know whether or not a given subgroup of G_K^L is topologically closed?), it gives us the important information that the Galois group of an infinite algebraic Galois extension always contains non-closed subgroups, as it is a general fact that an infinite compact topological group always contains a subgroup which is not topologically closed. We will now describe the nature of the finite topology on G_K^L, namely, we will show that if $(L : K)$ is an algebraic Galois extension then the finite topology coincides with the Krull topology as defined in (I.29.33).

(22.29) Theorem. *Let $(L : K)$ be an algebraic Galois extension.*

(a) *If $F(U) = F(V)$ where $U, V \leq G_K^L$ are subgroups of finite index then $U = V$.*

(b) *For any finite set $X \subseteq L$ the set*

$$N_X := \{\sigma \in G_K^L \mid \sigma(y) = y \text{ whenever } y \text{ is a conjugate of some } x \in X\}$$

is a normal subgroup of G_K^L of finite index.

(c) *If $U^\star$ denotes the Krull closure of a subgroup $U \leq G_K^L$ then $F(U) = F(U^\star)$.*

(d) *Two subgroups $U, V \leq G_K^L$ satisfy $F(U) = F(V)$ if and only if they have the same Krull closure.*

(e) *A subgroup $U \leq G_K^L$ is closed in the algebraic sense of (21.13)(a) if and only if U is Krull-closed.*

(f) *If N is a normal subgroup of G_K^L of finite index, then there is a finite subset $X \subseteq L$ with $N_X \subseteq N$.*

(g) *The finite topology on G_K^L coincides with the Krull topology.*

(h) *The group G_K^L, equipped with the Krull topology, is a compact and totally disconnected topological group.*

(i) *If a subgroup $U \leq G_K^L$ has finite index then it is closed in the sense of (21.13)(a).*

Proof. Let us denote by $\mathcal{N}$ the set of all normal subgroups of G_K^L of finite index; then the Krull closure of a subset $A \subseteq L$ is $A^\star = \bigcap_{N \in \mathcal{N}} NA$.

(a) Let $F(U) = F(V)$. Then $W := \langle U \cup V \rangle$, the group generated by U and V, has finite index again. Hence (22.19) implies that $[G_K^L : W] = [F(W) : K] = [F(U) : K] = [G_K^L : U]$; together with the obvious inclusion $U \subseteq W$ this implies $U = W$ which means $V \subseteq U$. Analogously, one obtains $U \subseteq V$ so that in fact $U = V$.

(b) This follows immediately from (I.26.7), applied to the canonical action of G_K^L on L given by $\sigma \star x = \sigma(x)$; note that the orbit of an element $x \in L$ consists exactly of the conjugates of x.

(c) The inclusion $F(U^\star) \subseteq F(U)$ is trivial. Conversely, let $x \in F(U)$ and let $G_x := \{\sigma \in G_K^L \mid \sigma(x) = x\}$. Then $U^\star = \bigcap_{N \in \mathcal{N}} NU \subseteq N_{\{x\}} U \subseteq G_x G_x \subseteq G_x$. This means that $\sigma(x) = x$ for all $\sigma \in U^\star$. Since $x \in F(U)$ was arbitrarily chosen, this shows that $F(U) \subseteq F(U^\star)$.

(d) If $U^\star = V^\star$ then $F(U) = F(U^\star) = F(V^\star) = F(V)$, using part (c). Conversely, let $F(U) = F(V)$ and let $N \in \mathcal{N}$. Then NU and NV are subgroups of G_K^L of finite index with $F(NU) = F(N) \cap F(U) = F(N) \cap F(V) = F(NV)$ so that $NU = NV$ by part (a). But then $U^\star = \bigcap_{N \in \mathcal{N}} NU = \bigcap_{N \in \mathcal{N}} NV = V^\star$.

(e) If U is closed in the sense of (21.13)(a), i.e., if $U = G_{F(U)}^L$, then $F(U) = F(U^\star)$ by part (c) and hence $U^\star \subseteq G_{F(U^\star)}^L = G_{F(U)}^L = U$ which implies $U = U^\star$. Suppose conversely that $U = U^\star$. Since $F = F \circ G \circ F$, we have $F(U) = F(G_{F(U)}^L)$. By part (d) this means that U and $G_{F(U)}^L$ have the same Krull closure, i.e., that $U^\star = (G_{F(U)}^L)^\star$. But $U^\star = U$ by hypothesis and $(G_{F(U)}^L)^\star = G_{F(U)}^L$ because $G_{F(U)}^L$ is closed in the sense of (21.13)(a) and hence Krull-closed by what we just proved; hence $U = G_{F(U)}^L$ which is what we wanted to show.

(f) We observe that $N = N^\star$ by the very definition of the Krull topology; hence $G_{F(N)}^L = N$ by part (e). Let $\sigma_1 N, \ldots, \sigma_r N$ be the different cosets of N in G_K^L where $\sigma_1 = \mathrm{id}$. If $i \neq j$ then σ_i does not coincide with σ_j on the fixed field $F(N)$; indeed, if $\sigma_i \sigma_j^{-1} \equiv \mathrm{id}$ on $F(N)$ then $\sigma_i \sigma_j^{-1} \in G_{F(N)}^L = N$ and hence $\sigma_i N = \sigma_j N$. Thus for $2 \leq i \leq n$ we can choose $\alpha_i \in F(N)$ with $\sigma_i(\alpha_i) \neq \alpha_i$. Let $X := \{\alpha_2, \ldots, \alpha_n\}$; we claim that then $N_X \subseteq N$. Suppose there is an element $\sigma \in N_X \setminus N$ to get a contradiction; then $\sigma \in \sigma_i N$ for some index i with $2 \leq i \leq n$. Consequently, $\sigma_i^{-1}\sigma \in N$ and hence $(\sigma_i^{-1}\sigma)(\alpha_i) = \alpha_i$ because $\alpha_i \in F(N)$, which implies that $\sigma(\alpha_i) = \sigma_i(\alpha_i) \neq \alpha_i$. On the other hand $\sigma(\alpha_i) = \alpha_i$ because $\sigma \in N_X$. This contradiction shows that in fact $N_X \subseteq N$.

(g) Since a topology is uniquely determined by the closure operator, it is enough to show that $\overline{A} = A^\star$ for all subsets $A \subseteq L$; here $\overline{A}$ denotes the closure of A with respect to the finite topology on G_K^L whereas $A^\star$ denotes the closure with respect to the Krull topology.

Let $\sigma \in A^\star$ and let $X \subseteq L$ be finite; we have to find an element $\tau \in A$ with $\tau \equiv \sigma$ on X. Now $\sigma \in \bigcap_{N \in \mathcal{N}} NA \subseteq N_X A$, say $\sigma\tau^{-1} \in N_X$ with $\tau \in A$. Conversely, let $\sigma \in \overline{A}$ and let $N \in \mathcal{N}$; we have to find an element $\tau \in A$ with $\sigma\tau^{-1} \in N$ because then $\sigma \in NA$. By (f) we can find a finite subset $X \subseteq L$ with $N_X \subseteq N$; let Y be the set of all conjugates of the elements of X. Since $\sigma \in \overline{A}$, there is an element $\tau \in A$ with $\tau \equiv \sigma$ on Y; but then $\sigma\tau^{-1} \in N_X \subseteq N$.

(h) Since the Krull topology coincides with the finite topology, we know by (22.28)(b) that G_K^L is a compact topological group. Being Hausdorff, the group G_K^L is totally disconnected by (I.29.33)(g).

(i) This was already established in (22.19). ∎

Exercises

Problem 1. Show that the extensions $(\mathbb{Q}(\sqrt[4]{2}) : \mathbb{Q}(\sqrt{2}))$ and $(\mathbb{Q}(\sqrt{2}) : \mathbb{Q})$ are of Galois type, but that $(\mathbb{Q}(\sqrt[4]{2}) : \mathbb{Q})$ is not. Hence the property of being a finite Galois extension is not transitive.

Problem 2. Let K be a field of characteristic 2, let $a_1 \ldots, a_n$ be pairwise different elements of K and let $\sqrt{a_i}$ be an element whose square is a_i. Show that $L := K(\sqrt{a_1}, \ldots, \sqrt{a_n}) = K(\alpha)$ where $\alpha := \sqrt{a_1} + \cdots + \sqrt{a_n}$.
   Hint. If we had $L \neq K(\alpha)$ then there existed an element $\sigma \in G_{K(\alpha)}^L$ different from the identity.

Problem 3. Show that a finite separable extension $(L : K)$ has only a finite number of intermediate fields.

Problem 4. Show that an algebraic extension $(L : K)$ is a Galois extension if and only if every K-embedding of L into some algebraic closure of L is in fact an automorphism of L.

Problem 5. Consider the polynomial $f(x) := x^4 - 3 \in \mathbb{Q}[x]$ and let $\alpha \in \mathbb{C}$ be a root of f.
   (a) Decompose f into irreducible factors over $\mathbb{Q}(\alpha)$.
   (b) Find a splitting field L of f over $\mathbb{Q}$, find the degree of L over $\mathbb{Q}$ and determine the Galois group $G_{\mathbb{Q}}^L$.

Problem 6. Find all intermediate fields of the extension $(\mathbb{Q}(\sqrt{3}, \sqrt[4]{2}) : \mathbb{Q})$.

Problem 7. Let L be the splitting field of $(x^2 - 3)(x^3 - 2)$. Determine the Galois group $G_{\mathbb{Q}}^L$ and find all intermediate fields of the extension $(L : \mathbb{Q})$.

Problem 8. For $\alpha = \sqrt{6}$ and $\alpha = \sqrt[3]{3}$ let $L := \mathbb{Q}(i + \alpha)$.
   (a) Show that $L = \mathbb{Q}(i, \alpha)$.
   (b) Find $[L : \mathbb{Q}]$.
   (c) Find the minimal polynomial of $i + \alpha$ over $\mathbb{Q}$.
   (d) Find a vector space basis of L over $\mathbb{Q}$.
   (e) Determine the Galois group $G_{\mathbb{Q}}^L$ and its subgroup lattice.
   (f) Find out whether or not $(L : \mathbb{Q})$ a Galois extension.
   (g) Find all intermediate fields of the extension $(L : \mathbb{Q})$.

Problem 9. In each of the cases $\alpha = \sqrt{5 + 2\sqrt{5}}$, $\alpha = i\sqrt[6]{3}$ and $\alpha := (1 + i)\sqrt[4]{5}$, answer the following questions.
   (a) What is the minimal polynomial of α over $\mathbb{Q}$?
   (b) Is $(\mathbb{Q}(\alpha) : \mathbb{Q})$ a Galois extension?
   (c) What is the Galois group of $(\mathbb{Q}(\alpha) : \mathbb{Q})$?
   (d) What are the intermediate fields of $(\mathbb{Q}(\alpha) : \mathbb{Q})$?

Problem 10. Show that $\mathbb{Q}(i, \sqrt[4]{5})$ is the splitting field of $f(x) = x^4 - 5$ and prove that $G_\mathbb{Q}^L$ is cyclic of order 4.

Problem 11. Let L be the splitting field of $f(x) = (x^2 - 2)(x^2 - 5)(x^2 - 7)$ over $\mathbb{Q}$.
  (a) Find all intermediate fields of the extension $(L : \mathbb{Q})$.
  (b) Find the Galois groups G_Z^L for the intermediate fields $Z = \mathbb{Q}(\sqrt{10}, \sqrt{14})$, $Z = \mathbb{Q}(\sqrt{10}, \sqrt{35})$, $Z = \mathbb{Q}(\sqrt{10} + \sqrt{14})$ and $Z = \mathbb{Q}(\sqrt{10} + \sqrt{70})$.

Problem 12. For each of the polynomials $f(x) = x^3 - 3$, $f(x) = x^4 - 4$, $f(x) = x^5 - 5$ and $f(x) = x^4 + 2x^2 - 6$, find the splitting field L of f over $\mathbb{Q}$, determine its Galois group $G_\mathbb{Q}^L$ and list all intermediate fields of the extension $(L : \mathbb{Q})$.

Problem 13. Let L be the splitting field of $x^5 - 2$ over $\mathbb{Q}$.
  (a) Show that $[L : \mathbb{Q}] = 20$ and that $L = \mathbb{Q}(\sqrt[5]{2} + e^{2\pi i/5})$.
  (b) Show that $G_\mathbb{Q}^L = \langle \sigma, \tau \mid \sigma^5 = \tau^4 = \mathrm{id}, \tau\sigma\tau^{-1} = \sigma^2 \rangle$.
  (c) Find all intermediate fields of the extension $(L : \mathbb{Q})$.

Problem 14. Let L be the splitting field of $f(x) = x^7 - 6$ over $\mathbb{Q}$ and let $\varepsilon = e^{2\pi i/7}$.
  (a) Show that $[L : \mathbb{Q}] = 42$.
  (b) Show that $L = \mathbb{Q}(\sqrt[7]{6} + \varepsilon)$.
  (c) Show that $G_\mathbb{Q}^L = \langle \sigma, \tau \mid \sigma^7 = \tau^6 = \mathrm{id}, \tau\sigma\tau^{-1} = \sigma^3 \rangle$.
  (d) Show that the intermediate fields of $(L : \mathbb{Q})$ are exactly the fields $\mathbb{Q}(\sqrt{-7})$, $\mathbb{Q}(\varepsilon + \bar{\varepsilon})$, $\mathbb{Q}(\varepsilon)$, $\mathbb{Q}(\alpha_i)$, $\mathbb{Q}(\alpha_i, \sqrt{-7})$ and $\mathbb{Q}(\alpha_i, \varepsilon + \bar{\varepsilon})$ where $\alpha_1, \ldots, \alpha_7$ are the roots of f. (Hence there are 24 intermediate fields.)

Problem 15. Let L be the splitting field of $f(x) = x^5 + x + 1$. Show that the proper intermediate fields of $(L : \mathbb{Q})$ are exactly the fields $\mathbb{Q}(\sqrt{-3})$, $\mathbb{Q}(\sqrt{-23})$, $\mathbb{Q}(\sqrt{69})$, $\mathbb{Q}(\sqrt{-3}, \sqrt{69})$, $\mathbb{Q}(\alpha_1, \alpha_2, \alpha_3)$, $\mathbb{Q}(\alpha_i)$, $\mathbb{Q}(\alpha_i, \sqrt{-3})$ and $\mathbb{Q}(\alpha_i, \sqrt{69})$ where $\alpha_1, \alpha_2, \alpha_3$ are the roots of $g(x) = x^3 - x^2 + 1$.

Problem 16. Suppose that $\alpha, \beta \in \mathbb{C}$ satisfy the equations $\alpha^5 = 7$ and $\beta^4 + \beta^3 + \beta^2 + \beta + 1 = 0$.
  (a) Show that $(\mathbb{Q}(\alpha) : \mathbb{Q})$ is not a Galois extension.
  (b) Show that $(\mathbb{Q}(\beta) : \mathbb{Q})$ is a Galois extension whose Galois group is abelian.
  (c) Show that $(\mathbb{Q}(\alpha, \beta) : \mathbb{Q})$ is a normal extension and determine its degree.
  (d) Determine the Galois group of $\mathbb{Q}(\alpha, \beta)$ over $\mathbb{Q}(\beta)$.

Problem 17. (a) Let $\mathbb{Q} \subseteq Z_1, Z_2 \subseteq \mathbb{C}$ where $Z_1 = \mathbb{Q}(\sqrt[3]{2})$ and $Z_2 = \mathbb{Q}(\varepsilon\sqrt[3]{2})$ with $\varepsilon = e^{2\pi i/3}$. Show that $[Z_1 Z_2 : Z_2]$ is not a divisor of $[Z_1 : K]$.
  (b) Suppose that $K \subseteq Z_1, Z_2 \subseteq L$ are fields and that $(Z_1 : K)$ is a finite Galois extension. Show that $[Z_1 Z_2 : Z_2]$ is a divisor of $[Z_1 : K]$.

Problem 18. Let L be the splitting field of an irreducible separable polynomial $f \in K[x]$. Show that if G_K^L is abelian then $[L : K] = \deg f$.

Hint. Show that $L = K(\alpha)$ where α is any root of f.

Problem 19. Let $(L : \mathbb{Q})$ be a finite Galois extension such that $G_{\mathbb{Q}}^L$ is abelian. Show that $f(x) = x^5 - 2$ is irreducible over L.

Problem 20. Let $\mathbb{Q} \subseteq L \subseteq \mathbb{C}$ be fields such that $(L : \mathbb{Q})$ is a Galois extension.

(a) Show that $[L : L \cap \mathbb{R}] \leq 2$.

(b) Show that $(L \cap \mathbb{R} : \mathbb{Q})$ is not necessarily a Galois extension.

(c) Show that if $(L \cap \mathbb{R} : \mathbb{Q})$ is a Galois extension then $G_{L \cap \mathbb{R}}^L$ is contained in the center of $G_{\mathbb{Q}}^L$.

Problem 21. Let $(L : K)$ be a finite Galois extension and let $U \leq G_K^L$.

(a) Show that $\mathrm{tr}_U = \mathrm{tr}_{F(U)}^L$.

(b) Show that $F(U) = K(\alpha)$ if and only if $\sigma(\alpha) = \alpha$ for all $\sigma \in U$ and $\sigma(\alpha) \neq \alpha$ for all $\sigma \in G_K^L \setminus U$.

Problem 22. Let $(L : K)$ be a field extension. Show that if $V \subseteq L^n$ is the zero set of a family of polynomials in $L[x]$ then $V \cap K^n$ is the zero set of a family of polynomials in $K[x]$. (In the language of algebraic geometry this means that if $V \subseteq L^n$ is an L-variety then $V \cap K^n$ is a K-variety.)

Hint. Prove the result first for finite Galois extensions, then for arbitrary finite extensions, then for purely transcendental extensions of finite transcendence degree and finally in the general case. (This is a good example in which a general theorem which holds for arbitrary field extensions can be established by proving it for special types of field extensions and then using the structure theory of field extensions to derive the general result.)

Problem 23. (a) Let $(L : K)$ be a finite Galois extension and let Z_1, Z_2 be two intermediate fields of $(L : K)$. Show that the following statements are equivalent:

(1) $Z_2 = \sigma(Z_1)$ for some $\sigma \in G_K^L$;

(2) $G_{Z_1}^L$ and $G_{Z_2}^L$ are conjugate subgroups of G_K^L;

(3) there are elements $\alpha_1, \alpha_2 \in L$ having the same minimal polynomial over K such that $Z_1 = K(\alpha_1)$ and $Z_2 = K(\alpha_2)$.

Show that if these conditions are satisfied then $[Z_1 : K] = [Z_2 : K]$.

(b) Let $\varepsilon := e^{2\pi i/3}$. Verify the claim in part (a) in the example $K = \mathbb{Q}$, $L = \mathbb{Q}(\varepsilon, \sqrt[3]{2})$, $Z_1 = \mathbb{Q}(\sqrt[3]{2})$ and $Z_2 = \mathbb{Q}(\varepsilon\sqrt[3]{2})$.

Problem 24. Let $K = \mathbb{Q}(t)$ be the rational function field in one variable over $\mathbb{Q}$. Show that $f(x) = x^n - t$ is irreducible in $K[x]$. Determine the Galois group G_K^L where L is a splitting field of f over K.

Problem 25. Let $A : V \to V$ be an endomorphism of a finite-dimensional K-vector space V and let p be the minimal polynomial of A (in the sense of (I.10.16)).

(a) Show that for each vector $v \in V$ there is a unique monic polynomial p_v of minimal degree such that $p_v(A)v = 0$ and that each other polynomial q with $q(A)v = 0$ is divisible by p_v.

(b) Show that if $\deg p = d$ then there is a vector $v \in V$ such that $v, Av, \ldots, A^{d-1}v$ are linearly independent.

Hint. Let $p = p_1^{d_1} \cdots p_k^{d_k}$ be the prime decomposition of p in $K[x]$. Since $(p/p_i)(A) \neq \mathbf{0}$ there is a vector v_i with $(p/p_i)(A)v_i \neq 0$. Show that if $u_i := (p/p_i^{d_i})v_i$ then $p_{u_i} = p_i^{d_i}$. Now let $v := u_1 + \cdots + u_k$ and argue that $p_v = p$ so that $\deg p_v = d$; then conclude that $v, Av, \ldots, A^{d-1}v$ are linearly independent.

Problem 26. Find a normal basis of $\mathbb{Q}(i)$ over $\mathbb{Q}$.

Problem 27. Let $\varepsilon = e^{2\pi i/n}$. Show that the numbers ε^k where k is coprime with n form a normal basis of $(\mathbb{Q}(\varepsilon) : \mathbb{Q})$ if and only if n is square-free.

Problem 28. Let $(L : K)$ be a finite Galois extension and let $\alpha \in L$. Show that $\mathrm{tr}_K^L(\alpha) = 0$ if and only if α is a sum of elements of the form $\sigma(\beta) - \beta$ with $\sigma \in G_K^L$ and $\beta \in L$. **Hint.** Express α in terms of a normal basis $\{\sigma(\beta) \mid \sigma \in G_K^L\}$ of $(L : K)$.

Problem 29. Let Z_1 and Z_2 be intermediate fields of a finite Galois extension $(L : K)$. Show that $G_{Z_1 \cap Z_2}^L = \langle G_{Z_1}^L \cup G_{Z_2}^L \rangle$ and $G_{Z_1 Z_2}^L = G_{Z_1}^L \cap G_{Z_2}^L$.

Problem 30. Find a field L with subfields Z_1 and Z_2 such that $(L : Z_1)$ and $(L : Z_2)$ are finite Galois extensions whereas $(L : Z_1 \cap Z_2)$ is an infinite extension.

Problem 31. Let $L \subseteq \mathbb{C}$ be the splitting of all polynomials $x^2 + px + q$ with $p, q \in \mathbb{Q}$.

(a) Show that $L = \mathbb{Q}(\{i\} \cup \{\sqrt{p} \mid p \in P\})$ where P is the set of all prime numbers.

(b) Show that $(L : \mathbb{Q})$ is an infinite Galois extension. Moreover, show that the Galois group $G_{\mathbb{Q}}^L$ is abelian with $\sigma^2 = \mathrm{id}$ for all $\sigma \in G_{\mathbb{Q}}^L$.

(c) Show that if Z is an intermediate field of $(L : \mathbb{Q})$ with $[Z : \mathbb{Q}] < \infty$ then $(Z : \mathbb{Q})$ is a Galois extension, and $[Z : \mathbb{Q}]$ is a power of 2.

(d) Show that $G_{\mathbb{Q}}^L$ has the same cardinality as the power set of $\mathbb{N}$ and possesses an uncountable number of subgroups of index 2 only countably many of which are closed in the Krull topology.

Problem 32. Show that if $(L : K)$ is a finite Galois extension then the Krull topology on G_K^L is the discrete topology. (Thus no additional information in finite Galois theory can be expected from topological considerations.) Conversely, show that if $(L : K)$ is an algebraic Galois extension such that the Krull topology on G_K^L is discrete then $[L : K] < \infty$.

Problem 33. Let $(L : K)$ be an algebraic Galois extension and let Z be an intermediate field. Show that the subspace topology induced on G_Z^L by the Krull topology of G_K^L is the Krull topology of G_Z^L.

Problem 34. Let $(L : K)$ be an algebraic Galois extension and let $U \leq G_K^L$. Moreover, let $\mathfrak{Z}$ be the family of all intermediate fields of $(L : K)$ such that $(Z : K)$ is a finite Galois extension. Show that the following conditions are equivalent:
(1) U is dense in G_K^L;
(2) if $Z \in \mathfrak{Z}$ then each $\sigma \in G_K^Z$ can be extended to an element of U;
(3) if $Z \in \mathfrak{Z}$ then $G_K^Z \cong U/U \cap G_Z^L$.

Problem 35. Let $(L : K)$ be an algebraic Galois extension.
(a) Show that a subgroup U is stable if and only if $Z := F(U)$ is a Galois extension of K.
(b) Let Z be a stable intermediate field of $(L : K)$. Is in this case the isomorphism $G_K^Z \cong G_K^L/G_Z^L$ even an isomorphism of topological groups?

Problem 36. Let $(L : K)$ be an algebraic Galois extension with Galois group $G := G_K^L$. Let U be the Krull closure of the commutator subgroup G' in G and let $F(U)$ be the fixed field of U. Show that $F(U)$ is the maximal Galois extension of K with abelian Galois group which is contained in L.

Problem 37. Show that $(\sigma \bullet f)(x) := \sigma\big(f(\sigma^{-1} \star x)\big)$ in the proof of (22.22) defines indeed an automorphic action.

Problem 38. Let K be a field and let $L = K(x_1, x_2, \ldots)$ be the rational function field in an infinite number of variables. Show that G_K^L is not closed in L^L (with the finite topology); hence (22.28)(d) is wrong for transcendental extensions.

Problem 39. Show that if $(L : K)$ is a Galois extension with an infinite Galois group G_K^L then $[L : K] = \infty$.
Hint. Choose $\alpha_1, \ldots, \alpha_n \in L$ such that $Z := K(\alpha_1, \ldots, \alpha_n)$ is a Galois extension of K with $[Z : K] \geq n$. Then each element of G_K^Z can be extended to an element of G_K^L so that $|G_K^L| \geq |G_K^Z|$.

Problem 40. Let L be a field, let $\Gamma \leq \mathrm{Aut}\, L$ be a group which is compact as a subspace of L^L (with the finite topology) and let $K := F(\Gamma)$. Show that $(L : K)$ is a Galois extension. (This generalizes (21.25)(d).)
Hint. For each element $\alpha \in L$ the set $\{\sigma(\alpha) \mid \sigma \in G\}$ is finite.

Problem 41. Let $(L : K)$ be an algebraic Galois extension and let $\mathcal{N}$ be the set of all normal subgroups of G_K^L of finite index. Show that $\bigcap_{N \in \mathcal{N}} N = \{\mathrm{id}\}$.

Problem 42. Let $(L : K)$ be an algebraic Galois extension. Show that if $U \leq G_K^L$ is a subgroup of G_K^L then its Krull closure is $U^* = G_{F(U)}^L$.

Problem 43. Let $(L : K)$ be an algebraic Galois extension. Show that the sets

$$\mathcal{A} := \{gG_K^Z \mid g \in G, K \subseteq Z \subseteq L \text{ with } [Z : K] < \infty\} \quad \text{and}$$
$$\mathcal{B} := \{gG_K^Z \mid g \in G, K \subseteq Z \subseteq L \text{ with } [Z : K] < \infty \text{ and } (Z : K) \text{ Galois}\}$$

are bases for the Krull topology on $G := G_K^L$.

Problem 44. Let L be an algebraic closure of a finite field K. Show that G_K^L has no elements of finite order except the identity element.

Problem 45. For any prime number p, let C_p be an algebraic closure of $\mathbb{Z}_p$ and let G_p be the Galois group of $(C_p : \mathbb{Z}_p)$.
(a) Show that G_p is abelian.
(b) Let $Z := \{x \in C_p \mid x^{p^{2^m}} \text{ for some } m \in \mathbb{N}\}$ and let $\sigma : Z \to Z$ be the Frobenius automorphism of Z, given by $\sigma(x) = x^p$. Show that $\sigma^{2^k} \to \text{id}$ as $k \to \infty$ with respect to the finite topology on $G_{\mathbb{Z}_p}^Z$.
(c) Show that $G_p \cong G_q$ for all primes p, q.

Problem 46. Let p be a prime number and let $\overline{K}$ be an algebraic closure of $K := \mathbb{Z}_p$. Consider each Galois field $\text{GL}(p^n)$ as sitting inside $\overline{K}$ (so that $\text{GF}(p^n) = \{x \in \overline{K} \mid x^{p^n} = x\}$); this gives an ascending chain

$$K = \text{GF}(p) \subseteq \text{GF}(p^2) \subseteq \text{GF}(p^3) \subseteq \cdots \subseteq \overline{K}.$$

Let $L := \bigcup_{n=1}^{\infty} \text{GF}(p^n)$. Show that $(L : K)$ is an infinite Galois extension and that $G_K^L \cong (R, +)$ where R is the ring of all rational numbers m/p^n with $m \in \mathbb{Z}$ and $n \in \mathbb{N}_0$. Moreover, show that this isomorphism is a topological isomorphism if R is equipped with the topology induced by the p-adic valuation $|\cdot|_p$ which is defined by

$$|x|_p := \begin{cases} 0, & \text{if } x = 0; \\ p^{-e(p)}, & \text{if } x = \pm p^{e(p)} \prod_{\substack{q \text{ prime,} \\ q \neq p}} q^{e(q)}. \end{cases}$$

(This definition makes sense because each nonzero rational number has a unique representation of the form $\pm \prod_i p_i^{e(p_i)}$ where the product runs over all prime numbers and where the exponents $e(p_i)$ are integers of which only finitely many are nonzero.)

23. The Galois group of a polynomial

WE introduced the concept of a group as a mathematical means to describe symmetries; in fact, we learned to distinguish different types of symmetry in geometrical patterns by looking in each case at the underlying group of transformations leaving invariant the given pattern. However, when the notion of a group was shaped and formulated historically, this was not done to describe geometrical patterns – for obvious reasons: One does not need an abstract mathematical instrument to describe facts which are obvious to one's bare eyes. On the other hand, if it comes to describing "hidden" symmetries such a mathematical tool might be very useful, and this was recognized for the first time in the theory of polynomials.† Mathematicians like Joseph-Louis de Lagrange (1736-1813), Niels Hendrik Abel (1802-1829) and, above all, Evariste Galois (1811-1832) were well aware of the fact that each polynomial f possesses an underlying hidden symmetry, namely, the symmetry between those roots of f which cannot be algebraically distinguished.

To see what this means, consider a polynomial $f \in K[x]$ and suppose that $\alpha_1, \ldots, \alpha_n$ are the different roots of f (in some extension of K). Of course, the symmetric group Sym_n acts as the group of permutations of these roots. Now the basic idea of Galois was to single out the subgroup of those permutations which preserve the arithmetic relations between the roots; it turns out that this subgroup – today called the "Galois group" of the polynomial f – measures in some sense (to be clarified in a later section) how difficult it is to solve the equation $f(x) = 0$ explicitly. The precise definition of the Galois group of a polynomial is as follows.

(23.1) Definition. *Let $f \in K[x]$ be a polynomial with the different roots $\alpha_1, \ldots, \alpha_n$ in some extension of K. The **Galois group** $\mathrm{Gal}_K(f)$ of f over K is the set of all permutations $\sigma \in \mathrm{Sym}_n$ with the following property: If $\Phi(\alpha_1, \ldots, \alpha_n) = 0$ for a polynomial $\Phi \in K[x_1, \ldots, x_n]$ in n variables, then also $\Phi(\alpha_{\sigma(1)}, \ldots, \alpha_{\sigma(n)}) = 0$.*

IT is important to observe that the Galois group depends on the chosen base-field K; this dependence is reflected in the notation $\mathrm{Gal}_K(f)$ instead of simply $\mathrm{Gal}(f)$. Clearly, if $K \subseteq L$ then $\mathrm{Gal}_K(f) \supseteq \mathrm{Gal}_L(f)$.

(23.2) Example. Consider $f(x) = x^2 + 1 \in \mathbb{Q}[x]$. The roots of f are $\pm i \in \mathbb{C}$. Let K be any intermediate field of $(\mathbb{C} : \mathbb{Q})$. We claim that

$$\mathrm{Gal}_K(f) = \begin{cases} \{\mathrm{id}\}, & \text{if } i \in K; \\ \{\mathrm{id}, (12)\} = \mathrm{Sym}_2, & \text{if } i \notin K. \end{cases}$$

Suppose $i \notin K$ and let $\Phi \in K[x, y]$ be a polynomial with $\Phi(i, -i) = 0$; we want to show that then also $\Phi(-i, i) = 0$. The polynomial $h(x) := \Phi(x, -x)$ has i as a root. Hence h is divisible by the minimal polynomial of i over K, namely $x^2 + 1$, and thus has also $-i$ as a root. But this means exactly $\Phi(-i, i) = 0$.

Suppose $i \in K$. Then the polynomial $\Phi(x, y) := x - i \in K[x, y]$ has $(i, -i)$ as a root, but not $(-i, i)$. This shows that $(12) \notin \mathrm{Gal}_K(f)$.

---

† Elementary particle physics is another field in which group theory is successfully applied to describe "hidden symmetries".

THE following theorem provides – at least in principle – a way of determining the Galois group of any given polynomial f without knowing the roots of f.

(23.3) Theorem. *Let K be a field and let $f \in K[x]$. Then the following steps constitute an algorithm to find the Galois group $\mathrm{Gal}_K(f)$.*

(a) Determine the number n of distinct roots of f and call these roots $\alpha_1, \dots, \alpha_n$.

(b) Define a polynomial F in $n+1$ variables $z, u_1, \dots, u_n$ by

$$F(z, u_1, \dots, u_n) := \prod_{\sigma \in \mathrm{Sym}_n} \left(z - (u_{\sigma(1)}\alpha_1 + \dots + u_{\sigma(n)}\alpha_n) \right) .$$

This polynomial has coefficients in K and can be determined without prior knowledge of the roots $\alpha_1, \dots, \alpha_n$.

(c) Decompose F into its irreducible factors in the unique factorization domain $K[z, u_1, \dots, u_n]$, say

$$F = F_1 \cdots F_r .$$

(d) If we define an action of Sym_n on $K[z, u_1, \dots, u_n]$ by $(\sigma \star P)(z, u_1, \dots, u_n) := P(z, u_{\sigma^{-1}(1)}, \dots, u_{\sigma^{-1}(n)})$ and if F_i is any factor of F, then

$$\mathrm{Gal}_K(f) = \{\sigma \in \mathrm{Sym}_n \mid \sigma \star F_i = F_i\} ;$$

i.e., whether or not an element $\sigma \in \mathrm{Sym}_n$ belongs to the Galois group of f can be decided by simply checking whether or not σ leaves one factor of F (and hence all factors of F) invariant.

Proof. (a) If $f' \neq 0$ then the number n is simply the degree of $f/\gcd(f, f')$, and the greatest common divisor of f and f' can be found by applying the Euclidean algorithm. If $f' = 0$ then necessarily $\mathrm{char}\, K = p \neq 0$ and $f(x) = g_1(x^p)$ for some $g_1 \in K[x]$. If $g_1' = 0$ then we can write $g_1(x) = g_2(x^p)$ and hence $f(x) = g_2(x^{p^2})$ with some $g_2 \in K[x]$. Continuing in this way we can write $f(x) = g(x^{p^n})$ where $g \in K[x]$ is such that $g' \neq 0$. Then the number of different roots of g can be determined by the above argument; but f and g have the same number of distinct roots because the roots of f are exactly the p^n-th roots of roots of g, and $\alpha^{p^n} = \beta^{p^n}$ if and only if $\alpha = \beta$.

(b) Since $u_{\sigma^{-1}(1)}\alpha_1 + \dots + u_{\sigma^{-1}(n)}\alpha_n = u_1\alpha_{\sigma(1)} + \dots + u_n\alpha_{\sigma(n)}$ for all $\sigma \in \mathrm{Sym}_n$, the coefficients of F are obviously symmetric polynomials in $\alpha_1, \dots, \alpha_n$ and hence can be expressed by the coefficients of f (which lie in K).

(c) Since $K[z, u_1, \dots, u_n]$ is a unique factorization domain, we know that there *is* always such a decomposition. However, for suitable fields K (for example for all fields obtained from $\mathbb{Q}$ or the fields $\mathbb{Z}_p$ by adjoining a finite number of elements) there is also an algorithm which allows one to *find* such a decomposition in a finite number of steps. (See (8.11) through (8.17).)

(d) Let F_i be a factor of F. By suitably numbering the roots of f, we can assume that F_i contains the factor $z - (u_1\alpha_1 + \dots + u_n\alpha_n)$. Now, clearly, if $\sigma \star F_i = F_i$ then σ permutes the linear factors of F_i over $L := K(\alpha_1, \dots, \alpha_n)$ so that in particular $\sigma \star \left(z - (u_1\alpha_1 + \dots + u_n\alpha_n) \right) = z - (u_{\sigma^{-1}(1)}\alpha_1 + \dots + u_{\sigma^{-1}(n)}\alpha_n)$ is again a linear factor of F_i. Conversely, if σ transforms $z - (u_1\alpha_1 + \dots + u_n\alpha_n)$ to a linear factor λ of F_i then $\sigma \star F_i$ is an irreducible factor of $\sigma \star F = F$ which has the factor λ in common with F_i; this implies $\sigma \star F_i = F_i$. Hence $\sigma \star F_i = F_i$ if and only if $u_1\alpha_1 + \dots + u_n\alpha_n$ and

$u_{\sigma^{-1}(1)}\alpha_1 + \cdots + u_{\sigma^{-1}(n)}\alpha_n = u_1\alpha_{\sigma(1)} + \cdots + u_n\alpha_{\sigma(n)}$ have the same minimal polynomial over $K(u_1, \ldots, u_n)$ (namely F_i), i.e., are conjugates over $K(u_1, \ldots, u_n)$. By (13.13) this is the case if and only if there is an element $\varphi \in G_{K(u_1,\ldots,u_n)}^{L(u_1,\ldots,u_n)}$ with

$$u_1\alpha_{\sigma(1)} + \cdots + u_n\alpha_{\sigma(n)} = \varphi(u_1\alpha_1 + \cdots + u_n\alpha_n) = u_1\varphi(\alpha_1) + \cdots + u_n\varphi(\alpha_n)$$

which means $\varphi(\alpha_i) = \alpha_{\sigma(i)}$ for $1 \leq i \leq n$. Observing that each element of $G_{K(u_1,\ldots,u_n)}^{L(u_1,\ldots,u_n)}$ is the unique extension of an element of G_K^L, we recognize that $\sigma \star F_i = F_i$ if and only if there is an element $\bar{\sigma} \in G_K^L$ with $\bar{\sigma}(\alpha_i) = \alpha_{\sigma(i)}$ for all i, which means exactly that $\sigma \in \mathrm{Gal}_K(f)$. ∎

THE method just described is not of much practical value, because the execution of the algorithm leads to abominable computations except in the simplest cases; however, the theorem assures us that we can in principle (assuming unlimited patience) determine the Galois group of any given polynomial. On the other hand, (23.3) yields the following corollary which will be used later on to obtain concrete partial information on the Galois group of a given polynomial.

(23.4) Theorem. *Let R be a unique factorization domain and let P be a prime ideal of R. Let K and $\overline{K}$ be the quotient fields of R and $\overline{R} := R/P$, respectively. Let $f \in R[x]$ be a monic polynomial and let $\overline{f} \in \overline{R}[x]$ be obtained from f by reducing the coefficients modulo P. If both f and $\overline{f}$ have no multiple roots † , then these roots can be ordered in such a way that $\mathrm{Gal}_{\overline{K}}(\overline{f}) \subseteq \mathrm{Gal}_K(f) \subseteq \mathrm{Sym}_n$.*

REMARK. This statement is stronger than just saying that $\mathrm{Gal}_{\overline{K}}(\overline{f})$ is isomorphic to a subgroup of $\mathrm{Gal}_K(f)$; there is in fact an embedding of $\mathrm{Gal}_{\overline{K}}(\overline{f})$ into $\mathrm{Gal}_K(f)$ which preserves the permutation structure.

Proof. Let $\alpha_1, \ldots, \alpha_n$ be the roots of f. Find a decomposition of

$$F(z, u_1, \ldots, u_n) := \prod_{\sigma \in \mathrm{Sym}_n} \left(z - (u_{\sigma(1)}\alpha_1 + \cdots + u_{\sigma(n)}\alpha_n)\right)$$

into irreducible factors as in (23.3), say $F = F_1 \cdots F_r$. By (8.2)(c) these factors can actually be chosen in $R[z, u_1, \ldots, u_n]$. Reducing modulo P, we obtain a factorization $\overline{F} = \overline{F_1} \cdots \overline{F_r}$ (in which the factors $\overline{F_i}$ need not be reducible). Then $G := \mathrm{Gal}_K(f)$ consists of those $\sigma \in \mathrm{Sym}_n$ with $\sigma \star F_1 = F_1$, whereas $\sigma \star F_1 \in \{F_2, \ldots, F_n\}$ for $\sigma \in \mathrm{Sym}_n \backslash G$. Consequently,

$(\star) \quad \sigma \star \overline{F_1} = \overline{F_1}$ if $\sigma \in \mathrm{Sym}_n$ and $\sigma \star \overline{F_1} \in \{\overline{F_2}, \ldots, \overline{F_n}\}$ if $\sigma \in \mathrm{Sym}_n \backslash G$.

Let λ be an irreducible factor of $\overline{F_1}$ in $\overline{K}[z, u_1, \ldots, u_n]$. Then $\overline{G} := \mathrm{Gal}_{\overline{K}}(\overline{f})$ consists of those $\sigma \in \mathrm{Sym}_n$ with $\sigma \star \lambda = \lambda$. Since then $\lambda = \sigma \star \lambda$ is an irreducible factor of $\sigma \star \overline{F_1}$, this yields $\sigma \star \overline{F_1} = \overline{F_1}$ which in view of $(\star)$ implies that $\sigma \in G$. ∎

---

† Clearly, if $\overline{f}$ has no multiple roots then the same is automatically true for f.

CONCRETE applications of (52.4) will be given below; let us first gather more information on Galois groups. The discussion will be enormously facilitated by relating the concept of the Galois group of a polynomial with that of the Galois group of a field extension because this allows us to use the structure theory of field extensions acquired up to this point to study polynomials. In fact, we could have defined the Galois group of a polynomial once the notion of a splitting field was available to us, but we would not have been able to obtain very concrete results without the machinery of field theory developed in the meantime. This is the reason why we did not follow the historical approach of studying polynomials as basic objects, but followed the "Dedekind-Artin linearization of Galois theory" in which field extensions are the basic objects.

LET us now show how the two notions of Galois groups defined in (21.2) and (23.1) are related.

(23.5) Theorem. *Let $f \in K[x]$ be a polynomial with coefficients in a field K and let $L = K(\alpha_1, \ldots, \alpha_n)$ be a splitting field of f where $\alpha_1, \ldots, \alpha_n$ are the pairwise different roots of f. Then each element $\sigma \in G_K^L$ determines a unique element $\overline{\sigma} \in \mathrm{Sym}_n$ such that $\sigma(\alpha_i) = \alpha_{\overline{\sigma}(i)}$ for $1 \leq i \leq n$, and the mapping*

$$\theta : \begin{array}{ccc} G_K^L & \to & \mathrm{Sym}_n \\ \sigma & \mapsto & \overline{\sigma} \end{array}$$

is an injective group homomorphism whose image is exactly $\mathrm{Gal}_K(f)$. Hence $\mathrm{Gal}_K(f)$ and G_K^L are isomorphic.

Proof. The crucial observation is that any element $\sigma \in G_K^L$ permutes the roots of f. Indeed, if $f(x) = a_0 + a_1 x + \cdots + a_n x^n$ and $f(\alpha) = 0$ (i.e., $\alpha = \alpha_i$ for some i), then

$$0 = \sigma(0) = \sigma\big(f(\alpha)\big) = \sigma(a_0 + a_1\alpha + \cdots + a_n\alpha^n)$$
$$= a_0 + a_1\sigma(\alpha) + \cdots + a_n\sigma(\alpha)^n = f\big(\sigma(\alpha)\big)$$

because the automorphism σ leaves invariant all the coefficients a_i which lie in K. So $\sigma(\alpha)$ is again a root of f; i.e., $\sigma(\alpha) = \alpha_j$ for some j. Also, if $\sigma(\alpha_i) = \sigma(\alpha_j)$ then $i = j$ because $\sigma : L \to L$ is a bijection. Hence the mapping θ is well-defined. We have to show that θ is an injective group homomorphism with $\mathrm{im}\,\theta = \mathrm{Gal}_K(f)$. First of all, $\mathrm{id}(\alpha_i) = \alpha_i$ for all i, hence $\overline{\mathrm{id}} = \mathrm{id}$; also, $(\sigma\tau)(\alpha_i) = \sigma(\alpha_{\overline{\tau}(i)}) = \alpha_{\overline{\sigma}\,\overline{\tau}(i)}$ so that $\overline{\sigma\tau} = \overline{\sigma}\,\overline{\tau}$. This shows that θ is a group homomorphism. Now if $\sigma \in \ker\theta$ then $\sigma(\alpha_i) = \alpha_i$ for all i; but then σ is the identity on $K(\alpha_1, \ldots, \alpha_n) = L$. This shows that θ is one-to-one. Let us prove finally that $\mathrm{im}\,\theta = \mathrm{Gal}_K(f)$. To show that $\mathrm{im}\,\theta \subseteq \mathrm{Gal}_K(f)$ observe the following: If

$$\Phi(x_1, \ldots, x_n) = \sum_{(k_1, \ldots, k_n)} a_{k_1, \ldots, k_n} x_1^{k_1} \cdots x_n^{k_n} \in K[x_1, \ldots, x_n]$$

satisfies $\Phi(\alpha_1, \ldots, \alpha_n) = 0$, then, for any element $\sigma \in G_K^L$, we have

$$0 = \sigma(0) = \sigma\big(\Phi(\alpha_1, \ldots, \alpha_n)\big) = \sigma\Big(\sum_{(k_1, \ldots, k_n)} a_{k_1, \ldots, k_n} \alpha_1^{k_1} \cdots \alpha_n^{k_n} \Big)$$

$$= \sum_{(k_1, \ldots, k_n)} a_{k_1, \ldots, k_n} \alpha_{\overline{\sigma}(1)}^{k_1} \cdots \alpha_{\overline{\sigma}(n)}^{k_n} = \Phi(\alpha_{\overline{\sigma}(1)}, \ldots, \alpha_{\overline{\sigma}(n)}) \ .$$

To prove conversely that $\text{Gal}_K(f) \subseteq \text{im}\,\theta$ we have to show that every element $\tau \in \text{Gal}_K(f)$ (which can be considered as a permutation of the set $\{\alpha_1, \ldots, \alpha_n\}$) extends to an automorphism σ of L so that we have $\tau = \overline{\sigma}$. An arbitrary element of $L = K(\alpha_1, \ldots, \alpha_n)$ has the form

$$(\star) \qquad \sum_{(k_1,\ldots,k_n)} a_{k_1,\ldots,k_n} \alpha_1^{k_1} \cdots \alpha_n^{k_n}$$

with coefficients $a_{k_1,\ldots,k_n} \in K$. Define $\sigma : L \to L$ by

$$\sigma\Big(\sum_{(k_1,\ldots,k_n)} a_{k_1,\ldots,k_n} \alpha_1^{k_1} \cdots \alpha_n^{k_n} \Big) := \sum_{(k_1,\ldots,k_n)} a_{k_1,\ldots,k_n} \alpha_{\tau(1)}^{k_1} \cdots \alpha_{\tau(n)}^{k_n} \, .$$

Now an element of L may have different representations of the form $(\star)$ but the hypothesis that $\Phi(\alpha_1, \ldots, \alpha_n) = 0$ implies $\Phi(\alpha_{\tau(1)}, \ldots, \alpha_{\tau(n)}) = 0$ guarantees that σ is well-defined. Then σ is obviously an automorphism of L with $\sigma|_K = \text{id}_K$, i.e. an element of G_K^L, such that $\sigma(\alpha_i) = \alpha_{\tau(i)}$ for all i; i.e., $\tau = \overline{\sigma}$. ∎

BEARING this result in mind, we will often identify G_K^L with $\text{Gal}_K(f)$; it will always be clear from the context whether we think of an element of a Galois group as a field automorphism or as a permutation.

BEFORE we look at various examples, let us see a first example how properties of polynomials or field extensions are reflected in properties of the corresponding Galois groups.

(23.6) Proposition. *Suppose the polynomial $f \in K[x]$ has no multiple roots. Then f is irreducible if and only if $\text{Gal}_K(f)$ acts transitively on the set of roots of f.*

Proof. Let L be a splitting field of f. If f is irreducible then for any two roots α, α' of f there is an isomorphism $\sigma : K(\alpha) \to K(\alpha')$ which leaves K pointwise fixed and maps α to α'; by (13.5)(b) this isomorphism can be extended to an element of G_K^L. This shows that the action of G_K^L on the roots of f is transitive.

Suppose conversely that $\text{Gal}_K(f)$ acts transitively on the set of roots of f. Let $p \in K[x]$ be an irreducible factor of f and let α be a root of p. If β is any root of f, then by assumption there is an element $\sigma \in G_K^L$ with $\sigma(\alpha) = \beta$. But then $p(\beta) = p(\sigma(\alpha)) = \sigma(p(\alpha)) = \sigma(0) = 0$ so that β is a root of p. This shows that every root of f is a root of p; since f has no multiple roots, this implies $f = p$ (up to a multiplicative constant) so that f is irreducible.

(23.7) Example. Let us determine the Galois group of the polynomial $p(x) = (x^2 - 2)(x^2 - 3) \in \mathbb{Q}[x]$; we simply write $\text{Gal}(p)$ instead of $\text{Gal}_\mathbb{Q}(p)$. The roots of p are

$$\alpha_1 = \sqrt{2}\,, \quad \alpha_2 = -\sqrt{2}\,, \quad \alpha_3 = \sqrt{3}\,, \quad \alpha_4 = -\sqrt{3}\,.$$

Intuitively, the roots α_1 and α_2 cannot be distinguished from each other by arithmetic relations over $\mathbb{Q}$; whenever $f(\sqrt{2}) = 0$ for some $f \in \mathbb{Q}[x]$, then also $f(-\sqrt{2}) = 0$. So we expect that the transposition (12) belongs to $\mathrm{Gal}(p)$. The same holds for (34). On the other hand, α_1 and α_2 satisfy the equation $x^2 - 2 = 0$, but α_3 and α_4 do not. Hence an element of $\mathrm{Gal}(p)$ has to leave invariant the sets $\{1,2\}$ and $\{3,4\}$. Thus the Galois group of p will be

$$\mathrm{Gal}(p) \;=\; \{\mathrm{id}, (12), (34), (12)(34)\} \;.$$

Since the splitting field of p is $\mathbb{Q}(\sqrt{2}, \sqrt{3})$ this is what we expect from (21.4). Let us prove a bit more strictly that $(13) \notin \mathrm{Gal}(p)$, but $(12) \in \mathrm{Gal}(p)$. The polynomial $\Phi(x_1, x_2, x_3, x_4) := (x_1^2 - 2)(x_3^2 - 3)$ in 4 variables satisfies $\Phi(\alpha_1, \alpha_2, \alpha_3, \alpha_4) = f(\sqrt{2}, -\sqrt{2}, \sqrt{3}, -\sqrt{3}) = 0$, but $\Phi(\alpha_3, \alpha_2, \alpha_1, \alpha_4) = \Phi(\sqrt{3}, -\sqrt{2}, \sqrt{2}, -\sqrt{3}) = -1 \neq 0$; whence $(13) \notin \mathrm{Gal}(p)$, as follows from the definition of $\mathrm{Gal}(p)$. Now suppose that Φ is a polynomial in 4 variables with rational coefficients such that $\Phi(\alpha_1, \alpha_2, \alpha_3, \alpha_4) = 0$; we claim that this implies $\Phi(\alpha_2, \alpha_1, \alpha_3, \alpha_4) = 0$. Indeed, the polynomial $h(x) := \Phi(x, -x, \alpha_3, \alpha_4) \in \mathbb{Q}(\sqrt{3})[x]$ satisfies $h(\sqrt{2}) = \Phi(\alpha_1, \alpha_2, \alpha_3, \alpha_4) = 0$, hence is divisible by the minimal polynomial of $\sqrt{2}$ over $\mathbb{Q}(\sqrt{3})$ which is $x^2 - 2$. But then also $-\sqrt{2}$ is a root of h; i.e., $0 = h(-\sqrt{2}) = \Phi(\alpha_2, \alpha_1, \alpha_3, \alpha_4)$.

(23.8) Example. Consider the polynomials $f(x) = (x^2 + 1)(x^2 + 2)$ and $g(x) = x^4 - 2x^2 + 9$ in $\mathbb{Q}[x]$. Here f has the roots $\pm i, \pm i\sqrt{2}$, whereas g has the roots $\pm(\sqrt{2} \pm i)$. Hence f and g have the same splitting field, namely $L = \mathbb{Q}(i, \sqrt{2})$, and consequently the same abstract Galois group; in fact, $\mathrm{Gal}_{\mathbb{Q}}(f)$ and $\mathrm{Gal}_{\mathbb{Q}}(g)$ are both isomorphic to $G_{\mathbb{Q}}^L$. However, there is no inner automorphism of Sym_4 that would carry $\mathrm{Gal}_{\mathbb{Q}}(f)$ into $\mathrm{Gal}_{\mathbb{Q}}(g)$, as can be seen by considering the cycle-structures of the elements of both Galois groups. Let us do this.

The Galois group $G_{\mathbb{Q}}^L$ has four elements, given by the possible combinations of signs in the assignments $i \mapsto \pm i$, $\sqrt{2} \mapsto \pm\sqrt{2}$; this can be verified in a way analogous to the procedure in example (21.4). Let us number the roots of f and g as

$$(\alpha_1, \alpha_2, \alpha_3, \alpha_4) \;=\; (i, -i, i\sqrt{2}, -i\sqrt{2}) \quad \text{and}$$
$$(\beta_1, \beta_2, \beta_3, \beta_4) \;=\; (\sqrt{2} + i, \sqrt{2} - i, -\sqrt{2} + i, -\sqrt{2} - i) \;.$$

Checking how the elements of $G_{\mathbb{Q}}^L$ permute the roots α_i and β_i, we find that

$$\mathrm{Gal}_{\mathbb{Q}}(f) \;=\; \{\mathrm{id}, (12), (34), (12)(34)\} \quad \text{and}$$
$$\mathrm{Gal}_{\mathbb{Q}}(g) \;=\; \{\mathrm{id}, (12)(34), (13)(24), (14)(23)\} \;.$$

Observe that $\mathrm{Gal}_{\mathbb{Q}}(f)$ does not act transitively on the roots of f whereas $\mathrm{Gal}_{\mathbb{Q}}(g)$ acts transitively on the roots of g. According to (23.6), this reflects the fact that f is reducible whereas g is irreducible.

THE following corollary is sometimes useful to determine the Galois group of a product of polynomials.

(23.9) Proposition. *Let K be a field and let $f, g \in K[x]$ be polynomials all of whose irreducible factors are separable. Let Z_1 and Z_2 be the splitting fields of f and g in some fixed algebraic closure $\overline{K}$ of K. If $Z_1 \cap Z_2 = K$ then $\mathrm{Gal}_K(fg) \cong \mathrm{Gal}_K(f) \times \mathrm{Gal}_K(g)$.*

Proof. Using (22.17)(b) we see that $\mathrm{Gal}_K(fg) \cong G_K^{Z_1 Z_2} \cong G_K^{Z_1} \times G_K^{Z_2} \cong \mathrm{Gal}_K(f) \times \mathrm{Gal}_K(g)$. ∎

(23.10) Dedekind's Reciprocity Theorem. *Let K be a field and $f, g \in K[x]$ irreducible separable polynomials. Let α be a root of f and β a root of g.*

(a) If σ and τ are elements of $\mathrm{Gal}_K(fg)$ then $\sigma(\alpha)$ and $\tau(\alpha)$ are conjugates over $K(\beta)$ if and only if $\sigma^{-1}(\beta)$ and $\tau^{-1}(\beta)$ are conjugates over $K(\alpha)$.

(b) Suppose that $f = f_1 \cdots f_m$ is a prime factorization of f over $K(\beta)$ whereas $g = g_1 \cdots g_n$ is a prime factorization of g over $K(\alpha)$. Then $m = n$, and the g_i can be ordered such that

$$\frac{\deg f_1}{\deg g_1} = \frac{\deg f_2}{\deg g_2} = \cdots = \frac{\deg f_m}{\deg g_m} \ .$$

Proof. Let L be a splitting field of fg so that $\mathrm{Gal}_K(fg) \cong G_K^L$.

(a) The elements $\sigma(\alpha)$ and $\tau(\alpha)$ are conjugates over $K(\beta)$ if and only if there is an element $\Theta \in G_{K(\beta)}^L$ with $\Theta\big(\sigma(\alpha)\big) = \tau(\alpha)$, i.e., $\tau^{-1}\Theta\sigma \in G_{K(\alpha)}^L$. Analogously, the elements $\sigma^{-1}(\beta)$ and $\tau^{-1}(\beta)$ are conjugates over $K(\alpha)$ if and only if there is an element $\overline{\Theta} \in G_{K(\alpha)}^L$ with $\overline{\Theta}\big(\sigma^{-1}(\beta)\big) = \tau^{-1}(\beta)$, i.e., $\tau\overline{\Theta}\sigma^{-1} \in G_{K(\beta)}^L$. Setting $\overline{\Theta} := \tau^{-1}\Theta\sigma$ if Θ is given and $\Theta := \tau\overline{\Theta}\sigma^{-1}$ if $\overline{\Theta}$ is given, we see that these two conditions are equivalent.

(b) We define a mapping $\varphi : \{1, \ldots, m\} \to \{1, \ldots, n\}$ as follows: Pick an element $\sigma \in G_K^L$ such that $\sigma(\alpha)$ is a root of f_i; then let $\varphi(i)$ be the unique index k such that $\sigma^{-1}(\beta)$ is a root of g_k. We claim that φ is well-defined and is a bijection.

First, since $\mathrm{Gal}_K(f)$ acts transitively on the roots of f we can find $\sigma \in \mathrm{Gal}_K(f)$ such that $\sigma(\alpha)$ is a root of f_i, and σ can be extended to an element of G_K^L. Then since G_K^L permutes the roots of g we have $g_k\big(\sigma^{-1}(\beta)\big) = 0$ for some k. Moreover, part (a) says that $\sigma(\alpha)$ and $\tau(\alpha)$ are roots of the same f_i if and only if $\sigma^{-1}(\beta)$ and $\tau^{-1}(\beta)$ are roots of the same g_k. Hence φ does not depend on the particular choice of σ and is therefore unambiguously defined and injective. Consequently, $m \leq n$. Exchanging the roles of f and g, we see that $m = n$ and that φ is a bijection.

Finally, let $N := |\mathrm{Gal}_{K(\alpha)}(fg)| = |G_{K(\alpha)}^L|$ be the cardinality of the stabilizer of α. For any root $\overline{\alpha}$ of f, the set $\{\sigma \in G_K^L \mid \sigma(\alpha) = \overline{\alpha}\}$ is a coset of this stabilizer and hence has cardinality N, due to problem 10 in section [I.26]. Thus $\{\sigma \in G_K^L \mid \sigma(\alpha)$ is a root of $f_i\}$ has cardinality $N \cdot \deg f_i$ for each $1 \leq i \leq n$[†] whereas $G_K^L = \{\sigma \in G_K^L \mid \sigma(\alpha)$ is

---

[†] Let $d := \deg f_i$ and let $\alpha_1, \ldots, \alpha_d$ be the roots of f_i. Then the set in question is the disjoint union of the d sets $\{\sigma \in G_K^L \mid \sigma(\alpha) = \alpha_k\}$ $(1 \leq k \leq d)$ each of which has cardinality N.

a root of f} has cardinality $N \cdot \deg f$. Using part (a), we obtain

$$\frac{\deg f_i}{\deg f} = \frac{|\{\sigma \in G_K^L \mid \sigma(\alpha) \text{ is a root of } f_i\}|}{|G_K^L|}$$
$$= \frac{|\{\sigma \in G_K^L \mid \sigma^{-1}(\alpha) \text{ is a root of } g_{\varphi(i)}\}|}{|G_K^L|} = \frac{\deg g_{\varphi(i)}}{\deg g}$$

which shows that if we arrange the g_i, such that $\varphi(i) = i$ for all i then we have the desired equation. ∎

DEDEKIND'S Reciprocity Theorem yields the following corollary which gives us some information on how a polynomial which is irreducible over a field K can split over an extension of K.

(23.11) Proposition. *Let $(L : K)$ be a finite Galois extension. If $f \in K[x]$ is irreducible over K and has a prime factorization $f = f_1 \cdots f_m$ over L then $\deg f_1 = \cdots = \deg f_m$.*

Proof. Let β be a primitive element of L over K and let $g \in K[x]$ be its minimal polynomial. Moreover, let α be a root of f, let g_i be an irreducible factor of g over $K(\alpha)$ and let β_i be a root of g_i. Then β_i is a conjugate of β over K which implies that $K(\beta_i) = K(\beta) = L$. Consequently, $\deg g_i = [K(\beta_i, \alpha) : K(\alpha)] = [L : K(\alpha)]$. This shows that all irreducible factors g_i of g over $K(\alpha)$ have the same degree. But then (23.10)(b) yields immediately the claim. ∎

LET us now determine Galois groups in concrete cases. The following result will turn out to be helpful.

(23.12) Proposition. *Let $f \in K[x]$ be a polynomial with distinct roots $\alpha_1, \ldots, \alpha_n$ and let $\Delta := \prod_{i<j} (\alpha_i - \alpha_j)$ so that $\Delta^2 = D(f)$ is the discriminant of f. Moreover, let L be a splitting field of f and let $G := \mathrm{Gal}_K(f) \cong G_K^L$. Then the fixed field of $G \cap \mathrm{Alt}_n$ is $K(\Delta)$.*

Proof. Let $\sigma \in G$. Then $\sigma \in \mathrm{Alt}_n$ if and only if $\sigma(\Delta) = \Delta$, i.e., if $\sigma = \mathrm{id}_{K(\Delta)}$. This shows that $G \cap \mathrm{Alt}_n = G_{K(\Delta)}^L$. Applying F on both sides and using that $F \circ G = \mathrm{id}$, we obtain $F(G \cap \mathrm{Alt}_n) = K(\Delta)$. ∎

(23.13) Theorem. *Let $f \in K[x]$ be an irreducible separable polynomial of degree 3.[†] Let $D(f)$ be the discriminant of f and let $G = \mathrm{Gal}_K(f)$ be the Galois group of f. Then*

$$G = \begin{cases} \mathrm{Alt}_3 \cong \mathbb{Z}_3, & \text{if } D(f) \text{ is a square in } K, \\ \mathrm{Sym}_3, & \text{otherwise.} \end{cases}$$

Proof. The only subgroups of Sym_3 acting transitively on $\{1,2,3\}$ are Alt_3 and Sym_3; hence G must be one of these by (23.6). If $\alpha_1, \alpha_2, \alpha_3 \in L$ are the roots of f, let

$$\Delta := (\alpha_1 - \alpha_2)(\alpha_1 - \alpha_3)(\alpha_2 - \alpha_3) ;$$

then $\sigma \in \mathrm{Alt}_3$ if and only if $\sigma(\Delta) = \Delta$ due to (23.12).

If $G \cong \mathrm{Alt}_3$, then $\sigma(\Delta) = \Delta$ for all $\sigma \in G_K^L$, so $\Delta \in F(G_K^L) = K$. Hence $D(f) = \Delta^2$ is a square in K. Conversely, if $D(f) = k^2$ for some $k \in K$, then $k = \pm\Delta$ so that Δ is an element of K. But then for all $\sigma \in G_K^L$ we have $\sigma(\Delta) = \Delta$, i.e., $\sigma \in \mathrm{Alt}_3$. ∎

(23.14) Examples. (a) The discriminant of $x^3 + px + q$ is $-4p^3 - 27q^2$. For example, if $f(x) = x^3 - x \pm 1$ then $D(f) = -23$; hence $\mathrm{Gal}_K(f)$ if and only if -23 is a square in K. Thus $\mathrm{Gal}_{\mathbb{Q}}(f) = \mathrm{Sym}_3$ whereas $\mathrm{Gal}_{\mathbb{Q}(i\sqrt{23})}(f) = \mathrm{Alt}_3$. Also, if $g(x) = x^3 - 3x + 1$ then $D(g) = 81$; hence $\mathrm{Gal}_K(g) = \mathrm{Alt}_3$ for any base-field K.

(b) The discriminant of a "pure" cubic $g(x) = x^3 \pm q$ is $-27q^2 = -3 \cdot (3q)^2$. Hence

$$\mathrm{Gal}_K(g) = \begin{cases} \mathrm{Sym}_3, & \text{if } -3 \text{ is not a square in } K; \\ \mathrm{Alt}_3, & \text{if } -3 \text{ is a square in } K. \end{cases}$$

For example, $g(x) = x^3 - 10$ has Galois group Sym_3 over $\mathbb{Q}$ or over $\mathbb{Q}(\sqrt{2})$, but has Galois group Alt_3 over $\mathbb{Q}(i\sqrt{3})$.

ANALOGOUSLY, we can determine the Galois groups of quartic polynomials. To prepare the discussion, we introduce subgroups of Sym_4 as follows:

$$U_1 = \{(1234), (13)(24), (1432), \mathrm{id}\} ,$$
$$U_2 = \{(1243), (14)(23), (1342), \mathrm{id}\} ,$$
$$U_3 = \{(1324), (12)(34), (1423), \mathrm{id}\} ,$$
$$V = \{\mathrm{id}, (12)(34), (13)(24), (14)(23)\} ,$$
$$W_1 = \{\mathrm{id}, (12), (34), (12)(34), (13)(24), (14)(23), (1324), (1423)\} ,$$
$$W_2 = \{\mathrm{id}, (13), (24), (12)(34), (13)(24), (14)(23), (1234), (1432)\} ,$$
$$W_3 = \{\mathrm{id}, (14), (23), (12)(34), (13)(24), (14)(23), (1243), (1342)\} .$$

[†] If char $K \neq 3$ then the separability of f is automatic.

(23.15) Theorem. *Let* $f \in K[x]$ *be an irreducible separable polynomial of degree 4 and let* φ *be its resolvent cubic. If* Z *is a splitting field of* φ *over* K *and if* $m := [Z : K]$ *then*

$$\mathrm{Gal}_K(f) = \begin{cases} \mathrm{Sym}_4, & \text{if } m = 6; \\ \mathrm{Alt}_4, & \text{if } m = 3; \\ V, & \text{if } m = 1; \\ W_1, W_2 \text{ or } W_3, & \text{if } m = 2 \text{ and } f \text{ is irreducible over } Z; \\ U_1, U_2 \text{ or } U_3, & \text{if } m = 2 \text{ and } f \text{ is reducible over } Z. \end{cases}$$

Remark. *The ambiguity between* U_1, U_2 *and* U_3 *and the ambiguity between* W_1, W_2 *and* W_3 *is unavoidable because these groups are conjugate under an inner automorphism of* Sym_4 *induced by simply renumbering the roots of* f. *Moreover, the condition whether or not* f *is reducible over* Z *can be effectively checked because the splitting field* Z *of* φ *can be determined by Cardano's formula.*

Proof. Since f is irreducible the Galois group $G := \mathrm{Gal}_K(f)$ is a transitive subgroup of Sym_4. Now it is straightforward (although tedious) to check that the only transitive subgroups of Sym_4 are Sym_4, Alt_4 and the groups V, U_i and W_i defined above.

Let L be a splitting field of f over K so that $K \subseteq Z \subseteq L$; since f is separable, we know that $(L : K)$ is a Galois extension. Now the decisive property of the resolvent cubic is that its splitting field Z is the fixed field of those elements of G which lie in V, i.e., that $Z = F(G \cap V)$. Applying the main theorem of Galois theory, we see that $G_Z^L = G_{F(G \cap V)}^L = G \cap V$; this implies that $G_K^Z \cong G_K^L / G_Z^L = G/(G \cap V)$ so that $m = [Z : K] = |G_K^Z| = |G|/|G \cap V|$. Checking all the transitive subgroups of Sym_4, we find that

$$\frac{|G|}{|G \cap V|} = \begin{cases} 6, & \text{if } G = \mathrm{Sym}_4; \\ 3, & \text{if } G = \mathrm{Alt}_4; \\ 1, & \text{if } G = V; \\ 2, & \text{if } G \in \{U_1, U_2, U_3, W_1, W_2, W_3\}. \end{cases}$$

If G is one of the W_i then $\mathrm{Gal}_Z(f) \cong G_Z^L = G \cap V = V$ is still transitive on the roots of f which by (23.6) implies that f is still irreducible over the larger field Z. If G is one of the U_i then $\mathrm{Gal}_Z(f) \cong G_Z^L = G \cap V$ has only two elements and hence does not act transitively on the roots of f; hence (23.6) implies that in this case f is reducible over Z. ∎

(23.16) Examples. Consider an irreducible separable quartic polynomial of the special form $f(x) = x^4 + a$ over a field K. One easily computes that the resolvent cubic of f is $\varphi(x) = x^3 - 4ax = x(x^2 - 4a)$ and hence has the splitting field $Z = K(\sqrt{a})$.

For example, let us determine the Galois groups of $f(x) = x^4 - 5$ over $K_1 := \mathbb{Q}$, $K_2 := \mathbb{Q}(i\sqrt{5})$ and $K_3 := \mathbb{Q}(i)$; clearly f is irreducible over each of these fields. For $i = 1, 2, 3$ let $Z_i := K_i(\sqrt{-5})$.

(a) Since $[Z_1 : K_1] = [\mathbb{Q}(\sqrt{-5}) : \mathbb{Q}] = 2$ and since f is irreducible over $Z_1 = \mathbb{Q}(\sqrt{-5})$, theorem (23.15) implies that $\mathrm{Gal}_{K_1}(f)$ is one of the groups W_1, W_2, W_3.

(b) Since $[Z_2 : K_2] = [\mathbb{Q}(\sqrt{-5}) : \mathbb{Q}(\sqrt{-5})] = 1$, theorem (23.15) implies that

$\mathrm{Gal}_{K_2}(f) = V$.

(c) Since $[Z_3 : K_3] = [\mathbb{Q}(i, \sqrt{-5}) : \mathbb{Q}(i)] = 2$ and since f is reducible over $Z_3 = \mathbb{Q}(i, \sqrt{-5}) = \mathbb{Q}(i, \sqrt{5})$, theorem (23.15) implies that $\mathrm{Gal}_{K_3}(f)$ is one of the groups U_1, U_2, U_3.

NEXT, we identify the Galois groups of certain polynomials of prime degree. For such polynomials the following result is useful.

(23.17) Lemma. *Suppose that $f \in K[x]$ is irreducible of prime degree p and has no multiple roots. Then $\mathrm{Gal}_K(f)$ contains a p-cycle.*

Proof. $\mathrm{Gal}_K(f)$ acts transitively on the set $\{1, \ldots, p\}$; hence p divides $\mathrm{Gal}_K(f)$ by (I.26.6). Then Sylow's theorems guarantee the existence of an element of order p in $\mathrm{Gal}_K(f)$; this element must be a p-cycle, due to (I.22.4)(b). ∎

(23.18) Proposition. *If $f \in \mathbb{Q}[x]$ is irreducible of prime degree p and has exactly two non-real roots, then $\mathrm{Gal}_\mathbb{Q}(f)$ is the full symmetric group Sym_p.*

Proof. Let $\alpha_1, \ldots, \alpha_p$ be the (distinct) roots of f and let $L = \mathbb{Q}(\alpha_1, \ldots, \alpha_p)$ be the splitting field of f in $\mathbb{C}$. Then the complex conjugation $z \mapsto \bar{z}$ is an element of $G_\mathbb{Q}^L$ which permutes the two non-real roots and fixes the other roots, and hence corresponds to a transposition $\tau \in \mathrm{Gal}_\mathbb{Q}(f)$. On the other hand $\mathrm{Gal}_\mathbb{Q}(f)$ contains a p-cycle by lemma (23.17). Now σ and τ generate all of Sym_p by (I.22.5)(d); whence $\mathrm{Gal}_\mathbb{Q}(f) = \mathrm{Sym}_p$. ∎

(23.19) Examples. (a) Both $f(x) = x^5 - 2x^4 + 2 \in \mathbb{Q}[x]$ and $g(x) = x^5 - 6x + 3$ have the same Galois group $\mathrm{Gal}_\mathbb{Q}(f) = \mathrm{Gal}_\mathbb{Q}(g) = \mathrm{Sym}_5$. In fact, f and g are irreducible by Eisenstein's criterion and have prime degree 5. We show that each of the polynomials f and g has exactly 3 real roots; then the claim follows immediately from (23.18).

We have $f'(x) = 5x^4 - 8x^3 = x^3(5x - 8)$; so f increases between $-\infty$ and 0, decreases between 0 and 8/5 and increases again between 8/5 and $+\infty$. Since $f(0) = 2 > 0$ and $f(8/5) = -8192/3075 < 0$, this, together with the fact that $f(x) \to \pm\infty$ as $x \to \infty$, shows that f has exactly 3 real roots.

Similarly, we have $g'(x) = 5x^4 - 6$ so that g increases between $-\infty$ and $a := -\sqrt[4]{6/5}$, decreases between a and $b := \sqrt[4]{6/5}$ and increases again between b and ∞. Since $g(a) > 0$ and $g(b) < 0$, this, together with the fact that $g(x) \to \pm\infty$ as $x \to \infty$, shows that g has exactly 3 real roots. ∎

(23.20) Example. Let p be an arbitrary prime number. Choose $a \in \mathbb{C} \setminus \mathbb{R}$ such that $2\,\mathrm{Re}\,a = a + \bar{a}$ and $|a|^2 = a\bar{a}$ lie in $\mathbb{Z}$. Let $a_1 := a$ and $a_2 := \bar{a}$ and choose pairwise distinct integers $a_3, \ldots, a_p \in \mathbb{Z}$. Then $f(x) := (x - a_1)(x - a_2)(x - a_3) \cdots (x - a_p)$ has exactly two non-real roots. However, we cannot apply (23.18) directly because f will not be irreducible in general. But we will construct from f a polynomial g which satisfies the hypotheses of (23.18).

Choose $\varepsilon > 0$ so small that the ε-balls around the r_i are disjoint and choose a prime number q so large that $g(x) := f(x) + (1/q)$ has exactly one root in each of these ε-balls. (This is possible because the roots of a polynomial depend continuously on its coefficients.)

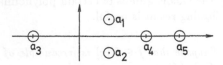

Since the non-real roots of g arise in conjugate pairs, we conclude that g also has exactly two non-real roots. Moreover, $q^2 f(x) + q$ is irreducible by Eisenstein's criterion; but then $g(x) = \left(q^2 f(x) + q\right)/q^2$ is also irreducible.

KNOWING from (22.9)(c) that each extension of finite fields is of Galois type with cyclic Galois group, we can now give a useful corollary of theorem (23.4).

(23.21) Theorem (Dedekind). *Let $f \in \mathbb{Z}[x]$ be a monic polynomial and let p be a prime number such that $\overline{f} \in \mathbb{Z}_p[x]$ (obtained from f by reducing the coefficients modulo p) has no multiple roots. Let $\overline{f} = \overline{f_1} \cdots \overline{f_r}$ be the prime factorization of $\overline{f}$ in $\mathbb{Z}_p[x]$ where $n_i = \deg \overline{f_i}$. Then $\mathrm{Gal}_\mathbb{Q}(f)$ contains a permutation of the form $\sigma = \sigma_1 \cdots \sigma_r$ where σ_i is an n_i-cycle.*

Proof. We note first that $\mathrm{Gal}_{\mathbb{Z}_p}(\overline{f})$ is cyclic; let $\overline{\sigma}$ be a generator. The orbits of the action of $\mathrm{Gal}_{\mathbb{Z}_p}(\overline{f})$ are the r different sets of roots of the polynomials $\overline{f_1}, \ldots, \overline{f_r}$. Then clearly the cycle decomposition of $\overline{\sigma}$ has the form $\overline{\sigma} = \overline{\sigma_1} \cdots \overline{\sigma_r}$ where $\overline{\sigma_i}$ is an n_i-cycle. By (23.4) the group $\mathrm{Gal}_\mathbb{Q}(f)$ possesses an element σ with the same cycle-structure. ∎

SINCE the Galois group $\mathrm{Gal}_{\mathbb{Z}_p}$ is always cyclic, the transition from $\mathrm{Gal}_\mathbb{Q}(f)$ to $\mathrm{Gal}_{\mathbb{Z}_p}(\overline{f})$ inevitably causes a loss of information. However, one can apply Dedekind's theorem with different prime numbers p to obtain more information.

(23.22) Example. Let $f(x) = x^5 - x - 1 \in \mathbb{Z}[x]$. Modulo $p = 2$, we have $\overline{f}(x) = (x^2 + x + 1)(x^3 + x^2 + 1) \in \mathbb{Z}_2[x]$; hence $\mathrm{Gal}_\mathbb{Q}(f)$ contains a product of a 2-cycle and a 3-cycle and therefore both a 2-cycle and a 3-cycle. Modulo $p = 3$, the polynomial $\overline{f}(x) = x^5 - x - 1 \in \mathbb{Z}_3[x]$ is irreducible by (13.21); hence $\mathrm{Gal}_\mathbb{Q}(f)$ contains a 5-cycle.[†] Since any 2-cycle and any 5-cycle together already generate Sym_5 due to (I.22.5)(d), we conclude that $\mathrm{Gal}_\mathbb{Q}(f) = \mathrm{Sym}_5$. Note that this cannot be obtained from (23.18) because f has exactly one real root.

---

† One can also easily verify by trial and error that $x^5 - x - 1$ is irreducible over $\mathbb{Z}_5$. This fact will also follow from a general result, namely the Artin-Schreier theorem to be proved later.

WE now show, using an argument of van der Waerden, that every symmetric group occurs as the Galois group of a polynomial over $\mathbb{Q}$.

(23.23) Proposition. *Let $n \in \mathbb{N}$. Then there is a polynomial $f \in \mathbb{Z}[x]$ of degree n with $\mathrm{Gal}_{\mathbb{Q}}(f) = \mathrm{Sym}_n$.*

Proof. In (13.19)(b) we proved that for any prime number p there are irreducible polynomials in $\mathbb{Z}_p[x]$ of arbitrary degree. Using this fact, we choose polynomials $f_1, f_2, f_3 \in \mathbb{Z}[x]$ of degree n which meet the following specifications:
  (1) f_1 is irreducible over $\mathbb{Z}_2$;
  (2) the irreducible factors of f_2 over $\mathbb{Z}_3$ have degrees 1 and $n-1$;
  (3) the irreducible factors of f_3 over $\mathbb{Z}_5$ have degrees 2 and $n-2$ if n is odd, but degrees $2, p, q$ with p, q odd if n is even.
  Now let $f := -15f_1 + 10f_2 + 6f_3 \in \mathbb{Z}[x]$. This polynomial has degree n and satisfies

$$f \equiv f_1 \bmod 2 \, , \quad f \equiv f_2 \bmod 3 \, , \quad f \equiv f_3 \bmod 5 \, .$$

The first condition implies that $\mathrm{Gal}_{\mathbb{Z}_2}(f) = \mathrm{Gal}_{\mathbb{Z}_2}(f_1)$ acts transitively on $\{1, \ldots, n\}$; hence the more so does $\mathrm{Gal}_{\mathbb{Q}}(f)$. The second condition implies that $\mathrm{Gal}_{\mathbb{Z}_3}(f) = \mathrm{Gal}_{\mathbb{Z}_3}(f_2)$ and hence $\mathrm{Gal}_{\mathbb{Q}}(f)$ contains an $(n-1)$-cycle, and the third condition implies that $\mathrm{Gal}_{\mathbb{Z}_5}(f) = \mathrm{Gal}_{\mathbb{Z}_5}(f_3)$ and hence $\mathrm{Gal}_{\mathbb{Q}}(f)$ contains a 2-cycle. Then (I.22.5)(e) implies that $\mathrm{Gal}_{\mathbb{Q}}(f) = \mathrm{Sym}_n$. ∎

TO conclude this section, we want to introduce the notion of a *general polynomial*. This notion arises naturally when one considers the problem of solving a polynomial equation $f(x) = 0$. The best possible solution of this problem is to find a formula which assigns to a polynomial

$$f(x) \; = \; x^n - a_1 x^{n-1} + a_2 x^{n-2} + \cdots + (-1)^{n-1} a_{n-1} x + (-1)^n a_n$$

with arbitrary coefficients $a_k \in K$† its roots $\alpha_1, \ldots, \alpha_n$ as functions of $a_1, \ldots, a_n$. For example, if $f(x) = x^2 + px + q$ then $x_{1,2} = \frac{1}{2}(-p \pm \sqrt{p^2 - 4q})$ by the quadratic formula. In every concrete example (such as $f(x) = x^2 + 2x + 3$) we could then find the roots by plugging the particular coefficients into the general formula (here $x_{1,2} = -1 \pm \sqrt{1-3} = -1 \pm i\sqrt{2}$). Now since $a_1, \ldots, a_n$ should stand for *arbitrary* elements of K it is natural to treat these coefficients as variables (or indeterminates). Thus one arrives at the following definition.

(23.24) Definition. *The **general polynomial** of degree n over a field K is*

$$f(x) \; = \; x^n - a_1 x^{n-1} + a_2 x^{n-2} + \cdots + (-1)^{n-1} a_{n-1} x + (-1)^n a_n \; \in \; K(a_1, \ldots, a_n)[x]$$

where $K(a_1, \ldots, a_n)$ is the rational function field over K in n variables $a_1, \ldots, a_n$.

---

† The signs attached to the coefficients are just for convenience.

LET us show how the introduction of the notion of a general polynomial, together with some Galois theory, allows us to give a new proof of the main theorem on symmetric polynomials.

(23.25) **Main Theorem on Symmetric Polynomials.** *Let $s_1, \ldots, s_n$ be the elementary symmetric polynomials in n variables $x_1, \ldots, x_n$ over a field K.*

(a) *$s_1, \ldots, s_n$ are algebraically independent; i.e., if $p(s_1, \ldots, s_n) = 0$ for some $p \in K[x_1, \ldots, x_n]$ then $p = 0$.*

(b) *Every symmetric rational function can be expressed as a rational function in $s_1, \ldots, s_n$.*

(c) *Every symmetric polynomial in $x_1, \ldots, x_n$ can be written as a polynomial in $s_1, \ldots, s_n$.*

Proof. (a) Let $f(x) = x^n - x_1 x^{n-1} + x_2 x^{n-2} - + \cdots + (-1)^n x_n$ be the general polynomial over K and let $\alpha_1, \ldots, \alpha_n$ be its roots $\big($in some extension of $K(x_1, \ldots, x_n)\big)$ so that $f(x) = (x - \alpha_1) \cdots (x - \alpha_n)$. If there is a relation $0 = p(s_1, \ldots, s_n) = p(\sum_i x_i, \sum_{i<j} x_i x_j, \ldots, x_1 x_2 \cdots x_n)$ then this relation must remain true if we plug in the values $\alpha_1, \ldots, \alpha_n$ for the variables $x_1, \ldots, x_n$; this gives

$$0 = p\Big(\sum_i \alpha_i, \sum_{i<j} \alpha_i \alpha_j, \ldots, \alpha_1 \alpha_2 \cdots \alpha_n\Big) = p(x_1, x_2, \ldots, x_n) .$$

But this means that p is the zero polynomial.

(b) Let S be the subfield of $L := K(x_1, \ldots, x_n)$ consisting of all symmetric rational functions. Then clearly $K(s_1, \ldots, s_n) \subseteq S$; we want to show that equality holds. Since Sym_n acts as a group of automorphisms on $L = K(x_1, \ldots, x_n)$ with fixed field S we have

(1) $$n! \leq |\mathrm{Sym}_n| \leq |G_S^L| .$$

(From (21.25)(e) we know that $G_S^L \cong \mathrm{Sym}_n$ but that is not even needed here.) On the other hand, $L = K(x_1, \ldots, x_n)$ is the splitting field of

$$p(x) = (x - x_1) \cdots (x - x_n) = x^n - s_1 x^{n-1} + s_2 x^{n-2} + \cdots + (-1)^n s_n \in K(s_1, \ldots, s_n)[x]$$

so that $[L : K(s_1, \ldots, s_n)] \leq n!$ by (13.5)(d); hence the more so $[L : S] \leq n! < \infty$. But then $|G_S^L| \leq [L : S]$ by (21.17). Using (1), this gives

$$n! \, [S : K(s_1, \ldots, s_n)] \leq |G_S^L| \cdot [S : K(s_1, \ldots, s_n)]$$
$$\leq [L : S] \cdot [S : K(s_1, \ldots, s_n)]$$
$$= [L : K(s_1, \ldots, s_n)] \leq n!$$

which implies that $[S : K(s_1, \ldots, s_n)] \leq 1$ and hence $S = K(s_1, \ldots, s_n)$; but this is the claim.

(c) Let p be a symmetric polynomial in $x_1, \ldots, x_n$. Then $p \in K(s_1, \ldots, s_n)$ which is the quotient field of $K[s_1, \ldots, s_n]$. Now $K[x_1, \ldots, x_n]$ (and hence p) is integral over $K[s_1, \ldots, s_n]$. But $K[s_1, \ldots, s_n]$ is isomorphic to $K[x_1, \ldots, x_n]$ by part (a) and hence is a unique factorization domain, thus integrally closed in its quotient field. This shows that $p \in K[s_1, \ldots, s_n]$ which is what we wanted to show. ∎

AS our last result in this section we exhibit the Galois group of a general polynomial.

(23.26) Theorem. *The Galois group of the general polynomial of degree n over a field K is the full symmetric group* Sym_n.

Proof. Let $L = K(a_1, \ldots, a_n)$ be the rational function field over K and let $f(x) = x^n + \sum_k (-1)^k a_k x^{n-k} \in L[x]$ be the general polynomial of degree n over K. Let $x_1, \ldots, x_n$ be the roots of f (in some extension of L). Then $a_i = s_i(x_1, \ldots, x_n)$; hence $L = K(s_1, \ldots, s_n)$, and a splitting field of f is given by $L(x_1, \ldots, x_n) = K(s_1, \ldots, s_n, x_1, \ldots, x_n) = K(x_1, \ldots, x_n)$. Consequently,

$$\mathrm{Gal}_L(f) \cong G_{K(s_1,\ldots,s_n)}^{K(x_1,\ldots,x_n)} \cong G_S^{K(x_1,\ldots,x_n)} \cong \mathrm{Sym}_n$$

where we used (21.25)(e) in the last equation. ∎

AN alternative formulation and proof of (23.26) is as follows.

(23.27) Theorem. *If $a_0, \ldots, a_{n-1}$ are algebraically independent over K then the Galois group of $f(x) = x^n + a_{n-1}x^{n-1} + \cdots + a_1 x + a_0$ over K is the full symmetric group* Sym_n.

Proof. Let $\alpha_1, \ldots, \alpha_n$ be the roots of f and suppose that there is an algebraic relation $g(\alpha_1, \ldots, \alpha_n)$ between these roots. Then $G := \prod_{\sigma \in \mathrm{Sym}_n} \sigma \star g$ is symmetric, hence can be written as a polynomial in the elementary symmetric polynomials $s_1, \ldots, s_n$. Consequently, $0 = G(\alpha_1, \ldots, \alpha_n) = P\big(s_1(\alpha_i), \ldots, s_n(\alpha_i)\big) = P(a_0, a_1, \ldots, a_{n-1})$. Thus $P = 0$, consequently $G = 0$, hence $g = 0$. This shows that there is no relation between the roots which, by the very definition (23.1) of the Galois group of a polynomial, implies that $\mathrm{Gal}_K(f) = \mathrm{Sym}_n$. ∎

Exercises

Problem 1. Find the Galois groups of the following polynomials over $\mathbb{Q}$.
(a) $x^8 - 1$
(b) $(x^2 - 2)(x^3 - 2)(x^3 - 3)$
(c) $(x^2 - 5)(x^2 - 20)$
(d) $(x^2 - 2)(x^2 - 5)(x^3 - x - 1)$
(e) $(x^3 - 2)(x^2 - 3)$
(f) $(x^3 - 2)(x^2 - 5)$

Problem 2. Let $p_1, \ldots, p_n$ be distinct prime numbers. Find the Galois group of $(x^2 - p_1)(x^2 - p_2) \cdots (x^2 - p_n)$ over $\mathbb{Q}$.

Problem 3. Show that if K is a field with char $K \neq 0$ and if $f \in K[x]$ is a non-constant polynomial then $f/\gcd(f, f')$ has exactly the same roots as f, but all with multiplicity 1.

Problem 4. Let L be a splitting field of $f \in K[x]$ and let $f(x) = (x - \alpha_1)^{n_1} \cdots (x - \alpha_k)^{n_k}$ where $\alpha_1, \ldots, \alpha_k$ are the pairwise different roots of f in L. Let $b_0, \ldots, b_k \in L$ be the coefficients of $g(x) := (x - \alpha_1) \cdots (x - \alpha_k)$ and let $Z := K(b_0, \ldots, b_k)$. Show that $\mathrm{Gal}_K(f) = \mathrm{Gal}_Z(g)$. (This shows that if we can determine the Galois group of a polynomial without multiple roots then we can also determine the Galois group of an arbitrary polynomial.)

Problem 5. Find the Galois groups of $x^4 + 2$ over $\mathbb{Q}$ and over $\mathbb{Q}(i)$.

Problem 6. Let L be the splitting field of $p(x) = x^4 - 7 \in \mathbb{Q}[x]$. Show that L contains $\sqrt{7}$ and i; then determine the Galois groups $G_{\mathbb{Q}}^L$, $G_{\mathbb{Q}(\sqrt{7})}^L$, $G_{\mathbb{Q}(i\sqrt{7})}^L$ and $G_{\mathbb{Q}(i)}^L$.

Problem 7. Let $\varepsilon = e^{2\pi i/3}$.
(a) Check that $f(x) = (x^2 - 2)(x^3 - 5)$ has the roots $\pm\sqrt{2}$ and $\varepsilon^k \sqrt[3]{5}$ with $0 \leq k \leq 2$.
(b) Check that $g(x) = x^6 - 6x^4 - 10x^3 + 12x^2 - 60x + 17$ has the roots $\pm\sqrt{2} + \varepsilon^k \sqrt[3]{5}$ with $0 \leq k \leq 2$.
(c) Show that f and g have the same splitting field over $\mathbb{Q}$, namely $L = \mathbb{Q}(\sqrt{2}, \sqrt[3]{5}, \varepsilon)$.
(d) Show that $\mathrm{Gal}_{\mathbb{Q}}(f) \leq \mathrm{Sym}_5$ and $\mathrm{Gal}_{\mathbb{Q}}(g) \leq \mathrm{Sym}_6$ are isomorphic.

Problem 8. Let $L := \mathbb{Q}(i\sqrt[4]{2})$ and $M := \mathbb{Q}(i\sqrt{2}, \alpha)$ where $\alpha^2 = 1 + i$.
(a) Show that L is the splitting field of $f(x) = x^4 - 2$ whereas M is the splitting field of $g(x) = x^4 - 2x^2 + 2$.
(b) Show that the fields L and M are not isomorphic.
(c) Show that the Galois groups $G_{\mathbb{Q}}^L$ and $G_{\mathbb{Q}}^M$ are isomorphic.
(d) Find all intermediate fields of the extensions $(L : \mathbb{Q})$ and $(M : \mathbb{Q})$.

Problem 9. Let α be a root of $f(x) = x^3 + x + 1 \in \mathbb{Z}_2[x]$. Show that $(\mathbb{Z}_2(\alpha) : \mathbb{Z}_2)$ is a Galois extension with Galois group $G_{\mathbb{Z}_2}^{\mathbb{Z}_2(\alpha)} \cong \mathbb{Z}_3$.

Hint. Check that α^2 and $\alpha^2 + \alpha$ are also roots of f.

Problem 10. Let $f(x) = x^p - x - 1$ where p is a prime.

(a) Show that f is irreducible over $\mathbb{Q}$.

(b) Show that if α is a root of f over $\mathbb{Z}_p$ then f splits over $\mathbb{Z}_p(\alpha)$. Show that the Galois group of f over $\mathbb{Z}_p$ is $(\mathbb{Z}_p, +)$.

Problem 11. Suppose that $f \in K[x]$ is irreducible and has n different roots $\alpha_1, \ldots, \alpha_n$. Show that if $\mathrm{Gal}_K(f)$ is abelian, then $|\mathrm{Gal}_K(f)| = n$.

Hint. $\mathrm{Gal}_K(f)$ acts transitively on $\{\alpha_1, \ldots, \alpha_n\}$.

Problem 12. Let $f \in K[x]$ be an irreducible separable polynomial. Show that $|\mathrm{Gal}_K(f)|$ is divisible by $\deg f$.

Problem 13. Let $K \subseteq L$ be a field extension and let $f \in K[x]$ be a polynomial with distinct roots $\alpha_1, \ldots, \alpha_n$ (in some splitting field). Show that $\mathrm{Gal}_L(f) = \mathrm{Gal}_K(f)$ if and only if $K(\alpha_1, \ldots, \alpha_n) \cap L = K$.

Problem 14. (a) Let $f \in \mathbb{Q}[x]$ be an irreducible polynomial of degree 4 and let $\alpha \in \mathbb{C}$ be a root of f. Show that if $\mathrm{Gal}_{\mathbb{Q}}(f)$ has more than 8 elements then α is not constructible.

(b) Show that if $\alpha \in \mathbb{C}$ satisfies the equation $\alpha^4 + \alpha + 1 = 0$ then α is not constructible. (Compare with problem 72 in section 12.)

Problem 15. Let $(L : K)$ be a finite Galois extension and let $f \in K[x]$ be an irreducible polynomial. Let f_1 and f_2 be two irreducible monic divisors of f in $L[x]$. Show that there is an element $\sigma \in G_K^L$ such that $f_2 = \sigma \star f_1$. (Consequently, $\deg f_2 = \deg f_1$.)

Problem 16. Show that an algebraic extension $(L : K)$ is normal if and only if all factors of an irreducible polynomial $f \in K[x]$ in the prime decomposition over L have the same degree.

Problem 17. Suppose that a finite Galois extension $(L : K)$ and elements $\alpha, \beta \in L$ are given. Let $\alpha = \alpha_1, \ldots, \alpha_m$ be the conjugates of α and $\beta = \beta_1, \ldots, \beta_n$ the conjugates of β. Assume that $[K(\alpha, \beta) : K] = mn$.

(a) Show that for any $1 \le i \le m$ and $1 \le j \le n$ there is an element $\sigma \in G_K^L$ with $\sigma(\alpha) = \alpha_i$ and $\sigma(\beta) = \beta_j$.

(b) Show that all of the elements $\alpha_i + \beta_j$ are conjugates of $\alpha + \beta$ whereas all elements $\alpha_i\beta_j$ are conjugates of $\alpha\beta$.

(c) Assume that $\alpha_i - \alpha_j \ne \beta_r - \beta_s$ whenever $i \ne j$ and $r \ne s$. Show that in this case $[K(\alpha + \beta) : K] = mn$ and hence $K(\alpha, \beta) = K(\alpha + \beta)$.

(d) Assume that $\alpha_i/\alpha_j \ne \beta_r/\beta_s$ whenever $i \ne j$ and $r \ne s$. Show that in this case $[K(\alpha\beta) : K] = mn$ and hence $K(\alpha, \beta) = K(\alpha\beta)$.

Problem 18. Let K be a field and let $f \in K[x]$ be a separable irreducible polynomial of prime degree p. Moreover, let $\alpha_1 \neq \alpha_2$ be two roots of f.

(a) Show that if N is a normal closure of $\big(K(\alpha_1 - \alpha_2) : K\big)$ and if $p \neq \operatorname{char} K$ then f splits over N.

(b) Show that if N is a normal closure of $\big(K(\alpha_1/\alpha_2) : K\big)$ and if f does not have the form $f(x) = c(x^p - k)$ with $c, k \in K$ then f splits over N.

Problem 19. Let $\alpha, \beta \neq 0$ be separable elements over a field K and let p be a prime number. Suppose that $[K(\alpha) : K] = p$ and $[K(\beta) : K] = n < p$.

(a) Show that if $p \neq \operatorname{char} K$ then $[K(\alpha + \beta) : K] = pn$ and hence $K(\alpha, \beta) = K(\alpha + \beta)$.

(b) Show that if $\alpha^p \notin K$ then $[K(\alpha\beta) : K] = pn$ and hence $K(\alpha, \beta) = K(\alpha\beta)$.

(c) Show that if $\beta^n \in K$ then $[K(\alpha\beta) : K] = pn$ and hence $K(\alpha, \beta) = K(\alpha\beta)$.

Problem 20. Let p be a prime number and let K a field with $\operatorname{char} K \neq p$. Show that if α, β are roots of irreducible polynomials $x^p - a$ and $x^p - b$ over K and if $[K(\alpha, \beta) : K] = p^2$ then $[K(\alpha + \beta) : K] = p^2$ and hence $K(\alpha, \beta) = K(\alpha + \beta)$.

Problem 21. Find the Galois groups of the following cubic polynomials over $\mathbb{Q}$.

(a) $x^3 - 3x + 1$

(b) $x^3 - 2$

(c) $x^3 - 4x + 1$

Problem 22. Let $L \subseteq \mathbb{C}$ be the splitting field of $f(x) = x^3 + x + 1 \in \mathbb{Q}[x]$.

(a) Show that $\operatorname{Gal}_{\mathbb{Q}}(f) = \operatorname{Sym}_3$.

(b) Show that there are exactly three intermediate fields of $(L : \mathbb{Q})$ which have degree 3 over $\mathbb{Q}$, namely the fields $\mathbb{Q}(\alpha)$ where α is a root of f.

(c) Show that there is a unique intermediate field of $(L, \mathbb{Q})$ which has degree 2 over $\mathbb{Q}$, namely $\mathbb{Q}(i\sqrt{31})$.

Problem 23. Show that $f(x) = 1 + x + (x^2/2!) + (x^3/3!)$ is irreducible over $\mathbb{Q}$ and has exactly one real root. Moreover, show that the Galois group of f over $\mathbb{Q}$ is Sym_3.

Problem 24. Let $f \in K[x]$ be a separable cubic polynomial with Galois group Sym_3 and let α, β, γ be the roots of f in some splitting field L. Show that the intermediate fields of the extension $(L : K)$ are $K(\Delta)$, $K(\alpha)$, $K(\beta)$, $K(\gamma)$, K and L.

Problem 25. (a) Let $f \in K[x]$ be a cubic polynomial where $\operatorname{char} K \neq 2$. Show that if $D(f)$ is a square in K then f is either irreducible or else splits over K.

(b) Let K be an arbitrary field and let $f(x) = x^3 - 3x + 1$. Show that f is either irreducible or else splits over K.

Problem 26. Find the Galois groups of the following quartic polynomials, each considered as an element of $\mathbb{Q}[x]$.

(a) $x^4 - 4$

(b) $x^4 + 4x^2 + 2$

(c) $x^4 - 4x^2 + 5$

(d) $x^4 - 2$

(e) $(x^2 + 1)(x^2 - 3)$

(f) $x^4 - 5x^2 + 6$

(g) $x^4 - 10x^2 + 4$

(h) $x^4 + 2x^2 + x + 3$

(i) $x^4 + 3x^3 + 3x - 2$

(j) $x^4 + 2x^2 + x + 3$

In each case find the splitting field $L \subseteq \mathbb{C}$ and determine all intermediate fields of the extension $(L : \mathbb{Q})$.

Problem 27. Let $K \subseteq \mathbb{R}$. Show that an irreducible quartic polynomial $f \in K[x]$ has exactly two real roots then $\mathrm{Gal}_K(f)$ equals Sym_4 or Alt_4.

Problem 28. Which groups occur as Galois groups of polynomials of the form $f(x) = x^4 + px^2 + q$ over a field of characteristic $\neq 2, 3$?

Problem 29. Show that $f(x) = x^5 - 20x^3 + 9x + 1$ has 5 real roots and that $\mathrm{Gal}_\mathbb{Q}(f) = \mathrm{Sym}_5$.

Problem 30. Show that if p and q are odd integers with $5|q$ then the Galois group of $f(x) = x^5 + 5px + 5q$ over $\mathbb{Q}$ is Sym_5.

Problem 31. Show that every polynomial of the form $f(x) = x^5 - kx + 1 \in \mathbb{Z}[x]$ with $k \geq 3$ has exactly three real roots and has the Galois group Sym_5.

Problem 32. Let p and q be different primes with $q \geq 2$ or $p \geq 13$. Show that the Galois group of $f(x) = x^5 - pqx + p$ is Sym_5.

Problem 33. Let p be a prime and let $f \in \mathbb{Z}_p[x] = x^4 - 10x^2 + 5$. Show that f is not irreducible then f either decomposes into four linear factors or else into two irreducible quadratic factors.

Problem 34. Let $f(x) = x^4 + x + 1$. Show that the Galois group of f over $\mathbb{Q}$ contains an element of order 3.

Problem 35. (a) Suppose that $f \in \mathbb{Z}[x]$ is an irreducible polynomial of degree n such that $\text{Gal}_{\mathbb{Q}}(f)$ does not contain any n-cycles. Show that f is reducible modulo p for every prime p.

(b) Find the Galois group of $f(x) = x^4 - 10x^2 + 1$ over $\mathbb{Q}$ and show that f factors into two quadratic factors over $\mathbb{Z}_p$, for any prime number p.

Problem 36. Find a polynomial $f \in \mathbb{Q}[x]$ of degree 6 such that $\text{Gal}_{\mathbb{Q}}(f) = \text{Sym}_6$.

Problem 37. Find polynomials in $\mathbb{Q}[x]$ with Galois groups $\mathbb{Z}_4$, $\mathbb{Z}_3$, $\mathbb{Z}_{12}$, Alt_4.

Problem 38. Let G be a finite group. Show that there is an irreducible polynomial $f \in \mathbb{Q}[x]$ such that G is isomorphic to a subgroup of $\text{Gal}_{\mathbb{Q}}(f)$.

24. Roots of unity and cyclotomic polynomials

IN this section we will study the roots of polynomials of the form $x^n - 1$. These solutions are called *roots of unity* and will be seen in subsequent chapters to play an important role in the general problem of solving polynomial equations.

(24.1) Definition. *Let K be a field.*

(a) *Every solution of the equation $x^n = 1$ (in some extension of K) is called an n-th* **root of unity**.

(b) *The set of all n-th roots of unity in K is denoted by $U(n, K)$ so that*

$$U(n, K) := \{ \varepsilon \in K \mid \varepsilon^n = 1 \} .$$

Moreover, we denote by $U(K) := \bigcup_{n \in \mathbb{N}} U(n, K)$ the set of all roots of unity in K.

(c) *An n-th root of unity ε is called* **primitive** *if if it is not an m-th root of unity for any $m < n$, i.e., if $\varepsilon^n = 1$, but $\varepsilon^m \neq 1$ for $1 \leq m \leq n - 1$.*

(24.2) Examples. (a) If n is odd then $U(n, \mathbb{Q}) = U(n, \mathbb{R}) = \{1\}$; if n is even then $U(n, \mathbb{Q}) = U(n, \mathbb{R}) = \{\pm 1\}$.

(b) For every $n \in \mathbb{N}$ we have $U(n, \mathbb{C}) = \{ e^{2k\pi i/n} \mid 1 \leq k \leq n \}$. Exactly $\varphi(n)$ of these n-th roots are primitive, namely the numbers $e^{2k\pi i/n}$ where k is coprime with n.

(c) If $K = \mathbb{Z}_{19}$ then $U(3, K) = \{1, 7, 11\}$; here 7 and 11 are primitive cubic roots.

(d) If $d \mid n$ (say $n = de$) then $U(d, K) \subseteq U(n, K)$. Indeed, $\varepsilon^d = 1$ implies $\varepsilon^n = (\varepsilon^d)^e = 1^e = 1$.

(e) If $\operatorname{char} K = p$ and if $n = p^m d$ with $p \nmid d$ then $U(n, K) = U(d, K)$. Here the inclusion $\supseteq$ is clear by part (d); let us prove the converse inclusion. If $\varepsilon^n = 1$ then $0 = \varepsilon^n - 1 = \varepsilon^{p^m d} - 1^{p^m} = (\varepsilon^d - 1)^{p^m}$ which implies that $\varepsilon^d - 1 = 0$, i.e., $\varepsilon^d = 1$. Thus if n is divisible by p, then K does not possess primitive n-th roots of unity.

(f) If $(L : K)$ is a field extension then $U(n, K) = U(n, L) \cap K$.

OBVIOUSLY, each $U(n, K)$ is a finite subgroup of $(K^\times, \cdot)$ (its order being at most n) and hence cyclic by (4.14). It is also clear that $\bigcup_{n \in \mathbb{N}} U(n, K)$ is exactly the set of those elements of $(K^\times, \cdot)$ which have finite order. A primitive n-th root of unity is an element of order exactly n. As example (24.2)(e) shows, primitive roots of unity need not exist for any order. The following result clarifies the situation.

(24.3) Proposition. *Let K be a field and let $n \in \mathbb{N}$.*

(a) *The following statements are equivalent:*

(1) *K possesses a primitive n-th root of unity;*

(2) *K possesses $\varphi(n)$ primitive n-th roots of unity where φ is Euler's function;*

(3) *K possesses n different n-th roots of unity, i.e., $|U(n, K)| = n$;*

(4) *the generators of $U(n)$ are exactly the primitive n-th roots of unity.*

(b) *The following statements are equivalent:*

(1) *There is a primitive n-th root in some extension of K;*

(2) *the polynomial $x^n - 1$ has no multiple roots;*

(3) *$\operatorname{char} K$ does not divide n.*

Proof. (a) Let us show first that (1) implies (2), (3) and (4). If ε is a primitive n-th root of unity then $U(n,K)$ consists of the n pairwise distinct elements ε^k where $1 \leq k \leq n$, and ε^k is primitive if and only if k is coprime with n, i.e., if ε^k generates $U(n,K)$. Conversely, if $|U(n,K)| = n$ then $U(n,K)$ is cyclic of order n; hence (3) implies (1). The implications (4)$\Longrightarrow$(1) and (2)$\Longrightarrow$(1) are trivial.

(b) Due to part (a), condition (1) holds if and only if $f(x) = x^n - 1$ has n different roots, which is the case if and only if f and f' are coprime. Since $f'(x) = nx^{n-1}$ this happens if and only if $f' \neq 0$, i.e., if char $K{\not|}n$. ∎

(24.4) Proposition. *Let K be a field.*

(a) *For all $n \in \mathbb{N}$ we have $\sum_{\varepsilon \in U(n,K)} \varepsilon = 0$.*

(b) *If $m, n \in \mathbb{N}$ are coprime then $U(mn, K) = U(m, K)U(n, K)$. More precisely, the multiplication map*

$$\mu : \begin{array}{ccc} U(m,K) \times U(n,K) & \to & U(mn,K) \\ (a,b) & \mapsto & ab \end{array}$$

is a group isomorphism.

Proof. (a) Due to (24.2)(e) we may assume without loss of generality that char $K{\not|}n$; in this case $U(n,K) = \{\omega^k \mid 0 \leq k \leq n-1\}$ where ω is a primitive n-th root of unity. Then $s := \sum_{\varepsilon \in U(n,K)} \varepsilon = \sum_{k=0}^{n-1} \omega^k$ so that $\omega s = \sum_{k=0}^{n-1} \omega^{k+1} = \omega_{k=0}^{n-1}\omega^k = s$, hence $0 = \omega s - s = (\omega - 1)s$. Since $\omega \neq 1$ this implies $s = 0$.

(b) Clearly, μ is a well-defined group homomorphism. To prove the isomorphism property, we can assume that char K divides neither m nor n. For char $K = 0$ this is trivial, and if char $K = p > 0$ we can write $m = p^r d$ and $n = p^s e$ with $p{\not|}d$ and $p{\not|}e$ to obtain $U(m, K) = U(d, K)$, $U(n, K) = U(e, K)$ and $U(mn, K) = U(p^{r+s}de, K) = U(de, K)$ due to (24.2)(e). Under this assumption, we can find primitive m-th and n-th roots of unity ε^m and ε^n in some extension L of K. By (24.2)(f) we can assume that $K = L$ so that $\varepsilon_m, \varepsilon_n \in K$. We claim that then $\varepsilon_m \varepsilon_n$ is a primitive (mn)-th root of unity (which implies that μ is surjective). Clearly, $(\varepsilon_m \varepsilon_n)^{mn} = 1$, and if $(\varepsilon_m \varepsilon_n)^k = 1$ then $1 = (\varepsilon_m \varepsilon_n)^{km} = \varepsilon_n^{km}$ so that km and hence k is divisible by n, say $k = \lambda n$. But then $1 = (\varepsilon_m \varepsilon_n)^{\lambda n} = \varepsilon_m^{\lambda n}$ so that m divides λn and hence λ, say $\lambda = \lambda' m$. Consequently, $k = \lambda n = \lambda' mn$ is a multiple of mn.

Since $|U(mn, K)| = mn = |U(m, K) \times U(n, K)|$ due to (24.3), the surjectivity of μ implies already that μ is an isomorphism. ∎

LET us now try to find the minimal polynomial of a root of unity over a field K. Of course, if ε is an n-th root of unity then ε is a root of $x^n - 1$, but this polynomial is always reducible (unless $n = 1$) because it is divisible by $x - 1$; namely, $x^n - 1 = (x - 1)(x^{n-1} + \cdots + x + 1)$. However, to find the explicit factorization of $x^n - 1$ into irreducible factors is not a trivial task. Quite often the observation is helpful that for n odd the polynomial $x^n + 1$ is divisible by $x + 1$; in fact, $x^n + 1 = (x + 1)(x^{n-1} - x^{n-2} + - \cdots + 1)$.

(24.5) Example. We have $x^{12}-1 = (x^6-1)(x^6+1) = (x^3-1)(x^3+1)((x^2)^3+1) = (x-1)(x^2+x+1)(x+1)(x^2-x+1)(x^2+1)(x^4-x^2+1)$; thus over any field K we have the factorization

$$(\star) \qquad x^{12}-1 \;=\; (x-1)(x+1)(x^2+x+1)(x^2+1)(x^4-x^2+1) \;.$$

Whether or not this decomposition can be refined further depends strongly on the base-field K.

(a) If $K = \mathbb{Q}$ then $(\star)$ is the prime factorization of $x^{12}-1$.

(b) The same is true for $K = \mathbb{R}$.

(c) If $K = \mathbb{Q}(i)$ then $x^2+1 = (x+i)(x-i)$; all other factors in $(\star)$ are irreducible.

(d) If $K = \mathbb{C}$ then of course all factors can be decomposed into linear factors; for example, $x^2 \pm x + 1 = (x \pm \frac{1-i\sqrt{3}}{2}) \cdot (x \pm \frac{1+i\sqrt{3}}{2})$.

(e) If $K = \mathbb{Z}_2$ then $x^2+1 = (x+1)^2$ and $x^4-x^2+1 = (x^2+x+1)^2$; the other factors in $(\star)$ are irreducible.

(f) If $K = \mathbb{Z}_{11}$ then $x^4-x^2+1 = (x^2+5x+1)(x^2-5x+1)$; all the other factors in $(\star)$ are irreducible.

(24.6) Definition. *Suppose the field K possesses a primitive n-th roots of unity. Then the polynomial $\Phi_n(x) := \prod_\varepsilon (x-\varepsilon)$ where ε ranges over all primitive n-th roots of unity is called the n-th **cyclotomic polynomial** † over K. Clearly, its degree is $\varphi(n)$.*

(24.7) Proposition. *Suppose the field K admits primitive n-th roots of unity for $1 \le n \le N$ (in some extension of K).*

(a) *The decomposition $x^N - 1 = \prod_{n|N} \Phi_n$ holds.*

(b) *We have $\Phi_1(0) = -1$ and $\Phi_n(0) = 1$ for $2 \le n \le N$.*

(c) *The coefficients of Φ_N lie in the prime field of K.*

Proof. (a) We have $x^N - 1 = \prod_\varepsilon (x-\varepsilon)$ where ε ranges over all N-th roots of unity. Since each such root is primitive for exactly one divisor of N, the claim follows.

(b) We have $\Phi_1(x) = x-1$ and $\Phi_2(x) = x+1$; hence $\Phi_1(0) = -1$ and $\Phi_2(0) = 1$. Let $N > 2$. Using part (a) and the fact that $\Phi_n(0) = 1$ for $1 < n < N$ by induction hypothesis, we see that

$$-1 \;=\; x^N - 1|_{x=0} \;=\; \underbrace{\Phi_1(0)}_{=-1} \cdot \underbrace{\prod_{\substack{n|N, \\ 1<n<N}} \Phi_n(0)}_{=1} \cdot \Phi_N(0) \;=\; -\Phi_N(0)$$

which implies that $\Phi_N(0) = -1$.

(c) Again we proceed by induction on N, the case $N = 1$ being trivial because $\Phi_1(x) = x-1$. Using part (b), we can write $x^N - 1 = \Phi_N(x)g(x)$ where g has coefficients in the prime field by induction hypothesis. ∎

---

† The Greek word "cyclotomic" roughly means "circle-dividing" and was chosen due to the fact that the n-th roots of unity in $\mathbb{C}$, i.e., the numbers $e^{2k\pi i/n}$, form the vertices of a regular n-gon and hence divide the unit circle into n equal parts.

FORMULA (24.7)(a) allows us to determine the polynomials Φ_n inductively; namely, Φ_n can be found by dividing $x^n - 1$ by $\prod_d \Phi_d$ where d ranges through the proper divisors of n. However, this has the big disadvantage that Φ_d has to be known for all proper divisors of n before Φ_n can be calculated. In the sequel, we will present a more effective way of determining cyclotomic polynomials.

(24.8) Proposition. *Let K be a field of characteristic 0.*

(a) *If $m, n \in \mathbb{N}$ then $\Phi_{mn}(x)$ divides $\Phi_n(x^m)$. If every prime divisor of m is also a prime divisor of n then $\Phi_{mn}(x) = \Phi_n(x^m)$. In particular, if p is a prime divisor of n then $\Phi_{pn}(x) = \Phi_n(x^p)$.*

(b) *If $p_1, \ldots, p_s$ are different prime numbers and if $r_1, \ldots, r_s \in \mathbb{N}$ then*

$$\Phi_{p_1^{r_1} \cdots p_s^{r_s}}(x) = \Phi_{p_1 \cdots p_s}\left(x^{p_1^{r_1 - 1} \cdots p_s^{r_s - 1}}\right).$$

(c) *If $n \geq 3$ is odd then $\Phi_{2n}(x) = \Phi_n(-x)$.*

(d) *If $n \in \mathbb{N}$ and if p is a prime number with $p \nmid n$ then $\Phi_n(x)\Phi_{np}(x) = \Phi_n(x^p)$.*

Proof. (a) If α is a root of Φ_{mn}, i.e., a primitive (mn)-th root of unity, then α^m is a primitive n-th root of unity, i.e., a root of Φ_n. Hence every root of Φ_{mn} is a root of $\Phi_n(x^m)$; since Φ_{mn} has only simple roots this implies that $\Phi_{mn}(x)$ divides $\Phi_n(x^m)$. Clearly, both Φ_{mn} and $\Phi_n(x^m)$ have the leading coefficient 1. If in addition every prime divisor of m is also a prime divisor of n then $\varphi(mn) = m \cdot \varphi(n)$ which means exactly that $\deg \Phi_{mn} = \deg \Phi_n(x^m)$. Thus $\Phi_{mn}(x) = \Phi_n(x^m)$ in this case.

(b) Apply part (a) with $n := p_1 \cdots p_s$ and $m := p_1^{r_1 - 1} \cdots p_s^{r_s - 1}$.

(c) If α is a root of Φ_{2n}, i.e., a primitive $(2n)$-th root of unity, then $-\alpha$ is a primitive n-th root of unity, i.e., a root of Φ_n. This shows that every root of Φ_{2n} is a root of $\Phi_n(-x)$; since Φ_{2n} has only simple roots this implies that Φ_{2n} divides $\Phi_n(-x)$. On the other hand, we claim that Φ_{2n} and $\Phi_n(-x)$ have the same degree and the same leading coefficient; once this is established we know that the two polynomials must coincide. First of all, we have $\deg \Phi_{2n} = \varphi(2n) = \varphi(2)\varphi(n) = \varphi(n) = \deg \Phi_n = \deg \Phi_n(-x)$. Second, since the leading coefficient of Φ_n is 1, the leading coefficient of $\Phi_n(-x)$ is $(-1)^{\varphi(n)} = 1$ which equals the leading coefficient of Φ_{2n}.

(d) Part (a) shows that $\Phi_{np}(x)$ divides $\Phi_n(x^p)$. But $\Phi_n(x)$ also divides $\Phi_n(x^p)$: Namely, if α is a root of Φ_n, i.e., a primitive n-th root of unity, then so is α^p, which means that α is a root of $\Phi_n(x^p)$; since Φ_n has only simple roots this implies that Φ_n divides $\Phi_n(x^p)$. Now Φ_{np} and Φ_n have no root in common, hence are coprime. Since both of these polynomials divide $\Phi_n(x^p)$, we conclude that $\Phi_n \cdot \Phi_{np}$ divides $\Phi_n(x^p)$. On the other hand, $\Phi_n \cdot \Phi_{np}$ and $\Phi_n(x^p)$ have the same leading coefficient (namely 1) and also the same degree because

$$\deg(\Phi_n \Phi_{np}) = \deg \Phi_n + \deg \Phi_{np} = \varphi(n) + \varphi(np) = \varphi(n) + \varphi(n)\varphi(p)$$
$$= \left(1 + \varphi(p)\right)\varphi(n) = p \cdot \varphi(n) = p \cdot \deg \Phi_n = \deg \Phi_n(x^p).$$

These facts together clearly imply the claim. ∎

THIS result allows one in fact to compute the cyclotomic polynomials Φ_n in a rather effective way. First, part (b) reduces the problem to the case that n has no multiple prime divisors; by part (c) we may even assume that only odd prime divisors occur. Then part (d) allows one to calculate Φ_n for $n = p_1 p_2 \cdots p_s$ in the following inductive way:

$$\Phi_{p_1 p_2}(x) = \frac{\Phi_{p_1}(x^{p_2})}{\Phi_{p_1}(x)}$$

$$\Phi_{p_1 p_2 p_3}(x) = \frac{\Phi_{p_1 p_2}(x^{p_3})}{\Phi_{p_1 p_2}(x)}$$

$$\vdots$$

$$\Phi_{p_1 \cdots p_s}(x) = \frac{\Phi_{p_1 \cdots p_{s-1}}(x^{p_s})}{\Phi_{p_1 \cdots p_{s-1}}(x)}$$

THE computation of Φ_n can be further simplified by observing that the coefficients of cyclotomic polynomials are symmetrically ordered, a fact which we are going to prove now.

(24.9) Proposition. *Let $n \geq 2$.*
(a) $\Phi_n(x) = x^{\varphi(n)} \Phi_n(x^{-1})$.
(b) *If $\Phi_n(x) = \sum_{i=0}^{\varphi(n)} a_i x^i$ then $a_i = a_{\varphi(n)-i}$ for all i.*

Proof. (a) Let $g(x) := x^{\varphi(n)} \Phi_n(x^{-1})$ (this is just the reciprocal polynomial of Φ_n). If α is a root of Φ_n, i.e., a primitive n-th root of unity, then so is α^{-1}; hence every root of Φ_n is a root of g which implies that Φ_n divides g. On the other hand, the degree of g is $\deg g = \varphi(n) = \deg \Phi_n$, and the leading coefficient of g is $\Phi_n(0) = 1$ and hence equals the leading coefficient of Φ_n. These observations show that $\Phi_n = g$.

(b) Using part (a), we obtain

$$\sum_{i=0}^{\varphi(n)} a_i x^i = \Phi_n(x) = x^{\varphi(n)} \Phi_n(x^{-1})$$

$$= x^{\varphi(n)} \sum_{i=0}^{\varphi(n)} a_i x^{-i} = \sum_{i=0}^{\varphi(n)} a_i x^{\varphi(n)-i} = \sum_{j=0}^{\varphi(n)} a_{\varphi(n)-j} x^j .$$

A comparison of coefficients now yields the claim. ∎

We now want to prove that over the field $\mathbb{Q}$ of rational numbers all cyclotomic polynomials are irreducible. The crucial part of the proof is contained in the following lemma.

(24.10) Lemma. *Let ε be a primitive n-th root of unity over $\mathbb{Q}$ and let $f \in \mathbb{Q}[x]$ be its minimal polynomial over $\mathbb{Q}$. Moreover, let p be a prime number with $p \nmid n$. If u is a root of f, then so is u^p.*

Proof. The polynomial f has coefficients in $\mathbb{Z}$ by (15.4). Let $\Phi_n \in \mathbb{Z}[x]$ be the n-th cyclotomic polynomial over $\mathbb{Q}$. Then f divides Φ_n in $\mathbb{Q}[x]$, hence also in $\mathbb{Z}[x]$ by (8.2)(c); say $\Phi_n = fg$ where $g \in \mathbb{Z}[x]$ is necessarily monic. Now if $f(u) = 0$, then $\Phi_n(u) = 0$ so that u is a primitive n-th root of unity and hence so is u^p; this implies that $\Phi_n(u^p) = 0$, hence $f(u^p) = 0$ or $g(u^p) = 0$. Suppose that $f(u^p) \neq 0$ to obtain a contradiction; then u is a root of the polynomial $\widehat{g}(x) := g(x^p)$ which is monic and lies in $\mathbb{Z}[x]$. Since f is the minimal polynomial of u, this implies that f divides $\widehat{g}$ in $\mathbb{Q}[x]$, hence also in $\mathbb{Z}[x]$ by (8.2)(c); say $\widehat{g} = fh$ where $h \in \mathbb{Z}[x]$ is necessarily monic. Now g^p and $\widehat{g}$ induce the same polynomial on $\mathbb{Z}_p[x]$; namely, if $g(x) = a_0 + a_1 x + \cdots + a_n x^n$ then reading the coefficients modulo p we obtain

$$
\begin{aligned}
g(x)^p &= (a_0 + a_1 x + \cdots + a_n x^n)^p = a_0^p + a_1^p x^p + \cdots + a_n^p x^{np} \\
&= a_0 + a_1 x^p + \cdots + a_n (x^p)^n = g(x^p) = \widehat{g}(x) \, .
\end{aligned}
$$

Hence if we denote for each polynomial $F \in \mathbb{Z}[x]$ the polynomial induced on $\mathbb{Z}_p$ by $\overline{F}$, then $\overline{g}^p = \overline{f} \cdot \overline{h}$. This implies that $\overline{f}$ and $\overline{g}$ have a proper common factor.[†] Consequently, $\overline{\Phi}_n = \overline{f}\overline{g}$ has a multiple root. But this is impossible because $\overline{\Phi}_n$ is a divisor of $x^n - 1$ in $\mathbb{Z}_p[x]$ and $x^n - 1$ has only simple roots. ∎

(24.11) Theorem. *All cyclotomic polynomials over $\mathbb{Q}$ are irreducible.*

Proof. Let $\Phi_n \in \mathbb{Z}[x]$ be the n-the cyclotomic polynomial over $\mathbb{Q}$, let ε be a primitive n-th root of unity and let $f \in \mathbb{Q}[x]$ be the minimal polynomial of ε over $\mathbb{Q}$; we want to show that $f = \Phi_n$. To do this, it is clearly enough to prove prove that each primitive n-th root of unity is a root of f (because then $\Phi_n \mid f$ whereas $f \mid \Phi_n$ is clear anyway). Each primitive n-th root of unity has the form ε^k where k is coprime with n. If $k = p_1 p_2 \cdots p_r$ is the factorization of k in (not necessarily pairwise different) prime numbers, then $p_i | n$ for all n. Hence applying lemma (24.10) repeatedly, we first conclude from $f(\varepsilon) = 0$ that $f(\varepsilon^{p_1}) = 0$, thence that $f(\varepsilon^{p_1 p_2}) = 0$, and so on. Finally, we arrive at the equation $f(\varepsilon^{p_1 p_2 \cdots p_r}) = 0$, i.e., $f(\varepsilon^k) = 0$. ∎

(24.12) Theorem. *Let K be a field and let $G = \mathrm{Gal}_K(x^n - 1)$ be the Galois group of $x^n - 1$ over K.*
  (a) *If* char $K = 0$ *then G is isomorphic to a subgroup of $(\mathbb{Z}_n^\times, \cdot)$.*
  (b) *If $K = \mathbb{Q}$ then G is isomorphic to $(\mathbb{Z}_n^\times, \cdot)$.*
  (c) *If* char $K = p > 0$ *then G is isomorphic to a subgroup of $(\mathbb{Z}_m^\times, \cdot)$ where $n = p^r m$ with $p \nmid m$.*
  In any case G is isomorphic to a subgroup of an abelian group and hence abelian.

  [†] Otherwise there would be polynomials $\alpha, \beta \in \mathbb{Z}_p[x]$ with $\alpha \overline{f} + \beta \overline{g} = 1$ and hence $\beta \overline{g} = 1 - \alpha \overline{f}$ so that $\overline{f} \cdot \overline{h} \beta^p = (\beta \overline{g})^p = (1 - \alpha \overline{f})^p$; but in this last equation the left-hand side is divisible by $\overline{f}$ whereas the right-hand side is not. (Clearly $\deg \overline{f} = \deg f \geq 1$ so that $\overline{f}$ is not a constant.)

Proof. We want to treat the cases (a) and (c) simultaneously; to do so, we observe that if char $K = p$ and m is chosen as indicated, then $x^n - 1$ and $x^m - 1$ have the same splitting field due to (24.2)(e) and hence isomorphic Galois groups. Therefore, nothing is lost if we assume that $m = n$. Hence we may assume in either case that char $K \nmid n$; then the splitting field of $x^n - 1$ over K is $K(\varepsilon)$ where ε is any primitive n-th root of unity. The elements of $G_K^{K(\varepsilon)}$ permute the primitive n-th roots of unity, i.e., the elements ε^k where $[k] \in \mathbb{Z}_n^\times$. Thus for any $\sigma \in G_K^{K(\varepsilon)}$ there is an integer $k = k(\sigma)$ relatively prime to n such that $\sigma(\varepsilon) = \varepsilon^k = \varepsilon^{k(\sigma)}$; here k is uniquely determined modulo n. This defines a mapping

$$\mu : \begin{array}{ccc} G_{\mathbb{Q}}^{\mathbb{Q}(\varepsilon)} & \to & \mathbb{Z}_n^\times \\ \sigma & \mapsto & [k(\sigma)] \end{array}.$$

We claim that μ is an injective group homomorphism. First of all, $\mathrm{id}(\varepsilon) = \varepsilon$ so that $\mu(\mathrm{id}) = [1]$. Also, if $k_1 = k(\sigma_1)$ and $k_2 = k(\sigma_2)$, then

$$(\sigma_1 \sigma_2)(\varepsilon) = \sigma_1\big(\sigma_1(\varepsilon)\big) = \sigma_1(\varepsilon^{m_2}) = \sigma_1(\varepsilon)^{m_2} = (\varepsilon^{m_1})^{m_2} = \varepsilon^{m_1 m_2} ;$$

this shows $\mu(\sigma_1 \sigma_2) = \mu(\sigma_1)\mu(\sigma_2)$. Let us check next that σ is one-to-one. If $\sigma \in \ker \mu$, then $\sigma(\varepsilon) = \varepsilon$ which implies that σ is the identity on all of $K(\varepsilon)$. This proves (a) and (c).

To prove (b) we have to show that μ is also onto if $K = \mathbb{Q}$. Here we use the fact that Φ_n is irreducible over $\mathbb{Q}$ to conclude that $\mathrm{Gal}\,\Phi_n \cong G_{\mathbb{Q}}^{\mathbb{Q}(\varepsilon)}$ acts transitively on the roots of Φ_n, which clearly implies that μ is onto. So we have proved that

$$\boxed{G_{\mathbb{Q}}^{\mathbb{Q}(\varepsilon)} \cong \mathbb{Z}_n^\times, \text{ if } \varepsilon \text{ is a primitive } n\text{-th root of unity.}}$$

In particular, $[\mathbb{Q}(\varepsilon) : \mathbb{Q}] = |G_{\mathbb{Q}}^{\mathbb{Q}(\varepsilon)}| = |\mathbb{Z}_n^\times| = \varphi(n)$ where φ is Euler's function. $\blacksquare$

(24.13) Example. Let us determine the Galois group of $x^{12} - 1 \in \mathbb{Q}[x]$. The splitting field of $x^{12} - 1$ (as well as the splitting field of Φ_{12}) is $L := \mathbb{Q}(\varepsilon)$ where $\varepsilon := e^{2\pi i/12}$. The primitive 12-th roots of unity are the numbers ε^k where k is relatively prime to 12, i.e., ε^k where $k = 1, 5, 7, 11$. Then (24.12) shows that

$$G_{\mathbb{Q}}^L = \{\sigma_k \mid k = 1, 5, 7, 11\}$$

where σ_k is defined by the condition $\sigma_k(\varepsilon) = \varepsilon^k$; we have $G_{\mathbb{Q}}^L \cong \mathbb{Z}_{12}^\times$. The multiplication table is as follows:

$$
\begin{array}{c|cccc}
 & \sigma_1 & \sigma_5 & \sigma_7 & \sigma_{11} \\
\hline
\sigma_1 & \sigma_1 & \sigma_5 & \sigma_7 & \sigma_{11} \\
\sigma_5 & \sigma_5 & \sigma_1 & \sigma_{11} & \sigma_7 \\
\sigma_7 & \sigma_7 & \sigma_{11} & \sigma_1 & \sigma_5 \\
\sigma_{11} & \sigma_{11} & \sigma_7 & \sigma_5 & \sigma_1
\end{array}
$$

The subgroup lattice of $G_{\mathbb{Q}}^L$ is shown in the following diagram.

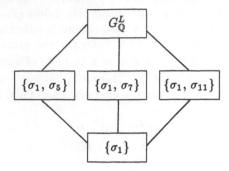

It is easy to verify that the corresponding fixed fields are

$$F(\{\sigma_1, \sigma_5\}) = \mathbb{Q}(i) ,$$
$$F(\{\sigma_1, \sigma_7\}) = \mathbb{Q}(i\sqrt{3}) = \mathbb{Q}(\sqrt{-3}) ,$$
$$F(\{\sigma_1, \sigma_{11}\}) = \mathbb{Q}(\sqrt{3}) .$$

Since the compositum of any two of these three intermediate fields of $(L : \mathbb{Q})$ is all of L, we conclude that ε can be expressed in terms of square roots; in fact, $\varepsilon = \varepsilon^{-3}\varepsilon^4 = e^{-2\pi i/4}e^{2\pi i/3} = (-1 + i\sqrt{3})/(2i) = \frac{1}{2}(i + \sqrt{3})$.

(24.14) Example. Let us determine the Galois group of $x^{15} - 1 \in \mathbb{Q}[x]$. The splitting field of $x^{15} - 1$ (as well as the splitting field of Φ_{15}) is $L := \mathbb{Q}(\varepsilon)$ where $\varepsilon := e^{2\pi i/15}$. The primitive 15-th roots of unity are the numbers ε^k where k is relatively prime to 15, i.e., ε^k where $k = 1, 2, 4, 7, 8, 11, 13, 14$. Then (24.12) shows

$$G_{\mathbb{Q}}^L = \{\sigma_k \mid k = 1, 2, 4, 7, 8, 11, 13, 14\}$$

where σ_k is defined by the condition $\sigma_k(\varepsilon) = \varepsilon^k$; we have $G_{\mathbb{Q}}^L \cong \mathbb{Z}_{15}^{\times} \cong \mathbb{Z}_2 \times \mathbb{Z}_4$. The following diagram shows the subgroup lattice of $G_{\mathbb{Q}}^L$.

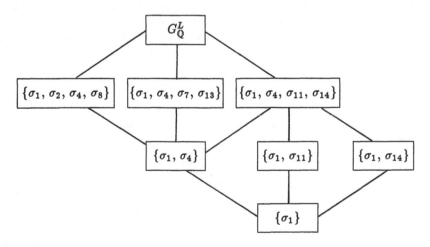

We have

$$\{\sigma_1, \sigma_4, \sigma_{11}, \sigma_{14}\} \cong \mathbb{Z}_2 \times \mathbb{Z}_2 \ ,$$
$$\{\sigma_1, \sigma_2, \sigma_4, \sigma_8\} \cong \mathbb{Z}_4 \quad \text{and}$$
$$\{\sigma_1, \sigma_4, \sigma_7, \sigma_{13}\} \cong \mathbb{Z}_4 \ .$$

IN the special case that n is a prime number we can use the fact that $\mathbb{Z}_n^\times$ is cyclic to determine the structure of the splitting field of $x^n - 1$.

(24.15) Theorem. *Let p be a prime number and $\varepsilon \in \mathbb{C}$ be a primitive p-th root of unity so that $L = \mathbb{Q}(\varepsilon)$ is a splitting field of $x^p - 1 \in \mathbb{Q}[x]$. Moreover, let λ be a primitive root modulo p, i.e., a generator of the multiplicative group $\mathbb{Z}_p^\times$.*

(a) The degree of L over $\mathbb{Q}$ is $N := p-1$, and the elements $\varepsilon_i := \varepsilon^{\lambda^i}$ $(0 \le i \le N-1)$ form a basis of L over $\mathbb{Q}$.

(b) The Galois group G_K^L is generated by the automorphism of L given by $\sigma(\varepsilon) = \varepsilon^\lambda$. This automorphism satisfies $\sigma(\varepsilon_i) = \varepsilon_{i+1}$ and $\sigma^k(\varepsilon_i) = \varepsilon_{i+k}$ for all i, k where the indices may be read modulo N.

(c) Let d be a divisor of N and let $e := N/d$. Then there is a unique intermediate field Z_d such that $[Z_d : \mathbb{Q}] = d$, and this field is given by

$$Z_d \ = \ \mathbb{Q}(\omega_0) \ = \ \mathbb{Q}(\omega_1) \ = \ \cdots \ = \ \mathbb{Q}(\omega_{d-1})$$

*where $\omega_k := \varepsilon_k + \varepsilon_{k+d} + \varepsilon_{k+2d} + \cdots + \varepsilon_{k+(e-1)d}$ is the sum of those ε_i with $i \equiv k \mod d$. Following Gauß, the quantities ω_k are called the e-**periods** of $(L : \mathbb{Q})$ because each ω_k consists of exactly e summands.*

Proof. (a) We have $[L : \mathbb{Q}] = \varphi(p) = p - 1$ by (24.12). By definition of λ the elements $\varepsilon, \varepsilon^\lambda, \varepsilon^{\lambda^2}, \ldots, \varepsilon^{\lambda^{N-1}}, \varepsilon^{\lambda^N} = 1$ are up to order just the elements $\varepsilon, \varepsilon^2, \ldots, \varepsilon^{N-1}, \varepsilon^N = 1$, and to prove that these form a basis of L over $\mathbb{Q}$ it is enough to prove that they are linearly independent over $\mathbb{Q}$. So suppose that $a_1\varepsilon + a_2\varepsilon^2 + \cdots + a_N\varepsilon^N = 0$ with $a_i \in \mathbb{Q}$; then division by ε yields $a_1 + a_2\varepsilon + \cdots + a_N\varepsilon^{N-1} = 0$ which shows that ε satisfies an equation of degree $\le N - 1 = p - 2$ over $\mathbb{Q}$. Since the minimal polynomial of ε over $\mathbb{Q}$ has degree $p - 1$, this implies that $a_1 = \cdots = a_N = 0$.

(b) The first claim follows immediately from (24.12). Also, $\sigma(\varepsilon_i) = \sigma(\varepsilon^{\lambda^i}) = \sigma(\varepsilon)^{\lambda^i} = (\varepsilon^\lambda)^{\lambda^i} = \varepsilon^{\lambda^{i+1}} = \varepsilon_{i+1}$; hence a k-fold application of σ to ε_i gives $\sigma^k(\varepsilon_i) = \varepsilon_{i+k}$. Finally, $\varepsilon_{i+N} = \varepsilon^{\lambda^{i+N}} = \varepsilon^{\lambda^i \lambda^N} = \varepsilon^{\lambda^i} = \varepsilon_i$.

(c) By the main theorem of Galois theory, the intermediate fields of $(L : K)$ are in 1-1-correspondence with the subgroups of G_K^L. Since G_K^L is cyclic of order N, there is a unique subgroup $U = U_d$ of index d for each divisor d of N, and the corresponding intermediate field is $Z_d = F(U_d)$. Now $U_d = \{\sigma^d, \sigma^{2d}, \ldots, \sigma^{ed} = \mathrm{id}\}$ where $e = N/d$. We now observe that $\omega_0, \ldots, \omega_{d-1}$ are invariant under the powers of σ^d, but not invariant under any other powers of σ (this follows easily from the facts that $\sigma^k(\varepsilon_i) = \varepsilon_{i+k}$ and that $(\varepsilon_0, \ldots, \varepsilon_{N-1})$ is a basis of L over $\mathbb{Q}$). Consequently, $F(U_d) = \mathbb{Q}(\omega_0) = \cdots = \mathbb{Q}(\omega_{d-1})$. ∎

(24.16) Example. Let us determine all intermediate fields of the splitting field of $x^{17} - 1$ over $\mathbb{Q}$. Here $p = 17$, hence $N = 16 = 2^4$. Hence there are 3 proper intermediate fields, one each of order $2, 4$ and 8. Since 3 is a generator of $\mathbb{Z}_{17}^{\times}$, we have $\varepsilon_0 = \varepsilon$, $\varepsilon_1 = \varepsilon^3$, $\varepsilon_2 = \varepsilon^9$, and so on.

(a) Let $d = 2$. Then there are two 8-periods, namely

$$
\begin{aligned}
\alpha_0 &= \varepsilon_0 + \varepsilon_2 + \varepsilon_4 + \varepsilon_6 + \varepsilon_8 + \varepsilon_{10} + \varepsilon_{12} + \varepsilon_{14} \\
&= \varepsilon + \varepsilon^{-8} + \varepsilon^{-4} + \varepsilon^{-2} + \varepsilon^{-1} + \varepsilon^8 + \varepsilon^4 + \varepsilon^2
\end{aligned}
$$

and

$$
\begin{aligned}
\alpha_1 &= \varepsilon_1 + \varepsilon_3 + \varepsilon_5 + \varepsilon_7 + \varepsilon_9 + \varepsilon_{11} + \varepsilon_{13} + \varepsilon_{15} \\
&= \varepsilon^3 + \varepsilon^{-7} + \varepsilon^5 + \varepsilon^{-6} + \varepsilon^{-3} + \varepsilon^7 + \varepsilon^{-5} + \varepsilon^6 .
\end{aligned}
$$

It is easy to check that $\alpha_0 + \alpha_1 = -1$ and $\alpha_0 \alpha_1 = -4$; hence α_0 and α_1 are root of the equation

$$
(1) \qquad\qquad\qquad y^2 + y - 4 = 0 .
$$

On the other hand, the roots of this equation are $\frac{1}{2}(-1 \pm \sqrt{17})$; hence $Z_2 = \mathbb{Q}(\alpha_0) = \mathbb{Q}(\alpha_1) = \mathbb{Q}(\sqrt{17})$.

(b) Now let $d = 4$. There are four 4-periods, namely

$$
\begin{aligned}
\beta_0 &= \varepsilon_0 + \varepsilon_4 + \varepsilon_8 + \varepsilon_{12} = \varepsilon + \varepsilon^{-4} + \varepsilon^{-1} + \varepsilon^4 , \\
\beta_1 &= \varepsilon_1 + \varepsilon_5 + \varepsilon_9 + \varepsilon_{13} = \varepsilon^3 + \varepsilon^5 + \varepsilon^{-3} + \varepsilon^{-5} , \\
\beta_2 &= \varepsilon_2 + \varepsilon_6 + \varepsilon_{10} + \varepsilon_{14} = \varepsilon^{-8} + \varepsilon^{-2} + \varepsilon^8 + \varepsilon^2 \quad \text{and} \\
\beta_3 &= \varepsilon_3 + \varepsilon_7 + \varepsilon_{11} + \varepsilon_{15} = \varepsilon^{-7} + \varepsilon^{-6} + \varepsilon^7 + \varepsilon^6 .
\end{aligned}
$$

Since $\beta_0 + \beta_2 = \alpha_0$, $\beta_0 \beta_2 = -1$ and $\beta_1 + \beta_3 = \alpha_1$, $\beta_1 \beta_3 = -1$, the numbers β_0, β_2 and β_1, β_3 are roots of the equations

$$
(2) \quad y^2 - \alpha_0 y - 1 = 0 ; \qquad (2') \quad y^2 - \alpha_1 y - 1 = 0 ,
$$

respectively. On the other hand, the roots of these equations are $\frac{1}{2}(\alpha_i \pm \sqrt{\alpha_i^2 + 4})$. Thus $Z_4 = \mathbb{Q}(\beta_0) = \mathbb{Q}(\beta_1) = \mathbb{Q}(\beta_1) = \mathbb{Q}(\beta_3) = Z_2(\sqrt{\alpha_0^2 + 4}) = Z_2(\sqrt{\alpha_1^2 + 4})$.

(c) Next, let $d = 8$. There are eight 2-periods; let us just write down two of them, namely $\gamma_0 = \varepsilon_0 + \varepsilon_8 = \varepsilon + \varepsilon^{-1}$ and $\gamma_4 = \varepsilon_4 + \varepsilon_{12} = \varepsilon^4 + \varepsilon^{-4}$. Then $\gamma_0 + \gamma_4 = \beta_0$ and $\gamma_0 \gamma_4 = \varepsilon^5 + \varepsilon^3 + \varepsilon^{-3} + \varepsilon^{-5} = \beta_1$; hence γ_0 and γ_4 are roots of the equation

$$
(3) \qquad\qquad\qquad y^2 - \beta_0 y + \beta_1 = 0
$$

which allows us again to explicitly obtain Z_8 as a quadratic extension of Z_4. Finally, multiplying the equation $\gamma_0 = \varepsilon + \varepsilon^{-1}$ by ε, we see that $\varepsilon \gamma_0 = \varepsilon^2 + 1$; hence ε satisfies the quadratic equation

$$
(4) \qquad\qquad\qquad y^2 - \gamma_0 y + 1 = 0
$$

over Z_8. This shows that the 17-th roots of unity can be obtained from $\mathbb{Q}$ by successively solving quadratic equations.

WE now exploit the fact that $[\mathbb{Q}(\varepsilon) : \mathbb{Q}] = \varphi(n)$ if ε is a primitive n-th root of unity to obtain some information on the number of roots of unity in an algebraic number field.

(24.17) Theorem. *Let K be an algebraic number field, i.e., a finite field extension of $\mathbb{Q}$. If K contains a primitive n-th root of unity then $\varphi(n) \leq [K : \mathbb{Q}]$. In particular, K contains only a finite number of roots of unity, and $U(K)$ is a finite cyclic group.*

Proof. Let ε be a primitive n-th root of unity in K. Then $\mathbb{Q}(\varepsilon) \subseteq K$ and hence $\varphi(n) = [\mathbb{Q}(\varepsilon) : \mathbb{Q}] \leq [K : \mathbb{Q}] < \infty$. Since φ takes each value only a finite number of times only a finite number of natural numbers n can satisfy this condition; this shows that $U(K)$ is a finite group. The fact that this group is cyclic is then a consequence of (4.14). ∎

(24.18) Example. *Let $K = \mathbb{Q}(\varepsilon)$ where ε is a primitive p-th root of unity, p being an odd prime number. Then K contains all $(2p)$-th roots of unity, but no others.*

Proof. The group $U(K)$ of all roots of unity in K is a finite cyclic group by (24.17); let $|U(K)| = m$. Obviously, all powers of $-\varepsilon$ and hence all $(2p)$-th roots of unity are contained in $U(K)$. Since $-\varepsilon \in U(K)$ has order $2p$ we have $2p \mid m$ by Lagrange's theorem. Let α be a generator of $U(K)$, i.e., a primitive m-th root of unity. Writing $m = p^r q$ with $r \geq 1$ and $p \nmid q$ we obtain

$$[\mathbb{Q}(\alpha) : \mathbb{Q}] = \varphi(m) = \varphi(p^r q) = p^{r-1}(p-1)\varphi(q) .$$

On the other hand we have $\mathbb{Q}(\alpha) \subseteq K = \mathbb{Q}(\varepsilon)$ and hence

$$[\mathbb{Q}(\alpha) : \mathbb{Q}] \leq [\mathbb{Q}(\varepsilon) : \mathbb{Q}] = \varphi(p) = p-1 .$$

Thus $p^{r-1}\varphi(q) \leq 1$ which implies that $r = 1$ and $q = 2$, i.e., that $m = 2p$. ∎

WE are now in a position to derive Kummer's lemma which played a key role in the proof of theorem (9.23) but was used without proof in section 9.

(24.19) Kummer's lemma. *Let $\varepsilon = e^{2\pi i/p}$ where p is an odd prime. Then every unit in $\mathbb{Z}[\varepsilon]$ is a product of a real unit and a power of ε.*

Proof. Let u be an arbitrary unit in $\mathbb{Z}[\varepsilon]$. Then there is a polynomial $f(x) = a_0 + a_1 x + \cdots + a_{p-2} x^{p-2} \in \mathbb{Z}[x]$ such that $u = f(\varepsilon)$. Hence $\bar{u} = f(\bar{\varepsilon}) = f(\varepsilon^{-1}) = f(\varepsilon^{p-1})$ is also a unit which implies that $v := u/\bar{u}$ is a unit. Now there are $p-1$ embeddings $\sigma_1, \ldots, \sigma_{p-1}$ of $\mathbb{Q}(\varepsilon)$ into $\mathbb{C}$, given by $\sigma_k(\varepsilon) = \varepsilon^k$. For each such embedding we obtain

$$\sigma_k(v) = \frac{\sigma_k(u)}{\sigma_k(\bar{u})} = \frac{\sigma_k\big(f(\varepsilon)\big)}{\sigma_k\big(f(\varepsilon^{p-1})\big)} = \frac{f\big(\sigma_k(\varepsilon)\big)}{f\big(\sigma_k(\varepsilon)^{p-1}\big)}$$

$$= \frac{f(\varepsilon^k)}{f(\varepsilon^{k(p-1)})} = \frac{f(\varepsilon^k)}{f(\varepsilon^{-k})} = \frac{f(\varepsilon^k)}{f(\varepsilon^k)}$$

and hence $|\sigma_k(v)| = 1$. Then (15.14)(b) implies that $\sigma_k(v)$ is a root of unity. Since $\mathbb{Q}(\varepsilon)$ contains only $(2p)$-th roots of unity, due to (24.18), we have

$$v = \pm \varepsilon^k$$

for some index $1 \leq k \leq p-1$. We claim that the sign must be "+". In fact, suppose that $v = -\varepsilon^k$, i.e., that $u = -\varepsilon^k \overline{u}$. Consider the ideal $I := \langle 1 - \varepsilon \rangle$ of $\mathbb{Z}[\varepsilon]$; all congruences that will occur in the sequel are modulo the ideal I. We have $\varepsilon \equiv 1$, hence $\varepsilon^k \equiv 1$ and therefore

$$u = a_0 + a_1 \varepsilon + \cdots + a_{p-2}\varepsilon^{p-2} \equiv a_0 + a_1 + \cdots + a_{p-2} := m \in \mathbb{Z}.$$

Then on the one hand $\overline{u} = -\varepsilon^{-k}u \equiv -u \equiv -m$, on the other hand $\overline{u} = f(\varepsilon^{p-1}) = a_0 + a_1 \varepsilon^{p-1} + \cdots + a_{p-2}\varepsilon^{(p-2)(p-1)} \equiv a_0 + a_1 + \cdots + a_{p-2} = m$ so that $2m \equiv 0$. This means that $2m$ is divisible by $1 - \varepsilon$ in $\mathbb{Z}[\varepsilon]$, i.e., that $2m \in \mathbb{Z} \cap \mathbb{Z}[\varepsilon](1-\varepsilon)$. Due to the formula $(\star)$ in the proof of (15.15) this implies that $2m \in \mathbb{Z}p$ so that p divides m. On the other hand we have $p = (1-\varepsilon)(1-\varepsilon^2)\cdots(1-\varepsilon^{p-1})$ due to (14.23); hence $1 - \varepsilon$ divides m in $\mathbb{Z}[\varepsilon]$ which means that $m \equiv 0$ and hence that $u \equiv 0$, i.e., that $u \in I$. Since u is a unit this forces that I is all of $\mathbb{Z}[\varepsilon]$, but this is impossible because $1 - \varepsilon$ is not a unit in $\mathbb{Z}[\varepsilon]$. This contradiction shows that we must have $v = \varepsilon^k$, i.e., $u = \varepsilon^k \overline{u}$. Let

$$s := \begin{cases} (p+k)/2 & \text{if } k \text{ is odd;} \\ (2p+k)/2 & \text{if } k \text{ is even.} \end{cases}$$

Then $2s \equiv k$ modulo p, hence $\varepsilon^k = \varepsilon^{2s}$ so that $u = \varepsilon^k \overline{u} = \varepsilon^{2s}\overline{u}$. This implies that

$$\frac{u}{\varepsilon^s} = \frac{\overline{u}}{\varepsilon^{-s}} = \frac{\overline{u}}{\overline{\varepsilon}^s}$$

is its own conjugate and hence is a real number. Since u is the product of this real number with ε^s the claim follows. ∎

TO conclude this section, we present three applications of roots of unity and cyclotomic polynomials, one in algebra, one in geometry and one in number theory. Let us start with the application in algebra, namely Wedderburn's celebrated result that a finite division ring is automatically commutative.†

---

† In view of problem 17 in section [I.5], this implies that for a finite plane geometry Desargues' theorem implies Pappus' theorem.

(24.20) Theorem (Wedderburn 1905). *Every finite division ring is a field.*

Proof (Artin). Suppose that R is a finite division ring which is not commutative. Then for each element $a \in R$ the ring

$$C(a) := \{x \in R \mid ax = xa\} \subseteq R$$

is again a division ring, and

$$K := \bigcap_{a \in R} C(a) = \{x \in R \mid xa = ax \text{ for all } a \in R\}$$

is even a field. Let $q := |K|$. Then $q > 1$ (since $0, 1 \in K$) and $q < |R|$ (since $K \subsetneq R$ by assumption). Viewing R as a vector space over the field K we see that $|R| = q^n$ for some $n > 1$. Now let $a \neq 0$. Viewing $C(a)$ as a vector space over K, we see that $|C(a)| = q^{n(a)}$ for some number $n(a) \in \mathbb{N}$; on the other hand, $C(a)^\times = C(a) \setminus \{0\}$ is the centralizer of a in the abelian group $R^\times$, and $K^\times$ is the center of this group. Then the class equation (I.27.2)(b) reads

$$|R^\times| = |K^\times| + \sum_{i=1}^{r} [R^\times : C(a_i)^\times]$$

where $\{a_1, \ldots, a_r\}$ is a set of representatives of those conjugacy classes with more than one element. Then $|C(a_i)^\times| = q^{n(a_i)} - 1$ is a proper divisor of $|R^\times| = q^n - 1$ which means that $n(i) := n(a_i)$ is a proper divisor of n. Thus the class equation can be more explicitly written as

$$q^n - 1 = q - 1 + \sum_{i=1}^{r} \frac{q^n - 1}{q^{n(i)} - 1} .$$

Let $\Phi_n \in \mathbb{Z}[x]$ be the n-th cyclotomic polynomial. Then $\Phi_n(q)$ divides $q^n - 1$ and divides $\sum_{i=1}^{r} \left((q^n - 1)/(q^{n(i)} - 1) \right)$, hence also divides $q - 1$. On the other hand, $\Phi_n(q) = \prod_\varepsilon (q - \varepsilon)$ where ε ranges through the primitive n-th roots of unity. Each such root has the form $\varepsilon = a + bi$ with $a < 1$ and $a^2 + b^2 = 1$. Then

$$|q - \varepsilon|^2 = |q - a - bi|^2 = (q - a)^2 + b^2 = q^2 - 2aq + a^2 + b^2$$
$$= q^2 - 2aq + 1 > q^2 - 2q + 1 = (q - 1)^2$$

which implies $|q - \varepsilon| > q - 1$. But then

$$|\Phi_n(q)| = \prod_\varepsilon |q - \varepsilon| > \prod_\varepsilon (q - 1) = (q - 1)^{\varphi(n)}$$

which is impossible since $\Phi_n(q)$ divides $q - 1$. ∎

AS a geometric application, we want to characterize those numbers $n \in \mathbb{N}$ for which a regular polygon can be constructed by using a ruler and a compass only. Before we can tackle this problem, we need a lemma.

(24.21) Lemma. *Let $\alpha := 2\pi/n$.*

(a) $\mathbb{Q} \subseteq \mathbb{Q}(\cos \alpha) \subseteq \mathbb{Q}(e^{i\alpha})$.

(b) $[\mathbb{Q}(e^{i\alpha}) : \mathbb{Q}(\cos \alpha)] = \begin{cases} 1, & \text{if } n = 1, 2; \\ 2, & \text{if } n \geq 3. \end{cases}$

(c) $(\mathbb{Q}(\cos \alpha) : \mathbb{Q})$ *is a Galois extension of degree* $\varphi(n)$ *if* $n = 1, 2$ *and of degree* $\frac{1}{2}\varphi(n)$ *if* $n \geq 3$.

Proof. (a) $\cos \alpha = \frac{1}{2}(e^{i\alpha} + e^{-i\alpha}) = \frac{1}{2}\left(e^{i\alpha} + (e^{i\alpha})^{-1}\right) \in \mathbb{Q}(e^{i\alpha})$.

(b) Let $p(x) := (x - e^{i\alpha})(x - e^{-i\alpha}) = x^2 - 2(\cos \alpha)x + 1 \in \mathbb{Q}(\cos \alpha)$. Then the degree $[\mathbb{Q}(e^{i\alpha}) : \mathbb{Q}(\cos \alpha)]$ is 2 if p if irreducible over $\mathbb{Q}(\cos \alpha)$ and 1 otherwise. Now the quadratic polynomial p is reducible over $\mathbb{Q}(\cos \alpha)$ if and only if its two roots $e^{i\alpha}$ and $e^{-i\alpha}$ are contained in $\mathbb{Q}(\cos \alpha) \subseteq \mathbb{R}$. Since $e^{\pm i\alpha} = \cos \alpha \pm i \sin \alpha$ this happens if and only if $0 = \sin \alpha = \sin \frac{2\pi}{n}$, i.e., if and only if $n = 1$ or $n = 2$.

(c) In (24.12) we proved that $(\mathbb{Q}(e^{i\alpha}) : \mathbb{Q})$ is a Galois extension of degree $\varphi(n)$ and that the Galois group $G_{\mathbb{Q}}^{\mathbb{Q}(e^{i\alpha})}$ is abelian so that every subgroup of $G_{\mathbb{Q}}^{\mathbb{Q}(e^{i\alpha})}$ is normal. Consequently, by the main theorem of Galois theory, every intermediate field $\mathbb{Q} \leq Z \leq \mathbb{Q}(e^{i\alpha})$ is of Galois type over $\mathbb{Q}$. This holds in particular for $Z := \mathbb{Q}(\cos \alpha)$. Then the degree formula

$$\underbrace{[\mathbb{Q}(e^{i\alpha}) : \mathbb{Q}]}_{= \varphi(n)} = \underbrace{[\mathbb{Q}(e^{i\alpha}) : \mathbb{Q}(\cos \alpha)]}_{= 1 \text{ or } 2} \cdot [\mathbb{Q}(\cos \alpha) : \mathbb{Q}]$$

proves the last claim. ∎

WE can now completely classify those natural numbers n for which a regular n-gon is constructible with a ruler and a compass. This classification is due to Carl Friedrich Gauß. According to his own records, Gauß discovered the constructibility of regular 17-gons on March 29, 1796, a few days before his 19-th birthday. This early success made him decide on a career as a mathematician rather than a philologue. A bronze statue was erected in Göttingen after his death whose pedestal has the shape of a regular 17-gon.

(24.22) Theorem. *Let $n \in \mathbb{N}$. A regular n-gon is constructible with a ruler and a compass if and only if $\varphi(n)$ is a power of 2. This is the case if and only if*

$$n = 2^m p_1 \cdots p_k$$

with $m \geq 0$ and pairwise different primes $p_1, \ldots, p_k$ of the form $p_i = 2^{m_i} + 1$.

Proof. The regular n-gon is constructible with ruler and compass if and only if $\cos \frac{2\pi}{n}$ is constructible from $\mathbb{Q}$. By (22.11) this is the case if and only if the degree of the smallest Galois extension of $\mathbb{Q}$ containing $\cos \frac{2\pi}{n}$ is a power of 2. Due to (24.21), this holds if and only if $\varphi(n)$ is a power of 2.

To prove the last claim, let us write down the prime decomposition of n, i.e.

$$n = 2^m p_1^{r_1} \cdots p_k^{r_k} \quad (m \geq 0, \ r_i \geq 1)$$

with pairwise different odd primes p_i. Now let $c := \begin{cases} 1, & m = 0, \\ 2^{m-1}, & m \geq 1. \end{cases}$ Using the properties of Euler's function, we obtain

$$\varphi(n) \;=\; c(p_1 - 1)p_1^{r_1 - 1} \cdots (p_k - 1)p_k^{r_k - 1} \;.$$

This number is a power of 2 if and only if all the $r_i = 1$ and $p_i - 1$ is a power of 2 for all i. ∎

(24.23) Remark. A number of the form $2^m + 1$ can be a prime only if m is a power of 2; namely, if $m = ab$ with b odd then $2^m + 1 = (2^a)^b + 1$ is divisible by $2^a + 1$. A prime number of the form $2^{2^k} + 1$ is called a **Mersenne prime**. Hence (24.22) can be reformulated as follows: A regular n-gon is constructible if and only if $n = 2^m p_1 \cdots p_k$ where $m \geq 0$ and where $p_1, \ldots, p_k$ are pairwise different Mersenne primes.

AS an application in number theory, we want to give a new proof of the quadratic reciprocity law (9.8) by using roots of unity.

RECALL that the Legendre symbol for an odd prime number p is defined as

$$\left(\frac{x}{p}\right) \;:=\; \begin{cases} 0, & \text{if } x = 0, \\ 1, & \text{if } x \neq 0 \text{ is a square in } \mathbb{Z}_p, \\ -1, & \text{if } x \neq 0 \text{ is not a square in } \mathbb{Z}_p \end{cases}$$

and is given by $\left(\frac{x}{p}\right) = x^{(p-1)/2}$ (where x on the right-hand side is interpreted as an element of $\mathbb{Z}_p$). We first give new proofs of (9.8)(a).

(24.24) Lemma. *Let p be an odd prime number. Then*

$$\left(\frac{1}{p}\right) = 1; \quad \left(\frac{-1}{p}\right) = \begin{cases} 1, & \text{if } p \equiv 1\,(4), \\ -1, & \text{if } p \equiv -1\,(4); \end{cases} \quad \left(\frac{2}{p}\right) = \begin{cases} 1, & \text{if } p \equiv \pm 1\,(8), \\ -1, & \text{if } p \equiv \pm 5\,(8). \end{cases}$$

Proof. The first statement is trivial, the second follows immediately from writing $\left(\frac{-1}{p}\right) = (-1)^{(p-1)/2}$. To prove the third statement, let C be an algebraic closure of $\mathbb{Z}_p$ and let $\varepsilon \in C$ be a primitive 8-th root of unity (so that $\varepsilon^4 = -1$). Then $y := \varepsilon + \varepsilon^{-1}$ satisfies $y^2 = \varepsilon^2 + 2 + \varepsilon^{-2} = 2 + \varepsilon^{-2}(\varepsilon^4 + 1) = 2 + \varepsilon^{-2} \cdot 0 = 2$ so that $\left(\frac{2}{p}\right) = 2^{\frac{p-1}{2}} = y^{p-1}$. Also, since we are in characteristic p, we have $y^p = (\varepsilon + \varepsilon^{-1})^p = \varepsilon^p + \varepsilon^{-p}$. If $p \equiv \pm 1\,(8)$ then $\varepsilon^8 = 1$ implies that $\varepsilon^p = \varepsilon^{\pm 1}$ and $\varepsilon^{-p} = \varepsilon^{\mp 1}$; therefore, $y^p = \varepsilon^{\pm 1} + \varepsilon^{\mp 1} = y$ which shows that $1 = y^{p-1} = \left(\frac{2}{p}\right)$. On the other hand, if $p \equiv \pm 5\,(8)$ then $\varepsilon^8 = 1$ implies that $\varepsilon^p = \varepsilon^{\pm 5}$ and $\varepsilon^{-p} = \varepsilon^{\mp 5}$. Since $\varepsilon^4 = -1$ we have $\varepsilon^5 = -\varepsilon$ and $\varepsilon^{-5} = -\varepsilon^{-1}$; therefore, $\varepsilon^p = -\varepsilon^{\pm 1}$ and $\varepsilon^{-p} = -\varepsilon^{\mp 1}$ which implies $y^p = -(\varepsilon^{\pm 1} + \varepsilon^{\mp 1}) = -y$ so that $-1 = y^{p-1} = \left(\frac{2}{p}\right)$. ∎

WE can also give a very elegant new proof of the fact that in the multiplicative group of a finite field there are as many squares as nonsquares.

(24.25) Lemma. *If K is a finite field then $(K^\times)^2$ has index 2 in $K^\times$.*

Proof. Let Ω be an algebraic closure of K. Define a homomorphism $\varphi : \Omega^\times \to \Omega^\times$ by $\varphi(x) = x^{(q-1)/2}$. Then $\ker \varphi = (K^\times)^2$. In fact, if $\xi \in (K^\times)^2$, say $\varphi = k^2$, then $\varphi(\xi) = a^{q-1} = 1$ which shows that $\xi \in \ker \varphi$. Conversely, if $\xi \in \ker \varphi$ let $a \in \Omega$ with $a^2 = \xi$. Then $a^{q-1} = \xi^{(q-1)/2} = \varphi(\xi) = 1$ which shows that $a \in K$ and hence $\xi \in (K^\times)^2$. Since $x^{(q-1)/2} - 1$ has only simple roots in Ω we have $|\ker \varphi| = \frac{q-1}{2} = \frac{1}{2}|K^\times|$. ∎

(24.26) Theorem. *Let $p \ne q$ be odd prime numbers and let C be an algebraic closure of $\mathbb{Z}_p$. In C, pick a primitive q-th root of unity ε.*

(a) The **Gauß sum** *$y := \sum_{x \in \mathbb{Z}_q}(\frac{x}{q})\varepsilon^x$ (which is in element of C) satisfies $y^2 = (-1)^{(q-1)/2} \cdot q$ and $y^{p-1} = (\frac{p}{q})$; in particular, both y^2 and y^{p-1} lie in the prime field $\mathbb{Z}_p$ of C.*

*(b) (**Quadratic Reciprocity Law.**) $(\frac{q}{p}) = (-1)^{\frac{p-1}{2} \cdot \frac{q-1}{2}}(\frac{p}{q})$.*

Proof. (a) Substituting $u = xy^{-1}$, we obtain

$$y^2 = \sum_{x,y \in \mathbb{Z}_q^\times}(\frac{xy}{q})\varepsilon^{x+y} = \sum_{u,y \in \mathbb{Z}_q^\times}(\frac{uy^2}{q})\varepsilon^{y(u+1)} = \sum_{u,y \in \mathbb{Z}_q^\times}(\frac{u}{q})\varepsilon^{y(u+1)}$$

$$= \sum_{y \in \mathbb{Z}_q^\times}(\frac{-1}{q}) + \sum_{u \ne -1}(\frac{u}{q})\sum_{y \in \mathbb{Z}_q^\times}\varepsilon^{y(u+1)}$$

Now $\sum_{y \in \mathbb{Z}_q^\times}(\frac{-1}{q}) = (q-1) \cdot (\frac{-1}{q})$, whereas for each $u \ne -1$ the sum $\sum_{y \in \mathbb{Z}_q^\times}\varepsilon^{y(u+1)}$ equals the sum of all primitive q-th roots of unity which is -1 due to (24.4)(a). Hence the above expression for y^2 becomes

$$y^2 = (q-1)(\frac{-1}{q}) - \sum_{u \ne -1}(\frac{u}{q}) = q \cdot (\frac{-1}{q}) - \sum_{\text{all } u}(\frac{u}{q}) = q \cdot (\frac{-1}{q}) \, ;$$

here we used that $\sum_{u \in \mathbb{Z}_q^\times}(\frac{u}{q}) = 0$ due to the fact that there are as many squares as non-squares in $\mathbb{Z}_q^\times$. Thus the claim for y^2 is established. To get the claim for y^{p-1}, we substitute $v = xp$ and observe that $(\frac{p^{-1}}{q}) = (\frac{p^{-1}}{q})(\frac{p^2}{q}) = (\frac{p}{q})$ to obtain

$$y^p = \sum_{x \in \mathbb{Z}_q}(\frac{x}{q})\varepsilon^{xp} = \sum_{v \in \mathbb{Z}_q}(\frac{vp^{-1}}{q})\varepsilon^v = (\frac{p^{-1}}{q})\sum_{v \in \mathbb{Z}_q}(\frac{v}{q})\varepsilon^v = (\frac{p}{q})y \, ;$$

division by y on both sides yields the claim for y^{p-1}.

(b) By part (a) we have $(\frac{p}{q}) = y^{p-1} = (y^2)^{\frac{p-1}{2}}$. Since $y^2 \in \mathbb{Z}_p$, this expression equals $(\frac{y^2}{p}) = (\frac{(-1)^{\frac{q-1}{2}}q}{p})$; due to the multiplicativity of the Legendre symbol this is $(\frac{-1}{p})^{\frac{q-1}{2}}(\frac{q}{p}) = (-1)^{\frac{p-1}{2} \cdot \frac{q-1}{2}}(\frac{q}{p})$. ∎

APART from the intrinsic beauty of this theorem, we can deduce also a structural result which is most easily formulated by introducing the following terminology.

(24.27) Definition. *A field extension $(L:K)$ is called a* **cyclotomic extension** *if L is obtained from K by adjoining roots of unity.*

WITH this terminology, the following corollary of (24.26) holds.

(24.24) Proposition. *Every quadratic extension of $\mathbb{Q}$ is contained in a cyclotomic extension of $\mathbb{Q}$.*

Proof. Every quadratic extension of $\mathbb{Q}$ can be written as $\mathbb{Q}(\sqrt{n})$ where $n \in \mathbb{Z}$ is a square-free number. If $n = \pm q_1 \cdots q_r$ is the prime factorization of n, then $\mathbb{Q}(\sqrt{n}) \subseteq \mathbb{Q}(\sqrt{q_1}, \ldots, \sqrt{q_r}, i)$; hence it is enough to prove that each extension of the form $\mathbb{Q}(\sqrt{q})$ with q prime is contained in a cyclotomic extension. For $q = 2$ we observe that $\varepsilon := (1 + i)/\sqrt{2}$ is an 8-th root of unity, and clearly $\mathbb{Q}(\sqrt{2}) \subseteq \mathbb{Q}(\varepsilon, i)$. If q is odd then (24.26)(a) shows that $q = (-1)^{(q-1)/2}y^2$ where $y \in \mathbb{Q}(\varepsilon)$ for some primitive q-th root of unity ε. Hence $\mathbb{Q}(\sqrt{q}) \subseteq \mathbb{Q}(\varepsilon)$ if $\frac{q-1}{2}$ is even and $\mathbb{Q}(\sqrt{q}) \subseteq \mathbb{Q}(\varepsilon, i)$ if $\frac{q-1}{2}$ is odd. ∎

THIS last proposition is a special case of a famous theorem of Kronecker and Weber which states that whenever $(Z : \mathbb{Q})$ is an algebraic extension such that $\operatorname{Aut} Z = G_{\mathbb{Q}}^{Z}$ is abelian then Z is contained in a cyclotomic extension of $\mathbb{Q}$.

Exercises

Problem 1. Let ε be a primitive p-th root of unity and let α be such that $\alpha^{p^{n-1}} = \varepsilon$. Show that α is a primitive p^n-th root of unity.

Problem 2. Let K be a field and let K_n be obtained from K by adjoining all n-th roots of unity (in some fixed algebraic closure of K). Show that if n is odd then $K_n = K_{2n}$.

Problem 3. For each number $n \in \{\pm 1, \pm 2, \pm 3, \pm 5\}$ find all roots of unity lying in $\mathbb{Q}(\sqrt{n})$.

Problem 4. Is it true or not true that whenever ε is a primitive n-th roots of unity then the roots of unity in $\mathbb{Q}(\varepsilon)$ are exactly the numbers $\pm \varepsilon^k$?

Problem 5. Find all natural numbers n with the property that the primitive n-th roots of unity are linearly independent over $\mathbb{Q}$.
Hint. Check the cases $n = 4, 6, 9, 15$ to get an idea.

Problem 6. Let $\varepsilon = e^{2\pi i/n}$. Prove the following statements!
(a) $(x - \varepsilon)(x - \varepsilon^2) \cdots (x - \varepsilon^n) = x^n - 1$.
(b) $1 + \varepsilon + \varepsilon^2 + \cdots + \varepsilon^{n-1} = 0$
(c) $1 \cdot \varepsilon \cdot \varepsilon^2 \cdots \varepsilon^{n-1} = (-1)^{n-1}$
(d) $(1 + \varepsilon)(1 + \varepsilon^2) \cdots (1 + \varepsilon^{n-1}) = \begin{cases} 0 & \text{if } n \text{ is even;} \\ 1 & \text{if } n \text{ is odd.} \end{cases}$

Problem 7. Compute all cyclotomic polynomials Φ_n for $1 \le n \le 20$.

Problem 8. Show that $\Phi_n(x) = \prod_{d|n}(x^{n/d} - 1)^{\mu(d)}$ where μ is the Möbius function.

Problem 9. Let $m, n > 1$ be coprime natural numbers and let U be the set of all primitive m-th roots of unity. Show that $\Phi_{mn}(x) = \prod_{u \in U} \Phi_n(ux)$.

Problem 10. Show that if $m, n \in \mathbb{N}$ are coprime then Φ_{mn} is a greatest divisor of $\Phi_m(x^n)$ and $\Phi_n(x^m)$ in $\mathbb{Q}[x]$.

Problem 11. (a) Show that if $p \ne q$ are prime numbers then the only coefficients of Φ_{pq} are 0, 1 and -1. (**Hint.** Write $\Phi_{pq}(x) = (1 - x)\Phi_q(x^p)(1 - x^q)^{-1}$ as an identity of formal power series.)
(b) Show that the coefficient of x^{41} and of x^7 in the expansion of Φ_{105} is -2.

Problem 12. Let Φ_n be the n-th cyclotomic polynomial and let ε be a primitive n-th root of unity. Show that

$$\Phi_n(1) = \begin{cases} 0, & \text{if } n = 1; \\ p, & \text{if } n \text{ is a power of the prime } p; \\ 1, & \text{if } n \text{ is neither } 1 \text{ nor a prime power.} \end{cases}$$

Conclude that $1 - \varepsilon$ is a unit in the ring $\mathbb{Z}[\varepsilon]$ if n is not a prime power, but a prime element if p is a prime power.

Problem 13. (a) Let n be a natural number and let p be a prime with $p \nmid n$. Let ℓ be the order of p in $\mathbb{Z}_n^\times$. Show that Φ_n can be decomposed into $\varphi(n)/\ell$ irreducible factors of degree ℓ over $\mathbb{Z}_p$.

(b) Find a natural number n such that Φ_n is reducible modulo p for any prime number p.

Problem 14. Let m and n be natural numbers with greatest common divisor d. Show that Φ_m decomposes into $\varphi(d)$ irreducible factors of equal degree over $\mathbb{Q}(e^{2\pi i/n})$.

Problem 15. For any $n \in \mathbb{N}$ let $\mathbb{Q}_n$ be the splitting field of the cyclotomic polynomial $\Phi_n \in \mathbb{Q}[x]$ so that $\mathbb{Q}_n = \mathbb{Q}(e^{2\pi i/n})$. Prove the following statements!

(a) The compositum of two cyclotomic fields is given by $\mathbb{Q}_n \mathbb{Q}_m = \mathbb{Q}_{\mathrm{lcm}(m,n)}$.

(b) $\mathbb{Q}_m \subseteq \mathbb{Q}_n$ if and only if $m \mid n$ or else $m \mid 2n$ with n odd.

(c) The intersection of two cyclotomic fields is given by $\mathbb{Q}_m \cap \mathbb{Q}_n = \mathbb{Q}_{\gcd(m,n)}$.

Hint. The groups $A := G_{\mathbb{Q}_m}^{\mathbb{Q}_m \mathbb{Q}_n}$, $B := G_{\mathbb{Q}_n}^{\mathbb{Q}_m \mathbb{Q}_n}$ and $C := G_{\mathbb{Q}_{\gcd(m,n)}}^{\mathbb{Q}_m \mathbb{Q}_n}$ are all subgroups of the abelian group $G := G_{\mathbb{Q}}^{\mathbb{Q}_m \mathbb{Q}_n}$. It is enough to show that $C = AB$ because then $\mathbb{Q}_{\gcd(m,n)} = F(C) = F(AB) = F(A) \cap F(B) = \mathbb{Q}_m \cap \mathbb{Q}_n$ where F denotes the formation of the fixed field.

(d) $\mathbb{Q}_\infty := \bigcup_{n \in \mathbb{N}} \mathbb{Q}_n$ is a subfield of $\mathbb{C}$, and $\mathrm{Aut}\, \mathbb{Q}_\infty$ is abelian.

Problem 16. Show that $\mathbb{Q}(e^{2\pi i/n}) \cap \mathbb{R} = \mathbb{Q}(\cos(2\pi/n))$ for all natural numbers n. For which values of n is this an extension of $\mathbb{Q}$ of degree 2?

Problem 17. Let $(L : \mathbb{Q})$ be a Galois extension such that $G_{\mathbb{Q}}^L$ is cyclic of order 4. Show that L cannot contain all fourth roots of unity.

Problem 18. Let K be the field obtained from $\mathbb{Q}$ by adjoining all roots of unity.

(a) Show that if $(N : \mathbb{Q})$ is a finite normal field extension with non-abelian Galois group then $N \not\subseteq K$.

(b) Show that $\sqrt[4]{2} \notin K$.

Hint. Let N be the splitting field of $x^4 - 2$ and use part (a).

Problem 19. Let $n \in \mathbb{N}$ and let $p_1, \ldots, p_k$ be pairwise different prime numbers. Show that $[\mathbb{Q}(\sqrt[n]{p_1}, \ldots, \sqrt[n]{p_k}) : \mathbb{Q}] = n^k$.

Hint. For the special case $n = 2$ see problem 33(c) in section 12. Show next that if n is even and ε is a primitive n-th root of unity then

$$
\begin{aligned}
&[\mathbb{Q}(\sqrt[n]{p_1}, \ldots, \sqrt[n]{p_k}) : \mathbb{Q}(\sqrt{p_1}, \ldots, \sqrt{p_k})] \\
&= [\mathbb{Q}(\sqrt[n]{p_1}, \ldots, \sqrt[n]{p_k}, \varepsilon) : \mathbb{Q}(\sqrt{p_1}, \ldots, \sqrt{p_k}, \varepsilon)] \\
&= \left(\frac{n}{2}\right)^k .
\end{aligned}
$$

Problem 20. Let K be a field and let L be the splitting field of $x^n - 1$ over K.
(a) Show that $(L : K)$ is a Galois extension.
(b) Show that if K is a finite field with q elements and if n is coprime with q then $[L : K]$ is the minimal number k such that n divides $q^k - 1$.
(c) Show that if n is a power of an odd prime then G_K^L is cyclic.
Hint. Use problem 26 in section 4.

Problem 21. For $n = 5, 7, 8$ and 11 let $L = \mathbb{Q}(e^{2\pi i/n})$ be the splitting field of $x^n - 1$ over $\mathbb{Q}$.
(a) Determine the Galois group $G_\mathbb{Q}^L$ and its subgroup lattice.
(b) Find all intermediate fields of the extension $(L : \mathbb{Q})$.

Problem 22. (a) Show that if p is a prime number of the form $p = 2^k + 1$ then the extension $(\mathbb{Q}(e^{2\pi i/p}) : \mathbb{Q})$ contains exactly one minimal intermediate field $Z \neq \mathbb{Q}$.
(b) Find all minimal intermediate fields of the extension $(\mathbb{Q}(e^{\pi i/10}) : \mathbb{Q})$.

Problem 23. Let $L := \mathbb{Q}(e^{\pi i/3}, e^{\pi i/5}, e^{2\pi i/15})$. Show that $(L : \mathbb{Q})$ is a Galois extension. Is the Galois group $G_\mathbb{Q}^L$ abelian?

Problem 24. Let $\varepsilon \in \mathbb{C}$ be a primitive p-th root of unity and let $\omega_0, \ldots, \omega_{d-1}$ be the (n/d)-periods of $(\mathbb{Q}(\varepsilon) : \mathbb{Q})$. Show that $\omega_0, \ldots, \omega_{d-1}$ form a basis of Z_d over $\mathbb{Q}$.

Problem 25. Express $\cos \frac{2\pi}{5}$ in terms of square roots of rational numbers.

Problem 26. (a) Show that the solutions of the equations (1), (2), (2′) and (3) in example (24.16) are all real.
(b) Prove that the regular 17-gon can be constructed with a ruler and a compass alone.
(c) Show that $\cos \frac{2\pi}{17}$ equals

$$
-\frac{1}{16} + \frac{1}{16}\sqrt{17} + \frac{1}{16}\sqrt{34 - 2\sqrt{17}} + \frac{1}{8}\sqrt{17 + 3\sqrt{17} - \sqrt{34 - 2\sqrt{17}} - 2\sqrt{34 + 2\sqrt{17}}} .
$$

(This formula was first given by Gauß in his *Disquisitiones Arithmeticae*.)

Problem 27. Which is the smallest number n such that an angle of n degrees can be constructed with a ruler and a compass only?

Problem 28. (a) For $k \in \mathbb{N}_0$ let $F_k := 2^{2^k} + 1$ be the k-th Fermat number. Show that $F_{k+1} - 2 = (F_k - 2)F_k$ and conclude that $F_m - 2 = \prod_{k<m} F_k$ for all m. (This shows that the Fermat numbers are pairwise coprime.)

(b) Let $n := 4,294,967,295 = 2^{32} - 1$. Show that a regular n-gon can be constructed by using a ruler and a compass only.

Problem 29. Let $p_1, \ldots, p_n$ be different odd primes and let $\varepsilon \in \mathbb{C}$ be a primitive $(8p_1 \cdots p_n)$-th root of unity. Show that $(x^2 + 1)(x^2 - 2)(x^2 - p_1) \cdots (x^2 - p_n)$ splits over $\mathbb{Q}(\varepsilon)$.

Problem 30. (a) Find a complex number α such that $(\mathbb{Q}(\alpha) : \mathbb{Q})$ is a Galois extension with $G_{\mathbb{Q}}^{\mathbb{Q}(\alpha)} \cong (\mathbb{Z}_{22}, +)$.

(b) Find a complex number α such that $(\mathbb{Q}(\alpha) : \mathbb{Q})$ is a Galois extension with $G_{\mathbb{Q}}^{\mathbb{Q}(\alpha)} \cong (\mathbb{Z}_{11}, +)$.

Problem 31. Let G be an arbitrary finite abelian group. Show that there is a Galois extension $(L : \mathbb{Q})$ such that $G_{\mathbb{Q}}^{L} \cong G$.

25. Pure equations and cyclic extensions

WE now turn to the problem of solving polynomial equations which is exactly the problem Galois had in mind when he developed his theory. Before we discuss arbitrary polynomial equations $f(x) = 0$ (which will be done in the next section), we will modestly start by only discussing "pure equations", i.e., equations of the form $x^n - a = 0$. It will turn out that the treatment of this special case gives us already the main tool to tackle polynomial equations in general.

(25.1) Definition. *Let K be a field. Every equation of the form $x^n - a = 0$, i.e., $x^n = a$, with $a \in K^\times$ is called a* **pure equation**, *and every solution of the equation $x^n = a$ (in some extension of K) is called an* **n-th root** *of a.*

SOLVING pure equations is closely tied up with finding roots of unity. In fact, if x_1, x_2 are roots of a pure polynomial $x^n - a$, i.e., satisfy the equation $x^n = a$, then $\varepsilon := x_2/x_1$ is a root of the equation $x^n = 1$; conversely, if $x_1^n = a$ and $\varepsilon^n = 1$ then $(\varepsilon x_1)^n = a$. This means that the roots of the equation $x^n = a$ are exactly the numbers εx_0 where x_0 is one particular root and where ε is an n-th root of unity.

WE now want to apply Galois theory to study pure equations. The idea is that the Galois group $\mathrm{Gal}_K(f)$ of a polynomial $f \in K[x]$ should reflect the properties of f; hence since polynomials of the form $f(x) = x^n - a$ are very special polynomials, we expect that their Galois groups are of a special type. This is indeed the case; under certain assumptions on the base-field K it will turn out that f is a "pure polynomial" if and only if its Galois group is cyclic. The next theorem proves the first half of this claim.

(25.2) Theorem. *Suppose the field K is such that $x^n - 1$ splits over K. Consider the pure polynomial $f(x) = x^n - a$ with $a \in K^\times$.*

(a) An extension L of K is a splitting field of f if and only if $L = K(\alpha)$ where α is any root of f.

(b) Let L be a splitting field of f. Then G_K^L is cyclic, and if α is any root of f then $|G_K^L|$ is the smallest number d such that $\alpha^d \in F(G_K^L)$; in particular $d \mid n$.

Proof. (a) If α is any root of f, then the totality of roots is given by the numbers $\varepsilon \alpha$ where $\varepsilon^n = 1$. Since all these elements ε lie in K, this shows that $L = K(\alpha)$ is a splitting field of f.

(b) If $\mathrm{char}\, K = p$ let m be the natural number such that $n = p^r m$ with $p \nmid m$; if $\mathrm{char}\, K = 0$ let $m = n$. We have $L = K(\alpha)$ by part (a), and the roots of f are the numbers $\varepsilon \alpha$ with $\varepsilon \in U(n, K) = U(m, K)$. Let ω be a primitive m-th roots of unity. Each $\sigma \in G_K^L$ maps α to another root of f, i.e., to an element $\omega^k \alpha$ where $k = k(\sigma)$ is uniquely determined modulo m. This gives a mapping

$$\kappa : \begin{array}{ccc} G_K^L & \to & (\mathbb{Z}_m, +) \\ \alpha & \mapsto & [k(\sigma)] \end{array} .$$

We have $\mathrm{id}(\alpha) = \omega^0 \alpha$ and hence $[k(\mathrm{id})] = [0]$; also, if $\sigma(\alpha) = \omega^r$ and $\tau(\alpha) = \omega^s$ then $(\sigma \circ \tau)(\alpha) = \sigma\big(\tau(\alpha)\big) = \sigma(\omega^s \alpha) = \omega^s \cdot \sigma(\alpha) = \omega^s \omega^r \alpha = \omega^{r+s} \alpha$ so that $[k(\sigma \circ \tau)] =$

$[k(\sigma)] + [k(\tau)]$. This shows that κ is a group homomorphism. Clearly κ is one-to-one because if $[k(\sigma)] = [0]$ then $\sigma(\alpha) = \alpha$ which implies that σ is the identity on $K(\alpha) = L$. Consequently, G_K^L is isomorphic to a subgroup of the cyclic group $(\mathbb{Z}_m, +)$ and hence is cyclic itself of an order which divides m and hence n.

Let $d := |G_K^L|$. If σ is a generator of G_K^L, then $u := \sigma(\alpha)/\alpha$ is a primitive d-th root of unity, and $\sigma(\alpha^d) = \sigma(\alpha)^d = (u\alpha)^d = u^d \alpha^d = \alpha^d$ so that $\alpha^d \in F(\langle\sigma\rangle) = F(G_K^L)$. Conversely, if $\alpha^e \in F(G_K^L)$ then $\alpha^e = \sigma(\alpha^e) = \sigma(\alpha)^e = (u\alpha)^e = u^e \alpha^e$ so that $u^e = 1$ which implies that e is divisible by d. This shows that d is the smallest number such that $\alpha^d \in F(G_K^L)$. ∎

TO prove the converse of (25.2) we need the following result on cyclic Galois extensions which has its name from the fact that its multiplicative version occurred as Theorem 90 in Hilbert's 1893 report on algebraic number theory.

(25.3) Hilbert's Theorem 90. *Let $(L : K)$ be a finite Galois extension with cyclic Galois group G_K^L, say $G_K^L = \langle\sigma\rangle$, and let $\alpha \in L$.*
  (a) *(Additive version.)* $\mathrm{tr}_K^L(\alpha) = 0$ *if and only if* $\alpha = \beta - \sigma\beta$ *for some $\beta \in L$.*
  (b) *(Multiplicative version.)* $N_K^L(\alpha) = 1$ *if and only if* $\alpha = \beta/\sigma(\beta)$ *for some* $\beta \in L^\times$.

Proof. Let $n = [L : K] = |G_K^L|$ so that $\sigma^n = \mathrm{id}$.
  (a) If $\alpha = \beta - \sigma\beta$ then $\mathrm{tr}_K^L(\alpha) = \sum_{k=0}^{n-1}\sigma^k(\alpha) = \sum_{k=0}^{n-1}\sigma^k(\beta) - \sum_{k=1}^{n}\sigma^k(\beta) = \beta - \sigma^n(\beta) = 0$. Conversely, suppose that $\mathrm{tr}_K^L(\alpha) = 0$. Since $\mathrm{tr}_K^L \not\equiv 0$ by (14.29), we can find an element ε with $\mathrm{tr}_K^L(\varepsilon) = 1$. For $1 \le k \le n - 1$ let

$$\beta_k := \sigma^k(\varepsilon)\sum_{i=0}^{k}\sigma^i(\alpha) = \sigma^k(\varepsilon)\big(\alpha + \sigma(\alpha) + \sigma^2(\alpha) + \cdots + \sigma^k(\alpha)\big) ;$$

we claim that then $\beta := \beta_0 + \beta_1 + \cdots + \beta_{n-2}$ has the desired property. Indeed, $\beta - \sigma(\beta) = \sum_{k=0}^{n-2}\beta_k - \sum_{k=0}^{n-2}\sigma(\beta_k) = \sum_{k=-1}^{n-3}\beta_{k+1} - \sum_{k=0}^{n-2}\sigma(\beta_k) = \beta_0 - \sigma(\beta_{n-2}) + \sum_{k=0}^{n-3}\big(\beta_{k+1} - \sigma(\beta_k)\big)$. Since obviously $\sigma(\beta_k) = \beta_{k+1} - \alpha\sigma^{k+1}(\varepsilon)$ and since $\beta_{n-1} = \sigma^{n-1}(\varepsilon)\mathrm{tr}_K^L(\alpha) = \sigma^{n-1}(\varepsilon) \cdot 0 = 0$, this becomes

$$\beta - \sigma(\beta) = \alpha\varepsilon + \alpha\sigma^{n-1}(\varepsilon) + \sum_{k=0}^{n-3}\alpha\sigma^{k+1}(\varepsilon)$$

$$= \alpha\big(\varepsilon + \sigma(\varepsilon) + \cdots + \sigma^{n-1}(\varepsilon)\big) = \alpha \cdot \mathrm{tr}_K^L(\varepsilon) = \alpha .$$

  (b) If $\alpha = \beta/\sigma(\beta)$ then

$$N_K^L(\alpha) = \alpha \cdot \sigma(\alpha) \cdot \sigma^2(\alpha)\cdots\sigma^{n-1}(\alpha) = \frac{\beta \cdot \sigma(\beta) \cdot \sigma^2(\beta)\cdots\sigma^{n-1}(\beta)}{\sigma(\beta) \cdot \sigma^2(\beta)\cdots\sigma^{n-1}(\beta) \cdot \sigma^n(\beta)} = 1$$

because $\sigma^n(\beta) = \beta$. Conversely, suppose that $N_K^L(\alpha) = 1$, i.e., that $\alpha \cdot \sigma(\alpha) \cdot \sigma^2(\alpha)\cdots\sigma^{n-1}(\alpha) = 1$. Then it is straightforward to check that for each choice of $\gamma \in L$ the element

$$\beta := \gamma + \alpha\sigma(\gamma) + \alpha\sigma(\alpha)\sigma^2(\gamma) + \cdots + \alpha\sigma(\alpha)\cdots\sigma^{n-2}(\alpha)\sigma^{n-1}(\gamma)$$

$$= \sum_{k=0}^{n-1}\left[\alpha\sigma(\alpha)\cdots\sigma^{k-1}(\alpha)\right]\sigma^k(\gamma)$$

satisfies the equation $\alpha \cdot \sigma(\beta) = \beta$. Hence if we can choose $\gamma \in L$ such that the corresponding element β is nonzero, we can divide by $\sigma(\beta)$ to obtain $\alpha = \beta/\sigma(\beta)$ as desired. But such a choice for γ is possible because $1, \sigma, \sigma^2, \ldots, \sigma^{n-1}$ are linearly independent by Dedekind's theorem (21.18) so that $\sum_{k=0}^{n-1}\left[\alpha\sigma(\alpha)\cdots\sigma^{k-1}(\alpha)\right]\sigma^k$ is not the zero mapping. ∎

AS a nice application of Hilbert's theorem, let us give a second proof of (22.9)(d).

(25.4) Proposition. *Let $K \subseteq L$ be finite fields. Then the norm function $N_K^L : L \to K$ is surjective.*

Proof. Since $N_K^L(0) = 0$ it is enough to show that $N := N_K^L : L^\times \to K^\times$ (which is a homomorphism of multiplicative groups) is surjective. Since $\operatorname{im} N \cong L^\times / \ker N$, we have $|L^\times| = |\ker N| \cdot |\operatorname{im} N|$. Analogously, the homomorphism $\Phi : L^\times \to L^\times$ given by $\beta \mapsto \beta/\sigma(\beta)$ satisfies $|L^\times| = |\operatorname{im} \Phi| \cdot |\ker \Phi|$. A comparison of these two equations yields

$$(\star) \qquad\qquad |\ker N| \cdot |\operatorname{im} N| = |\operatorname{im} \Phi| \cdot |\ker \Phi| .$$

Now $\ker \Phi = \{\beta \in L^\times \mid \sigma(\beta) = \beta\} = F(G_K^L) \setminus \{0\} = K^\times$ because $F(G_K^L) = K$ by (22.9)(a) (every extension of finite fields is a Galois extension) and $\operatorname{im} \Phi = \ker N$ by (25.3)(b). Hence $(\star)$ becomes

$$(\star\star) \qquad\qquad |\operatorname{im} \Phi| \cdot |\operatorname{im} N| = |\operatorname{im} \Phi| \cdot |K^\times|$$

so that $|\operatorname{im} N| = |K^\times|$ and hence $\operatorname{im} N = K^\times$. ∎

WE can now prove a converse of theorem (25.2).

(25.5) Theorem. *Suppose that K contains a primitive n-th root of unity ε. Let $(L : K)$ be a Galois extension such that G_K^L is cyclic of order $n := [L : K]$. Then L is the splitting field of some irreducible polynomial of the form $x^n - a$ with $a \in K^\times$.*

Proof. Let σ be a generator of G_K^L. Since $\varepsilon \in K$ we have $N_K^L(\varepsilon) = \varepsilon^n = 1$; hence by Hilbert's Theorem 90 there is an element $\beta \in L$ with $\varepsilon = \beta/\sigma(\beta)$ so that

$$(\star) \qquad\qquad \sigma(\beta) = \varepsilon^{-1}\beta .$$

Then $\sigma^m(\beta) = \varepsilon^{-m}\beta$ for all m because $\varepsilon \in K$; hence β has the n different conjugates $\varepsilon^m\beta$ with $0 \leq m \leq n-1$. Consequently, $\mathrm{id}_{K(\beta)}, \sigma|_{K(\beta)}, \ldots, \sigma^{n-1}|_{K(\beta)}$ are pairwise different elements of $G_K^{K(\beta)}$ so that

$$[K(\beta):K] \geq n = |G_K^L| = [L:K] = [L:K(\beta)] \cdot [K(\beta):K]$$

and hence $[L:K(\beta)] \leq 1$, i.e., $L = K(\beta)$. Moreover, $(\star)$ implies that $\sigma(\beta^n) = \sigma(\beta)^n = \varepsilon^{-n}\beta^n = \beta^n$ so that $\beta^n \in F(\langle\sigma\rangle) = F(G_K^L) = K$, say $\beta^n = a \in K$. We have shown that $L = K(\beta)$ where β is a root of a pure polynomial $x^n - a \in K[x]$; then (25.2)(a) gives the claim. Finally, $\deg(x^n - a) = n = [K(\beta):K]$ which implies that $x^n - a$ is the minimal polynomial of β over K. ∎

(25.6) Remark. The crucial point in the above proof was to find an element $\beta \in L$ with $\sigma(\beta) = \varepsilon^{-1}\beta$. This was done with Hilbert's Theorem 90. An inspection of the proof of Hilbert's theorem shows that β can be chosen in the form

$$\beta = \sum_{k=0}^{n-1}\left(\varepsilon \cdot \sigma(\varepsilon)\cdots\sigma^{k-1}(\varepsilon)\right)\sigma^k(\gamma) = \sum_{k=0}^{n-1}\varepsilon^k\sigma^k(\gamma)$$

with a suitable element γ. This observation gives a way to formulate the proof of (25.5) without invoking Hilbert's Theorem 90, as follows.

Let σ be a generator of G_K^L. Then the automorphisms $\mathrm{id}, \sigma, \ldots, \sigma^{n-1}$ are pairwise different, hence linearly independent by Dedekind's theorem. In particular, the mapping

$$\mathrm{Lag} := \mathrm{Lag}_\varepsilon := \mathrm{id} + \varepsilon\sigma + \cdots + \varepsilon^{n-1}\sigma^{n-1} : L \to L$$

is not identically zero, for any n-th root of unity ε, and if γ is chosen such that $\beta := \mathrm{Lag}(\gamma) \neq 0$ then

$$\begin{aligned}
\sigma(\beta) &= \sigma\left(\gamma + \varepsilon\sigma(\gamma) + \cdots + \varepsilon^{n-1}\sigma^{n-1}(\gamma)\right) \\
&= \sigma(\gamma) + \varepsilon\sigma^2(\gamma) + \cdots + \varepsilon^{n-1}\sigma^{n-1}(\gamma) = \frac{1}{\varepsilon}\beta.
\end{aligned}$$

As was shown in the proof of (25.5), this implies $\beta^n \in K$. Hence each of the mappings Lag_ε with $\varepsilon \in U(n,K)$ (which are called the **Lagrange resolvents** of $(L:K)$) has the property that $\mathrm{Lag}_\varepsilon(\gamma)^n \in K$ for all $\gamma \in L$. Moreover, since $\sum_{\varepsilon \in U(n)}\varepsilon = 0$ by (24.4)(a), we have $\sum_\varepsilon \mathrm{Lag}_\varepsilon = n \cdot \mathrm{id}$ where the sum runs over $U(n,K)$. These properties of the Lagrange resolvents will be used later on.

THEOREMS (25.2) and (25.5) state basically that cyclic Galois groups correspond to pure equations, under the crucial hypothesis that the base field contains primitive roots of unity of the appropriate order. The necessity of this hypothesis can be seen by observing that the Galois group of the pure polynomial $x^3 - 2 \in \mathbb{Q}[x]$ (namely Sym_3) is not cyclic (not even abelian); this is due to the fact that there are no cubic roots of unity in $\mathbb{Q}$.

In particular, theorems (25.2) and (25.5) are not applicable if $\mathrm{char}\, K = p$ and $p \mid n$ because the no primitive n-th roots of unity over K exist. However, in the special case $n = p$, there is a certain analogue to these theorems.

(25.7) Theorem (Artin-Schreier). *Let K be a field of prime characteristic p.*

(a) *An extension L of K is a splitting field of a polynomial of the form $x^p - x - a \in K[x]$ if and only if $L = K(\alpha)$ where α is any root of f; in this case $(L : K)$ is a Galois extension.*

(b) *Let L be the splitting field of a polynomial of the form $f(x) = x^p - x - a$ with $a \in K$. Then either $L = K$ or else f is irreducible over K and G_K^L is cyclic of oder p.*

(c) *Conversely, if $(L : K)$ is a Galois extension such that G_K^L is cyclic of order p, then L is a splitting field of a polynomial of the form $f(x) = x^p - x - a$.*

Proof. (a) Let $\alpha \in L$ be a root of f. Then for any $k \in \mathbb{Z}_p$ (which is the prime field of K) we have

$$
\begin{aligned}
f(\alpha + k) &= (\alpha + k)^p - (\alpha + k) - a = \alpha^p + k^p - \alpha - k - a \\
&= \underbrace{\alpha^p - \alpha - a}_{= f(\alpha) = 0} + \underbrace{k^p - k}_{= 0} = 0 ;
\end{aligned}
$$

hence f has the p different roots $\alpha, \alpha + 1, \ldots, \alpha + p - 1$. This shows that all irreducible factors of f are separable; hence $(L : K)$ is a Galois extension, and $L = K(\alpha)$.

(b) If one root of f lies in K then all do, so that $L = K$ in this case. Suppose that f has no root in K. If $\sigma \in G_K^L$ then σ permutes the roots of f; hence $\sigma(\alpha) = \alpha + k$ for some $k \in \mathbb{Z}_p$. Hence we have a map

$$
\Phi : \begin{array}{ccc} G_K^L & \to & (\mathbb{Z}_p, +) \\ \sigma & \mapsto & \sigma(\alpha) - \alpha \end{array}
$$

which is easily recognized to be a group isomorphism. In fact, if $\sigma(\alpha) = \alpha + k$ and $\tau(\alpha) = \alpha + l$ then $\Phi(\tau \circ \sigma) = \tau(\sigma(\alpha)) - \alpha = \tau(\alpha + k) - \alpha = \tau(\alpha) + k - \alpha = l + k = \Phi(\tau) + \Phi(\sigma)$ which shows that Φ is a group homomorphism. If $\sigma \in \ker \Phi$ then $\sigma(\alpha) = \alpha$ so that σ is the identity on $K(\alpha) = L$; this shows that Φ is one-to-one. Finally $G_K^L \neq \{\mathrm{id}\}$ because $(L : K)$ is Galois; hence $|G_K^L| = p$. This shows that Φ must be surjective. Hence G_K^L is isomorphic to a cyclic group of order p and thus itself cyclic of order p. Moreover, $[K(\alpha) : K] = p = |G_K^L| = [L : K]$ which implies that $L = K(\alpha)$.

(c) Let $G_K^L = \langle \sigma \rangle$. We have $\mathrm{tr}_K^L(1) = [L : K] \cdot 1 = p \cdot 1 = 0$ because $\mathrm{char}\, K = p$. Hence by (25.3)(a) there is an element $\alpha \in L$ with $1 = \sigma(\alpha) - \alpha$ so that $\sigma(\alpha) = \alpha + 1$. Then $\sigma^k(\alpha) = \alpha + k$ for $1 \leq k \leq p - 1$ (and even for all $k \in \mathbb{N}$ if we interpret k on the right-hand side as an element of $\mathbb{Z}_p \subseteq L$). These elements are pairwise distinct; hence $[K(\alpha) : K] \geq p = [L : K]$ which implies $L = K(\alpha)$. Also,

$$
\sigma(\alpha^p - \alpha) = \sigma(\alpha)^p - \sigma(\alpha) = (\alpha + 1)^p - (\alpha + 1) = \alpha^p - \alpha
$$

so that $\alpha^p - \alpha \in F(\langle \sigma \rangle) = F(G_K^L) = K$; hence $a := \alpha^p - \alpha$ lies in K, and α is a root of the polynomial $f(x) := x^p - x - a$. It was shown in part (a) already that then $L = K(\alpha)$ is a splitting field of f. ∎

WITH the results collected about cyclic extensions we can now state exactly when a pure polynomial over an arbitrary field is irreducible.

(25.8) Theorem. *Let K be an arbitrary field and let $f(x) = x^n - a$ with $a \in K^\times$. Then f is reducible if and only if at least one of the following two conditions holds:*

(1) there is a prime divisor q of n such that a is a q-th power in K, i.e., $a = \lambda^q$ for some $\lambda \in K$;

(2) n is divisible by 4, and $a = -4\lambda^4$ for some $\lambda \in K$.

Proof. Suppose (1) holds. Let $n = qm$ and $a = \lambda^q$ for some $\lambda \in K$. Then $x - \lambda$ divides $x^q - \lambda^q = x^q - a$; hence $x^m - \lambda$ divides $(x^m)^q - a = x^n - a$ so that $x^n - a$ is reducible.

Suppose (2) holds. Let $n = 4m$ and $a = -4\lambda^4$ for some $\lambda \in K$. Then

$$x^4 - a = x^4 + 4\lambda^4 = (x^2 - 2\lambda x + 2\lambda^2)(x^2 + 2\lambda x + 2\lambda^2)$$

is reducible over K; hence then so is $x^n - a = x^{4m} - a = (x^m)^4 - a$.

We have shown that any of the conditions (1) and (2) implies the reducibility of f. Suppose conversely that f is reducible. To show that (1) or (2) holds, we first treat the special case that $n = p$ is a prime number. What we have to establish in this special case is the following statement:

$(\star)$ | If the polynomial $x^p - a$ is reducible in $K[x]$ then it has a root in K.

Let α be a root of f in some splitting field L of f. If char $K = p$ then $f(x) = x^p - a = x^p - \alpha^p = (x - \alpha)^p$. Hence the only factors of f in L (up to units) are the polynomials $(x - \alpha)^i$ where $1 \leq i \leq p - 1$; thus if f has a proper factor in $K[x]$ then $\alpha \in K$. If char $K \neq p$ then there are p different roots of unity over L, and the roots of f are exactly the numbers $\varepsilon\alpha$ where ε ranges over these roots of unity. Let K' be obtained from K by adjoining a primitive p-th root of unity (so that K' is a splitting field of $x^p - 1$ over K). Since f is reducible over K then the more so over K'. Hence the minimal polynomial g of α over K' has a degree less than $\deg f = p$; on the other hand $\deg g = [K'(\alpha) : K'] = |G_{K'}^{K'(\alpha)}|$ divides p by (25.2)(b) and thus must be 1 so that $g(x) = x - \alpha$. This shows that $\alpha \in K'$. Now $G_K^{K'}$ is isomorphic to a subgroup of the cyclic group $\mathbb{Z}_p^\times$ by (24.12) and hence cyclic itself, say $G_K^{K'} = \langle\sigma\rangle$. Since σ permutes the roots of f, we have $\sigma(\alpha) = u\alpha$ for some p-th root of unity u. If $u = 1$ then $\alpha \in F(\langle\sigma\rangle) = F(G_K^{K'}) = K$. If $u \neq 1$ then u is a primitive root of unity. If in this case $\sigma(u) = u$ then $u \in F(\langle\sigma\rangle) = F(G_K^{K'}) = K$ so that $K' = K$ which implies again that $\alpha \in K$, and we are done in this case also. Finally, if $\sigma(u) \neq u$, then both u and $\sigma(u)/u$ are primitive p-th roots of unity; thus there is an exponent $1 \leq k \leq p - 1$ such that $u = (\sigma(u)/u)^k$. Then $\lambda := \alpha/u^k$ satisfies

$$\sigma(\lambda) = \sigma\left(\frac{\alpha}{u^k}\right) = \frac{\sigma(\alpha)}{\sigma(u)^k} = \frac{u\alpha}{u \cdot u^k} = \frac{\alpha}{u^k} = \lambda$$

and hence lies in $F(\langle\sigma\rangle) = F(G_K^{K'}) = K$; on the other hand $\lambda^p = \alpha^p/(u^p)^k = \alpha^p/1 = \alpha^p = a \in K$. This shows that f has a p-th root in K (namely λ).

Now we want to show in general that the reducibility of $x^n - a$ implies (1) or (2); to do so, we proceed by induction on n. Since the case that n is a prime has already been settled, the induction beginning is clear. Let $n \geq 4$, let q be a prime divisor of n and set $m := n/q$; if n is not a power of 2 we choose q to be odd. Now if α is a root of f in some splitting field, then α^q is a root of $x^m - a$. If $x^m - a$ is reducible, then conditions (1) and (2) hold for m (by induction hypothesis), hence the more so for n, and we are done. So suppose that $x^m - a$ is irreducible. Then $x^m - a$ is the minimal polynomial of α^q over K so that $[K(\alpha^q) : K] = m$. If $g(x) := x^q - \alpha^q$ were irreducible over $K' := K(\alpha^q)$, then we would have $[K(\alpha) : K(\alpha^q)] = [K'(\alpha) : K'] = q$ and hence $[K(\alpha) : K] = [K(\alpha) : K(\alpha^q)] \cdot [K(\alpha^q) : K] = q \cdot m = n$; but this would imply that f is the minimal polynomial of α over K and hence irreducible, contradicting our hypothesis. Hence g is reducible over K'. Applying $(\star)$, we see that g has a root, say β, in K' so that $\beta^q = \alpha^q$. Then

$$(\star\star) \qquad N_K^{K'}(\beta)^q = N_K^{K'}(\beta^q) = N_K^{K'}(\alpha^q) = -a(-1)^m$$

by (14.20)(a) because $x^m - a$ is the minimal polynomial of $\alpha^q \in K'$ over K. If n is not a power of 2 then we chose q as an *odd* prime divisor of n; we claim that in this case a has a q-th root in K so that condition (1) holds. In fact, if m is odd, then $(\star\star)$ reads $N_K^{K'}(\beta)^q = a$ so that $N_K^{K'}(\beta)$ is a q-th root of a in K; if m is even then (due to the fact that q is odd) equation $(\star\star)$ can be rewritten as $\left(-N_K^{K'}(\beta)\right)^q = a$ so that $-N_K^{K'}(\beta)$ is a q-th root of a in K.

We are left with the case that $n = 2^s$ with $s \geq 2$. Then $q = 2$ and $m = 2^{s-1}$ so that $(\star\star)$ reads $-a = N_K^{K'}(\beta)^2$ which shows that $-a$ is a square in K; let us write

$$-a = k^2$$

for short. Let i be a primitive 4-th root of unity; then

$$(\star\star\star) \qquad x^n - a = x^{2^s} - a = (x^{2^{s-1}} + ik)(x^{2^{s-1}} - ik) .$$

If $i \in K$ then the equation $-a = k^2$ becomes $a = (ik)^2$ so that a is a square in K; hence in this case condition (1) is satisfied. It remains to consider the case that $i \notin K$. Suppose that in this case $x^{2^{s-1}} + ik$ is irreducible over $K(i)$; then so is its conjugate $x^{2^{s-1}} - ik$, and hence $(\star\star\star)$ is the prime decomposition of f in $K(i)[x]$. Since $K[x]$ and $K(i)[x]$ are both unique factorization domains and since $x^n - a$ is reducible in $K[x]$ by hypothesis, we conclude that $(\star\star\star)$ is a decomposition in $K[x]$, contradicting our assumption that $i \notin K$. This contradiction shows that $x^{2^{s-1}} + ik$ is reducible over $K(i)$. We can apply the induction hypothesis to this polynomial and thus know that (1) or (2) holds. Now (1) says that ik is a square in $K(i)$, and if (2) holds then $ik = -4\lambda_0^4$ for some $\lambda_0 \in K(i)$, i.e., $ik = (2i\lambda_0^2)^2$; hence in each of the two cases ik is a square in $K(i)$. Writing $ik = (\lambda + i\mu)^2 = \lambda^2 - \mu^2 + 2i\lambda\mu$, we conclude that $\lambda^2 = \mu^2$ and $2\lambda\mu = k$, hence $k^2 = 4\lambda^2\mu^2 = 4\lambda^4$ and consequently $a = -k^2 = -4\lambda^4$ which is condition (2). ∎

SUPPOSE that K is a field which contains a primitive n-th root of unity. Then the splitting field L of a pure polynomial $f(x) = x^n - a \in K[x]$ is $L = K(\alpha)$ where α is *any* root of f, and we can unambiguously write $L = K(\sqrt[n]{a})$. We also saw that the field extensions $(L : K)$ of this form, i.e., the ones that are obtained from K by adjoining an n-th root, can be characterized in terms of their Galois groups. Namely, in (25.2) and (25.5) we established the following equivalence:

$$\boxed{L = K(\sqrt[n]{a}) \text{ for some } a \in K^\times \iff G_K^L \text{ is cyclic of order } \leq n .}$$

We now ask – following Ernst Eduard Kummer (1810-1893) – whether or not a similar characterization exists for those fields which are obtained from K by adjoining *several* n-th roots. More precisely, for any set $A \subseteq K^\times$ we denote by $\sqrt[n]{A}$ the set of all n-th roots of elements of A (in some fixed algebraic closure $\overline{K}$ of K). Then we try to characterize the fields of the form $K(\sqrt[n]{A})$ in terms of their Galois groups; i.e., we try to establish an equivalence of the following form:

$$L = K(\sqrt[n]{A}) \text{ for some } A \subseteq K^\times \iff G_K^L \text{ satisfies ??? .}$$

Before we do so, let us observe two things. First, $\left(K(\sqrt[n]{A}) : K\right)$ is always a Galois extension, because $K(\sqrt[n]{A})$ is the splitting field of the family of polynomials $\{x^n - a \mid a \in A\}$, and each member of this family is separable due to the existence of a primitive n-th root of unity. Second, we have $K(\sqrt[n]{A}) = K(\sqrt[n]{B})$ where B is the subgroup of $(K^\times, \cdot)$ generated by A and $(K^\times)^n := \{k_1 \cdots k_n \mid k_i \in K^\times\}$. Hence it is no loss of generality to assume that A is a subgroup of $K^\times$ containing $(K^\times)^n$. Now before we can present Kummer's solution of characterizing fields of the form $K(\sqrt[n]{A})$ we have to introduce the notion of a *pairing* of abelian groups. (See (I.25.28) through (I.25.32) for the concept of the character group of an abelian group which is involvded in the following definition.)

(25.9) Definition. *A **pairing** of two abelian groups G and H is a mapping $\varphi : G \times H \to \mathbb{C}^\times$ which is multiplicative in each component. This is tantamount to saying that*

$$\varphi_1 : \begin{matrix} G & \to & H^* \\ g & \mapsto & \varphi(g, \cdot) \end{matrix} \quad and \quad \varphi_2 : \begin{matrix} H & \to & G^* \\ h & \mapsto & \varphi(\cdot, h) \end{matrix}$$

*are group homomorphisms. Such a pairing is called **nondegenerate** if both φ_1 and φ_2 are injective, i.e., if $\varphi(g, h) = 1$ for all $h \in H$ implies that $g = e_G$ and if $\varphi(g, h) = 1$ for all $g \in G$ implies that $h = e_H$.*

(25.10) Proposition. *Let $\varphi : G \times H \to \mathbb{C}^\times$ be a nondegenerate pairing of abelian groups. Then G is finite if and only if H is, and in this case $\varphi_1 : G \to H^*$ and $\varphi_2 : H \to G^*$ are isomorphisms. Consequently, $G \cong H$.*

Proof. Suppose without loss of generality that G is finite (if H is finite a completely analogous argument can be used). Since φ is nondegenerate the mappings φ_1 and φ_2 are injective; also, $G^* \cong G$ and $H^* \cong H$ by (I.25.32)(a). Consequently,

$$|G| = |\varphi_1(G)| \le |H^*| = |H| = |\varphi_2(H)| \le |G^*| = |G|$$

so that $|G| = |H|$. This implies that φ_1 and φ_2 are in fact isomorphisms; hence $G \cong H^* \cong H$. ∎

FINALLY, the following terminology will allow us an easy formulation of Kummer's results.

(25.11) Definition. *A group G is called **of exponent n** if $g^n = e$ for all $g \in G$.*

(25.12) Theorem (Kummer Correspondence). *Let K be a field containing a primitive n-th root of unity and fix an algebraic closure $\overline{K}$ of K. Then a field extension $(L : K)$ has the form $L = K(\sqrt[n]{A})$ if and only if G_K^L is abelian of exponent n. More precisely, there are inverse bijections*

$$
\boxed{\begin{array}{c} \text{subgroups } A \text{ of } K^\times \\ \text{with } (K^\times)^n \subseteq A \end{array}} \;\;\begin{array}{c} F \\ \rightleftarrows \\ G \end{array}\;\; \boxed{\begin{array}{c} \text{intermediate fields } L \text{ of } (\overline{K} : K) \\ \text{with } G_K^L \text{ abelian of exponent } n \end{array}}
$$

given by

$$
F(A) := K(\sqrt[n]{A}) \quad and \quad G(L) := (L^\times)^n \cap K^\times .
$$

Here $(K(\sqrt[n]{A}) : K)$ is finite if and only if the quotient group $A/(K^\times)^n$ is finite, and in this case $G_K^{K(\sqrt[n]{A})} \cong A/(K^\times)^n$ so that $[K(\sqrt[n]{A}) : K] = [A : (K^\times)^n]$.

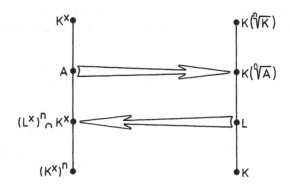

Proof. Let A be a subgroup of $K^\times$ containing $(K^\times)^n$ and let $L := K(\sqrt[n]{A})$. If $\alpha \in \sqrt[n]{A}$, say $\alpha^n = a \in A \subseteq K$, and if $\sigma \in G_K^L$ then $\sigma(\alpha)^n = \sigma(\alpha^n) = \sigma(a) = a = \alpha^n$; hence $\sigma(\alpha) = \varepsilon\alpha$ where ε is an n-th root of unity. To emphasize the dependence of ε on σ and on α we write

$$
(1) \qquad\qquad \sigma(\alpha) = \varepsilon(\sigma, \alpha)\,\alpha \quad \text{where } \sigma \in G_K^L \text{ and } \alpha \in \sqrt[n]{A} .
$$

Now let $\alpha, \beta \in \sqrt[n]{A}$ and $\sigma, \tau \in G_K^L$. Then $\varepsilon(\sigma, \alpha\beta)\alpha\beta = \sigma(\alpha\beta) = \sigma(\alpha)\sigma(\beta) = \varepsilon(\sigma, \alpha)\alpha \cdot \varepsilon(\sigma, \beta)\beta$; consequently,

$$
(2) \qquad\qquad \varepsilon(\sigma, \alpha\beta) = \varepsilon(\sigma, \alpha) \cdot \varepsilon(\sigma, \beta) .
$$

Also, $\varepsilon(\tau\sigma, \alpha) = (\tau\sigma)(\alpha) = \tau\big(\sigma(\alpha)\big) = \tau\big(\varepsilon(\sigma, \alpha)\alpha\big) = \varepsilon(\sigma, \alpha)\tau(\alpha) = \varepsilon(\sigma, \alpha)\varepsilon(\tau, \alpha)\alpha$. Using the fact that $K^\times$ is abelian, this shows that

$$
(3) \qquad\qquad \varepsilon(\tau\sigma, \alpha) = \varepsilon(\sigma, \alpha)\varepsilon(\tau, \alpha) = \varepsilon(\tau, \alpha)\varepsilon(\sigma, \alpha) = \varepsilon(\sigma\tau, \alpha) .
$$

Conditions (1) and (3) show that $\tau\sigma = \sigma\tau$ on $\sqrt[n]{A}$ and hence on all of L. Moreover, we have $\sigma^n(\alpha) = \varepsilon(\sigma, \alpha)^n\alpha = \alpha$ so that $\sigma^n = \mathrm{id}$ on $\sqrt[n]{A}$ and hence on L. This shows that if $L = K(\sqrt[n]{A})$ then G_K^L is abelian of exponent n. On the other hand, if L is such that G_K^L is abelian of exponent n, then clearly $(L^\times)^n \cap K^\times$ is a subgroup of $K^\times$ containing $(K^\times)^n$. This shows that the mappings F and G are well-defined.

Next we want to establish the following condition:

$$(4) \qquad\qquad\qquad \text{if } \alpha^n = \beta^n \text{ then } \varepsilon(\sigma,\alpha) = \varepsilon(\sigma,\beta) \ .$$

In fact, if $\beta^n = \alpha^n$ then $\beta = u\alpha$ for some n-th root of unity u and hence $\varepsilon(\sigma,\beta) = \sigma(\beta) = \sigma(u\alpha) = u \cdot \sigma(\alpha) = u \cdot \varepsilon(\sigma,\alpha)\alpha = \varepsilon(\sigma,\alpha)\beta$. The results proved up to now show that there is a pairing

$$(5) \qquad \begin{array}{ccc} G_K^L \times A & \to & U(n,K) \\ (\sigma,a) & \mapsto & \varepsilon(\sigma,\alpha) \end{array} \quad \text{where } \alpha \text{ is any element with } \alpha^n = a.$$

This pairing is nondegenerate in the first variable: If $\varepsilon(\sigma,a) = 1$ for all a then $\sigma = \mathrm{id}$ by (1). On the other hand, if $\alpha^n = a$ where $\varepsilon(\sigma,a) = 1$ for all $\sigma \in G_K^L$ then (1) shows that $\alpha \in F(G_K^L) = K$ and hence $a \in (K^\times)^n$. Thus (5) yields a nondegenerate pairing

$$(6) \qquad \begin{array}{ccc} G_K^L \times A/(K^\times)^n & \to & U(n,K) \\ \big(\sigma, a \cdot (K^\times)^n\big) & \mapsto & \varepsilon(\sigma, \sqrt[n]{a}) \end{array}$$

where $\sqrt[n]{a}$ means any element α with $\alpha^n = a$. Then (25.10) implies that G_K^L is finite if and only if $A/(K^\times)^n$ is finite and that $G_K^L \cong A/(K^\times)^n$ in this case. It remains to show that F and G are inverse bijections.

First, if L is any intermediate field of $(\overline{K} : K)$ and if A is any subgroup of $K^\times$ containing $(K^\times)^n$, then trivially

$$(7) \qquad\qquad\qquad (F \circ G)(L) \subseteq L \quad \text{and} \quad (G \circ F)(A) \supseteq A$$

and consequently

$$(8) \qquad\qquad\qquad F \circ G \circ F = F \quad \text{and} \quad G \circ F \circ G = G \ .$$

Let us now prove that

$$(9) \qquad\qquad\qquad G \circ F = \mathrm{id} \ ,$$

i.e., that $G\big(F(A)\big)(A) = A$ for all A. As was stated above, the inclusion $\supseteq$ is trivial. The converse inclusion $\subseteq$ is clear if $[F(A) : K] < \infty$ because then

$$\left| \frac{A}{(K^\times)^n} \right| = [F(A) : K] = [(F \circ G \circ F)(A) : K] = \left| \frac{G\big(F(A)\big)}{(K^\times)^n} \right|$$

due to (25.10) which together with the inclusion $A \subseteq G\big(F(A)\big)$ implies that $A = G\big(F(A)\big)$. Now let A be arbitrarily chosen and let $\alpha \in G\big(F(A)\big)$; we want to show that $\alpha \in A$. Since $K(\sqrt[n]{a}) \subseteq K(\sqrt[n]{(G \circ F)(A)}) = (F \circ G \circ F)(A) = F(A) = K(\sqrt[n]{A})$ there are elements $a_1, \ldots, a_r \in A$ with $\sqrt[n]{\alpha} \in K(\sqrt[n]{a_1}, \ldots, \sqrt[n]{a_r})$. Let A_0 be the subgroup of $K^\times$ generated by $\{a_1, \ldots, a_r\}$ and $(K^\times)^n$; then $\sqrt[n]{\alpha} \in K(\sqrt[n]{A_0}) = F(A_0)$ and hence $\alpha \in G\big(F(A_0)\big)$. But since $A_0/(K^\times)^n$ is finite we have $G\big(F(A_0)\big) = A_0 \subseteq A$ so that $\alpha \in A$ as desired.

What remains is to establish that

$$(10) \qquad\qquad F \circ G = \mathrm{id} \, ,$$

i.e., that $F(G(L)) = L$ for all L. Only the inclusion $\supseteq$ needs still to be proved. We consider three cases. *First case:* If G_K^L is a finite cyclic group then $L = K(\alpha)$ where $\alpha^n \in K$, due to (25.5) and (25.2). But then $\alpha = \sqrt[n]{\alpha^n} \in \sqrt[n]{G(L)} \subseteq F(G(L))$ and consequently $L = K(\alpha) \subseteq F(G(L))$. *Second case:* If $[L : K]$ is finite then G_K^L is a finite abelian group, hence is a direct product of cyclic subgroups which by the Galois correspondence can be written as $G_K^{L_1}, \ldots, G_K^{L_r}$ with intermediate fields $K \subseteq L_i \subseteq L$. But then $G_K^L = \prod_{i=1}^r G_K^{L_i} = G_K^{L_1 \cdots L_r}$ so that L equals the compositum $L = L_1 \cdots L_r$ of the L_i. Since $L_i \subseteq F(G(L_i))$ for all i by the first case and since $G \circ F$ preserves inclusions, this gives

$$
\begin{aligned}
L = L_1 \cdots L_r &\subseteq F(G(L_1)) \cdots F(G(L_r)) \\
&\subseteq F(G(L)) \cdots F(G(L)) = F(G(L)) \, .
\end{aligned}
$$

Third (general) case: Given the intermediate field L, let $\alpha \in L$; then $\alpha \in L_0$ for some intermediate field L_0 of $(L : K)$ with $[L_0 : K] < \infty$. Then $G_K^{L_0}$ is a subgroup of G_K^L, hence also abelian of exponent n. Consequently, $L_0 \subseteq F(G(L_0))$ by the second case so that $\alpha \in L_0 \subseteq F(G(L_0)) \subseteq F(G(L))$. Since α was an arbitrary element of L this proves the claim. ∎

THERE is also a Kummer correspondence in characteristic p which is in relation to (25.12) in the same way that (25.7) is in relation with (25.2) and (25.5). (See problem 17 below.)

Exercises

Problem 1. Let K be a field and let $a \in K^\times$. Show that if char $K \nmid n$ then the splitting field of $x^n - a$ contains a primitive n-th root of unity.

Problem 2. Let $a \in \mathbb{Z} \setminus \{0, \pm 1\}$ be square-free. Find the Galois groups of $x^4 - a$ and of $x^n - a$ (n odd) over $\mathbb{Q}$.

Problem 3. Let t be a variable. Find the Galois groups of $x^4 - t$ over $\mathbb{C}(t)$ and over $\mathbb{R}(t)$.

Problem 4. Let $a \in \mathbb{Z} \setminus \{0, \pm 1\}$ be a square-free number.
(a) Let p be a prime number and let $K \subseteq \mathbb{C}$ be the splitting field of $x^p - a$. Show that $[K : \mathbb{Q}] = p(p-1)$.
(b) Let $p_1, \ldots, p_r$ be distinct odd prime numbers and let $K \subseteq \mathbb{C}$ be the splitting field of $f(x) := (x^{p_1} - a) \cdots (x^{p_r} - a)$. Show that $[K : \mathbb{Q}] = p_1(p_1 - 1) \cdots p_r(p_r - 1)$.
(c) Let $p_1, \ldots, p_r$ be distinct odd prime numbers and let $K \subseteq \mathbb{C}$ be the splitting field of $f(x) := (x^2 - a)(x^{p_1} - a) \cdots (x^{p_r} - a)$. Moreover, let $m := 2p_1 \cdots p_r$. Show that

$$[K : \mathbb{Q}] = \begin{cases} p_1(p_1 - 1) \cdots p_r(p_r - 1), & \text{if } \sqrt{a} \in \mathbb{Q}(e^{2\pi i/m}); \\ 2p_1(p_1 - 1) \cdots p_r(p_r - 1), & \text{if } \sqrt{a} \notin \mathbb{Q}(e^{2\pi i/m}). \end{cases}$$

Problem 5. Let $f \in K[x]$ be irreducible of prime degree $p \neq$ char K and let L be a splitting field of f. Show that if there are two different roots $\alpha \neq \beta$ of f with $\beta \in K(\alpha)$ then $L = K(\alpha)$ and G_K^L is cyclic of order p.

Problem 6. Let $f \in K[x]$ be an irreducible separable polynomial where K is a field with char $K \neq 2$. Show that if $\mathrm{Gal}_K(f)$ is cyclic then $D(f)$ is a square in K if and only if $|\mathrm{Gal}_K(f)|$ is odd.

Problem 7. Let K be a field of characteristic $p \neq 0$ and let $a, b \in K \setminus \{0\}$.
(a) Show that $f(x) = x^p - b^{p-1}x - a$ is either irreducible or splits over K.
(b) Show that $g(x) = ax^p + b^{p-1}x^{p-1} - 1$ is either irreducible or splits over K.
Proof. Substitute $x = by$ in part (a) and $x = 1/y$ in part (b).

Problem 8. Let K be a field. Show that $x^{2^n} + 1$ is irreducible in $K[x]$ for all $n \in \mathbb{N}$ if and only if none of the elements -1, 2, -2 is a square in K.

Problem 9. Let m and n be coprime natural numbers. Use (25.8) to prove that $x^{nm} - a$ is irreducible if and only if both $x^m - a$ and $x^n - a$ are irreducible. Compare with the proof outlined in problem 27 of section 12.

Problem 10. Let K be a field of characteristic p and let $f \in K[x]$ be a separable polynomial.

(a) Show that if α is a root of a polynomial $x^p - x - a \in K[x]$ then $\mathrm{Gal}_{K(\alpha)}(f)$ either equals $\mathrm{Gal}_K(f)$ or else is a normal subgroup of index p.

(b) Show that if $U \leq \mathrm{Gal}_K(f)$ is a normal subgroup of index p then there is a root α of a polynomial $x^p - x - a \in K[x]$ such that $U = \mathrm{Gal}_{K(\alpha)}(f)$.

Problem 11. Let K be a field of characteristic $p \neq 0$ and suppose that $a \in K$ is such that $f(x) := x^p - x - a$ is irreducible in $K[x]$.

(a) Show that if α is a root of f (in some extension of K) then $g(x) := x^p - x - a\alpha^{p-1}$ is irreducible over $K(\alpha)$. Conclude that if β is a root of g then $[K(\alpha,\beta) : K] = p^2$.

Hint. To prove the irreducibility of g over $K(\alpha)$ it is enough to show that g has no root in $K(\alpha)$. Let $\xi := \sum_{i=0}^{p-1} k_i \alpha^i$ be an arbitrary element of $K(\alpha)$ and calculate $g(\xi)$.

(b) Explain how one can construct for any given natural number n an extension L of K such that $[L : K] = p^n$.

Problem 12. Let $\overline{\mathbb{Q}}$ be an algebraic closure of $\mathbb{Q}$ and let $\alpha \in \overline{\mathbb{Q}} \setminus \mathbb{Q}$. Suppose that K is an intermediate field of $(\overline{\mathbb{Q}} : \mathbb{Q})$ which is maximal with the property of not containing α. Prove that if $(L : K)$ is a finite extension of K then G_K^L is cyclic.

Problem 13. Let $\overline{K}$ be an algebraic closure of a field K. Show that if Z is the fixed field of some element $\sigma \in G_K^{\overline{K}}$ and if $(L : Z)$ is a finite extension, then G_Z^L is cyclic.

Problem 14. Let $(L : K)$ be a Galois extension in characteristic $p \neq 0$ such that G_K^L is cyclic of order p^m with $m \geq 1$. Let σ be a generator of G_K^L. Prove the following statements!

(a) If $\beta \in L$ satisfies $\mathrm{tr}_K^L(\beta) = 1$ then $\beta^p - \beta = \sigma(\alpha) - \alpha$ for some $\alpha \in L$.

(b) If α is as in (a) then $f(x) := x^p - x - \alpha$ is irreducible over L.

(c) If θ is a root of f then $(L(\theta) : K)$ is a Galois extension whose Galois group is cyclic of order p^{m+1}. One can find a generator σ^* of $G_K^{L(\theta)}$ extending σ and satisfying $\sigma^*(\theta) = \theta + \beta$.

Problem 15. Let A be a subgroup of the abelian group G. Then every character φ of G/A induces a character $(\mathrm{ind}\,\varphi)$ of G via $(\mathrm{ind}\,\varphi)(g) := \varphi(gA)$, and every character χ of G can be restricted to a character $(\mathrm{res}\,\chi) := \chi|_A$ of A. Show that the sequence

$$\{1\} \;\to\; (G/A)^\star \;\overset{\mathrm{ind}}{\to}\; G^\star \;\overset{\mathrm{res}}{\to}\; A^\star \;\to\; \{1\}$$

is exact (which means that the image of each of the homomorphisms in this chain equals the kernel of the following homomorphism). Conclude in particular that each element of $A^\star$ can be extended to an element of $G^\star$.

Problem 16. Let G be a finite abelian group. Show that

$$\begin{array}{ccc} G^\star \times G & \to & \mathbb{C}^\times \\ (\chi, g) & \mapsto & \chi(g) \end{array}$$

is a nondegenerate pairing.

Hint. To prove the nondegeneracy in the second variable, use that if $g \neq e$ in G then every element of $\langle g \rangle^\star$ can be extended to an element of $G^\star$.

Problem 17. Let K be a field of prime characteristic p and let $\overline{K}$ be a fixed algebraic closure of K.

(a) Let A be a subgroup of $(K, +)$ containing $\{x^p - x \mid x \in K^\times\}$. Let $F(A)$ be the intermediate field of $(\overline{K} : K)$ which is obtained from K by adjoining all roots of the polynomials $x^p - x - a$ with $a \in A$. Show that the Galois group $G_K^{F(A)}$ is abelian of exponent p.

(b) Let L be an intermediate field of $(\overline{K} : K)$ such that G_K^L is abelian of exponent p. Show that $G(L) := \{x \in K^\times \mid x = y^p - y \text{ for some } y \in L^\times\}$ is a subgroup of $(K, +)$ which contains $\{x^p - x \mid x \in K^\times\}$.

(c) Show that the mappings F and G defined in parts (a) and (b) constitute a 1-1-correspondence between the subgroups of $(K, +)$ containing $\{x^p - x \mid x \in K\}$ and the intermediate fields of $(\overline{K} : K)$ whose Galois group over K is abelian of exponent p. Moreover, show that such an intermediate fields has finite degree if and only if $\{x^p - x \mid x \in K^\times\}$ has finite index in the corresponding subgroup, and that the degree and the index coincide in this case.

26. Solvable equations and radical extensions

ONE main goal in classical algebra was to find explicit solutions of polynomial equations. Here explicitly solving an equation $f(x) = 0$ with $f \in K[x]$ means writing down the roots of f as "radicals", i.e., as expressions of the form

$$(\star) \qquad \sqrt[n]{\cdots + \sqrt[m]{\cdots} + \sqrt[k]{\cdots} + \cdots + \cdots}$$

where only elements of the base-field K occur under the root signs. In other words, the goal is to obtain the roots of f from the coefficients of f by performing rational operations (addition, subtraction, multiplication and division) and by extracting roots.

(26.1) Example. To find the roots of $f(x) = x^4 - 10x^2 + 1 \in \mathbb{Q}[x]$, we substitute $u := x^2$ and solve the equation $u^2 - 10u + 1 = 0$ which yields $u = 5 \pm 2\sqrt{6}$; hence the roots of f are $\pm\sqrt{5 \pm 2\sqrt{6}}$ where all four combinations of signs are possible. Noting that $(\sqrt{3} \pm \sqrt{2})^2 = 3 \pm 2\sqrt{6} + 2 = 5 \pm 2\sqrt{6}$ and hence $\sqrt{3} \pm \sqrt{2} = \sqrt{5 \pm 2\sqrt{6}}$, we see that the roots of f can also be written in the form $\pm(\sqrt{3} \pm \sqrt{2})$.

AS was pointed out in section 20, there are general formulas for equations up to degree 4 (under certain conditions on the base-field). Such formulas are highly desirable, because they allow one to find the roots of any given polynomial by simply plugging its coefficients into a general expression determined once and for all (such as the well-known quadratic formula). However, at the beginning of the 19th century it was realized (by Abel and Ruffini) that no such formulas do exist in degrees greater than 4. On the other hand, certain polynomials of arbitrarily high degrees allow explicit solutions in the above sense, for example all cyclotomic polynomials. Hence we want to investigate the conditions under which a polynomial equation can be "solved by radicals". To be able to apply the field-theoretical methods developed before (and also to clarify the problem), we will focus on the domains in which one can find "explicit solutions", namely, on "radical extensions" of the coefficient field of the polynomial in question.

(26.2) Definitions. (a) *A field extension* $(L : K)$ *is called a* **radical extension** *if* L *can be obtained from* K *by successively adjoining roots of existing elements, i.e., if* L *has the form* $K(\alpha_1, \ldots, \alpha_m)$ *where some power of* α_i *lies in* $K(\alpha_1, \ldots, \alpha_{i-1})$ *for* $1 \leq i \leq m$. *This means exactly that* α_i *satisfies a pure equation over* $K(\alpha_1, \ldots, \alpha_{i-1})$.

(b) *An element* α *which is algebraic over* K *is said to be* **expressible by radicals** *over* K *if* α *is contained in some radical extension of* K.

(c) *A polynomial equation* $f(x) = 0$ *with* $f \in K[x]$ *is called* **solvable by radicals** *or simply* **solvable** *if all roots of* f *are expressible by radicals.*

(26.3) Remarks. (a) Every radical extension is a finite algebraic extension; this is immediately clear from the definition.

(b) By inserting further elements α_i, if necessary, we can arrange in (26.2)(a) that for each i a *prime* power of α_i lies in $K(\alpha_1, \ldots, \alpha_{i-1})$. This follows immediately from the observation that $K(\alpha^{p_1 p_2 \cdots p_r}) \subseteq K(\alpha^{p_1}, \alpha^{p_1 p_2}, \ldots, \alpha^{p_1 p_2 \cdots p_n})$.

(c) A chain of fields $K = K_1 \subseteq K_2 \subseteq \cdots \subseteq K_{m-1} \subseteq K_m = L$ is called a **radical chain** if for each i there is an element $\alpha_i \in K_{i+1}$ such that $K_{i+1} = K_i(\alpha_i)$ and $\alpha_i^{n_i} \in K_i$ for some $n_i \in \mathbb{N}$ (so that α_i is an n_i-th root of an element of K_i). Then $(L : K)$ is a radical extension if and only if there is a radical chain starting with K and ending with L.

(26.4) Examples. (a) As was shown in example (26.1), the splitting field L of $x^4 - 10x^2 + 1 \in \mathbb{Q}[x]$ is $\mathbb{Q}(\sqrt{5 \pm 2\sqrt{6}}) = \mathbb{Q}(\sqrt{3} \pm \sqrt{2})$. Using (12.4)(c) we see that $L = \mathbb{Q}(\sqrt{2}, \sqrt{3}) = \mathbb{Q}(\sqrt{2} + \sqrt{3}) = \mathbb{Q}(\sqrt{5 + 2\sqrt{6}})$, hence also $L = \mathbb{Q}(\sqrt{2}, \sqrt{3}) = \mathbb{Q}(\sqrt{2}, \sqrt{6}) = \mathbb{Q}(\sqrt{3}, \sqrt{6})$. Thus there are various different radical chains which exhibit L as a radical extension of $\mathbb{Q}$. Of course, we can also find radical chains like

$$\mathbb{Q} \subseteq \mathbb{Q}(i) \subseteq \mathbb{Q}(\frac{1+i}{\sqrt{2}}) \subseteq \mathbb{Q}(\frac{1+i}{\sqrt{2}}, \sqrt{3})$$

which do not end with L itself but with a field strictly containing L.

(b) If $K = \mathbb{R}$ or $\mathbb{C}$, then every polynomial equation $f(x) = 0$ with $f \in K[x]$ is solvable by radicals; this is an immediate consequence of the fundamental theorem of algebra.

(c) Suppose that the field K contains a primitive n-th root of unity. Then each Galois extension $(L : K)$ with cyclic Galois group G_K^L is a radical extension; indeed, by (25.5) and (25.2)(a) there is an element $\alpha \in L$ such that $L = K(\alpha)$ and $\alpha^n \in K$ where $n = [L : K]$.

(d) If $(L : K)$ is finite and purely inseparable, then $(L : K)$ is a radical extension; this is an immediate consequence of (14.15)(c).

TO see that there are indeed polynomial equations which are not solvable by radicals, we give the following example.

(26.5) Example. The general quadratic equation over $\mathbb{Z}_2$ is not solvable by radicals. More precisely, consider the rational function field $K := \mathbb{Z}_2(p, q)$ in two variables p, q over $\mathbb{Z}_2$; then the splitting field of $f(x) = x^2 + px + q \in K[x]$ is not contained in a radical extension of K. The proof of this statement is outlined in problem 1 below.

LET us collect some basic properties of radical extensions.

(26.6) Proposition. (a) *If $(L : K)$ is a radical extension and if $K \subseteq Z \subseteq L$ is an intermediate field, then $(L : Z)$ is also a radical extension.*

(b) *Suppose $Z_1, \ldots, Z_r$ are intermediate fields of a field extension $(L : K)$. If all the $(Z_i : K)$ are radical extensions, then $(Z_1 Z_2 \cdots Z_r : K)$ is a radical extension too. In short, finite composita of radical extensions are radical again.*

(c) *If $(L : K)$ is a radical extension and if N is a normal closure of $(L : K)$ then $(N : K)$ is also a radical extension.*

(d) *Every separable radical extension is contained in a radical extension of Galois type.*

Proof. (a) If $L = K(a_1, \ldots, a_n)$ with $a_i^{n_i} \in K(a_1, \ldots, a_{i-1})$, then the more so $L = Z(a_1, \ldots, a_n)$ where $a_i^{n_i} \in K(a_1, \ldots, a_{i-1}) \subseteq Z(a_1, \ldots, a_{i-1})$.

(b) It is clearly enough to treat the case $r = 2$, for then the general case will follow by induction. Let $K \subseteq K(\alpha_1) \subseteq \cdots \subseteq K(\alpha_1, \alpha_2, \ldots, \alpha_m) = Z_1$ and $K \subseteq K(\beta_1) \subseteq \cdots \subseteq K(\beta_1, \beta_2, \ldots, \beta_n) = Z_2$ be radical chains for Z_1 and Z_2, respectively. Then clearly

$$K \subseteq K(\alpha_1) \subseteq \cdots \subseteq K(\alpha_1, \ldots, \alpha_m) \subseteq K(\alpha_1, \ldots, \alpha_m, \beta_1)$$
$$\subseteq \cdots \subseteq K(\alpha_1, \ldots, \alpha_m, \beta_1, \ldots, \beta_n) = Z_1 Z_2$$

is a radical chain for $Z_1 Z_2$.

(c) By (13.18)(b) the normal closure N of $(L : K)$ can be written as $N = L_1 \cdots L_r$ where each L_i is K-isomorphic with L and hence a radical extension of K; thus (c) follows from (b).

(d) Let $(L : K)$ be a separable radical extension and let N be a normal closure of $(L : K)$. Then $(N : K)$ is radical by part (c) and separable by (14.12)(c), hence of Galois type by (22.1). ∎

THE importance of parts (c) and (d) is that they make available to us the theory of normal and separable field extensions for the investigation of radical extensions. As a first application, let us return to the definition of solvable equations. We called a polynomial equation $f(x) = 0$ solvable if *all* of its roots can be expressed in terms of radicals. We might as well have chosen the apparently weaker condition that at least *one* of its roots can be expressed in terms of radicals. We now show that these two conditions coincide under the natural condition that f is irreducible.

(26.7) Theorem. *Let $f \in K[x]$ be irreducible. Then the following two statements are equivalent:*
(1) *there is a radical extension of K containing a root of f;*
(2) *there is a radical extension of K over which f splits.*

Proof. Since (2)$\Longrightarrow$(1) is trivial it is enough to prove the implication (1)$\Longrightarrow$(2). Let $(L : K)$ be a radical extension such that L contains a root of f. Let N be a normal closure of $(L : K)$. Then f splits over N by (13.10)(2), and $(N : K)$ is a radical extension by (26.6)(c). ∎

BEFORE we proceed any further, we should ask ourselves whether the (intuitive) notion of explicitly solving a polynomial equation is really captured in the (mathematically precise) concept of forming radical extensions. This formalization of our intuitive idea is not completely satisfactory in one point, namely, the ambiguity of the root sign $\sqrt[m]{a}$: an equation $x^m - a = 0$ has in general m different solutions, and if we obtain the solutions of a polynomial equation in the form

$$(\ast) \qquad \sqrt[n]{\cdots + \sqrt[m]{\cdots} + \sqrt[k]{\cdots} + \cdots + \cdots}$$

it is not clear which of the m possibilities for an occurring radical should be chosen (and whether all possible choices for the different radicals occurring in the expression

($\star$) may be combined). Let us introduce a special name for the type of situation which is most desirable in this respect.

(26.8) Definition. *Let $f \in K[x]$ be a polynomial. The equation $f(x) = 0$ is called* **strongly solvable by radicals** *if it is solvable by radicals and if the following condition is satisfied: If one choice for the radicals in an expression* ($\star$) *yields a root of f, then all choices do (with the natural understanding that if a radical $\sqrt[n]{a}$ occurs more than once it must be given the same value each time).*

(26.9) Examples. (a) In example (26.1) we wrote the solutions of the equation $x^4 - 10x^2 + 1 = 0$ in the form $\pm(\sqrt{3} \pm \sqrt{2})$ with the implicit understanding that $\sqrt{3}$ and $\sqrt{2}$ are the *positive* solutions of the equations $u^2 = 3$ and $u^2 = 2$, respectively. However, there is no purely algebraic way of distinguishing between $+\sqrt{a}$ and $-\sqrt{a}$ for $a > 0$. Therefore, we might also write down the solutions of the equation $x^4 - 10x^2 + 1 = 0$ as $x = \sqrt{3} + \sqrt{2}$ with the understanding that *any* of the two possibilities for $\sqrt{3}$ and any of the two possibilities for $\sqrt{2}$ may be chosen.

(b) The equation $x^2 + x + 1 = 0$ over $\mathbb{Z}_2$ is solvable, but not strongly solvable by radicals. Indeed, each root ε of $f(x) = x^2 + x + 1$ is a primitive cubic root of unity, hence satisfies $\varepsilon^3 = 1$ and can thus be written as $\varepsilon = \sqrt[3]{1}$. But this would also allow for the solution $\varepsilon = 1$ which is not a root of f. Hence writing $\varepsilon = \sqrt[3]{1}$ does not yield a solution of the equation $x^2 + x + 1 = 0$ in the strong sense. The fact that there is no such solution whatsoever is outlined in problem 1 below.

FORTUNATELY, there is an easily verified criterion for strong solvability.

(26.10) Definitions. (a) *A chain of fields $K = K_1 \subseteq K_2 \subseteq \cdots \subseteq K_{m-1} \subseteq K_m = L$ is called a* **strong radical chain** *if for each i there is an element $\alpha_i \in K_{i+1}$ such that $K_{i+1} = K_i(\alpha_i)$ and such that α_i is a root of an irreducible pure polynomial $f(x) = x^{p_i} - k_i$ of prime degree over K_i.*

(b) *A field extension $(L : K)$ is called a* **strong radical extension** *if there is a strong radical chain starting with K and ending with L.*

(26.11) Examples. (a) Let n be a prime number. Suppose that K contains a primitive n-th root of unity and that $(L : K)$ is a Galois extension such that G_K^L is cyclic of order n. Then $(L : K)$ is a strong radical extension by (25.5).

(b) If $L = K(\alpha)$ where α is purely inseparable over K, then $(L : K)$ is a strong radical extension. To see this we recall the following fact which was proved in (14.15)(c): If β is purely inseparable over a field F of characteristic p, there is a natural number n with $\beta^{p^n} \in F$, and if n is chosen minimal with this property, then $x^{p^n} - \beta^{p^n}$ is irreducible over F. Applying this fact to the fields $F = (\alpha^{p^i})$ and the elements $\beta = \alpha^{p^{i-1}}$, we see that

$$ K \subseteq K(\alpha^{p^{n-1}}) \subseteq K(\alpha^{p^{n-1}}, \alpha^{p^{n-2}}) \subseteq \cdots \subseteq K(\alpha^{p^{n-1}}, \alpha^{p^{n-2}}, \ldots, \alpha^p, \alpha) $$

is a strong radical chain.

(c) If $(L : Z)$ and $(Z : K)$ are strong radical extensions, then so is $(L : K)$; simply concatenate two strong radical chains for $(L : Z)$ and $(Z : K)$ to obtain one for $(L : K)$.

(d) Every finite purely inseparable extension $(L : K)$ is a strong radical extension. In fact, we can write $L = K(\alpha_1, \ldots, \alpha_n)$ where each α_i is purely inseparable over K and hence over $K(\alpha_1, \ldots, \alpha_{i-1})$; then we can apply parts (c) and (d) to obtain the claim.

THE relevance of the concepts just introduced becomes clear in the next proposition.

(26.12) Proposition. *Suppose a splitting field of a polynomial $f \in K[x]$ is contained in a strong radical extension of K. Then the equation $f(x) = 0$ is strongly solvable by radicals.*

Proof. There is a strong radical chain $K = K_1 \subseteq \cdots \subseteq K_m = R$ such that R contains a splitting field L of f. If $K_{i+1} = K_i(\alpha_i)$ where α_i is a root of an irreducible polynomial $x^{p_i} - k_i$ over K_i, then two different choices α_i and α_i' for $\sqrt[p_i]{k_i}$ will yield conjugate elements; i.e., there is an isomorphism $\sigma_i : K_i(\alpha_i) \to K_i(\alpha_i')$ with $\sigma_i|_{K_i} = \mathrm{id}_{K_i}$ and $\sigma_i(\alpha_i) = \alpha_i'$. This isomorphism can by (13.5)(b) be extended to an automorphism $\sigma : \overline{K} \to \overline{K}$ of an algebraic closure $\overline{K}$ of K which is the identity on $K \subseteq K_i$. Hence any two choices for the values of a radical occurring in the expression $(\star)$ yields elements which can be transformed into each other by an element of $G_K^{\overline{K}}$; since $G_K^{\overline{K}}$ permutes the roots of f, this gives the claim. ∎

HAVING this in mind, we observe that the problem of the ambiguity of roots is mainly a problem concerning roots of unity. In fact, two different choices for the value of $\sqrt[m]{\alpha}$, i.e., any two solutions x_1, x_2 of the equation $x^m = \alpha$, are related by $x_2 = \varepsilon x_1$ where ε is an m-th root of unity; hence if we can tell apart the different m-th roots of unity, we can also tell apart the different values of $\sqrt[m]{\alpha}$ for any $\alpha \neq 0$.

Now it would indeed be very unsatisfactory to write down the solutions of $x^m - 1 = 0$ simply as $x = \sqrt[m]{1}$ and leave it at that. Fortunately, if the characteristic of the basefield is not too bad, one can always express m-th roots of unity by radicals of order lower than m, which will show inductively that all roots of unity can be distinguished by writing them as radical expressions in the strong sense.

(26.13) Example. We want to write down the sixth roots of unity over $\mathbb{Q}$. Clearly, writing $\sqrt[6]{1}$ or even $\sqrt[12]{1}$ is not satisfactory. Noting that $x^6 - 1 = (x^3 - 1)(x^3 + 1) = (x - 1)(x^2 + x + 1)(x + 1)(x^2 - x + 1)$, we see that the sixth roots of unity are $+1$ (primitive first root of unity), -1 (primitive second root of unity), $\frac{1}{2}(-1 \pm \sqrt{-3})$ (primitive third roots of unity) and $\frac{1}{2}(1 \pm \sqrt{-3})$ (primitive sixth roots of unity). Here the radical expressions $\frac{1}{2}(\pm 1 \pm \sqrt{-3})$ are not problematic, because any possible choice for $\sqrt{-3}$ (i.e., any root of $x^2 + 3 = 0$) will lead to a solution.

(26.14) Proposition. *Let K be a field, let $U(n)$ be the set of all n-th roots of unity over K and let K_n be obtained from K by adjoining $\bigcup_{m \leq n} U(m)$. If $\mathrm{char}\, K = 0$ or if $\mathrm{char}\, K$ is greater than the largest prime factor of n, then $(K_n : K)$ is a strong radical extension.*

Remark. Example (26.9)(b) shows that the condition concerning the characteristic of K cannot be dropped.

Proof. We proceed by induction on n. For $n = 1$ and $n = 2$ the claim is trivial because $U(1) = \{1\}$ and $U(2) = \{\pm 1\}$ are contained in K. Suppose the claim is proved for all numbers $< n$. We distinguish three different cases.

(a) Let $n = p$ be a prime number. By the condition on the characteristic of K, there is a primitive n-th root of unity ε. Then $(K_n : K_{n-1}) = (K_{n-1}(\varepsilon) : K_{n-1})$ is a Galois extension whose Galois group $G_{K_{n-1}}^{K_n}$ is cyclic of order d where d is a divisor of $n - 1$, due to (25.2)(b). Since K_{n-1} contains a primitive d-th root of unity, example (26.11)(a) is applicable and shows that $(K_n : K_{n-1})$ is a strong radical extension. Since also $(K_{n-1} : K)$ is a strong radical extension by induction hypothesis, example (26.11)(c) gives the claim in this case.

(b) Let $n = p^k$ be a prime power with $k \geq 2$. Now if ε is a primitive n-th root of unity then $\varepsilon_0 := \varepsilon^p$ is a p^{k-1}-st root of unity and hence lies in K_{n-1}. This means that ε is a root of the polynomial $f(x) = x^p - \varepsilon_0 \in K_{n-1}[x]$. If f is irreducible then $(K_n : K_{n-1}) = (K_{n-1}(\varepsilon) : K_{n-1})$ is a strong radical extension; since also $(K_{n-1} : K)$ is a strong radical extension, then so is $(K_n : K)$ by (26.11)(c), and we are done. If f is reducible, then f has a root ε_1 in K_{n-1} due to (25.8)($\star$) so that $\varepsilon_1^p = \varepsilon_0 = \varepsilon^p$. This implies that $\varepsilon = u\varepsilon_1$ where $u \in U(p)$. Since then both u and ε_1 lie in K_{n-1}, we have $\varepsilon \in K_{n-1}$ and hence $K_n = K_{n-1}(\varepsilon) = K_{n-1}$ in this case, and the induction hypothesis yields the claim.

(c) Let n be a number with at least two different prime factors. Then we can write $n = n_1 n_2$ where $n_1, n_2 < n$ are coprime. Hence $U(n) = U(n_1)U(n_2)$ by (24.4)(b) and consequently $K_n = K_{n-1}$ so that $(K_n : K) = (K_{n-1} : K)$ is a strong radical extension by induction hypothesis.

THE possibility of adjoining roots of unity will be very helpful to study strong radical extensions. For example, we will exploit this possibility by applying the following lemma.

(26.15) Lemma. (a) *Let $K \subseteq Z_1, Z_2 \subseteq L$ be fields such that $(Z_1 : K)$ and $(Z_2 : K)$ are strong radical extensions. If either $\operatorname{char} K = 0$ or if $\operatorname{char} K$ is greater than the largest prime divisor of $[Z_1 : K] \cdot [Z_2 : K]$, then $(Z_1 Z_2 : K)$ is a strong radical extension too.*

(b) *Suppose that L is the splitting field of a polynomial $f \in K[x]$ and that either $\operatorname{char} K = 0$ or that $\operatorname{char} K$ is greater than the largest prime divisor of $[L : K]$. Then L is contained in a strong radical extension of K if and only if each root of f is contained in a strong radical extension of K.*

Proof. (a) Let $K \subseteq K(\alpha_1) \subseteq \cdots \subseteq K(\alpha_1, \ldots, \alpha_m) = Z_1$ and $K \subseteq K(\beta_1) \subseteq \cdots \subseteq K(\beta_n) = Z_2$ be strong radical chains where each of the elements α_i and β_i satisfies an irreducible equation $x^{p_i} - k_i$ over the previous field in the corresponding chain. Then the occurring exponents p_i are exactly the prime divisors of $[Z_1 : K]$ and $[Z_2 : K]$; let p be the maximum of these exponents. To exhibit $(Z_1 Z_2 : K)$ as a radical extension, we cannot simply concatenate the above two chains as in the proof of (26.6)(b) because a polynomial can lose the property of irreducibility if the base-field is extended. However, the following trick allows to overcome this difficulty.

Let K' be obtained from K by adjoining all roots of unity of order $\leq p$; then $(K' : K)$ is a strong radical extension by (26.14). Now

$$K' \subseteq K'(\alpha_1) \subseteq \cdots \subseteq K'(\alpha_1, \ldots, \alpha_m) \subseteq K'(\alpha_1, \ldots, \alpha_m, \beta_1)$$
$$\subseteq \cdots \subseteq K'(\alpha_1, \ldots, \alpha_m, \beta_1, \ldots, \beta_n) = Z_1 Z_2$$

is indeed a strong radical chain. In fact, either the irreducible equation $x^{p_i} - k_i$ satisfied by an element α_i or β_i is still irreducible over the previous field in this chain, or the polynomial $x^{p_i} - k_i$ completely splits over this previous field (in which two successing fields in our chain coincide).

(b) If each root α_i is contained in a strong radical extension Z_i of K, then $[Z_i : K]$ divides $[L : K]$, and L is contained in the compositum $Z_1 \cdots Z_n$ which is again a strong radical extension of K by part (a). The converse implication is trivial. ∎

IT was Galois' brilliant idea to correlate – in modern terminology – the problem of solving a polynomial equation $f(x) = 0$ over a field K with the structure of the group $\text{Gal}_K(f)$. The key result to realize this idea was already proved in the previous section; namely, theorem (25.5) essentially states that if $(L : K)$ is a Galois extension with cyclic Galois group G_K^L then L can be obtained form K by adjoining a root of a pure equation; in particular, $(L : K)$ is a radical extension.

Even though theorem (25.5) only deals with the case that G_K^L is cyclic, it can be used in a more general setting as follows. Suppose G_K^L has a normal subgroup N such that G_K^L/N is cyclic. Let $Z := F(N)$ be the fixed field of N. Then the main theorem of Galois theory yields $G_K^Z \cong G_K^L/G_Z^L = G_K^L/N$, so that G_K^Z is cyclic, say of order n. But then (25.5) tells us that we can obtain Z from K by adjoining an n-th root, i.e., by adjoining an element $\alpha \in Z$ such that $\alpha^n \in K$. It is a natural question whether or not one can iterate this process. This will work if G_K^L possesses a chain

$$\{\text{id}\} = N_1 \trianglelefteq N_2 \trianglelefteq \cdots \trianglelefteq N_k = G_K^L$$

where each factor group N_{i+1}/N_i is cyclic. But as was shown in (I.28.12)(c), this means precisely that G_K^L is a solvable group! (It was exactly this observation that led to the notion of solvability of groups.) By stepwise ascending from one fixed field $F(N_i)$ to the next fixed field $F(N_{i-1})$, we finally reach L by successively adjoining roots. These observations constitute already an outline of the proof of the following theorem.

(26.16) Theorem. *Let $(L : K)$ be a normal field extension such that G_K^L is solvable. If* char K *does not divide* $[L : K]$, *then L is contained in a radical extension of K. If* char $K = 0$ *or if* char K *is greater than the largest prime divisor of* $[L : K]$, *then L is even contained in a strong radical extension of K.*

REMARK. Examples (26.5) and (26.9)(b) show that the conditions on the characteristic of K cannot be dropped. Problem 16 below shows that one also cannot dispose of the normality of $(L : K)$.

Proof. Let us first reduce the problem to the case that $(L : K)$ is of Galois type. By (22.2)(c) we can write $L = SP$ where $(S : K)$ is a Galois extension and where $(P : K)$ is purely inseparable. Then $(P : K)$ is automatically a (strong) radical extension by (26.4)(d) and (26.11)(d), respectively. Moreover, G_K^S is solvable as the homomorphic image of the solvable group G_K^L under the restriction map $R : G_K^L \to G_K^S$, and char K is 0 or greater than the greatest prime divisor of $[S : K]$. Hence if (26.16) is established for Galois extensions, then we can conclude that S is contained in a (strong) radical extension of K. But then so is $L = SP$, due to (26.6)(b) and (26.15)(a).

Hence we may assume that $(L : K)$ is of Galois type. Since G_K^L is solvable, theorem (I.28.12)(c) shows that there is a chain of subgroups

$$G_K^L = N_0 \unrhd N_1 \unrhd N_2 \unrhd \cdots N_m = \{\mathrm{id}\}$$

such that each quotient groups N_i/N_{i+1} is cyclic of order p_i where p_i is a prime divisor of n. Invoking the Galois correspondence for $(L : K)$ and writing $Z_i := F(N_i)$ for the fixed field of N_i, then we obtain the chain of fields

$$K = Z_0 \subseteq Z_1 \subseteq Z_2 \subseteq \cdots \subseteq Z_m = L$$

where each $(Z_{i+1} : Z_i)$ is a Galois extension whose Galois group $G_{Z_i}^{Z_{i+1}}$ is isomorphic to N_i/N_{i+1} and hence cyclic of order p_i.

We would like to apply (25.5) to the extensions $(Z_{i+1} : Z_i)$, but unfortunately this is not possible in a straightforward way because Z_i does not contain a primitive p_i-th root of unity in general. Therefore, we extend each field Z_i to a field Z_i' by adjoining all roots of unity of orders $\leq \max\{p_1, \ldots, p_m\}$. Then $(K' : K) = (Z_0' : Z_0)$ is a (strong) radical extension by (26.14).

Since Z_{i+1} is of Galois type over Z_i, we know that Z_{i+1} is the splitting field of a polynomial $f \in Z_i[x]$ whose irreducible factors are separable. But then Z_{i+1}' is the splitting field of $f(x)(x^{p_i} - 1) \in Z_i[x] \subseteq Z_i'[x]$ whose irreducible factors are separable again because char $K \neq p_i$. Hence $(Z_{i+1}' : Z_i')$ is a Galois extension again. The restriction map $R : G_{Z_i'}^{Z_{i+1}'} \to G_{Z_i}^{Z_{i+1}}$ (which is obviously an injective group homomorphism) shows that $G_{Z_i'}^{Z_{i+1}'}$ is isomorphic to a subgroup of $G_{Z_i}^{Z_{i+1}}$; hence $|G_{Z_i'}^{Z_{i+1}'}| = [Z_{i+1}' : Z_i']$ divides $|G_{Z_i}^{Z_{i+1}}| = [Z_{i+1} : Z_i] = p_i$ which implies that either $[Z_{i+1}' : Z_i'] = 1$ (in which case $Z_{i+1}' = Z_i'$) or else $[Z_{i+1}' : Z_i'] = p_i$ (in which case $(Z_{i+1}' : Z_i')$ is a strong radical extension of oder p_i by (25.5)). Thus in each case $(Z_{i+1}' : Z_i')$ is a strong radical extension. Consequently, (26.11)(c) implies that $(L' : K')$ is a strong radical extension.

As we observed above, $(K' : K)$ is also a (strong) radical extension; hence so is $(L' : K)$ by (26.11)(c). ∎

THE converse statement is also true, even without any condition on the base field. This is proved in theorem (26.18) below; (26.17) covers the most important special case (to which the general case will be traced back).

(26.17) Theorem. *Let $(L : K)$ be a radical extension of Galois type. Then G_K^L is a solvable group.*

Proof. Let $L = K(a_1, \ldots, a_m)$ where $a_i^{n_i} \in K(a_1, \ldots, a_{i-1}) =: K_i$ for all i. We may assume that char K does not divide any of the n_i. For char $K = 0$ this is clear; suppose char $K = p$ and $n_i = p^r s$ with $p \nmid s$. Then $(a_i^s)^{p^r} = a_i^{n_i} \in K_i$; hence a_i^s is purely inseparable over K_i by (14.15)(c). But L is separable over K by (22.1), hence over K_i; thus a_i^s is also separable over K_i. Due to (14.15)(a) this implies that $a_i^s \in K_i$ so that we can replace n_i by s. Now let $n := n_1 \cdots n_m$; since char K does not divide n, there is a primitive n-th root of unity ε in some extension of K. Then $L(\varepsilon) = K(a_1, \ldots, a_m, \varepsilon)$ is a radical extension of K which is of Galois type; indeed, if L is the splitting field of $p \in K[x]$, then $L(\varepsilon)$ is the splitting field of $p(x)(x^n - 1)$, and $x^n - 1$ has no multiple roots because char $K \nmid n$. Now for all i, the extension $\big(L(\varepsilon) : K_i(\varepsilon)\big)$ is of Galois type by the main theorem. But then $\big(K_{i+1}(\varepsilon) : K_i(\varepsilon)\big)$ is of Galois type too because $K_{i+1}(\varepsilon)$ contains a root of a polynomial of the form $x^{n_i} - k_i$ with $k_i \in K_i$ and hence is the splitting field of this polynomial because of the existence of all p_i-th roots of unity in $K_{i+1}(\varepsilon)$. Then $G_{K_{i+1}(\varepsilon)}^{L(\varepsilon)} \unlhd G_{K_i(\varepsilon)}^{L(\varepsilon)}$ by the main theorem of Galois theory, and $G_{K_i(\varepsilon)}^{L(\varepsilon)}/G_{K_{i+1}(\varepsilon)}^{L(\varepsilon)} \cong G_{K_i(\varepsilon)}^{K_{i+1}(\varepsilon)}$ which is cyclic by (25.2)(b). So the chain of normal subgroups

$$\{\mathrm{id}\} = G_{L(\varepsilon)}^{L(\varepsilon)} = G_{K_m(\varepsilon)}^{L(\varepsilon)} \unlhd G_{K_{m-1}(\varepsilon)}^{L(\varepsilon)} \unlhd \cdots \unlhd G_{K_1(\varepsilon)}^{L(\varepsilon)} \unlhd G_K^{L(\varepsilon)}$$

shows that $G_K^{L(\varepsilon)}$ is solvable.

Now $(L : K)$ and $\big(L(\varepsilon) : K\big)$ are of Galois type; so $G_L^{L(\varepsilon)} \unlhd G_K^{L(\varepsilon)}$ and $G_K^L \cong G_K^{L(\varepsilon)}/G_L^{L(\varepsilon)}$, by the main theorem of Galois theory. This shows that G_K^L is a homomorphic image of the solvable group $G_K^{L(\varepsilon)}$, hence solvable itself. ∎

(26.18) Theorem. *Let $(L : K)$ be a field extension such that L is contained in some radical extension of K. Then G_K^L is solvable.*

Proof. The proof consists in reducing the claim to the situation in (26.17). Suppose L is contained in the radical extension R of K. We first let $K_0 := F(G_K^L)$, then choose a normal closure N of $(R : K_0)$ and finally let $K_1 := F(G_{K_0}^N)$.

We claim first that $(N : K_1)$ is a radical extension of Galois type; then (26.17) will imply that $G_{K_1}^N$ is solvable. To wit: $(R : K)$ is radical, hence $(R : K_0)$ is radical by (26.4)(a), thus $(N : K_0)$ is radical by (26.4)(c), and consequently $(N : K_1)$ is radical by (26.4)(a). Also, using the Galois correspondence of $(N : K_0)$, we see that $F(G_{K_1}^N) = (F \circ G \circ F)(G_{K_0}^N) = F(G_{K_0}^N) = K_1$ which shows that $(N : K_1)$ is of Galois type.

Now observe that $G_{K_1}^N = (G \circ F \circ G)(K_0) = G(K_0) = G_{K_0}^N$; hence $G_{K_0}^N$ is solvable. The restriction map

$$R: \begin{array}{ccc} G_{K_0}^N & \to & G_{K_0}^L \\ \sigma & \mapsto & \sigma|_{K_0} \end{array}$$

is a well-defined surjective homomorphism. In fact, $(L : K_0)$ is of Galois type because the Galois correspondence of $(L : K_0)$ yields $F(G_{K_0}^L) = (F \circ G \circ F)(G_K^L) = F(G_K^L) = K_0$; hence L is stable under $G_{K_0}^N$. This shows that R is well-defined. On the other hand, R is surjective because $(N : L)$ is normal.

Thus $G_{K_0}^L$ is a homomorphic image of the solvable group $G_{K_0}^N$ and hence solvable itself. But $G_{K_0}^L = (G \circ F \circ G)(K) = G_K^L$; thus G_K^L is solvable as desired. ∎

THE most important fact contained in the above results is the following statement.

(26.19) Theorem. *Let $f \in K[x]$ where* $\operatorname{char} K = 0$. *Then the equation* $f(x) = 0$ *is (strongly) solvable by radicals if and only if the Galois group* $\operatorname{Gal}_K(f)$ *is solvable.*

Proof. The statement is clearly contained in (26.16) and (26.18). ∎

THIS theorem allows us to immediately show that certain equations are *not* solvable by radicals.

(26.20) Example. The general polynomial of degree n over $\mathbb{Q}$ is solvable by radicals if and only if $n \le 4$. Indeed, the Galois group of f is Sym_n by (23.26), and Sym_n is solvable if and only if $n \le 4$.

AS was outlined in the remarks preceding (23.24), the result expressed in (26.20) means that there are general formulas to solve equations of degree ≤ 4 (a fact which we know already from our discussion in section 20), but that there are no such general formulas for degrees ≥ 5. However, it might still be true that for each *individual* polynomial $f \in \mathbb{Q}[x]$ of degree $n \ge 5$ there is an *individual* way of expressing its roots by radicals. However, not even this is true, as the following example shows.

(26.21) Example. Suppose that $f \in \mathbb{Q}[x]$ is irreducible of prime degree $p \ge 5$ and has exactly two non-real roots. Then $\operatorname{Gal}_{\mathbb{Q}}(f)$ equals Sym_p by (23.18) and hence is not a solvable group. Thus for example the equation $x^5 - 2x^4 + 2 = 0$ is not solvable by radicals.

TO be able to prove the non-solvability of a polynomial equation in some other concrete examples, we deduce the following criterion for solvability.

(26.22) Proposition. *Let* $\operatorname{char} K = 0$ *and let* $f \in K[x]$ *be irreducible of prime degree p. Let L be the splitting field of f and let* $\alpha_1, \ldots, \alpha_p \in L$ *be the roots of f. Then the equation* $f(x) = 0$ *can be solved by radicals if and only if* $L = K(\alpha_i, \alpha_j)$ *for any two roots* α_i, α_j.

Proof. By theorem (26.19), the solvability of the equation $f(x) = 0$ is equivalent to the solvability of the Galois group $\operatorname{Gal}_K(f)$. Since f is irreducible over K the group $\operatorname{Gal}_K(f)$ acts transitively on $\{1, \ldots, p\}$. So we can exploit the characterization of solvable transitive subgroups of Sym_p obtained in (I.28.23); namely, a transitive subgroup G of Sym_p is solvable if and only if no element $\sigma \neq \operatorname{id}$ in G has more than one fixed point.

Suppose $G_K^L \cong \mathrm{Gal}_K(f)$ is solvable. Let i, j be arbitrary. Then every element of $G_{K(\alpha_i, \alpha_j)}^L \le G_K^L$ has the two fixed points α_i and α_j, hence must be the identity. This shows $G_{K(\alpha_i, \alpha_j)}^L = \{\mathrm{id}\}$. Now $(L : K(\alpha_i, \alpha_j))$ is a Galois extension. Therefore, $K(\alpha_i, \alpha_j) = F(G_{K(\alpha_i, \alpha_j)}^L) = F(\{\mathrm{id}\}) = L$.

Conversely, let $L = K(\alpha_i, \alpha_j)$ for any i, j. Suppose that $\sigma \in \mathrm{Gal}_K(f)$ has at least two fixed points α_i and α_j. Then, since $L = K(\alpha_i, \alpha_j)$, all elements of L are fixed under σ; whence $\sigma = \mathrm{id}$. By the above-mentioned characterization (I.28.23), the group $\mathrm{Gal}(f)$ is solvable. ∎

IN some sense, theorem (26.19) does not *answer* the question when an equation $f(x) = 0$ is solvable; it simply *translates* the problem to the equivalent question when the Galois group $\mathrm{Gal}_K(f)$ is solvable. If this group-theoretical question is not easier to answer than the original question, then our results might be considered quite nice from an aesthetical viewpoint, but they are of no practical value. The same is true for proposition (26.22). It gives a necessary and sufficient field-theoretical condition for the solvability of the equation $f(x) = 0$, but how should one check this condition for a given polynomial? However, in (26.23) below, we will exploit proposition (26.22) to give a satisfying answer in concrete examples.

(26.23) Proposition. *Suppose $f \in \mathbb{Q}[x]$ is irreducible of prime degree $p \ge 5$. If the equation $f(x) = 0$ is solvable by radicals, then f has either exactly one real root or else p real roots. In other words, if f has at least two real roots and one non-real root, then the equation $f(x) = 0$ is not solvable by radicals.*

Proof. Note that f has at least one real root because the degree of f is an odd number. Suppose that $\alpha_1, \alpha_2 \in \mathbb{R}$ and $\alpha_3 \in \mathbb{C} \backslash \mathbb{R}$ are roots of f. Then $\alpha_3 \notin \mathbb{Q}(\alpha_1, \alpha_2) \subsetneq \mathbb{R}$. So the claim follows immediately from (26.22). ∎

(26.24) Examples. (a) Let p, q be prime numbers with $p \ge 5$. Then no root of the polynomial

$$f(x) \;=\; x^p - 2qx + q$$

is expressible by radicals. To see this, we note first that f is irreducible by Eisenstein's criterion. Also, since $f'(x) = px^{p-1} - 2q$ has exactly two real roots, Rolle's theorem shows that f has at most three real roots; on the other hand f has at least three real roots by the intermediate value theorem, because $f(x) \to -\infty$ as $x \to -\infty$, $f(0) = q > 0$, $f(1) = 1 - q < 0$ and $f(x) \to \infty$ as $x \to \infty$. Consequently, f has exactly three real roots, and (26.23) gives the claim.

(b) The situation encountered in (26.21) is also covered by (26.23).

(c) Suppose $f \in \mathbb{Q}[x]$ is irreducible of prime degree p such that $p - 1$ is divisible by 4 and $D(f) < 0$. Then f has at least one real root because $\deg f = p$ is odd and at least one non-real root, because $D(f) < 0$. If f had exactly one real root (and hence $p - 1$ non-real roots) then (4.29)(c) would imply $D(f) > 0$; hence f has at least two real roots. Hence (26.23) shows that the equation $f(x) = 0$ is not solvable by radicals.

UNTIL now we only applied theorem (26.19) in a negative way: We exhibited for certain polynomials f the non-solvability of the corresponding Galois group and concluded that the equation $f(x) = 0$ is not solvable by radicals. What about applying (26.19) in a positive way? In other words, if the Galois group of a polynomial f turns out to be solvable then we know that the equation $f(x) = 0$ is solvable by radicals – but how can we actually find the roots of f and write them down as radical expressions? In principle, the answer to this question is already contained in the proof of (26.16), and the following description merely elaborates the decisive steps.

(26.25) Theorem. *The following steps constitute an algorithm to express the roots of a polynomial $f \in K[x]$ in terms of radicals.*

(a) *Determine the Galois group $G = \mathrm{Gal}_K(f)$ of f (which we assume to be solvable) and write down a sequence $G = G_0 \trianglerighteq G_1 \trianglerighteq G_2 \trianglerighteq \cdots \trianglerighteq G_{r-1} \trianglerighteq G_r = \{\mathrm{id}\}$ such that G_{i-1}/G_i is cyclic of prime order p_i.*

(b) *For $1 \le i \le r-1$ choose $\sigma_i \in G_{i-1}$ such that the coset $[\sigma_i] = \sigma_i G_i$ generates G_{i-1}/G_i.*

(c) *Let $\alpha_1, \ldots, \alpha_n$ be the (unknown) roots of f. For all indices $1 \le i \le r$ and all roots of unity $\varepsilon_i \in U(p_i)$, consider the Lagrange resolvents*

$$\mathrm{Lag}_{(i,\varepsilon_i)} := \mathrm{id} + \varepsilon_i \sigma_i + \cdots + \varepsilon_i^{p_i-1} \sigma_i^{p_i-1}$$

and form the expressions $\alpha(\varepsilon_i, \ldots, \varepsilon_r)$ which are inductively defined as follows:

$$\alpha(\varepsilon_r) := \mathrm{Lag}_{(r,\varepsilon_r)}(\alpha_1)^{p_r} , \quad \alpha(\varepsilon_{i-1}, \ldots, \varepsilon_r) := \mathrm{Lag}_{(i-1,\varepsilon_{i-1})}\big(\alpha(\varepsilon_i, \ldots, \varepsilon_r)\big)^{p_{i-1}}$$

so that

$$\alpha(\varepsilon_i, \ldots, \varepsilon_r) := \mathrm{Lag}_{(i,\varepsilon_i)}\left(\cdots \left(\mathrm{Lag}_{(r-1,\varepsilon_{r-1})}\big(\mathrm{Lag}_{(r,\varepsilon_r)}(\alpha_1)^{p_r}\big)^{p_{r-1}}\right) \cdots\right)^{p_i} .$$

(d) *The expression $\alpha(\varepsilon_1, \ldots, \varepsilon_r)$ is an element of K and can be calculated from the coefficients of f (without any prior knowledge about the roots of f).*

(e) *The root α_1 of f can be expressed by radicals as follows:*

$$(\star) \qquad \alpha_1 = \frac{1}{p_r} \sum_{\varepsilon_r} \sqrt[p_r]{\frac{1}{p_{r-1}} \sum_{\varepsilon_{r-1}} \sqrt[p_{r-1}]{\cdots \frac{1}{p_1} \sum_{\varepsilon_i} \sqrt[p_1]{\alpha(\varepsilon_1, \ldots, \varepsilon_r)}}}$$

where $\sum_{\varepsilon_i}$ denotes summation over all elements $\varepsilon_i \in U(p_i)$.

(f) *Plug all possible values of the radical expression $(\star)$ into the original expression $(\star)$ and discard extraneous solutions.*

Proof. (a) The Galois group of f can (in principle) be determined by the method described in (23.3), and a chain of subgroups as indicated exists by (I.28.12)(c).

(b) Such an element σ_i exists because G_{i-1}/G_i has prime order p_i and hence is cyclic.

(c) For $1 \le i \le r$ let $K_i := F(G_i)$ be the fixed field of G_i; then $K = K_0 \subseteq K_1 \subseteq K_2 \subseteq \cdots \subseteq K_{r-1} \subseteq K_r = L$ where $[K_i : K_{i-1}] = p_i$. Since $G_{i-1}/G_i = G^L_{K_{i-1}}/G^L_{K_i} \cong G^{K_i}_{K_{i-1}}$, we can identify σ_i with a generator of $G^{K_i}_{K_{i-1}}$. In (25.6) it was shown that the

Lagrange resolvents $\mathrm{Lag}_{(i,\varepsilon_i)} : K_i \to K_i$ have the property that $\mathrm{Lag}_{(i,\varepsilon_i)}(\alpha)^{p_i} \in K_{i-1}$ whenever $\alpha \in K_i$. This implies that $\alpha(\varepsilon_i,\ldots,\varepsilon_r) \in K_{i-1}$ for all i; in particular, $\alpha(\varepsilon_1,\ldots,\varepsilon_r) \in K_0 = K$.

(d) We know that $\alpha(\varepsilon_1,\ldots,\varepsilon)$ lies in the fixed field $K_0 = F(G_0)$ of $G_0 = G$. If $G = \mathrm{Sym}_n$ this means that $\alpha(\varepsilon_1,\ldots,\varepsilon_r)$ is a symmetric expression in $\alpha_1,\ldots,\alpha_n$ and hence can be expressed in terms of the elementary symmetric polynomials in $\alpha_1,\ldots,\alpha_n$, i.e., in the coefficients of f; thus we are done in this case.

If $G_0 \subsetneqq \mathrm{Sym}_n$, we denote by $\beta_1,\ldots,\beta_k \in K$ the different elements of the set $\{\sigma(\alpha(\varepsilon_1,\ldots,\varepsilon_r)) \mid \sigma \in \mathrm{Sym}_n\}$ and define the **resolvent polynomial** $q \in K[x]$ as

$$q(y) := (y - \beta_1)\cdots(y - \beta_k) = b_0 + b_1 y + \cdots + b_k y^k .$$

The coefficients of q are symmetric in $\beta_1,\ldots,\beta_k$; since Sym_n merely permutes $\beta_1,\ldots,\beta_k$, they are also symmetric in $\alpha_1,\ldots,\alpha_n$, hence can be determined from the coefficients of f. Since $\alpha(\varepsilon_1,\ldots,\varepsilon_r)$ is one of the β_i's, we only have to find all roots of q in the base-field K; each of these is considered a possible value for $\alpha(\varepsilon_1,\ldots,\varepsilon_r)$. (For the elimination of extraneous solutions see part (f) below.)

(e) From our discussion on Lagrange resolvents in (25.6) we know that $(1/p_i)\sum_{\varepsilon_i} \mathrm{Lag}_{(i,\varepsilon_i)}$ is the identity on K_i. Since $\alpha(\varepsilon_i,\ldots,\varepsilon_r) \in K_i$ for all i, we find first that

$$(1) \quad \alpha_1 = \frac{1}{p_r}\sum_{\varepsilon_r} \mathrm{Lag}_{(r,\varepsilon_r)}(\alpha_1) = \frac{1}{p_r}\sum_{\varepsilon_r} \sqrt[p_r]{\mathrm{Lag}_{(r,\varepsilon_r)}(\alpha_1)^{p_r}} = \frac{1}{p_r}\sum_{\varepsilon_r} \sqrt[p_r]{\alpha(\varepsilon_r)}$$

and then since $\alpha(\varepsilon_r) \in K_{r-1}$ also

$$(2) \qquad\qquad \alpha(\varepsilon_r) = \frac{1}{p_{r-1}}\sum_{\varepsilon_{r-1}} \mathrm{Lag}_{(r-1,\varepsilon_{r-1})}(\varepsilon_r)$$

$$= \frac{1}{p_{r-1}}\sum_{\varepsilon_{r-1}} \sqrt[p_{r-1}]{\mathrm{Lag}_{(r-1,\varepsilon_{r-1})}(\alpha(\varepsilon_r))^{p_{r-1}}} = \frac{1}{p_{r-1}}\sum_{\varepsilon_{r-1}} \sqrt[p_{r-1}]{\alpha(\varepsilon_{r-1},\varepsilon_r)} ;$$

substituting (2) into (1) we obtain

$$\alpha_1 = \frac{1}{p_r}\sum_{\varepsilon_r} \sqrt[p_r]{\frac{1}{p_{r-1}}\sum_{\varepsilon_{r-1}} \sqrt[p_{r-1}]{\alpha(\varepsilon_{r-1},\varepsilon_r)}} .$$

Continuing in this way, we finally arrive at the formula $(\star)$.

(f) Since the numbering of the roots of f is completely arbitrary, we proved that each root of f can be written in the form $(\star)$; this does *not* say, however, that every expression of this form is indeed a root. Each time a root is taken, there is an ambiguity; in general we expect a nonzero field element to have n different n-th roots, a point which was touched upon in (20.2) and before (26.8) already. Now reading the expression $(\star)$ from the innermost root outward, we see that we always have to take a p_i-th root of some previously determined quantity; but to take a p_i-th root there are in general p_i possibilities, and we did not specify at all which root to take. Also, if $G \neq \mathrm{Sym}_n$ so that we have to introduce a resolvent polynomial, there are different possibilities for $\alpha(\varepsilon_1,\ldots,\varepsilon_r)$. Thus the process introduces extraneous answers that must be eliminated by working backwards from $\alpha(\varepsilon_1,\ldots,\varepsilon_r)$ to the roots of f via the formula $(\star)$. In general, considering all expressions of the form $(\star)$ (with all possible choices for $\alpha(\varepsilon_1, \dots, \varepsilon_r)$), we get a relatively large (but finite!) number of candidates for the roots of f; which of these are in fact roots has to be checked by plugging them into the original equation $f(x) = 0$,[†] discarding the ones which do not satisfy this equation. This trial-and-error-method is not very elegant, but since formula $(\star)$ allows only for a finite number of candidates, it always leads to finding all roots. Moreover, if n roots are found we know that there cannot be any others, and the process can be stopped. ∎

WE now illustrate (and also clarify) the algorithm described in (26.25) by applying it in concrete cases; in particular, we want to recover – in the systematical way devised by Galois theory – the formulas given in section 20 to solve polynomial equations of low degree. The reader should not be surprised by the relative complexity of the calculations involved; as always in mathematics, 'abstract' theorems are usually rather elegant and hide the tedious (but fruitful!) computations which are involved in concrete examples.

(26.26) Example: The general quadratic equation. Consider the general quadratic polynomial $f(x) = x^2 + px + q$ over a field K_0, i.e., $f \in K[x]$ where $K = K_0(p, q)$ consists of all rational functions in two indeterminates p and q. Let α_1 and α_2 be the roots of f in some splitting field L of f and let us follow the instructions given in (26.25).

(a) We have $G = \mathrm{Sym}_2 = \{\mathrm{id}, (12)\}$; hence $r = 1$, $G_0 = \{\mathrm{id}, (12)\}$, $G_1 = \{\mathrm{id}\}$ and $p_1 = 2$.

(b) A generator of $G_0 \cong G_0/G_1$ is $\sigma_1 = (12)$.

(c) Since the second roots of unity are ± 1, the Lagrange resolvents needed are $\mathrm{Lag}_{(1,1)} = \mathrm{id} + \sigma_1$ and $\mathrm{Lag}_{(1,-1)} = \mathrm{id} - \sigma_1$; hence we have to form the elements $\alpha(1) = \mathrm{Lag}_{(1,1)}(\alpha_1)^2 = (\alpha_1 + \alpha_2)^2$ and $\alpha(-1) = \mathrm{Lag}_{(1,-1)}(\alpha_1)^2 = (\alpha_1 - \alpha_2)^2$.

(d) Since $\alpha_1 + \alpha_2 = -p$ and $\alpha_1\alpha_2 = q$ by Vieta's formula, we obtain

$$\alpha(1) = (\alpha_1 + \alpha_2)^2 = p^2 \quad \text{and}$$
$$\alpha(-1) = (\alpha_1 - \alpha_2)^2 = (\alpha_1 + \alpha_2)^2 - 4\alpha_1\alpha_2 = p^2 - 4q .$$

(e) The solution formula $(\star)$ yields

$$\alpha = \frac{1}{2} \sum_{\varepsilon_1 \in \{\pm 1\}} \sqrt{\alpha(\varepsilon_1)} = \frac{1}{2}(\sqrt{\alpha(1)} + \sqrt{\alpha(-1)}) = \frac{1}{2}(\sqrt{p^2} + \sqrt{p^2 - 4q}) .$$

(f) If $\sqrt{p^2 - 4q}$ denotes any fixed choice of one of the two square roots of $p^2 - 4q$, the above formula can be written as $\alpha = \frac{1}{2}(\pm p \pm \sqrt{p^2 - 4q})$. It is easily checked that the two candidates $\frac{1}{2}(p \pm \sqrt{p^2 - 4q})$ do *not* satisfy the original equation $x^2 + px + q = 0$, whereas the two other candidates do. Hence we obtain $\alpha = \frac{1}{2}(-p \pm \sqrt{p^2 - 4q})$ for the roots of f; this is the well-known quadratic formula, of course.

[†] We should point out that it is often not clear at first glance whether or not a given radical expression satisfies the equation $f(x) = 0$; think of equations like $\sqrt[3]{10 + \sqrt{108}} - \sqrt[3]{-10 + \sqrt{108}} = 2$.

(26.27) Example: The general cubic equation. Consider the general cubic polynomial $f(x) = x^3 - s_1 x^2 + s_2 x - s_3$ over a field K_0, i.e., $f \in K[x]$ where $K = K_0(s_1, s_2, s_3)$ consists of all rational functions in three indeterminates s_1, s_2 and s_3. Let α_1, α_2 and α_3 be the roots of f in some splitting field L of f and let us follow the instructions given in (26.25).

(a) We have $G = \text{Sym}_3$ by (23.26); a composition series is given by

$$\text{Sym}_3 \; \trianglerighteq \; \text{Alt}_3 \; \trianglerighteq \; \{\text{id}\} \; .$$

Hence $r = 2$, $p_1 = |\text{Sym}_3 \,/\, \text{Alt}_3| = 2$ and $p_2 = |\text{Alt}_3 \,/\{\text{id}\}| = 3$.

(b) A generator of G_0/G_1 is induced by $\sigma_1 = (12)$, whereas a generator of $G_1 \cong G_1/G_2$ is $\sigma_2 = (123)$.

(c) Since there are two square roots of unity ± 1 and three cubic roots of unity $1, \omega, \omega^2$ involved (where ω denotes a primitive cubic root of unity), we need the Lagrange resolvents

$$\begin{aligned}
\text{Lag}_{(2,1)} &= \text{id} + \sigma_2 + \sigma_2^2, \\
\text{Lag}_{(2,\omega)} &= \text{id} + \omega\sigma_2 + \omega^2\sigma_2^2, \\
\text{Lag}_{(2,\omega^2)} &= \text{id} + \omega^2\sigma_2 + \omega\sigma_2^2, \\
\text{Lag}_{(1,1)} &= \text{id} + \sigma_1, \\
\text{Lag}_{(1,-1)} &= \text{id} - \sigma_1
\end{aligned}$$

where $\sigma_1 = (12)$, $\sigma_2 = (123)$ and $\sigma_2^2 = (132)$; note that we consequently identify permutations of the set $\{\alpha_1, \alpha_2, \alpha_3\}$ with the corresponding automorphisms of $L = K(\alpha_1, \alpha_2, \alpha_3)$. Hence we have to form the elements

$$\begin{aligned}
\alpha(1) &= \text{Lag}_{(2,1)}(\alpha_1)^3 &=& \quad \left((\text{id} + \sigma_2 + \sigma_2^2)(\alpha_1)\right)^3 &=& \quad (\alpha_1 + \alpha_2 + \alpha_3)^3 \\
\alpha(\omega) &= \text{Lag}_{(2,\omega)}(\alpha_1)^3 &=& \quad \left((\text{id} + \omega\sigma_2 + \omega^2\sigma_2^2)(\alpha_1)\right)^3 &=& \quad (\alpha_1 + \omega\alpha_2 + \omega^2\alpha_3)^3 \\
\alpha(\omega^2) &= \text{Lag}_{(2,\omega^2)}(\alpha_1)^3 &=& \quad \left((\text{id} + \omega^2\sigma_2 + \omega\sigma_2^2)(\alpha_1)\right)^3 &=& \quad (\alpha_1 + \omega^2\alpha_2 + \omega\alpha_3)^3
\end{aligned}$$

and for each $\varepsilon \in \{1, \omega, \omega^2\}$ the elements

$$\begin{aligned}
\alpha(\;1, \varepsilon) &= \text{Lag}_{(1,\;1)}\big(\alpha(\varepsilon)\big)^2 &=& \quad \left((\text{id} + \sigma_1)(\alpha(\varepsilon)\right)^2 \quad \text{and} \\
\alpha(-1, \varepsilon) &= \text{Lag}_{(1,-1)}\big(\alpha(\varepsilon)\big)^2 &=& \quad \left((\text{id} - \sigma_1)(\alpha(\varepsilon)\right)^2
\end{aligned}$$

which concluded step (c) in our algorithm.

(d) This is the messiest part of the algorithm: We have to express all the terms $\alpha(\varepsilon_1, \varepsilon_2)$ with $\varepsilon_1 \in \{\pm 1\}$ and $\varepsilon_2 \in \{1, \omega, \omega^2\}$ by the coefficients of f. This can be done directly (without introducing a resolvent polynomial) because G is the full symmetric group Sym_3. First of all, we have $\alpha_1 + \alpha_2 + \alpha_3 = s_1$; hence we have immediately

$$\boxed{\alpha(1,1) \;=\; 4s_1^6 \quad \text{and} \quad \alpha(-1,1) \;=\; 0 \,.}$$

Further, we calculate

$$\begin{aligned}
(\alpha_1 + \omega\alpha_2 + \omega^2\alpha_3)^3 =\; & (\alpha_1^3 + \alpha_2^3 + \alpha_3^3) \\
& + 3\omega(\alpha_1\alpha_2^2 + \alpha_2\alpha_3^2 + \alpha_3\alpha_1^2) \\
& + 3\omega^2(\alpha_1\alpha_3^2 + \alpha_3\alpha_2^2 + \alpha_2\alpha_1^2) \\
& + 6\alpha_1\alpha_2\alpha_3 \; ;
\end{aligned}$$

exchanging α_1 and α_2 in this formula yields

$$(\alpha_2 + \omega\alpha_1 + \omega^2\alpha_3)^3 = (\alpha_1^3 + \alpha_2^3 + \alpha_3^3)$$
$$+ 3\omega(\alpha_2\alpha_1^2 + \alpha_1\alpha_3^2 + \alpha_3\alpha_2^2)$$
$$+ 3\omega^2(\alpha_2\alpha_3^2 + \alpha_3\alpha_1^2 + \alpha_1\alpha_2^2)$$
$$+ 6\alpha_1\alpha_2\alpha_3 \ .$$

Consequently,

$$(\alpha_1 + \omega\alpha_2 + \omega^2\alpha_3)^3 + (\alpha_2 + \omega\alpha_1 + \omega^2\alpha_3)^3$$

(1)
$$= 2(\alpha_1^3 + \alpha_2^3 + \alpha_3^3) + 3(\omega + \omega^2)\sum_{i\neq j}\alpha_i\alpha_j^2 + 12\alpha_1\alpha_2\alpha_3$$
$$= 2(\alpha_1^3 + \alpha_2^3 + \alpha_3^3) - 3\sum_{i\neq j}\alpha_i\alpha_j^2 + 12\alpha_1\alpha_2\alpha_3$$
$$= 2(3s_3 - 3s_1s_2 + 3s_1^3) - 3(s_1s_2 - 3s_3) + 12s_3$$
$$= 27s_3 - 9s_1s_2 + 2s_1^3$$

and

$$(\alpha_1 + \omega\alpha_2 + \omega^2\alpha_3)^3 - (\alpha_2 + \omega\alpha_1 + \omega^2\alpha_3)^3$$

(2)
$$= 3(\omega^2 - \omega)(\alpha_1\alpha_3^2 + \alpha_3\alpha_2^2 + \alpha_2\alpha_1^2 - \alpha_2\alpha_3^2 - \alpha_3\alpha_1^2 - \alpha_1\alpha_2^2)$$
$$= 3(\omega^2 - \omega)(\alpha_1 - \alpha_2)(\alpha_1 - \alpha_3)(\alpha_2 - \alpha_3) \ .$$

Exchanging the roles of ω and ω^2 in the above equations (which is possible, of course, since both choices of a primitive cubic root of unity are equivalent), we find that

(3) $\qquad (\alpha_1 + \omega^2\alpha_2 + \omega\alpha_3)^3 + (\alpha_2 + \omega^2\alpha_1 + \omega\alpha_3)^3 = 27s_3 - 9s_1s_2 + 2s_1^3$

and

(4) $\ (\alpha_1 + \omega^2\alpha_2 + \omega\alpha_3)^3 - (\alpha_2 + \omega^2\alpha_1 + \omega\alpha_3)^3 = 3(\omega - \omega^2)(\alpha_1 - \alpha_2)(\alpha_1 - \alpha_3)(\alpha_2 - \alpha_3) \ .$

Squaring (1) and (3) we see that

$$\boxed{\alpha(1, \omega) = \alpha(1, \omega^2) = (27s_3 - 9s_1s_2 + 2s_1^3)^2 \ ;}$$

squaring (2) and (4) and using the fact that $1 + \omega + \omega^2 = 0$ and hence $(\omega^2 - \omega)^2 = \omega^4 - 2\omega^3 + \omega^2 = \omega - 2 + \omega^2 = -3$, we see that

$$\boxed{\alpha(-1, \omega) = \alpha(-1, \omega^2) = -27 \cdot D(f) \ ;}$$

where $D(f) = s_1^2 s_2^2 + 18 s_1 s_2 s_3 - 4 s_2^3 - 4 s_1^3 s_3 + 27 s_3^2$ denotes the discriminant of f.

(e) The solution formula $(\star)$ (multiplied by 3 on both sides) yields

(★)
$$3\alpha = \sum_{\varepsilon_2\in\{1,\omega,\omega^2\}} \sqrt[3]{\frac{1}{2}\sum_{\varepsilon_1\in\{\pm 1\}}\sqrt{\alpha(\varepsilon_1,\varepsilon_2)}}$$
$$= \sum_{\varepsilon_2\in\{1,\omega,\omega^2\}} \sqrt[3]{\frac{1}{2}\left(\sqrt{\alpha(1,\varepsilon_2)} + \sqrt{\alpha(-1,\varepsilon_2)}\right)}$$
$$= \sqrt[3]{\frac{1}{2}\left(\sqrt{\alpha(1,1)} + \sqrt{\alpha(-1,1)}\right)}$$
$$+ \sqrt[3]{\frac{1}{2}\left(\sqrt{\alpha(1,\omega)} + \sqrt{\alpha(-1,\omega)}\right)}$$
$$+ \sqrt[3]{\frac{1}{2}\left(\sqrt{\alpha(1,\omega^2)} + \sqrt{\alpha(-1,\omega^2)}\right)} \ .$$

A comparison of (2) and (4) shows that the square-roots of $\alpha(-1,\omega)$ and $\alpha(-1,\omega^2)$ occurring in this formula must have different signs. Letting $U := \frac{1}{2}(27s_3 - 9s_1 s_2 + 2s_1^3)$ and $V := -27 \cdot D(f)$ and plugging the results of (d) into $(\star)$, we get

$$(\star) \qquad 3 \cdot \alpha = \sqrt[3]{\pm s_1^3} + \sqrt[3]{\pm U + \sqrt{V}} + \sqrt[3]{\pm U - \sqrt{V}} .$$

(f) Since we have two choices for the sign of U, two choices for taking the square root of V and three choices for taking the cubic roots of $\pm s_1^3$ and of $\pm U \pm \sqrt{V}$, the above formula will yield extraneous solutions which have to be eliminated by trial and error. To obtain one solution of the equation $f(x) = 0$, we are completely free in our choice of the roots (with the natural restriction that we assign the same value to the square-root of V and the cubic root of $\pm U \pm \sqrt{V}$ at each occurrence), due to strong solvability. However, the cubic equation $f(x) = 0$ has three different roots, and to distinguish these we have to evaluate the radicals in the formula $(\star\star)$ in a consistent way; for example, we must have $\alpha_1 + \alpha_2 + \alpha_3 = s_1$ (this equation is an easily checked necessary criterion for consistency). A possible set of solutions is

$$
\begin{aligned}
3\alpha_1 &= s_1 + \sqrt[3]{U + \sqrt{V}} + \sqrt[3]{U - \sqrt{V}} \\
3\alpha_2 &= s_1 + \omega\sqrt[3]{U + \sqrt{V}} + \omega^2\sqrt[3]{U - \sqrt{V}} \\
3\alpha_3 &= s_1 + \omega^2\sqrt[3]{U + \sqrt{V}} + \omega\sqrt[3]{U - \sqrt{V}}
\end{aligned}
$$

which is just Cardano's solution.

(26.28) Example: A special cubic equation. Let us now present an example of a polynomial f whose Galois group is solvable but is not a full symmetric group so that a resolvent polynomial has to be constructed to find the roots of f. It is easy to verify that the polynomial

$$f(x) = x^3 - 3x + 1$$

has the discriminant $D(f) = 81$; hence by (23.13) the Galois group of f equals Alt_3 and is therefore cyclic of order 3.

(a) We have $G = \mathrm{Alt}_3$; hence $r = 1$, $G_0 = \{\mathrm{id}, (123), (132)\}$, $G_1 = \{\mathrm{id}\}$ and $p_3 = 2$.

(b) A generator of $G_0 \cong G_0/G_1$ is $\sigma_1 = (123)$.

(c) Let ω be a primitive cubic root of unity; then the Lagrange resolvents needed are $\mathrm{Lag}_{(1,1)} = \mathrm{id} + \sigma_1 + \sigma_1^2$, $\mathrm{Lag}_{(1,\omega)} = \mathrm{id} + \omega\sigma_1 + \omega^2\sigma_1^2$ and $\mathrm{Lag}_{(1,\omega^2)} = \mathrm{id} + \omega^2\sigma_1 + \omega\sigma_1^2$. First of all, we obtain $\alpha(1) = \mathrm{Lag}_{(1,1)}(\alpha_1)^3 = (\alpha_1 + \alpha_2 + \alpha_3)^3 = 0^3 = 0$. Second, observing that $\sigma_1(\alpha_1) = (123) \star \alpha_1 = \alpha_2$ and $\sigma_1^2 \star \alpha_1 = (132) \star \alpha_1 = \alpha_3$, we see that $\alpha(\omega) = (\alpha_1 + \omega\alpha_2 + \omega^2\alpha_3)^3$ equals

$$\underbrace{\alpha_1^3 + \alpha_2^3 + \alpha_3^3}_{=\,-3} + 3\omega\underbrace{(\alpha_1\alpha_2^2 + \alpha_2\alpha_3^2 + \alpha_3\alpha_1^2)}_{=:\,B} + 3\omega^2\underbrace{(\alpha_1\alpha_3^2 + \alpha_3\alpha_2^2 + \alpha_2\alpha_1^2)}_{:=\,A} + 6\underbrace{\alpha_1\alpha_2\alpha_3}_{=\,-1}$$

so that $\alpha(\omega) = -9 + 3\omega A + 3\omega^2 B$. Analogously, we obtain $\alpha(\omega^2) = -9 + 3\omega B + 3\omega^2 A$. In summary, we have

$$
\begin{aligned}
\alpha(1) &= 0, \\
\alpha(\omega) &= -9 + 3\omega A + 3\omega^2 B, \\
\alpha(\omega^2) &= -9 + 3\omega B + 3\omega^2 A.
\end{aligned}
$$

(d) Using the results from (c) and the fact that $\omega + \omega^2 = -1$, we can calculate $\alpha(\omega) + \alpha(\omega^2)$ and $\alpha(\omega)\alpha(\omega^2)$ and obtain the resolvent polynomial

$$q(y) = (y - \alpha(\omega))(y - \alpha(\omega^2)) = y^2 - (\alpha(\omega) + \alpha(\omega^2))y + \alpha(\omega)\alpha(\omega^2)$$
$$= y^2 + (18 + 3(A + B))y + 9(9 + 3(A + B) + (A + B)^2 - 3AB) .$$

Now we observe that $A + B$ and AB can are symmetric polynomials in α_1, α_2 and α_3 and hence are known elements of the base-field K; in fact, we have $A + B = \sum_{i \neq j} \alpha_i \alpha_j^2 = 3a_0 - a_1 a_2 = 3$, $(A - B)^2 = D(f) = 81$ and $4AB = (A + B)^2 - (A - B)^2 = 9 - 81 = -72$ and hence $AB = -18$. Consequently, we have

$$q(y) = y^2 + 27y + 9^3 = y^2 + 27y + 27^2 .$$

Using the quadratic formula, we see that the two roots of this polynomials are $\frac{27}{2}(-1 \pm \sqrt{-3})$, i.e., the numbers 27ω and $27\omega^2$. On the other hand, by the very definition of q, the roots of q are $\alpha(\omega)$ and $\alpha(\omega^2)$; hence either $\alpha(\omega) = 27\omega$ and $\alpha(\omega^2) = 27\omega^2$ or vice versa.

(e) The solution formula $(\star)$ in (26.25) yields

$(\star)$

$$\begin{aligned}
\alpha &= \frac{1}{3} \sum_{\varepsilon_1 \in \{1, \omega, \omega^2\}} \sqrt[3]{\alpha(\varepsilon_1)} \\
&= \frac{1}{3}\left(\sqrt[3]{\alpha(1)} + \sqrt[3]{\alpha(\omega)} + \sqrt[3]{\alpha(\omega^2)}\right) \\
&= \frac{1}{3}\left(\sqrt[3]{27\omega} + \sqrt[3]{27\omega^2}\right) \\
&= \sqrt[3]{\omega} + \sqrt[3]{\omega^2} .
\end{aligned}$$

(f) For each of the two cubic roots occurring in the formula $(\star)$ there are three possibilities; if ξ is one of them, then the others are $\omega\xi$ and $\omega^2\xi$. (In $\mathbb{C}$ we can write these three possibilities as $e^{2\pi i/9}$, $e^{8\pi i/9}$ and $e^{14\pi i/9}$.) Combining all possibilities, we get 9 candidates for the roots of f, but only three of these candidates are indeed roots; they can be written as $\alpha_1 = \xi_1 + \xi_1^{-1}$, $\alpha_2 = \xi_2 + \xi_2^{-1}$ and $\alpha_3 = \xi_3 + \xi_3^{-1}$ where ξ_1, $\xi_2 = \omega\xi_1$ and $\xi_3 = \omega^2\xi_1$ are the different solutions of $\xi^3 = \omega$. More concisely, we can write

$$\begin{aligned}
\alpha_1 &= \omega^{1/3} + \omega^{-1/3}, \\
\alpha_2 &= \omega^{4/3} + \omega^{-4/3}, \\
\alpha_3 &= \omega^{7/3} + \omega^{-7/3} = \omega^{2/3} + \omega^{-2/3} ;
\end{aligned}$$

in exponential notation this reads

$$\alpha_1 = 2\cos\frac{2\pi}{9}, \qquad \alpha_2 = 2\cos\frac{8\pi}{9}, \qquad \alpha_3 = 2\cos\frac{14\pi}{9}$$

which shows that all three roots are real numbers between -2 and 2. (The case that all three roots of a cubic polynomial are real is called *casus irreducibilis*.)

(26.29) Example: The general quartic equation. Consider the general quartic polynomial $f(x) = x^4 - s_1 x^3 + s_2 x^2 - s_3 x + s_4$ over a field K_0, i.e., $f \in K[x]$ where $K = K_0(s_1, s_2, s_3, s_4)$ consists of all rational functions in four indeterminates s_1, s_2, s_3, s_4. Let $\alpha_1, \alpha_2, \alpha_3, \alpha_4$ be the roots of f in some splitting field L of f. We do not intend to go fully through the lengthy calculations which are necessary to express the roots of f by radicals, but we only want to point out in principle how to proceed, following the algorithm (26.25).

(a) The Galois group G of f is Sym_4 by (23.26). We consider the composition series $G_0 \trianglerighteq G_1 \trianglerighteq G_2 \trianglerighteq G_3 \trianglerighteq G_4$ given by

$$\mathrm{Sym}_4 \trianglerighteq \mathrm{Alt}_4 \trianglerighteq \underbrace{\{\mathrm{id}, (12)(34), (13)(24), (14)(23)\}}_{=: \, U} \trianglerighteq \{\mathrm{id}, (12)(34)\} \trianglerighteq \{\mathrm{id}\} \, .$$

Hence $r = 4$; the factor groups G_{i-1}/G_i have orders $p_1 = 2$, $p_2 = 3$, $p_3 = 2$ and $p_4 = 2$; this means that we can find radical expressions for the roots of f by first taking a square root (indeed, a square root of $D(f)$), then a cubic root, then a square root, and finally another square root.

(b) As generators $[\sigma_i]$ for G_{i-1}/G_i for $1 \le i \le 4$ we can choose

$$\sigma_4 := (12)(34) \, , \quad \sigma_3 := (13)(24) \, , \quad \sigma_2 := (123) \, , \quad \sigma_1 := (12) \, .$$

(c) Starting with $\alpha = \alpha_1$, we obtain

$$\alpha(1) = \big((\mathrm{id} + \sigma_4)(\alpha_1)\big)^2 = (\alpha_1 + \alpha_2)^2 \quad \text{and}$$
$$\alpha(-1) = \big((\mathrm{id} - \sigma_4)(\alpha_1)\big)^2 = (\alpha_1 - \alpha_2)^2 \, ;$$

these two elements are invariant under G_3, hence belong to $K_3 = F(G_3)$. Next, we determine the four elements

$$
\begin{aligned}
\alpha(1,1) &= \big((\mathrm{id} + \sigma_3)\alpha(1)\big)^2 &= \big((\alpha_1 + \alpha_2)^2 + (\alpha_3 + \alpha_4)^2\big)^2 \\
\alpha(-1,1) &= \big((\mathrm{id} - \sigma_3)\alpha(1)\big)^2 &= \big((\alpha_1 + \alpha_2)^2 - (\alpha_3 + \alpha_4)^2\big)^2 \\
\alpha(1,-1) &= \big((\mathrm{id} + \sigma_3)\alpha(-1)\big)^2 &= \big((\alpha_1 - \alpha_2)^2 + (\alpha_3 - \alpha_4)^2\big)^2 \\
\alpha(-1,-1) &= \big((\mathrm{id} - \sigma_3)\alpha(-1)\big)^2 &= \big((\alpha_1 - \alpha_2)^2 - (\alpha_3 - \alpha_4)^2\big)^2
\end{aligned}
$$

which are invariant under G_2, hence lie in $F(G_2) = K_2$. This is the point where we stop our presentation. If we wanted to continue, we would choose a primitive cubic root of unity ω and then calculate for each of the quantities $\alpha(\varepsilon_3, \varepsilon_4)$ with $\varepsilon_3, \varepsilon_4 \in \{\pm 1\}$ the three elements

$$
\begin{aligned}
\alpha(1, \varepsilon_3, \varepsilon_4) &= \big((\mathrm{id} + \sigma_2 + \sigma_2^2)\alpha(\varepsilon_3, \varepsilon_4)\big)^3 \\
\alpha(\omega, \varepsilon_3, \varepsilon_4) &= \big((\mathrm{id} + \omega\sigma_2 + \omega^2\sigma_2^2)\alpha(\varepsilon_3, \varepsilon_4)\big)^3 \\
\alpha(\omega^2, \varepsilon_3, \varepsilon_4) &= \big((\mathrm{id} + \omega^2\sigma_2 + \omega\sigma_2^2)\alpha(\varepsilon_3, \varepsilon_4)\big)^3
\end{aligned}
$$

which altogether gives 12 elements $\alpha(\varepsilon_2, \varepsilon_3, \varepsilon_4)$ in $F(G_1) = K_1$ where $\varepsilon_2 \in \{1, \omega, \omega^2\}$ and $\varepsilon_3, \varepsilon_4 \in \{\pm 1\}$. For each of these we would then determine the two elements

$$\alpha(1, \varepsilon_2, \varepsilon_3, \varepsilon_4) = \big((\mathrm{id} + \sigma_1)\alpha(\varepsilon_2, \varepsilon_3, \varepsilon_4)\big)^2 \quad \text{and}$$
$$\alpha(-1, \varepsilon_2, \varepsilon_3, \varepsilon_4) = \big((\mathrm{id} - \sigma_1)\alpha(\varepsilon_2, \varepsilon_3, \varepsilon_4)\big)^2$$

which finally lie in $F(G_0) = K_0 = K$. Since G_0 is all of Sym_4, these last elements (24 altogether) are symmetric expressions in $\alpha_1, \alpha_2, \alpha_3, \alpha_4$ and hence can be determined from the coefficients of f.

READERS with good nerves and great patience are invited to continue the above calculations. We will instead give an alternate way of finding the various radical extensions which arise in solving the general quartic equation. Note that the algorithm described in (26.25) can be called a "top-to-bottom"-method to explicitly find the roots $\alpha_1, \ldots, \alpha_n$ of a polynomial $f \in K[x]$. Given a composition series $G_0 \trianglerighteq G_1 \trianglerighteq \cdots \trianglerighteq G_r$ of $G = \mathrm{Gal}_K(f)$ and the corresponding chain $K_0 \subseteq K_1 \subseteq \cdots \subseteq K_r$ of fixed fields, we start with the largest field $K_r = K(\alpha_1, \ldots, \alpha_n)$, then work our way down through $K_{r-1}, \ldots, K_2, K_1$, applying Lagrange resolvents, to finally arrive at elements down in the base-field $K_0 = K$ from which the roots of f can be obtained by successively taking roots.

We will show below how to go the reverse way "from bottom to top", i.e., how to ascend from the base-field $K = K_0$ over K_1, K_2 and K_3 to K_4 by successively adjoining roots of quantities.

(26.31) Proposition. *The following steps lead to the splitting field of the general quartic polynomial $f \in K[x]$ by successively adjoining roots.*
   (a) *Starting with $K_0 = F(G_0) = F(\mathrm{Sym}_4)$, we find that*

$$K_1 := F(G_1) = F(\mathrm{Alt}_3) = K\Big(\prod_{i<j}(\alpha_i - \alpha_j)\Big) = K(\sqrt{D(f)})$$

whe.e $D(f)$ is the discriminant of f.
   (b) *It is easy to check that the fixed field of $G_2 = U$ is $K_2 = F(U) = K(u_1, u_2, u_3)$ where*

$$u_1 = -\frac{1}{2}(\alpha_1 + \alpha_2)(\alpha_3 + \alpha_4),$$

$$u_2 = -\frac{1}{2}(\alpha_1 + \alpha_3)(\alpha_2 + \alpha_4),$$

$$u_3 = -\frac{1}{2}(\alpha_1 + \alpha_4)(\alpha_2 + \alpha_3).$$

The elements u_1, u_2, u_3 can be identified by finding the roots of the cubic polynomial

$$\varphi(x) := (x - u_1)(x - u_2)(x - u_3) = x^3 - \sigma_1 x^2 + \sigma_2 x - \sigma_3$$

where

$$\sigma_1 := -s_2,$$

$$\sigma_2 := \frac{1}{4}(s_2^2 + s_1 s_3 - 4s_4),$$

$$\sigma_3 := -\frac{1}{8}(s_1 s_2 s_3 - s_1^2 s_4 - s_3^2).$$

Letting $U := \frac{1}{2}(27\sigma_3 - 9\sigma_1\sigma_2 + 2\sigma_1^3)$ and $V := -27 \cdot D(\varphi) = -27 \cdot D(f)^\dagger$, example (26.27) shows that

$$
\begin{aligned}
3u_1 &= \sigma_1 + \sqrt[3]{U + \sqrt{V}} + \sqrt[3]{U - \sqrt{V}}, \\
3u_2 &= \sigma_1 + \omega\sqrt[3]{U + \sqrt{V}} + \omega^2\sqrt[3]{U - \sqrt{V}}, \\
3u_3 &= \sigma_1 + \omega^2\sqrt[3]{U + \sqrt{V}} + \omega\sqrt[3]{U - \sqrt{V}}.
\end{aligned}
$$

$\dagger$ It is easy to check that

$$(u_1 - u_2)(u_1 - u_3)(u_2 - u_3) = -(\alpha_1 - \alpha_2)(\alpha_1 - \alpha_3)(\alpha_1 - \alpha_4)(\alpha_2 - \alpha_3)(\alpha_2 - \alpha_4)(\alpha_3 - \alpha_4)$$

so that $D(\varphi) = D(f)$. (See problem 7 in section 20.)

(c) *Next, the fixed field of $G_3 = \{\mathrm{id}, (12)(34)\}$ is*

$$K_3 := F(G_3) = K_2(P) = K_2(Q) \quad \text{where} \quad P := \alpha_1 + \alpha_2 \quad \text{and} \quad Q := \alpha_3 + \alpha_4 \; ;$$

the equality $K_2(P) = K_2(Q)$ is obvious because $P + Q = s_1 \in K \subseteq K_2$. Since

$$(x - P)(x - Q) = (x - (\alpha_1 + \alpha_2))(x - (\alpha_3 + \alpha_4)) = x^2 - s_1 x - 2u_1$$

we have

$$P, Q = \frac{1}{2}(s_1 \pm \sqrt{s_1^2 + 8u_1}) \; .$$

(d) *Finally, the splitting field of f is*

$$K_4 = F(\{\mathrm{id}\}) = K_3(\alpha_1) = K_3(\alpha_2) = K_3(\alpha_3) = K_3(\alpha_4) \; .$$

Since

$$(x - \alpha_1)(x - \alpha_2) = x^2 - (\alpha_1 + \alpha_2)x + \alpha_1\alpha_2$$

$$= x^2 - Px + \frac{(s_1 - 2P)s_4}{(s_1 - P)(s_2 + 2u_1) - s_3}$$

we have

$$\alpha_{1,2} = \frac{P}{2} \pm \sqrt{\frac{P^2}{4} - \frac{(s_1 - 2P)s_4}{(s_1 - P)(s_2 + 2u_1) - s_3}} \; .$$

Proof. (a) This is an immediate consequence of (23.12).

(b) We observe that u_1, u_2, u_3 are exactly the different elements of the set $\{\sigma \star u_1 \mid \sigma \in \mathrm{Sym}_4\}$, i.e., the elements of Sym_4 simply permute u_1, u_2, u_3; hence $\sigma \star \varphi = \varphi$ for all $\sigma \in \mathrm{Sym}_4$. This shows that the coefficients of φ are symmetric polynomials in $\alpha_1, \alpha_2, \alpha_3, \alpha_4$, hence expressible as polynomials in s_1, s_2, s_3, s_4. The given values for the coefficients of φ can be obtained by a straightforward (if somewhat tedious) computation. (See problem 13 below.)

Let us remark that the factor $-\frac{1}{2}$ in the definition of u_1, u_2, u_3 is just for convenience; the reason for this choice is as follows. We saw in (20.3) that in characteristic $\neq 2$ every quartic polynomial can be reduced to the form $f(x) = x^4 + px^2 + qx + r$; then φ becomes

$$\varphi(x) = x^3 + px^2 + \frac{1}{4}(p^2 - 4r)x - \frac{1}{8}q^2 = \frac{1}{8}(8x^3 + 8px^2 + (2p^2 - 8r)x - q^2) \; .$$

Up to the factor $1/8$, this is exactly Ferrari's resolvent cubic! This is why we chose the normalization factor $-1/2$ in the definition of u_1, u_2, u_3; otherwise we would have obtained a slightly different resolvent cubic.

Another possible choice for u_1, u_2, u_3 is

$$u_1 = \alpha_1\alpha_2 + \alpha_3\alpha_4, \quad u_2 = \alpha_1\alpha_3 + \alpha_2\alpha_4, \quad u_3 = \alpha_1\alpha_4 + \alpha_2\alpha_3 \; ;$$

this choice would have led to

$$\varphi(x) = x^3 - s_2 x^2 + (s_1 s_3 - 4s_4)x - (s_1^2 s_4 - 4s_2 s_4 + s_3^2) \; ;$$

if $f(x) = x^4 + px^2 + qx + r$, this reads $\varphi(x) = x^3 - px^2 - 4rx + (4pr - q^2)$.

(c) It is obvious that $K_3(\alpha_1) = K_3(\alpha_2)$ and that $K_3(\alpha_3) = K_3(\alpha_4)$ because $\alpha_1 + \alpha_2 = P \in K_3$ and $\alpha_3 + \alpha_4 = Q \in K_3$. However, the equality $K_3(\alpha_1) = K_3(\alpha_3)$ is not so obvious. To prove it, we note first that

$$\alpha_1\alpha_2 = \frac{(s_1 - 2P)s_4}{(s_1 - P)(s_2 + 2u_1) - s_3} \in K_3 \quad \text{and}$$

$$\alpha_3\alpha_4 = \frac{(s_1 - 2Q)s_4}{(s_1 - Q)(s_2 + 2u_1) - s_3} \in K_3 .$$

But then

$$\underbrace{\pm\sqrt{D(f)}}_{\in K_1 \subseteq K_3} = (\alpha_1 - \alpha_2)\underbrace{(\alpha_1 - \alpha_3)(\alpha_1 - \alpha_4)}_{= \alpha_1^2 - Q\alpha_1 + \alpha_3\alpha_4}\underbrace{(\alpha_2 - \alpha_3)(\alpha_2 - \alpha_4)}_{= \alpha_2^2 - Q\alpha_2 + \alpha_3\alpha_4}(\alpha_3 - \alpha_4)$$

so that $\alpha_3 - \alpha_4 \in K_3(\alpha_1, \alpha_2) = K_2(\alpha_1)$. But then also $\alpha_3 = \frac{1}{2}(\alpha_3 - \alpha_4 + Q)$ belongs to $K_3(\alpha_1)$. ∎

Exercises

Problem 1. Let $f(x) = x^2 + px + q$ be the general quadratic polynomial over $\mathbb{Z}_2$, i.e., $f \in K[x]$ where $K = \mathbb{Z}_2(p, q)$.

(a) Let $K \subseteq Z \subseteq L$ be fields such that L is obtained from Z by adjoining a p-th root of unity where p is a prime number. Show that if f has no root in Z, then f has also no root in L.

(b) Let $K \subseteq Z \subseteq L$ be such that L is obtained from Z by adjoining a p-th root of an element of K where p is a prime number. Show that if f has no root in Z, then f has also no root in L.

(c) Show that the equation $f(x) = 0$ is not solvable by radicals.

(d) Modify the above argument to show that the equation $x^2 + x + 1 = 0$ over $\mathbb{Z}_2$ is not strongly solvable by radicals.

Problem 2. Let $K \subseteq Z_1$, $Z_2 \subseteq L$ be fields. Prove the following statements!

(a) If $(Z_1 : K)$ is a radical extension then so is $(Z_1 Z_2 : Z_2)$.

(b) If $(Z_1 Z_2 : Z_2)$ and $(Z_2 : K)$ are radical extensions then so is $(Z_1 Z_2 : K)$.

Problem 3. Let K be a field of characteristic zero.

(a) Let $L = K(\alpha)$ where $\alpha^n \in K$. Suppose that $(R : K)$ is a radical Galois extension R. Consider the polynomial

$$p(x) := \prod_{\sigma \in G_K^R} \left(x^n - \sigma(\alpha^n) \right)$$

to show that there is a radical Galois extension of K containing L.

(b) Use part (a) to show that every radical extension of K is contained in a radical Galois extension of K. (This was proved in (26.6)(d) with a different argument.)

Problem 4. (a) Express $\mathbb{Q}(e^{2\pi i/5})$, $\mathbb{Q}(e^{2\pi i/7})$ and $\mathbb{Q}(e^{2\pi i/8})$ as strong radical extensions.

(b) Find all subfields of $\mathbb{Q}(e^{2\pi i/15})$ and exhibit them as radical extensions of $\mathbb{Q}$.

(c) Find a subfield of $\mathbb{Q}(e^{2\pi i/7})$ which is not a radical extension of $\mathbb{Q}$.

(d) Show that $\left(\mathbb{Q}(e^{2\pi i/3}, e^{2\pi i/7}) : \mathbb{Q} \right)$ is a strong radical extension but that $\left(\mathbb{Q}(e^{2\pi i/7}) : \mathbb{Q} \right)$ is not.

Problem 5. Decide for each of the following quintic polynomials $f \in \mathbb{Q}[x]$ whether or not the equation $f(x) = 0$ is solvable by radicals. If possible, find a radical extension L of $\mathbb{Q}$ over which f splits.

(a) $f(x) = x^5 - 2x^4 + 2$

(b) $f(x) = x^5 - 4x^2 + 2$

(c) $f(x) = x^5 - 4x + 2$

(d) $f(x) = x^5 + 5x - 10$

Problem 6. Decide for each of the following polynomials $f \in \mathbb{Q}[x]$ whether or not the equation $f(x) = 0$ is solvable by radicals. If possible, find a radical extension L of $\mathbb{Q}$ over which f splits.
   (a) $f(x) = x^6 - 6x^2 + 3$
   (b) $f(x) = x^7 - 10x^5 + 15x + 5$
   (c) $f(x) = x^7 - 10x^5 + 15x + 5$

Problem 7. Let $K \subseteq \mathbb{R}$ and let $f \in K[x]$ be a polynomial of prime degree p such that the equation $f(x) = 0$ is solvable by radicals over K.
   (a) Show that if $p \equiv 1\,(4)$ then $D(f) > 0$.
   (b) Show that if $p \equiv 3\,(4)$ then $D(f) < 0$ if and only if g has a single real root.

Problem 8. Write down a polynomial $f \in \mathbb{Q}[x]$ such that some, but not all of the roots of f are expressible as radicals over $\mathbb{Q}$.

Problem 9. Let $f \in K[x]$ be an irreducible cubic polynomial where the field K does not contain a primitive cubic root of unity. Show that there is no radical extension of f containing the roots of f.

Problem 10. Let $f \in \mathbb{Q}[x]$ be an irreducible cubic polynomial. Show that there is no radical extension L of $\mathbb{Q}$ contained in $\mathbb{R}$ such that f splits over L (even if f has three real roots).

Problem 11. Let $\xi \in \mathbb{C}$ be a primitive 9-th root of unity; then the discussion in (26.28) showed that the roots of $f(x) = x^3 - 3x + 1$ are given by

$$\alpha_1 = \xi + \xi^{-1}, \quad \alpha_2 = \xi^4 + \xi^{-4} \quad \text{and} \quad \alpha_3 = \xi^7 + \xi^{-7} = \xi^2 + \xi^{-2}.$$

Show that $\mathbb{Q}(\alpha_1, \alpha_2, \alpha_3)$ is strictly contained in $\mathbb{Q}(\xi)$.

Problem 12. Go through the steps as in (26.27), (26.28) and (26.29) to find the solutions of the following equations.
   (a) $x^3 + x^2 - 2x - 1 = 0$.
   (b) $x^3 - 7x + 5 = 0$
   (c) $x^4 + 4x + 2 = 0$

Problem 13. Prove the assertion in (26.31)(b). (Compare with problem 7 in section 20.)

Problem 14. Let $K = \mathbb{Z}_3(t)$ be the rational function field over $\mathbb{Z}_3$.
   (a) Show that $f(x) = x^6 + x^3 + t$ is irreducible over K.
   (b) Show that if α is a root of f then $\big(K(\alpha) : K\big)$ is normal but not separable.
   (c) Is the equation $f(x) = 0$ solvable by radicals?

Problem 15. Let K be a field of characteristic $p \neq 0$ and let $f \in K[x]$ be a polynomial all of whose irreducible factors are separable. Show that if $\mathrm{Gal}_K(f)$ is solvable then a field over which f splits can be constructed by successively adjoining roots of polynomials $x^q - a$ with a prime $q \neq p$ and a already constructed or of the form $x^p - x - a$ with a already constructed.

Problem 16. Suppose that $\alpha \in \mathbb{C}$ satisfies the equation $\alpha^5 - 6\alpha + 3 = 0$. Show that the extension $(\mathbb{Q}(\alpha) : \mathbb{Q})$ is not normal. Moreover, show that $G_{\mathbb{Q}}^{\mathbb{Q}(\alpha)}$ is trivial (and hence solvable) but that $\mathbb{Q}(\alpha)$ is not contained in a radical extension of $\mathbb{Q}$.

Problem 17. (a) Show that $\mathbb{Q}(\sqrt{3}\sqrt[3]{5}) = \mathbb{Q}(\sqrt{3}, \sqrt[3]{5}) =: L$ and determine the degree of L over $\mathbb{Q}$.
(b) Find the minimal polynomial p of $\sqrt{3}\sqrt[3]{5}$ over $\mathbb{Q}$.
(c) Determine the Galois group $G_{\mathbb{Q}}^L$. Is $(L : \mathbb{Q})$ a Galois extension?
(d) Find the normal closure of $(L : \mathbb{Q})$ and determine its degree over $\mathbb{Q}$.
(e) Is the equation $p(x) = 0$ solvable by radicals?

27. Epilogue: The idea of Lie theory as a Galois theory for differential equations

IN this epilogue we want to briefly describe the brilliant conception of the Norwegian mathematician Marius Sophus Lie (1842-1899) of transferring Galois' theory of polynomial equations to an analogous theory of differential equations. In both theories the fundamental underlying idea is that of exploiting hidden symmetries between the (a priori unknown) solutions of an equation (be it an algebraic or a differential equation) to solve this equation; in both cases the mathematical tool to describe these "hidden symmetries" is the concept of a group.

Let us recapitulate Galois' approach to the study of polynomial equations. Suppose that a polynomial equation $f(x) = 0$ over a field K has n different solutions $\alpha_1, \ldots, \alpha_n$ which are parameterized by the underlying set $\{1, \ldots, n\}$. The group Sym_n acts on these solutions via $\sigma \star \alpha_i = \alpha_{\sigma_i}$. Now Galois' crucial idea was to study those permutations σ whose effect on the solutions α_i cannot be detected by algebraic means (which means that $\alpha_{\sigma(i)}$ and α_i satisfy the same algebraic equations over K); these permutations form a group which was called the *Galois group* of the polynomial f.

Now consider a differential equation $y' = f(x, y)$, i.e., $y'(x) = f(x, y(x))$, where, say, f is continuously differentiable. † The solutions of this differential equation are curves $x \mapsto \big(x, y(x)\big)$ in $\mathbb{R}^2$, called the *integral curves* of the equation $y' = f(x, y)$. Through each point (x_0, y_0) in $\mathbb{R}^2$ there is a unique maximal integral curve, due to the existence and uniqueness theorem for differential equations. Thus if we denote by $\varphi_{(x_0, y_0)}$ the unique integral curve through (x_0, y_0) then the solutions of the equation $y' = f(x, y)$ are parameterized by the points (x_0, y_0) of the underlying set $\mathbb{R}^2$. Now the group of all "permutations" of this underlying set, i.e., the group of all diffeomorphisms σ of $\mathbb{R}^2$, acts on these solutions via $\sigma \star \varphi_{(x_0, y_0)} = \varphi_{\sigma(x_0, y_0)}$. In analogy to what Galois had done, Lie studied those diffeomorphisms σ whose effect on the solutions $\varphi_{(x_0, y_0)}$ cannot be detected by analytical means. To see what this means we just observe that a diffeomorphism of $\mathbb{R}^2$ is simply a coordinate transformation $\sigma : (x, y) \mapsto (x^\star, y^\star)$, and to say that the effect of such a transformation cannot be detected is tantamount to saying that the differential equation in the new coordinates $(x^\star, y^\star)$ is the same as that in the old coordinates (x, y); i.e., if $dy/dx = f(x, y)$ then $dy^\star/dx^\star = f(x^\star, y^\star)$. It is easily verified that these coordinate transformations form a group which we call the *Lie group* of the differential equation $y' = f(x, y)$.

(27.1) Definition. *The **Lie group** of a differential equation $dy/dx = f(x, y)$ is the group of those diffeomorphisms $\sigma : (x, y) \mapsto (x^\star, y^\star)$ such that $dy^\star/dx^\star = f(x^\star, y^\star)$. These are just those diffeomorphisms which map solutions of the equation $y' = f(x, y)$ to other solutions of this equation; namely, the integral curve of (x, y) is mapped to the integral curve through $(x^\star, y^\star) = \sigma(x, y)$.*

---

† The theory to be described can be extended in a straightforward way to arbitrary ordinary and partial differential equations on manifolds, but the case of ordinary equations of one variable suffices completely to convey the basic idea, which is all we want to do in this epilogue.

(27.2) Example. Consider a differential equation of the special form $y' = f(x)$. In this case the integral curve $(x, y(x))$ through a point (x_0, y_0) is simply given by $y(x) = y_0 + \int_{x_0}^{x} f(t)\mathrm{d}t$. This means that the integral curves can all be obtained from each other by parallel translations in the y-direction.

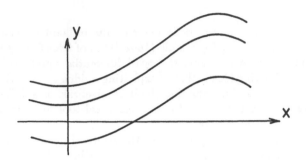

Thus any differential equation $y' = f(x)$ is invariant under the group of all translations $\Phi_t : \mathbb{R}^2 \to \mathbb{R}^2$ given by $\Phi_t(x, y) = (x, y + t)$. In fact, if $(x^*, y^*) := (x, y + t)$ then $\mathrm{d}y^*/\mathrm{d}x^* = \mathrm{d}(y + t)/\mathrm{d}x = \mathrm{d}y/\mathrm{d}x = f(x) = f(x^*)$.

ONE technical but important remark has to be made.

(27.3) Remark. To be useful in practice, definition (27.1) has to be modified. Namely, we will have to admit not only diffeomorphisms, i.e., coordinate transformations which are defined on all of $\mathbb{R}^2$, but also *local diffeomorphisms*, i.e., coordinate transformations which are only defined on some open subset of $\mathbb{R}^2$. (For example, transformation to polar coordinates defines a diffeomorphism between $\mathbb{R}^2 \setminus \{(x, 0) \mid x \geq 0\}$ and $(0, \infty) \times (0, 2\pi)$.) This gives rise to the notion of a *local group* . In some examples we will hint at the relevance of this concept, but for the purpose of this brief description we prefer not to deal with all the technical consequences of the existence of transformations which are not everywhere defined.

ONE of Lie's great inspirations was to study the (local) group G of those transformations of $\mathbb{R}^2$ which leave invariant a given differential equation $y' = f(x, y)$ by studying its "one-dimensional" subgroups. More precisely, he considered *one-parameter groups* of diffeomorphisms, i.e., families $(\Phi_t)_{t \in \mathbb{R}}$ of diffeomorphisms such that $\Phi_0 = \mathrm{id}$ and $\Phi_s \circ \Phi_t = \Phi_{s+t}$ for all $s, t \in \mathbb{R}$. (Again, it is not necessary to require that all the mappings Φ_t be defined on all of $\mathbb{R}^2$; it is enough to assume that for each open subset Ω of $\mathbb{R}^2$ there is a number $\varepsilon > 0$ such that Φ_t is defined on Ω for $|t| < \varepsilon$. Then one requires that $\Phi_s \circ \Phi_t = \Phi_{s+t}$ whenever both sides are defined.) With each such (local) one-parameter group, Lie considered its *infinitesimal generator* which indicates the direction in which (Φ_t) spreads out from the identity element $\mathrm{id} = \Phi_0$.

(27.4) Definition. *The* **infinitesimal generator** *of a (local) one-parameter group*

$$\Phi : \begin{array}{ccc} \mathbb{R} & \to & \text{Diff } \mathbb{R}^2 \\ t & \mapsto & \Phi_t \end{array}$$

of diffeomorphisms of $\mathbb{R}^2$ *is the linear differential operator*

$$U := \varphi \frac{\partial}{\partial x} + \psi \frac{\partial}{\partial y} \quad \text{where} \quad \begin{pmatrix} \varphi(x,y) \\ \psi(x,y) \end{pmatrix} := \frac{d}{dt}\Big|_{t=0} \Phi_t(x,y) \ .$$

NOTE that applying the differential operator U to a function $f : \mathbb{R}^2 \to \mathbb{R}$ has just the effect of taking the directional derivative of f in the direction of the transformation Φ; in fact, the chain rule yields

$$(Uf)(x,y) = \varphi(x,y)\frac{\partial f}{\partial x}(x,y) + \psi(x,y)\frac{\partial f}{\partial y}(x,y) = \frac{d}{dt}\Big|_{t=0} f\big(\Phi_t(x,y)\big)$$

which shows that $(Uf)(x,y)$ is the rate of change of f at the point (x,y) along the curve $t \mapsto \Phi_t(x,y)$.

(27.5) Proposition. *For all analytic functions* $f : \mathbb{R}^2 \to \mathbb{R}$ *we have* $f \circ \Phi_t = e^{tU}f$.

Proof. We fix a point $(x,y) \in \mathbb{R}^2$ and write the function $h : \mathbb{R} \to \mathbb{R}$ given by

$$h(t) := f\big(\Phi_t(x,y)\big)$$

as a Taylor series at $t_0 = 0$. This requires knowledge of the values $h^{(k)}(0)$ for all $k \geq 0$. Now

$$\begin{aligned}
(U^k f)(x,y) &= \frac{\partial}{\partial t_1} \cdots \frac{\partial}{\partial t_k} f\big(\Phi_{t_k} \circ \cdots \circ \Phi_{t_1}(x,y)\big)|_{t_1 = \cdots = t_k = 0} \\
&= \frac{\partial}{\partial t_1} \cdots \frac{\partial}{\partial t_k} f\big(\Phi_{t_k + \cdots + t_1}(x,y)\big)|_{t_1 = \cdots = t_k = 0} \\
&= \frac{d^k}{dt^k} f\big(\Phi_t(x,y)\big)|_{t=0} = h^{(k)}(0) \ ;
\end{aligned}$$

hence

$$(f \circ \Phi_t)(x,y) = \sum_{k=0}^{\infty} \frac{t^k}{k!} h^{(k)}(0) = \sum_{k=0}^{\infty} \frac{t^k}{k!} (U^k f)(x,y) = (e^{tU}f)(x,y) \ .$$

■

(27.6) Examples. (a) If $\Phi_t \begin{pmatrix} x \\ y \end{pmatrix} = \begin{pmatrix} x \\ y \end{pmatrix} + t \begin{pmatrix} a \\ b \end{pmatrix}$ (where $a,b \in \mathbb{R}$ are fixed numbers) then

$$\begin{pmatrix} \varphi(x,y) \\ \psi(x,y) \end{pmatrix} = \begin{pmatrix} a \\ b \end{pmatrix} \quad \text{so that} \quad U = a\frac{\partial}{\partial x} + b\frac{\partial}{\partial y} \ .$$

(b) If $\Phi_t(x,y) = \begin{pmatrix} \cos t & -\sin t \\ \sin t & \cos t \end{pmatrix} \begin{pmatrix} x \\ y \end{pmatrix}$ then

$$\begin{pmatrix} \varphi(x,y) \\ \psi(x,y) \end{pmatrix} = \begin{pmatrix} 0 & -1 \\ 1 & 0 \end{pmatrix} \begin{pmatrix} x \\ y \end{pmatrix} = \begin{pmatrix} -y \\ x \end{pmatrix} \quad \text{so that} \quad U = -y\frac{\partial}{\partial x} + x\frac{\partial}{\partial y}.$$

(c) If $\Phi_t(x,y) = \begin{pmatrix} e^{\alpha t} & 0 \\ 0 & e^{\beta t} \end{pmatrix} \begin{pmatrix} x \\ y \end{pmatrix}$ then

$$\begin{pmatrix} \varphi(x,y) \\ \psi(x,y) \end{pmatrix} = \begin{pmatrix} \alpha & 0 \\ 0 & \beta \end{pmatrix} \begin{pmatrix} x \\ y \end{pmatrix} = \begin{pmatrix} \alpha x \\ \beta y \end{pmatrix} \quad \text{so that} \quad U = \alpha x\frac{\partial}{\partial x} + \beta y\frac{\partial}{\partial y}.$$

(d) If $\Phi_t(x,y) = \big(x+t,\ xy/(x+t)\big)$ then

$$\begin{pmatrix} \varphi(x,y) \\ \psi(x,y) \end{pmatrix} = \begin{pmatrix} 1 \\ -y/x \end{pmatrix} \quad \text{so that} \quad U = \frac{\partial}{\partial x} - \frac{y}{x}\frac{\partial}{\partial y}.$$

(e) If $\Phi_t(x,y) = \big(x/(1-ty),\ y/(1-ty)\big)$ then

$$\begin{pmatrix} \varphi(x,y) \\ \psi(x,y) \end{pmatrix} = \begin{pmatrix} xy \\ y^2 \end{pmatrix} \quad \text{so that} \quad U = xy\frac{\partial}{\partial x} + y^2\frac{\partial}{\partial y}.$$

(Note that in the last two examples the mappings Φ_t are not defined on all of $\mathbb{R}^2$ but are only local diffeomorphisms.)

(27.7) **Exercise.** Check in the above examples that $\Phi_s \circ \Phi_t = \Phi_{s+t}$ whenever both sides are defined.

(27.8) **Exercise.** Let Φ_t be a one-parameter group of diffeomorphisms of $\mathbb{R}^2$ and suppose that the orbits of Φ_t are level sets of a function p so that $\{\Phi_t(x_0,y_0) \mid t \in \mathbb{R}\} = \{(x,y) \in \mathbb{R}^2 \mid p(x,y) = p(x_0,y_0)\}$ for all $(x_0,y_0) \in \mathbb{R}^2$. Show that for a function $f : \mathbb{R}^2 \to \mathbb{R}$ the following conditions are equivalent:
  (1) f is invariant under Φ_t, i.e., $f\big(\Phi_t(x,y)\big) = f(x,y)$ for all $(x,y) \in \mathbb{R}^2$ and all $t \in \mathbb{R}$;
  (2) f is constant on the orbits of Φ_t;
  (3) there is a function $g : \mathbb{R} \to \mathbb{R}$ such that $f(x,y) = g\big(p(x,y)\big)$ for all $(x,y) \in \mathbb{R}^2$.

SINCE it is our goal to find the one-parameter groups which leave invariant the family of integral curves of a given differential equation, we now ask ourselves in general for the conditions under which a family of curves $w(x,y) = C$ (where C parameterizes the family) is invariant under a one-parameter group of diffeomorphisms.

(27.9) Examples. The following pictures show three families of curves in $\mathbb{R}^2$.

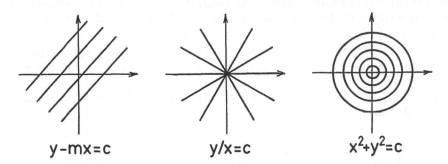

$$y-mx=c \qquad y/x=c \qquad x^2+y^2=c$$

(a) The family $\omega(x,y) = y - mx = C$ (where m is fixed) is obviously invariant under all translations and hence under all one-parameter groups

$$\Phi_t(x,y) \;=\; \begin{pmatrix} x \\ y \end{pmatrix} + t \begin{pmatrix} a \\ b \end{pmatrix} \quad \text{with} \quad \begin{pmatrix} a \\ b \end{pmatrix} \in \mathbb{R}^2 \;.$$

(b) The family $\omega(x,y) = y/x = C$ is obviously invariant under all rotations and all dilatations, i.e., under all one-parameter groups

$$\Phi_t(x,y) \;=\; \begin{pmatrix} \cos(t\varphi) & -\sin(t\varphi) \\ \sin(t\varphi) & \cos(t\varphi) \end{pmatrix} \quad \text{and}$$

$$\Psi_t(x,y) \;=\; \begin{pmatrix} e^{ta} & 0 \\ 0 & e^{ta} \end{pmatrix} \begin{pmatrix} x \\ y \end{pmatrix}$$

where $\varphi, a \in \mathbb{R}$.

(c) The family $\omega(x,y) = x^2 + y^2 = C$ is obviously invariant under the one-parameter groups Φ_t and Ψ_t named in part (b).

(27.10) Theorem. *Let Φ_t be a one-parameter group of diffeomorphisms of $\mathbb{R}^2$ and let $\mathfrak{F}$ be a family of curves which are indexed by a parameter C, say $\omega(x,y) = C$. Then the following conditions are equivalent:*

(1) The family $\mathfrak{F}$ is invariant under the one-parameter group Φ_t.

(2) There is a family of functions $g_t : \mathbb{R} \to \mathbb{R}$ with $\omega \circ \Phi_t = g_t \circ \omega$.

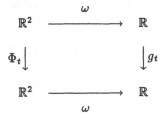

(3) There is a function $g : \mathbb{R} \to \mathbb{R}$ with $U\omega = g \circ \omega$.

Here the one-parameter group Φ_t leaves invariant each individual curve in the family $\mathfrak{F}$ (and not just the family $\mathfrak{F}$ as a whole) if and only if $g \equiv 0$ in condition (3).

Proof. (1)$\Longrightarrow$(2). Fix $t \in \mathbb{R}$. By hypothesis, the diffeomorphism Φ_t maps the curve $\omega(x,y) = C$ to another curve in the family $\mathfrak{F}$, say $\omega(x,y) = C_t$. Define $g_t : \mathbb{R} \to \mathbb{R}$ such that $g_t(C) = C_t$ for all parameter values C of the family $\mathfrak{F}$; then $g_t(\omega(x,y)) = g_t(C) = C_t = \omega(\Phi_t(x,y))$.

(2)$\Longrightarrow$(1). If $\omega \circ \Phi_t = g_t \circ \omega$ and if $\omega(x,y) = C$ then $\omega(\Phi_t(x,y)) = g_t(\omega(x,y)) = g_t(C)$; hence Φ_t maps the curve $\{(x,y) \in \mathbb{R}^2 \mid \omega(x,y) = C\}$ to the curve $\{(x,y) \in \mathbb{R}^2 \mid \omega(x,y) = g_t(C)\}$.

(2)$\Longrightarrow$(3). By (27.5) and by condition (2) we have $(e^{tU}\omega)(x,y) = \omega(\Phi_t(x,y)) = g_t(\omega(x,y))$; hence

$$(U\omega)(x,y) \;=\; \frac{\mathrm{d}}{\mathrm{dt}}\Big|_{t=0}(e^{tU}\omega)(x,y) \;=\; \frac{\mathrm{d}}{\mathrm{dt}}\Big|_{t=0} g_t(\omega(x,y)) \;=:\; g(\omega(x,y)) \,.$$

(3)$\Longrightarrow$(2). Using the product rule and the chain rule we can easily check that

$$U(fg) \;=\; (Uf)g + f(Ug) \quad \text{and}$$
$$U(h \circ \omega) \;=\; (h' \circ \omega) \cdot (U\omega)$$

for all functions $f, g, \omega : \mathbb{R}^2 \to \mathbb{R}$ and $h : \mathbb{R} \to \mathbb{R}$. Then the equation $(U\omega) = g \circ \omega$ in (3) implies that

$$U^2\omega \;=\; U(g \circ \omega) \;=\; (g' \circ \omega) \cdot (g \circ \omega),$$
$$U^3\omega \;=\; U(g' \circ \omega) \cdot (g \circ \omega) + (g' \circ \omega) \cdot U(g \circ \omega),$$

and so on; inductively, we see that for all $k \in \mathbb{N}$ the quantity $(U^k\omega)(x,y)$ depends only on the value $\omega(x,y)$. Consequently, for any fixed value of t, the value $\omega(\Phi_t(x,y)) = (e^{tU}\omega)(x,y) = \sum_{k=0}^{\infty}(t^k/k!)(U^k\omega)(x,y)$ depends only on $\omega(x,y)$, hence can be written as $g_t(\omega(x,y))$ with a well-defined function $g_t : \mathbb{R} \to \mathbb{R}$. $\blacksquare$

(27.11) Examples. (a) Let $\omega(x,y) = y - mx$ where $m \in \mathbb{R}$ is fixed. The one-parameter group Φ_t given in (27.9)(a) has the infinitesimal generator

$$U \;=\; a\frac{\partial}{\partial x} + b\frac{\partial}{\partial y} \,.$$

Hence $(U\omega)(x,y) = -ma + b$ which is a constant and hence trivially depends only on $\omega(x,y)$, in accordance with (27.10)(3). Note that $U\omega \equiv 0$ if and only if $b - ma = 0$. In view of (27.10) this means that the translation by $(a,b)^T$ leaves invariant every individual curve of the form $y - mx = C$ if and only if $b - ma = 0$, a fact which is immediately clear from a geometrical point of view.

(b) Let $\omega(x,y) = y/x$. The one-parameter groups Φ_t and Ψ_t given in (27.9)(b) have the infinitesimal generators

$$U \;=\; -y\varphi\frac{\partial}{\partial x} + x\varphi\frac{\partial}{\partial y} \quad \text{and} \quad V \;=\; ax\frac{\partial}{\partial x} + ay\frac{\partial}{\partial y} \,.$$

Hence $(U\omega)(x,y) = \varphi((y^2/x^2) + 1) = \varphi(\omega(x,y)^2 + 1)$ and $(V\omega)(x,y) = 0$. In view of (27.10) this means that the one-parameter groups Φ_t leave the family of curves $y/x = C$

invariant as a whole, whereas the one-parameter groups Ψ_t even leave invariant each individual member of this family.

(c) Let $\omega(x,y) = x^2 + y^2$ and let Φ_t and Ψ_t as in part (b). Then $(U\omega)(x,y) = 0$ and $(V\omega)(x,y) = 2a(x^2 + y^2) = 2a\omega(x,y)$. In view of (27.10) this means that the one-parameter groups Ψ_t leave the family of curves $x^2 + y^2 = C$ invariant as a whole, whereas the one-parameter groups Φ_t even leave invariant each individual member of this family.

WE now want to pursue our original idea of exploiting symmetries in solving a differential equation.

(27.12) Theorem. *Suppose that the family of the integral curves $\omega(x,y) = C$ of a differential equation*

$$(\star) \qquad \qquad \frac{dy}{dx} = \frac{\eta(x,y)}{\xi(x,y)}$$

(or $\xi dy - \eta dx = 0$ in differential notation) is invariant under a one-parameter group Φ_t with infinitesimal generator U. Let $g : \mathbb{R} \to \mathbb{R}$ be a function with $U\omega = g \circ \omega$.† Assume that $g \not\equiv 0$ and let G be an antiderivative of $1/g$. Then the solution of the differential equation $(\star)$ is implicitly given by the equation

$$(\star\star) \qquad \qquad G\big(\omega(x,y)\big) = const.$$

Proof. The rate of change of a function $f : \mathbb{R}^2 \to \mathbb{R}$ in the direction of the integral curves $x \mapsto \big(x, y(x)\big)$ of $(\star)$ is given by

$$\frac{d}{dx} f\big(x, y(x)\big) = \frac{\partial f}{\partial x} + \frac{\partial f}{\partial y} \cdot y'(x) = \frac{\partial f}{\partial x} + \frac{\eta(x,y)}{\xi(x,y)} \frac{\partial f}{\partial y} \; ;$$

hence the differential operator

$$Z := \xi \frac{\partial}{\partial x} + \eta \frac{\partial}{\partial y} = \left\langle \begin{pmatrix} \xi \\ \eta \end{pmatrix}, \begin{pmatrix} \partial/\partial x \\ \partial/\partial y \end{pmatrix} \right\rangle$$

is up to the scalar factor $\xi(x,y)$ just the directional derivative in the direction of the integral curves of $(\star)$. In particular, we have $Zf = 0$ for any function f which is constant along the integral curves of $(\star)$. Since ω and hence also $\Omega := G \circ \omega$ is constant along the integral curves of $(\star)$ we have

$$(1) \qquad \qquad \xi \frac{\partial \Omega}{\partial x} + \eta \frac{\partial \Omega}{\partial y} = Z\Omega = 0 \; .$$

Moreover, we have

$$(2) \qquad \begin{aligned} \varphi \frac{\partial \Omega}{\partial x} + \psi \frac{\partial \Omega}{\partial y} &= U\Omega = U(G \circ \omega) = (G' \circ \omega)(U\omega) \\ &= (G' \circ \omega)(g \circ \omega) = (G'g) \circ \omega = 1 \circ \omega = 1 \; . \end{aligned}$$

---

† Such a function exists by (27.10)(3).

Equations (1) and (2) together can be written as a linear system

$$\begin{pmatrix} \xi & \eta \\ \varphi & \psi \end{pmatrix} \begin{pmatrix} \partial\Omega/\partial x \\ \partial\Omega/\partial y \end{pmatrix} = \begin{pmatrix} 0 \\ 1 \end{pmatrix}$$

whose solution is easily seen to be

$$\frac{\partial\Omega}{\partial x} = -\frac{\eta}{\xi\psi - \eta\varphi}, \quad \frac{\partial\Omega}{\partial y} = \frac{\xi}{\xi\psi - \eta\varphi}$$

or, in differential notation,

$$d\Omega = \frac{\xi dy - \eta dx}{\xi\psi - \eta\varphi}$$

which implies that Ω is constant along the integral curves of $(\star)$; this is what we wanted to show. (Using the language of the theory of differential equations, we have shown that $1/(\xi\psi - \eta\varphi)$ is an integrating factor for the differential equation $(\star)$.) ∎

(27.13) Example. Consider a differential equation of the form

$$\frac{y - xy'}{x + yy'} = F(x^2 + y^2) .$$

We claim that the integral curves of such an equation are invariant under all rotations about the origin in $\mathbb{R}^2$ which, according to the proof of (27.12), means that there is an integrating factor which only depends on $x^2 + y^2$. To verify this claim we consider the integral curve through a fixed point (x, y). Let α be the elevation angle of this curve at (x, y) so that $y' = \tan\alpha$. Moreover, let β be the angle between the integral curve and the straight line connecting the points $(0, 0)$ and (x, y).

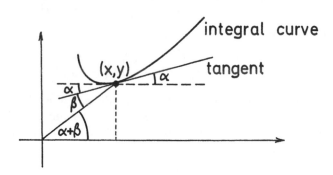

As the above drawing shows, we have

$$\frac{y}{x} = \tan(\alpha + \beta) = \frac{\tan\alpha + \tan\beta}{1 - \tan\alpha \cdot \tan\beta} = \frac{y' + \tan\beta}{1 - y' \tan\beta} .$$

Consequently, $y - yy' \tan\beta = xy' + x \tan\beta$, hence $y - xy' = (x + yy') \tan\beta$ and therefore

$$\tan\beta = \frac{y - xy'}{x + yy'} = F(x^2 + y^2).$$

This shows that all integral curves intersect a fixed circle $x^2 + y^2 = r^2$ at the same angle; this implies the claim. ∎

(27.14) Exercise. Solve the differential equation

$$y' = \frac{y + x(x^2 + y^2)}{x - y(x^2 + y^2)} \; .$$

Hint. This equation can be written in the form discussed in (27.13).

QUITE often it is clear from the provenance or the special form of a differential equation that its solutions are invariant under a certain group of transformations. However, there is also the following purely arithmetical invariance criterion.

(27.15) Theorem. *Consider a differential equation*

$(\star)$ $$\frac{dy}{dx} = \frac{\eta(x,y)}{\xi(x,y)}$$

with the associated differential operator $Z = \xi(\partial/\partial x) + \eta(\partial/\partial y)$ and let (Φ_t) be a one-parameter group with infinitesimal generator U. Then the family of integral curves of $(\star)$ is invariant under (Φ_t) if and only if there is a function $\lambda : \mathbb{R}^2 \to \mathbb{R}$ such that

$$[U, Z] := U \circ Z - Z \circ U = \lambda Z \; .$$

Proof. Suppose that the integral curves $\omega(x,y) = C$ of $(\star)$ are invariant under (Φ_t). Then $Z\omega = 0$ because applying Z means, up to a scalar factor, just taking the derivative in the direction of the integral curves. Then also $[U, Z]\omega = U(Z\omega) - Z(U\omega) = U(0) - Z(g \circ \omega) = 0 - (g' \circ \omega) \cdot Z\omega = 0 - (g' \circ \omega) \cdot 0 = 0$. The two equations $Z\omega = 0$ and $[U, Z]\omega = 0$ can be written in components and then rewritten as a linear system

$$\begin{pmatrix} Z_1 & Z_2 \\ [U, Z]_1 & [U, Z]_2 \end{pmatrix} \begin{pmatrix} \partial\omega/\partial x \\ \partial\omega/\partial y \end{pmatrix} = \begin{pmatrix} 0 \\ 0 \end{pmatrix} \; .$$

We conclude that the coefficient matrix cannot be invertible; hence its two rows are linearly independent which means that $[U, Z] = \lambda Z$ for some λ.

Conversely, if $[U, Z] = \lambda Z$ then $0 = \lambda \cdot Z\omega = [U, Z]\omega = U(Z\omega) - Z(U\omega) = U(0) - Z(U\omega) = -Z(U\omega)$. Since Z is, up to a scalar factor, just the directional derivative in the direction of the integral curves $\omega(x,y) = C$ of $(\star)$ this means that $U\omega$ is constant along these integral curves. Consequently, $(U\omega)(x,y)$ depends only on $\omega(x,y)$; hence there is a function g such that $U\omega = g \circ \omega$. By (27.10)(3) this gives the claim. ∎

AT this point we want to conclude our considerations. We have seen how the fundamental idea of solving an equation by exploiting symmetries in the solutions of this equation was transferred from polynomial equations to differential equations. We have developed the theory far enough to obtain one or two nontrivial results – delving any further into this topic must be left to books on differential equations and Lie theory.

Bibliography

ALGEBRA IN GENERAL

Garrett Birkhoff, Thomas C. Bartee, *Modern Applied Algebra*; McGraw-Hill 1970
Israel N. Herstein, *Topics in Algebra*; Blaisdell 1964
Thomas W. Hungerford, *Algebra*; Springer 1974
Nathan Jacobson, *Lectures in Abstract Algebra* (3 volumes);
    van Nostrand 1951/1953/1964
Serge Lang, *Algebra*; Addison-Wesley 1965
Bartel L. van der Waerden, *Algebra* (2 volumes); Frederick Ungar 1970

RINGS AND FIELDS

N. J. Divinsky, *Rings and Radicals*; University of Toronto Press 1965
M. Gray, *A Radical Approach to Algebra*; Addison-Wesley 1970
Israel N. Herstein, *Noncommutative Rings*; Mathematical Association of America
    (Carus Mathematical Monographs) 1968
Israel N. Herstein, *Topics in Ring Theory*; University of Chicago Press 1965
Irving Kaplansky, *Fields and Rings*; The University of Chicago Press 1969
Paul McCarthy, *Algebraic Extensions of Fields*; Blaisdell 1966
Neal H. McCoy, *Theory of Rings*; Macmillan 1964
Masayoshi Nagata, *Field Theory*; Marcel Dekker 1977
D. G. Northcott, *Lessons on Rings, Modules and Multiplicities*;
    Cambridge University Press 1968
Abraham Robinson, *Numbers and Ideals*; Holden-Day 1965
David Sharpe, *Rings and factorization*; Cambridge University Press 1987

GALOIS THEORY

Emil Artin, *Galois Theory*; Notre Dame 1944
Harold M. Edwards, *Galois Theory*; Springer 1984
Lisl Gaal, *Classical Galois Theory with Examples*; Chelsea Publishing Company 1971
Irving Kaplansky, *An Introduction to Differential Algebra*; Hermann 1957
Jean-Pierre Tignol, *Galois Theory of algebraic equations*; Longman 1987

COMMUTATIVE ALGEBRA

Michael F. Atiyah, I. G. Mac Donald, *Introduction to Commutative Algebra*;
    Addison-Wesley 1969
Irving Kaplansky, *Commutative Rings*; Allyn and Bacon 1970
Max D. Larsen, Paul J. McCarthy, *Multiplicative Theory of Ideals*;
    Academic Press 1971
D.G. Northcott, *Ideal Theory*; Cambridge University Press 1965
Oscar Zariski, Pierre Samuel, *Commutative Algebra* (2 volumes); Springer 1958/1960

ALGEBRAIC GEOMETRY

William Fulton, *Algebraic Curves*; Benjamin 1969
William Fulton, *An Introduction to Algebraic Geometry*; Benjamin 1969
Wolfgang Gröbner, *Algebraische Geometrie I*; Bibliographisches Institut 1968
Robin Hartshorne, *Algebraic Geometry*; Springer 1977
W.E. Jenner, *Rudiments of Algebraic Geometry*; Oxford University Press 1963
Keith Kendig, *Elementary Algebraic Geometry*; Springer 1977
Ernst Kunz, *Introduction to Commutative Algebra and Algebraic Geometry*;
    Birkhäuser, 1985
Miles Reid, *Undergraduate Algebraic Geometry*; Cambridge University Press 1988
Igor R. Shafarevich, *Basic Algebraic Geometry*; Springer 1977

NUMBER THEORY

Senon I. Borewicz, Igor R. Šafarevič, *Number Theory*; Academic Press 1966
J. W. S. Cassels, A. Fröhlich (eds.), *Algebraic Number Theory*; Thompson 1967
Harvey Cohn, *A Classical Invitation to Algebraic Numbers and Class Fields*; Springer
Fred Wayne Dodd, *Number theory in the quadratic field with golden section unit*;
    Polygonal Publishing House 1983
Harold M. Edwards, *Fermat's Last Theorem*; Springer 1977
Neal Koblitz, *A Course in Number Theory and Cryptography*; Springer 1987
Serge Lang, *Algebraic Number Theory*; Addison-Wesley 1970
Daniel A. Marcus, *Number Fields*; Springer 1977
Paulo Ribenboim, *Algebraic Numbers*; Wiley-Interscience 1972
Paulo Ribenboim, *13 Lectures on Fermat's Last Theorem*; Springer 1979
Jean-Pierre Serre, *A Course in Arithmetic*; Springer 1973
Ian Stewart, David Tall, *Algebraic Number Theory*; Chapman and Hall 1979
Edwin Weiss, *Algebraic Number Theory*; McGraw-Hill 1963

Index

† Terms are referenced by section number. P, problem; A-E, Appendices A-E.